MOLECULAR BIOLOGY
INTELLIGENCE
UNIT

Tetrapyrroles
Birth, Life and Death

Martin J. Warren, PhD
Protein Science Group
Department of Biosciences
University of Kent
Canterbury, Kent, UK

Alison G. Smith, PhD
Department of Plant Sciences
University of Cambridge
Cambridge, UK

LANDES BIOSCIENCE
AUSTIN, TEXAS
USA

SPRINGER SCIENCE+BUSINESS MEDIA
NEW YORK, NEW YORK
USA

TETRAPYRROLES: BIRTH, LIFE AND DEATH

Molecular Biology Intelligence Unit

Landes Bioscience
Springer Science+Business Media, LLC

ISBN: 978-0-387-78517-2 Printed on acid-free paper.

Springer Science+Business Media, LLC, 233 Spring Street, New York, New York 10013, USA
http://www.springer.com

Please address all inquiries to the Publishers:
Landes Bioscience, 1002 West Avenue, Austin, Texas 78701, USA
Phone: 512/ 637 6050; FAX: 512/ 637 6079
http://www.landesbioscience.com

Printed in the United States of America.

9 8 7 6 5 4 3 2 1

Library of Congress Cataloging-in-Publication Data

Tetrapyrroles : birth, life, and death / edited by Martin J. Warren, Alison G. Smith.
    p. ; cm.
  Includes bibliographical references and index.
  ISBN 978-0-387-78517-2 (alk. paper)
  1. Tetrapyrroles. 2. Chlorophyll. 3. Porphyrins. 4. Heme. I. Warren, Martin J. II. Smith, Alison G.
  [DNLM: 1. Tetrapyrroles. QD 401 T348 2009]
  QP670.T48 2009
  572'.46--dc22

                              2008005850

About the Editors...

MARTIN J. WARREN, PhD, is Professor of Biochemistry at the University of Kent, Canterbury, UK. For the past twenty years he has studied tetrapyrrole biosynthesis, including the synthesis of the macrocyclic ring structure and its transformation into heme, siroheme and vitamin B_{12}. He has recently been awarded a Biotechnology and Biological Science Research Council (BBSRC) professional fellowship to continue his work on metabolic engineering of complex biosynthetic pathways. He received his PhD from the University of Southampton, Hants, UK, and did his postdoctoral work at Texas A&M, College Station, Texas, USA.

About the Editors...

ALISON SMITH, PhD, is Professor of Plant Biochemistry in the Department of Plant Sciences, University of Cambridge, UK. She obtained a BSc in Biochemistry from the University of Bristol, and a PhD in Plant Biochemistry from the University of Cambridge. Her research is focused on metabolism in plants, microbes and algae, in particular that for tetrapyrrole biosynthesis and vitamin metabolism. She is known for her work on the identification and characterisation of the enzymes involved in haem biosynthesis in higher plants and in determining the subcellular location and regulation of the pathway. More recently, she was instrumental in establishing the widespread interaction in which algae obtain vitamin B_{12} from bacteria in exchange for fixed carbon.

CONTENTS

Preface xv

1. **An Historical Introduction to Porphyrin and Chlorophyll Synthesis** 1
   Michael R. Moore
   Historical Introduction to Porphyrins and Porphyrias 1
   Structure 4
   The Early Chemical Era 9
   The Biochemical Descriptive Era 10
   Early Description of Porphyria 11
   Classification of Porphyrias 12
   Enzymes 14
   Acute Porphyria 15
   Nonacute Porphyrias 17
   Porphyria in Animals 18
   Porphyrin Synthesis in the Animal Kingdom 19
   The Harderian Gland 19
   Phototherapy and Cancer 19
   Retrospective Diagnoses 20
   Ephemera: Porphyrinurias 21

2. **Biosynthesis of 5-Aminolevulinic Acid** 29
   Dieter Jahn and Dirk W. Heinz
   Condensation of Succinyl-CoA and Glycine
      into Aminolevulinic Acid 29
   Transfer RNA-Dependent Aminolevulinic Acid Formation 30

3. **5-Aminolaevulinic Acid Dehydratase, Porphobilinogen Deaminase
   and Uroporphyrinogen III Synthase** 43
   Heidi L. Schubert, Peter T. Erskine and Jonathan B. Cooper
   5-Aminolaevulinic Acid Dehydratase 43
   Porphobilinogen Deaminase 58
   Uroporphyrinogen III Synthase 61

4. **Transformation of Uroporphyrinogen III into Protohaem** 74
   Johanna E. Cornah and Alison G. Smith
   Uroporphyrinogen III Decarboxylase 76
   Coproporphyrinogen III Oxidase/Dehydrogenase 77
   Protoporphyrinogen IX Oxidase 79
   Ferrochelatase 81
   Organization of Pathway 82

5. **Inherited Disorders of Haem Synthesis: The Human Porphyrias** 89
   Michael N. Badminton and George H. Elder
   Overview 89
   Molecular Genetics and Pathogenesis 93
   Mechanisms of Disease 96
   New Approaches to Management 97

6. **Heme Degradation: Mechanistic and Physiological Implications** 101
Angela Wilks
   Evolution and Biological Function of Heme Oxygenase 101
   Sequence and Structural Conservation within the Heme
      Oxygenase Enzymes 103
   Crystallographic Studies 103
   Mechanism of Heme Oxygenase 107
   Biliverdin Reduction to Bilirubin 110
   Biliverdin IXα Reductase 110
   Biliverdin IXβ Reductase 113

7. **Regulation of Mammalian Heme Biosynthesis** 116
Amy E. Medlock and Harry A. Dailey
   Regulation of Heme Biosynthesis by ALA Synthase 117
   Regulation at Sites Other Than ALAS 121

8. **Tetrapyrroles in Photodynamic Therapy** 128
David I. Vernon and Ian Walker
   Brief History 128
   Singlet Oxygen: The Cytotoxic Agent 129
   Singlet Oxygen Targets 129
   Light Delivery and Requirement 130
   Photodynamic Damage 130
   Mechanisms of Tumour and Cellular Uptake................... 131
   Tetrapyrroles in Photodynamic Therapy 131
   Haematoporphyrin Derivative (HpD) and Photofrin 131
   The Ideal Properties of a Photosensitiser 132
   Second Generation Photosensitisers................................ 133
   5,10,15,20 Tetrakis (meso-hydroxyphenyl) Chlorin
      (m-THPC, Foscan, temoporfin) 133
   5,10,15,20 Tetrakis (meso-hydroxyphenyl) Bacteriochlorin
      (m-THPBC) 134
   Benzoporphyrin Derivative (BPD, Verteporfin) 135
   Tin Ethyl Etiopurpurin (SnEt2, Purlytin, Rostaporfin) 136
   Mono-L-Aspartyl Chlorin e6 (Npe6, MACE, Talaporfin) 136
   Palladium-Bacteriopheophorbide (TOOKAD, WST009) 137
   2-[1-hexyloxyethyl]-2-Devinyl Pyropheophorbide-a
      (HPPH, Photochlor) 138
   Phthalocyanines.. 139
   Lutetium Texaphyrin (Lu-tex, Motexafin Lutetium) 139
   5-Aminolaevulinic Acid 140
   Clinical ALA-PDT 141
   Other Applications of ALA-PDT 141
   Aminolaevulinic Acid Esters 141
   Photodetection of Tumours 142

9. **Heme Transport and Incorporation into Proteins** 149
   Linda Thöny-Meyer
   Localization of Heme in Prokaryotes 150
   Bacterial Heme Transport 151
   Heme Transport in Eukaryotes 153
   Unassisted Heme Transport 155
   Heme-Protein Assembly 155

10. **Heme and Hemoproteins** 160
    Andrew W. Munro, Hazel M. Girvan, Kirsty J. McLean,
    Myles R. Cheesman and David Leys
    The Heme Synthetic Pathway 160
    Structural Variations of the Heme Cofactor 162
    Heme Iron Coordination in Hemoproteins 165
    Diversity of Hemoprotein Form and Function 168
    Spectroscopic Analysis of Hemoproteins 171
    Novel Aspects and Future Prospects 175

11. **Novel Heme-Protein Interactions—Some More Radical**
    Than Others 184
    Ann Smith
    Heme as a Sensor: Interactions of Heme with Proteins
       That Lead to Recognition of Gaseous Molecules:
       Oxygen, Carbon Monoxide and Nitric Oxide 186
    Heme Binding to Ion Channels 187
    Novel Low-Spin Heme-Protein Interactions 188
    Heme-Binding Proteins That Are Protective, Preventing
       Heme-Mediated Oxidative Stress 189
    Heme Transport across Enterocytes and Proof of Principle ... 193
    Interactions of Heme with Transcription Factors 195
    Heme-Protein Interactions and the Control of Circadian Rhythms ... 196
    Novel Heme-Protein Interactions for the Control
       of Intracellular Heme Levels 199
    Relationships between ATP Concentrations, Oxygen Tension
       and Heme Transporters, Many of Which Also Interact
       with Porphyrins 201

12. **Synthesis and Role of Bilins in Photosynthetic Organisms** 208
    Nicole Frankenberg-Dinkel and Matthew J. Terry
    Structure and Spectral Properties of Protein-Bound Bilins 208
    Synthesis of Biliverdin IXα by Heme Oxygenases 210
    Biosynthesis of Bilins by Ferredoxin-Dependent Bilin Reductases 212
    Assembly of Phycobiliproteins and Phytochromes 216
    The Roles of Bilins in Photosynthetic Organisms 216

13. **Phytochromes: Bilin-Linked Photoreceptors in Bacteria and Plants ... 221**
Matthew J. Terry and Alex C. McCormac
The Phytochromes—A Diverse Family of Photoreversible
Photoreceptors 222
Phytochrome Photosensory Domains 224
Physiological Roles of Phytochrome-Like Proteins in Prokaryotes 226
Phytochrome Function in Flowering Plants 227
Specific Roles for Specific Phytochromes 227
Phytochrome Mode of Action 229
Phytochrome Regulation of Tetrapyrrole Synthesis 229

14. **Biosynthesis of Chlorophyll and Bacteriochlorophyll 235**
Derren J. Heyes and C. Neil Hunter
The Insertion of the Central Magnesium Ion 237
Methylation of Ring C 238
The Missing Link in Chlorophyll Biosynthesis: The Formation
of the Isocyclic Ring E of Protochlorophyllide 239
Reduction of the 8-Vinyl Group 240
Two Routes for the Reduction of Pchlide 241
POR: A Light-Driven Enzyme 241
DPOR: A Multi-Subunit Enzyme 243
The Steps Unique to Bacteriochlorophyll Biosynthesis 244
The Final Steps: Addition and Reduction of the Phytol Tail 245

15. **Regulation of Tetrapyrrole Synthesis in Higher Plants 250**
Matthew J. Terry and Alison G. Smith
Regulation of the Plant Tetrapyrrole Pathway—At the Heart
of Plant Metabolism? 252
Turning on the Tap—Regulation of the Synthesis of the Initial
Precursor, ALA 253
Decision Time at the Branchpoints 256
Regulation of the Chlorophyll Branch 258

16. **Regulation of the Late Steps of Chlorophyll Biosynthesis 263**
Wolfhart Rüdiger
Light Regulation via NADPH: Protochlorophyllide
Oxidoreductase (POR) 264
Regulation of Chlorophyll b Biosynthesis 267
A New Role for Carotenoids? 269

17. **Chlorophyll Breakdown 274**
Bernhard Kräutler
Chlorophyll Breakdown in Higher Plants 274
Early Steps 275
Cleavage of the Chlorophyll Macroring 276
The Arrival at Colorless and Nonfluorescent Chlorophyll
Breakdown Products 278

Breakdown Beyond the Stage of Colorless Tetrapyrrolic Catabolites ... 280
Chlorophyll Catabolites from Other Sources 282

18. **Vitamin B$_{12}$: Biosynthesis of the Corrin Ring** 286
Ross M. Graham, Evelyne Deery and Martin J. Warren
The First Common Step: Production of Precorrin-2 287
The Aerobic Pathway 288
Production of Hydrogenobyrinic Acid 291
Proteins of Unknown Function 293
The Anaerobic Pathway 293
Production of Cobalt-Precorrin-6A 295

19. **Conversion of Cobinamide into Coenzyme B$_{12}$** 300
Jorge C. Escalante-Semerena, Jesse D. Woodson, Nicole R. Buan
and Carmen L. Zayas
Attachment of 5-Deoxyadenosine, the Upper (*Coβ*) Ligand
of Coenzyme B$_{12}$ 300
The Nucleotide Loop Assembly (NLA) Pathway 304

20. **The Regulation of Cobalamin Biosynthesis** 317
Jeffrey G. Lawrence
The Complexity of Cobalamin 317
Cobalamin in Context: Regulating a Branch Point 319
Operon Induction and Physiological Significance 321
Operon Repression and mRNA Binding 323
The Synergy of Cobalamin Transport and Synthesis 324

21. **Coenzyme B$_{12}$-Catalyzed Radical Isomerizations** 330
Dominique Padovani and Ruma Banerjee
Structural Insights into the B$_{12}$-Dependent Isomerases 331
Co-C Bond Activation in B$_{12}$-Dependent Isomerases 333
Radical Flights: Conformational Changes at Play 336
Rearrangement Reactions Catalyzed by B$_{12}$-Dependent Enzymes 338

22. **Biosynthesis of Siroheme and Coenzyme F$_{430}$** 343
Martin J. Warren, Evelyne Deery and Ruth-Sarah Rose
Siroheme Biosynthesis 343
Coenzyme F$_{430}$ Biosynthesis 346

23. **Role of Coenzyme F$_{430}$ in Methanogenesis** 352
Evert C. Duin
Methanogenesis 353
Free Factor 430 354
Name That Signal 355
EPR Signals in Whole Cells 358
Structure of MCR and the Nickel Site 360
Oxidation State of Nickel in the MCRox1 Form 362
Activation and Inactivation of Methyl-Coenzyme M Reductase 363

Prelude to the Catalytic Mechanism 366
Catalytic Mechanism 367
Anaerobic CH4 Oxidation 370
MCR Is Still a Mystery 370

24. **The Role of Siroheme in Sulfite and Nitrite Reductases** 375
M. Elizabeth Stroupe and Elizabeth D. Getzoff
   Siroheme-Containing Sulfite and Nitrite Reductases Represent
      a Single Enzyme Class 375
   Diversity between the SiR and NiR Enzymes 377
   Symmetry Defines Homology between the Assimilatory
      and Dissimilatory Enzymes 378
   SiRs and NiRs Have Multiple Redox Centers and Intricate
      Spectroscopic Features 378
   X-Ray Crystallographic Structures Support the Spectroscopic Data ... 378
   Siroheme Is at the Heart of the Six-Electron Reduction of Sulfite
      to Sulfide or Nitrite to Ammonia 379
   Siroheme Anchors the Transformation of Sulfite to Sulfide
      or Nitrite to Ammonia 381
   The Siroheme Tetrapyrrole Shows Significant Departure
      from Planarity 382
   Siroheme's Structural and Electronic Characteristics Control
      Anion Interactions 386
   A Possible π Cation Radical Intermediate 387

25. **The Role of Heme d_1 in Denitrification** 390
Stuart J. Ferguson
   Structure of Cytochrome cd_1 391
   Mechanism of Nitrite Reduction 394
   Insights into d_1 heme Chemistry from Model Compound Studies 395
   Biosynthesis of d_1 heme 396

Index 401

EDITORS

Martin J. Warren
Protein Science Group
Department of Biosciences
University of Kent
Canterbury, Kent, UK
Email: m.j.warren@kent.ac.uk
Chapters 18, 22

Alison G. Smith
Department of Plant Sciences
University of Cambridge
Cambridge, UK
Email: as25@cam.ac.uk
Chapters 4, 15

CONTRIBUTORS

Note: Email addresses are provided for the corresponding authors of each chapter.

Michael N. Badminton
Department of Medical Biochemistry
  and Immunology
School of Medicine
Cardiff University
Heath Park, Cardiff, UK
Email: badmintonmn@cardiff.ac.uk
Chapter 5

Ruma Banerjee
Department of Biological Chemistry
University of Michigan
Ann Arbor, Michigan, USA
Email: rbanerje@umich.edu
Chapter 21

Nicole R. Buan
Department of Bacteriology
University of Wisconsin, Madison
Madison, Wisconsin, USA
Chapter 19

Myles R. Cheesman
Department of Chemical Sciences
  and Pharmacy
University of East Anglia
Norwich, UK
Chapter 10

Jonathan B. Cooper
Laboratory for Protein Crystallography
Centre for Amyloidosis and Acute
  Phase Proteins
UCL Department of Medicine
London, UK
Email: j.b.cooper@soton.ac.uk
Chapter 3

Johanna E. Cornah
Department of Plant Sciences
University of Cambridge
Cambridge, UK
Chapter 4

Harry A. Dailey
Biomedical and Health Sciences Institute
Paul D. Coverdell Center
University of Georgia
Athens, Georgia, USA
Email: hdailey@uga.edu
Chapter 7

Evelyne Deery
Protein Science Group
Department of Bioscience
University of Kent
Canterbury, Kent, UK
Chapters 18, 22

Evert C. Duin
Department of Chemistry
   and Biochemistry
Auburn University
Auburn, Alabama, USA
Email: duinedu@auburn.edu
Chapter 23

George H. Elder
Department of Medical Biochemistry
   and Immunology
School of Medicine
Cardiff University
Heath Park, Cardiff, UK
Chapter 5

Peter T. Erskine
Laboratory for Protein Crystallography
Centre for Amyloidosis and Acute
   Phase Proteins
UCL Department of Medicine
London, UK
Chapter 3

Jorge C. Escalante-Semerena
Department of Bacteriology
University of Wisconsin, Madison
Madison, Wisconsin, USA
Email: escalante@bact.wisc.edu
Chapter 19

Stuart J. Ferguson
Department of Biochemistry
University of Oxford
Oxford, UK
Email: stuart.ferguson@bioch.ox.ac.uk
Chapter 25

Nicole Frankenberg-Dinkel
Physiology of Microorganisms
Ruhr-University Bochum
Bochum, Germany
Email: nicole.frankenberg@rub.de
Chapter 12

Elizabeth D. Getzoff
The Scripps Research Institute
La Jolla, California, USA
Email: edg@scripps.edu
Chapter 24

Hazel M. Girvan
Faculty of Life Sciences
University of Manchester
Manchester Interdisciplinary Biocentre
Manchester, UK
Chapter 10

Ross M. Graham
School of Medicine and Pharmacology
University of Western Australia
Fremantle Hospital
Fremantle, Australia
Chapter 18

Dirk W. Heinz
Department of Structural Biology
German Research Center
   for Biotechnology
Braunschweig, Germany
Chapter 2

Derren J. Heyes
Department of Life Sciences
Manchester Interdisciplinary Biocentre
University of Manchester
Manchester, UK
Chapter 14

C. Neil Hunter
University of Sheffield
Sheffield, UK
Email: c.n.hunter@sheffield.ac.uk
Chapter 14

Dieter Jahn
Institute of Microbiology
Technical University Braunschweig
Braunschweig, Germany
Email: d.jahn@tu-bs.de
Chapter 2

Bernhard Kräutler
Institute of Organic Chemistry
University of Innsbruck
Innsbruck, Austria
Email: bernhard.kraeutler@uibk.ac.at
Chapter 17

Jeffrey G. Lawrence
Department of Biological Sciences
University of Pittsburgh
Pittsburgh, Pennsylvania, USA
Email: jlawrenc@pitt.edu
Chapter 20

David Leys
Department of Life Sciences
University of Manchester
Manchester Interdisciplinary Biocentre
Manchester, UK
Chapter 10

Alex C. McCormac
School of Biological Sciences
University of Southampton
Southampton, UK
Email: mjt@soton.ac.uk
Chapter 13

Kirsty J. McLean
Department of Life Sciences
University of Manchester
Manchester Interdisciplinary Biocentre
Manchester, UK
Chapter 10

Amy E. Medlock
Biomedical and Health Sciences Institute
Paul D. Coverdell Center
University of Georgia
Athens, Georgia, USA
Chapter 7

Michael R. Moore
National Research Centre
    for Environmental Toxicology
University of Queensland
Brisbane, Queensland, Australia
Email: m.moore@uq.edu.au
Chapter 1

Andrew W. Munro
Department of Life Sciences
University of Manchester
Manchester Interdisciplinary Biocentre
Manchester, UK
Email: andrew.munro@manchester.ac.uk
Chapter 10

Dominique Padovani
Department of Biological Chemistry
University of Michigan
Ann Arbor, Michigan, USA
Chapter 21

Ruth-Sarah Rose
Protein Science Group
Department of Biosciences
University of Kent
Canterbury, Kent, UK
Chapter 22

Wolfhart Rüdiger
Department Biologie I, Botanik
Universität München
München, Germany
Email: ruediger@lrz.uni-muenchen.de
Chapter 16

Heidi L. Schubert
Department of Biochemistry
University of Utah
Salt Lake City, Utah, USA
Chapter 3

Ann Smith
School of Biological Sciences
University of Missouri, Kansas City
Kansas City, Missouri, USA
Email: smithan@umkc.edu
Chapter 11

M. Elizabeth Stroupe
The Scripps Research Institute
La Jolla, California, USA
Chapter 24

Matthew J. Terry
School of Biological Sciences
University of Southampton
Southampton, UK
Email: mjt@soton.ac.uk
Chapters 12, 13, 15

Linda Thöny-Meyer
Laboratory of Biomaterials, Empa
Swiss Federal Laboratories for Materials
    Testing and Research
St. Gallen, Switzerland
Email: linda.thoeny@empa.ch
Chapter 9

David I. Vernon
Institute of Molecular and Cellular
    Biology
Department of Biological Sciences
University of Leeds
Leeds, UK
Email: d.i.vernon@leeds.ac.uk
Chapter 8

Ian Walker
Institute of Molecular and Cellular
    Biology
Department of Biological Sciences
University of Leeds
Leeds, UK
Chapter 8

Angela Wilks
Department of Pharmaceutical Sciences
School of Pharmacy
University of Maryland
Baltimore, Maryland, USA
Email: awilks@rx.umaryland.edu
Chapter 6

Jesse D. Woodson
Department of Bacteriology
University of Wisconsin, Madison
Madison, Wisconsin, USA
Chapter 19

Carmen L. Zayas
Department of Bacteriology
University of Wisconsin, Madison
Madison, Wisconsin, USA
Chapter 19

PREFACE

Excluding the biological polymers such as proteins, lipids and nucleic acids, modified tetrapyrroles are the biological molecules that have had the greatest impact on the evolution of life over the past 4 billion years. They are involved in a wide variety of fundamental processes that underpin central primary metabolism in all kingdoms of life, from photosynthesis to methanogenesis. Moreover, and as an added attraction to those of us who are fortunate enough to work in this research area, they bring colour into the world and it is for this reason that these compounds have been appropriately dubbed the 'pigments of life'. To understand how and why these molecules have been so universally integrated into the life processes one has to appreciate the chemical properties of the tetrapyrrole scaffold and, where appropriate, the chemical characteristics of the centrally chelated metal ion. In this book we have tried to address why these molecules are employed in nature, how they are made and what happens to them after they have finished their usefulness.

It was about five years ago that we organised a conference on tetrapyrroles where we wanted to integrate the synthesis of this remarkable collection of molecules with their biological function. At the time, several papers were reporting on the metabolism of modified tetrapyrroles when their function was over, and it was apparent that we had assembled speakers that covered the birth (biogenesis), life (function) and death (breakdown) of tetrapyrroles. We found the conference very stimulating not least because it covered a vast area of biochemistry but also because it integrated many seemingly disparate topics. We also felt it would make an interesting and comprehensive academic account. This book has been in the pipeline for a long time but the collection of chapters by the leading figures in the world of tetrapyrrole science has made the wait worthwhile.

Tetrapyrroles have been at the heart of so many fascinating discoveries in biology, chemistry and medicine. They are responsible for maintaining the ecological balance of oxygen in the atmosphere, for the production of a trillion tonnes of methane gas per year, for modulating some of the most complex chemistry observed in nature, and possibly even for turning America's last king mad. We hope that the following chapters will not only act as an important reference source of material but also highlight the multidisciplinarity of the topic, where the application of a range classical and biophysical techniques is required to make substantive progress in the area. We have enjoyed reading this collection of work and hope that it stimulates further research into the biochemistry of modified tetrapyrroles.

Martin J. Warren, PhD
Alison G. Smith, PhD

Acknowledgements

The editors would like to thank all those in the tetrapyrrole community who have made this such a vibrant and stimulating research area. We would also like to thank the Biochemical Society which originally sponsored a symposium on tetrapyrroles—their birth, life and death, which stimulated the composition of this book. Finally, we would like to thank the authors who contributed to this work for their time and patience—we got there in the end.

CHAPTER 1

An Historical Introduction to Porphyrin and Chlorophyll Synthesis

Michael R. Moore*

Historical Introduction to Porphyrins and Porphyrias

Introduction

Porphyrins are the extroverts of chemistry. Bright purple and fluorescent, they are used biologically in the processes of energy capture and utilization. Porphyrins are the key to life. It has been suggested that abiotic formation of porphyrins, in particular uroporphyrinogen would have provided the first pigments necessary for the eventual synthesis of the chlorophylls. This would have facilitated the emergence of simple photosynthetic organisms in primordial earth through enhanced efficiency of energy capture and utilisation (Fig. 1).

Communication

The development of the science of porphyrins and chlorophyll has progressed at a breakneck pace over the last century. A key to the development of medicine and science in the 20th century has been the improving levels of communication on an international basis. However, by the 1960s it was becoming increasingly easy for scientists to communicate both in person and electronically. At the same time many national groupings established between the middle and end of the 20th Century have grown and consolidated the interrelationships between scientists and physicians in the study of haem and chlorophyll synthesis. As an illustration of this continuing communication, Table 1 shows the sequence of meetings of the Tetrapyrrole Discussion Group, (TPDG) in the United Kingdom from its inception in 1977.

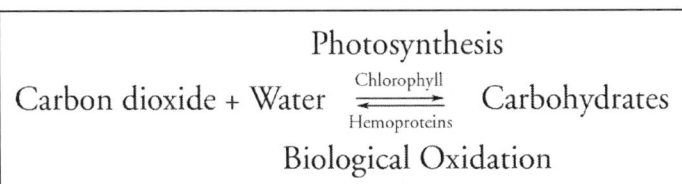

Figure 1. Carbohydrate synthesis and degradation using chlorophyll and hemoproteins.

*Michael R. Moore—National Research Centre for Environmental Toxicology, University of Queensland, 39 Kessels Road, Coopers Plains, Brisbane, Queensland 4108, Australia. Email: m.moore@uq.edu.au

Tetrapyrroles: Birth, Life and Death, edited by Martin J. Warren and Alison G. Smith.
©2009 Landes Bioscience and Springer Science+Business Media.

Table 1. The Tetrapyrrole Discussion Group (TPDG) a list of meetings since inception

Year	Date	Site	Organiser
1977	5 July	University College, Cardiff	George Elder/ Stan Brown, Francesco De Matteis
1978	7 April	Middlesex Hospital, London	Francesco De Matteis
1979	29-30 March	University of Leeds	Stan Brown
1980	3-4 January	University of Bristol	Trevor Griffiths, Owen Jones
1980	25-26 September	Western Infirmary, University of Glasgow	Michael Moore
1981	10 April	Queen Mary College, London	Ray Bonnett
1982	23 April	University of Southampton	Peter Jordan
1983	21-22 March	University College, Cardiff	Tony Jackson, George Elder
1984	6 January	Royal Free Hospital, London	Barbara Billing
1984	16-17 July	University of Leeds	Stan Brown
1985	19-20 September	University of Edinburgh	Jenny Houghton
1986	April	The Royal Society	Albert Neuberger
1987	20-23 July	University of Southampton Medical School '10th Anniversary Meeting'	Peter Jordan
1988	11-12 April	University of Leeds	Jenny Houghton
1989	4-7 April	University of Glasgow 'A Century of Porphyria'	Michael Moore
1989	December	The Royal Society	Francesco De Matteis
1990	December	Queen Mary and Westfield College, London	Ray Bonnett
1992	6-7 April	University of Southampton	Peter Jordan
1993	5-6 January	University of Cambridge	Alison Smith
1993	15-16 September	University of Leeds	Jenny Houghton
1994	11-12 July	Queen Mary and Westfield College	Ray Bonnett, MartinWarren
1995	10-12 April	University of Leicester	C-K Lim, Andy Smith
1996	11-12 January	University of Southampton	Peter Shoolingin-Jordan
1996	10-11 September	University of York	Timmins
1997	30 June-1 July	Queen Mary and Westfield College '20th Anniversary Meeting'	Ray Bonnett
1998	23-24 March	University of Cambridge	Alison Smith
1998	7 December	The Royal Society	Martin Warren
1999	1-2 September	University of Essex	Ross Boyle
2001	4-5 January	University of Leicester	Andy Smith
2001	13-14 September	University of Cardiff	Mike Badminton
2002	8-10 April	Heriott-Watt University	Martin Warren
2003	9-10 January	University of Leeds	David Vernon
2003	28-30 August	Lund University, Sweden	Lars Hederstedt, Mats Hansson
2004	16-17 September	University of Cambridge	Alison Smith
2005	1-2 July	University of Hull	Ross Boyle

The porphyrias provide examples of the derangement of the pathway that synthesizes these tetrapyrroles. From the outset, the name "porphyria" described not the diseases but the lustrous purple-red crystalline porphyrins:

πορφυρου

[porphuros or purple].

Again the importance of this topic is reflected in the number of international meetings organized in the last half-century (Table 2). Chapter 5 describes recent advances in this field.

Table 2. A chronology of international porphyrin and porphyria meetings post-1955

Year	Date	Site	Organiser(s)	Background
1955	8-10 Feb	London	Rimington	CIBA Foundation
1959	31 Aug-4 Sept	Canberra	Lemberg, Legge	IUB/AAS Hematin Enzymes
1963	18-24 Sept	Cape Town	Eales	
1968*	8 April	London	Rimington, Neuberger	UCH/ Biochem Soc Symp. 28.
1968*		Crystal Mt	Tschudy	1st Gordon Conference on
to		to	to	Tetrapyrroles GRCs every even
2006		Newport	Ann Smith	year thereafter
1970	2-6 Dec	Cape Town	Eales	
1970				GRC
1972	3-6 April	New York	Adler	NY Acad Sci
1972				GRC
1973*	28-30 June	Marburg	Doss	
1974				GRC
1975*	19-20 Feb	London	Neuberger, Kenner	Royal Society
1975*	1-4 May	Freiburg	Doss	
1975		New York	Adler	NY Acad Sci
1975*	17-19 July	Helsinki	Koskelo & Tenhunen	
1976*	July	Wolfeboro	GRC	
1977	June28-July 1	Marburg	Doss	400th Anniversary Phillips U.
1977*	Sept	Cardiff	Brown, De Matteis	1st Meeting of TPDG Elder/Moore
1978*	July	Wolfeboro	GRC	
1978		Kyoto	Sano	
1979*	29 Oct-1 Nov	Buenos Aires	Batlle	CIPYP
1980*	4-8 Aug	Wolfeboro	Troxler	GRC
1982	23-28 May	Tokyo	Mascaro	Dermatology
1982	1-6 Aug	Wolfeboro	Marks	GRC
1984*	5 Nov	Minneapolis	Pierach	Watson Symposium
1984*	6-10 Aug	New London	Smith (K)	GRC
1985	19-22 June	Paris	Nordmann	
1986*	28 July-1 Aug	Wolfeboro	Sassa	GRC
1986*	27-29 Oct	Rye Brook NY	Silbergeld	NY Acad Sci
1987*	12-13 Nov	London Ont	Haust	
1987*	7-10 Oct	Rome	Topi	
1988*	25-29 July	Wolfeboro	Correia	GRC
1989*	4-7 April	Glasgow	Moore	Century of Porphyria TPDG - Goldberg Festschrift
1990*	23-27 July	Wolfeboro	Beale	GRC
1992*	30 April-4 May	Papendal	Wilson, De Rooij	
1992	Kushner	GRC		
1992*	22-24 Oct	Göttingen	Seubert	Ippen Festschrift
1993*	16-17 March	Oslo	Moan	Rimington 90th Birthday
1993*	22-26 Nov	Melbourne	Blake	
1994	24-29 July	Wolfeboro	McDonagh	GRC
1995*	28 June-2 July	Helsinki	Tenhunen, Mustajoki	
1996*	February	Cape Town	Meissner, Kirsch	International association for the study of liver disease (IASLD)
1996*	14-19 July	Henniker	Bloomer	GRC
1998	12-17 July	Newport	Lagarias, Shoolingin-Jordan	GRC

Table continued on next page

Table 2. Continued

Year	Date	Site	Organiser(s)	Background
1999*	July	Hamburg	Martasek	
2000*	16-21 July	Newport	Ortiz de Montellano	GRC
2000*	10-13 Sept	Paris	Deybach festschrift	Millenium, Nordmann
2000		Dijon	Guillard	ISPP-1
2002	8-10 April	Edinburgh	Heathcote, Warren	Biochem Soc, TPDG
2002	June	Kyoto	Ogoshi	ISPP-2
2002*	14-19 July	Newport	Ferreira	GRC
2003	21-25 Sept	Prague	Martasek	
2004		New Orleans	Smith, Kadish, Sessler	ISPP-3
2004	July	USA	O'Brian	GRC
2005	Feb	Cape Town	Meissner	Kirsch festschrift
2006	July	USA	Smith (A)	GRC
2007	April	Rotterdam	De Rooij	Wilson Festschrift

GRC = Gordon Research Conference

Descriptions in Antiquity

Although the first verifiable cases of porphyria were identified in the middle of the 19th Century, some of the writings of the ancients seem to show clear descriptions of attacks of acute porphyria (Table 3). Hippocrates, 460 to 370BC, gives a splendid description of a woman from Thasus who suffered from many of the features that we now recognise to be associated with the acute attack of porphyria.

The allusions to porphyrins, around 1840, by Lecanu[1] and other workers, preceded the first clinical presentations of porphyria by Schultz[2] and Baumstark[3] by 30 years. The porphyrias belong to that larger group of diseases described by Garrod[4] in 1923 as "inborn errors of metabolism." They demonstrate a unique combination of neurological and dermatological features, which show characteristic variations from one condition to another, the reasons for which may be sought in enzymic change within the haem biosynthetic pathway. The history of various aspects of the porphyrias and porphyrin metabolism has been recorded by Florkin and Stotz[5] and Moore[6] (Table 3).

Structure

The basic porphyrin nucleus is a unique biological structure. It consists of a macrocycle of four pyrrole rings linked by four methene bridges. The normal biological intermediate is not this highly conjugated porphyrin, but the hexa-hydro porphyrin, the porphyrinogen in which each of the methene bridges is reduced.

An important feature of this complex ring structure is its metal-binding capability. The most commonly bound metals are iron and magnesium. In this form the metalloporphyrins reach their true apotheosis. Haem, an iron-containing complex usually bound to various proteins, is central to all biological oxidations. Haemoproteins are also used as oxygen carriers. The chlorophylls are the magnesium-porphyrin compounds, which are central in solar energy utilization in the biosphere.

As well as the systematic formation of porphyrins by biological systems, abiotic synthesis of porphyrins has been described in which a primitive chemical system has produced porphyrin-like compounds through the high entropy of their formation.[7] Such synthesis was important in the ontogenesis of terrestrial life, since it would have facilitated the emergence of life forms on primordial earth by increasing the efficiency of oxido-reductive processes as well as of energy capture.[8] Porphyrins are found in fossil life forms[9] and have even been identified in rocks from the moon.[10]

Table 3. Selected chronology of studies of porphyrins

Year	Researcher	Studies
460-370BC	Hippocrates	Described a case akin to menstrually-related acute porphyria
1831	Laennec	Anaemia of lead poisoning
1837	Lecanu	Acid extraction of hemin from blood
1841	Scherer	Hemin extraction—iron free
1844	Mülder	Purple-red fluid devoid of iron
1864	Hoppe-Seyler	Named hemoglobin
1867	Thudichum	Spectrum and red fluorescence of porphyrins
1871	Hoppe-Seyler	Prepared porphyrins from blood and showed they were pyrroles
1874	Schultz	Congenital porpyria—pemphigus leprosus)
	Baumstark	Urinary porphyrin pigments—an error of biosynthesis
1879	Hoppe-Syeler	Chlorophyll tetrapyrrole, phylloporphyrin, structurally similar to heme
1980	McMunn	Salicylate-induced porphyrinogenesis
1883	Soret	Light absorption around 400nm of hemin
	Krukenberg	Red fluorescence of eggshells
1884	McMunn	Discovery of cytochromes (myohematin)
	Brouwier	Chocolate-colored bones in abbatoir carcasses
1888	Nencki, Sieber	Hematoporphyrin dicarboxylic
1889	Stokvis	Sulfonal-induced acute porphyria
	McMunn	Porphyrins in echinoderms and the black slug
1890	Ranking, Pardington	Described the symptoms of acute porphyria
1891	Geill, Fehr	Induction of acute porphyria in women by sulfonal
	Copeman	Neurosis in female porphyrics
1892	Church	Turacin shown to be a porphyrin
1893	Barnes	Named photosynthesis (carbonic acid and chlorophyll)
1895	Stokvis	Animal porphyria
	Binnendijk	Excess porphyrins in urine of lead-poisoned patient
1896	Saillet	Coproporphyrin (urospectrine)
1903	Zaleski	Mesoporphyrin
	Erbslöh	Axonal degeneration in acute porphyria
1904	Laidlaw	Preparation of protoporphyrin
1906	Willstätter	Chlorophyll characterization commenced
	Dobrschansky	Barbiturate-induced porphyria
1908	Hausmann	Porphyrin photosensitization of mice
1911	Günther	First classification of porphyria
1912	Küster	Correct tetrapyrrolic structure of heme
1913	Meyer-Betz	Hematoporphyrin sensitization of skin
1915	Willstätter	Nobel prize for chlorophyll work
1923	Kämmerer	Porphyrin from putrefying meat—protoporphyrin
	Garrod	Porphyria—inborn error of metabolism
1924	Fischer and Kögl	Ooporphyrin (protoporphyrin) in eggshells
	Policard	Preferential uptake of hematoporphyrin by rat sarcoma
1925	Keilin	Rediscovers cytochromes
	Derrien and Turchini	Oviduct porphyrin deposition on eggshells
		Intense red fluorescence of harderian gland
1929	Fischer and Zeile	Hemin synthesis confirming Küster's structure
		Similarities of side chains of heme and chlorophyll

Table continued on next page

Table 3. Continued

Year	Researcher	Studies
1929	Borst and Königsdörffer	Autopsy study of Petry
1930	Fischer	Nobel prize for porphyrin work
1931	Sachs	Chromogen giving red coloration wth Ehrlich's reagent
	Körbler	Porphyrin uptake by tumors
1932	Grotepass	Urinary coproporphyrin III in lead poisoning
	Hijmans van den Bergh	Lead increase in erythrocyte protoporphyrin
1933	Watson	Stercobilins
	Dhéré et al	Studies of microbial porphyrins
1935	Franke, Fikenstscher	Doubling of urinary coproporphyrin excretion after ethanol
1936	Dobriner	First case of what would be called hereditary coproporphyria
	Rimington	Congenital porphyria in cattle
1937	Waldenström	Review, naming of porphyria cutanea tarda and classification of acute porphyria
	Turner	Porphyrins in fox squirrels
	Guildemeister	Albumin binding of porphyrins
1939	Rimington, Hemmings	Coproporphyrinuria following use of sulfanilamide
	Waldenström	Isolation of porphobilinogen
1941	Watson, Schwartz	Simple test for urinary porphobilinogen
1944	Strong	Harderian gland porphyrin and carcinoma in mice
	Clare and Stephens	Congenital porphyria in pigs
1945	Barnes	Variegate porphyria in south Africa
1946	Shemin, Rittenberg	Isotopic study of porphyrin biosynthesis
	Corwin, Erdman	IR investigation of NH tautomerism
1948	Folkers, Smith	Isolation and characterization of vitamin B_{12}
1949	Pauling, Itano	Sickle cell hemoglobin
	Lemberg, Legge	Description of biosynthetic sequence
1950	Neuberger et al, Shemin et al	Uroporphyrin first formed biosynthetically
	Rimington, Sveinsson	Spectral absorption of porphyrins
1951	Watson et al	Erythropoietic and hepatic classification of porphyria
	Shemin, Wittenberg	Requirement for succinyl-CoA
1952	Westall	Isolation of porphobilinogen
1953	Shemin, Russell, Neuberger, Scott	Role of ALA in biosynthesis
	Kosenow, Treibs	Erythropoietic protoporphyria
1955	Goldberg, Rimington	Allylisopropylacetamide-induced experimental porphyria
	Hodgkin et al	X-ray structure of vitamin B12
	Bogorad	Reduced porphyrins—porphyrinogens are biosynthetic intermediates
	Scott	Photosensitizing effect of ALA
	Shemin	The succinate-glycine cycle
	Berger, Goldberg	Named hereditary coproporphyria
	Tio	Porphyria cutanea tarda associated with adenoma
	Schwartz et al, Rassmussen-Taxdal et al	Use of hematoporphyrin derivative on cancer
1956	Goldberg et al	Ferrochelatase—the iron-incorporating enzyme
	Waldenström	Recognition that each porphyria can be localized to a defect in one enzyme

Table continued on next page

Table 3. Continued

Year	Researcher	Studies
1957	Cam	Turkish porphyria—hexachlorobenzene-induced
	Lascelles, Schulman, Richert	ALA synthase—requirement for pyridoxal phosphate
1957	Haeger	ALA in urine in lead poisoning
1958	Granick	ALA synthase the primary control point of heme biosynthesis
1959	Solomon, Figge	3,5-Dicarbethoxy-1,4-dihydrocollidine experimental porphyria
	Becker and Bradley	First application of NMR to porphyrin structure
1961	Calvin	Nobel prize for photosynthesis work
	Magnus et al, and Langhof et al	Named and studied erythropoietic protoporphyria
	Lipson et al	Use of purified hematoporphyrin derivative
	Gajdos et al	Role of ATP in control of porphyrin synthesis
1962	Perutz and Kendrew	Nobel prize for X-ray crystal structure of hemoglobin
1964	Illis	Porphyria and the etiology of werewolves
	Heide et al	Hemopexin, a heme-binding γ-globulin
	Dagg et al	Similarities between acute porphyria and lead poisoning
1965	Woodward	Nobel prize for total synthesis of chlorophyll and vitamin B12
1966	Macalpine and Hunter	"Royal malady"
1969	Pinol-Aguade et al	Hepatoerythropoietic porphyria
	Gajdos et al	Homozygous hereditary coproporphyria
	Romeo and Levin	Uro'gen III cosynthase and congential porphyria
1970	Strand et al	Acute intermittent porphyria related to deficiency of PBG deaminase
	Koskelo et al	Binding of ^{14}C-porphyrins by hemopexin and albumin
1971	Bonkowsky et al	Hematin use in acute porphyria
	Moore et al	Ethanol inhibition of ALA dehydratase
1973	Beale et al	ALA synthesis from glutamate, through DOVA
1974	Jackson et al	Protoporphyrinogen oxidation enzymically controlled
1975	Bonkowsky et al, Bottomley et al, De Goeij et al	Ferrochelatase and erythropoietic protoporphyria
	Beale et al	C_5 Pathway, glutamate $\rightarrow$ ALA
1976	Elder	Hydroxymethylbilane synthase as secondary control stage
	Wetterberg	International review of drugs in porphyria
	Brodie et al, Elder et al	Coproporphyrinogen oxidase and hereditary coproporphyria
	Kushner et al	Uro'gen decarboxylase and PCT
1977	Brodie et al	Control of acute porphyrias by PBG deaminase
1979	Jordan et al, Battersby et al	Mechanism of formation of uroporphyrinogen III from hydroxymethylbilane (preuroporphyrinogen)
	Bird et al	First description of plumboporphyria (ALA dehydratase deficiency)
	Moore et al	Acute and nonacute classification of porphyria
	Tephly et al	DDC forms *N*-alkylated porphyrins
1980	Meisler et al	Gene for porphobilinogen deaminase on chromosome 11
	Brenner and Bloomer	Protoporphyrinogen oxidase and variegate porphyria
1981	Mustajoki, Anderson et al	Immunological variants of PBG deaminase—CRIM positive and CRIM negative
1982	Day et al	Concurrent (dual) variegate porphyria and PCT
	Nordmann et al	Description of harderoporphyria

Table continued on next page

Table 3. Continued

Year	Researcher	Studies
1983	Eiberg et al	Gene for ALA dehydratase on chromosome 9
	Grandchamp et al	Gene for coproporphyrinogen oxidase on chromosome 3
	Batlle et al	Use of ALA dehydratase replacement therapy
1984	Kordic et al	Homozygous variegate porphyria
	Grandchamp et al	cDNA for PBG deaminase mRNA
	De Verneuil et al	Gene for uroporphyrinogen decarboxylase on chromosome 1
	Youngs	Chester Porphyria
1986	Deybach et al	Homozygous erythropoietic protoporphyria
1987	Jordan and Warren	Dipyrromethane co-factor for PBG Deaminase
	Siepker	FAD cofactor for protoporphyrinogen oxidase
1989	Riddle et al	Two forms of ALA Synthase
1990	Bishop et al	ALA-S2 and X-linked Sideroblastic anaemia—location Xp11.21
1991	Whitcombe et al	Ferrochelatase 18q22
1991	Astrin et al	Uroporphyrinogen co-synthase 10q25.3
1992	Cotteral	X-linked sideroblastic anaemia—due to ALA-S2
1993	Norton	Chester Porphyria on 11q
1994	Dailey et al	Two iron, two sulphur clusters in ferrochelatase
1994	Cacheux et al	Coproporphyrinogen oxidase 3q12
1995	Taketani et al and Roberts et al	Protoporphyrinogen oxidase 1q22-23
1996	Meissner et al	R59W mutation in protoporphyrinogen oxidase—South African variegate porphyria
1998	Röhl, Warren	"Purple Secret" further evaluation of the 'Royal Malady'

Chlorophyll

Prior to 1893, the process by which plants reduce carbon dioxide to organic matter was termed assimilation. However, in that year, Barnes[11] proposed that the biological process for the synthesis of complex carbon compounds in the presence of chlorophyll under the influence of light should be designated as photosynthax or photosynthesis. The history of this process is described by Gest.[12]

The naming of chlorophyll in 1818 was by Pelletier and Caventou[13] from the Greek

χλóροσ θύλλον

[Colour of Leaf]

On our planet chlorophyll synthesis is an extremely conspicuous process. Rüdiger[14] noted that the surge in chlorophyll synthesis in the spring on our planet is the clearest evidence of the presence of life on earth. Elucidation of the structure of chlorophyll by Fischer and Orth[15] occurred in 1940 and a verification of this structure by its total synthesis by Woodward[16] in 1960.

In chlorophyll the four pyrrole rings are linked into a tetrapyrrole with a magnesium atom at the centre of the structure. To form the first of these pyrrole rings there is an alternative route for the synthesis of 5-aminolaevulinate, which starts from glutamate through a three-step pathway, which is called the C5 pathway; in contrast to the synthesis from glycine and succinyl-CoA. This was demonstrated by Beale et al.[17] The overall mechanisms are described by von Wettstein et al.[18]

In the subsequent stages of the pathway, living systems have developed two ways of transforming uroporphyrinogen III into metal complex macrocycles. One is oxidative and leads to the chlorophylls and to haem, which the obligatory acceptor for the enzymic oxidation is oxygen. The second method is nonoxidative and makes use of methylation. This pathway is found in anaerobic organisms and for compounds like cyanocobalamin develops a macrocycle, which is structured to hold the cobalt ion.[19] Recent work on the biosynthetic pathways of haem, chlorophylls and corrins are described in other chapters in this book.

The Early Chemical Era

Early Studies

The history of the porphyrins begins with the work of Lecanu, Berzelius,[20] Scherer,[21] and Mülder.[22] Scherer added concentrated sulfuric acid to dried and powdered blood and washed the precipitate free of iron. He thus had shown that the red coloration of blood was not due to iron. In Mülder's study, he described a "purple-red fluid" without any iron, which he named "Eisenfreises hämatin" (iron-free haematin). This red substance was called "cruentine" by Thudichum[23] in his report to the Privy Council of Great Britain in 1867. He defined its spectrum, and noted that, "it fluoresced with a splendid blood-red color."

Contemporaneously, the first tentative efforts were being made to understand the part played by the tetrapyrroles in living organisms. In 1871, Hoppe-Seyler[24] found that "iron-free haematin" was a mixture of two substances, the main constituent of which he called "hämatoporphryin." Three years later, Schultz[2] published the clinical details of a case of so-called "Pemphigus Leprosus" for his doctoral thesis. The patient was a 33-year old weaver who had suffered from skin photosensitivity from the age of 3 months. His spleen was enlarged and he passed a wine-red urine; this urine was investigated by Baumstark[3] who named two pigments derived from it—"urorubrohaematin" and "urofuscohaematin." The importance of his observations was in his interpretation that the source of the porphyrin pigments was from an error of biosynthesis. The case was a description of congenital porphyria. This was the first association of this class of pigments in urine with a disease in humans. The autopsy records intense red-brown discoloration of the skeleton, a feature of this disease in both animals and humans.

In 1880, MacMunn,[25] who later discovered the cytochromes in 1884, described a dark pigment excreted in the urine of a patient who had been taking sodium salicylate. MacMunn called this pigment "urohaematin," but later, he renamed it "urohaematoporphyrin" because "it bears a very striking resemblance to haematoporphyrin." Hoppe-Seyler[26] studied the porphyrin in chlorophyll and rediscovered the property of red fluorescence first seen by Thudichum. He named it "phylloporphyrin." The term "porphyrin" was used by others such as Church[27] in his description of the porphyrin from Turaco feathers (turacin). Finally, the major spectroscopic feature of porphyrins, the strong absorptions lying around 400 nm was described for haemoglobin in 1883 by Soret.[28] This absorption band for porphyrins is still called the Soret band.

Biochemical Developments

A true biology of the porphyrins could not at that time be formulated because the exact route to their biosynthesis, together with their relationship within intermediary metabolism, were lacking. Many erroneous views were propounded during this period concerning the origins and interrelationships of the porphyrins, views, which obfuscated and retarded progress in the biological sphere. It was thought, for example, that porphyrins arose as degradation products of haemoproteins, such as haemoglobin, by removal of iron and the protein moiety and that protoporphyrin, so formed, was "detoxicated" by progressive carboxylation providing, in uroporphyrin, a more hydrophylic molecule for urinary excretion. These mistaken assumptions inevitably confused any understanding of the porphyrias, although Günther's[29] clinical classification in 1911 helped to resolve some of the misunderstanding. One may also admire Baumstark's foresight in 1874. He believed the urinary pigments of Schultz's 1874 case arose by an error in the biosynthesis of haemoglobin and not by any fault in its degradation, a concept that was not confirmed for another 50 years.

The chemical excreted in the porphyrias remained a matter of debate. The urinary pigment had been thought to be haematoporphyrin.[30-32] Garrod[33] showed that the absorption spectra of the urinary porphyrins were being masked by other chromophores in the urine, but it was not until 1915 that Fischer[34] showed that "urineporphyrin" was quite discrete from "haematoporphyrin." Nencki and his coworker, Sieber[35] contributed to the knowledge of the time by showing that haematoporphyrin was a dicarboxylic porphyrin. Saillet[36] prepared "urospectrine" in 1896 from urine which was subsequently named "coproporphyrin"[37] and also showed the presence of this compound in urine as a colorless chromogen (probably "coproporphyrinogen"). "Protoporphyrin" was also prepared unknowingly at this time by Laidlaw.[38] The correct structure of haem was first proposed by Küster[39] in 1912 but subsequently rejected by other workers, of greater stature, such as Willstätter and Fischer. Following

Figure 2. Richard Willstätter—Nobel laureate in   Figure 3. Hans Fischer Nobel laureate in 1930
chemistry 1915 (© The Nobel Foundation).   (© The Nobel Foundation).

the separation studies of Willstätter and coworkers[40,41] for which he was awarded a Nobel Prize (Fig.
2), Fischer began a series of studies, which continued for 30 years until his death in 1945. During this
time, he was awarded the Nobel prize for chemistry in 1930 (Fig. 3).

Mathias Petry and Hans Fischer

Mathias Petry was one of Günther's cases of congenital haematoporphyria. Petry became both
laboratory aide and source of porphyrins for Fischer. Fischer worked with him until Petry's death in
January 1925 when Fischer undertook a chemicopathological autopsy, which he published under
the name of "Porphyrinurie".[42] Borst and Königsdörffer[43] published the extensive pathological
autopsy. Laidlaw,[38] Fischer,[34] and Schumm[44] differentiated the naturally occurring porphyrin of
haem itself from haematoporphyrin and the name "protoporphyrin" was suggested for this substance by Fischer[45]

The Biochemical Descriptive Era

It was during the 1930s that the next generation started their work. During this time, some key
names and prominent contributions emerged: the discovery of Ehrlich's positive chromogen in the
urine of patients with acute porphyria in attack by Sachs.[46] Waldenstrom studied Sachs' Ehrlich's
positive chromogen, which, together with Vahlquist, he named "porphobilinogen"[47] in 1939

Haem Biosynthesis

The next milestone in the development of this subject rested upon the emergence of a systematic
biochemical description of the pathway of haem biosynthesis. The names most commonly associated with this are Shemin and Neuberger (Figs. 4, 5). Their work encompassed the early description
of how "5N-glycine was incorporated into haem by humans and animals.[48,49] This led first, to the
realization that ALA was a precursor of porphyrins[50,51] and at the same time, that the monopyrrole,
porphobilinogen, was, indeed, the precursor of uro-, copro-, protoporphyrin, and haem.[52]

Figure 4. David Shemin.   Figure 5. Albert Neuberger.

A "pyrrolic intermediate" was postulated as a vital step in the biosynthesis of the tetrapyrrole ring. The decisive contribution came in 1952 when the substance called porphobilinogen, excreted by patients suffering from acute intermittent porphyria, was shown to be a monopyrrole.[53,54] It was shown to give rise, enzymatically, to uroporphyrinogen when incubated with a haemolysate of chicken red cells.[52]

Early Description of Porphyria

After sulfonal was introduced as a hypnotic by Kast[55] in 1888, Stokvis[56] reported that an elderly woman who had taken sulfonal excreted dark-red urine and later died. He considered that the pigment in this urine, producing coloration resembling port wine, was similar to, but not identical with, haematoporphyrin. Harley[57] (1890) reported a fatal case of an unusual form of nervous disturbance associated with dark-red urine in a 27 year-old woman who had been given sulfonal and presented many of the neurological features of porphyria. Ranking and Pardington[58] (1890) described two women who excreted "haematoporphyrin" and who exhibited the gastrointestinal and neuropsychiatric manifestations of acute intermittent porphyria. The terms "porphyria" and "porphyrinuria" emerged gradually and slowly replaced "haematoporphyria" and "haematoporphyrinuria." Sometimes sulfonal or the allied drugs, tetronal and trional, had been taken for variable periods prior to the onset of symptoms. Other cases had no obvious relationship to drugs and, presumably, were precipitated by other unknown causes. When barbiturates were introduced into medicine in 1903, it was not long until a report of an acute porphyric attack, precipitated by diethyl barbituric acid, was reported.[59]

Drugs and Chemical Porphyria

The earliest evidence that we have for the drug-related development of porphyria occurred when the drug sulfonal induced the first published example of an attack of acute porphyria by Stokvis.[56] The ability of certain compounds to increase porphyrin synthesis in experimental animals has been used as a means of examining the processes of haem biosynthesis.[60] Experimentation with animals added much to this area of knowledge but confusingly, some drugs produced different patterns of porphyrin excretion in different species due to differences in response and susceptibility. Some investigators maintained that drugs merely precipitated attacks in patients already suffering from genetic acute porphyria, whilst others disputed this.

Allyl Compounds

Schmid and Schwartz[61] (1952) found that the hypnotic compound Sedormid (allylisopropyl acetylurea) could produce a porphyria in laboratory rodents which was akin to acute intermittent porphyria in humans. Allylisopropylacetamide (AIA), a structural analogue of Sedormid was soon found to be equally effective porphyrinogenic agents[62] (Fig. 6).

DDC

A chance observation by Solomon and Figge[63] (1959) revealed that the substituted dihydropyridine, 3,5-diethoxycarbonyl-1,4 dihydro collidine (DDC) caused an experimental porphyria. Tephly et al (1981),[64] De Matteis and Marks (1983)[65] and others studied the chemical events following experimental administration of the porphyrinogenic drug, DDC. It has been shown that the N-alkylated protoporphyrins are the immediate inhibitors of ferrochelatase and that they originate from the haem of the cytochrome P450, DDC being the methyl donor.

Fungicides

Hexachlorobenzene, has also been shown to cause experimental porphyria in humans. An outbreak of cutaneous porphyria in Turkey[66] was related to the ingestion of seed-wheat treated with hexachlorobenzene as a fungicide, which initiated the work, which continues to the present into the effects of polyhalogenated hydrocarbons on haem biosynthesis.[67,68] Another therapeutic fungicide, griseofulvin, was also shown to be porphyrinogenic.[69]

The means by which each of these compounds, AIA, DDC, and hexachlorobenzene, influence the biosynthetic pathway are of interest since they provide a useful pointer to the diversity of influence of chemical compounds upon the processes of control of haem synthesis and provide the rationale for development of the lists of drugs contraindicated for use in acute porphyria.

Classification of Porphyrias

In his papers of 1911 and 1922, Günther[29,70] was the first to classify the diseases of porphyrin metabolism. In the first of these he quoted 14 cases from the literature in which acute symptoms of porphyria arose spontaneously, "haematoporphyria acuta," and 56 cases of "haematoporphyria acuta toxica," in which the symptoms were associated with the ingestion of sulfonal, trional, or veronal. He also defined and named, for the first time, the very rare condition, congenital porphyria, "haematoporphyria congenital" in which the predominating symptoms were due to skin photosensitivity. In Glasgow in 1898, McCall-Anderson[71] had described two brothers, both of whom had solar sensitivity and excreted "haematoporphyrin" in the urine. Meyer-Betz[72] was the first of Fischer's coworkers. He injected 200 mg of haematoporphyrin into his own veins in 1913. Knowing the marked photodynamic effect of this substance on mice, paramecia, and erythrocytes, from the studies of Hausmann,[73,74] he intended to remain indoors and expose himself very cautiously to light. However, he was a practicing physician and soon after the injection he received an urgent message from a sick patient and he felt obligated to make a house call. It was a sunny day and though he strove to avoid direct exposure to the sun, he was unsuccessful. Shortly thereafter, he developed an extreme solar urticaria of the hands and face. This experiment revealed the potent photosensitizing influence of this particular porphyrin.

In Günther's second work in 1922,[70] he elaborated on his first thesis, quoting further cases. He noted the possibility that acute haematoporphyria might be hereditary and suggested that people liable to develop acute or congenital haematoporphyria had a "porphyrism" with certain notable physical and mental characteristics—neurosis, insomnia, dark hair, and pigmented skin. In a survey of the clinical features of acute haematoporphyria he described a triad of symptoms which were commonly present, namely abdominal pain, constipation, and vomiting.

Figure 6. Sam Schwartz (courtesy of Studio Minne-apolis).

Figure 7. Cecil Watson (courtesy of Dublin Pro-ductions).

Hepatic and Erythropoietic Classification

Waldenström (1937)[75] made a clinical survey of 103 cases of acute porphyria found in Sweden. He reviewed some previously published cases of chronic haematoporphyria (Günther's classification) in which light sensitivity occurred some years after birth, at times associated with abdominal pain. For these cases, he substituted the name "porphyria cutanea tarda". This classification was further extended by Watson[76] in 1960, Goldberg and Rimington in 1962[62] and Eales[77] in 1979 (Figs. 7-9).

Classification into Acute and Nonacute Porphyria

The general aim in most classifications was to provide a clinical basis consistent with the known biochemical features. The full elucidation had however, to await the complete description of each of these diseases as a specific enzymic disorder of the haem biosynthetic pathway, a process that evolved over the years between 1955 and 1980.

The clinical manifestations of the porphyrias vary enormously. The traditional classification of the diseases as either hepatic or erythropoietic, dependent upon the primary site (or what was thought to be the primary site) of overproduction of the porphyrins, is inadequate. For that reason the classification into the acute and nonacute types of porphyria based upon the main clinical presentation, offers a more satisfactory means of subdivision of this group of diseases.

The major feature of these diseases is that they may be provoked into an acute attack with a neuropsychiatric or neurovisceral syndrome associated with increases in production and urinary excretion of the porphyrin precursors ALA and PBG. The reasons for these changes have been sought in a number of theories which presently have settled upon an excess of ALA, a deficiency of haem, or a combination of these two.[78] The neuropsychiatric or neurovisceral syndrome is not present in the nonacute porphyrias, which consist of congenital erythropoietic porphyria, erythropoietic protoporphyria, and PCT. The primary presenting feature in the nonacute porphyrias is skin photosensitivity.

Figure 8. Abe Goldberg.   Figure 9. Len Eales.

Enzymes

With the description of the pathway sequence, the time was ripe for the elucidation of the catalytic steps—the enzymic description of the biosynthetic pathway.

Thereafter, each of the stages of the pathway was exhaustively examined. The most important point in this sequence is the first one, the formation of ALA by ALA synthase. Granick provided the evidence from his studies that this was indeed the control point of the biosynthetic pathway[79] (Fig. 10). The last enzyme of the sequence to be described was protoporphyrinogen oxidase.[80] It was clear however, that additional control points would have to be sought in the biosynthetic sequence. One such enzymic control point was demonstrated at porphobilinogen deaminase (PBG-D).[67,81] It was at this site in the pathway that the formation of uroporphyrinogen III was eventually elucidated by Jordan[82] and Battersby et al.[83]

Enzymic Classification

The most important feature of current levels of understanding of the porphyrias is that the metabolic disorder can, in all cases, be localized to one specific enzyme within the haem biosynthetic pathway.[81,84]

Thus, in the acute porphyrias the expression is largely hepatic. In acute intermittent porphyria, the defect has been shown to lie at the level of porphobilinogen deaminase (EC 4.3.1.8). In hereditary coproporphyria the defect lies at the level of coproporphyrinogen oxidase (EC 1.3.3.3), in variegate porphyria at protoporphyrinogen oxidase (EC 1.3.3.4), and in the exceptionally rare "plumboporphyria" at ALA dehydratase (EC 4.2.1.24). The resultant overproduction of coproporphyrinogen and protoporphyrin respectively in these diseases can account for their photocutaneous manifestations.

In the nonacute porphyrias, the expression is both hepatic and erythropoietic. The deficient enzymes are in congenital erythropoietic porphyria, uroporphyrinogen cosynthetase (EC 4.2.1.75), erythropoietic protoporphyria, ferrochelatase (EC 4.99.1.1); and PCT, uroporphyrinogen decarboxylase (EC 4.1.1.37). A classification based upon these enzymic lesions may have merit but

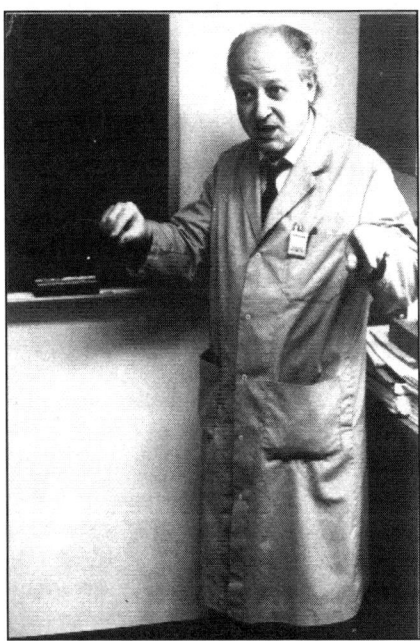

Figure 10. Sam Granick (courtesy of S. Sassa).

Figure 11. Claude Rimington (courtesy of Greta Rimington).

there are logistical problems in the application of such a classification, not the least of which is the difficulty in reproducible measurement of enzyme activities and the nonavailability of biopsy material upon which the estimations might be carried out.

Molecular Genetics

Molecular genetics have allowed a more fundamental recognition of the nature of the genetic defects in these diseases. The early concept that was shown to be true for *Staphylococcus aureus* was that the genes would all lie on one chromosome and be cotransducible; that is unfortunately not so. In humans the early investigations in this sphere have shown clearly that they are not clustered in the human genome, but dispersed among different chromosomes. From the work thus established, cDNA probes have been synthesized for these proteins, which are facilitating further investigation of the molecular defect and of the familial association of these diseases.

Acute Porphyria

Waldenström's studies in Sweden were greatly aided by the presence in his patients' urines of porphobilinogen, which forms a red color with Ehrlich's aldehyde reagent (an acidic solution of paradimethyl-amino-benzaldehyde). As early as 1890, Harley had noted that the urine of a case of sulfonal-induced acute porphyria contained a chromogen which, when oxidized, became a red pigment. Such a substance in the urine of a patient with acute porphyria gave a red coloration insoluble in chloroform, with Ehrlich's aldehyde reagent, and which was therefore not urobilinogen (Sachs, 1931). Waldenström (1937) showed that this was not only excreted in the urine of every one of his patients with acute porphyria, but that some apparently healthy relatives of these patients also excreted it. The idea of a "latent porphyria" was thus conceived. Later, Waldenström and Vahlquist (1939) considered that the Ehrlich-reacting chromogen, which they named porphobilinogen and partly purified, was a dipyrromethane.

The liver and kidney of fatal cases of acute porphyria contained porphobilinogen[85,86] which was isolated from the urine of a patient with acute porphyria in University College Hospital in London,[53] Cookson and Rimington[54] showed it to be a monopyrrole and Haeger[87] found that two-thirds of patients with latent or manifest acute intermittent porphyria excreted excess 5-aminolevulinic acid (ALA) in addition to increased porphobilinogen (Fig. 11).

Concurrent Porphyria

Concurrent porphyria has been defined as two differing types of porphyria occurring in the same individual.[88] A large kindred in Chester have been described, with excretion and enzymic patterns of both acute intermittent porphyria and variegate porphyria (both PBG-deaminase and protoporphyrinogen oxidase with low activity). In another study, Day et al[89] reported 25 patients with a variant of variegate porphyria combined with porphyria cutanea tarda which they called "dual porphyria." Studies performed in six of these dual porphyria patients by Meissner et al[90] showed that erythrocyte uroporphyrinogen decarboxylase activity was reduced as was protoporphyrinogen oxidase activity. Such parallel inheritance need not imply that there is further genetic disadvantage since the effects of multiple enzyme inhibition in the pathway are not necessarily additive, as is shown by the example of lead poisoning.[91] This is not the case for homozygous porphyrias in which the dermatological and other features are usually severe.[92]

Neuropathy

The acute porphyrias have in common the features of acute abdominal pain, limb weakness, and neuropsychiatric presentation. This symptomatology can be explained on a neurogenic basis. Despite advances in genetics and biochemistry, the link between these biochemical abnormalities and the neuropathological and clinical manifestations remains unclear. The earliest case of acute porphyria examined pathologically was that of Campbell[93] in 1898 who failed to find any nervous system abnormality. However, in 1903, Erbslöh[94] described features of axonal demyelination in the femoral nerve from a porphyric patient who had died after treatment with sulfonal. Mason[95] and coworkers (1933) showed that the most characteristic lesions were seen in the nervous system and affected peripheral nerves and sympathetic ganglia.

Psychiatric Aspects

The acute attack may reveal a variety of psychiatric manifestations, including anxiety, depression, and frank psychosis. The literature on the psychiatric features is sparse. The earliest cases were described by Copeman[96] in 1891 and by Campbell in 1898. It is clear that a frequent misdiagnosis of hysteria is made, although Brugsch[97] recognized it as a distinct psychosis in his review of the literature. The available data show that the psychopathology is related to affective neurotic, rather than psychotic, features and a truly schizophreniform presentation has not been observed.[84] Tishler[98] and coworkers in 1985 in 3867 psychiatric inpatients found that 0.2% (8 patients) who experienced episodic psychosis and depression had confirmed acute intermittent porphyria. In general terms, the psychiatric phenomena may be expected in up to 70% of acute attacks.

Monopyrroles in Porphyria and Other Disorders

It is intriguing that disturbances of porphyrin metabolism have been observed in schizophrenics and that excesses of a monopyrrole-hydroxyhaemopyrrole lactam are present in urine of patients with acute porphyria and some psychiatric disorders.[95,96,97,99,100,101] Prior to its discovery in the urine of porphyric patients an association between "mauve factor" excretion and psychiatric illness had been made. Typically the factor was shown to be present in up to 50% of endogenous psychoses and 70% of acute schizophrenia.

Mauve factor was identified using a chromatographic method that suggested it was an alkyl pyrrole. Mass spectrometric analysis of a single component showed a molecular ion that was suggestive of a mono-ethyl dimethyl pyrrole that led to the hypothesis that the "mauve factor" was essentially due to kryptopyrrole. Subsequent studies by Graham et al showed the identification to be incorrect and in fact due to the heterologous pyrrole, haemopyrrole in its lactam form.[101,102]

Nonacute Porphyrias

Congenital Porphyria

Congenital porphyria is one of the rarest of the porphyrias, but was the first recorded case of porphyria in the literature, probably because of the severity and dramatic nature of its symptoms. The case was described by Schultz in 1874. His patient was a 33 year-old weaver who had suffered from photosensitivity of the skin from the age of three months. His urine was wine red and contained pigments which resembled Hoppe-Seyler's acid haematoporphyrin. At autopsy, his bones were found to be dark brown in color. Further case reports were made by other authors.[71,103,104] Günther[29] named the disease haematoporphyria congenital in 1911.

In subsequent autopsies of two cases of congenital porphyria, the porphyrins were found to be concentrated in the bone marrow rather than in the liver. Congenital porphyria was therefore renamed "porphyria erthyropoietica". In subsequent patients examined two different types of normoblast were found, an abnormal type which exhibited marked porphyrin fluorescence, apparently in the nuclei (fluorocytes), and other normal normoblasts without such fluorescence. It was then thought that these abnormal normoblasts which carried the trait representing the inborn error of metabolism.[105]

Porphyria Cutanea Tarda

Porphyria cutanea tarda (PCT) is the most common form of porphyria. Patients present with cutaneous photosensitivity but do not experience attacks of neurovisceral dysfunction. PCT differs from the other porphyrias in that there is no clear pattern of inheritance in the majority of cases although in some, familial transmission can clearly be established. Waldenström[106] introduced the term "porphyria cutanea tarda" (PCT) in 1957.

PCT is usually classified into two main types based on the relative importance of inherited and acquired factors.[107] The majority of patients may be classified as **sporadic**. These patients have no family history of the disorder and its development appears largely related to chronic alcohol ingestion or the use of the contraceptive pill. Erythrocyte URO-D activity is normal in these patients although hepatic URO-D activity is depressed. In the sporadic or toxic disease polyhalogenated hydrocarbons have been shown to be the etiological agent in some cases. The most common precipitant is however, alcohol ingestion, which is known to disturb porphyrin metabolism in normal subjects.[108] Orten et al[109] studied the urinary excretion of porphyrins and precursors in chronic alcoholics and noted increased coproporphyrin excretion but no significant increase in excretion of uroporphyrin, ALA, or PBG.

Toxic Porphyria

If one excludes the porphyria caused by sulfonal, toxic porphyria in humans only became clearly established after 1957 when there was a disastrous outbreak of porphyria among Turkish peasants who had inadvertently ingested the fungicide, hexachlorobenzene, with their wheat and bread.[66,110] In this Turkish population, Günther's postulate of "toxic porphyria" in his early classification was fully substantiated.

Hepatoerythropoietic Porphyria

Hepatoerythropoietic porphyria is a homozygous form of familial PCT first reported by Pinol-Aguade[111] in 1969. This topic is well reviewed by Smith (1986).[112] All have exhibited severe mutilating photosensitivity from birth. In addition, there are hepatic changes and there is usually a mild normochromic normocytic anemia. URO-D activity examined in erythrocytes and fibroblasts is markedly reduced to less than 10% of normal.

All the above forms of PCT present with a similar clinical picture of cutaneous photosensitivity and an identical porphyrin excretion pattern due to reduced hepatic URO-D activity. However the mechanism of the enzyme defect appears to vary, genetic factors being most important in the familial type and acquired factors in the sporadic or toxic type.

Erythropoietic Protoporphyria

Erythropoietic protoporphyria is principally characterized by acute solar photosensitivity. The nature of the disease was first clearly established in 1961 by Magnus et al[113] and Langhof et al[114] although there had been earlier reports.[115] Despite its late description this is a relatively common condition although its prevalence has not been precisely calculated. Onset may occasionally be much later and not necessarily include many of the described features of this disease.[116]

Porphyria in Animals

Hepatic Porphyria

Since acute intermittent porphyria is the most common of the porphyrias affecting humans, one might expect that cases of this condition would have been encountered in animals but, up to the present, no clear example of acute intermittent porphyria in animals has been reported. This might be explained by the difficulties in diagnosis of the disease, should it occur in animals.

Congenital Porphyria in Cattle

Several reports described the finding of chocolate or brown-colored bones in abattoir carcasses. The first such report was by Brouwier[117] in 1884, followed a year later by one from Tappeiner[118] of red-brown bones in swine. Mosselman and Hebrant[119] were convinced that it was formed from haemoglobin, while Ingier[120] thought it to be melanin associated with another pigment derived from chlorophyll. The similarity between this condition in animals and human congenital porphyria was recognized by Schmey,[121] who proposed for it the name "osteohaemochromatosis" instead of the misleading "ochronosis" with which it had previously been known. That the pigment in the bone was a porphyrin seems to have been first clearly realized by Möller-Sorensen[122] in 1920. Credit for the suggestion of its hereditary nature should probably go to Witte.[123]

Studies in South Africa

In 1936, a herd of grade short horn cattle was discovered in South Africa in which no fewer than 13 cases of congenital porphyria were seen.[124] Excessive quantities of uroporphyrin I and coproporphyrin I were found in the organs and body fluids and there was definite evidence of photosensitization. The urine was wine red in color and was rich in uroporphyrin and coproporphyrin; the faeces, erythrocytes, and plasma were also rich in coproporphyrin I. The entire skeleton was deep brown in color and afforded large quantities of uroporphyrin I.[62] The inheritance was relatively easy to trace as autosomal recessive.

Porphyria in Pigs

As with cattle, dark-colored bones had been noticed in pig carcasses long before living cases of congenital porphyria were diagnosed.[118,120,121,125] The first living cases were observed in New Zealand and reported by Clare and Stephens.[126] In Denmark, what appeared to be congenital porphyria in pigs appeared in the Thisted district during 1951 and 1954.[127] Affected animals had discolored teeth displaying pinkish-red fluorescence in ultraviolet light. A closer study of the teeth showed that the porphyrin was not evenly distributed, but was mainly located in the dentine layer just below the enamel.[128]

Cats and Dogs

In 1964, a report appeared of a young kitten whose deciduous teeth were brownish with red fluorescence in ultraviolet light and whose urine had been blood-colored since the cat was two months old; it was otherwise normal. When the permanent teeth erupted they were lighter in color and devoid of fluorescence. A littermate and some kittens from a former litter of the mother also had discolored teeth, suggesting a dominant inheritance.[129] The absence of skin photosensitivity, dominant inheritance, and decrease of porphyrin deposition with age is similar to the porphyria found in pigs.

In studies in Cape Town it has been shown that many dogs have abnormally high excretion of coproporphyrin in urine, which may relate to their carnivorous, rather than omnivorous, eating patterns. Owen[130] et al (1962) described a young dog with permanent teeth showing a transient pink color in ultraviolet light, but both disappeared after several months.

Porphyrin Synthesis in the Animal Kingdom

Free porphyrins in varying amounts are found in widely, although somewhat erratically, distributed in most living organisms. Microbial porphyrins represent a useful source of material for the examination of porphyrin synthesis,[131] for example *Rhodobacter sphaeroides* was used by Lascelles[132] in her studies of the control of porphyrin synthesis.

Among the mammals, a few genera appear to produce much more porphyrin than others; these belong to the family of rodents. The rat produces a relatively large quantity of protoporphyrin, by synthesis in the harderian glands. Squirrels also produce much porphyrin and their bones have a pale-brown color due to deposition of uroporphyrin and fluoresce pale red in ultraviolet light. This is most marked in the American fox squirrel, *Sciurus niger*.[133] Only *Tamias striatus* has this in common with *S. niger*[134] Levin and Flyger[135] found reduced activity of the enzyme uroporphyrinogen III cosynthetase in haemolysates and tissue extracts from fox squirrels as compared with grey squirrels. The animals appear in every way normal without untoward symptoms accompanying their high porphyrin production and must therefore, be regarded as physiological examples of excess porphyrin synthesis. Analogous to this are the molluscs that deposit quantities of uroporphyrin in their shells,[136] or among other higher forms of animal life, the group of birds known as the touracos, or plantain eaters, who utilize the copper complex of uroporphyrin III, called turacin, for the deep-red areas of pigmentation in their flight feathers.[27,137,138] Porphyrins are used in the coloration of eggs,[139] probably by deposition within the oviduct.[140]

The Harderian Gland

Johann Jakob Harder (1656-1711) found and named the retro-orbital gland in the deer, which he named glandula nova lachrymalis in his publication of 1694.[141] Since then, many investigators have found that this gland is present not only in some mammals but also in amphibians, reptiles and birds. The precise purpose of the gland remains an enigma and has probably not been studied in greater depth because of its absence in primates.[142]

The gland has a remarkable capacity to synthesis porphyrins.[140] In particular, the presence of many of the porphyrins of the latter part of the biosynthetic pathway are found in the gland in rodents in such quantities that intermediates such as tri-carboxylic haemoporphyrin can be isolated and characterised in the gland.[143,144,145] Numerous studies have shown that porphyrin production in the gland is profoundly influenced by the presence of steroids.[146,147]

The harderian gland in the golden hamster is an extremely rich source of porphyrins.[148] It has been extensively studied in this rodent. Although the female gland is arguably the richest natural source of porphyrins known, the male gland contains little porphyrin, possibly because the rate-limiting enzyme, 5-aminolevulinic acid (ALA) synthase is more active in the female than the male.[149] All of these, including the role of hormone replacement,[144] emphasize the role that steroids play in porphyrin synthesis and the activities of the enzymes of the biosynthetic pathway.

Phototherapy and Cancer

The early foundation of photochemotherapy may be sought in the work of the 1903 Nobel laureate Neils Finsen. Thereafter, Hausmann in 1908 and 1911[73,74] showed that haematoporphyrin photosensitized both paramecia and mice. Normal human tissue also reacts in a pathological fashion when saturated with porphyrins and exposed to light, as was so clearly demonstrated by Meyer-Betz in 1913.[72] It is a simple step to conclude that this photoreaction might be usefully employed in the destruction of pathological tissue. This was aided by the early description in 1924 by Policard[150] of the preferential uptake by rat sarcoma of haematoporphyrin. That injected porphyrins accumulated in tumors were observed by Körbler.[151] The work in 1955 of both Schwartz[152] and coworkers and Rassmussen-Taxdal et al[153] is of interest since it pointed the way for future work in humans.

Haematoporphyrin Derivative

The foundation of the present use of haematoporphyrin derivative is based on the work of Lipson et al[154] who showed that some components of haematoporphyrin derivative (HPD) were better localized in malignant tissue than "crude" haematoporphyrin and in the development of lasers. Such chemical treatment of the haematoporphyrin produces a complex mixture of substances, such as mono- and diacetates of haematoporphyrin, protoporphyrin, deuteroporphyrin, and, in particular, a dihaematoporphyrin ether or ester.[155,156,157]

5 Aminolevulinate

In the second half of the twentieth century it became obvious that 5 Aminolevulinate (ALA) could be converted to porphyrins. In his early study in 1955 Scott[158] showed that, when given to humans, a dose of ALA gave rise to a transient but marked photosensitivity with erythema of the exposed skin resembling mild sunburn reaching a maximum between 9 and 12 hours. The transient nature of the ALA induced photoreaction meant that it too was considered as a potential candidate for use in phototherapy. By bypassing the rate limiting stage in haem biosythesis, through the use of ALA with consequent overproduction of porphyrins 'in situ', would allow such tissues to become photosensitized and subject to photodestruction, given the appropriate dose of photoactivating light. As a consequence of this, ALA based photodynamic therapy has taken its place in the treatment of cancer with considerable successes in the succeeding years.[159,160]

Mechanism of Effect

The fundamental principle behind this therapy is that since tumor tissue will preferentially accumulate porphyrins, these may be used for both identification and therapy of neoplastic tissue. Unfortunately, other metabolically active tissue, such as liver and kidney, also accumulate porphyrins, but being remote from light sources these are relatively safe from the photodestructive effects of therapy. The red fluorescence seen in ultraviolet light of tumors after injection of various porphyrins, has proved of help to the surgeon in the localization of neoplasms during an operation. A serious side effect of therapy is continued light photosensitivity of exposed skin for a considerable time after treatment.

This area of study is one of which work continues to seek better localizers and/or sensitizers, based either on specific synthesis of porphyrin derivatives or through the use of other naturally occurring porphyrins or synthetic porphyrins and in the development of better light sources. At the present time, the clinical interest in this subject has been stimulated by technological developments. These relate principally to the development of better laser light sources such as gold vapor lasers and other forms of tunable lasers and endoseopic probes.

Retrospective Diagnoses

Royal Malady

The history of the porphyrias is naturally only truly reliable from the time at which there was concurrent medical observation and scientific mensuration. Any studies prior to the well-documented works at the end of the last century are, therefore, liable to be steeped in anecdotal inaccuracy. It is however, of interest to consider the hypothesis pro-pounded by MacAlpine and Hunter in 1964 that porphyria, possibly variegate porphyria, was present in the Royal Houses of Stuart and Hanover in the United Kingdom.[161] The ability to carry out such investigations depended not only on the inevitable extensive documentation of royalty, but also on the very precise descriptions conveyed to us over time by their physicians.

Of these, one of the more remarkable is that of Sir Theodore Turquet de Mayerne, physician to James, VI and I. He described one acute episode, following a hunting trip:

> "On his return he passed blood red urine.... He also told me that he quite frequently passed water, red like Alicante wine but without attendant pain...."

Mary Queen of Scots had suffered similarly to her son from acute attacks of abdominal colic, described as

> "...he labored under painful colic from flatus (an affliction from which his mother also suffered)..."

The history of the "Royal Malady" took a further intriguing turn following the studies of Röhl, Warren and Hunt[162] in 1998. These authors pursued a scientific/medical detective investigation, as a consequence of which, they discovered the bones of Princess Feodora of Reuss in Poland. Following exhumation, bone samples were obtained for DNA analysis. The authors were similarly able to obtain skeletal samples from Theodora's mother, Princess Charlotte of Saxe-Meiningen.

Mitochondrial DNA can be used to establish family relationships provided that there is a continuous maternal line of descent. Czarina Alexandra and Princess Charlotte were first cousins and importantly both were daughters of Queen Victoria's daughters, Princess Alice and Princess Victoria respectively. All should consequentially have inherited the same mitochondrial DNA from Queen Victoria. Such analysis proved that the bones from Princess Feodora's grave did not derive from the Romanov line. However those from Princess Charlotte's grave did derive from that line. As a consequence of the DNA analysis the workers were able to show that there was a novel protoporphyrinogen oxidase gene in Princess Charlotte's bones. Finally the workers found that there was convincing evidence that Prince William of Gloucester had variegate porphyria. Prince William was tragically killed in an aircraft accident in 1972 and in consequence DNA evaluation has never been possible, but the clinical evidence was sufficiently robust to convince Sir Abraham Goldberg and Dr Geoffrey Dean of its merit.

The historical implications of these observations are profound and, if true could imply that the loss by Britain of the American Colonies could be ascribed to a genetic disease[163] and, potentially that Kaiser Wilhelm, whose mother was Princess Victoria could have suffered from the psychiatric features of Variegate Porphyria.[164]

Vincent Van Gogh

Retrospective diagnosis of porphyria continues to be a matter of interest to many workers throughout the world. Wilfred Arnold[165] studied the history of Vincent van Gogh in some detail, his descent into psychiatric illness and the reasons for its causation. Many of the features he exhibited were consistent with acute porphyria precipitated by chemicals such as the solvents and alcohol to which he was exposed. No objective evidence of acute porphyria in his family has yet been discovered. However the subjective evidence is highly suggestive of this condition.

Werewolves

A bizarre suggestion has been that persons with congenital porphyria or hepatoerythropoietic porphyria or other of the homozygous porphyrias were the werewolves or vampires of legend. Lycanthropy (magical transformation of human to wolf) certainly did not take place, but the subjects' skin mutilation, hypertrichosis, and desire to eschew light exposure may have led the superstitious to this conclusion.[166] The medieval descriptions of werewolves included: "pale yellowish excoriated skin"—explicable by haemolytic anemia and pruritis; reddish teeth—erythrodontia; and "habitation in isolated regions, such as Central European valleys"—familial association and inbreeding in such areas. It is not easy to explain either fear of garlic or lust for blood. The effects of garlic might have related to the oxidative metabolism of diallyl disulfide. In this the needs of the patient should be considered. The press hysteria in the United States following these disclosures, established an inaccurate and unjustified perception of porphyria in the public mind and did considerable harm to the porphyric patient.[167]

Ephemera: Porphyrinurias

The heterogeneous group of diseases best described as porphyrinurias are those in which the disturbances of porphyrin metabolism have been brought about by endogenous and exogenous factors other than the genetic ones linked to the porphyrias. The porphyrin normally excreted in excess is coproporphyrin. This category includes lead poisoning, hereditary tyrosinemia,[168] ethanol abuse,[108] myocardial infarction,[169,170] and the effects of drugs like carbamazepine[171] and many other compounds such as polyhalogenated hydrocarbons.[172] Coproporphyrinuria has also been reported in a number of conditions such as liver disease and hepatocellular carcinoma.[37,173] Porphyrins are probably hepatotoxic and may even be associated with the development of neoplasms.[174] When Lithner and Wetterberg[175] carried out a retrospective study of 20 years of the relationship between hepatocellular carcinoma and acute intermittent porphyria, they found this to be significant.

In addition to those porphyrinurias, there are changes in the pattern of porphyrin isomer excretion in the hyperbilirubinemias of the Dubin-Johnson syndrome and in the Rotor syndrome. In these two conditions the quantity of porphyrin synthesized and excreted is not necessarily in excess of normal upper limits, but they are associated with excess production of series I isomer porphyrin.[176-178] A key worker in this area was Peniti Koskelo (Fig. 12).

Figure 12. Peniti Koskelo.

Lead

The connection between lead and porphyrin biosynthesis is reputed to have been first made by Binnendjik (cited by Stokvis, 1895), but the connection between this metal and anemia had been made very much earlier by Lannaec[179] in 1831. This was confirmed by Garrod in 1892[33] who observed abnormal porphyrin excretion in the urine of a patient with lead poisoning and then in 1895 by Stokvis[31] who found that lead-poisoned rabbits excreted excess urinary porphyrins. Other effects of lead on the haemopoietic system were reported before the end of the century by Behrend,[180] who observed stippled basophils in the blood of a patient with lead poisoning.

Lead Porphyrinemia and Porphyrinuria

The crude compound that Garrod identified as haematoporphyrin was shown to be coproporphyrin III by Duesberg.[181] Liebig[182] had suggested that it was produced by the action of lead on the bone marrow. By 1932 Grotepass.[183] had demonstrated elevated coproporphyrin in urine in lead poisoning.

The increased concentration of a free porphyrin in blood was identified by Hijmans van den Bergh et al[184] in 1932 as protoporphyrin IX located in the erythrocytes of subjects dosed with lead. In 1958, finally, diminution of ALA dehydratase activity (ALA-D) was identified as a means of lead assessment.[185]

Arsenic

In 1960, studies of hair taken from Napoleon Bonaparte by the University of Glasgow showed that his hair contained considerable concentrations of this metal. Subsequent activation analysis has also found increased arsenic in his hair and would potentially confirm this diagnosis.

Of greatest interest here is arsenic's impact on haem biosynthesis and in particular upon uroporphyrin decarboxylase, especially when one juxtaposes this information against the recent findings by Warren (personal communication) that George III had substantial levels of arsenic in his hair (100 times background). The reasons for such high levels could be sought in the use of arsenic both as a medicine and as a hair powder.

Woods and Fowler[186] showed haem biosynthetic changes associated with arsenic and subsequent studies have confirmed that not only is there increased excretion of porphyrins[187] with decreased activity of uroporphyrinogen decarboxylase which might be used as a biomarker of arsenic exposure.[188,189] Concordant with the hypothesis that As could have contributed to exacerbation of acute porphyria in George III is the finding of precipitation of variegate porphyria in a patient from southern USA who consumed "moonshine" contaminated with lead and arsenic.[190]

Pseudoporphyria

Poh-Fitzpatrick[191] pointed out in 1986 that the term "pseudoporphyria" had been used to describe a bullous dermatosis associated with a number of dermatological conditions that bear some resemblance to porphyria, often induced by many drugs but which were not porphyrias! The term "pseudoporphyria" should not be used to describe them, but only to describe conditions in which alterations of porphyrin metabolism can be found, such as the bullous dermatosis of haemodialysis.[192]

Mycosis Porphyria

In addition to hereditary porphyria in cattle, an acquired form has been described, attributable to a fungal infection.[193] Similar to this was the case reported by Lim et al[194] of a 24 year-old man who had increased faecal porphyrin excretion, resembling that seen in variegate porphyria. The abnormal faecal porphyrins were shown to be the result of excessive consumption of brewer's yeast, which was shown to have a high porphyrin content.

References

1. Lecanu LR. Études chimiques sur le sang humain (cited by Berzelius, 1840), 4° Paris No. 395. 1837.
2. Schultz JH. Ein fall von pemphigus leprosus, complicirt durch lepra visceralis. Griefswald: Inaugural Diss, 1874.
3. Baumstark F. Zwei pathologische Harnfarbstoff. Pflugers Arch ges Physiol 1874; 9:568-584.
4. Garrod AE. Inborn Errors of Metabolism. 2nd ed. London: Hodder and Stoughton, 1923.
5. Biosynthesis of tetrapyrroles and of corrinoids. In: Florkin M, Stotz EH, eds. Comprehensive Biochemistry: A History of Biochemistry. Amsterdam: Elsevier, 1979:193-238.
6. Moore MR, McColl KEL, Goldberg A. The porphyries. Diabet Metab 1979; 5:323-336.
7. Hodgson GW, Baker BL. Porphyrin abiogenesis from pyrrole and formaldehyde under simulated geochemical conditions. Nature 1967; 216:29-32.
8. Mercer-Smith JA, Mauzerall DC. Photochemistry of porphyrins-a model for the origin of photosynthesis. Photochem Photobiol 1984; 39:397-405.
9. Bonnett R, Czechowski F. Metalloporphyrins in coal-gallium porphyrins in bituminous coal. J Lab Chem Sci 1984; 1:125-132.
10. Hodgson GW. Cosmochemical evolution of large organic molecules-illustrative laboratory simulations for porphyrins. Ann NY Acad Sci 1972; 194:86-97.
11. Barnes CR. On the food of green plants. Bot Gaz 1893; 18:403-411.
12. Gest HO. History of the word photosynthesis and evolution of its definition. Photosynthesis Research 2002; 73:7-10.
13. Pelletier F, Caventou JB. Ann Chim Phys 1818; 9(2):194; (J Pharm 1818; 3:486).
14. Rüdiger W. Chlorophyll metabolism: From outer space down to the molecular level. Phytochemistry 1997; 46:1151-1167.
15. Fischer H, Orth H. Die chemie des Pyrrols. Leipzig: Adademische Verlagsgesellschaft, 1940.
16. Woodward RB. The total synthesis of chlorophyll. Pure Appl Chem 1960; 2:383-404.
17. Beale SI, Gough SP, Granick S. Biosynthesis of δ-aminolevulinic acid from the intact carbon skeleton of glutamic acid in greening barley. Proc Natl Acad Sci USA 1975; 72:2719-2723.
18. Von Wettstein D, Gough S, Kannangara CG. Chlorophyll Biosynthesis. The Plant Cell 1995; 7:1039-1057.
19. Battersby AR. Biosynthesis of the pigments of life. Proc R Soc Lond B 1985; 225:1-26.
20. Berzelius JJ. Lehrbuch der Chemie. Arnoldische Buchbandlung. Leipzig: Dresden, 1840:67-69.
21. Scherer J. Chemische-physiologische untersuchungen. Ann Chem Phar 1841; 40:1-64.
22. Mülder GH. Über eisenfreises hämatin. J Prakt Chem 1844; 32:186-197.
23. Thudichum JLW. Report on researches intended to promote an improved chemical identification of disease. 10th Report of The Medical Officer. London: Privy Council, H.M.S.O., 1867:152-233, (App. 7).
24. Hoppe-Seyler F. Das hämatin. Tubinger Med Chem Untersuchungen 1871; 4:523-533.
25. MacMunn CA. Further researches into the colouring-matters of human urine, with an account of their artificial production from bilirubin, and from haematin. Proc Roy Soc Lond Ser B 1880; 31:206-237.
26. Hoppe-Seyler F. Über das chlorophyll der pflanzen. Hoppe-Seyler's Z Physiol Chem 1879; 205:193-197.
27. Church AH. Researches in Turacin an animal pigment containing copper 11. Philos Trans Roy Soc Lond Ser A 1892; 183:511-530.
28. Soret JL. Récherches sur l'absorption des rayons ultraviolets par diverses substances. Arch Sci Phys Nat 1883; 10:430-485.
29. Günther H. Die hämatoporphyrie. Deutsch Arch Klin Med 1911; 105:89-146.
30. Hammarsten O. Tva fall af hamatoporphyrin i urinen. Uppsala Lak For Forh 1891; 26:259-288.
31. Stokvis BJ. Zur pathogenese der hamatoporphyrinurie. Z Klin Med 1895; 28:1-9.
32. Salkowski E. Über vorkommen und nachweis des hämatoporphyrins in härn. Hoppe-Seyler's Z Physiol Chem 1891; 15:286-309.
33. Garrod A. On the occurrence and detection of haematoporphyrin in the urine. J Physiol 1892; 13:598-620.

34. Fischer H. Uber das urinporphyrin. 1. Mitteilung. Hoppe-Seyler's Z Physiol Chem 1915; 95:34-60.
35. Nencki M, Sieber N. Über das hämatoporphyrin. Arch Exp Pathol Pharmacol 1888; 24:430-446.
36. Saillet H. De l'urospectrine (ou urohematoporphyrine urinale normlae). Rev Med 1896; 16:542-552.
37. Fischer H, Zerweck W. Über uroporphyrinogen heptamethylester und eine neue uberfuhrung von uro-koproporphyrin. Hoppe-Seyler's Z Physiol Chem 1924; 137:242-264.
38. Laidlaw PP. Some observations on blood pigments. J Physiol 1904; 31:464-472.
39. Küster W. Beitrage zur Kenntnis des Bilirubins und Hämins. Hoppe Seyler's Z Physiol Chem 1912; 82:463-483.
40. Willstätter R, Mieg W. Untersuchungen über das chlorophyll. Liebigs Ann Chem 1906; 350:1-47.
41. Willstätter R, Stoll A. Untersuchungen über chlorophyll. Methoden und ergebnisse. Berlin: Springer, 1913:424.
42. Fischer H, Hilmer H, Lindner F et al. Chemische befunde bei einem fall von porphyrie (Petry). Hoppe-Seyler's Z Physiol Chem 1925; 150:44-101.
43. Borst M, Königsdörffer H. Untersuchungen uber porphyrie mit besonderer berucksichtigung der Porphyrie Congenita. Leipzig: Hirzel, 1929.
44. Schumm O. Uber die naturlichen porphyrine. Z Physiol Chem 1924; 126:169-202.
45. Fischer H, Orth H. Chemie des Pyrrols. Leipzig: Acad Verlag, 1937.
46. Sachs P. Ein fall von akuter porphyrie mit hochgradiger muskelatrophied. Klin Wschr 1931; 10:1123-1125.
47. Waldenström J, Vahlquist BC. Studien uber die entstehung der roten harnpigmente (uroporphyrin und porphobilin) bein der akuten porphyrie aus ihrer farblosen vorstufe (porphobilinogen). Hoppe-Seyler's Z Physiol Chem 1939; 260:189-209.
48. Shemin D, Russell CS, Abramsky T. The succinate glycine cycle. 1. The mechanism of pyrrole synthesis. J Biol Chem 1955; 215:613.
49. Muir HM, Neuberger A. The biogenesis of porphyrins. 2. The origin of the methyne carbon atoms. Biochem J 1950; 47:97-104.
50. Shemin D, Russell CS. Succinate glycine cycle. J Amer Chem Soc 1953; 75:4873-4875.
51. Neuberger A, Scott JJ. Aminolaevulinic acid and porphyrin synthesis. Nature 1953; 172:1093-1094.
52. Falk JE, Dresel EIB, Rimington C. Porphobilinogen as a porphyrin precursor and interconversion of porphyrins in a tissue system. Nature 1953; 172:292.
53. Westall RG. Isolation of porphobilinogen from the urine of a patient with acute porphyria. Nature 1952; 170:614.
54. Cookson GH, Rimington C. Porphobilinogen. Biochem J 1954; 57:476-484.
55. Kast A. Über die art der darreichung und verordnung des sulfonals. Ther Mh 1888; 11:316-319.
56. Stokvis BJ. Over twee zeldsame kleurstoffen in urine van zicken. Ned Tijds Geneesk 1889; 13:409-417.
57. Harley V. Two fatal cases of an unusual form of nerve disturbance associated with red urine, probably due to defective tissue oxidation. Brit Med J 1890; 11:1169-1170.
58. Ranking JE, Pardington GL. Two cases of haematoporphyrin in the urine. Lancet 1890; ii:607-609.
59. Dobrschansky M. Einiges uber malonal. Wiener Med Presse 1906; 47:2145.
60. De Matteis F. Disturbances of liver metabolism caused by drugs. Pharm Rev 1967; 19:523-550.
61. Schmid R, Schwartz S. Experimental porphyria: Hepatic type produced by sedormid. Proc Soc Exp Biol Med 1952; 81:685-689.
62. Goldberg A, Rimington C. Diseases of Porphyrin Metabolism. Springfield: Thomas, 1962:I:11.
63. Solomon HM, Figge FHJ. Disturbance in porphyrin metabolism caused by feeding diethyl-1,4-dihydro-2,4,6-trimethylpyridine-3,5-dicarboxylate. Proc Soc Exp Biol Med 1959; 100:583-586.
64. Tephly TR, Coffman BL, Ingall G et al. Identification of N-methylprotoporphyrin IX in livers of untreated mice and mice treated with 3,5-diethoxycarbonyl-1,4-dihydrocollidine-source of the methyl group. Arch Biochem Biophys 1981; 212:120-126.
65. De Matteis F, Marks GS. The effect of N-methylprotoporphyrin and succinylacetone on the regulation of haem biosynthesis in checken hepatocytes in culture. FEBS Lett 1983; 159:127-131.
66. Cam C. Report on few cases of congenital porphyria. Nester (Instanbul) 1957; 1:2-6.
67. Elder GH. Acquired disorders of haem synthesis. Essays Med Biochem 1976; 2:75-114.
68. Strik JJTWA, Koeman JH. Chemical Porphyria in Man. Amsterdam: Elsevier North-Holland, 1979.
69. De Matteis F, Rimington C. Disturbance of porphyrin metabolism caused by griseofulvin in mice. Brit J Dermatol 1963; 75:91-104.
70. Günther H. Die bedeutung der hamatoporphyrinurie in der physiologie und pathologie. Ergebnisse der Allgemienen Pathol Anat 1922; 20:608-764.
71. McCall-Anderson T. Hydroa aestivale in two brothers, complicated with the presence of hematoporphyrin in the urine. Brit J Dermatol 1898; 10:1-4.

72. Meyer-Betz F. Untersuchungen über die biologische (photodynamische) wirkung des hämatoporphyrins und anderen derivate des blut und gallenfarbstoffs. Deutsch Arch Klin Med 1913; 112:476-503.
73. Hausmann W. Uber die sensibilisierende wirkung tierischer farbstoffe. Anwendung Biochem Z 1908; 14:275-283.
74. Hausmann W. Die sensibilisierende wirkung des hamatoporphyrins. Biochem Z 1911; 30:276-316
75. Waldenström J. Studien uber porphyries. Acta Med Scand Suppl 1937; 82:120.
76. Watson CJ. The problem of porphyria-some facts and questions. N Engl J Med 1960; 263:1205-1215.
77. Eales L. Clinical chemistry of the porphyries. In: Dolphin D, ed. The Porphyrias. New York: Academic Press, 1979:665-793.
78. Yeung-Laiwah AC, Moore MR, Goldberg A. Pathogenesis of acute porphyria. Quart J Med 1987; 63:377-392.
79. Granick S. Porphyrin synthesis in erythrocytes. 1. Formation of 5-aminolaevulinic acid in erythrocytes. J Biol Chem 1963; 232:1101-1117.
80. Jackson AH, Games DE, Couch P et al. Conversion of coproporphyrinogen III to protoporphyrin IX. Enzyme 1974; 17:81-87.
81. Brodie MJ, Moore MR, Goldberg A. Enzyme abnormalities in the porphyrias. Lancet 1977; ii:699-701.
82. Jordan PM, Burton G, Nordlov H et al. Preuroporphyrinogen-a substrate for uroporphyrinogen III cosynthetase. J Chem Soc Chem Comm 1979; 204-205.
83. Battersby AR, Fookes CJR, Matcham GWJ et al. Biosynthesis of the pigments of life: Formation of the macrocycle. Nature 1980; 285:17-21.
84. Waldenström J. Studies on the incidence and heredity of acute porphyria in Sweden. Acta Genet 1956; 6:122-131.
85. Prunty FTG. Acute porphyria-some properties of porphobilinogen. Biochem J 1945; 39:446-451.
86. Gray CH. Porphyria. Arch Intern Med 1950; 85:459-470.
87. Haeger G. Urinary 5-aminolaevulinic acid and porphobilinogen in different types of porphyria. Lancet 1958; ii:606-607.
88. Moore MR, McColl KEL, Rimington C et al. Disorders of Porphyrin Metabolism. New York: Plenum, 1987.
89. Day RS, Eales L, Meissner D. Coexistent variegate porphyria and porphyria cutanea tarda. N Engl J Med 1982; 307:36-41.
90. Meissner PN, Sturrock ED, Moore MR et al. Protoporphyrinogen oxidase, porphobilinogen-deaminase and uroporphyrinogen decarboxylase in variegate porphyria. Biochem Soc Trans 1985; 13:203-204.
91. Moore MR, Goldberg A. Health implications of the hemopoietic effects of lead. In: Mahaffey K, ed. Dietary and Environmental Lead: Human Health Effects. Amsterdam: Elsevier, 1985.
92. Hift RJ, Meissner PN, Todd G et al. Homozygous variegate porphyria: An evolving clinical syndrome. Postgrad Med J 1993; 69:781-786.
93. Campbell K. A case of haematoporphyrinuria. J Men Sci 1898; 44:305-313.
94. Erbslöh W. Zur pathologie und pathologischen anatomie der toxischer polyneuritis nach sulfonalgebrauch. Deutsch Z Nervenheik 1903; 23:197-204.
95. Mason VR, Courville C, Ziskind E. The porphyrins in human disease. Medicine 1933; 12:355-439.
96. Copeman SM. Porphyrinuria. Proc Pathol Soc Lond Lancet 1891; i:197.
97. Brugsch J. Porphyrine. J Ambrosius Barth, Leipzig 1959.
98. Tishler PV, Woodward B, O'Connor J et al. High prevalence of intermittent acute porphyria in a psychiatric in-patient population. Amer J Psych 1985; 142:1430-1436.
99. Irvine DG, Wetterberg L. Kryptopyrrole-like substance in acute intermittent porhyria. Lancet 1972; 2:1201.
100. Graham DJM, Brodie MJ, McColl KEL et al. Quantitation of 3-ethyl-5-hydroxy-4,5-dimethyl-Δ_3-pyrrolin-2-one in the urine of patients with acute intermittent porphyria. Eur J Clin Invest 1979; 9:40-53.
101. Graham DJM, Moore MR, Thompson GG et al. The effect of 4-ethyl-5hydroxy-3,5-dimethyl- Δ_3 pyrrolin-2-one on porphyrin synthesis in the rat. Biochem Soc Trans 1976; 4:1089-1091.
102. Moore MR, Graham DJM. Monopyrroles in porphyria psychosis and lead exposure. Internat J Biochem 1980; 12:827-832.
103. Gagey PL. Un cas d'hémoglobinurie en cours d'un xeroderma pigmentosum. Thèse de Paris, 1896.
104. Vollmer E. Über hereditare syphilis und haematoporphyrinurie. Arch Dematol and Syphilogy (Berlin) 1903; 65:221-234.
105. Schmid R, Schwartz S, Watson CJ. Porphyrin content of bone marrow and liver in the various forms of porphyria. Arch Intern Med 1954; 93:167-190.
106. Waldenström J. The porphyries as inborn errors of metabolism. Amer J Med 1957; 22:758-773.
107. Tio TH. Beschouwingen over de porphyria cutanea tarda. Doctoral Thesis. Amsterdam: 1956.

108. Moore MR, McColl KEL, Goldberg A. The effects of alcohol on porphyrin biosynthesis and metabolism. In: Rosalki SB, ed. Clinical Biochemistry of Alcoholism. Oxford: Churchill Livingstone: 1984:161-187.
109. Orten JM, Doehr SA, Bond C et al. Urinary excretion of porphyrins and porphyrin intermediates in human alcoholics. Quart J Studies on Alcohol 1963; 24:598-609.
110. Cetingil AI, Ozen MA. Toxic porphyria. Blood 1960l; 16:1002-1010.
111. Pinol-Aguade J, Castells A, Indocochea A et al. A case of biochemically unclassifiable hepatic porphyria. Brit J Dermatol 1969; 81:270-275.
112. Smith SG. Hepatoerythropoietic porphyria. Sem Dermatol 1986; 5:125-137.
113. Magnus IA, Jarret A, Prankerd TAJ et al. Erythropoietic protoporphyria: A new porphyria syndrome with solar urticaria due to protoporphyrinaemia. Lancet 1961; ii:448-451.
114. Langhof H, Müller H, Rietschel I. Untersuchungen zur familiaren protoporphyrinamischen lichurticaria. Arch Klin Exp Dermatol 1961; 212:506-518.
115. Kosenow W, Treibs A. Lichtuberempfindlichkeit und porphyrinamie. Z Kinderkeikunde 1953; 73:82-92.
116. Murphy GM, Hawk JLM, Magnus IA. Late-onset erythropoietic protoporphyria with unusual cutaneous features. Arch Dermatol 1985; 121:1309-1312.
117. Brouwier L. Quelques observations recueilles dans le service d'inspection de l'abbatoir de Liege: Alteration des os. Echo Vet (Liege) 1884; 271-273.
118. Tappeiner H. Untersuchung pigmentirter knochen vom schweine. Sitzungserber Ges Morph Physiol München 1885; 1:38-41.
119. Mosselman D, Hebrant A. Coloration abnormale du squelette chez une bête de boucherie. Ann Med Vet 1898; 47:201-206.
120. Ingier A. Ochronose bei tieren. Zieglers Beirage zur Pathol Anat 1911; 51:199-208.
121. Schmey M. Über ochronose bei mensch und tier. Frankfurter Z Pathol 1913; 12:218-328.
122. Mö1ler-Sorenson A. On haemochromatosis ossium (Ochronose) has husdyrene. In: Den KGL, ed. Veterinaer-Og Landbohojskoles Aarsskrift. Copenhagen: 1920:122-139.
123. Witte H. Ein fall von ochronose bei einem bullen und einem von ihre stammenden kalbe. Z Fleisch U Milchhyg 1914; 24:334.
124. Rimington C. Some cases of congenital porphyrinuria in cattle-chemical studies upon living animals and postmortem material. Onderstepoort J Vet Sci Animal Ind 1936; 7:567-609.
125. Poulson V. Om ochronotiske tilstande hos mennesker og. dyr. (on ochronotic states in man and animals). Med. Diss., University of Copenhagen, 1910.
126. Clare T, Stephens EH. Congenital porphyria in pigs. Nature (London) 1944; 153:252-253.
127. Jørgensen SK. Congenital porphyria in pigs. Brit Vet J 1959; 115:160-175.
128. Jørgensen SK, With TK. Congenital porphyria in animals other than man. In: Rook AJ, Walton GS, eds. Comparative Physiology and Pathology of the Skin. Oxford: Blackwell, 1965:317-331.
129. Tobias G. Congenital porphyria in a cat. J Amer Vet Med Assoc 1964; 145:462-463.
130. Owen LN, Stevenson DE, Keilin J. Abnormal pigmentation and fluorescence in canine teeth. Res Vet Sci 1962; 3:139-146.
131. Vannotti A. Porphyrins—Their Biological and Chemical Importance. London: Hilger and Watts, 1954.
132. Lascelles J. Tetrapyrrole Biosynthesis and Its Regulation. New York: Benjamin, 1964.
133. Rimington C. Porphyrins. Endeavour 1955; 14:126-135.
134. Turner WJ. Studies on Porphyria. 1. Observations on the fox squirrel (Sciurus niger). J Biol Chem 1937; 118:519-530.
135. Levin EY, Flyger V. Uroporphyrinogen, III. Cosynthetase activity in the fox squirrel (Sciurus niger). Science 1971; 174:59-60.
136. Kennedy GY. Pigments of marine invertebrates. Adv Mar Biol 1979; 16:309-381.
137. Church AH. Researches in turacin, an animal pigment containing copper. Phil Trans R Soc 1869; 159:627-636.
138. Fischer H, Hilger J. Zur kenntnis der naturlichen porphyrine. 8 mitteilung. Uber das vorkommen von uroporphyrin (als kuppfersalz, Turacin) in den turakusvogeln und den nachweis von koproporphyrin in der hefe. Hoppe-Seyler's Z Physiol Chem 1924; 138:49-67.
139. With TK. Porphyrins in egg shells. Biochem J 1973; 137:597-598.
140. Derrien E, Turchini J. Sur les fluoresences rouges de certains tissus on secreta animaux en lumiere ultraviolette-nouvelles observations de fluorescences rouges chez les animaux. C R Soc BW (Paris) 1925; 92:1028-1032.
141. Harder JJ. Glandula nova lacrymalis una cum ducta excretoria in cervis et damis detecta. Acta Eruditorum Lipsiae 1694; 11:49-52.
142. Brownscheidle CM, Niewenhuis RJ. Ultrastructure of the Harderian gland in male albino rats. Anat Rec 1978; 190:735-754.

143. Kennedy GY, Jackson AH, Kenner GW et al. Isolation, structure and synthesis of a tricarboxylic porphyrin from the Harderian glads of the rat. FEBS Lett 1970; 6:9-12.
144. Spike RC, Johnston HS, McGadey J et al. Quantitative studies on the effects of hormones on structure and porphyrin biosynthesis in the Harderian gland of the female golden hamster. I. The effects of ovariectomy and androgen administration. J Anat 1985; 142:59-72.
145. Spike RC, Johnston HS, McGadey J et al. Quantitative studies on the effects of hormones on structure and porphyrin biosynthesis in the Harderian gland of the female golden hamster. II. The time course of changes after ovariectomy. J Anat 1986; 145:67-77.
147. Payne AP, McGadey J, Moore MR et al. Androgenic control of the Harderian gland in the male golden hamster. J Endocr 1977; 75:73-82.
148. Mindegaard J. Studier verrend porphyrinsynthese og porfyrinholdige biologiske pigmentinkorns oprindelse og ultrastruktur, Afdelingen for biokemi og erneering. Danmarks Tekniske Hojskole 1976.
148. Woolley GW, Worley J. Sexual dimorphism in the Harderian gland of the hamster (Cricetus auratus). Anat Rec 1954; 118:416-417.
149. Thompson GG, Hordovatzi X, Moore MR et al. Sex differences in haem biosynthesis and porphyrin content in the Harderian gland of the golden hamster. Int J Biochem 1984; 16:849-852.
150. Policard A. Étude sur les aspects offerts par des tumeurs experimentales examinées a la lumiere de Wood. C R Seances Soc Biol 1924; 91:1423-1424.
151. Körbler J. Untersuchung von krebsgewebe im fluoreszenzerregenden licht. Strahlentherapie 1931; 41:510-518.
152. Schwartz S, Absolon K, Vermund H. Some relationships of porphyrins, X-rays and tumors. Univ Minnesota Med Bull 1955; 27:7-13.
153. Rassmussen-Taxdal DS,Ward GE, Figge FHJ. Fluorescence of human lymphatic and cancer tissues following high doses of intravenous hematoporphyrin derivative. Cancer 1955; 8:78-81.
154. Lipson RL, Baldes EJ, Olsen AM. The use of a derivative of hematoporphyrin in tumor detection. J Nat Cancer Inst 1961; 26:1-11.
155. Kessel D. Components of hematoporphyrin derivatives and their tumor-localizing capacity. Cancer Res 1982; 42:1703-1706.
156. Dougherty TJ, Potter WR, Weishaupt KR. An overview of the status of photoradiation therapy. In: Doiron TJ, Gomer CJ, eds. Porphyrin Localisation and Treatment of Tumors. New York: Alan R. lass, 1984.
157. Kessel D, Cheng ML. Biological and biophysical properties of the tumor-localizing component of hematoporphyrin derivative. Cander Res 1985; 45:3053-3057.
158. Scott JJ. The Metabolism of δ-aminolevinlate acid in CIBA Foundation Symposium on Porphyrin Biosynthesis and Metabolism. In: Wolstenholme GEW, Millar ECP, eds. London: J. and A. Churchill Ltd., 1955.
159. Peng Q, Warloe T, Berg K et al. 5-Aminolevulinic Acid-Based Photodynamic Therapy: Clinical Research and Future Challenges. Cancer 1997; 12(79):2282-2308.
160. Brown JE, Brown SB, Vernon DI. Photodynamic therapy—New light on cancer treatment. JSDC 1999; 115:249-253.
161. MacAlpine I, Hunter R. George the III and the 'Mad Business.' London: Penguin, 1969.
162. Röhl JCG, Warren M, Hunt D. Purple Secret: Genes. 'Madness' and the Royal Houses of Europe 1998.
163. Ware M. Porphyria-A royal malady. London: British Medical Association, 1968.
164. Cox TM, Jack N, Lofthouse S et al. King George 111 and porphyria: An elecental hypothesis and investigation. Lancet 2005; 366:332-335.
165. Arnold WN. Vincent Van Gogh: Chemicals, Crises, and Creativity. Boston: Birkhäuser, 1992.
166. Illis L. On porphyria and the aetiology of werewolves. Proc Roy Soc Med 1964; 57:23-26.
167. Dresser N. Vampires are in the headlines, but patients pay the price. Cal Folk Soc Modesto Calif 1986.
168. Gentz J, Johansson S, Lindblad B et al. Excretion of delta-aminolaevulinic acid in hereditary tyrosinaemia. Clin Chim Acta 1969; 23:257-263.
169. Koskelo P. Studies of urinary coproporphyrin excretion in acute coronary diseases. Ann Med Intern Fenniae 1956.
170. Koskelo P, Heikkila J. Urinary excretion of porphyrin precursors in myocardial infarction. Acta Med Scand 1965; 178:681.
171. Yeung-Laiwah AC, Rapeport WG et al. Carbamazepine-induced nonhereditary acute porphyria. Lancet 1983; i:790-792.
172. Marks GS. The effects of chemicals on hepatic heme biosynthesis. TIPS 1981; 2:59-61.
173. Udagawa M, Horie Y, Hirayama C. Aberrant porphyrin metabolism in hepatocellular carcinoma. Biochem Med 1984; 31:131-139.

174. Tio TH, Leijnse B, Jarrett A et al. Acquired porphyria from a liver tumour. Clin Sci 1957; 16:517-527.
175. Lithner F, Wetterberg L. Hepatocellular carcinoma in patients with acute intermittent porphyria. Acta Med Scand 1984; 215:272-274.
176. Koskelo P, Toivonen I, Adlercreutz H. Urinary coproporphyrin isomer distribution in the Dubin-Johnson syndrome. Clin Chem 1967; 13:1006-1008.
177. Ben Ezzer J, Rimington C, Shani M et al. Abnormal excretion of the isomers of urinary coproporphyrin by patients with Dubin-Johnson syndrome in Israel. Clin Sci 1971; 40:17-30.
178. Cohen C, Kirsch RE, Moore MR. Porphobilinogen-deaminase and the synthesis of porphyrin isomers in Dubin-Johnson syndrome. S Afr Med J 1986; 70:36-39.
179. Lannaec RTM. Traite sur l'auscultation mediate. 4th ed. Chaude Paris 1831.
180. Behrend B. Uber endoglobulare einschlusse volker blutkorperchen. Deutsch Med Wschr 1899; 25:254.
181. Duesberg R. Über die anämien - 1. Porphyrie und erythropoese. Arch Exp Pathol 1931; 162:249.
182. Liebig NS. Uber die experimentelle Bleihämatoporphyrie. Arch Exp Pathol Pharmacol 1927; 125:16-27.
183. Grotepass W. Zur kenntnis des im harn auftretenden porphyrins bei bleivergiftung. Hoppe-Seyler's Z Physiol Chem 1932; 205:193-197.
184. Van den Bergh HAA, Grotepass W, Revers FE. Beitrag über das porphyrin in Blut und Galle. Klin Wschr 1932; 11:1534-1536.
185. Gibson KD, Neuberger A, Scott JJ. The purification and properties of delta-aminolaevulinic acid dehydratase. Biochem J 1955; 61:618-629.
186. Woods JS, Fowler BA. Effects of chronic arsenic exposure on hematopoietic function in adult mammalian liver. Environ Health Perspect 1977; 19:209-13.
187. Fowler BA, Mahaffey KR. Interactions among lead, cadmium and arsenic in relation to porhyrin excretion patterns. Environ Health Perspect 1978; 25:87-90.
188. Ng JC, Qi L, Moore MR. Porphyrin profiles in blood and urine as a biomarker for exposure to various arsenic species. Cell Mol Biol (Noisy-le-grand) 2002; 48:111-23.
189. Wang JP, Qi L, Zheng B et al. Porphyrins as early biomarkers for arsenic exposure in animals and humans. Cell Mol Biol (Noisy-le-grand) 2002; 48:835-43.
190. Hughes GS, Davis L. Variegate porphyria and heavy metal poisoning from ingestion of 'moonshine'. South Med J 198; 76:1027-9.
191. Poh-Fitzpatrick M.B. Porphyria, pseudoporphyria, pseudopseudoporphyria. Arch Dermatol 1986; 122:403-404.
192. Topi GC, D'Alessandro GL, De Costanza F et al. Porphyria and pseudoporphyria in hemodialized patients. Int J Biochem 1980; 12:963-967.
193. Kaneko J, Cornelius CE. Clinical Biochemistry of Domestic Animals. 2nd ed. Vol. 1. New York: Academic Press, 1970.
194. Lim CK, Rideout JM, Peters TJ. Pseudoporphyria associated with consumption of brewer's yeast. Brit Med J 1984; 298:1640.

CHAPTER 2

Biosynthesis of 5-Aminolevulinic Acid

Dieter Jahn* and Dirk W. Heinz

Abstract

5-Aminolevulinic acid (ALA) is the general precursor of all known tetrapyrroles. Currently, two different biosynthetic routes for ALA formation are known. Humans, animals, fungi and the α-group of the proteobacteria employ the one-step-condensation of succinyl-coenzyme A and glycine catalyzed by pyridoxal 5′-phosphate-dependent ALA synthase. In plants, algae, archaea and all other bacteria ALA is formed by two enzymes. The initial substrate glutamyl-tRNA is synthesized by glutamyl-tRNA synthetase and supplied both to protein and to tetrapyrrole biosynthesis. During the first committed step of ALA synthesis a NADPH-dependent glutamyl-tRNA reductase reduces glutamyl-tRNA to form glutamate-1-semialdehyde. The aldehyde is subsequently transaminated by glutamate-1-semialdehyde-2,1-aminomutase to yield ALA. Evidence for metabolic channeling of the reactive aldehyde between glutamyl-tRNA reductase and the aminomutase is outlined based on the structures of both enzymes. The enzymatic mechanisms deduced from biochemical investigations and recently solved crystal structures are described for all participating enzymes.

Introduction

Tetrapyrroles, such as hemes, chlorophylls, vitamin B_{12} and coenzyme F_{430} are essential metabolites for almost all living organisms. The common precursor for all these tetrapyrroles is 5-aminolevulinic acid (ALA). In nature it is synthesized by two alternative, unrelated biosynthetic routes: Mammals, fungi and the α-group of the proteobacteria use the so-called 'Shemin pathway', the condensation of succinyl coenzyme A and glycine catalyzed by 5-aminolevulinic acid synthase (ALAS) for ALA formation.[1] In plants, archaea and most bacteria ALA formation starts from the C_5-skeleton of glutamate.[2] This C_5-pathway involves two enzymes. NADPH-dependent glutamyl-tRNA reductase (GluTR) catalyzes the reduction of glutamyl-tRNA to glutamate-1-semialdehyde (GSA). In the subsequent reaction pyridoxamine 5′-phosphate (PMP)-dependent glutamate-1-seminaldehyde-2,1-aminomutase (GSAM) transaminates GSA to form ALA.[3] This chapter focuses on the structure and function of the enzymes belonging to both pathways.

Condensation of Succinyl-CoA and Glycine into Aminolevulinic Acid

5-Aminolevulinic Acid Synthase (ALAS)

5-Aminolevulinic acid synthase (ALAS) (EC 2.3.1.37) catalyzes the condensation of glycine and succinyl coenzyme A to produce ALA, carbon dioxide and free coenzyme A (Fig. 1). The enzyme is found in nonplant eukaryotes such as humans, animals and fungi. The α-group of the proteobacteria including *Rhodobacter*, *Agrobacterium*, *Rhizobium* and *Rickettsia* species are the only prokaryotic representatives carrying ALAS. Phylogenetically, mitochondria also belong into this bacterial group.[4]

*Corresponding Author: Dieter Jahn—Institute of Microbiology, Technical University Braunschweig, Spielmannstrasse 7, D-38106, Braunschweig, Germany. Email: d.jahn@tu-bs.de

Tetrapyrroles: Birth, Life and Death, edited by Martin J. Warren and Alison G. Smith. ©2009 Landes Bioscience and Springer Science+Business Media.

ALAS requires pyridoxal 5′-phosphate (PLP) as cofactor. The enzyme belongs to the α-oxoamine synthase subfamily of PLP-dependent enzymes which typically catalyze the condensation of an amino acid and a carboxylic acid coenzyme A thioester in combination with the decarboxylation of the amino acid. ALAS was simultaneously described in 1958 by Neuberger and coworkers for avian preparations and Shemin and his group for bacterial cell free extracts.[5,6] Over the years ALAS has been purified from various bacterial and eukaryotic sources.[7] Mammals carry two ALAS isoforms which are encoded by two different genes localized on two distinct chromosomes. One gene encodes the housekeeping isoform (ALAS1/ALAS-H), the other gene the erythroid-specific enzyme (ALAS2/ALAS-E). The latter is required to sustain the extra need for heme during hemoglobin formation in the erythrozytes.[4,7] Unlike mammals, most α-proteobacteria use only one form of ALAS. The only known exception is *Rhodobacter sphaeroides* which also contains two isoforms of ALAS encoded by *hemA* and *hemT*.[8] ALAS is a functional homodimer in which residues from both subunits contribute to the active site. Based on the solved crystal structure of the closely related *E. coli* 8-amino-7-oxononanoate synthase a model for the structure of *R. sphaeroides* ALAS was recently proposed.[9]

Catalytic Mechanism

The mechanistic and steric course of the ALAS reaction has been intensively investigated by various groups (Fig. 1). Elegant radiolabeling studies using *R. sphaeroides* ALAS provided the basis for a model of the catalytic mechanism.[4,7] This model was continually improved by using stopped-flow and quenched-flow kinetic analyses.[10] Catalysis begins, as in other PLP-dependent enzymes, with a transaldimination reaction between the substrate glycine and the internal aldimine formed between the aldehyde group of PLP and an active site lysine residue (Fig. 1). After formation of the external aldimine, the pro-R proton of glycine gets removed. The resulting transient quinonoid form is formed in the presence of the second substrate.[11,12] During reaction of this quinonoid intermediate with succinyl coenzyme A the coenzyme A group is cleaved off and an aldimine to α-amino-β-ketoadipate is formed. The quinonoid form is then stabilized by the cleavage of the C_α-carboxylate bond. Protonation at the C-5 position of the aldimine leads to the formation and dissociation of ALA from the active site of ALAS.[12] In the intensively investigated murine ALAS Lys313 forms the Schiff base linkage with the PLP cofactor.[4] The same residue is proposed to function as the catalytic base for proton abstraction from the external aldimine during quinonoid intermediate formation.[11] Arg439 was found to be important for glycine binding.[13] A glycine-rich loop comprising Gly142 and Gly144 as well as Tyr 124 were proposed to be involved in PLP cofactor binding.[13,14] Asp279 was charaterized as an enhancer of the electron withdrawing capacity of PLP via the stabilization of protonated form of the pyridinium ring nitrogen.[14]

Transfer RNA-Dependent Aminolevulinic Acid Formation

The precursor of the C_5-pathway for ALA formation is glutamyl-tRNA (Fig. 2). The same glutamyl-tRNA simultaneously participates in protein and tetrapyrrole biosynthesis.[3] To reflect the importance of glutamyl-tRNA formation to the biosynthetic pathway of ALA, some properties of glutamyl-tRNA synthetase (GluRS) and its substrate tRNA^{Glu} are briefly described below.

tRNA^{Glu}

The investigation of barley chloroplast, *Chlamydomonas reinhardtii* and *Chlorella* extracts indicated that a tRNA-like molecule is essential for the synthesis of ALA from glutamate.[15] Sequence analysis of the active RNA species from barley identified the compound as tRNA^{Glu}.[16] The sequenced barley tRNA^{Glu} contained a total of ten modified nucleotides and an UUC anticodon. For *C. reinhardtii* and *Synechocystis* the same tRNA^{Glu} was demonstrated to support tetrapyrrole and protein biosynthesis.[3]

A variety of in vitro and in vivo approaches established a set of nucleotides, so called identity elements, of *Escherichia coli* tRNA^{Glu} required for the recognition by its cognate glutamyl-tRNA synthetase (Fig. 3A). These identity elements consist of U34, U35, C36 and A37 in the anticodon loop, G1:C72, U2:A71 and C4:G69 in the acceptor stem, U11:A24, U13:G22::A46 and C12:G23::C9 in the augmented D helix formed of the D-stem helix with several neighbouring residues and the variable loop (N:N and N::N denote secondary and tertiary nucleotide base pairings). Moreover, A46 and the absence of nucleotide 47 (D47) in the short variable loop were found to stabilize

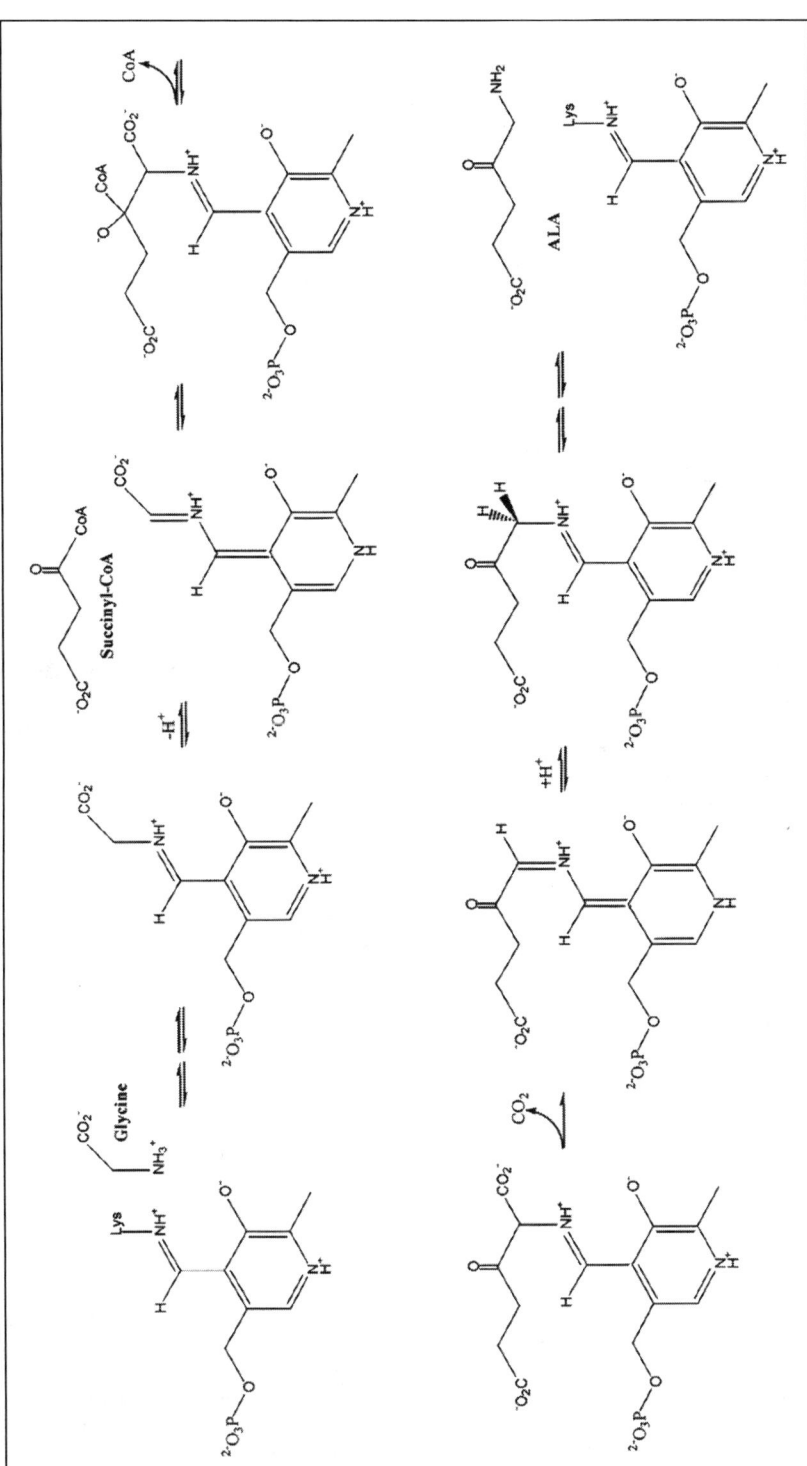

Figure 1. Pyridoxal 5′-phosphate-dependent reaction mechanism of 5-aminolevulinic acid synthase (ALAS).

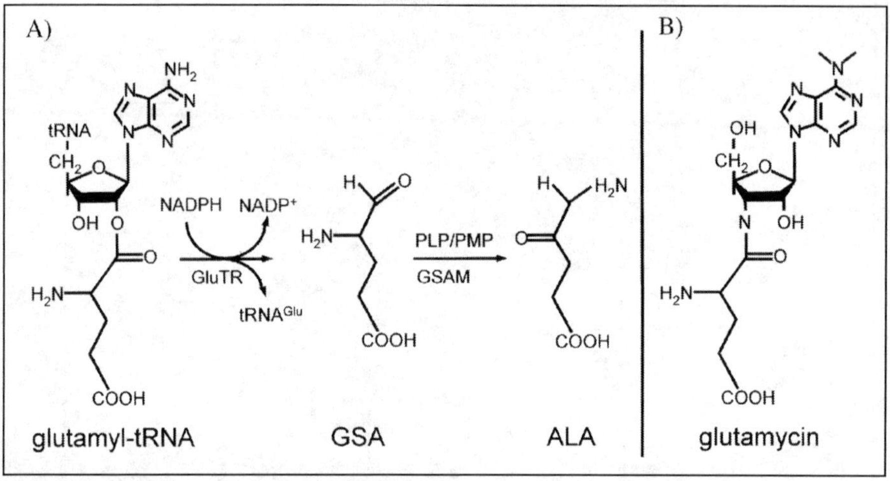

Figure 2. A) The C5-pathway for 5-aminolevulinic acid (ALA) synthesis: NADPH-dependent glutamyl-tRNA-reductase (GluTR) reduces tRNA-bound glutamate to glutamate-1-semialdehyde (GSA). Glutamate-2,1-semialdehyde aminomutase (GSAM) then transaminates GSA to ALA. B) The inhibitor glutamycin is an analogue of the 3'-end of aminoacylated tRNA^{Glu}.

U13:G22::A46, a tertiary base pairing essential for tRNA-GluRS interaction.[17,18] The importance of the 2-thio groups in methylaminomethyl-2-thiouridine (mnm5s2U) in the wobble position U34 for the recognition of tRNA^{Glu} by *E. coli* GluRS in vitro was challenged by in vivo experiments with hypomodified tRNA^{Glu}.[19,20]

The identity elements of glutamyl-tRNA for GluTR were mainly subject to theoretical considerations. The isolation of an *Euglena gracilis* plastidic tRNA^{Glu} gene mutant impaired in chlorophyll biosynthesis but capable of protein biosynthesis revealed the only verified glutamyl-tRNA identity element in position C56.[21]

Glutamyl-tRNA Synthetase

Glutamyl-tRNA synthetase (GluRS, glutamic acid:tRNA^{Glu} ligase, EC 6.1.1.17) esterifies glutamate with the 2'-terminal hydroxy group of tRNA (Fig. 4A). A two step mechanism including the activation of the amino acid by ATP and its subsequent transfer to the tRNA is generally accepted[22] (Fig. 4A):

1. GluRS + Glu + ATP (+ tRNA^{Glu}) → GluRS + Glu-AMP + PP$_i$ (+ tRNA^{Glu})
2. GluRS + Glu-AMP + tRNA^{Glu} → GluRS + Glu-tRNA^{Glu} + AMP

GluRS is one out of ten class I aminoacyl-tRNA synthetases (Fig. 3A). Its ATP-binding domain bears a variation (HVGG) of the characteristic HIGH motif. GluRS is closely related to glutaminyl-tRNA synthetase.[22] Some bacteria, most archaea and organelles lack glutaminyl-tRNA synthetase. In these cases GluRS glutamylates both tRNA^{Glu} and tRNA^{Gln}, and the misacylated Glu-tRNA^{Gln} is converted to the required Gln-tRNA^{Gln} (glutaminyl-tRNA) by an amidotransferase.[23]

Biochemical and Structural Characterization of GluRS

GluRS has been purified from various bacteria, green algae, plants and mammalian cells. Bacterial and green algal enzymes were found to be active as monomers with a M$_r$ of ~50,000. *E. coli* and *B. subtilis* GluRS forms a complex with adenylosuccinate AMP-lyase (EC 4.3.2.2.). This complex protects GluRS against heat denaturation, increases its affinity for glutamate and ATP and coordinates purine metabolism and protein biosynthesis. Chloroplast, mitochondrial and wheat cytoplasmic GluRSs are dimeric. The K$_M$ of GluRS for ATP is in the range of 10^{-4} and 10^{-5} M, for glutamate about 10^{-5} M, and for tRNA^{Glu} about 10^{-7} M. As observed for other aminoacyl-tRNA synthetases, GluRS activity is Mg^{2+}-dependent. Furthermore, the *E. coli* enzyme is Zn^{2+}-dependent. Only bacterial and plastidic enzymes provide the appropriate precursor for tetrapyrrole biosynthesis.[22]

The crystal structure of monomeric *Thermus thermophilus* GluRS has been solved for both the uncomplexed protein[24] and in complex with tRNA^{Glu}.[25] The enzyme consists of five distinct domains arranged to form an elongated, slightly curved molecule (Fig. 3A). The Rossmann-type nucleotide-binding fold of domain 1 is conserved in all class-I tRNA synthetases. It contains the two characteristic ATP-binding motifs, HIGH (here HVGG) and KMSKS. The adjacent regions of domains 2 and 3 are also structurally conserved. Located on either side of the ATP-binding pocket of domain 1, they create a conspicuous groove that accommodates the tRNA acceptor arm. The remaining conserved α-helices of domain 3 interact with both the D-stem region at the concave elbow angle of the tRNA as well as the anticodon stem. Structural features unique to each synthetase are the major determinants of specificity. Regions belonging to domains 2 and 3 create the glutamate recognition pocket. Domains 4 and 5 are structurally unique to GluRS in their entirety. They are reserved for tRNA^{Glu}-anticodon recognition and discrimination. Only U35 is recognized by a planar main-chain segment (Thr444-O and Pro445-N). Arg417 and Arg435 are crucial for anticodon recognition. They bind the phosphate and base of C34, respectively. Arg358 hydrogen bonds C36 in a Watson-Crick-like pattern.

Glutamyl-tRNA Reductase

The discovery that tRNA^{Glu} is involved in the biosynthesis of ALA prompted the formulation of a novel two-step-pathway (Fig. 1): In the first step, tRNA^{Glu}-bound glutamate is reduced to GSA, which is transaminated to ALA in a second step. The enzymes catalyzing these first committed steps of tetrapyrrole formation are glutamyl-tRNA reductase (GluTR) and glutamate-1-aldehyde-1,2-aminomutase (GSAM).[3]

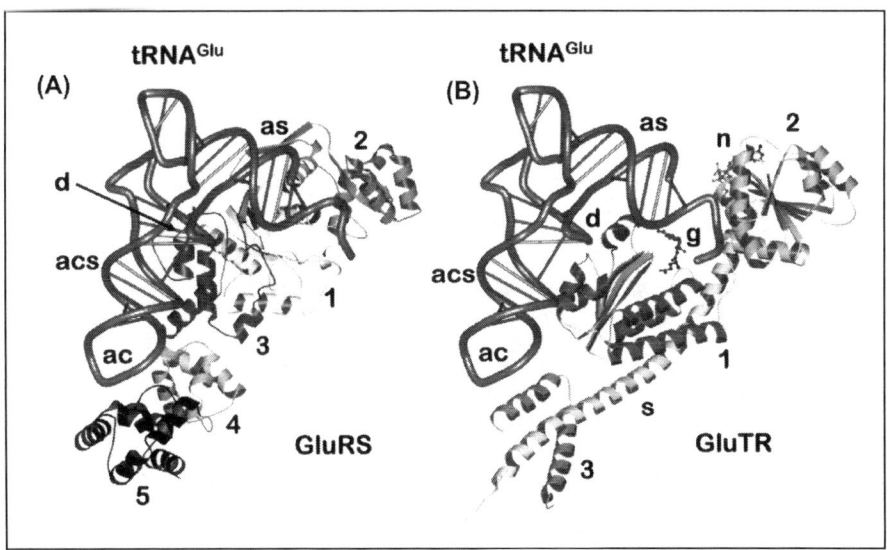

Figure 3. A ribbon diagram of the complexes of (A) tRNA^{Glu}/glutamyl-tRNA synthetase (B) tRNA^{Glu}/GluTR monomer (modeled). The complexes have been oriented to provide the equivalent view of the tRNA^{Glu}. Individual domains are numbered. as, acceptor stem of tRNA^{Glu}; ac, anticodon; acs, anticodon stem; d, D-stem; g, glutamycin; n, NADPH (modeled); s, spinal helix. Note the similarity in binding modes in both proteins. Generated using *MOLSCRIPT* (Kraulis, 1991) rendered with *POVRAY* (www.povray.org) as implemented in *GL_RENDER* (www.hhmi.swmed.edu/external/Doc/Gl_render).

Biochemical and Structural Characterization of GluTR

Initial characterizations of bacterial and plant GluTRs resulted in controversy regarding their relative molecular masses, cofactor requirements, and potential catalytic.[26,27] In retrospect, the extended structure of GluTR[28,29] solubility problems due to aggregation and nonspecific disulfide bridge formation may partly explain these discrepancies.[30]

Finally, the biochemical and crystallographic investigation of recombinant GluTR from the extreme thermophilic archaeon *Methanopyrus kandleri* and *E. coli* established the catalytic mechanism and its structural basis.[28-30] Spectroscopic and biochemical analysis of both GluTRs demonstrated that neither heme, flavins, nor metal ions are required for catalysis. Treatment of *M. kandleri* and *E. coli* GluTR with iodoacetamide and 5,5'-dithiobis(2-nitrobenzoic acid) abolished enzyme activity indicating the involvement of a nucleophilic cysteine in catalysis. All cysteines of both GluTRs were individually replaced by serines. Only mutants C48S (*M. kandleri*) and C50S (*E. coli*) were completely inactive indicating that these cysteines represent the active site nucleophiles. These residues correspond to the only cysteines conserved throughout all known GluTRs. Interestingly, both enzymes efficiently turn over glutamyl-tRNA in the absence of NADPH, liberating glutamate. The rate of this esterase activity is comparable to GluTR reductase activity in the presence of NADPH. Similar esterase activities are also observed for other enzymes such as GAPDH, thiol proteinases and aldehyde dehydrogenases that form a covalent acyl-enzyme intermediate involving an active site cysteinyl residue. Corresponding GluTR mutants C48S/C50S did not possess any esterase activity. Based on these observations a catalytic mechanism for GluTR was proposed (Fig. 4B). The sulfhydryl group of Cys48 (Cys50 in *E. coli*) nucleophilically attacks the α-carbonyl group of tRNA-bound glutamate forming an enzyme-bound thioester intermediate with the concomitant release of tRNA^{Glu} (Fig. 6). The thioester reaction intermediate was isolated and visualized for *E. coli* GluTR.[30] Hydride transfer from NADPH to the thioester-bound glutamate produces glutamate-1-semialdehyde while in the absence of NADPH, a water molecule takes its place, hydrolyzing the reactive thioester bond to release free glutamate.[28-30]

The reduction and inhibition capabilities of various NADPH analogues were also investigated. NADH, lacking the 2'-phosphoryl group, does not substitute for NADPH, nor does it inhibit *M. kandleri* GluTR. Interestingly, reduced β-nicotinamide mononucleotide corresponding to removal of the adenosine phosphate moiety does, by contrast, inhibit NADPH-binding. Removal of the adenine amino group (reduced nicotinamide hypoxanthine dinucleotide phosphate) resulted in the inhibition of GluTR. Surprisingly, both 2'-NADPH and 3'-NADPH are equally active. Thus all the major determinants of NADPH are required for efficient recognition and utilization by GluTR.[28]

The crystal structure of *M. kandleri* GluTR revealed an extended V-shaped dimer[29] (Fig. 5). Each monomer consists of three distinct domains arranged along an extended curved 'spinal' α-helix. The N-terminal domain consists of two subdomains: a small $\beta\alpha\beta\beta\alpha\alpha\beta$ and a larger subdomain containing three α-helices. These three helices roughly align with the spinal helix to form a 4-helix bundle. This first domain is linked via a short loop to a second carrying a classical NAD(P)H-binding fold. Domain 2 is followed by the 110-Å spinal helix of 18 α-helical turns. At the N-terminus this spinal helix reinforces the loop between domain 1 and 2, then passes domain 1, and C-terminally extends into and forms part of the dimerization domain. This third domain consists of a symmetric, six-helix bundle, three helices deriving from each of two interacting monomers. The crystal structure of GluTR was solved in complex with glutamycin[28,29] (Fig. 5). The ester bond linking the 3' end of tRNA^{Glu} and glutamate is by necessity labile. To generate a stable substrate analogue representing the last adenosine residue of the tRNA the bridging oxygen was replaced by an imino group. This analogue was synthesized from the structurally related puromycin aminonucleoside and named glutamycin. *M. kandleri and E. coli* GluTR are competitively inhibited by glutamycin.[28,30] The inhibitor binds within a deep pocket between the subdomains of domain 1 and is specifically recognized by an array of strictly conserved amino acids (Figs. 2B,4,5). The bidentate salt-bridge between the carboxylate group of glutamycin and Arg50 at the bottom of the pocket represents the most discriminating interaction. This resembles the carboxylate-recognition mode of aspartyl-tRNA synthetase and presumably of glutamyl-tRNA synthetases.[25] GluTR specifically recognizes the α-amino group and less specifically the ribose moiety of the inhibitor.

The observed structure of the enzyme-inhibitor complex supports the proposed catalytic mechanism for the enzyme[28] (Figs. 4,5). To obtain crystals of GluTR the oxidation-sensitive Cys48 had

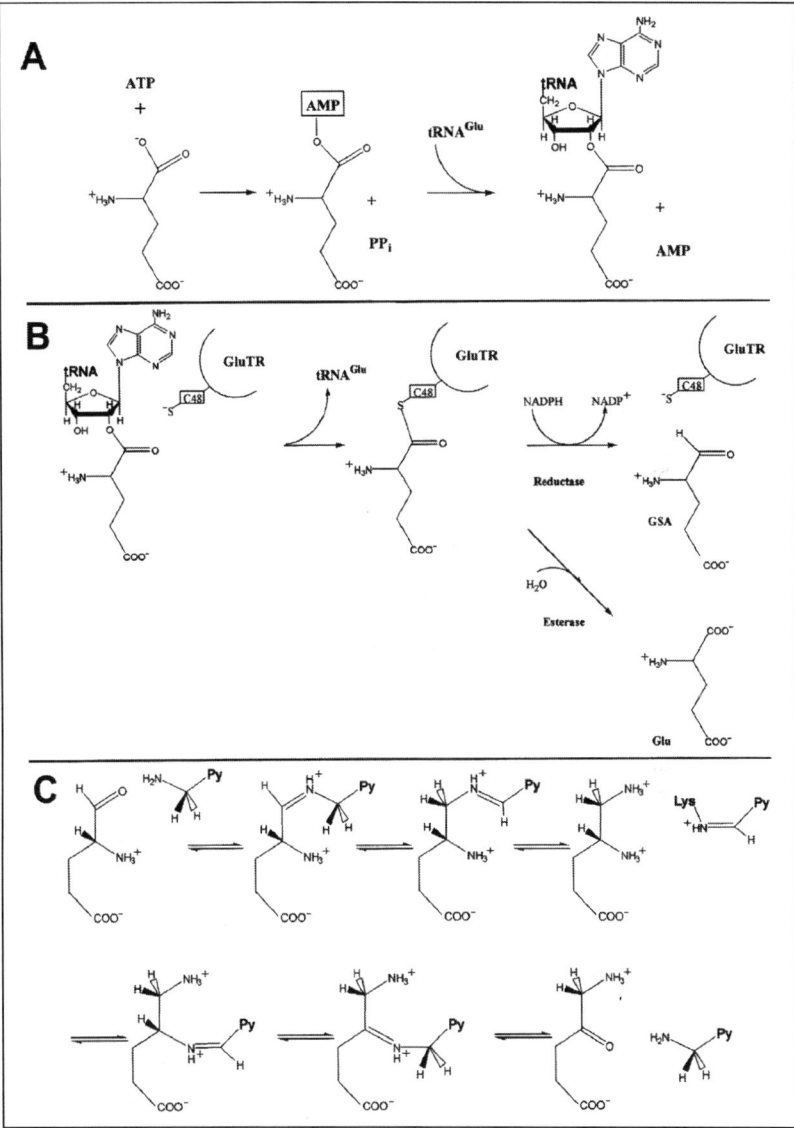

Figure 4. The chemical reactions catalyzed by (A) glutamyl-tRNA synthetase (GluRS), (B) GluTR and (C) GSAM. A) GluRS specifically binds ATP and glutamate, allowing the amino acid carboxyl oxygen of glutamate to displace pyrophosphate from ATP. In a second step the 2'/3'-OH of glutamyl-tRNA attacks the carbonyl carbon of glutamate-AMP displacing AMP and generating glutamyl-tRNA^{Glu}. B) The thiol group of the catalytic cysteine 48 of GluTR initiates a nucleophilic attack at the tRNA-bound carbonyl carbon of glutamyl-tRNA, covalently binding glutamate to the enzyme and displacing glutamyl-tRNA. In a second step hydride transfer from NADPH releases glutamate-1-semialdehyde (GSA) from GluTR, restoring the latter. In the absence of NADPH, GluTR functions as an esterase, transferring a hydroxyl group from water, producing glutamate. C) GSA from GluTR is transferred to GSAM. Here the pyridoxamine-5-phosphate (PMP) binds the aldehyde of GSA forming an aldimine. Proton transfer and Schiff' base formation producing the 5-pyridoxyl phosphate form of GSAM, releases the intermediate 4,5-diaminovalerate (DAVA). Aldimine formation at the former amino group of GSA, proton transfer and release of PMP produces the final product ALA.

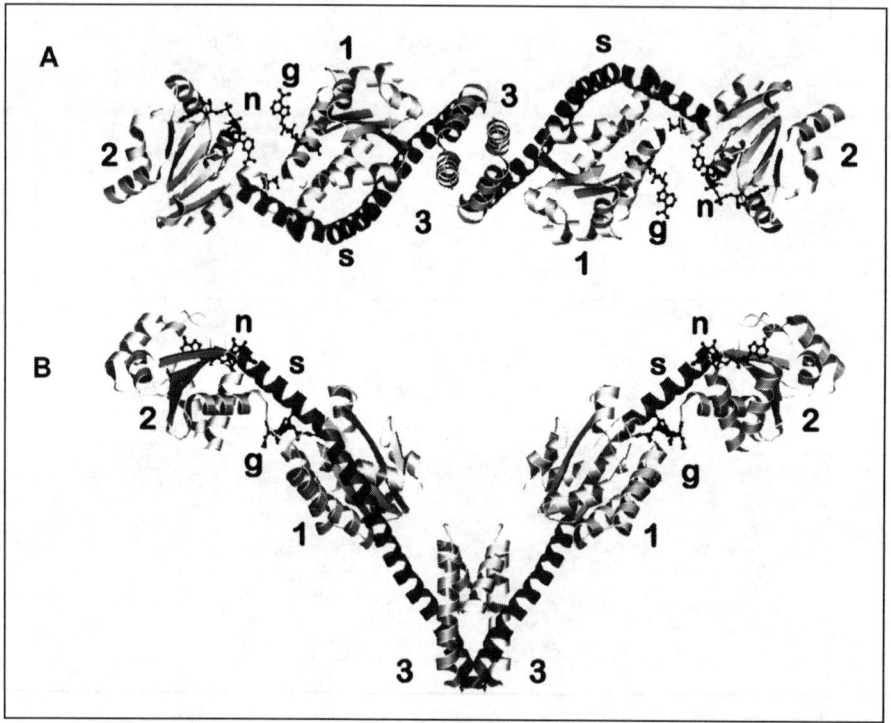

Figure 5. A schematic diagram of the *Methanopyrus kandleri* GluTR dimer viewed (A) perpendicular to and (B) along the 2-fold-axis. GluTR is composed of a catalytic domain (1), an NADPH-binding domain (2) and a dimerization domain (3) linked by a 'spinal' α-helix (s). Glutamycin (black) is recognized by the catalytic domain. Abbreviations and software as in Figure 2, except rendering by *RASTER3D* (Merrit and Murphy, 1994).

to be replaced by serine resulting in an inactivate enzyme.[29] In the crystal structure Ser48 (Cys48) is located at the edge of the glutamate-binding pocket. Its hydroxyl (sulfhydryl) group is in close proximity to the α-carbonyl carbon atom of glutamycin ready to nucleophilically attack the activated α-carboxylate of glutamyl-tRNA. This attack would lead to the observed covalent thioacyl intermediate and release of tRNA^Glu. In a second step the thioacyl intermediate would be reduced to the product GSA by hydride transfer from NADPH. The canonical NADPH-binding site in domain 2 is not occupied in the crystal structure. However, a reliable position for NADPH may be inferred from structurally related NAD(P)H-binding domains. The resulting distance between the nicotinamide moiety and the glutamate-binding pocket of 21 Å indicates that domain 2 must rotate around the end of the spinal helix to close the active site and to place NADPH close to the thioacyl-bound glutamate for hydride transfer to occur. The currently observed open structure of GluTR may therefore be described as a 'preactive' state. To what extent the individual steps of glutamyl-tRNA binding, tipping of domain 2, opening of the NADPH binding pocket and NADPH-binding occur in concert or consecutively is presently not clear.

The glutamate-binding pocket of GluTR clearly establishes the location of the 3'-terminal nucleotide of the glutamyl-tRNA substrate. Placing the acceptor arm of the *E. coli* glutamyl-tRNA structure[25] close to the active site pocket using glutamycin as a guide leads to a striking shape complementarity between the concave elbow region of the tRNA and the catalytic domain of GluTR. In this enzyme-substrate model glutamyl-tRNA may thus be moved immediately adjacent to the GluTR, placing the acceptor arm into the cleft between the catalytic and the NADPH-binding domains (Fig. 3B). The elbow region of the tRNA and in particular the D-stem would interact with

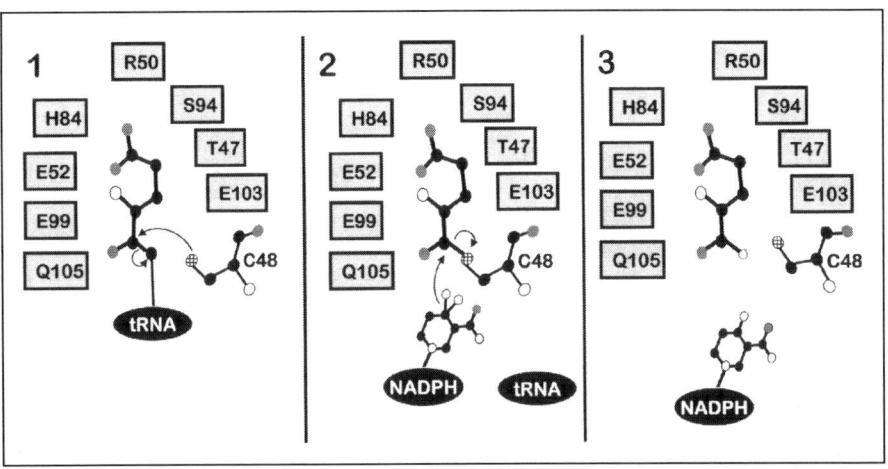

Figure 6. Catalytic mechanism of GluTR. Conserved residues responsible for specific substrate recognition through an intricate hydrogen-bonding network are indicated. The reactive cysteine residue nucleophilically attacks the aminoacyl bond of glutamyl-tRNA (1). An enzyme-localized thioester intermediate is formed with the release of free tRNA^{Glu} (2). The thioester is reduced by hydride transfer from NADPH leading to GSA (3).

the catalytic domain, while the anticodon would come close to the α-helical dimerization domain. Overall this hypothetical enzyme-substrate model reveals a large degree of surface complementarity between tRNA and GluTR. Surprisingly the regions of the tRNA interacting with the reductase are similar to those of the GluRS-tRNA complex: the acceptor arm lies within a large furrow between domains 1 and 2, allowing a multitude of specific interactions. The elbow region and the anticodon stem interact with domain 2 (domain 3 in GluRS). The anticodon itself appears to interact with domain 3, possibly allowing its specific recognition. Interestingly in both GluRS and GluTR, the (proposed) anticodon-recognition region contains a high number of arginines, though these are not highly conserved. For GluRS the tRNA complex was required to finally establish which arginines are responsible for anticodon recognition.[25]

The gene for GluTR, *hemA*, was cloned from various bacteria.[3] Its expression was intensively studied in *E. coli*, *S. typhimurium* and *P. aeruginosa*. The level of transcriptional regulation is low and influenced by oxygen tension, cellular heme concentration and the presence of alternative electron acceptors for anaerobic growth like nitrate.[31,32] Changes in cellular GluTR concentration were explained by heme-induced proteolytic degradation of the enzyme.[33] The N-terminal region of *S. typhimurium* GluTR was found to bind heme causing structural rearrangements and leading to proteolytic turnover.[34] In plants two genes encoding GluTR, *HEMA1* and *HEMA2*, have been cloned from *Arabidopsis thaliana*,[35] cucumber[36] and soybean.[37] Expression of *A. thaliana* HEMA1 is light-regulated through phytochrome photoreceptors. HEMA2, expressed only in the roots of seedlings, is not regulated by light.[38] In cucumber HEMA2 transcription is similarly independent of light. Transcription, however, occurs in all tissues.[36] Barley contains at least three HEMA genes.[39] Two HEMA genes are light-regulated, one follows circadian growth conditions.[39] This is discussed in more detail in Chapter 15.

Glutamate-1-Semialdehyde-2,1-Aminomutase

The two-step synthesis of ALA from glutamyl-tRNA requires the exchange of amino and oxo functionalities between carbon 1 and 2 of GSA. This reaction is catalyzed by glutamate-1-semialdehyde-2,1-aminomutase (GSAM, EC 5.4.3.8). The reaction differs from a classical aminotransferase reaction by its intramolecular nature. Nevertheless, GSAM represents a typical aminotransferase in structure and catalysis.

Biochemistry and Structure of GSAM

GSAM has been purified from various bacterial, green algal, and plant sources and was mainly found to be mainly dimeric. Two principle catalytic routes are thinkable. Reaction of the pyridoxal 5′-phosphate (PLP) form of GSAM results in the formation of a dioxovalerate intermediate. The pyridoxamine 5′-phosphate (PMP) gives rise to diaminovalerate (DAVA) as a reaction intermediate. Recombinant GSAMs from *E. coli*, *Synechococcus* sp. and other sources were found to catalyze both reactions.[40,41] Kinetic investigations of *Synechococcus* sp. and pea GSAM eventually identified DAVA as the true intermediate.[40,42,43] The active site lysine responsible for Schiff′s base formation with PMP was identified for various GSAMs.[41,44] The half-reactions of the PMP-dependent amino transfer are outlined in Figure 4C. Spectroscopic analyses, including stop flow experiments, identified the various proposed reaction intermediates.[41,45-47] GSAM catalyzes an anomalous enantiomeric reaction discriminating between (S)-GSA and (R)-GSA. Interestingly, (R)-GSA is a substrate for the first half-reaction but the resulting (R)-DAVA is either inactive or a poor substrate for the second half-reaction.[43,45] GSAMs are subject to inhibition by gabaculine, an inhibitor of GABA transaminase and other aminotransferases. Other inhibitors, 4-aminohex-5-ynoate, 4-aminohex-5-enoate and enantiomers of diaminopropyl sulfate have been synthesized.[48]

GSAM is a member of a large family of structurally related, PLP-dependent proteins that include aminotransferases, racemases, decarboxylases, synthetases and mutases. GSAM shares the prototypical aspartate aminotransferase fold and is structurally particularly similar to dialkylglycine decarboxylase and 4-aminobutyrate aminotransferase. Mechanistically GSAM differs from aminotransferases in that it exchanges amino and oxo functionalities within a single substrate molecule. It is a dimeric protein with an overall ellipsoidal shape when viewed along the dimer axis (Fig. 7).[49] The monomers interact through a large, convoluted interface. Morphologically distinct domains are not apparent. Instead, three domains may be rationalized based on three β-sheets: a three-stranded, antiparallel β-sheet in domain 1, a seven-stranded, parallel β-sheet (one strand is antiparallel) in domain 2, and a four-stranded, antiparallel β-sheet in domain 3. All domains, as well as interdomain connections, additionally include α-helices and long loops of low secondary structure content. Both cofactor and substrate are bound at the monomer-monomer interface. Domain 2 of one monomer accommodates the PLP/PMP head group, while the phosphate is additionally coordinated by a loop belonging to domain 2 of the second monomer. Loops from all three domains participate in lining the substrate pocket. In the crystal structure, the GSAM dimer is observed to be imperfectly symmetric. A loop consisting of residues 159-172

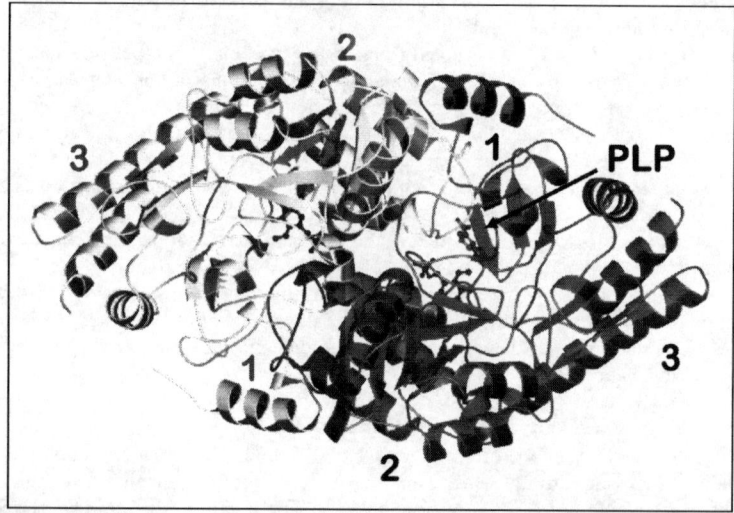

Figure 7. Ribbon diagram of dimeric GSAM.[49] One monomer is depicted in white the second in gray. Cofactors (PMP) and inhibitor (gabaculine) are shown as black ball and stick representations.

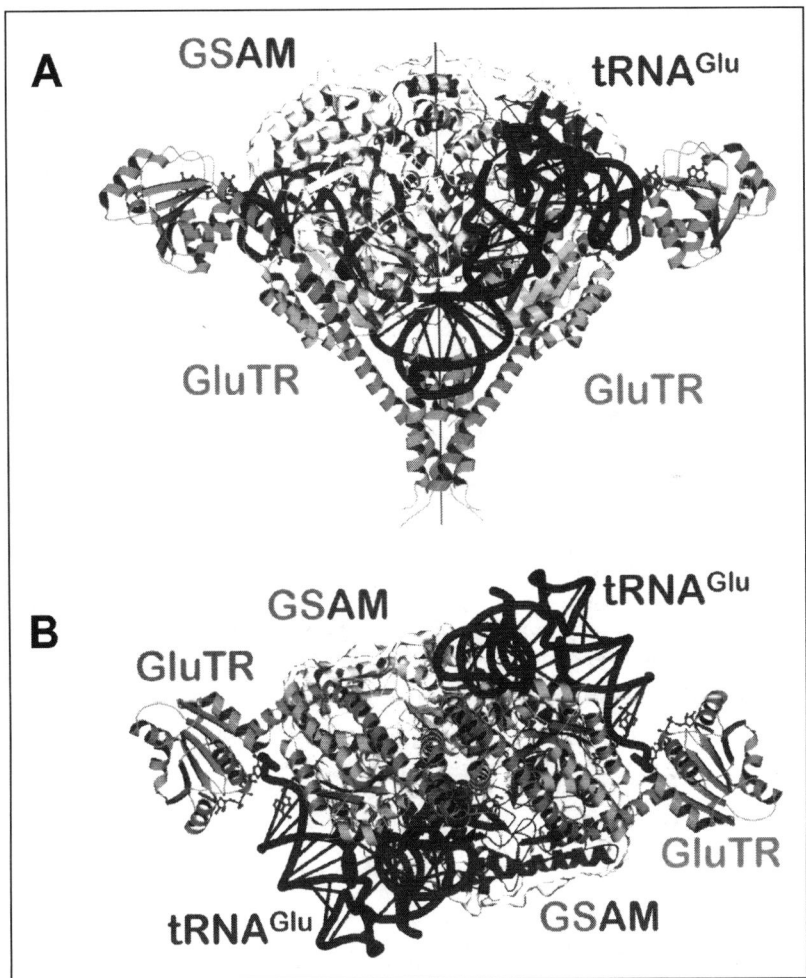

Figure 8. The proposed ternary complex of GluRS/tRNA^{Glu}/GSAM viewed (A) perpendicular to the common two-fold axis and (B) along this axis from the dimerization domain of GluTR (opposite direction to Fig. 4A). The ribbon depiction of GluTR is rendered in shades of green. tRNA^{Glu} is represented as a backbone model (purple), while GSAM is shown by a transparent surface covering a gray and white ribbon diagram. (Produced as Fig. 2 with the addition of using *GRASP*. Cofactor labels: n, NADPH; g, glutamycin; p, PMP; i, gabaculin (inhibitor). A color version of this figure is available online at www.eurekah.com.

and laterally covering the substrate pocket is partly disordered in one monomer, while it is well structured in the second. This loop is essential for proper enzyme function.[50] Presumably, GSAM oscillates between two conformational states in which one monomer is in the closed, active state (with ordered active-site lid), while the second is in a relaxed state, allowing product and substrate to diffuse out of and into the active site, respectively.

GSAMs are encoded by *hemL* in bacteria and by *gsa* in plants.[51-53] The plant *gsa* genes are light-regulated.[52,54] with the involvement of heterotrimeric G-protein, phospholipase C-catalyzed InsP$_3$ formation, Ca^{2+}-release and Ca^{2+}/CaM-dependent protein kinase activation of downstream signaling in *C. reinhardtii*.[55,56] Transcription of *hemL* from pathogenic *P. aeruginosa* and the soil bacterium *B. subtilis* is oxygen stress regulated.[32,53,57]

Metabolic Channeling of Glutamate-1-Semialdehyde

The tRNA-dependent formation of ALA found in plants and most bacteria requires the concerted action of the two enzymes GluTR and GSAM. Metabolically they are linked by the semialdehyde GSA (Fig. 2). The expected high chemical reactivity of GSA in the cellular medium mandates for a direct transfer of the aldehyde from GluTR to GSAM ensuring efficient formation of ALA. The three-dimensional shape of both GluTR and GSAM suggests an attractive solution to this metabolic problem. Placing the structure of the GSAM dimer from *Synechococcus sp.*[49] alongside the extended and similarly dimeric GluTR utilizing exactly the same symmetry axis immediately suggests that the open space delimited by the GluTR monomers could comfortably accommodate GSAM. The resulting model complex (Fig. 8) displays a striking degree of surface complementarity between both enzymes. The most important result of the proposed GluTR/GSAM complex is that the putative active site entrance of each GSAM monomer[49] is positioned opposite a depression in domain 1 of GluTR located opposite to the active site pocket. This depression and the glutamate recognition pocket are separated only by Arg50 and guarded by the conserved His84. The proposed complex may thus indicate that the GluTR-product GSA leaves the enzyme via this 'back door' of the glutamate recognition pocket and directly channels to the active site of GSAM, a distance of about 26 Å, without being exposed to the aqueous environment.

Both tRNA^{Glu} (see above) and GSAM can thus independently be docked onto GluTR, each in a single plausible position. Though separately docked, the model of the ternary complex of GluTR, tRNA^{Glu} and GSAM does not lead to steric clashes between GSAM and tRNA^{Glu}, instead GSAM could laterally extend the interaction surface of GluTR for the tRNA.

The proposed complex thus supports notions that major metabolic pathways in the cell might be organized as 'metabolons', where the respective enzymes closely interact with each other in an organized manner to ensure rapid and efficient synthesis of important metabolites.[26,27,29]

Acknowledgements

We thank Dr. Jürgen Moser, Dr. Wolf-Dieter Schubert and Dr. Stefan Schauer for their invaluble contibutions in the functional and structural characterization of GluTR. Financial support was granted by the Deutsche Forschungsgemeinschaft and Fonds der Chemischen Industrie.

References

1. Shemin D, Russell CS. Delta-aminolevulinic acid, its role in the biosynthesis of porphyrins and purines. J American Chem Soc 1953; 75:4873-4875.
2. Beale SI, Castelfranco PA. ^{14}C incorporation from exogenous compounds into -aminolevulinic acid by greening cucumber cotyledons. Biochem Biophys Res Commun 1973; 52:143-149.
3. Jahn D, Verkamp E, Söll D. Glutamyl-transfer RNA: A precursor of heme and chlorophyll biosynthesis. Trends Biochem Sci 1992; 17:215-218.
4. Ferreira GC. 5-Aminolevulinate synthase and mammalian heme biosynthesis. In: Ferreira GC, Moura JJG, Franco R, eds. Iron Metabolism. Weinheim, Germany: Wiley-VCH, 1999:15-29.
5. Kikuchi G, Kumar AM, Tamalge P et al. The enzymatic synthesis of d-aminolevulinic acid. J Biol Chem 1958; 233:1214-1219.
6. Gibson KD, Laver WG, Neuberger A. Initial steps in the biosynthesis of porphyrins. The formation of d-aminolevulinic acid from glycine and succinyl-CoA by particles of chicken erythrocytes. Biochem J 1958; 70:71-81.
7. Jordan PM. New comprehensive biochemistry; No. 19. In: Neuberger A, Van Deenen LLM, eds. Biosynthesis of Tetrapyrroles. Amsterdam: Elsevier, 1991.
8. Bolt EL, Kryszak L, Zeilstra-Ryalls J et al. Characterization of the Rhodobacter sphaeroides 5-aminolaevulinic acid synthase isoenzymes, HemA and HemT, isolated from recombinant Escherichia coli. Eur J Biochem 1999; 265:290-299.
9. Shoolingin-Jordan PM, Al-Daihan S, Alexeev D et al. 5-Aminolevulinic acid synthase: Mechanism, mutations and medicine. Biochim Biophys Acta 2003; 1647:361-366.
10. Zhang J, Ferreira GC. Transient state kinetic investigation of 5-aminolevulinate synthase reaction mechanism. J Biol Chem 2002; 277:44660-44669.
11. Hunter GA, Ferreira GC. Lysine-313 of 5-aminolevulinate synthase acts as a general base during formation of the quinonoid reaction intermediates. Biochemistry 1999; 38:3711-3718.
12. Hunter GA, Ferreira GC. Presteady-state reaction of 5-aminolevulinate synthase. Evidence for a rate-determining product release. J Biol Chem 1999; 274:12222-12228.

13. Tan D, Harrison T, Hunter GA et al. Role of arginine 439 in substrate binding of 5-aminolevulinate synthase. Biochemistry 1998; 37:1478-1484.
14. Gong J, Hunter GA, Ferreira GC. Aspartate-279 in aminolevulinate synthase affects enzyme catalysis through enhancing the function of the pyridoxal 5'-phosphate cofactor. Biochemistry 1998; 37:3509-3517.
15. Huang DD, Wang WY, Gough SP et al. delta-Aminolevulinic acid-synthesizing enzymes need an RNA moiety for activity. Science 1984; 225:1482-1484.
16. Schon A, Krupp G, Gough S et al. The RNA required in the first step of chlorophyll biosynthesis is a chloroplast glutamate tRNA. Nature 1986; 322:281-284.
17. Sekine S, Nureki O, Sakamoto K et al. Major identity determinants in the "augmented D helix" of tRNA(Glu) from Escherichia coli. J Mol Biol 1996; 256:685-700.
18. Sekine S, Nureki O, Tateno M et al. The identity determinants required for the discrimination between tRNA^{Glu} and tRNA^{Asp} by glutamyl-tRNA synthetase from Escherichia coli. Eur J Biochem 1999; 261:354-360.
19. Madore E, Florentz C, Giege R et al. Effect of modified nucleotides on Escherichia coli tRNA^{Glu} structure and on its aminoacylation by glutamyl-tRNA synthetase. Predominant and distinct roles of the mnm5 and s2 modifications of U34. Eur J Biochem 1999; 266:1128-1135.
20. Kruger MK, Sorensen MA. Aminoacylation of hypomodified tRNAGlu in vivo. J Mol Biol 1998; 284:609-620.
21. Stange-Thomann N, Thomann HU, Lloyd AJ et al. A point mutation in Euglena gracilis chloroplast tRNA(Glu) uncouples protein and chlorophyll biosynthesis. Proc Natl Acad Sci USA 1994; 91:7947-7951.
22. Freist W, Gauss DH, Soll D et al. Glutamyl-tRNA sythetase. Biol Chem 1997; 378:1313-1329.
23. Tumbula DL, Becker HD, Chang WZ et al. Domain-specific recruitment of amide amino acids for protein synthesis. Nature 2000; 407:106-110.
24. Nureki O, Vassylyev DG, Katayanagi K et al. Architectures of class-defining and specific domains of glutamyl-tRNA synthetase. Science 1995; 267:1958-1965.
25. Sekine S, Nureki O, Shimada A et al. Structural basis for anticodon recognition by discriminating glutamyl-tRNA synthetase. Nat Struct Biol 2001; 8:203-206.
26. Schubert WD, Moser J, Schauer S et al. Structure and function of glutamyl-tRNA reductase, the first enzyme of tetrapyrrole biosynthesis in plants and prokaryotes. Photosynth Res 2002; 74:205-215.
27. Moser J, Schubert WD, Heinz DW et al. Structure and function of glutamyl-tRNA reductase involved in 5-aminolaevulinic acid formation. Biochem Soc Trans 2002; 30:579-584.
28. Moser J, Lorenz S, Hubschwerlen C et al. Methanopyrus kandleri Glutamyl-tRNA Reductase. J Biol Chem 1999; 274:30679-30685.
29. Moser J, Schubert WD, Beier V et al. V-shaped structure of glutamyl-tRNA reductase, the first enzyme of tRNA-dependent tetrapyrrole biosynthesis. EMBO J 2001; 20:6583-6590.
30. Schauer S, Chaturvedi S, Randau L et al. Escherichia coli glutamyl-tRNA reductase. Trapping the thioester intermediate. J Biol Chem 2002; 277:48657-48663.
31. Krieger R, Rompf A, Schobert M et al. The Pseudomonas aeruginosa hemA promoter is regulated by Anr, Dnr, NarL and integration host factor. Mol Genet Genomics 2002; 267:409-417.
32. Schobert M, Jahn D. Regulation of heme biosynthesis in nonphototrophic bacteria. J Mol Microbiol Biotechnol 2002; 4:287-294.
33. Wang L, Elliott M, Elliott T. Conditional stability of the HemA protein (glutamyl-tRNA reductase) regulates heme biosynthesis in Salmonella typhimurium. J Bacteriol 1999; 181:1211-1219.
34. Wang L, Wilson S, Elliott T. A mutant HemA protein with positive charge close to the N terminus is stabilized against heme-regulated proteolysis in Salmonella typhimurium. J Bacteriol 1999; 181:6033-6041.
35. Kumar AM, Csankovszki G, Soll D. A second and differentially expressed glutamyl-tRNA reductase gene from Arabidopsis thaliana. Plant Mol Biol 1996; 30:419-426.
36. Tanaka R, Yoshida K, Nakayashiki T et al. Differential expression of two hemA mRNAs encoding glutamyl-tRNA reductase proteins in greening cucumber seedlings. Plant Physiol 1996; 110:1223-1230.
37. Sangwan I, O'Brian MR. Expression of a soybean gene encoding the tetrapyrrole-synthesis enzyme glutamyl-tRNA reductase in symbiotic root nodules. Plant Physiol 1999; 119:593-598.
38. McCormac AC, Fischer A, Kumar AM et al. Regulation of HEMA1 expression by phytochrome and a plastid signal during de-etiolation in Arabidopsis thaliana. Plant J 2001; 25:549-561.
39. Bougri O, Grimm B. Members of a low-copy number gene family encoding glutamyl-tRNA reductase are differentially expressed in barley. Plant J 1996; 9:867-878.
40. Smith MA, Grimm B, Kannangara CG et al. Spectral kinetics of glutamate-1-semialdehyde aminomutase of Synechococcus. Proc Natl Acad Sci USA 1991; 88:9775-9779.

41. Ilag LL, Jahn D. Activity and spectroscopic properties of the Escherichia coli glutamate 1-semialdehyde aminotransferase and the putative active site mutant K265R. Biochemistry 1992; 31:7143-7151.
42. Pugh CE, Harwood JL, John RA. Mechanism of glutamate semialdehyde aminotransferase. Roles of diamino- and dioxo-intermediates in the synthesis of aminolevulinate. J Biol Chem 1992; 267:1584-1588.
43. Friedmann HC, Duban ME, Valasinas A et al. The enantioselective participation of (S)- and (R)-diaminovaleric acids in the formation of delta-aminolevulinic acid in cyanobacteria. Biochem Biophys Res Commun 1992; 185:60-68.
44. Grimm B, Smith MA, von Wettstein D. The role of Lys272 in the pyridoxal 5-phosphate active site of Synechococcus glutamate-1-semialdehyde aminotransferase. Eur J Biochem 1992; 206:579-585.
45. Smith MA, Kannangara CG, Grimm B. Glutamate 1-semialdehyde aminotransferase: Anomalous enantiomeric reaction and enzyme mechanism. Biochemistry 1992; 31:11249-11254.
46. Brody S, Andersen JS, Kannangara CG et al. Characterization of the different spectral forms of glutamate 1-semialdehyde aminotransferase by mass spectrometry. Biochemistry 1995; 34:15918-15924.
47. Smith MA, King PJ, Grimm B. Transient-state kinetic analysis of Synechococcus glutamate 1-semialdehyde aminotransferase. Biochemistry 1998; 37:319-329.
48. Contestabile R, Jenn T, Akhtar M et al. Reactions of glutamate 1-semialdehyde aminomutase with R- and S-enantiomers of a novel, mechanism-based inhibitor, 2,3-diaminopropyl sulfate. Biochemistry 2000; 39:3091-3096.
49. Hennig M, Grimm B, Contestabile R et al. Crystal structure of glutamate-1-semialdehyde aminomutase: An alpha2-dimeric vitamin B6-dependent enzyme with asymmetry in structure and active site reactivity. Proc Natl Acad Sci USA 1997; 94:4866-4871.
50. Contestabile R, Angelaccio S, Maytum R et al. The contribution of a conformationally mobile, active site loop to the reaction catalyzed by glutamate semialdehyde aminomutase. J Biol Chem 2000; 275:3879-3886.
51. Grimm B. Primary structure of a key enzyme in plant tetrapyrrole synthesis: Glutamate 1-semialdehyde aminotransferase. Proc Natl Acad Sci USA 1990; 87:4169-4173.
52. Ilag LL, Kumar AM, Söll D. Light regulation of chlorophyll biosynthesis at the level of 5-aminolevulinate formation in Arabidopsis. Plant Cell 1994; 6:265-275.
53. Hungerer C, Troup B, Romling U et al. Cloning, mapping and characterization of the Pseudomonas aeruginosa hemL gene. Mol Gen Genet 1995; 248:375-380.
54. Vavilin DV, Vermaas WF. Regulation of the tetrapyrrole biosynthetic pathway leading to heme and chlorophyll in plants and cyanobacteria. Plant Physiol 2002; 115:9-24.
55. Herman CA, Im CS, Beale SI. Light-regulated expression of the gsa gene encoding the chlorophyll biosynthetic enzyme glutamate 1-semialdehyde aminotransferase in carotenoid-deficient Chlamydomonas reinhardtii cells. Plant Mol Biol 1999; 39:289-297.
56. Im CS, Beale SI. Identification of possible signal transduction components mediating light induction of the Gsa gene for an early chlorophyll biosynthetic step in Chlamydomonas reinhardtii. Planta 2000; 210:999-1005.
57. Chen L, Keramati L, Helmann JD. Coordinate regulation of Bacillus subtilis peroxide stress genes by hydrogen peroxide and metal ions. Proc Natl Acad Sci USA 1995; 92:8190-8194.

5-Aminolaevulinic Acid Dehydratase, Porphobilinogen Deaminase and Uroporphyrinogen III Synthase

Heidi L. Schubert, Peter T. Erskine and Jonathan B. Cooper*

Abstract

The three enzymes 5-aminolaevulinic acid dehydratase (ALAD, E.C.4.2.1.24; some times referred to as porphobilinogen synthase), porphobilinogen deaminase (EC 4.3.1.8; also known as hydroxymethylbilane synthase) and uroporphyrinogen III synthase (U3S; E.C.4.2.1.75) together convert 5-aminolaevulinic acid (ALA) into uroporphyrinogen III, from which all tetrapyrroles are synthesized. The X-ray structures of several ALADs have been determined showing that the enzyme forms a large homo-octameric structure with all eight active sites on the outer surface. Each subunit adopts the TIM barrel fold with an N-terminal arm which forms extensive inter-subunit interactions. The active site of each subunit is located in a pronounced cavity formed by loops at the C-terminal ends of the strands forming the TIM barrel. Current proposals for the catalytic mechanism involve the binding of both substrate moieties by formation of Schiff bases with two invariant active site lysine residues. Structural studies of porphobilinogen deaminase have shown that the enzyme has three domains, two of which show a strong structural resemblance to a number of periplasmic binding proteins. The reaction catalysed by uroporphyrinogen III synthase involves cyclization and ring inversion, predicted to proceed through a spirocyclic intermediate. X-ray structures of the enzyme from humans and a thermophilic bacterium have enabled models of the catalytic process to be proposed.

5-Aminolaevulinic Acid Dehydratase

Introduction

5-Aminolaevulinic acid dehydratase (ALAD, E.C.4.2.1.24), also referred to as porphobilinogen synthase, catalyzes one of the initial steps in the biosynthesis of tetrapyrroles involving the condensation of two 5-aminolaevulinic acid (ALA) molecules to form the pyrrole porphobilinogen (PBG)[1-5] (Fig. 1). The enzyme plays an essential role in tetrapyrrole biosynthesis by catalysing the formation of the pyrrole porphobilinogen from two molecules of 5-aminolaevulinic acid. Four porphobilinogen molecules are subsequently condensed in a reaction catalyzed by porphobilinogen deaminase to form the linear tetrapyrrole, preuroporphyrinogen (or 1-hydroxymethylbilane), which is cyclized and rearranged by uroporphyrinogen III synthase to give uroporphyrinogen III, the first macrocyclic tetrapyrrole in the pathway.[1-5] The three steps from ALA to uroporphyrinogen III are common to the biosynthesis of haem, chlorophyll, cobalamins and all other tetrapyrroles.

*Corresponding Author: Jonathan B. Cooper—School of Biological Sciences, University of Southampton, Bassett Crescent East, Southampto, SO16 7PX, UK. Email: j.b.cooper@soton.ac.uk

Tetrapyrroles: Birth, Life and Death, edited by Martin J. Warren and Alison G. Smith.
©2009 Landes Bioscience and Springer Science+Business Media.

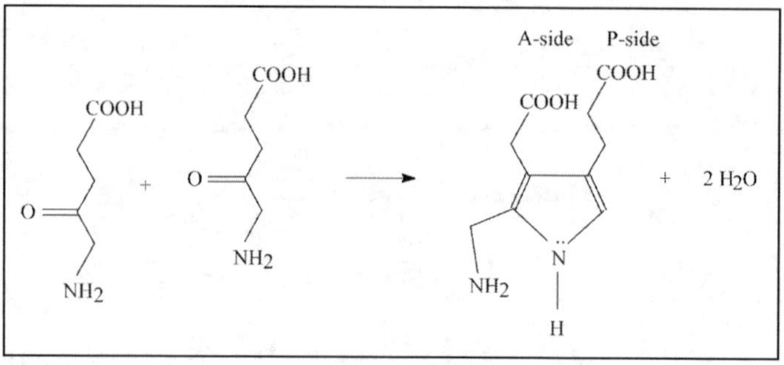

Figure 1. The Knorr-type condensation reaction catalyzed by 5-aminolevulinic acid dehydratase (ALAD) indicating the A- and P-sides of the product porphobilinogen.

Single turnover experiments have shown that the first substrate molecule to bind to the enzyme ultimately forms the 'propionate' half of the product PBG, whilst the second substrate molecule forms the 'acetate' half of PBG.[6] This has led to the widely used terminology of the 'P' and 'A' binding sites in the enzyme which bind the substrates forming the propionate and acetate halves of the product, respectively (see Fig. 1).

ALADs have been purified from a variety of sources including bovine liver,[7] human erythrocytes,[8] bacteria such as *E. coli*[9] and plants such as spinach.[10-11] There are differences between the ALAD enzymes in terms of their metal requirements, kinetic parameters, pH dependence, inactivation by inhibitors and susceptibility to oxidation. There is evidence that all ALADs require a metal ion for activity with animal enzymes using zinc but with plant enzymes requiring magnesium. Representatives of both classes exist in bacteria.

In humans, clinically significant deficiencies in ALAD activity arise from genetic mutations and from inhibition of the enzyme as a result of lead poisoning or type I tyrosinaemia.[12-16]

Whilst the human and bovine ALAD enzymes have been purified in high yield from blood and liver where the enzyme is present in high abundance,[7-8] the majority of recent studies have involved the production of recombinant enzyme in *E. coli*.[17] The enzyme activity can be assayed spectrophotometrically by the use of Ehrlich's reagent[18] which forms a colored adduct with the product porphobilinogen. The absorbance of this adduct is measured at 555 nm where its molar absorption coefficient is 6.02×10^4 $M^{-1}cm^{-1}$.

Gel-filtration studies have shown the majority of ALAD enzymes to be octameric although there are early reports that the plant enzyme is hexameric.[1] Further evidence that the enzyme forms octamers was provided by EM and solution scattering studies[19-20] and was finally confirmed by X-ray structural studies.[21] The sequences are now available for ALADs from dozens of species. All ALADs share a high degree of amino acid sequence identity and contain about 350 residues per subunit, although the plant enzymes are extended by about 100 residues at the N-terminus to form the transit peptide for targeting the enzyme to chloroplasts.[1-5] Studies on bovine and human ALAD showed that catalysis proceeds by the formation of a Schiff base link at the P-site between the 4-keto-group of substrate and an invariant lysine residue equivalent to Lys 247 in *E. coli* ALAD.[22] Later structural studies established the importance of a second invariant lysine residue (195 in *E. coli* ALAD).[21] Comparison of ALAD primary sequences reveals a strong degree of similarity which is most notable in the vicinity of these two active site lysine residues. Another conserved region of the ALAD sequence has been implicated in metal ion binding at the active site.[23]

The nature of the metal ligands in bovine ALAD has been studied using EXAFS which suggested that one of the enzyme's zinc ions has 4 cysteine ligands whereas the other is coordinated by more polar ligands.[24] Subsequent structural studies established that the catalytic zinc ion is coordinated tetrahedrally by only three cysteines allowing one ligand position to be occupied by a solvent molecule that is likely to be displaced by bound substrate during the reaction.[21]

Most of the zinc-dependent ALADs can bind 2 zinc ions per subunit and these sites have been classified variously as A and B or α and β[9,24-25] where the metal at the α site is essential for catalytic activity and the metal at the β site is thought to have more of a structural role. The catalytic α-site is now known to consist of three conserved zinc-binding cysteine residues. In contrast, ALADs from plants and some prokaryotes lack several of these characteristic cysteines and have a requirement for magnesium ions. For example, two of the conserved cysteines are replaced by aspartates, which are more appropriate for coordination of the essential Mg^{2+} ions.[23] Accordingly the magnesium dependent enzymes are less sensitive to oxidation than the zinc dependent ALADs.

X-Ray Structure

A number of high resolution X-ray crystal structures have been determined recently[21,26-27] (e.g., PDB codes 1aw5, 1b4e, 1b4k) confirming that the subunits of the enzyme associate to form compact octamers. Recently, evidence for hexameric quaternary forms of some ALADs has been found.[28-29] The monomers of the ALAD octamer are organized with 422 or D_4 point group symmetry with their active sites oriented towards the solvent region. The octamer of *E. coli* enzyme (PDB code 1b4e), which will be treated as the archetypal ALAD for the remainder of this chapter, has overall dimensions of 104 x 104 x 83 $Å^3$ consistent with those deduced from EM and solution scattering studies of the bovine enzyme.[19-20] Each subunit adopts the $(\alpha/\beta)_8$ or TIM-barrel fold with an N-terminal arm approximately 30 residues in length which forms extensive intersubunit interactions. The TIM-barrel fold is shared by a number of aldolases, a class of enzymes that are functionally related to ALAD. In ALAD the most extensive of the quaternary contacts are those between pairs of monomers which associate with their arms wrapped around each other to form compact dimers. These dimers further associate to form the octamer which possesses a solvent filled channel of 15-20 Å diameter passing through the centre.

Structure of the Monomer

The first 30 amino acids of *E. coli* ALAD form an extended arm-like structure pointing away from the compact α/β domain (Fig. 2A). The arm, which is of variable length in different ALADs, includes a region of distorted 3_{10}-helix (residues 8-12) followed by an α-helix (residues 15-21) denoted α1. The remaining tertiary structure of the ALAD monomer is dominated by the $(\alpha/\beta)_8$ or TIM-barrel formed by an 8-membered cylindrical β-sheet surrounded by 8 α-helices. In all enzymes adopting the $(\alpha/\beta)_8$ fold, the active site is located in an opening formed by the loops connecting the C-terminal ends of the parallel β-strands in the barrel with their ensuing α-helical segments.[30] In *E. coli* ALAD the TIM-barrel is formed from the following elements of secondary structure from N- to C-terminus: β1, α2, β4, α3, β5, α4, β6, α5, β7, α6, β8, α7, β9, α9, β10 and α10 as shown in Figure 2A.[26] The loop regions between these α and β segments are elaborated extensively at the active site end of the barrel where the regular alternation of helices and strands is broken by the insertion of several extra secondary structure elements. One example of this is the β-hairpin which lies between β1 and α2. Here the hydrogen bonds made by the strands of this hairpin (β2 and β3) mean that they are part of the main β-sheet of the molecule and provide an extension to it at the active site end of the barrel. Another elaboration on the basic TIM-barrel fold is the loop covering the active site (197-220) which includes a region of α-helix involving residues 202-208 (in *E. coli*) denoted α_{act}. There was very little electron density for this region of the yeast ALAD molecule when crystallized in the absence of inhibitor[21] but it is very well defined in the *E. coli* ALAD structure with the competitive inhibitor laevulinic acid bound. Finally, there is also a helix (α8) that lies between β9 and α9 in the primary sequence. This helix is important for quaternary interactions about a 2-fold axis within the dimer and will be discussed later.

The longest stretches of conserved sequence in ALADs are all involved in maintaining the structure of the active site which is formed mainly by the loop regions at the exposed end of the β-barrel. Several of these loops make quaternary contacts with neighboring subunits, implying that they may play a role in transmitting conformational change throughout the octamer. The residues in the loop covering the active site (197-220 in *E. coli* ALAD) are also very strongly conserved indicating that they have an important role in substrate binding during catalysis.

Structure of the Dimer

The N-terminal arm of ALAD (residues 1-30 in *E. coli*) adopts a conformation in which it partly wraps around the $(\alpha/\beta)_8$ barrel domain of a neighboring monomer to form a dimer (Fig. 2B).

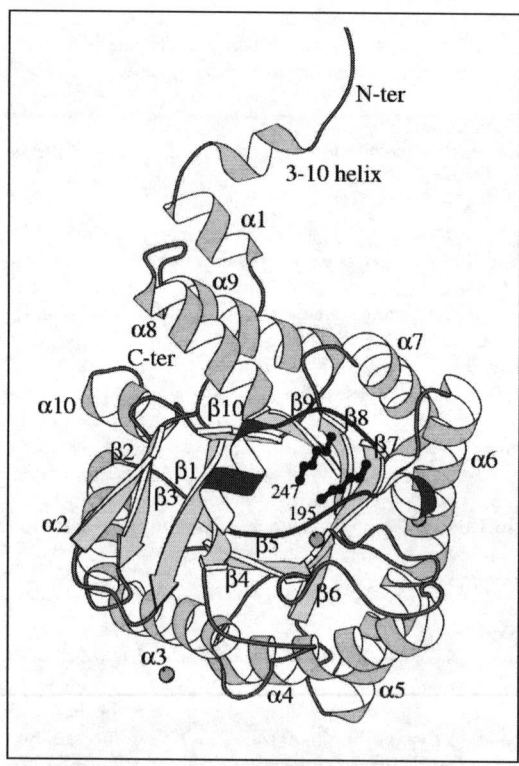

Figure 2A. This shows a ribbon diagram of the fold of the *E. coli* ALAD monomer (PDB code 1b4e). The dominant feature of the tertiary structure is the closed $(\alpha/\beta)_8$ or TIM-barrel with the active site located in a pronounced cavity (facing reader). The active site flap (197-220) is shaded and the enzyme's N-terminal arm region (residues 1-30), which forms extensive quaternary contacts, is shown above the barrel. The two adjacent lysines implicated in catalysis (195 and 247) are shown in dark grey. Lys 247 forms a Schiff base link with P-site ALA. Grey spheres indicate the locations of the zinc binding sites. The active site zinc ion held by three cysteines can be seen close to the two lysines in the centre of the TIM-barrel. The zinc ion on the right hand side (β-site) is involved in inter-subunit contacts and that on the lower left is involved in crystal contacts.

These interactions occur between monomers related by an intervening 2-fold axis so as to generate dimers resembling the number 69. The $(\alpha/\beta)_8$ domains of each monomer interact extensively about the 2-fold axis so that the two active sites are pointing approximately perpendicular to one another. This interaction is common to all ALADs which have been analyzed structurally and is in accord with cross-linking studies of bovine ALAD[20] which suggested that the strongest oligomeric interaction between ALAD monomers is dimer formation. The N-terminus of the arm region of each monomer interacts with helices $\alpha4$ and $\alpha5$ of the second subunit in the dimer. The arm region itself adopts helical conformations between residues 8-12 and 15-21 (in *E. coli* ALAD) and these regions interact with helix $\alpha6$ of the adjacent subunit in the dimer.

The principal barrel-barrel contacts in the dimer involve the helical segments $\alpha7$, $\alpha8$ and $\alpha9$. Helix $\alpha8$ is approximately perpendicular to the 2-fold axis of the dimer and pairs in an antiparallel manner with the equivalent helix in the neighboring monomer. Helix $\alpha8$ appears to be an elaboration of the classical $(\alpha/\beta)_8$ barrel, as mentioned above, perhaps to allow extensive nonpolar inter-subunit contacts within the ALAD dimers. Helix $\alpha7$ interacts with the equivalent helix in the neighboring monomer in a parallel manner as they are roughly aligned with the intervening 2-fold axis. Residues in helix $\alpha7$ and $\alpha9$ form several intra- and inter-molecular electrostatic interactions

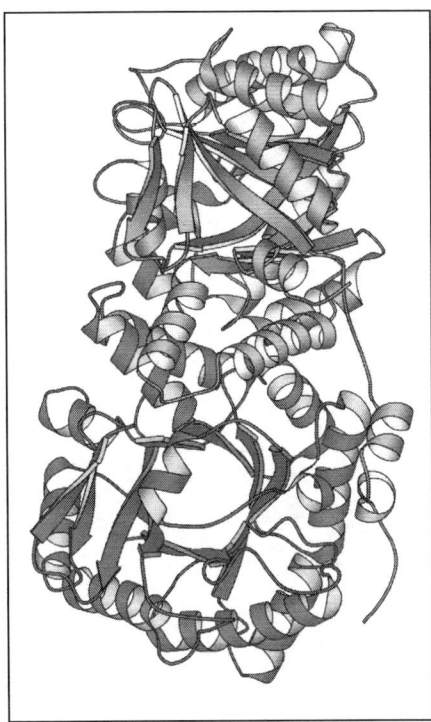

Figure 2B. Shows the formation of *E. coli* ALAD dimers. The two monomers (shaded differently) associate extensively about an intervening 2-fold axis and their arm regions wrap around the TIM-barrel domain of the neighboring subunit.

involving conserved or invariant residues indicating that these largely buried ionic interactions may be important for conserving the quaternary structures of all ALADs.

Structure of the Octamer

The dimers in the ALAD octamer have the gross appearance of ellipsoids with their long axes inclined slightly with respect to the 4-fold symmetry axis. The overall appearance of the octamer is shown in Figure 2C, where the dimers are colored separately. All dimers within the octamer interact with each other in an identical manner and these interactions are mediated principally by the arm regions that are on the surface of each dimer. In the dimer-dimer contacts, the two helical segments in the arm region (residues 8-12 and 15-21 in *E. coli* ALAD) associate with one end of a β-barrel in the neighboring dimer. This end of the β-barrel is essentially 'capped' by the arm region of the neighboring subunit leaving the other end of the barrel, which forms the active site, exposed on the surface of the octamer.

The Active Site

The loops at the exposed end of the β-barrel form a pronounced cavity, the base of which is dominated by two spatially adjacent lysine side chains numbered 195 and 247 in *E. coli* ALAD (Fig. 3). Both of these residues occur at the C-terminal ends of adjacent β-strands in the barrel, namely β7 and β8. The side chains of these two lysines emerge from a hydrophobic pocket formed by the side chains of several tyrosines and other highly conserved residues. The active site also contains a number of invariant polar residues such as Glu 123, Ser 273, Asp 118 and Ser 165 whose putative roles in substrate binding and catalysis are discussed later. Lys 247, which is known to form a Schiff base to the P-site substrate molecule, has a more hydrophobic environment than its neighbor (Lys 195).

The substrate binding cleft in ALAD has a number of aromatic residues which are predominantly tyrosines whose hydroxyl groups point towards the active site; these aromatic groups are almost totally invariant. The two phenylalanine side chains at positions 36 and 79 appear to form a hydrophobic patch against which nonpolar residues in the helical segment (α_{act}) of the loop covering the active site are packed. The hydroxyl groups of tyrosines 270 and 312 are in hydrogen-bonding contact and are close to the side chain -NH$_2$ group of Lys 247. As will be discussed later, these tyrosine residues are likely to have an important role in substrate binding since they interact with the inhibitor, laevulinic acid, covalently bound to Lys 247. The -NH$_2$ group of the other lysine, Lys 195, is approximately 5Å from a well defined zinc ion which is held by 3 cysteine side chains and a solvent molecule.

The Active Site Metal Ion (α-site)

The cysteine residues numbered 120, 122 and 130 in *E. coli* ALAD are located in the loop connecting β5 and α4 (Fig. 2A). This region of the molecule was previously identified as a potential metal binding region from primary sequence comparisons, since it possesses a number of likely zinc binding cysteine residues which are absent in the magnesium binding plant ALADs.[23] Two of the

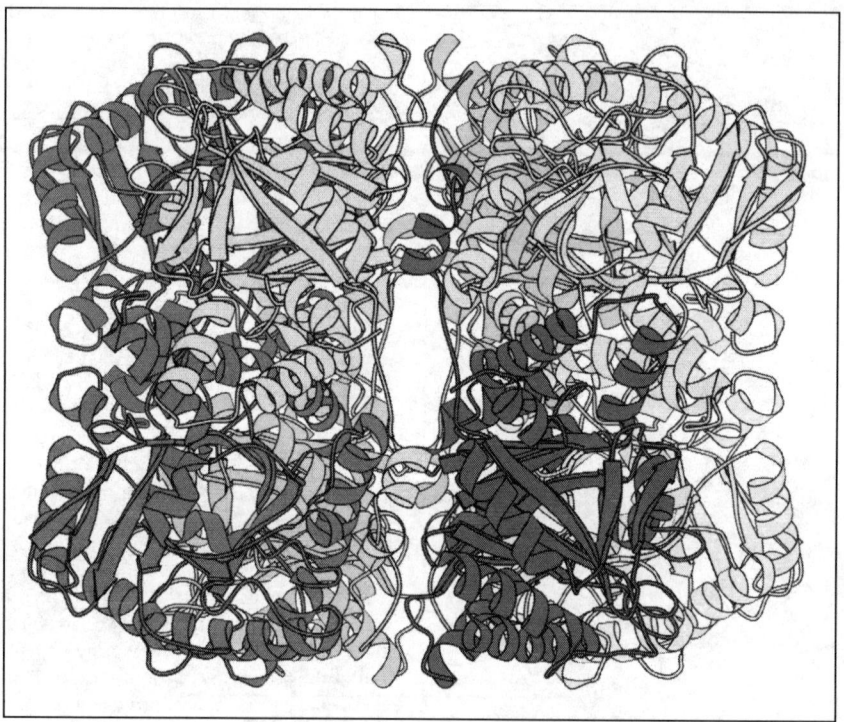

Figure 2C. Shows the organisation of the *E. coli* ALAD octamer. The dimers (shaded differently) are oriented with their long axes slightly inclined with respect to the central 4-fold axis of the octamer (vertical). Interactions between adjacent dimers are mediated principally by the arm regions. This figure was prepared using the program BOBSCRIPT.[68]

above three cysteine residues (122 and 130) are replaced by aspartate in plant ALADs and Cys 120 becomes Ala. The three cysteine side chains in *E. coli* ALAD along with a water molecule coordinate the zinc ion with tetrahedral geometry (see Fig. 3). There is good electron density for this solvent ligand which lies between the metal ion and the two active site lysines. The zinc ion is close to the side chain of the invariant Ser 165 which itself hydrogen-bonds to the carboxyl group of an invariant aspartate (Asp 118). These two residues interact indirectly with the zinc ion through a network of hydrogen bonded water molecules. The proximity of these interacting groups to the Schiff base lysine indicates that they most probably participate in substrate binding or catalysis. It is very unusual to find an active site zinc ion coordinated by three cysteine ligands since zinc shows a preponderance of imidazole and carboxyl ligands at the active sites of other enzymes such as the functionally related metalloaldolases.

The structure of magnesium-dependent ALAD from the pathogenic bacterium *Pseudomonas aeruginosa* has been solved with laevulinic acid bound[27] (PDB code 1b4k) revealing that this octameric enzyme effectively consists of four dimers. The monomers of each dimer differ from each other by having 'open' and 'closed' active site pockets. In the 'closed' subunit the active site is shielded by the active site loop which is in a well-defined conformation covering the bound laevulinic acid residue. In the 'open' subunit, the active site loop is partially disordered although laevulinic acid appears to be bound. The most surprising finding is that no metal ions could be detected in the active site of either subunit. However a hydrated magnesium ion was found at the second metal-binding site (β-site) of the 'closed' subunit, which is not a conserved feature of ALADs (see below). In the 'open' subunit, conformational differences in the side chains of amino acids preclude magnesium from binding to the β-site.

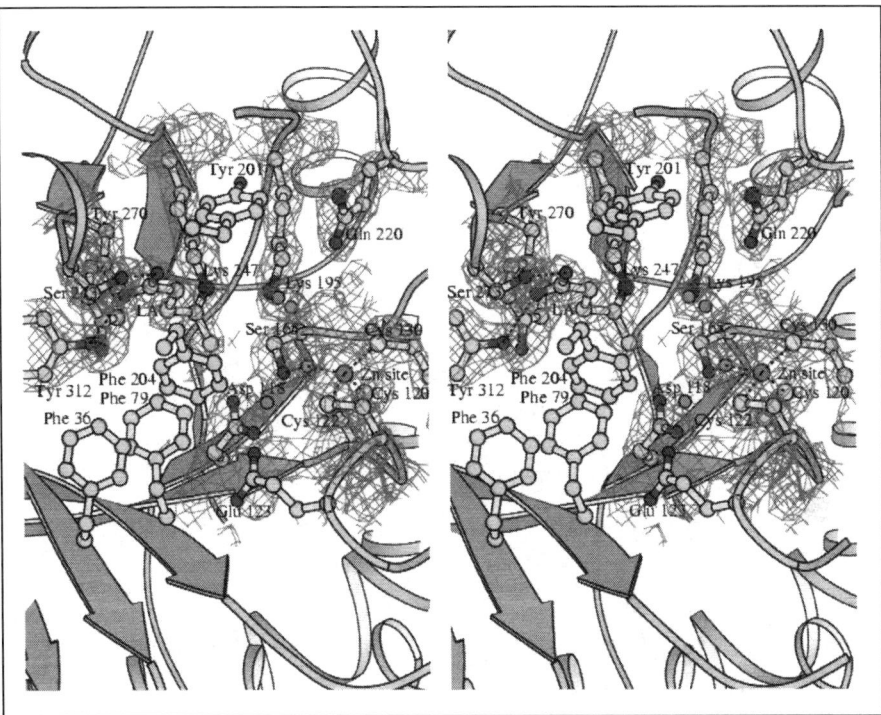

Figure 3. A stereo-view of the active site of *E. coli* ALAD showing the two lysines implicated in catalysis (247 and 195) in the top centre. Lysine 247 has formed a Schiff base link to the inhibitor laevulinic acid (LA) which is hydrogen bonded to Ser 273 and Tyr 312 (centre left). The zinc binding site consists of cysteines 120, 122 and 130 and a solvent molecule which coordinate the metal ion (shown on the right). The $2F_o$-F_c electron density at 2.0 Å resolution (contoured at 1.2σ) is shown for selected residues as grey lines. Phe 204 (shown below the levulinic acid moiety) is notable since it was disordered in the native yeast ALAD structure, i.e., its interaction with the inhibitor may contribute to ordering of the active site flap (197-220). The feature of electron density containing two solvent molecules just above Ser 165 and in front of Lys 195 may originate from a disordered laevulinic acid residue at the A-site. This figure was prepared using the program BOBSCRIPT.[68]

The Allosteric Metal Binding Site (β-site)

Adjacent to the side chain of Lys 195 in *E. coli* ALAD is a water-filled pocket which lies between the TIM-barrel domain and the N-terminal arm of the neighboring subunit in the dimer. This pocket contains a zinc ion which is coordinated octahedrally by one side chain oxygen of Glu 232 (in helix α_6) and five solvent molecules, presumed to be waters (Fig. 4). These water molecules are hydrogen-bonded to surrounding side chain atoms of Asp 169, Gln 171 and Asp 236. There have been many studies of the metal ion requirements of ALAD and there is experimental evidence for the enzyme having at least two distinct metal binding sites which contribute to its catalytic function.[24-25] While the closest side chain atom of Lys 195 is less than 8 Å from this zinc ion, the lysine's ε-NH_2 group is over 11 Å from the metal ion. Hence, it is difficult to envisage a direct catalytic role for this metal site. Nonetheless, it would be expected to contribute significantly to the stability of the octamer since the arm residue Arg 12 (from the neighboring subunit) forms a salt-bridge with the residue coordinating the metal ion (Glu 232). This would appear to be the most likely site at which Mg^{2+} could bind to this enzyme. The apparent activating properties of magnesium on *E. coli* ALAD may therefore stem from an effect on the active site flap which passes close to this metal ion or an effect on the enzyme structure close to the active site. In *E. coli*

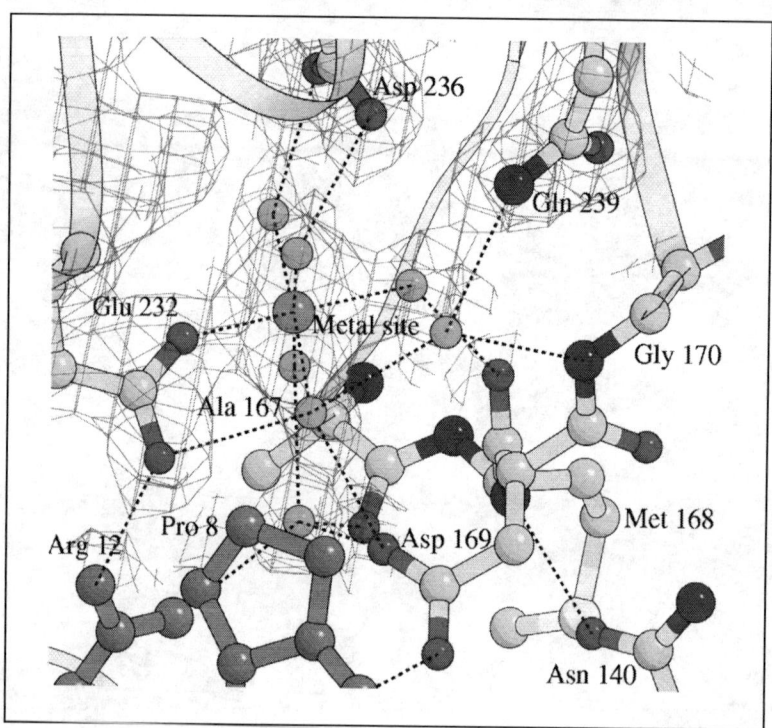

Figure 4. The allosteric metal binding site (β-site) of *E. coli* ALAD occupied by a penta-hydrated zinc ion bound to Glu 232. This glutamate side chain and five water molecules coordinate a zinc ion with approximate octahedral geometry. This site is at the subunit interface in the octamer. The shaded arginine and proline side chains (bottom left) originate from the N-terminal arm region of the adjacent monomer. The $2F_o$-F_c electron density at 2.0 Å resolution (contoured at 1.2σ) is shown for selected residues as grey lines. This figure was prepared using the program BOBSCRIPT.[68]

ALAD it has been shown that magnesium activates the enzyme even at optimal zinc concentrations.[9,17] However this effect is not observed with yeast ALAD and in the latter enzyme the β-site is disrupted by an arginine side chain (Arg 251) which forms a five-membered salt-bridge interaction involving Asp 180, Arg 185, Arg 251, Asp 252 and Asp 183. In *E. coli* ALAD several of these residues are replaced by smaller side chains thereby creating space for the second metal-binding site. The absence of this site (β-site) in yeast ALAD may account for the failure of magnesium ions to activate ALAD from this species at optimal zinc ion concentrations.[17] It is difficult to rationalize this intriguing difference between ALAD enzymes.

In the *P. aeruginosa* enzyme, the subunits within each dimer differ significantly at the catalytic- and metal binding sites (27). Only in the 'closed' monomer is magnesium bound to the enzyme at the β-site. The hydrated magnesium ion at this site interacts with an arginine side chain (Arg 181). In contrast, in the 'open' subunit, the absence of bound magnesium causes this arginine to adopt a different conformation in which it points towards the catalytic site and forms a salt-bridge with one of the putative active site metal ligands, Asp 139, thereby pulling the side chain of this residue away from the active site. Thus it is proposed that the side chain of Arg 181 acts as a conformational switch coupling the two metal binding sites. When the β-site is occupied by a metal ion, Arg 181 will be in a position that allows Asp 139 to fulfill its putative metal-binding role at the catalytic α-site. Thus occupation of the β-site leads to an increase in enzyme activity. However, it should be emphasized that the role of a metal ion at the active site of the *P. aeruginosa* enzyme has been questioned since EPR and NMR experiments using Mn^{2+} and ^{13}C-labelled PBG appear to exclude the involvement of a metal ion in substrate binding.[31] Similarly the structure of the plant-like ALAD from *C. vibrioforme* shows that it lacks a magnesium at the catalytic site[32] and there is evidence that some ALADs are completely metal independent.[28]

Substrate Binding

Definition of the P-Site

The X-ray structure of the competitive inhibitor laevulinic acid bound to the P-site of the enzyme has been analyzed for ALADs from a number of species. The inhibitor makes a Schiff-base to Lys 247 and its carboxyl group forms three hydrogen bonds, one with the side chain of Tyr 312 and two with Ser 273 involving its hydroxyl and main chain nitrogen (Fig. 3). These two residues are invariant in ALADs as are most of the residues which interact with the hydrophobic moiety of the laevulinic acid. One important aromatic residue belonging to the loop covering the active site is Phe 204. This residue packs against the laevulinic acid methylene groups and shields them from solvent. The active site loop was completely invisible in the native yeast ALAD structure, presumably due to its flexibility in the absence of substrate or inhibitor. Hence the hydrophobic interaction between P-site laevulinic acid or ALA and Phe 204 may be crucially important for ordering the remaining loop residues so that they can function in binding the second substrate.

The structures of yeast ALAD complexed with substrate (ALA) and three inhibitors: laevulinic acid, succinylacetone (SA) and 4-keto-5-aminolaevulinic acid (KAH) (Fig. 5), have been solved at high resolution[33] (PDB codes 1h7o, 1h7n, 1h7r and 1h7p, respectively). The structure of the complex with ALA was solved with data from crystals of the enzyme which had been purified solely by ammonium sulphate fractionation. Thus it would appear that ALA remains bound to the enzyme during initial extraction and purification and it was speculated that the bound ALA would normally be lost during the later stages of enzyme purification, most probably during

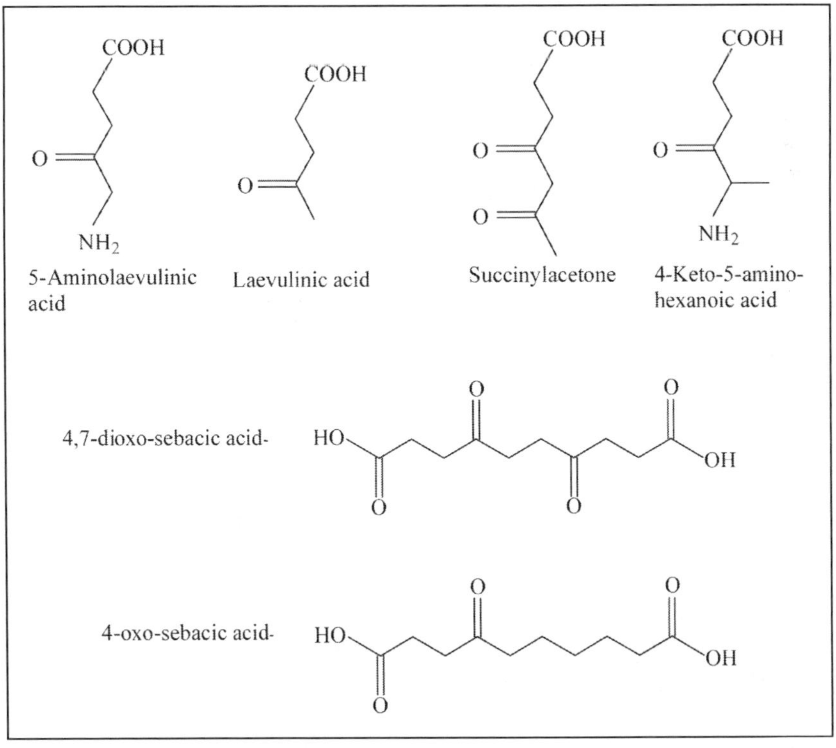

Figure 5. The formulae of substrate ALA together with various inhibitors of ALAD.

gel-filtration. All of these ligands form a Schiff base link with lysine 263 (equivalent to Lys 247 in *E. coli*) at the catalytic centre with the exception of KAH which appears to be linked covalently to the lysine but has become trapped as a carbinolamine intermediate.[33]

The substrate ALA differs from the inhibitor laevulinic acid by possession of a C-5 amino group which interacts with the side chains of Asp 131 and Ser 179 in the complex with yeast ALAD. These residues, which are equivalent to Asp 118 and Ser 165 in *E. coli* ALAD, are invariant, implying that they have an important role in catalysis. Both ALA and laevulinic acid bind to yeast ALAD in two slightly different conformations[33] but they interact with the same set of residues in each case.

All of the inhibitors induce a significant ordering of the flap covering the active site. Succinylacetone appears to be unique by inducing a number of conformational changes in loops covering the active site which may be important for understanding the cooperative properties of ALAD enzymes. The part of the active site flap most affected by conformational change upon SA binding is in the region of amino acids 225-235. These residues come close to the side chain of Lys 210 (Lys 195 in *E. coli*) at the catalytic centre of the enzyme. The tip of this side chain appeared to move about 2 Å from its position in the native enzyme upon SA binding but was largely unaffected by the binding of the other ligands. Thus it has been suggested that the large conformational changes in the active site flap which occur upon SA binding are a knock-on effect of a very local conformational change in Lys 210 caused by SA binding to Lys 263. The active site flap is also in contact with residues in the N-terminal arm region of a neighboring monomer. These residues are sandwiched between the active site flap and the base of a TIM barrel within an adjacent dimer. This suggests that conformational change in an active site in one dimer can be communicated to neighboring dimers via this contact point. The effects these conformational changes would have on the active sites of neighboring subunits are difficult to predict since the active sites in ALAD are far apart. It is reasonable to expect that changes in the conformation of the active site flap due to substrate binding in one subunit could transmit a conformational change to neighboring monomers via the subunit contacts described above. Subtle changes in relative subunit orientation may affect the flexibility or conformation of the active site flap since it forms part of the subunit interface. Thus substrate or inhibitor binding to one subunit may affect the catalytic turnover of neighboring monomers.

An intriguing result was obtained with 4-keto-5-amino-hexanoic acid which seems to form a stable carbinolamine intermediate with Lys 263.[33] It appears to define the structure of an intermediate of Schiff base formation which the substrate forms upon binding to the P-site of the enzyme. The intermediate seems to be held by hydrogen bonds between its 5-amino group and the side chains of Tyr 207 and Asp 131, and between its carbinolamine -OH group and the side chain of Lys 210.

Definition of the A-Site

Kinetic studies of ALAD have suggested that the K_M of the P-site is lower than that of the A-site (4.6 μM as against 66 μM for the *E. coli* enzyme).[34-35] This correlates with the finding that many substrate analogues studied up till now bind predominantly in the P-site and the adjacent A-site appears to be occupied only by water molecules, and a number of the side chains forming this site are disordered. ALADs typically have turnover numbers (k_{cat}) of around 1.0 per second at their optimal pH values which are usually in the pH range of 8.0 to 9.0.[17]

The interactions that an ALA molecule could make at the A-site of the enzyme can be predicted with reasonable confidence from the laevulinic acid complexes analyzed so far (Fig. 3). Adjacent to the P-site laevulinic acid is a solvent-filled pocket lined by the following highly conserved residues: Asp 118, Ser 165, Lys 195, Arg 205, Arg 216, Gln 220 and the zinc-binding cluster involving cysteines 120, 122 and 130 (*E. coli* numbering). There is good electron density for a solvent molecule bound datively to the zinc ion and this is within H-bonding distance of the amino group of P-side ALA. It has been suggested that this water molecule may be a zinc-bound hydroxide which abstracts a proton from C-3 of A-side ALA during formation of the C-C bond which eventually links both substrates. In addition, an electrostatic link may exist between the C-5 amino group of P-side substrate and the zinc-bound hydroxide. This putative hydroxide forms an H-bond with Ser 165 and forms an indirect H-bond with Asp 118 via a water molecule. Both of these residues have been implicated in the mechanism by site-directed mutagenesis studies[36] and their proximity to the C-5 amino group of P-side ALA indicates that they have a role in the catalytic mechanism. Further evidence that this may be the A-site of the enzyme is provided by the presence of an unusual elongated feature of electron density in the *E.*

coli structure which might represent an additional laevulinic acid molecule bound with low occupancy (Fig. 3).

It was suggested that dicarboxylic acids of appropriate length may be able to cross-link the carboxylic acid binding groups associated with the A and P substrate binding sites.[37-38] The structures of ALAD from *E. coli* and yeast have been analyzed in complex with the 10 carbon chain irreversible dicarboxylic acid inhibitors 4-oxosebacic acid and 4,7-dioxosebacic acid[39-40] (formulae are shown in Fig. 5) (PDB codes 1i8j, 1l6s, 1l6y, 1eb3, 1gjp). Both inhibitors bind by forming a Schiff base link with Lys 247 (*E. coli* numbering) at the active site. The most intriguing result of these studies is the novel finding that 4,7-dioxosebacic acid forms a second Schiff base with the enzyme involving Lys 195. It has been known for many years that P-side substrate forms a Schiff base (with Lys 247) but prior to work on the dicarboxylic acid inhibitors there has been no evidence that binding of A-side substrate involves formation of a Schiff base with the enzyme.

In the 4,7-dioxosebacic acid complexes analyzed it is notable that hydrogen bonds are formed between the A-site carboxyl group and the side chain of Gln 220 as well as the guanidinium groups of Arg 205 and Arg 216 (in *E. coli* numbering) (Fig. 6). Hence these arginine residues are very likely to form a salt bridge with the carboxylate of A-side substrate. These arginines (along with Gln 220) are strongly conserved residues. Interestingly, 4,7-dioxosebacic acid was found to have higher potency for the zinc-requiring ALADs than for the magnesium requiring enzymes.[41]

The structure of the substrate analogue 5-fluorolaevulinic acid bound to a mutant of *P. aeruginosa* ALAD has been determined at high resolution[42] (PDB code 1gzg). The mutation D139N converts one of the putative metal ligands into an asparagine residue that is less likely to coordinate metal ions. The same residue has been implicated in coupling the two metal binding sites (catalytic and allosteric; see above). In this remarkable complex, two inhibitor moieties are bound to the enzyme via Schiff base linkages with the two invariant lysines (K205 and K260) and a metal ion (probably sodium) is coordinated by the C-5 fluoro-groups of the two inhibitor molecules. Intriguingly, the

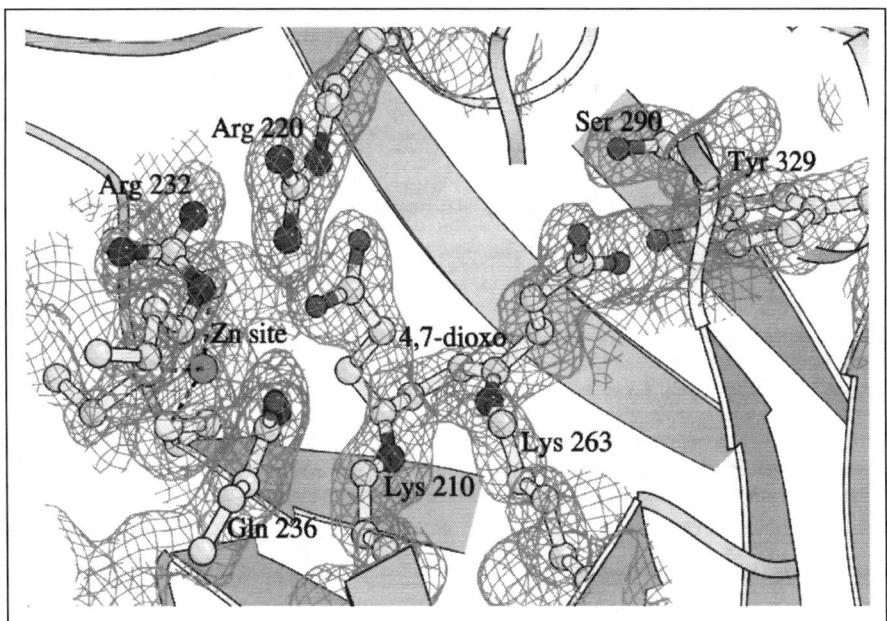

Figure 6. The refined electron density map for the inhibitor 4,7-dioxosebacic acid bound to yeast ALAD solved at 1.75 Å resolution and contoured at 1.5 rms. The P-site is on the right-hand side and the zinc-binding cysteines as well as the conserved arginines, which form the A-site, are on the left. This figure was prepared using the program BOBSCRIPT.[68]

other putative metal ligand Asp 131 is involved in binding the metal ion as are the invariant Ser 175 (165 in *E. coli*) and Asp 127 (118 in *E. coli*) residues. These findings are consistent with the idea that whilst the native enzyme may not bind a metal ion at the active site, a metal binding site is formed when the substrates bind. These findings may be applicable to all the plant-like ALAD enzymes and in vivo the metal ion involved is more likely to be potassium than sodium.

Catalytic Mechanism

The synthesis of PBG from two molecules of ALA requires the formation of a C-N bond and a C-C bond with the loss of two water molecules. Single-turnover experiments established that the substrate moiety to bind first forms a Schiff base link with the enzyme at the P-site.[6] It has been suggested that the enzyme may function by forming an additional Schiff base linking the A-site substrate to the enzyme.[38] This was proposed prior to determination of the X-ray structure which showed that the side chain of the Schiff base lysine (Lys 247) is spatially adjacent to the side chain of another invariant lysine (Lys 195). Further evidence for a double Schiff base mechanism comes from recent structural studies in which both of the invariant active site lysines have been observed forming Schiff bases with bound inhibitor molecule(s).[39-42]

In addition to the Schiff base(s) formed between the enzyme and substrate, another Schiff base is formed between the two substrate moieties. This involves the C-4 atom of the A-side substrate and the amino group of the P-side substrate, eventually forming the C-N bond linking both ALA molecules in the product. The C-C bond formed between the substrates results from nucleophilic attack by a stabilized carbanion at the C-3 position of the A-side substrate, on the carbonyl carbon of P-side substrate. The latter would be rendered highly electropositive by its Schiff base link to the enzyme.

The ε-NH$_2$ group of Lys 247 would have to be neutral for it to act as a nucleophile and condense with the first ALA molecule to bind to the enzyme. Of the two lysines at the active site, Lys 195 is more likely to be positively charged since its environment is more polar than that of Lys 247. A positive charge on Lys 195 would serve to lower the pK_a of Lys 247 and make it a more effective nucleophile for Schiff base formation with P-side ALA. The Schiff base formed by Lys 247 and ALA could then become protonated and act in a similar way on the pK_a of Lys 195 allowing it to nucleophilically attack the A-side ALA molecule. The central region of the A-side ALA would be in a position to interact with the zinc ion and its associated water molecule or hydroxide ion. These groups could facilitate the deprotonation reactions and assist stabilisation of the C-3 carbanion prior to the aldol condensation which forms the C-C bond linking the two ALA molecules. The C-N bond results from formation of an inter-substrate Schiff base.

The major issue of discussion in the field has been the order in which the bonds linking the two ALA molecules are formed, i.e., whether C-C bond formation occurs before or after C-N bond formation. The formation of a Schiff base between the A-side ALA moiety and Lys 195 would further help to labilize the C-3 hydrogen atoms of A-side ALA yielding a bound enamine (see Fig. 7). This could lead to nucleophilic attack of the enamine in the A-site on the C-4 of P-side ALA thus forming the first C-C bond linking the substrates. The inter-substrate C-N bond could then form. Whilst it is possible that the C-N bond could form first, thus yielding a Schiff base linking both substrates, this would imply that the Schiff base between A-side ALA and Lys 195 observed in recent studies[39-42] has no role in the mechanism other than perhaps to anchor the A-side substrate prior to catalysis. If instead it is assumed that the Schiff base linking A-side ALA with Lys 195 has a catalytic function (rather than just a passive binding role), then it is likely that inter-substrate C-C bond formation would occur before the C-N bond formation, as shown in Figure 7.

The final deprotonation shown in Figure 7 involves the C-5 of P-side ALA. This step, which has been shown to be stereospecific for the *pro-R* hydrogen,[43] could be catalyzed by Lys 247 in view of its proximity. In the reaction a base (B2 in Fig. 7) assists the deprotonation of the N-5 of P-side ALA. It is possible that the two proximal active site residues Asp 118 and Ser 165 may be involved in this process and site-directed mutagenesis[36] implicates both of these residues in the mechanism.

Recently, the structure of a putative intermediate resembling covalently bound product has also been determined by cocrystallising the yeast enzyme in the presence of added substrate ALA.[44] This pyrrole-like intermediate is covalently bound to the P-side lysine residue and its A-side amino group is datively bound to the active site zinc ion (Fig. 8). The P- and A-side carboxyl groups make essentially the same interactions as those of 4,7 dioxosebacic acid and other inhibitors, thus confirming the

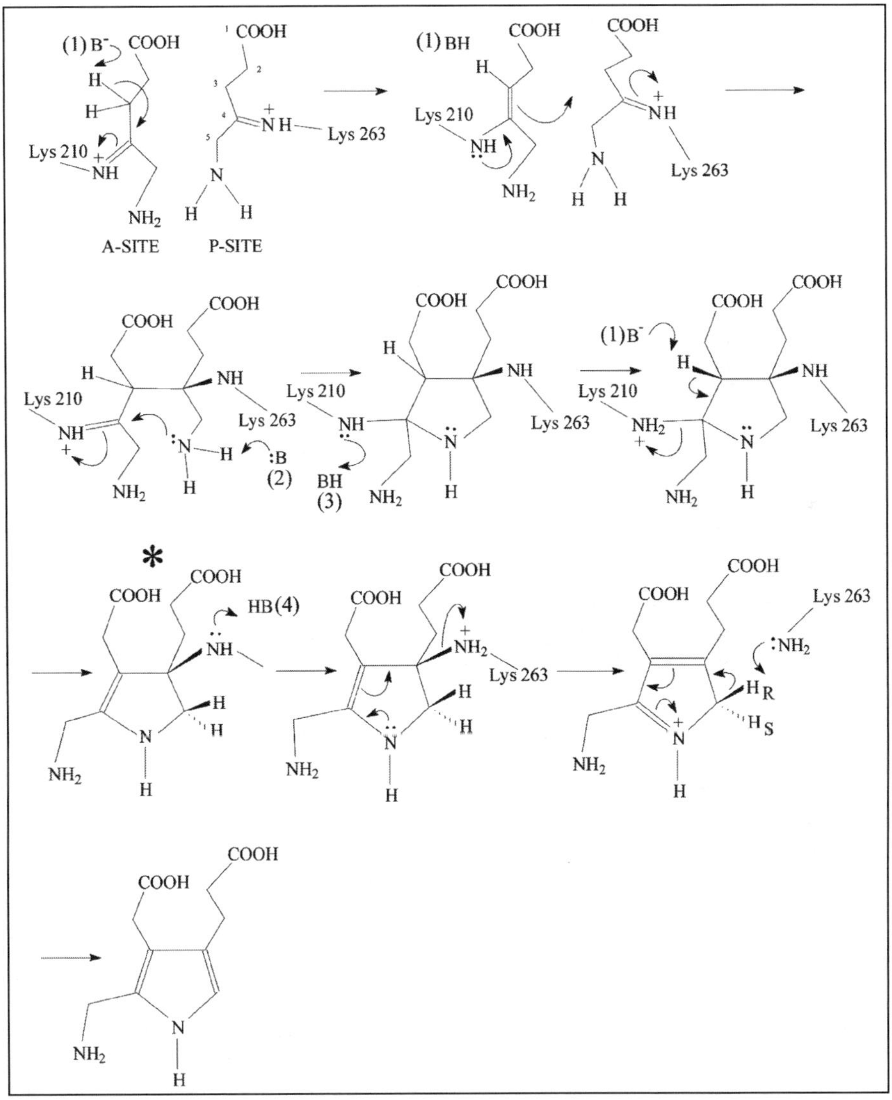

Figure 7. A proposed reaction mechanism of ALAD catalysis. Note that both ALA molecules are bound to the enzyme by Schiff bases to lysines 195 and 247 (*E. coli* ALAD numbering).

physiological relevance of these earlier inhibitor studies. It is likely that this intermediate (which is indicated in the proposed reaction pathway by an asterisk in Figure 7 represents a stable form of bound product that is awaiting displacement from the active site by an incoming substrate moiety or a conformational change within the octamer.

Human Alad Variants, Inhibitors and Porphyria

The hereditary deficiency of functional dehydratase in humans is associated with the genetic disease Doss or ALA dehydratase porphyria.[12] This is a rare homozygous disease with severe neurological symptoms that are thought to be due to the accumulation of 5-aminolaevulinate which structurally resembles the neurotransmitter GABA, and may have pharmacologically significant properties. The disease is inherited as an autosomal recessive trait and results in a hugely increased excretion of ALA and coproporphyrin III in the urine.[45-48] A more detailed consideration of porphyrias is given in Chapter 5.

Interestingly, it has been shown that the population has a very wide range of ALAD activity with 2% of individuals having less than 50% of the normal level.[45] The human ALAD gene consists of 12 exons and resides on chromosome 9q34. It has two promoter regions which generate different transcripts in erythroid and nonerythroid cells by alternative splicing, although the enzyme produced in both tissues is identical. Determination of the cDNA sequence of human ALAD[49] allowed several natural variants to be identified one of which, K59N, is estimated to be present in 10% of the population.[50] The lysine residue altered in this mutation is close to the carboxyl of Asp 52 i.e., could form a salt bridge with it in the wild-type enzyme. The absence of this lysine in ALADs from other species indicates that the loss of this ionic interaction in the K59N human ALAD mutant may not prevent the enzyme folding or functioning which is consistent with the natural abundance of this variant. The mutated residue occurs at the active site end of the TIM barrel at a point where the regular α/β alternation of the secondary structure is interrupted by a β-hairpin. The hairpin extends into the active site cavity and hence the K59N mutation may have indirect effects on ligand binding in the human enzyme.

Five other variants (F12L, G133R, R240W, A274T and V275M) associated with ALAD porphyria have been detected[51-53] and from the enzyme's structure it is apparent that these mutations probably have indirect structural effects on the ALAD molecule rather than direct effects on the catalytic apparatus. Glycine 133 is an invariant residue in ALADs and adopts a remarkably extended conformation with its ϕ and ψ angles almost exactly 180 degrees. This conformation causes some steric hindrance for all amino acids, except glycine. The substitution of Gly 133 with an arginine in human ALAD could substantially disrupt the local structure of the molecule particularly since the glycine appears to reside in a tight hydrophobic pocket. In the primary sequence this glycine is adjacent to one of the residues which form the best defined zinc binding site (Cys 132 in human ALAD) and so disrupting this region of the molecule may alter its metal binding competence and catalytic activity.

The other two porphyria related mutations (A274T and V275M) occur in one of the β-strands (β9) forming the closed $(\beta/\alpha)_8$ barrel and are situated towards the end that is buried by quaternary interactions. Both residues are in tightly packed hydrophobic environments. The side chain of residue 274 points into the central hydrophobic core of the barrel where ALAD, like all enzymes adopting this fold, has an abundance of small hydrophobic amino acids such as Leu, Ile and Val. Substitution of this residue with the more polar β-branched amino acid threonine may disrupt this region of the molecule. This part of the ALAD molecule is close to the cavity at the base of the TIM barrel which interacts with the arm region of an adjacent subunit by the capping interaction described above. Therefore the A274T substitution may also effect the quaternary interactions of the mutant ALAD enzyme. Accordingly, expression of the mutant allele in chinese hamster ovary cells resulted in unstable enzyme although its catalytic activity was around 50% of that of the wild-type enzyme.[51] The V275M human ALAD mutation affects a residue that resides in the toroidal hydrophobic milieu between the strands and helices of the barrel. A methionine residue substituted at position 275 in human ALAD could therefore be sterically disruptive to the local structure of the enzyme.

The mutation R240W was found in the other allele of the A274T patient and resulted in inactive enzyme when this allele was expressed in chinese hamster ovary cells.[51] Arg 240 occurs at a position where it can make an intra-molecular salt bridge to an invariant aspartate (172 in human ALAD). This ion-pair is buried beneath the arm region of the adjacent subunit in the vicinity of residues 16-18 which are conserved in many ALADs. The R240W mutation may therefore destabilise the human ALAD oligomer due to the loss of this buried ion-pair at the subunit interface. However the arginine is not strongly conserved in ALADs and is frequently replaced by both polar and non-polar amino acids although tryptophan never occurs here presumably because it would sterically disrupt inter-subunit contacts with the neighbouring monomer as may occur in the R240W mutant. However, heterologous expression of the mutant allele resulted in an inactive enzyme with apparently normal stability.[51] Interestingly this mutation occurs at what is the β-metal binding site in ALADs activated by magnesium. Recently a mutation in the N-terminal arm region of the molecule has been identified[53] which probably affects the extensive quaternary interactions made by this part of the molecule.

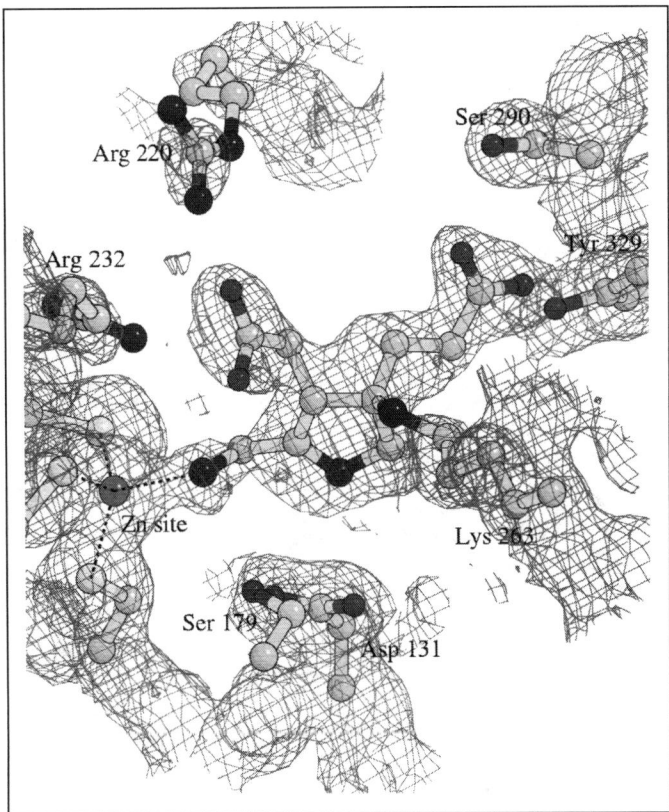

Figure 8. The electron density at 1.6 Å resolution of a putative reaction intermediate bound to yeast ALAD when the enzyme is cocrystallised with substrate ALA. The intermediate is covalently bound to the P-site lysine residue (Lys 263) on the right-hand side and the A-site with the bound zinc ion is shown on the left.

Elevated levels of ALA are also associated with the hereditary disease type I tyrosinaemia[14-16] which is thought to stem from the accumulation of succinylacetone, a breakdown product of tyrosine and a potent inhibitor of ALAD. Hereditary tyrosinaemia type I is a rare and lethal disease which affects sufferers in early life. The disease is characterized principally by raised plasma tyrosine, liver cirrhosis and renal tubule problems. The course of tyrosinaemia type I can vary from fulminating liver failure within a few months of birth to a more slowly developing form with later onset which progresses to liver malignancy within the first decades of life. Patients suffer from neurological crises and it has been found that peripheral axons of long nerves are subject to degeneration and secondary demyelination. Large amounts of succinylacetoacetate and succinylacetone are excreted by patients with type I tyrosinaemia. Succinylacetone is an effective inhibitor of ALAD having a K_i of 0.8 mM for the *E. coli* enzyme[17] and it is thought to be responsible for many of the symptoms of this disease.

Inhibition of ALAD by lead ions is one of the major manifestations of acute lead poisoning which often leads to neurological and psychotic disturbances. Anaemia is one of the main symptoms of lead poisoning and can be attributed to the inhibition of several enzymes in the tetrapyrrole biosynthetic pathway including ALAD. It has been shown that human ALAD binds lead tightly with a sub-picomolar inhibition constant[54] and the structure of the yeast enzyme complexed with lead shows that Pb^{2+} ions bind at the triple-cysteine site, displacing the catalytically essential zinc ion.[21,55-56] In addition it has been shown that the same site also binds mercury and platinum ions which may have implications for the pathology of poisoning by these heavy metals.[56]

Porphobilinogen Deaminase

The enzyme porphobilinogen deaminase (widely referred to as hydroxymethylbilane synthase or HMBS; EC 4.3.1.8) catalyses the third step of the haem biosynthesis pathway in which four molecules of the monopyrrole porphobilinogen are condensed to form a linear tetrapyrrole, preuroporphyrinogen (1-hydroxymethylbilane; see Fig. 9). One of the most common of the hereditary porphyrias is due to genetic lesions in the gene for this enzyme which give rise to the disease acute intermittent porphyria (AIP)[1-3] (see Chapter 5 for more details).

Purification of porphobilinogen deaminase (PBGD) from various species confirms that it is monomeric with an M_r in the range 34 to 44 kD.[1-3] The PBGD enzymes from most sources exhibit high thermal stability, pH optima in the range 8.0 - 8.5 and isoelectric points in the range 4.0 - 4.5. Isotopic labelling and single turnover studies showed that the pyrrole forming ring A (Fig. 9) is the first to bind to the enzyme followed by rings B, C and finally D.

Studies of the *E. coli* PBGD enzyme showed that it possessed a dipyrromethane cofactor (Fig. 10) which is linked to the enzyme by a thioether linkage to an invariant cysteine residue (Cys 242 in *E. coli* numbering).[57] This cofactor can be derived from two molecules of the normal substrate porphobilinogen or from cleavage of the product, preuroporphyrinogen.[58] The cofactor acts as a primer to which four porphobilinogen molecules are attached sequentially prior to cleavage of the link between the cofactor and the first substrate molecule on completion of the reaction. Thus the cofactor remains attached to the enzyme when the product is released.

The X-ray structure of the *E. coli* enzyme has been solved at 2.0 Å resolution.[59] The polypeptide of 313 amino acids is folded into three domains each of approximately the same size (Fig. 11). Part-way through the N-terminal domain (domain I) the polypeptide leaves to form domain II and then returns to complete the last strand of domain I. The two short connecting regions between these domains could act as hinge points. The general architecture of domains I and II shows a strong resemblance to a number of periplasmic binding proteins.

The dipyrromethane cofactor is attached to a loop on domain III and positioned at the mouth of a deep active site cleft formed between domains I and II. The propionate and acetate groups of the cofactor rings are involved in ion-pair interactions with a number of conserved arginine side chains, some of which have been shown to be mutated in carriers and sufferers of AIP. The cofactor ring furthest from the attachment point (outer pyrrole) can adopt two conformations depending on its oxidation state. Most of the interactions between the cofactor and enzyme side chains involve residues from domain II. In contrast, both pyrrole nitrogens form hydrogen bonds with a

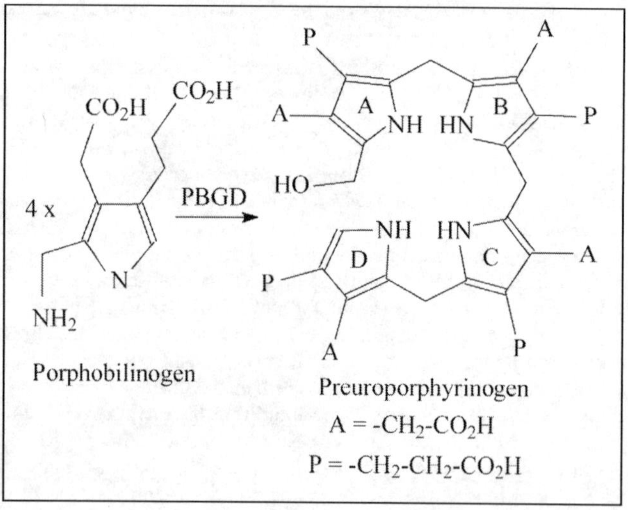

Figure 9. The reaction catalysed by porphobilinogen deaminase. Four molecules of the pyrrole porphobilinogen are condensed to form the linear tetrapyrrole preuroporphyrinogen (hydroxymethylbilane).

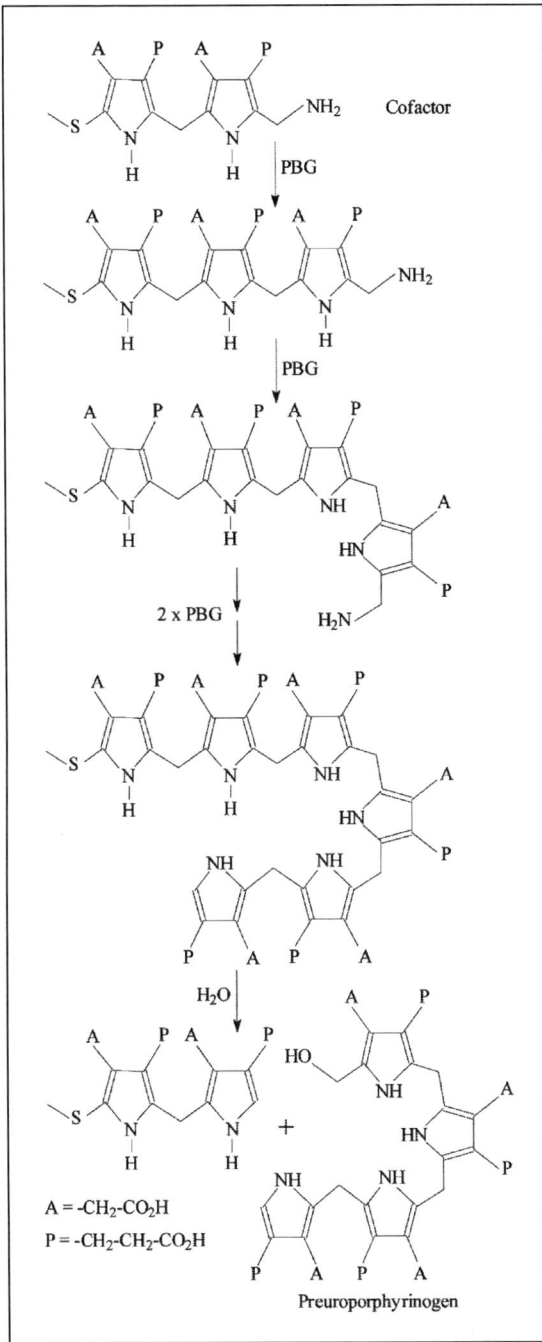

Figure 10. The catalytic cycle of porphobilinogen deaminase. The dipyrromethane cofactor is covalently attached to the enzyme by a thioether bond with a cysteine residue. Four substrates are added in sequence giving ES, ES_2, ES_3 and ES_4. Finally hydrolysis of the linkage between the substrate and the cofactor releases the product.

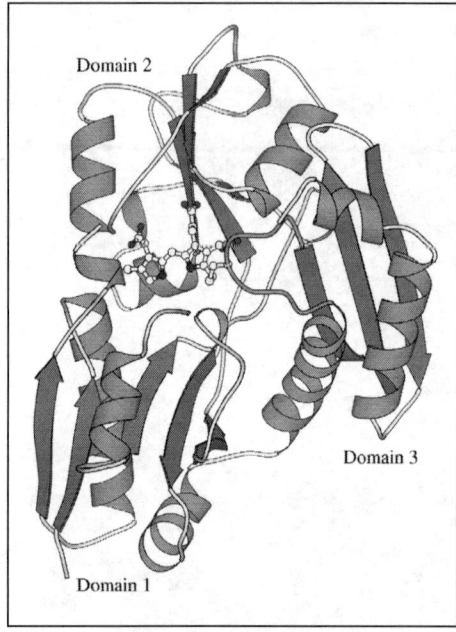

Figure 11. The X-ray structure of porphobilinogen deaminase showing the dipyrromethane cofactor located between domains 1 and 2. The cofactor is covalently attached to a conserved cysteine which resides in domain 3. This figure was prepared using the program BOBSCRIPT.[68]

domain I residue, Asp 84, which is invariant. This aspartate is clearly an important residue in catalysis. Accordingly the enzyme activity drops dramatically when this residue is mutated to glutamate. The X-ray structure of this mutant shows that the hydrogen bond normally formed between Asp 84 and the outer pyrrole nitrogen is disrupted. Asp 84 is thought to function in catalysis by deaminating porphobilinogen once it has bound to the cofactor yielding a reactive azafulvene intermediate which can condense with the next substrate molecule. Evidence that the conformation of PBGD changes during the tetrapolymerisation comes from the observation that the enzyme becomes increasingly susceptible to inactivation by N-ethylmaleimide (NEM) during the reaction. The susceptible cysteine is between domains 2 and 3. It may be that these conformational changes are due to the enzyme 'pulling' the chain of pyrroles through the active site cleft to allow each added pyrrole access to Asp 84 for subsequent condensation with the next substrate. Alternatively these conformational changes may be due to reorganisation of the polypyrrole as it is assembled allowing each newly incorporated pyrrole access to the enzyme's catalytic apparatus. The same catalytic machinery could also facilitate the final cleavage of the bond linking ring A with the cofactor thereby releasing the tetrapyrrole product. It is not clear at present how the enzyme manoeuvres the growing intermediates during assembly of the tetrapyrrole product. The sidechains of the conserved arginines appear to be involved since mutagenesis of these residues leads to loss of activity and the accumulation of different intermediates (ES_1, ES_2, ES_3 and ES_4) in different proportions depending on the nature of the mutation.[60]

The gene for human PBGD has been located on the long arm of chromosome 11 and it consists of 15 exons interrupted by 14 introns.[61-66] The gene encodes two isozymes whose expression is tissue-specific and controlled by alternate splicing. A housekeeping isozyme (44 kD, 361 residues) is expressed in all cells and its promoter lies in the 5' flanking region of the DNA. The transcript is made of exon 1 and exons 3-15. A promoter that is active only in erythroid cells resides within intron 1 and produces a transcript containing exons 2-15. The corresponding protein is 344 amino acids in length (42 kD) i.e., 17 residues shorter than the housekeeping gene at the N-terminus.

The sequences of human and *E. coli* PBGD are ~46% identical but more than 70% of the amino acids are similar. The residues which are identical in both enzymes are those that make up the active site cleft, the hydrophobic core of the domains and loop regions with strict conformational constraints e.g., tight turns involving conserved glycines. The human housekeeping enzyme is slightly longer at the N-terminal end and both forms of the human enzyme contain an insertion of 20 residues in domain III. The structure of the *E. coli* enzyme shows that the intron-exon boundaries of the human gene correspond to loop regions in the tertiary structure of PBGD.

Acute Intermittent Porphyria

The hereditary deficiency in activity of PBGD (usually by 50%) gives rise to the inducible disease acute intermittent porphyria (AIP) which ranks among the most common forms of porphyria with an incidence of 1-2 per 10,000.[62-66] Numerous different mutations affecting the erythroid

PBGD have been identified in AIP sufferers varying from single base changes to insertions and deletions which destroy the reading frame, as well as truncations and exon deletions.[67] In contrast, only around 8 different mutations affecting only the nonerythroid or housekeeping PBGD have been found in AIP sufferers.

X-ray crystallography of the *E. coli* enzyme has shown that many of the mutations affect crucial arginine residues which bind the side chain carboxylates of the cofactor and/or substrate. Some of these mutations are associated with the CRIM positive phenotype. Engineering equivalent mutations into the *E. coli* gene can lead to enzyme which either is unable to assemble the cofactor, is unable to bind substrate or accumulates intermediates of the tetrapyrrole assembly process: ES, ES_2 and ES_3.[60]

Uroporphyrinogen III Synthase

Uroporphyrinogen III (uro'gen III) is the last common precursor in the biosynthesis of all the tetrapyrrole cofactors, the so called pigments of life, heme, chlorophyll, cobalamin, siroheme and coenzyme F_{430}.[1,2,69-72] Conversion of individual pyrrole rings into the asymmetric cyclic tetrapyrrole involves two enzymes, porphobilinogen (PBG) deaminase and uro'gen III synthase (U3S), their coupling is so significant that U3S was once referred to as cosynthase. Despite the fact that U3S was first purified almost 50 years ago,[73] and that there has been substantial interest in the mechanism of cyclic tetrapyrrole formation, the product of porphobilinogen deaminase, and consequently the substrate for U3S, was not identified until 1979 with the identification of a hydroxylated linear tetrapyrrole, hydroxymethylbilane (HMB).[74-77] Originally, over twenty-five different mechanisms proposing the catalytic action of U3S were debated, but in 1961 Mathewson and Corwin proposed a mechanism that has been continually supported by experimental data ever since.[78] U3S is the second fastest enzyme in the heme biosynthetic pathway,[79] and competes effectively with the uncatalyzed autocyclization that occurs at one tenth the rate of uro'gen III formation.[77]

Catalytic Mechanism

HMB is predicted to bind to the enzyme in a circular conformation. Due to sp^3-hybridized bridge carbons between each pyrrole ring, both the linear tetrapyrrole substrate and the cyclic tetrapyrrole product are puckered molecules where the direction of pyrrole nitrogens can be described as up or down, with respect to the pseudo-plane of the final tetrapyrrole ring. The proposed mechanism suggests that the pyrrole nitrogens of the A, C and D rings will point towards a central point, while the orientation of the B ring is less restrained.[80] In order to facilitate the rearrangement of ring D, the enzyme must position it such that the covalent connection to ring C is close in space to the carbon containing the hydroxyl moiety on ring A.

Catalysis begins with rearrangement of ring A resulting in a positive charge on the pyrrole nitrogen, loss of the hydroxyl moiety as water, and formation of the first azafulvene intermediate (Fig. 12).[78,80-82] Subsequent bond rearrangements on ring D facilitate attack by the A ring azafulvene on the substituted α position of the D-ring pyrrole forming a spirocyclic pyrrolenine intermediate that is predicted to be the transition state. Collapse of the spiro intermediate occurs by breaking the bond between the D- and C-ring leaving a second azafulvene on ring C. At this point, the enzyme must alter the orientation of the D-ring such that the free α-position is in close proximity to the C-ring azafulvene. The final cyclic tetrapyrrole is formed through attack on the D-ring by the C-ring azafulvene with subsequent bond rearrangements returning the pyrrole rings to their minimum energy state.

The inherent symmetry of the mechanism containing two azafulvenes is attractive. The attack on ring D by ring A in the second step is repeated mechanistically by the attack on ring D by ring C in the fourth step. The mechanism also takes advantage of the inherent polarizablility of the pyrrole nitrogens on all three rings involved. The current proposal highlights the possibility that a single catalytic residue positioned near the pyrrole nitrogens of all three pyrrole rings (A, C and D) may be able to facilitate the reaction.

Support for this complex mechanism has been built up over the last thirty years and is summarized herein. Pioneering ^{13}C-NMR studies by Battersby[74,83] revealed that during cyclization the C15-C16 bond is broken and that the C16-C20 and C19-C15 bonds are formed (Fig. 12). Historically, the tetrapyrrole ring numbering starts with ring A carbon, such that the bridge carbons are considered carbons 5, 10, 15, and 20. The bond breakage and bond formation must happen sequentially since the free D ring pyrrole has never been isolated as an intermediate nor does it

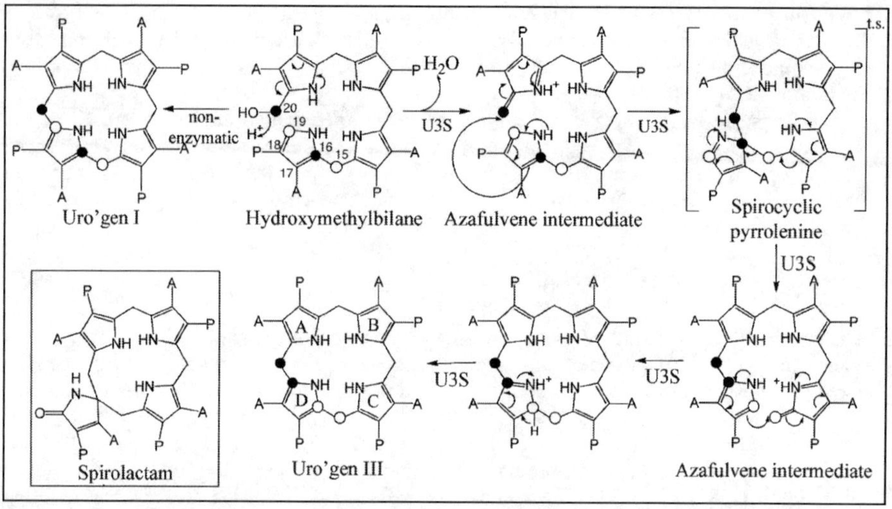

Figure 12. The enzymatic mechanism of uroporphyrinogen III synthase proposed for the catalytic formation of uro'gen III versus the nonenzymatic cyclization to uro'gen I. A spirocyclic intermediate has been proposed as the transition state, a hypothesis that is supported by the inhibitory effects of a synthetic spirolactam.[74,78] Filled and empty circles on ring carbons reflect early ^{13}C labeled NMR experiments and highlight the bond changes of the proposed mechanism.[70,79] A = $CH_2CO_2^-$, P = $CH_2CH_2CO_2^-$.

function as a substrate.[75] Interestingly, the linear asymmetric tetrapyrrole containing a D ring that has the acetate and propionate side chains reversed still functions as a substrate for the enzyme.[84] The enzyme presumably functions along the normal catalytic trajectory and flips the D ring of this asymmetric substrate to produce the symmetric tetrapyrrole uro'gen I. The result not only reduces the mechanistic importance of the D ring side chains, but also confirms that the asymmetric linear tetrapyrrole is not a normal catalytic intermediate.[84]

The existence of an azafulvene intermediate was shown by trapping an aminomethylbilane product by running the U3S reaction under cryogenic conditions in the presence of 200 mM ammonium chloride.[81] The aminomethylbilane is postulated to form by reaction of the first azafulvene on ring A with the free ammonium ion. Unfortunately, no amino-modified or reduced intermediate of ring C has been trapped to support the existence of the second azafulvene. This may indicate that either this portion of the catalytic reaction occurs deep inside an inaccessible enzyme active site, or that the reaction proceeds through an alternative mechanism.[81]

Early objections to the formation of a spirocyclic intermediate focused on the potential instability of the cyclic compound. To test this hypothesis two spirocyclic intermediates were synthesized, a spirolactam and a dinitrile pyrrolenine.[85] The structure of the dinitrile pyrrolenine was determined by X-ray crystallography confirming the viability and integrity of the proposed intermediate as well as its highly puckered conformation (Fig. 13).[85] Both compounds were isolated and tested as inhibitors of U3S. The spirolactam (Fig. 12), differs from the proposed spirocyclic intermediate by containing an amide in place of an imine and functions as a potent inhibitor of the enzyme with a K_i of 1-2 μM.[80,82,86] This value is approximately equal to the K_m for the HMB substrate,[87] and supports the theory that the spirolactam is mimicking the transition state intermediate.

The restrictions on pyrrole nitrogen and side chain modifications have been probed by testing the activity of U3S against a series of modified bilanes. The enzyme remains active against substrates that have the acetate and propionate groups switched on either the C or D ring.[84] The acetate and propionate side chains on ring D can also be replaced by a methyl and ethyl groups, respectively,[84,88] suggesting that the carboxylate moieties do not form specific contacts with the enzyme or play a role in the electrostatics of the mechanism. In contrast, reversal of the side chains of either the A or B rings result in a modified bilane that is no longer a substrate for U3S.[88] It has been suggested that the

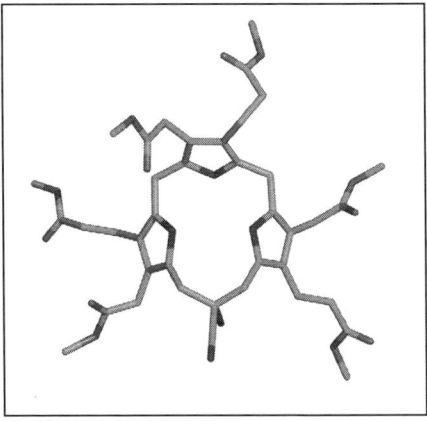

Figure 13. The structure of a dinitrile pyrrolenine determined by X-ray crystallography confirms the viability and integrity of the proposed intermediate of uroporphyrinogen III synthase as well as the highly puckered conformation of the intermediate (Fig. 12).

position and electron withdrawing potential of the acetate moiety on ring A may influence the rate and direction of the reaction.[80] Indeed it has been suggested that the direction in which the spirocyclic intermediate breaks down may be favored by the greater electron withdrawing effect on the acetate moiety proximate to the C ring rather than the weaker polarization of the propionate near the A ring.[80] In addition to potentially participating in the catalytic mechanism, these side chains may form specific interactions with the enzyme. Finally, methyl groups attached to the pyrrole nitrogens of the C and D rings also inhibit catalysis supporting the need for a polarizable nitrogen in those positions.[89]

Coupled Enzymatic Assay

The understanding of the coupling between PBG deaminase and U3S prompted the development of a quick and efficient coupled enzymatic assay to replace earlier thin layer chromatography separation of coproporphyrin products.[87,90] The assay relies on the fact that the enzyme catalyzed reaction to form uro'gen III is significantly faster than the uncatalyzed autocyclization of HMB to form uro'gen I. The use of proper controls permits the subtraction of background uro'gen I formation and eliminates the need for HPLC separation of the I and III isomers.[91] The original assays were analyzed spectrophotometrically by absorbance (A_{405}) or fluorescence (λ_{ex} 399-404 nm; λ_{em} 615-618 nm) to eliminate the need for the time consuming HPLC assays.[87] Modern assays once again incorporate HPLC analysis for quantitative determination of the I and III isomer ratios.[91]

Briefly, purified PBG deaminase is mixed with wild-type or mutant U3S and preincubated 1-2 minutes prior to the addition of porphobilinogen. The reaction is allowed to proceed for 1-2 minutes at 37°C in Tris buffer at pH 8.2. Production of uro'gen III is terminated by the addition of strong acid (either I_2/TCA or HCl). When using the I_2/TCA mixture the product is further oxidized with metabisulfite and acidified with HCl prior to absorbance measurements.[92] Alternatively, the acidified mixture can be oxidized by UV irradiation and monitored by HPLC analysis.[93] HPLC analysis provides direct quantitation of the I and III isomers, while the absorbance assay depends on a control experiment without U3S to estimate background levels of uro'gen I formation.

Structure

The tertiary structure of mammalian uro'gen III synthase has been determined by X-ray crystallography to 1.85 Å resolution (pdb code 1JR2).[93] The protein folds into two αβα sandwich domains connected by a long two-stranded β-sheet (Fig. 14A). The N-terminal two strands and one helix combine with the C-terminal three strands and four helices to form domain 1. The central portion of the polypeptide creates domain 2 composed of a four stranded β-sheet surrounded by four primary and three minor α-helices. In each domain the β-sheets are parallel and point towards the cleft in the center of the molecule between the two domains.

The conserved residues cluster at the cleft between the two domains.[93] U3S is the most diverse enzyme in the heme biosynthetic pathway with species variation ranging from 77% identical between human and mouse, to as low as 21% between human and *Fusobacterium nucleatum*. In contrast, the mouse and human uro'gen decarboxylase enzymes are 92% identical. Aligning 34 different U3S sequences reveals 12 residues that are identical in at least 33/34 sequences, Thr62, Ser63, Thr103, Glu127, Gly144, Gly159, Tyr168, Ser197, Pro198, Lys220, Gly225 and Thr228. These residues primarily occupy two clusters, one on each side of the central cleft (Fig. 14A). The distance between the two clusters is ~15-20 Å, but the size of the known structure of the dinitrile spiropyrrolenine is approximately 3-5 Å x ~12-14 Å, suggesting that a significant conformational change is required to form specific interactions with the substrate and catalyze ring closure.[93]

In support of a necessary conformational change, the recent crystal structure of U3S from an extremely thermophilic bacterium, *Thermus thermophilus* (Tt) Hb8, reveals a "closed" conformation[94] (Fig. 14C). Three independent molecules of the *T. thermophilus* U3S from two crystal forms exhibit two examples of "closed" enzymes. The most "closed" structure, from PDBcode 1WD7 can be aligned with the human U3S using domain 2, residues 39-170, with RMS deviation values of 2.5 Å. An amazing rotation of domain 1 is apparant such that equivalent residues within the two structures can be separated up to 24 Å apart. The rotation is not generated by the two domains clamping down on each other like a jaw, but rather a 60° clockwise rotation of domain 1 with respect to domain 2 (Fig. 14).

Aside from a few invariant residues involved in the hydrophobic core of the molecule, the only two invariant potentially catalytic residues are Tyr168 and Thr228 (Figs. 14 and 15, human numbering). These two residues are 4 and 9 Å distance from their Tt counterpart, but are separated from each other by a distance of 19 Å in the "open" human structure and only 13.8 Å in the "closed" Tt structure. This closure does not fully eliminate the gap between the two domains, but does reduce the separation such that now side chains from the two domains can physically interact with each other (Fig. 14). The addition of this "closed" conformation to the U3S story helps us appreciate the flexibility of the U3S enzyme, but a substrate/product/inhibitor-bound complex is still needed to provide the details of how the invariant residues might cooperate in the catalytic mechanism.

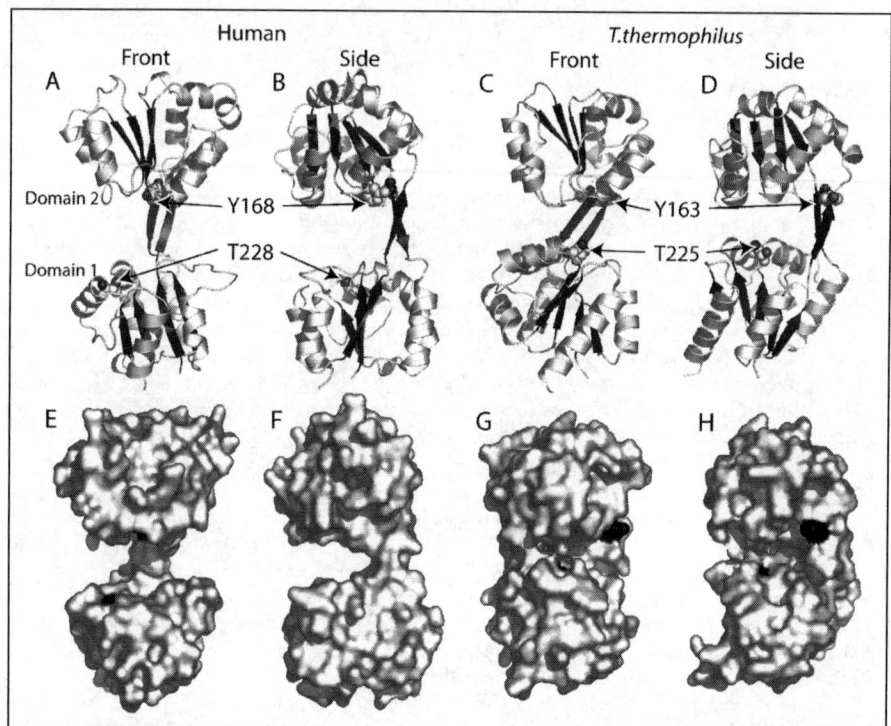

Figure 14. The structure of human U3S in comparison to *T. thermophilus* U3S. The two structures are aligned based on residues within domain 2; a large conformational change is observed in domain 1. Conserved residues, hY168(TtY163) and hT228 (TtT225) are positioned on either side of the main cleft. They are considerable closer to each other in the "closed" Tt crystal structure. A and E) Front view of hU3S in cartoon and surface showing the locations of hY168 and hT228 in small spheres and in black, respectively. B and F) side view of hU3S. C and G) Front view of TtU3S. D and H) TtU3S side view.

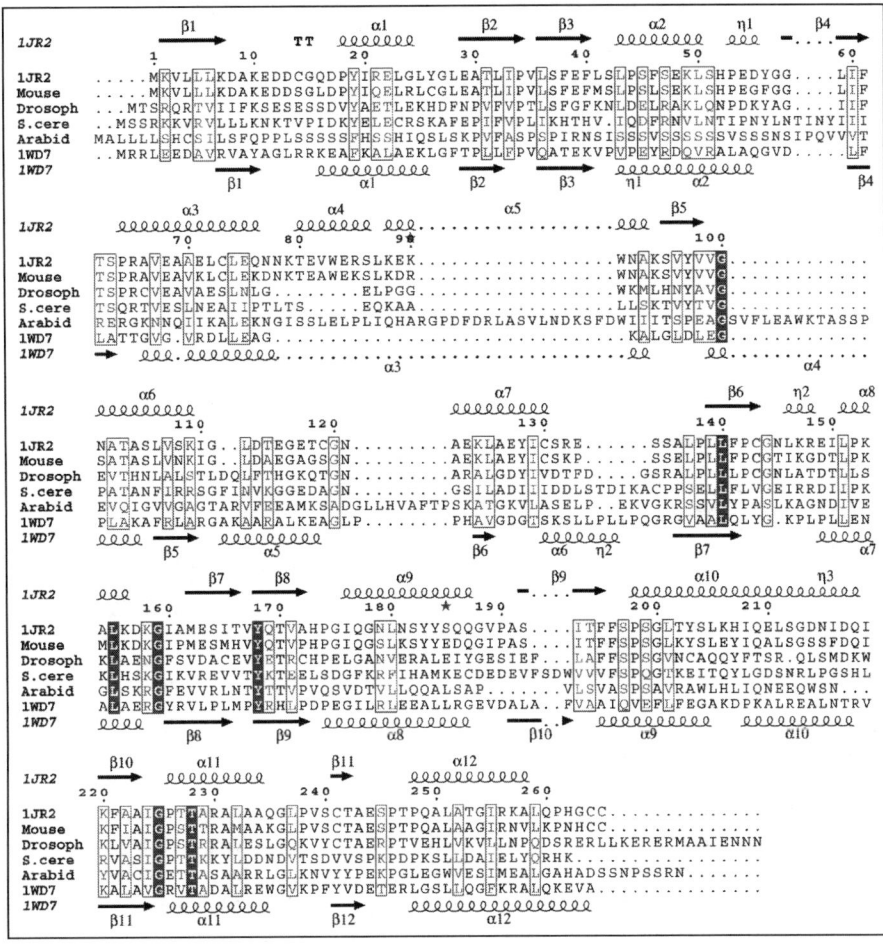

Figure 15. Alignment of several diverse U3S sequences highlighting invariant residues hY168 and hT228. Secondary structure elements of the human U3S (PDBcode 1JR2) and the *T thermophilus* U3S (PDBcode 1WD7) are also shown.

Site-specific point mutations have been made in several of the conserved residues and mutant enzymes were tested for their ability to catalyze the formation or uro'gen III (Table 1). Two different groups have reported mutational analysis on the human (h) and *Anacystis nidulans* (an) enzyme respectively, and though their results are strikingly different, the conclusions are similar.[92,93] The most controversial of mutations involves the hTyr168/anTyr166 mutations to Phe. The activity of the human mutation is reduced to 50% of wild-type, while the reported activity of the *A. nidulas* mutation is ~1%. While this seems like a disparate result, the results compare two different species and use slightly different methodology. Both mutations were among the most severe of those tested and strongly suggest a role in catalysis for this tyrosine. Human T103, equivalent with anT100, were also mutated to Ala and shown to reduce the catalytic activity to 63 and ~25% of wild type respectively. The only other significant disruption of catalysis occurred in the hThr228 to Ala mutation that only retained 32% of catalytic activity. Despite their conservation, many additional residues did not show significant loss of activity when mutated (Table 1).

In lieu of a ligand-bound crystal structure, the exact details of this complicated catalytic mechanism remain mysterious. The location of conserved residues, deleterious mutations and the known

Table 1. The effects of site-specific mutations in uroporphyrinogen III synthase

Mutant	Species	Relative Activity (%)
R15Q	*A.nidulans*	100
T62A	Human	97.8 ± 3.2
S63A	Human	97.1 ± 7.2
S68A	*A.nidulans*	100
R65A	Human	74.1 ± 3.4
T100A	*A.nidulans*	25
T103A	Human	63.5 ± 1.1
D112G	*A.nidulans*	50
E127A	Human	100.6 ± 10.6
Y166F	A.nidulans	<1
Y168F	Human	49.1 ± 1.3
S197A	Human	101.1 ± 3.7
K220A	Human	109.3 ± 4.3
T227A	Human	115.2 ± 9.3
T228A	Human	32.2 ± 3.1

Grouped mutations represent the analogous residue in the two species, human[89] and *A nidulans*.[88]

structure of potential inhibitors all suggest that the two structural domains must close upon the active site during catalysis to impose a rigid framework within which the enzyme could specifically facilitate catalysis. Whether the reaction is catalyzed by a specific general acid or general base remains to be seen.

Disease

Deficiency in U3S results in congenital erythropoietic porphyria (CEP; Chapter 20) a disease causing severe photosensitivity due to the accumulation of photocatalytic porphyrins in the tissues.[95-115] To date at least 36 different genomic mutations have been identified in patients exhibiting CEP, 4 mutations in the 5'untranslated region of the erythroid specific transcript, 21 missense mutations and 11 other mutations including splice defects, shifts in reading frame, and multiple nucleotide deletions (Table 2) (reviewed in ref. 96). The location of the missense mutations can be mapped to the structure of the enzyme.[93] Since the publication of that report, eight additional mutations have been reported[101] and the predicted structural effect of all the described mutations is reported in Table 2.

Conclusion

The three enzymes described in this chapter catalyse the conversion of 5-aminolaevulinic acid (ALA) to a cyclic tetrapyrrole: uroporphyrinogen III; the final step involving tetrapyrrole ring closure and a remarkable stereochemical rearrangement of the D pyrrole ring. The first enzyme, ALA dehydratase, is predominantly an octameric protein but the intriguing findings of hexameric forms of this enzyme and the suggestions that marked changes in the alternative quaternary structures provide a means of regulating the activity[29] are worthy of much further study. Other fascinating aspects, such as the metal ion dependency of the enzyme from different species and the recent findings that some appear to be completely metal-independent[28] along with the variable occurrence of an allosteric site, raise many questions for further research. Likewise, studies of substrate binding and the role of domain movements in both porphobilinogen deaminase and uroporphyrinogen III synthase will be important to define the conformational manoeuvring that takes place both in the enzyme and the substrate during catalysis.

Table 2. Structural and kinetic effects of known mutations in U3S

Exon/ Intron	Allele Designation	Nucleotide/ Codon Change	Structural Effect	Effect on Coding Sequence	Catalytic Activity	Ref.
Erythroid-specific promoter:						
	T70C		n/a	Altered a GATA-1 binding element	2.9[a]	114
	G76A		n/a		53.9[a]	114
	T86C		n/a		43.3[a]	114
	C90A		n/a	Altered a putative CP2 binding element	8.3[a]	114
Coding region:						
Exon 2	V3F	7/GGT→TGT	Larger residue inserted into hydrophobic core	Val→Phe	<1.0	111
	L4F	10/CTT→TTT	Larger residue inserted into hydrophobic core	Leu→Phe	1.8	115
	Y19C	56/TAT→TGT	Loss of buried packing interaction	Tyr→Cys	n/a	107
	21ΔG		n/a	Frameshift after cDNA nt 21 and premature termination after codon 22	n/a	108
Intron 2	IVS2+1	gt→at	n/a	5'donor splice site mutation; deletion of exon 2	n/a	107
Exon 4:	148Δ98		n/a	Deletion of 98 nt from cDNA nt 148-245; frameshift, resulting premature termination	n/a	112
	P53L	158/CCTCTT	Expected to disrupt packing	Pro→Leu	<1.0	115
	T62A	184/ACC→GCC	Solvent exposed to active site clef.	Thr→Ala	<1.0	113
	A66V	197/GCA→GTA	Larger residue inserted into hydrophobic core	Ala→Val	14.5	113
	A69T	205/GCA→ACA	Larger residue inserted into hydrophobic core	Ala→Thr	1.4	108
	C73R	217/TGT→CGT	Charged residue inserted into hydrophobic core	Cys→Arg	<1.0	115
	E81D	242/GAA→GAT	Distant from active site – solvent accessible	Glu→Asp, altered 5' splicing and exon 4 skipping	30.0	108
	V82F	243/GTC→TTC	Large hydrophobic exposed to solvent	Val→Phe	35.8	107

continued on next page

Table 2. Continued

Exon/Intron	Allele Designation	Nucleotide/Codon Change	Structural Effect	Effect on Coding Sequence	Catalytic Activity	Ref.
Exon 5:	V99A	296/GTT→GCT	Loss of packing interactions	Val→Ala	5.6	107
	A104V	311/GCT→GTT	Larger residue inserted into hydrophobic core	Ala→Val	7.7	107
Exon 7:	398insG		n/a	Frameshift after nt 398 in codon 133 and premature termination after codon 196	n/a	108
Exon 8:	H173Y	517/CAC→TAC	Backside of active site cleft	His → Tyr	<1.0	109
Intron 8:	IVS8$^{-23A\to G}$		n/a	Putative branchpoint mutation	n/a	109
Exon 9:	Q187P	558/CAG→CCG	Proline distorts chain direction	Gln → Pro	<1.0	109
	G188R	562/GGG→AGG	Glycine involved in loop, Phi/Psi angle restrictions	Gly → Arg	4.3	104
	G188W	562/GGG→TGG		Gly → Trp	1.7	108
	627Δ6ins39		n/a	Deletion of codons 210 and 211 and in-frame insertion of 13 codons in exon 9 after cDNA nt 627	n/a	108
	633insA		n/a	Frameshift, stop codon 28 codons downstream	n/a	107 110
	S212P	634/TCT→CCT	Disruption of helix, packing constraints	Ser → Pro	<1.0	111
	I219S	656/ATT→AGT	Hydrophilic residue into hydrophobic core	Ile → Ser	1.5	108
Intron 9	IVS9^{A+4}	gtaag→gtag	n/a	Deletion of exon 9 in some mRNAs	n/a	107
Exon 10:	660ins80		n/a	Frameshift at codon 221	n/a	112
	G225S	673/GGC→AGC	Tight turn	Gly → Ser	1.2	107
	T228M	683/ACG→ATG	Larger residue could disrupt packing	Thr → Met	<1.0	113
	P248Q	743/CCA→CAA	Tight turn, hydrophilic residue into hydrophobic packing environment	Pro → Gln	<1.0	109
	Q249X	683/CAA→TAA	Truncated protein, loss of helix 12	Glu → Ter	1.1	107
	672ins28		n/a	Insertion of 9 codons after codon 224, then a frameshift and elongation of the transcript by 49 codons, with termination after codon 314	<2.0	108

[a] Activity using Luciferase expression activity.

Acknowledgements

Heidi Schubert is supported by grants from the NIH, GM56775 and DK02794 and Peter Erskine was supported by BBSRC (UK) grant B13781.

References

1. Jordan PM. Biosynthesis of tetrapyrroles. New Comprehensive Biochemistry 1991; 19:1-65.
2. Jordan PM. Highlights in haem biosynthesis. Curr Opin Struc Biol 1994; 4:902-911.
3. Warren MJ, Scott AI. Tetrapyrrole assembly and modification into the ligands of biologically functional cofactors. TIBS 1990; 15:486-491.
4. Jaffe EK. Porphobilinogen synthase, the first source of heme's asymmetry. J Bioenerg Biomemb 1995; 27:169-179.
5. Jaffe EK. The porphobilinogen synthase family of metalloenzymes. Acta Crystallogr 2000; D56:115-128.
6. Jordan PM, Gibbs PNB. Mechanism of action of 5-aminolevulinate dehydratase from human erythrocytes. Biochem J 1985; 227:1015-1020.
7. Gibson K, Neuberger A, Scott JJ. The purification and properties of δ-aminolevulinic acid dehydratase. Biochem J 1955; 61:618-629.
8. Anderson PM, Desnick RJ. δ-Aminolevulinate dehydrase from human erythrocytes. J Biol Chem 1979; 254:6924-6930.
9. Spencer P, Jordan PM. Purification and characterisation of 5-aminolevulinic acid dehydratase from E. coli and a study of reactive thiols at the metal-binding domain. Biochem J 1993; 290:279-287.
10. Schneider HAW, Liedgens W. An evolutionary tree based on monoclonal antibody recognized surface-features of a plastid enzyme (5-aminolevulinate dehydratase). Z Naturforsch 1981; 36c:44-50.
11. Liedgens W, Lutz C, Schneider HAW. Molecular-properties of 5-aminolevulinic acid dehydratase from Spinacia-oleracea. Eur J Biochem 1983; 135:75-79.
12. Doss M, Von-Tieperman R, Schneider J et al. New types of hepatic porphyria with porphobilinogen synthase defect and intermittent acute clinical manifestation. Klin Wochenschr 1979; 57:1123-1127.
13. Brennan MJW, Cantrill RC. δ-Aminolevulinic acid is a potent agonist for GABA autoreceptors. Nature (Lond) 1979; 280:514-515.
14. Mitchell G, Larochelle J, Lambert M et al. Neurological crises in hereditary tyrosinemia. New England J Med 1990; 322:432-437.
15. Lindstedt S, Holme E, Lock EA et al. Treatment of hereditary tyrosinemia type-I by inhibition of 4-hydroxyphenylpyruvate dioxygenase. Lancet 1992; 340:813-817.
16. Lindblad B, Lindstedt S, Steen G. On the enzymic defects in hereditary tyrosinaemia. Proc Natl Acad Sci USA 1977; 74:4641-4645.
17. Senior N, Thomas PG, Cooper JB et al. Comparative studies of the 5-aminolevulinic acid dehydratase from P. sativum, E. coli and S. cerevisiae. Biochem J 1996; 320:401-412.
18. Mauzerall D, Granick S. The occurrence and determination of δ-aminolevulinic acid and porphobilinogen in urine. J Biol Chem 1956; 219:435-446.
19. Wu W, Shemin D, Richards KE et al. The quaternary structure of δ-aminolevulinic acid dehydratase from bovine liver. Proc Natl Acad Sci USA 1974; 71:1767-1770.
20. Pilz I, Schwarz E, Vuga M et al. Small angle X-ray scattering study of bovine porphobilinogen synthase. Biol Chem Hoppe-Seyler 1988; 369:1099-1103.
21. Erskine PT, Senior N, Awan S et al. X-ray structure of 5-aminolevulinate dehydratase, a hybrid aldolase. Nat Struct Biol 1997; 4:1025-1031.
22. Gibbs PNB, Jordan PM. Identification of a lysine at the active site of human 5-aminolevulinate dehydratase. Biochem J 1986; 236:447-451.
23. Boese QF, Spano AJ, Li J et al. 5-Aminolevulinic acid dehydratase in pea. Identification of an unusual metal-binding domain in the plant enzyme. J Biol Chem 1991; 266:17060-17066.
24. Dent A, Beyersmann D, Block C et al. Two different zinc sites in bovine 5-aminolevulinate dehydratase distinguished by extended X-ray absorption fine structure. Biochemistry 1990; 29:7822-7828.
25. Mitchell LW, Jaffe EK. Porphobilinogen synthase from Escherichia coli is a Zn(II) metalloenzyme stimulated by Mg(II). Arch Biochem Biophys 1993; 300:169-177.
26. Erskine PT, Norton E, Cooper JB et al. The X-ray structure of 5-aminolevulinic acid dehydratase from Escherichia coli complexed with the inhibitor levulinic acid at 2.0 Å resolution. Biochemistry 1999; 38:4266-4276.
27. Frankenberg N, Erskine PT, Cooper JB et al. High resolution crystal structure of a Mg^{2+}-dependent porphobilinogen synthase. J Molec Biol 1999; 289:591-602.

28. Bollivar DW, Clauson C, Lighthall R et al. Rhodobacter capsulatus porphobilinogen synthase, a high activity metal independent hexamer. BMC Biochem 2004; 5:17.
29. Breinig S, Kervinen J, Stith L et al. Control of tetrapyrrole biosynthesis by alternate quaternary forms of porphobilinogen synthase. Nat Struct Biol 2003; 10:757-763.
30. Branden C, Tooze J. Introduction to Protein Structure. New York: Garland, 1991.
31. Frankenberg N, Jahn D, Jaffe EK. Pseudomonas aeruginosa contains a novel type V porphobilinogen synthase with no required catalytic metal ions. Biochemistry 1999; 38:13976-13982.
32. Coates L, Beaven G, Erskine PT et al. The X-ray structure of the plant like 5-aminolaevulinic acid dehydratase from Chlorobium vibrioforme complexed with the inhibitor laevulinic acid at 2.6 Ångstrom resolution. J Molec Biol 2004; 342:563-570.
33. Erskine PT, Newbold R, Brindley AA et al. The X-ray structure of yeast 5-aminolaevulinic acid dehydratase complexed with substrate and three inhibitors. J Molec Biol 2001; 312:133-141.
34. Jarret C, Stauffer F, Henz ME et al. Inhibition of Escherichia coli porphobilinogen synthase using analogs of postulated intermediates. Chem Biol 2000; 7:185-196.
35. Neier R. A novel synthesis of porphobilinogen: Synthetic and biosynthetic studies. J Heterocyclic Chem 2000; 37:487-508.
36. Shoolingin-Jordan PM, Spencer P, Sarwar M et al. 5-Aminolaevulinic acid dehydratase: Metals, mutants and mechanism. Biochem Soc Trans 2002; 30:584-590.
37. Stauffer F, Zizzari E, Engeloch-Jarret C et al. Inhibition studies of porphobilinogen synthase from Escherichia coli differentiating between the two recognition sites. Chem Bio Chem 2001; 2:343-354.
38. Neier R. Chemical synthesis of porphobilinogen and studies of its biosynthesis. Adv Nitrogen Heterocycles 1996; 2:35-146.
39. Erskine PT, Coates L, Newbold R et al. The X-ray structure of yeast 5-aminolaevulinic acid dehydratase complexed with two diacid inhibitors. FEBS Lett 2001; 503:196-200.
40. Kervinen J, Jaffe EK, Stauffer F et al. Mechanistic basis for suicide inactivation of porphobilinogen synthase by 4,7-dioxosebacic acid, an inhibitor that shows dramatic species selectivity. Biochemistry 2001; 40:8227-8236.
41. Jaffe EK, Kervinen J, Martins J et al. Species-specific inhibition of porphobilinogen synthase by 4-oxosebacic acid. J Biol Chem 2002; 277:19792-19799.
42. Frere F, Schubert WD, Stauffer F et al. Structure of porphobilinogen synthase from Pseudomonas aeruginosa in complex with 5-fluorolevulinic acid suggests a double Schiff base mechanism. J Molec Biol 2002; 320:237-247.
43. Chaudhry AG, Jordan PM. Stereochemical studies on the formation of porphobilinogen. Biochem Soc Trans 1976; 4:760-761.
44. Erskine PT, Coates L, Butler D et al. X-ray structure of a putative reaction intermediate of 5-ALA-dehydratase. Biochem J 2003; 373:733-738.
45. Thunell S, Holmberg L, Lundgren J. Aminolevulinate dehydratase porphyria in infancy a clinical and biochemical-study. J Clin Chem Clin Biochem 1987; 25:5-14.
46. Nordmann Y, Puy H. Human hereditary hepatic porphyrias. Clin Chim Acta 2002; 325:17-37.
47. Gross U, Hoffmann GF, Doss MO. Erythropoietic and hepatic porphyries. J Inherit Metab Dis 2000; 23:641-661.
48. Thunell S. Porphyrins, porphyrin metabolism and porphyrias. Scand J Clin Lab Invest 2000; 60:509-540.
49. Wetmur JG, Bishop DF, Cantelmo C et al. Human δ-aminolevulinate dehydratase—Nucleotide-sequence of a full-length cDNA clone. Proc Natl Acad Sci USA 1986; 83:7703-7707.
50. Wetmur JG, Kaya AH, Plewinska M et al. Molecular characterization of the human δ-aminolevulinate dehydratase-2 (ALAD2) allele—Implications for molecular screening of individuals for genetic susceptibility to lead-poisoning. Am J Hum Genet 1991; 49:757-763.
51. Ishida N, Fujita H, Fukuda T et al. Cloning and expression of the defective genes from a patient with δ-aminolevulinate dehydratase porphyria. J Clin Invest 1992; 89:1431-1437.
52. Plewinska M, Thunell S, Holmberg L et al. δ-Aminolevulinate dehydratase deficient porphyria—identification of the molecular lesions in a severely affected homozygote. Am J Hum Genet 1991; 49:167-174.
53. Akagi R, Yasui Y, Harper P et al. A novel mutation of δ-aminolaevulinate dehydratase in a healthy child with 12% erythrocyte enzyme activity. Brit J Haematol 1999; 106:931-937.
54. Simons TJB. The affinity of human erythrocyte porphobilinogen synthase for Zn^{2+} and Pb^{2+}. Eur J Biochem 1995; 234:178-183.
55. Warren MJ, Cooper JB, Wood SP et al. Lead poisoning, haem synthesis and 5-aminolevulinic acid dehydratase. TIBS 1998; 23:217-221.

56. Erskine PT, Duke EMH, Tickle IJ et al. MAD analyses of yeast 5-aminolaevulinate dehydratase. Their use in structure determination and in defining the metal binding sites. Acta Crystallogr D 2000; D56:421-430.
57. Jordan PM, Warren MJ. Evidence for a dipyrromethane cofactor at the catalytic site of Escherichia-coli porphobilinogen deaminase. FEBS Lett 1987; 225:87-92.
58. Awan SJ, Siligardi G, Shoolingin-Jordan PM et al. Reconstitution of the holoenzyme form of Escherichia coli porphobilinogen deaminase from apoenzyme with porphobilinogen and preuroporphyrinogen: A study using circular dichroism spectroscopy. Biochemistry 1997; 36:9273-9282.
59. Louie GV, Brownlie PD, Lambert R et al. Structure of porphobilinogen deaminase reveals a flexible multidomain polymerase with a single catalytic site. Nature 1992; 359:33-39.
60. Shoolingin-Jordan PM. Structure and mechanism of enzymes involved in the assembly of the tetrapyrrole macrocycle. Biochem Soc Trans 1998; 26(3):326-336.
61. Wang AL, Arredondo-Vega FX, Giampietro PF et al. Regional gene assignment of human porphobilinogen deaminase and esterase-A4 to chromosome 11q23-11qter. Proc Natl Acad Sci USA 1981; 78:5734-5738.
62. Grandchamp B, Picat C, Mignotte V et al. Tissue-specific splicing mutation in acute intermittent porphyria. Proc Natl Acad Sci USA 1989; 86:661-664.
63. Chretien S, Dubart A, Beaupain D et al. Alternative transcription and splicing of the human porphobilinogen deaminase gene result either in tissue-specific or in housekeeping expression. Proc Natl Acad Sci USA 1988; 85:6-10.
64. Raich N, Romeo PH, Dubart A et al. Molecular-cloning and complete primary sequence of human-erythrocyte porphobilinogen deaminase. Nucl Acid Res 1986; 14:5955-5968.
65. Yoo HW, Warner CA, Chen CH et al. Hydroxymethylbilane synthase—Complete genomic sequence and amplifiable polymorphisms in the human gene. Genomics 1993; 15:21-29.
66. Deybach JC, Puy H. Porphobilinogen deaminase gene structure and molecular defects. J Bioenergetics and Biomembranes 1995; 27:197-205.
67. Wood S, Lambert R, Jordan PM. Molecular basis of acute intermittent porphyria. Molec Med Today 1995; 1:232-239.
68. Esnouf R. An extensively enhanced version of MolScript that includes greatly enhanced coloring capabilities. J Molec Graphics and Mod 1997; 15:132.
69. Beale SI, Weinstein JD. Biosynthesis of heme and chlorophylls. New York: McGraw-Hill, 1990.
70. Battersby AR, Leeper FJ. Biosynthesis of the pigments of life: Mechanistic studies on the conversion of porphobilinogen to uroporphyrinogen III. Chem Rev 1990; 90(7):1261-1274.
71. Battersby AR, Fookes CJ, Matcham GW et al. Biosynthesis of the pigments of life: Formation of the macrocycle. Nature 1980; 285(5759):17-21.
72. Jordan PM. The biosynthesis of 5-aminolevulinic acid and its transformation into coproporphyrinogen in animals and bacteria. In: Dailey HA, ed. Biosynthesis of heme and chlorophylls. New York: McGraw-Hill, 1990:55-122.
73. Bogorad L. The enzymic synthesis of porphyrins from porphobilinogen. II. Uroporphyrin III. J Biol Chem 1957; 233(2):510-515.
74. Battersby A, Fookes C, McDonald E et al. Biosynthesis of type-III porphyrins: Proof of intact enzymatic conversion of the head-to-tail bilane into uro'gen III by intramolecular arrangement. J Chem Soc Chem Commun 1978; 185-186.
75. Jordan PM, Burton G, Nordlov H et al. Preuroporphyrinogen: A substrate for uroporphyrinogen III cosynthase. J Chem Soc Chem Comm 1979; 204-205.
76. Burton G, Fagerness PE, Hosozawa S et al. ^{13}C N.M.R. Evidence for a new intermediate, preuroporphyrinogen, in the enzymatic transformation of prophobilinogen into uroporphyrins I and III. J Chem Soc Chem Comm 1979; 202-204.
77. Battersby AR, Fookes CJ, Gustafson-Potter KE et al. Proof by synthesis that unrearranged hydroxymethylbilane is the product from deaminase and the substrate for cosynthase in the biosynthesis of uro'gen III. J Chem Soc Chem Comm 1979; 1155-1158.
78. Mathewson J, Corwin A. Biosynthesis of pyrrolepigments: A mechanism for porphybilinogen polymerization. J Am Chem Soc 1961; 83:135-137.
79. Desnick RJ, Bishop DF. Preface. Enzyme 1982; 28:91-92.
80. Leeper FJ. The evidence for a spirocyclic intermediate in the formation of uroporphyrinogen III by cosynthase. Ciba Found Symp 1994; 180:111-123.
81. Pichon C, Atshaves BP, Stolowich NJ et al. Evidence for an intermediate in the enzymatic formation of uroporphyrinogen III. Bioorg Med Chem 1994; 2(3):153-68.

82. Stark WM, Hart GJ, Battersby ARJ. Synthetic studies on the proposed spiro intermediate for bio-synthesis of the natrual porphyrins: Inhibition of cosynthase. Chem Soc Chem Commun 1986; 465-467.
83. Battersby AR, McDonald E. Biosynthesis of porphyrins and corrins. Philos Trans R Soc Lond B Biol Sci 1976; 273(924):161-80.
84. Battersby AR, Fookes CJ, Matcham GWJ et al. Biosynthesis of natural porphyrins: Studies with isomeric hydroxymethylbilanes on the specificity and action of cosynthase. Angew Chem Int Ed Engl 1981; 20:293-295.
85. Stark WM, Baker MG, Raithby PR et al. The spiro intermediate proposed for biosynthesis of the natural porphyrins: Synthesis and properties of its macrocycle. J Chem Soc Chem Comm 1985; 1294.
86. Cassidy MA, Crockett N, Leeper FJ et al. Synthetic studies on the proposed spiro intermediate for biosynthesis of the natural porphyrins: The stereochemical probe. J Chem Soc Chem Commun 1991; 6:384-386.
87. Jordan PM. Uroporphyrinogen III cosynthetase: A direct assay method. Enzyme 1982; 28(2-3):158-169.
88. Battersby AR, Fookes CJ, Pandey PS. Linear tetrapyrroleic intermediates for biosynthesis of the natural porphyrins. Experiments with modified substrates. Tetrahedron 1983; 39:1919-1926.
89. Pichon C, Atshaves BP, Xue T et al. Studies on uro'gen III synthase with modified bilanes. Bioorg Med Chem 1994; 4(9):1105-1110.
90. Falk JE, Benson A. Separation of uroporphyrin esters I and III by paper chromatography. Biochem J 1953; 55:101-104.
91. Nordlov H, Jordan PM, Burton G et al. Improved separation of uroporphyrin isomers by high-performance liquid chromatography. J Chromatog 1980; 190(1):221-225.
92. Roessner CA, Ponnamperuma K, Scott AI. Mutagenesis identifies a conserved tyrosine residue im-portant for the activity of uroporphyrinogen III synthase from Anacystis nidulans. FEBS Lett 2002; 525(1-3):25-28.
93. Mathews MA, Schubert HL, Whitby FG et al. Crystal structure of human uroporphyrinogen III synthase. EMBO J 2001; 20(21):5832-9.
94. Mizohata E, Matsuura T, Sakai H et al. Crystal structure of uroporphyrinogen III synthase from thermus thermophilus Hb8 (PDB codes 1WCW and 1WD7) protein data bank.
95. Romeo G, Levin EY. Uroporphyrinogen 3 cosynthetase in human congenital erythropoietic por-phyria. Proc Natl Acad Sci USA 1969; 63(3):856-63.
96. Desnick RJ, Astrin KH. Congenital erythropoietic porphyria: Advances in pathogenesis and treat-ment. Br J Haematol 2002; 117(4):779-95.
97. Bensighoum M, Larou M, Lemeur M et al. The disruption of mouse uroporphyrinogen III syn-thase (uros) gene is fully lethal. Transgenics 1998; 2:275-280.
98. Xu W, Astrin KH, Desnick RJ. Molecular basis of congenital erythropoietic porphyria: Mutations in the human uroporphyrinogen III synthase gene. Hum Mutat 1996; 7(3):187-192.
99. Tsai SF, Bishop DF, Desnick RJ. Human uroporphyrinogen III synthase: Molecular cloning, nucle-otide sequence, and expression of a full-length cDNA. Proc Natl Acad Sci USA 1988; 85(19):7049-7053.
100. Aizencang G, Solis C, Bishop DF et al. Human uroporphyrinogen-III synthase: Genomic organiza-tion, alternative promoters, and erythroid-specific expression. Genomics 2000; 70(2):223-231.
101. Shady AA, Colby BR, Cunha LF et al. Congenital erythropoietic porphyria: Identification and expression of eight novel mutations in the uroporphyrinogen III synthase gene. Br J Haematol 2002; 117(4):980-987.
102. Piomelli S, Poh-Fitzpatrick MB, Seaman C et al. Complete suppression of the symptoms of con-genital erythropoietic porphyria by long-term treatment with high-level transfusions. N Engl J Med 1986; 314(16):1029-1031.
103. Poh-Fitzpatrick MB, Piomelli S, Seaman C et al. Congenital erythropoietic porphyria: Complete suppresion of symptoms by long-term high-level transfusion with deferoxamine infusion iron res-cue. In: Orfanos C, Stadler R, Gollnick H, eds. Dermatology in Five Continents. Berlin: Springer-Verlag, 1988:876-879.
104. Tezcan I, Xu W, Gurgey A et al. Congenital erythropoietic porphyria successfully treated by allo-geneic bone marrow transplantation. Blood 1998; 92(11):4053-4058.
105. Mazurier F, Geronimi F, Lamrissi-Garcia I et al. Correction of deficient CD34+ cells from periph-eral blood after mobilization in a patient with congenital erythropoietic porphyria. Mol Ther 2001; 3(3):411-417.

106. Moreau-Gaudry F, Mazurier F, Bensidhoum M et al. Metabolic correction of congenital erythropoietic porphyria by retrovirus-mediated gene transfer into Epstein-Barr virus-transformed B-cell lines. Blood 1995; 85(6):1449-1453.

107. Xu W, Warner CA, Desnick RJ. Congenital erythropoietic porphyria: identification and expression of 10 mutations in the uroporphyrinogen III synthase gene. J Clin Invest 1995; 95:905-912.

108. Shady AA, Colby BR, Cunha LF et al. Congenital erythropoietic poprphyria: identification and expession of eight novel mutations in the uroporphyrinogen III synthase gene. Brit J Haematol 2002; 117:980-987.

109. Fontanellas A, Bensidhoum M, Enriquez de Salamanca R et al. A systematic analysis of the mutations of the uroporphyrinogen III synthase gene in congenital erythropoietic porphyria. Eur J Hum Genet 1996; 4:274-282.

110. Bensidhoum M, Larou M, Lemeur M et al. The disruption of mouse uroporphyrinogen III synthase (uros) gene is fully lethal. Transgenics 1998; 2:275-280.

111. Takamura N, Hombrados I, Tanigawa K et al. Novel point mutation in the uroporphyrinogen III synthase gene causes congenital erythropoietic porphyria of a Japanese family. Am J Med Genet 1997; 70:299-302.

112. Boulechfar S, Da Silva V, Deybach JC et al. Heterogeneity of mutations in the uroporphyrinogen III synthase gene in congenital erythropoietic porphyria. Hum Genet 1992; 88:320-324.

113. Warner CA, Yoo HW, Roberts AG, Desnick RJ. Congenital erythropoietic porphyria: identification and expression of exonic mutations in the uroporphyrinogen III synthase gene. J Clin Invest 1992; 89:693-700.

114. Solis C, Aizencang GI, Astrin KH et al. Uroporphyrinogen III synthase erythroid promoter mutations in adjacent GATA1 and CP2 elements cause congenital erythropoietic porphyria. J Clin Invest 2001; 107:753-762.

115. Deybach JC, De Verneuil H, Boulechfar S et al. Point mutations in the uroporphyrinogen-III synthase gene in congenital erythropoietic porphyria (Gunther's disease). Blood 1990; 75:1763-1765.

CHAPTER 4

Transformation of Uroporphyrinogen III into Protohaem

Johanna E. Cornah and Alison G. Smith*

Abstract

Haem is an essential cofactor for virtually all organisms. It is made from the common tetrapyrrole progenitor, uroporphyrinogen III, by four sequential enzymes: uroporphyrinogen III decarboxylase, coproporphyrinogen III oxidase, protoporphyrinogen IX oxidase, and ferrochelatase. Each of the enzymes catalyses a remarkable reaction, with the first three required to carry out the same reaction at multiple sites on the substrate molecule. Now that the crystal structures are available for each of the proteins, the mechanisms of these essential enzymes are beginning to be elucidated. Despite the universality of haem synthesis however, there are differences between organisms. Firstly in many bacteria there are anaerobic forms of the two oxidases, which appear to have completely different origins from the aerobic forms found in eukaryotes. Secondly, in certain bacteria some or all of these enzymes are missing completely; either they are pathogenic and can take up haem from their host, or there are alternative, as yet uncharacterized, enzymes. Finally, within the eukaryotes, the subcellular distribution of the enzymes differs depending on the organism, which has ramifications for the regulation of the biosynthetic pathway.

Introduction

Uroporphyrinogen III (Fig. 1) is the first cyclic tetrapyrrole and the backbone for all the biologically active tetrapyrrole-derived molecules in living systems. The insertion of divalent metal ions and elaboration of the side chains of this macrocycle generate some of the most biologically important and widespread molecules on earth (Chapter 1). The most abundant is chlorophyll (and bacteriochlorophyll), characteristic of photosynthetic organisms. Haem is synthesized by the vast majority of all eukaryotes and prokaryotes, although there are some notable exceptions including certain obligate pathogens such as *Haemophilus influenzae*,[1] and the nematode worm *Caenorhabditis elegans*.[2] Further metabolism of haem leads to the production of bilins, linear tetrapyrroles that act as light-harvesting pigments or sensors in photosynthetic organisms (Chapters 12 and 13), and may play a signaling role in animals. All these molecules are derived from protoporphyrin IX, synthesized from uroporphyrinogen III by oxidation and decarboxylation reactions. Alternatively, uroporphyrinogen III is methylated for direction into the synthesis of sirohaem, and the corrins, coenzyme F_{430} and vitamin B_{12} (Chapters 18, 19 and 20).

This chapter considers recent advances in our understanding of the universal part of the pathway that converts uroporphyrinogen III into protohaem. This conversion is achieved by the action of four enzymes: uroporphyrinogen III decarboxylase (UROD), coproporphyrinogen III oxidase (CPO) (or CP dehydrogenase in some bacteria), protoporphyrinogen IX oxidase (PPO) and ferrochelatase (Fig. 1). The first two of these enzymes catalyse oxidative decarboxylations of the substituents of the porphyrin rings, PPO then oxidizes the colourless protoporphyrinogen IX to

*Corresponding Author: Alison G. Smith—Department of Plant Sciences, University of Cambridge, Downing Street, Cambridge CB2 3EA, UK. Email: as25@cam.ac.uk

Tetrapyrroles: Birth, Life and Death, edited by Martin J. Warren and Alison G. Smith.
©2009 Landes Bioscience and Springer Science+Business Media.

the fully conjugated protoporphyrin IX, and finally the insertion of ferrous iron is catalysed by ferrochelatase. These enzymes are remarkable in the reactions that they catalyse, and in the exquisite regulation that they must be under to avoid the uncontrolled accumulation of the pathway intermediates, which are toxic. Studies of the enzymes from four disparate groups of organisms have each provided a unique perspective on the pathway. Budding yeast (*Saccharomyces cerevisiae*) can respire anaerobically, and so mutants defective in haem synthesis, and therefore devoid of mitochondrial cytochromes, can be readily generated. These have provided a valuable resource for cloning of genes by functional complementation, not just of the endogenous yeast genes,[3] but also those from other organisms, and for testing that genes identified by genome sequencing are indeed functional.[4,5] The study of these enzymes in mammals has been driven in part by the importance of haem synthesis in erythrocytes, and by the fact that defects in the enzymes lead to the diseases known as porphyrias (Chapter 5). In plants, haem synthesis must take place concomitantly with that for chlorophyll, with the two chelatases competing for the protoporphyrin IX macrocycle as substrate. The subcellular location and organization of the enzymes in the plant cell may well play

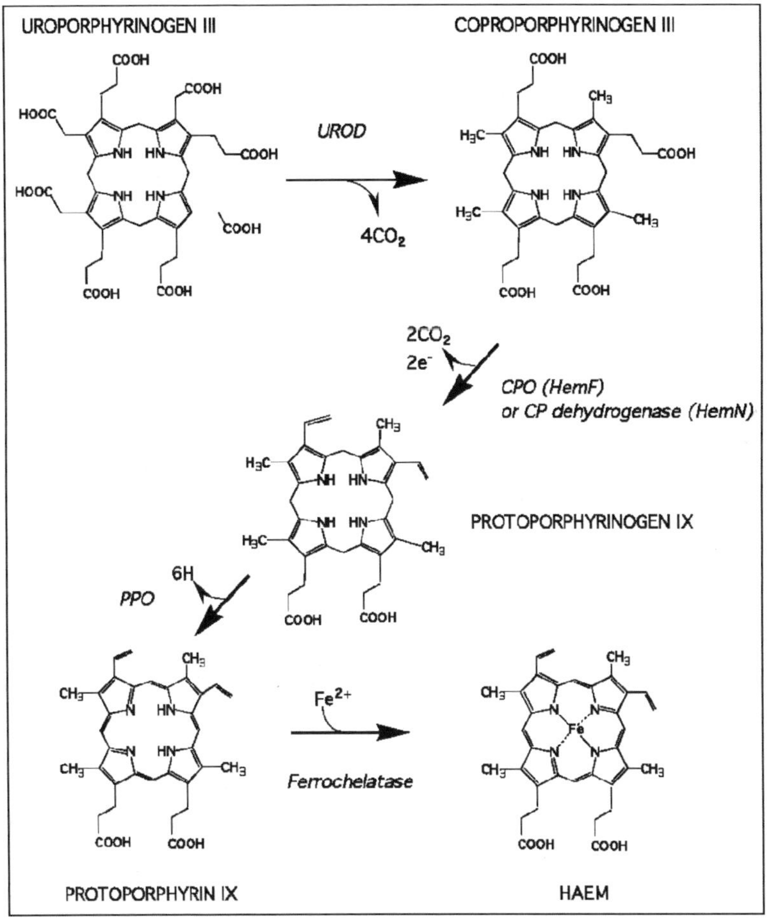

Figure 1. Interconversion of uroporphyrinogen III to haem, carried out by four enzymes. In eukaryotes these are uroporphyrinogen III decarboxylase (UROD), coproporphyrinogen III oxidase (CPO; encoded by *hemF*), protoporphyrinogen IX oxidase (PPO), and ferrochelatase. In many bacteria the second step is catalysed under anaerobic condition by an unrelated enzyme called CP dehydrogenase (encoded by *hemN*). There is also an anaerobic form of the third enzyme, which has not been well-characterised.

a role in the regulation of this branchpoint (Chapter 15). Lastly, the ability of some bacteria to live anaerobically means that the oxidation of coproporphryinogen III and protoporphyrinogen IX must be carried out by enzymes that do not require molecular oxygen. In fact it is becoming clear that there is much more variation in haem synthesis enzymes in prokaryotes than had previously been thought (reviewed by ref. 1).

All four enzymes have been purified from a variety of sources, and the genes encoding them have been cloned. Despite some differences in the pathways between organisms, the sequences of these enzymes show that there is a high level of functional conservation in bacteria, yeast, mammals and plants. Figure 2 shows the phylogenetic relationships between the enzymes from different sources, discussed in more detail in the sections on individual enzymes. However, it is possible to discern a common theme, namely that the enzymes form animals and yeast form a clade, as do the plant and algal enzymes, whereas those from bacteria are more variable. As well as generating sequence information, cloning of the genes has allowed overexpression of recombinant proteins, providing the opportunity to investigate the kinetic characteristics and the reaction mechanisms in some detail. This has been enhanced spectacularly in recent years by the publication of the crystal structures of each of the enzymes. At the same time, studies of the enzymes in vivo are beginning to provide an insight into the organisation and regulation of the pathway as a whole.

Uroporphyrinogen III Decarboxylase

Uroporphyrinogen III decarboxylase (UROD, EC 4.1.1.37), reviewed in detail by Shoolingin-Jordan,[6] catalyses the decarboxylation of the four acetate side chains of uroporphyrinogen III to generate coproporphyrinogen III (Fig. 1). In plants, bacteria and yeast, this enzyme must compete for uroporphyrinogen III with uroporphyrinogen III methylase, the first enzyme on the pathway to sirohaem and corrin biosynthesis (Chapters 18, 19 and 20), but little or nothing is known about the regulation of this branchpoint in any organism. UROD has been purified from many species including bacteria, yeast and humans, and the genes cloned. Sequence similarity between URODs from different organisms is of the order of 10% identity and 33% similarity, with a region at the N-terminus where 8 out of 10 residues are invariant. Phylogenetic analysis (Fig. 2A) shows that human, mouse, *Drosophila melanogaster* and yeast (*S. cerevisiae*) URODs cluster together with those from the eubacteria *E. coli* and *Bradyrhizobium japonicum*, whereas the plant enzymes form a distinct clade with that from the photosynthetic cyanobacterium *Synechocystis*, suggesting that the plant enzyme was derived from the endosymbiont that gave rise to chloroplasts. In the higher plant *Arabidopsis thaliana*, a second gene for UROD is present that does not cluster with the enzyme from tobacco and *Synechocystis*, suggesting that it evolved separately, or alternatively is a pseudogene. This latter suggestion is rendered more likely by the observation that in the completed genome sequence of rice, *Oryza sativa*, there is only a single gene for UROD.

The crystal structures of both the human[7] and tobacco enzymes[8] have been solved and found to be dimeric, as was the recombinant human protein, although that purified from yeast was reportedly monomeric.[9] Each protomer is a single domain containing a $(\beta\alpha)_8$ barrel, with an active site cleft containing a number of invariant polar residues. These include several arginines and a histidine, likely to be involved in binding the carboxyl groups of the substrate, and aspartate-86 and tyrosine-164 (human numbering) that may function in catalysis.

UROD is the only known decarboxylase without a requirement for prosthetic groups or cofactors. The reaction proceeds through intermediate porphyrinogens with 3, 2 and 1 acetate groups. Although under conditions of substrate excess the decarboxylations are found to occur in a random order, at physiological substrate concentrations the reaction occurs in an ordered fashion starting with the acetate on ring D followed by rings A, B and finally C.[10] The absence of repetitive motifs in the sequence of UROD, together with the fact that the chirality of each of the four acetate α-carbons is the same and conserved during the decarboxylation reactions,[6] provides evidence that there is a single active site on each subunit. The substrate and each of the three intermediates are quite different from one another, so it is remarkable that all four decarboxylations should be catalysed by one enzyme. Furthermore, to position ring A at the same active site as that for the decarboxylation of ring D, the first reaction intermediate would have to flip by 180°. One possible explanation for the reaction mechanism draws on the fact that the two funnel-shaped monomers are in head-to-head orientation, such that the two active sites are at the interface. This creates a single extended cleft that is more protected from the solvent, and could accommodate two substrate molecules. On the basis of this, it was proposed that ring D decarboxylation occurs on one monomer, then moves to the active site of the second for decarboxylation of ring A, without the need for flipping.[8] An alternative proposal has come from the crystal structure of human UROD in complex with the product coproporphyrinogen III.[11] In this structure the tetrapyrrole is moulded into a highly nonplanar domed conformation, such that the central NH groups of the pyrrole rings are precisely orientated to hydrogen bond with the Oδ1 atom of the invariant Asp86. The authors propose a catalytic mechanism in which protonation of a Cα atom of the pyrrole ring is stabilized by Asp86, the only negatively charged side chain in the active site. This protonated pyrrole ring would then act as an electron sink allowing decarboxylation to occur, in a manner analogous to pyridoxal phosphate dependent decarboxylases. There is little specific coordination of the peripheral substituents, and the propionate groups are partially exposed to solvent. The central binding of Asp86 with the pyrrole NH groups would thus allow identical interactions to be made with protein as the substrate and the intermediates bind and rebind after rotating 90° about the contact point with Asp86.[11]

Prokaryotic UROD is encoded by the *hemE*, which in many gram positive bacteria, such as *B. subtilis*, is in an operon with genes for protoporphyrinogen IX oxidase (*hemY*) and ferrochelatase (*HemH*).[12] The *hemE* gene, with considerable similarity to its eukaryotic homologue, is present in most prokaryotic genomes thus far sequenced, with notable exceptions being a number of pathogenic bacteria, where some, or all, of the haem biosynthesis genes are missing.[1] This presumably reflects the loss of many biosynthetic genes from these so-called degenerate genomes during their evolution into the pathogenic lifestyle. Haem is either taken up from the host, as in *Haemophilus influenzae*, or dispensed with altogether as in the spirochaete *Borrelia burgdorferi*. Interestingly many archaea do not have a recognizable *hemE* gene (nor indeed genes for the other three enzymes to haem), and yet they possess haem-dependent proteins.[1] It is unlikely given the ecological niches of these extremophiles that they have an available source of exogenous haem for uptake, and thus presumably have an alternative, as yet uncharacterized, route for haem biosynthesis. Indeed evidence for such an alternative comes from radiolabelled feeding studies with the sulphur-reducing bacterium *Desulfovibrio vulgaris*, which identified the conversion of uroporphyrinogen III into coproporphyrinogen via precorrin 2, an intermediate of the corrin/sirohaem pathway[13] (see Chapter 18).

Coproporphyrinogen III Oxidase/Dehydrogenase

The next step in the sequence catalyses the oxidative decarboxylation of ring A and B propionate groups to vinyl groups to generate protoporphyrinogen IX (Fig. 1), (reviewed in detail by Akhtar ref. 14). As for UROD, the enzyme thus carries out the same reaction at more than one site on the substrate, so any reaction mechanism must provide an explanation for this. In eukaryotes, the reaction is catalysed by CPO, requiring molecular oxygen to act as an electron acceptor and results in the release of 2 molecules of carbon dioxide per molecule of coproporphyringen III oxidized (CPO oxygen-dependent, EC 1.3.3.3). However, in anaerobic and facultatively-anaerobic bacteria there is another enzyme that converts coproporphyrinogen III to protoporphyrinogen IX, encoded by *hemN*, that does not require oxygen. The two forms of the enzyme share no sequence identity, indicating that they evolved independently of one another. In prokaryotes, the oxygen-dependent CPO (encoded by *hemF*) is restricted to contain subclasses of the *Proteobacteria* and cyanobacteria,[1] and the oxygen-independent form, HemN, represents the more common form of the enzyme. Since HemN does not use molecular oxygen, a more appropriate name for the enzyme is coproporphyrinogen III dehydrogenase (EC 1.3.99.22). In facultative bacteria like *E. coli* both genes are present, and the expression of the *hemN* gene is induced by oxygen limitation. HemN from *E. coli* has been overexpressed and purified under strict anaerobic conditions.[15] Maximal activity required SAM, NAD(P)H and cytoplasmic extracts of *E. coli*, and little or no activity was measurable without the extract, so it is likely that this provided the electron donors/acceptors for the reaction. The enzyme was a monomer of 52 kDa, and contained an oxygen-sensitive 4Fe-4S cluster that was essential for enzyme activity. Sequence similarity to other Fe.S proteins suggested that the enzyme was a member of the radical S-adenosyl methionine (SAM) family. In the reduced state, the Fe.S cluster of these enzymes transfers an electron to SAM, causing cleavage to produce a highly reactive 5′deoxyadenosyl radical. This can then abstract a hydrogen from the substrate. The crystal structure of *E. coli* HemN with SAM bound provided the first structural indication of the mechanism of radical SAM enzymes.[16] The protein has two domains, the largest one (residues 36-364) being a curved $(\beta\alpha)_6$ barrel enclosing the active site pocket, which bears some structural similarity to $(\beta\alpha)_8$ (TIM) barrels. The 4Fe.4S cluster is located here, coordinated by three cysteines, and the fourth Fe is coordinated by a juxtaposed SAM molecule. Intriguingly a second SAM is bound close by, and modeling the substrate coproporphyrinogen III into the structure led to speculation that this may be involved in oxidation of the second propionate side-chain, but further work is necessary to resolve this question.

Our understanding of the oxygen-dependent CPO is also incomplete, although we have considerable information about the protein and the expression of the gene in animals (see Chapter 7). The eukaryotic enzyme is a homodimer of subunits of about 35 kDa in yeast and plants,[5] and 50 kDa in humans.[17] Both the plant and mammalian enzymes are synthesized initially as precursor proteins with N-terminal extensions to direct import of the protein into the chloroplast and mitochondrion respectively (see section 6.1), but the mammalian enzyme has an additional 100 or so residues at the N-terminus of the mature protein. There is considerable sequence similarity throughout the protein, but particularly at the C-terminus, where there is a stretch of 35 amino acids of which 24 are identical between animals, bacteria, yeast and plants. Phylogenetic analysis (Fig. 2B) reveals a virtually congruent tree to that for UROD except that this time the cyanobacterial enzyme is quite dissimilar to CPO from all other organisms. In this case, plant CPO was presumably derived from the eukaryotic host nucleus rather than the chloroplast progenitor.

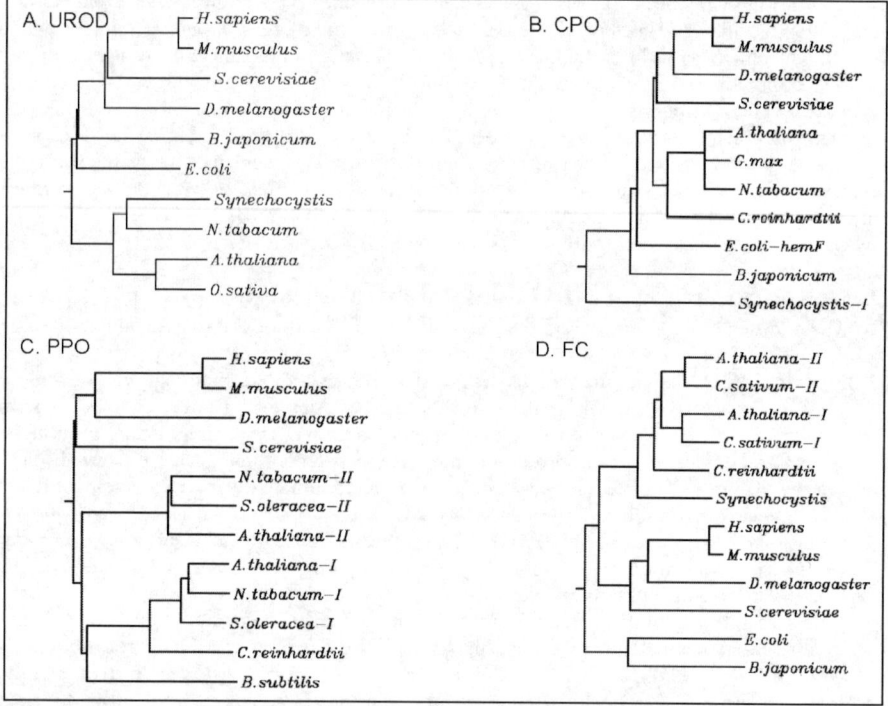

Figure 2. Phylogenetic trees for proteins for each of UROD, CPO, PPO and FC from representative organisms with complete genome sequences, together with some sequences mentioned in the text. *H. sapiens* = human; *M. musculus* = mouse; *D. melanogaster* = fruit fly; *S. cerevisiae* = budding yeast; *B. japonicum* = *Bradyrhizobium japonicum* (soil bacterium); *N. tabacum* = tobacco; *A. thaliana* = Arabidopsis; *O. sativa* = rice; *G. max* = soybean; *S. oleracea* = spinach; *C. reinhardtii* = *Chlamydomonas reinhardtii* (green alga); *B. subtilis* = *Bacillus subtilis*; *C. sativum* = cucumber. Note that that the only bacterial species to encode PPO is B. subtilis. Amino acid sequences were aligned using CLUSTALW (http://clustalw.genome.jp/). The sequences were trimmed to remove all insertions so that the sequences for each enzyme were the same length, and then used to draw Neighbor-joining trees.

The enzyme has an absolute requirement for molecular oxygen for activity, which led to the proposal of a metal ion involved in catalysis. Recombinant mouse CPO was reported to contain bound Cu^{2+} ions,[18] whereas purified HemF from *E. coli* was found to have 0.2 - 0.6 mol of manganese/ mol of protein, and the enzyme was stimulated by exogenous manganese.[19] Mutant forms in which highly conserved histidine residues were altered to leucine were inactive, and for two of these (His106Leu and His96Leu) the amount of manganese bound was reduced or abolished. However, the crystal structures of both the yeast[20] CPO, solved to 2 Å, and that from human, solved to 1.58Å,[17] show no evidence for a bound metal ion. Indeed even when crystals of the human enzyme were grown in the presence of transition metals Fe, Zn, Mn or Cu, none were found in the X-ray data. Moreover, the conserved histidines are too far apart (14-22 Å) in the structure to serve as metal-ion ligands.

The CPO protein from both yeast and humans is dimeric and adopts a novel 7-stranded antiparallel sheet flanked by helices, with a deep active site cleft, lined with conserved residues. In one crystal form of the yeast enzyme, the cleft is closed to bulk solvent by movement of a helix over the entrance. The enclosed cavity has the dimensions of the substrate, which could be modeled into the space, although the resolution of the current structure was insufficient to establish which residues are involved in substrate recognition. In the human enzyme, solved to higher resolution, a citrate molecule was identified bound in the active site, allowing residues Ser244, His258, Asn260, Arg262, Arg282 and Arg 332 to be assigned roles in substrate binding and decarboxylation. Furthermore His258 is positioned appropriately to be involved in catalysis, as previously predicted from mutation studies on mouse CPO.[18] No other cofactors appear to be required for CPO, and the reaction mechanism for the two oxidative decarboxylations is completely unknown, although information from the crystal structures has been used to propose schemes involving proton abstraction from the porphyrin ring, and direct reaction of the substrate with molecular oxygen.[17,21]

Nevertheless, it has been demonstrated conclusively that the decarboxylations of the propionates proceed in strict order, with the A ring proprionate first, followed by that on ring B. The propionate groups on rings C and D are not substrates, but they do appear to influence enzyme activity,[22] providing evidence for a model of the active site involving recognition of the proprionate on ring C. Above a certain threshold, increasing amounts of either the natural substrate or analogues where the propionates on rings C and D were modified resulted in a reduction in enzyme activity. This was found to be due to inhibition by the product protoporphyrinogen IX,[23] suggesting that this may be a means of regulation of CPO in vivo. Interestingly, over-expression of the oxygen-dependent CPO from *E. coli* and mouse can mediate the conversion of protoporphyrinogen IX to protoporphyrin IX in an *E. coli hemG* mutant defective in PPO activity.[24] This suggests that the enzyme is able to accept either copro- or protoporphyrinogen as substrate, but is presumably prohibited from doing so by strict substrate allocation within the cell (see below).

Protoporphyrinogen IX Oxidase

The penultimate step of protohaem synthesis is the conversion of the colourless protoporphyrinogen IX to the highly conjugated protoporphyrin IX (Fig. 1). This reaction can occur spontaneously in the light in the presence of oxygen, but is catalysed in vivo by the enzyme protoporphyrinogen IX oxidase (PPO; oxygen-dependent, EC 1.3.3.4), (reviewed in detail by Akhtar ref.14).

The reaction requires the removal of 6 electrons, and in eukaryotes and some bacteria, such as *B. subtilis*, molecular oxygen acts as the electron acceptor, and three hydrogen peroxide molecules are released per protoporphyrinogen IX oxidised. PPO requires a flavin cofactor for activity, and must be extracted from the membranous fraction of the cell using detergent; its activity in vitro is stimulated by the addition of fatty acids or phospholipids. It is a dimer of subunits of about 50 kDa, and when purified contains FAD, associated with a highly conserved dinucleotide-binding motif, found also in monoamine oxidases and phytoene desaturase.[25] The enzyme from different organisms shares limited overall sequence similarity, but a clear flavin-binding domain is conserved in all. Higher plants have two isoforms of the enzyme, found in chloroplasts and mitochondria respectively (see below),[26] with the chloroplast isoform clustering with the enzyme from the green alga *Chlamydomonas reinhardtii* (Fig. 2C). Interestingly, the two higher plant forms are as dissimilar to one another as they are to the animals/yeast clade, suggesting that they diverged from one another early on in the plant lineage.

In eukaryotes PPO is on the outer surface of the inner mitochondrial membrane, and in plant chloroplasts it is associated with both the thylakoid and envelope membranes.[27] However, the primary amino acid sequence of the enzyme from all sources reveals no obvious membrane spanning regions of the protein, leading to the proposal that membrane anchoring might be mediated by acylation of PPO with a fatty acid.[28] Further understanding of its association with the membrane has come from the recent crystal structure of the mitochondrial isoform PPO-II from tobacco, complexed with a phenyl-pyrazol inhibitor.[29] The structure reveals a dimer, with each protomer made up of three domains: an FAD-binding domain, a substrate-binding domain enclosing a narrow cavity, and finally an alpha-helical domain that encompasses most of the dimer interface, and is likely to constitute the membrane anchor. Helices α_4, α_5 and α_{11} contain apolar residues that would face the lipid part of the membrane, and would insert monotypically, essentially spanning just one half of the bilayer. Modeling the substrate into the active site demonstrated that the methylene bridge C_{20} between rings A and D is orientated towards the reactive N_5 atom of FAD. The narrow cavity precludes rotation of the substrate, so the authors propose that the other hydride abstractions occur from C_{20}, after imine-enamine tautomerisations. Beneath the active site there is a U-shaped channel leading into the membrane bilayer, which the authors suggest might provide a means to deliver the product, protoporphyrin IX, to the next enzyme ferrochelatase. It was possible to dock the structure of tobacco PPO-II with that of ferrochelatase from humans, such that there were considerable interactions between residues in the two enzymes.[29]

The location of the enzyme on a membrane has implications not just for the regulation of tetrapyrrole synthesis within cells (see below), but also for the mechanism of action of the light-dependent peroxidising herbicides, tetrahydrophthalimides, and diphenyl ethers such as acifluorfen. These inhibit PPO at nanomolar levels by competing with protoporphyrinogen IX at the substrate binding site.[30] The result is the accumulation of the product of PPO, protoporphyrin IX, which rapidly partitions into the plasma membrane causing lipid peroxidation. The explanation for this apparent contradiction is that inhibition of PPO leads to a build-up of its substrate protoporphyrinogen IX, which then accumulates in the cytosol.[31] Here it can be rapidly oxidized either nonenzymically or by a nonspecific oxidase in the plasma membrane or ER[32] and the protoporphyrin IX that results is then not accessible to ferrochelatase or Mg-chelatase.[31] The implication of this observation is that PPO has a role to play in the transport of protoporphyrin(ogen) IX into the matrix of the mitochondrion, where the active site of ferrochelatase is located (see below). These herbicides have their effect only in the light, because this is required for the lipid peroxidising effects of protoporphyrin IX. Nevertheless, oxygen-dependent PPO from all organisms is sensitive to these compounds, and this property has been used to probe the active site of PPO from yeast and human, and to test candidate drugs for the photodynamic therapy of cancer.[33]

As is the case for CPO, an oxygen-independent PPO, apparently unrelated to the oxygen-dependent enzyme, exists for the generation of haem and/or bacteriochlorophyll in many bacteria, but unlike CPO, only a single form of PPO is present in any one cell.[34] Whereas the gram-positive *B. subtilis* has a eukaryotic PPO, *E. coli* encodes no genes with sequence similarity to this enzyme. Instead, the so-called *hemG* mutant of *E. coli*, which had no measurable PPO activity, was rescued by a gene that encodes a protein of 21 kDa.[35] However, the overexpressed HemG protein did not have PPO enzyme activity, suggesting that this protein is a subunit of a larger complex. This may well be because of the need for the coupling of the oxygen-independent protoporphyrinogen oxidation to the respiratory chain. It has been shown that any compound that can serve as a terminal oxygen-acceptor will allow protoporphyrinogen oxidation,[34] in a reaction unaffected by herbicides such as acifluorfen. Interestingly, the *E. coli hemG* mutant can be complemented with clones encoding the oxygen-dependent PPO from tobacco and mouse,[26,36] revealing functional, if not sequence, conservation.

In a search for orthologues of *hemG* or *hemY* (the bacterial gene encoding oxygen-dependent PPO) in genomes of prokaryotes that contain all the other genes of haem synthesis, only 14 of 28 genera examined contained a recognisable PPO gene.[1] As the authors conclude, this represents a major gap in our understanding of haem biosynthetic pathway in prokaryotic systems. Although the reaction catalysed by PPO can proceed nonenzymically in vitro, PPO mutants in *E.coli*[35] and *B. subtilis*[12] are haem-defective, illustrating an absolute requirement for PPO activity. It is likely that either there is an unidentified PPO in prokaryotes, or another protein has PPO activity, as is clearly the case in higher plants treated with the PPO-inhibiting herbicides.[30,32] Alternatively, as explained above, it may be that this reaction can be catalysed by CPO when present in excess.[37]

Ferrochelatase

The terminal enzyme of protohaem synthesis, ferrochelatase (EC 4.99.1.1), reviewed in detail by Dailey,[38,39] is by far the most well-characterized of the haem-synthesis enzymes. It catalyses the incorporation of ferrous iron into protoporphyrin IX to generate protohaem, and ferrochelatases isolated from various sources all have very similar catalytic properties. This indicates that the major features of the reaction are conserved, although a notable difference is that whereas the enzyme from *B. subtilis* is a water soluble monomer,[40] eukaryotic ferrochelatases appear to be dimers.[41,42] These are associated with the mitochondrial membrane, and in plants, also with the chloroplast membranes[43,44] (discussed in more detail below). The enzyme requires no cofactors for activity, but in vitro the yeast enzyme has an absolute requirement for fatty acids to be active,[42] indicative of its association with the inner mitochondrial membrane. Genes encoding ferrochelatase have been isolated and cloned from many organisms. Like PPO, higher plants have two ferrochelatase isoforms, but in this case they are more closely related to each other than to the enzyme from other organisms, indicating that they evolved after the divergence of plants from animals and fungi (Fig. 2D).

Ferrochelatases can use a range of porphyrins and divalent metal ions as substrates. Among other variants of the substituents on rings A and B, both mesoporphyrin (where the vinyl groups at positions 2 and 4 are replaced with ethyl groups) and deuteroporphyrin (with H at positions 2 and 4) are good substrates, and the higher solubility in aqueous solution of deuteroporphyrin has meant that it is often used in enzyme assays in vitro.[45] In contrast, the propionate groups on rings C and D appear to be crucial for enzyme activity, possibly to allow correct orientation of the porphyrin macrocyle within the active site of the enzyme. Similarly, N-alkylporphyrins are not substrates, but are potent inhibitors, with K_i for N-methyl-protoporphyrin of 1^{-10} nM. These compounds have a distorted macrocycle structure, which has been proposed to act as a transition state analogue (see below).[39] In addition to ferrous iron, a range of bivalent cations can be utilized by ferrochelatase, including Co^{2+} and Zn^{2+}, although the specificity varies with the source of the enzyme. Ferric iron is not a substrate, and understanding the mechanism of reduced iron delivery to ferrochelatase in vivo is an important question.

Studies of ferrochelatase using a variety of approaches, including kinetic studies of the enzyme, investigation of metallation by catalytic antibodies, and resonance Raman spectroscopy, all support a model in which the enzyme causes distortion of the porphyrin to a nonplanar conformation, thus facilitating the insertion of the metal ion (reviewed in ref. 39). The results of solution studies of nonenzymatic metalation are consistent with a so-called "sitting atop complex" where the incoming metal sits on the porphyrin macrocycle, as the pyrrole protons leave from the opposite side. Assuming that the enzymic mechanism is similar, this would mean that the active site has residues involved in metal binding on one side, and those for the abstraction of protons from the pyrrole nitrogens on the other.

Support for these proposals has come from the X-ray crystal structure of ferrochelatase from three sources, *B. subtilis*, human and yeast.[40,41,46] The structures show a remarkable degree of structural and active site residue conservation, despite sequence identity of less than 10% between the human and bacterial proteins. The structure of the *B. subtilis* enzyme and each protomer of the dimeric eukaryotic enzyme are essentially similar. They are composed of two similar domains each containing a four-stranded parallel beta sheet and five or six alpha helices. The enzyme has an obvious cleft structure approximately 25 Å deep, which contains a catalytically important histidine residue conserved in all organisms. Most informatively, the *B. subtilis* enzyme has been crystallized with a tightly bound molecule of *N*-methyl-mesoporphyrin at the active site, in which the distorted macrocycle is clearly visible.[47] The *B. subtilis* ferrochelatase possesses a coordinated hydrated magnesium ion that may interact with the active site metal coordination site based on its location and the experimental observation of stimulation of metal ion chelation by magnesium. Interestingly, the human ferrochelatase has a Ni^{2+} ion coordinated in a similar position. The structure of the yeast enzyme has been solved with both Co^{2+} (a substrate) and Cd^{2+} (an inhibitor), and the metal binding site has been confirmed to be the invariant H235, E314 and S275 residues.[46]

The eukaryotic enzyme is a homodimer and evidence from yeast suggests the enzyme displays a clear asymmetry between the monomers with respect to the porphyrin binding cleft and the mode of metal binding.[46] The sites involved in dimerization contain a number of hydrophobic residues, which are found to be conserved in all eukaryotic ferrochelatase sequences. Animal ferrochelatases and that from fission yeast, *Schizosaccharomyces pombe*, but not those from *S. cerevisiae* or higher plants, contain an iron-sulphur cluster at the C-terminus of each monomer. An unusual cysteine binding motif is involved in Fe-S cluster binding. The role of the Fe-S cluster is at present unknown although several proposals have been made including a structural role, as an iron response element, or in antioxidative protection of ferrous iron.[39]

Uniquely among the eukaryotes, higher plants contain two ferrochelatase genes.[4,48] The proteins they encode show a high degree of sequence conservation (73 to 90% amino acid identity in mature proteins), but fall into two distinct classes based on amino acid similarity (Fig. 2D). Class II ferrochelatases are found exclusively in chloroplasts, and are likely to be involved in synthesis of haem for photosynthetic cytochromes.[49] They contain a C-terminal LHC motif, which is characteristic of the *Lhc* gene family encoding the light harvesting chlorophyll *a/b* binding proteins.[50] In these proteins the LHC element forms a membrane spanning helix with conserved residues involved in binding chlorophylls *a*, *b* and carotenoids, which are thought to have a role in photo-protection of the light-harvesting complexes. The presence of an LHC element in one of the ferrochelatases has been proposed to act either as a membrane anchor for the thylakoid membrane, or more functionally, as a way of providing photo-protection for the intermediates of porphyrin biosynthesis.[51] Class I ferrochelatases are likely to be involved in haem synthesis for haem proteins in mitochondria and other cellular locations such as the ER and peroxisomes.[49] However, both isoenzymes are found in chloroplasts, implying that there may be some spatial separation of the two activities within the organelle.

Another unique feature of ferrochelatase in photosynthetic organisms is that it must compete for its substrate, protoporphyrin IX, with Mg-chelatase, the first enzyme on the chlorophyll branch (Chapter 14).[52] Although both enzymes catalyse essentially similar reactions, namely the insertion of a divalent cation into protoporphyrin IX, Mg chelatase is quite different from ferrochelatase, comprising three nonidentical subunits (Chl H, Chl I and Chl D) and requiring ATP for activity.[53] An intriguing insight into the possible evolution of these enzymes has come from the study of cobaltochelatases, involved in vitamin B_{12} synthesis (Chapter 18), of which there are two kinds, one of which operates under aerobic conditions, and after ring contraction, whereas the other acts early on in the pathway before ring contraction.[54] The aerobic enzyme, for example found in *Pseudomonas denitrificans*, requires ATP for activity and is a trimeric enzyme where one of the subunits (CobN) has considerable sequence similarity to Chl H. In contrast, the other cobaltochelatase, such as that from *Salmonella typhimurium*, is monomeric and has no ATP requirement. Most remarkably, the crystal structure of the enzyme has revealed that it has essentially an identical fold to that of ferrochelatase.[54]

Organization of Pathway

The conversion of uroporphyrinogen III to haem in vivo is much more than the sum of the individual enzyme reactions, not least because of the need to regulate this pathway to avoid build-up of the intermediates. The importance of this is illustrated by the serious consequences when these enzymes are deficient. The accumulation of excess intermediates results in porphyrias in humans (Chapter 5), and severe photobleaching damage in bacteria[55] and plants.[30,56] The production of haem must also be coordinated with the production of the cognate apoproteins (Chapter 9). In photosynthetic organisms, the utilization of protoporphyrin IX for haem synthesis must compete with that for chlorophyll (Chapter 15). In this section, we consider the physical organization of the enzymes within the cell, and the role this might play in the function and regulation of the pathway.

Association with Membranes

The current view is that UROD is soluble, since the native enzyme from all organisms studied, and recombinant forms expressed in *E. coli*, can be purified from the soluble phase.[9] Similarly, the oxygen-dependent CPO in bacteria and plants is a soluble enzyme,[3] but it is loosely associated with mitochondrial membranes in animals and yeast.[17,18] Both PPO and ferrochelatase are also membrane-associated in most organisms; although the well-characterized *B. subtilis* ferrochelatase is a soluble protein,[40] this seems to represent a rare exception. In anaerobic organisms, the oxygen-independent coproporphyrinogen dehydrogenase (HemN) is of necessity membrane-associated in order to use components of the respiratory chain as electron acceptors,[15] and this may also be true of oxygen-independent PPO. However, oxygen-dependent oxidases require no other cofactor, and instead the membrane-association of the terminal three enzymes of haem synthesis in aerobic organisms may be due to the fact that haem, and indeed all the intermediates from uroporphyrinogen III, are poorly soluble in aqueous solution. Association with the membrane may facilitate the passage of substrates between enzymes, and the delivery of haem to the apoproteins to which it is bound.[57]

Neither PPO nor ferrochelatase are predicted to have trans-membrane domains, and their association with the membrane has been proposed to be peripheral, perhaps mediated via association with either lipid or other protein components of the membrane. In plants, import in vitro into isolated chloroplasts of radiolabelled precursors of both ferrochelatase-I and ferrochelatase-II, followed by fractionation of the chloroplasts, found that the processed proteins were present in in both thylakoid and envelope membranes.[44] However, it could be removed by proteolysis, indicating that the enzyme did not span the membrane bilayer. An attractive explanation comes from the crystal structure of the two enzymes. Human ferrochelatase has a putative membrane-binding "lip" domain in the dimer,[41] and a similar feature is observed in tobacco PPO-II,[29] enabling both these enzymes to be anchored monotypically. Modeling of the two structures has revealed that they could dock together, thus spanning the complete membrane, although this would not explain the arrangement in chloroplasts, where both enzymes are on the stromal side of the thylakoid membrane.[27,44] Nevertheless, the association of the two enzymes, together with the subcellular location of these enzymes to specific compartments within the cell (see below) has the potential to play an essential role in regulation of the pathway.

Subcellular Location

Figure 3 summarizes our current knowledge of the subcellular location of the enzymes of haem synthesis in eukaryotic cells. In each case, they are distributed between at least two subcellular compartments, the cytosol and the mitochondrion in yeast and animals, and the plastids and mitochondrion in higher plants.

In plants, chloroplasts contain the enzymes for the entire tetrapyrrole pathway, so they are capable of synthesis of all four major endproducts, haem, chlorophyll, sirohaem and phytochromobilin, and it is thought that all enzymes up to CPO are exclusively in chloroplasts (or plastids in nonphotosynthetic tissue).[58] CPO activity is not detectable in pea[31] or soybean[5] mitochondria, and the protein from *Arabidopsis* has been shown to be exclusively in the plastid by immunological studies and the targeting of a GFP-fusion protein.[5] In contrast, PPO and ferrochelatase activities are detectable in both plastids and plant mitochondria,[31,45] although estimations of the relative proportions of these enzymes suggest that haem synthesis in the mitochondrion represents a relatively minor fraction of the total.[45] Higher plants encode two genes for PPO and ferrochelatase, the products of which are targeted differently. Two cDNAs encoding PPO were isolated from tobacco by functional complementation of an *E. coli hemG* mutant.[26] The two cDNAs encoded proteins of similar size (approx 500 amino acids) but with low levels of sequence identity (27%). PPO-I was imported into plastids and PPO-II into mitochondria. This result suggested a simple scenario where each organelle contains a PPO enzyme encoded by a different gene. While this may be the case in tobacco, investigation of the subcellular location of spinach PPO isoforms suggests a more complex situation. The precursor of spinach PPO-I protein is targeted to chloroplasts, but the PPO-II precursor is targeted to both chloroplasts and mitochondria by the alternative use of two in-frame initiation codons.[59] A similarly complex picture is observed for the two isoforms of ferrochelatase. Both ferrochelatase-I and -II from *Arabidopsis*, barley and cucumber are imported into isolated chloroplasts in vitro,[48,60] and target GFP to chloroplasts in vivo (Fui-Ching Tan and AGS, unpublished), but the ferrochelatase-I isoforms are also imported into isolated pea mitochondria in vitro. However, the in vitro import system for these mitochondria lacks specificity,[61] so these results cannot be taken as firm evidence for the presence of ferrochelatase-I in plant mitochondria. Nevertheless, the targeting behaviour of the two isoforms is clearly different. Furthermore the presence of PPO in plant mitochondria argues strongly that ferrochelatase is also present to prevent the accumulation of phototoxic protoporphyrin IX in the organelle.[45]

Since plant mitochondria contain the terminal two enzymes only, protoporphyrinogen IX (or possibly protoporphyrin IX) is presumably exported from plastids and delivered to the mitochondrion for

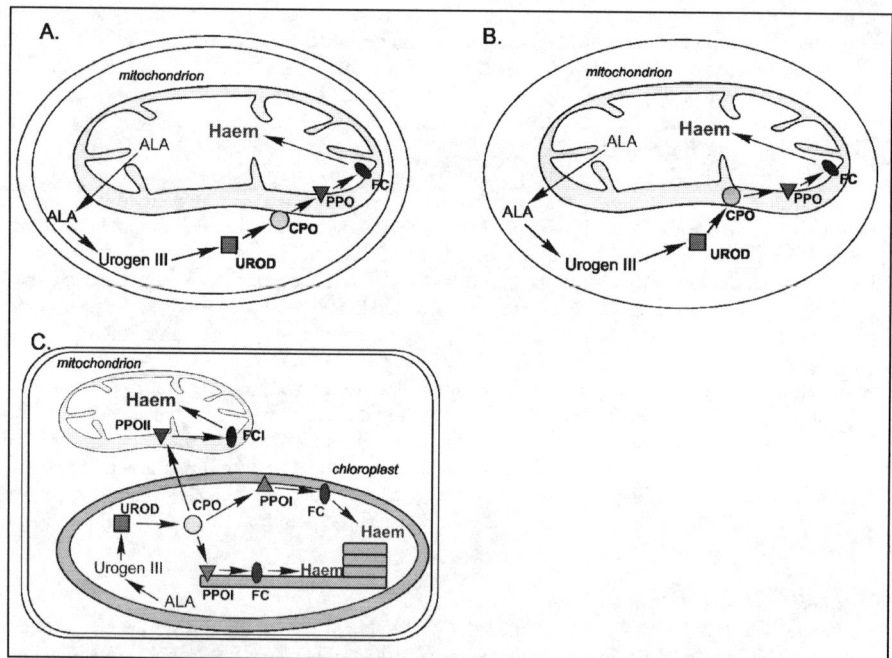

Figure 3. Models to illustrate our current understanding of the subcellular localisation of the enzymes of haem synthesis, including proposed membrane associations, in A) yeast, B) mammals, and C) plants. Enzymes are shown as square (UROD), circle (CPO), triangle (PPO) and oval (ferrochelatase; FC). Details of early steps of the pathway have been abbreviated. ALA = 5-aminolaevulinic acid; Urogen III = uroporphyrinogen III. Yeast and animal pathways differ only in the location of CPO. In plants, mitochondria have PPO-II and FCI isoforms, whereas chloroplasts contain PPO-I and both ferrochelatases. The chlorophyll branch of the pathway, which proceeds from PPO-I in the chloroplast, is omitted for clarity. Note that CPO acts as the major branch point for haem synthesis.

mitochondrial haem synthesis.[31] The capacity for this export has been demonstrated in isolated barley etioplasts, which export up to 50% of total protoporphyrinogen generated after feeding with excess ALA.[62] However, nothing else is known of this system, and no transport proteins have been identified either in chloroplasts or mitochondria.

The subcellular organisation of the pathway in yeast and animals is quite different in several respects (Fig. 3). ALA is synthesized from succinyl CoA and glycine in the mitochondrion, followed by the synthesis of coproporphyrinogen III in the cytosol (Chapter 7). Yeast CPO is a cytosolic protein associated with the mitochondrial outer membrane,[20] whereas in mammals CPO is in the intermembrane space,[17] but in both organisms, protoporphyrinogen IX is made on the cytosolic side of the inner mitochondrial membrane. Immunological studies indicated that in mammalian mitochondria, the active site of ferrochelatase was on the matrix side,[63] thus protoporphyrin IX produced by PPO must be delivered across the membrane. In this respect therefore, mitochondria from all three groups of eukaryotes have the same requirement. A role for PPO in the transport of protoporphyrin(ogen) IX across the plant mitochondrial membrane has been implicated in studies with herbicide inhibitors of PPO such as acifluoren[31] (see above), and it is possible that this enzyme plays the same role in yeast and animal mitochondria.

Evidence for Enzyme Complexes and Substrate Channeling

The final intermediates of haem biosynthesis, and haem itself, are poorly soluble and highly reactive. In addition, the substrate K_ms of the three terminal enzymes are in the micromolar range, which is above that expected to be present within a cell. The possibility of a complex between the terminal enzymes, which could channel the pathway intermediates from enzyme to enzyme, is an attractive hypothesis prevent their 'escape' or accumulation elsewhere in the cell/organelle. This would also provide the means to transport the intermediates across membranes given the distribution of the different enzymes. Evidence for substrate channeling between mammalian PPO and ferrochelatase was obtained from kinetic studies of solubilized mitochondrial membranes, although a stable membrane complex was considered unlikely to exist.[57] This was further supported by experiments with isolated murine mitochondria that accumulate protoporphyrin IX when supplied with coproporphyrinogen alone, but can produce Zn-protoporphyrin when supplied with exogenous Zn^{2+} (a nonendogenous substrate for ferrochelatase) in addition to coproporphyrinogen.[64] The idea of a complex between the enzymes has been lent further credence recently by ability to model interactions between the crystal structures of ferrochelatase and PPO-II[29] (see above).

Site-directed mutagenesis of conserved residues of unknown function in the *B. subtilis* ferrochelatase has revealed a site that may play a key role in substrate channeling.[65] Serine 54 is a conserved residue on the surface of the *B. subtilis* ferrochelatase molecule. The amino acid change Ser54Ala, reduced the growth rate of *B. subtilis* and resulted in the accumulation of coproporphyrin III in the growth medium, but in vitro the enzyme activity of the Ser54Ala mutant was as high as wild-type. This led the authors to propose that Ser54 is involved in interaction with another protein responsible for substrate reception or delivery of the enzymatic product. This hypothesis still has to be proved, but in combination with earlier work it provides evidence for some kind of enzyme association, if not a stable ternary complex, which suggests that substrate channeling is likely to occur.

The organization of the terminal enzymes for haem synthesis within the chloroplast in higher plants is also of interest, since ferrochelatase must compete with Mg-chelatase for protoporphyrin IX.[52] PPO activity has been found associated with both thylakoid and envelope membranes in spinach chloroplasts,[66] and, in spinach, PPO-I was specifically associated with thylakoids, whereas PPO-II was on the stromal side of the chloroplast inner envelope.[27] One possibility therefore is that there are two pools of protoporphyrin IX that can be utilized separately, perhaps by spatial separation of the two chelatases between the two membranes. However, both isoforms of ferrochelatase are found associated with envelope and thylakoid membranes after import into chloroplasts,[44] indicating that the regulation is more complex than this (Chapter 15). In the future these interactions may be explored using in vitro approaches such as immunological analysis of endogenous enzymes, or by testing the behaviour of recombinant enzymes using techniques such as the yeast two-hybrid system.

Iron Delivery to Ferrochelatase

One aspect of protohaem synthesis that is receiving more attention in recent years is the intimate relationship between iron and haem biosynthesis. In bacteria an iron-responsive protein has been identified that coordinates iron availability and the regulation of haem biosynthesis.[67] A direct interaction has been demonstrated in yeast. Ferrochelatase will only accept Fe^{2+} and not Fe^{3+} as substrate and therefore there needs to be a mechanism for the delivery of reduced iron to the active site of ferrochelatase. A study of the import of iron into yeast has revealed that simultaneous transport of iron and deuteroporphyrin across the inner mitochondrial membrane was necessary for haem synthesis, and iron preloaded into mitochondria could not serve as a substrate for ferrochelatase.[68] Iron uptake was driven by a membrane potential across the inner membrane, but import did not require ATP. More recently, the mitochondrial protein frataxin, has been implicated as an iron-chaperone for delivery of the metal to ferrochelatase.[69] Since this protein is also involved in Fe-S cluster assembly, this would provide the means to regulate metal delivery to the two iron-containing cofactors. Clearly, further investigation of the mechanisms of iron delivery in chloroplasts and mitochondria from other organisms is necessary to gain a comprehensive understanding of this regulation.

Conclusions

In this chapter we have summarized recent work on the enzymes involved in the conversion of uroporphyrinogen III to protohaem. For all four enzymes there is a great deal of functional conservation between organisms and across kingdoms. This is best illustrated with the remarkable conservation of catalytic residues in bacterial and human forms of ferrochelatase, as observed from the X-ray crystal structures. Indeed, the information gained from crystallographic studies illustrates the power of such work in finally defining enzyme mechanisms, but it also provides gratifying confirmation of the accuracy of earlier biochemical and molecular studies, carried out when crystallographic data were not available. We can look forward in the future to further insights as more refined structures of the enzymes, including those with substrate bound, become available. In addition, knowledge of their organization within the cellular context, and their interactions with each other, will enable a clearer understanding of the synthesis of haem in vivo.

Acknowledgements
We are grateful to the UK Biotechnology and Biological Sciences Research Council for funding.

References

1. Panek H, O'Brian MR. A whole genome view of prokaryotic haem biosynthesis. Microbiology 2002; 148:2273-2282.
2. Rao AU, Carta LK, Lesuisse E et al. Lack of heme synthesis in a free-living eukaryote. Proc Natl Acad Sci USA 2005; 102:4270-4275.
3. Labbe-Bois R. The ferrochelatase from Saccharomyces cerevisiae. Sequence, disruption, and expression of its structural gene HEM15. J Biol Chem 1990; 265:7278-7283.
4. Chow KS, Singh DP, Amanda RW et al. Two different genes encode ferrochelatase in Arabidopsis: Mapping, expression and subcellular targeting of the precursor proteins. Plant J 1998; 15:531-541.
5. Santana MA, Tan FC, Smith AG. Molecular characterisation of coproporphyrinogen oxidase from Glycine max and Arabidopsis thaliana. Plant Physiol Biochem 2002; 40:289-298.
6. Shoolingin-Jordan PM. The biosynthesis of coproporphyrinogen III. In: Kadish KM, Smith KM, Guilard R, eds. The Porphyrin Handbook. Vol 12. The Iron and Cobalt Pigments: Biosynthesis, Structure and Degradation. New York: Elsevier, 2003:33-74.
7. Whitby FG, Phillips JD, Kushner JP et al. Crystal structure of human uroporphyrinogen decarboxylase. EMBO J 1998; 17:2463-2471.
8. Martins BM, Grimm B, Mock HP et al. Crystal structure and substrate binding modeling of the uroporphyrinogen-III decarboxylase from Nicotiana tabacum—Implications for the catalytic mechanism. J Biol Chem 2001; 276:44108-44116.
9. Felix F, Brouillet N. Purification and properties of uroporphyrinogen decarboxylase from Saccharomyces cerevisiae. Yeast uroporphyrinogen decarboxylase. Eur J Biochem 1990; 188:393-403.
10. Luo J, Lim CK. Order of uroporphyrinogen III decarboxylation on incubation of porphobilinogen and uroporphyrinogen III with erythrocyte uroporphyrinogen decarboxylase. Biochem J 1993; 289:529-532.
11. Phillips JD, Whitby FG, Kushner JP et al. Structural basis for tetrapyrrole coordination by uroporphyrinogen decarboxylase. EMBO J 2003; 22:6225-6233.
12. Hansson M, Hederstedt L. Cloning and characterization of the Bacillus subtilis hemEHY gene cluster, which encodes protoheme IX biosynthetic enzymes. J Bacteriol 1992; 174:8081-8093.
13. Ishida T, Yu L, Akutsu H et al. A primitive pathway of porphyrin biosynthesis and enzymology in Desulfovibrio vulgaris. Proc Natl Acad Sci USA 1998; 95:4853-4858.
14. Akhtar M. Coproporphyrinogen III and protoporphyrinogen IX oxidases. In: Kadish KM, Smith KM, Guilard R, eds. The Porphyrin Handbook. Vol 12. The Iron and Cobalt Pigments: Biosynthesis, Structrue and Degradation. New York: Elsevier, 2003:75-92.
15. Layer G, Verfurth K, Mahlitz E et al. Oxygen-independent coproporphyrinogen-III oxidase HemN from Escherichia coli. J Biol Chem 2002; 277:34136-34142.
16. Layer G, Moser J, Heinz DW et al. Crystal structure of coproporphyrinogen III oxidase reveals cofactor geometry of Radical SAM enzymes. EMBO J 2003; 22:6214-6224.
17. Lee DS, Flachsova E, Bodnarova M et al. Structural basis of hereditary coproporphyria. Proc Natl Acad Sci USA 2005; 102:14232-14237.
18. Kohno H, Furukawa T, Tokunaga R et al. Mouse coproporphyrinogen oxidase is a copper-containing enzyme: Expression in Escherichia coli and site-directed mutagenesis. Biochim Biophys Acta 1996; 1292:156-162.
19. Breckau D, Mahlitz E, Sauerwald A et al. Oxygen-dependent coproporphyrinogen III oxidase (HemF) from Escherichia coli is stimulated by manganese. J Biol Chem 2003; 278:46625-46631.
20. Phillips JD, Whitby FG, Warby CA et al. Crystal structure of the oxygen-dependant coproporphyrinogen oxidase (Hem13p) of Saccharomyces cerevisiae. J Biol Chem 2004; 279:38960-38968.

21. Lash TD. The enigma of coproporphyrinogen oxidase: How does this unusual enzyme carry out oxidative decarboxylations to afford vinyl groups? Bioorg Med Chem Lett 2005; 15:4506-4509.
22. Lash TD, Kaprak TA, Shen L et al. Metabolism of analogues of coproporphyrinogen-III with modified side chains: Implications for binding at the active site of coproporphyrinogen oxidase. Bioorg Med Chem Lett 2002; 12:451-456.
23. Jones MA, He J, Lash TD. Kinetic studies of novel di- and tri-propionate substrates for the chicken red blood cell enzyme coproporphyrinogen oxidase. J Biochem (Tokyo) 2002; 131:201-205.
24. Narita S, Taketani S, Inokuchi H. Oxidation of protoporphyrinogen IX in Escherichia coli is mediated by the aerobic coproporphyrinogen III oxidase. Mol Gen Genet 1999; 261:1012-1020.
25. Dailey TA, Dailey HA. Identification of an FAD superfamily containing protoporphyrinogen oxidases, monoamine oxidases, and phytoene desaturase. Expression and characterization of phytoene desaturase of Myxococcus xanthus. J Biol Chem 1998; 273:13658-13662.
26. Lermontova I, Kruse E, Mock HP et al. Cloning and characterization of a plastidal and a mitochondrial isoform of tobacco protoporphyrinogen IX oxidase. Proc Natl Acad Sci USA 1997; 94:8895-8900.
27. Che FS, Watanabe N, Iwano M et al. Molecular characterization and subcellular localization of protoporphyrinogen oxidase in spinach chloroplasts. Plant Physiol 2000; 124:59-70.
28. Arnould S, Takahashi M, Camadro JM. Acylation stabilizes a protease-resistant conformation of protoporphyrinogen oxidase, the molecular target of diphenyl ether-type herbicides. Proc Natl Acad Sci USA 1999; 96:14825-14830.
29. Koch M, Breithaupt C, Kiefersauer R et al. Crystal structure of protoporphyrinogen IX oxidase: A key enzyme in haem and chlorophyll biosynthesis. EMBO J 2004; 23:1720-1728.
30. Matringe M, Camadro JM, Labbe P et al. Protoporphyrinogen oxidase as a molecular target for diphenyl ether herbicides. Biochem J 1989; 260:231-235.
31. Smith AG, Marsh O, Elder GH. Investigation of the subcellular location of the tetrapyrrole-biosynthesis enzyme coproporphyrinogen oxidase in higher-plants. Biochem J 1993; 292:503-508.
32. Lee HJ, Duke MV, Duke SO. Cellular Localization of protoporphyrinogen-oxidizing activities of etiolated barley (Hordeum vulgare L.) leaves (relationship to mechanism of action of protoporphyrinogen oxidase-inhibiting herbicides). Plant Physiol 1993; 102:881-889.
33. Fingar VH, Wieman TJ, McMahon KS et al. Photodynamic therapy using a protoporphyrinogen oxidase inhibitor. Cancer Res 1997; 57:4551-4556.
34. Dailey HA. Terminal steps of haem biosynthesis. Biochem Soc Trans 2002; 30:590-595.
35. Sasarman A, Letowski J, Czaika G et al. Nucleotide sequence of the hemG gene involved in the protoporphyrinogen oxidase activity of Escherichia coli K12. Can J Microbiol 1993; 39:1155-1161.
36. Dailey TA, Dailey HA, Meissner P et al. Cloning, sequence, and expression of mouse protoporphyrinogen oxidase. Arch Biochem Biophys 1995; 324:379-384.
37. Narita S, Taketani S, Inokuchi H. Oxidation of protoporphyrinogen IX in Escherichia coli is mediated by the aerobic coproporphyrinogen oxidase. Mol Gen Genet 1999; 261:1012-1020.
38. Dailey HA, Dailey TA, Wu CK et al. Ferrochelatase at the millennium: Structures, mechanisms and [2Fe-2S] clusters. Cell Mol Life Sci 2000; (57):1909-1926.
39. Dailey H, Dailey T. Ferrochelatase. In: Kadish KM, Smith KM, Guilard R, eds. The Porphyrin Handbook. Vol 12. The Iron and Cobalt Pigments: Biosynthesis, Structure and Degradation. New York: Elsevier, 2003:93-121.
40. Al-Karadaghi S, Hansson M, Nikonov S et al. Crystal structure of ferrochelatase: The terminal enzyme in heme biosynthesis. Structure 1997; 5:1501-1510.
41. Wu CK, Dailey HA, Rose JP et al. The 2.0 A structure of human ferrochelatase, the terminal enzyme of heme biosynthesis. Nat Struct Biol 2001; 8:156-160.
42. Grzybowska E, Gora M, Plochocka D et al. Saccharomyces cerevisiae ferrochelatase forms a homodimer. Arch Biochem Biophys 2002; 398:170-178.
43. Matringe M, Camadro JM, Joyard J et al. Localization of ferrochelatase activity within mature pea chloroplasts. J Biol Chem 1994; 269:15010-15015.
44. Roper JM, Smith AG. Molecular localisation of ferrochelatase in higher plant chloroplasts. Eur J Biochem 1997; 246:32-37.
45. Cornah JE, Roper JM, Singh DP et al. Measurement of ferrochelatase activity using a novel assay suggests that plastids are the major site of haem biosynthesis in both photosynthetic and nonphotosynthetic cells of pea (Pisum sativum L.). Biochem J 2002; 362:423-432.
46. Karlberg T, Lecerof D, Gora M et al. Metal binding to Saccharomyces cerevisiae ferrochelatase. Biochemistry 2002; 41:13499-13506.
47. Lecerof D, Fodje MN, Alvarez Leon R et al. Metal binding to Bacillus subtilis ferrochelatase and interaction between metal sites. J Biol Inorg Chem 2003; 8:452-458.

48. Suzuki T, Masuda T, Singh DP et al. Two types of ferrochelatase in photosynthetic and nonphotosynthetic tissues of cucumber—Their difference in phylogeny, gene expression, and localization. J Biol Chem 2002; 277:4731-4737.

49. Singh DP, Cornah JE, Hadingham S et al. Expression analysis of the two ferrochelatase genes in Arabidopsis in different tissues and under stress conditions reveals their different roles in haem biosynthesis. Plant Mol Biol 2002; 50:773-788.

50. Jansson S. A guide to the LHC genes and their relatives in Arabidopsis. Trends Plant Sci 1999; 4:236-240.

51. Jansson S, Andersson J, Kim SJ et al. An Arabidopsis thaliana protein homologous to cyanobacterial high-light-inducible proteins. Plant Mol Biol 2000; 42:345-351.

52. Cornah JE, Terry MJ, Smith AG. Green or red: What stops the traffic in the tetrapyrrole pathway? Trends Plant Sci 2003; 8:224-230.

53. Walker CJ, Willows RD. Mechanism and regulation of Mg-chelatase. Biochem J 1997; 327:321-333.

54. Schubert HL, Raux E, Wilson KS et al. Common chelatase design in the branched tetrapyrrole pathways of heme and anaerobic cobalamin synthesis. Biochemistry 1999; 38:10660-10669.

55. Nakahigashi K, Nishimura K, Miyamoto K et al. Photosensitivity of a protoporphyrin-accumulating, light-sensitive mutant (visA) of Escherichia coli K-12. Proc Natl Acad Sci USA 1991; 88:10520-10524.

56. Hu G, Yalpani N, Briggs SP et al. A porphyrin pathway impairment is responsible for the phenotype of a dominant disease lesion mimic mutant of maize. Plant Cell 1998; 10:1095-1105.

57. Ferreira GC, Andrew TL, Karr SW et al. Organization of the terminal two enzymes of the heme biosynthetic pathway. Orientation of protoporphyrinogen oxidase and evidence for a membrane complex. J Biol Chem 1988; 263:3835-3839.

58. Smith AG, Cornah JE, Roper JM et al. Compartmentation of tetrapyrrole metabolism in higher plants. In: Bryant JA, Burrell MM, Kruger NJ, eds. Plant Carbohydrate Metabolism. Oxford: BIOS Scientific Publishers, 1999:281-294.

59. Watanabe N, Che FS, Iwano M et al. Dual targeting of spinach protoporphyrinogen oxidase II to mitochondria and chloroplasts by alternative use of two in-frame initiation codons. J Biol Chem 2001; 276:20474-20481.

60. Chow KS, Singh DP, Roper JM et al. A single precursor protein for ferrochelatase-I from Arabidopsis is imported in vitro into both chloroplasts and mitochondria. J Biol Chem 1997; 272:27565-27571.

61. Cleary SP, Tan FC, Nakrieko KA et al. Isolated plant mitochondria import chloroplast precursor proteins in vitro with the same efficiency as chloroplasts. J Biol Chem 2002; 277:5562-5569.

62. Jacobs JM, Jacobs NJ. Porphyrin accumulation and export by isolated barley (Hordeum vulgare) plastids (effect of diphenyl ether herbicides). Plant Physiol 1993; 101:1181-1187.

63. Harbin BM, Dailey HA. Orientation of ferrochelatase in bovine liver mitochondria. Biochemistry 1985; 24:366-370.

64. Proulx KL, Woodard SI, Dailey HA. In situ conversion of coproporphyrinogen to heme by murine mitochondria: Terminal steps of the heme biosynthetic pathway. Protein Sci 1993; 2:1092-1098.

65. Olsson U, Billberg A, Sjovall S et al. In vivo and in vitro studies of Bacillus subtilis ferrochelatase mutants suggest substrate channeling in the heme biosynthesis pathway. J Bacteriol 2002; 184:4018-4024.

66. Matringe M, Camadro JM, Block MA et al. Localization within chloroplasts of protoporphyrinogen oxidase, the target enzyme for diphenylether-like herbicides. J Biol Chem 1992; 267:4646-4651.

67. Hamza I, Chauhan S, Hassett R et al. The bacterial irr protein is required for coordination of heme biosynthesis with iron availability. J Biol Chem 1998; 273:21669-21674.

68. Lange H, Kispal G, Lill R. Mechanism of iron transport to the site of heme synthesis inside yeast mitochondria. J Biol Chem 1999; 274:18989-18996.

69. Yoon T, Cowan JA. Frataxin-mediated iron delivery to ferrochelatase in the final step of heme biosynthesis. J Biol Chem 2004; 279:25943-25946.

CHAPTER 5

Inherited Disorders of Haem Synthesis:
The Human Porphyrias

Michael N. Badminton* and George H. Elder

Abstract

The porphyrias are inherited metabolic disorders resulting from partial deficiency of enzymes of the haem biosynthetic pathway. Each particular enzyme deficiency gives rise to increased levels of metabolites prior to the pathway blockage, which result in characteristic clinical features, and allow the individual conditions to be diagnosed. The acute neurovisceral attacks, which are common to ALA dehydratase deficiency porphyria (ADP), acute intermittent porphyria (AIP), variegate porphyria (VP) and hereditary coproporphyria are the result of neuronal damage by a mechanism which involves accumulation of ALA. Cutaneous manifestations comprise either bullous lesions (fragile skin, blisters), which are common to congenital erythropoietic porphyria (CEP), porphyria cutanea tarda (PCT), VP and HCP, or acute photosensitivity which is associated with erythropoietic protoporphyria (EPP). Skin lesions result from photoactivation of circulating porphyrins by light (400-410 nm) resulting in the generation of free oxygen radicals with the different presentations explained by the physicochemical properties of the individual porphyrins. All the porphyrias, apart from the sporadic form of PCT, are inherited either as autosomal dominant (AIP, HCP, PCT, VP, EPP) or autosomal recessive (ADP, CEP) disorders. The molecular genetics of each disorder has been fully characterised and investigation at the DNA level is now indispensable in management of families as it allows accurate presymptomatic testing, and genetic counselling. The mainstay of management remains prevention, either by avoiding sunlight in the case of the cutaneous porphyrias or, in the acute porphyrias, factors such as prescribed drugs which are known to precipitate acute attacks. For severely affected patients newly developed treatments include liver transplantation (in AIP) and bone marrow transplantation in CEP. Active research is underway to develop enzyme replacement for AIP and gene therapy for the erythropoietic porphyrias.

Overview

Of the more than five hundred inherited metabolic disorders described to date few are more widely recognized by the general public than the porphyrias. Public interest has been triggered by speculation of medical historians, widely publicised by the media, that porphyria was responsible for the madness of famous figures of the past such as King George III[1] and Vincent van Gogh.[2] Despite there being little scientific proof either for or against these theories[1] they contribute to the generally held misconception, even amongst the medical fraternity, that porphyria results in chronic psychiatric illness. Indeed, while media interest in porphyria stories may seem harmless to most scientists, it can result in stigmatisation, anxiety and stress among patients. This review aims to outline recent developments in our understanding of the pathogenesis, diagnosis and treatment of this heterogenous group of disorders.

*Corresponding Author: Michael N. Badminton—Department of Medical Biochemistry and Immunology, School of Medicine, Cardiff University, Heath Park, Cardiff CF14 4XN, UK. Email: badmintonmn@cardiff.ac.uk

Tetrapyrroles: Birth, Life and Death, edited by Martin J. Warren and Alison G. Smith.
©2009 Landes Bioscience and Springer Science+Business Media.

Table 1. Porphyria nomenclature and abbreviations

Porphyria		Enzyme/Gene	
Acute			
ALA dehydratase deficiency porphyria	ADP	ALA dehydratase	ALAD
Acute intermittent porphyria	AIP	Hydroxymethylbilane synthase	HMBS
Hereditary coproporphyria	HCP	Coproporphyrinogen oxidase	CPO
Variegate porphyria	VP	Protoporphyrinogen oxidase	PPOX
Nonacute			
Congenital erythropoietic porphyria	CEP	Uroporphyrinogen III synthase	UROS
Porphyria cutanea tarda	PCT	Uroporphyrinogen decarboxylase	UROD
Erythropoietic protoporphyria	EPP	Ferrochelatase	FECH

The porphyrias comprise a group of seven metabolic disorders in humans, each of which results from partial deficiency of a different enzyme of the haem biosynthetic pathway (Table 1, Fig. 1).[3] Each enzyme defect is inherited apart from UROD deficiency in the main type of PCT. The defects are partial as haem is essential for life and even in the autosomal recessive porphyrias residual enzyme activity is always measurable. Although no disease has yet been associated with defects in the ubiquitous ALAS1 gene; mutations in the erythroid specific ALAS (ALAS2) gene cause either X-linked sideroblastic anaemia[3] or, when deletions disrupt or delete the C-terminal region X-linked dominant protoporphyria[3b] (see "Other Autosomal Dominant Porphyrias; EPP). Each particular enzyme deficiency gives rise to increased levels of metabolites prior to the pathway blockage, and it is this accumulation of precursors that gives rise to the characteristic clinical features associated with the porphyrias, which may be acute neurovisceral attacks, bullous skin lesions, acute photosensitivity or a combination of these symptoms (Table 2). The tetrapyrrole pathway intermediate porphyrinogens are rapidly oxidised to the porphyrin equivalent which can be measured in plasma, urine, faeces or erythrocytes of symptomatic patients and the distinct and characteristic pattern of excretion allows the type of porphyria to be established in individual patients (Table 2).

All nucleated human cells synthesise haem using the pathway of haem biosynthesis shown in Figure 1. About 80% of total haem is made by erythroid cells in the bone marrow and used for formation of haemoglobin. In other cells haem is required for the action of many cellular proteins, including respiratory and other cytochromes, haem dependent oxidases (catalase, peroxidase etc.), myoglobin and neuroglobin. However the main site of nonerythroid haem synthesis is the liver, which accounts for about 10-15% of the body's total haem production; most hepatic haem being used for assembly of cytochrome P450 (CYP) enzymes. In all cells, the rate of haem synthesis is determined by the activity of ALAS of which there are two isoforms, ALAS1 and ALAS2, encoded by separate genes. In liver, ALAS1 activity is regulated through haem-mediated alteration of its rate of synthesis whereas the activity of the erythroid-specific isoform, ALAS2 is controlled mainly by iron (see Chapter 7).[4]

Clinical Features

Accumulation of haem precursors is associated with two types of illness: acute neurovisceral attacks that are always associated with accumulation of ALA, and skin lesions caused by photosensititisation by porphyrins (Fig. 1, Table 2) that may occur separately or in combination.

Acute neurovisceral attacks occur in a minority of patients who inherit the gene defect for an autosomal dominant acute porphyria (AIP, HCP or VP) and in ADP (Table 2). The peak occurrence of attacks is in the third decade and females are more likely to be affected than males. They are exceptionally rare before puberty and are also less likely to occur after the menopause. The acute attack presents with a combination of clinical features in which abdominal pain is an almost universal feature (Table 3).[5] In the early stages they may be mistaken for one of a number of more commonly occurring abdominal complaints including acute appendicitis and cholecystitis, and it is not

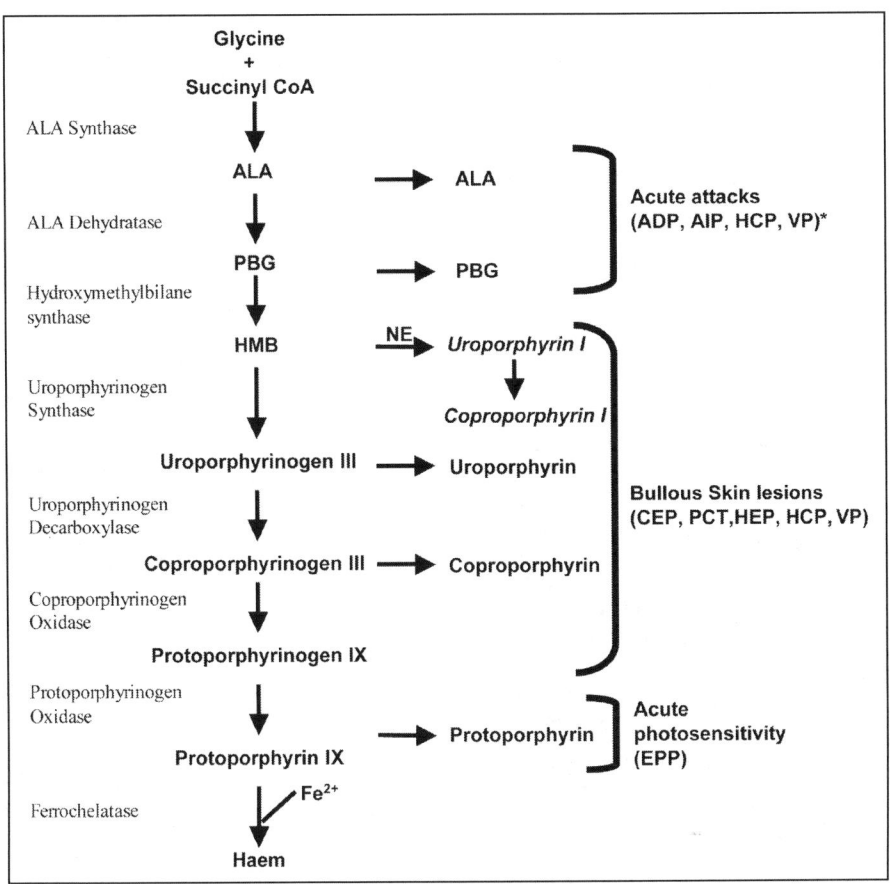

Figure 1. The haem biosynthetic pathway in mammals consists of eight enzymes, of which the first and last three are situated within the mitochondria. The type I isomers are produced following nonenzymatic (NE) cyclisation of HMB into uroporphyrinogen I. A commonly used alternative name for the enzyme hydroxymethylbilane synthase is porphobilinogen deaminase (PBG-D).* PBG exceeds ALA excretion in AIP, HCP and VP, but not in ADP where PBG is normal or slightly increased.

unusual for patients to have had at least one surgical procedure prior to the diagnosis being made. A recent study reported the average time from onset of symptoms to diagnosis as six years (range <1 to 49 years, n = 81).[6] Acute attacks may also be complicated by convulsions, which may be precipitated by a sudden fall in plasma sodium concentration, hypertension, tachycardia, and peripheral neuropathy. Psychiatric manifestations which occur during the attack invariably resolve when the patient is in remission, and even repetitive acute attacks do not cause long standing psychiatric disorders.[6] In the longer term, there is an increased risk of developing liver cancer[7] and renal disease leading to hypertension and kidney failure.[8]

Accumulation of porphyrins other than protoporphyrin, as occurs in CEP, PCT, HCP and VP, results in skin changes in sun-exposed areas described as bullous lesions. These lesions are identical in all four disorders, although they vary in severity and age of onset. These skin changes result from photoactivation of porphyrins with the production of reactive oxygen species that damage the cells of the dermis. The changes, which occur in sun-exposed areas, include fragile skin with trivial mechanical trauma leading to tearing or blisters. The majority of patients also have spontaneously occurring blisters (bullae) that are filled with clear fluid. These have a tendency to rupture resulting in lesions that heal slowly leaving scarring, milia (small white lesions) and areas of over or under

Table 2. **Characteristic biochemical findings in samples taken when patient is symptomatic**

Porphyria	Inheritance Prevalence of Overt Disease	Clinical	Increased Haem Precursors
ADP	AR *(Unknown)*	Acute	U: ALA RBC: Protoporphyrin (Zn)
AIP	AD *(1-2:100,000)*	Acute	U: *ALA, PBG* P: Peak[#] at 615-620 nm
CEP	AR *(< 1: 10⁶)*	Cutaneous	U: Uroporphyrin I, Coproporphyrin I
			F: Coproporphyrin I
			RBC: Protoporphyrin (Zinc and free),
			uroporphyrin I
			P: Peak[#] at 615-620 nm
PCT porphyrin	25% AD[1] *(1:25,000)*	Cutaneous (bullous)	U: Uroporphyrin, Heptacarboxylic
			F: Isocoproporphyrin, heptacarboxylic
			P: Peak[#] at 615-620 nm
HCP	AD *(Probably <1:250,000)*	Acute ± Cutaneous (bullous)	U: ALA PBG, Coproporphyrin III
			F: *Coproporphyrin III (isomer III:I ratio >1.4)*
			P: Peak[#] at 615-620 nm
VP	AD *(1:250,000)*	Acute ± Cutaneous (bullous)	U: PBG, ALA, Copro III
			F: *Protoporphyrin >Coproporphyrin III*
			P: *Peak[#] at 624- 627nm*
EPP	AD[1] *(1:100,000)*	Cutaneous (acute photosensitivity)	U: Normal RBC: Protoporphyrin (free)
			P: Peak[#] at 626-634 nm

AD, autosomal dominant; AR, autosomal recessive. U: Urine, F: Faeces, RBC: - erythrocyte, P: Plasma, Peak[#]: fluorescence emission. [1] See text. Precursors in italics are those that may also be found in presymptomatic or latent acute hepatic porphyria patients.

Table 3. **Summary of clinical features, diagnosis and management of the acute attack**

		Acute Attacks
Porphyrias		AIP,VP, HCP, ADP
Clinical Features:		Abdominal pain, vomiting, constipation, psychological symptoms, tachycardia, hypertension, peripheral neuropathy
Complications:		Seizures, hyponatraemia, respiratory paralysis
Diagnosis:		Increased urinary PBG (usually > 10-fold)
Treatment:	Specific	Intravenous human hemin (Orphan Europe), haematin (Abbott, USA)
	Supportive	Maintain fluid, calorie intake, Opiate analgesia, Antiemetic, laxative, anxiolytic

pigmentation. Another common finding is increased hair formation in nonandrogen sensitive sun-exposed skin. The skin lesions are usually worse in the summer months although many patients fail to correlate the lesions with sunlight.

The accumulation of protoporphyrin in skin leads to a distinct clinical presentation that is characteristic of only one porphyria, EPP. Presentation is usually in early childhood. Within minutes of exposure to sunlight patients suffer burning, stinging pain with subsequent redness (erythema) and swelling (oedema) in exposed areas of skin. The characteristic history should alert the clinician to

Table 4. *Some clinical conditions, other than porphyrias, that may be associated with abnormal porphyrin metabolism*

Sample	Increased Haem Precursors	Clinical Conditions
Urine	Coproporphyrin	Liver dysfunction, fever, drugs, alcohol
	ALA	Lead poisoning
Faeces	Protoporphyrin, other dicarboxylic porphyrins	Intestinal bleed, constipation, high protein intake
Plasma	Uroporphyrin	Renal failure
	coproporphyrin	Liver dysfunction
Erythrocytes	Zinc-protoporphyrin	Iron deficiency anaemia
		Lead poisoning

this disorder, as there is usually very little to find on examination by the time the patient gets to the doctor. Very occasionally there may be purpuric lesions following acute exposure. Long-term changes include skin thickening, especially over the knuckles, and small elliptical scars particularly on the face. A minority of patients develop liver dysfunction of whom a small proportion suffer severe life-threatening liver failure.[9]

Diagnosis

The diagnosis of porphyria, whatever the type, is made by analysing the pattern of haem precursor excretion in urine, faeces, plasma and erythrocytes. The properties of the different porphyrins determine where they accumulate and are excreted, and analysis of all four specimens may be needed to make a diagnosis. The required level of expertise is usually only available in specialist laboratories (for information see www.porphyria-europe.com). Clinical acumen alone is not sufficient as many other disorders can give rise to symptoms or signs similar to those found in the porphyrias. In addition, many clinical disorders result in secondary changes in porphyrin metabolism (Table 4) and without full porphyrin analysis may result in an erroneous diagnosis of porphyria being made.

The diagnosis of porphyria as a cause of acute symptoms rests on the demonstration of increased porphobilinogen excretion in the urine and should be part of the test repertoire in all acute hospitals. Screening tests may be used but a positive finding should always be confirmed by a quantitative assay based on the method of Mauzerall and Granick.[10] Measurement of urinary ALA in addition to PBG adds little further clinical information, and in our view is unnecessary as a front-line test. Having diagnosed an acute porphyria and initiated treatment, additional investigation is required to determine the type (Table 2).[11] Screening to establish which other family members are affected is best achieved by mutational analysis of the relevant gene.[12-14] Biochemical testing by enzyme or metabolite measurement has lower sensitivity and specificity; metabolite measurements are rarely abnormal before puberty. Due to the allelic heterogeneity of the acute porphyrias in most countries, mutational analysis usually requires prior investigation of an unequivocally affected family member. Ideally family screening should be undertaken in collaboration with a clinical genetics unit to ensure appropriate counselling and obtain informed consent. Testing children by parental request is considered ethically acceptable.

Investigation of patients with bullous skin lesions or acute photosensitivity is essential for diagnosing cutaneous porphyria and allowing the appropriate treatment to be undertaken. The key biochemical findings are specific enough to distinguish between each disorder (Table 2).

Molecular Genetics and Pathogenesis

Genes for all the human porphyrias have been characterized which has allowed molecular defects to be identified in all the porphyrias (Human Gene Mutation Database, www.hgmd.org) (Table 5). Molecular investigation has proved indispensable in the management of porphyria patients and their families by allowing accurate presymptomatic testing, preconceptual genetic counselling and, in rare circumstances, prenatal diagnosis.

Table 5. Molecular genetics of the porphyrias

Disorder	Gene	Chromosome	Gene Size (kb)	# Exons	Expression
ADP	*ALAD*	9q34	13	13	Ubiquitous and erythroid-specific mRNAs
AIP	*HMBS*	11q24.1-24.2	10	15	Ubiquitous and erythroid-specific isoenzymes
CEP	*UROS*	10q25.2-26.3	34	10	Ubiquitous and erythroid-specific mRNAs
PCT	*UROD*	1p34	3	10	Ubiquitous
HCP	*CPO*	3q12	14	7	Ubiquitous
VP	*PPOX*	1q21-23	5	13	Ubiquitous
EPP	*FECH*	18q21.3	45	11	Ubiquitous

Autosomal Dominant Acute Porphyrias

Enzyme activities in the autosomal dominant acute porphyrias (AIP, HCP, VP) are half-normal, reflecting expression of a normal gene *trans* to a mutation that abolishes or markedly decreases activity. The mutations all show incomplete penetrance with most affected individuals being clinically normal through life. Family studies indicate that about 10% of those who inherit the gene for one of these disorders develop symptoms but population surveys suggest a lower figure.[15] As yet, there is little information on genetic or other factors that determine penetrance. Endocrine factors and contact with drugs and other precipitants of symptoms contribute but by no means provide the sole explanation; the general genetic background is likely to be more important.[16] It has also been suggested that some mutations may be associated with a higher penetrance than others.[17]

With few exceptions, mutations in the autosomal dominant acute porphyrias are restricted to one or a few families. In AIP alone, over 200 mutations scattered throughout the gene have been identified (Human Gene Mutation Database, www.hgmd.org). Mutational analysis for presymptomatic diagnosis therefore necessitates a methodological approach that involves analysis of large parts of the gene including coding regions, splice sites and portions of 5' and 3' untranslated regions. In a few instances, a founder effect means that the majority of patients in a particular region carry an identical mutation that makes molecular testing much simpler. Examples include South Africa where the high prevalence of VP amongst the Afrikaner population is due to a missense mutation, R59W, the origin of which has been traced back to an immigrant from Holland who arrived in the Cape in the 17th century.[18] Similarly parts of Sweden are affected by a high incidence of AIP, resulting from a common mutation (W188X) in the *HMBS* gene.[19] All types of genetic defect have been described with approximately one third each of missense, frameshift and splice site mutations. In most cases mutations are inherited and can be demonstrated in family members, although de-novo mutations have been reported. The sensitivity of mutation detection by direct sequencing is greater than 96% for AIP and VP (S. Whatley, M Badminton, G Elder, unpublished data) and for those families in which no mutation is identified gene-tracking using intragenic polymorphisms may be helpful. Rare variants of each of the autosomal dominant acute porphyrias in which a deleterious mutation is inherited from each parent have been described.[20,21] Clinical presentation varies but symptoms usually start in childhood and tend to be more severe than in the parent disorder, with additional features, notably chronic neurological disease and skeletal abnormalities.

Other Autosomal Dominant Porphyrias

Half-normal enzyme activity is also inherited in an autosomal dominant pattern in two other disorders: EPP and familial (type 2) PCT. These conditions also show low clinical penetrance. However, in contrast to the autosomal dominant acute porphyrias where enzyme activities are the same whether or not symptoms are present, activities in EPP and in familial PCT are decreased to below half-normal in individuals with symptoms while remaining half-normal in their asymptomatic but affected relatives. This additional decrease is found in all tissues in EPP but is restricted to the liver in familial PCT.

EPP

Recent evidence has shown that in most EPP families clinical expression requires coinheritance of a low expression allele, carried by approximately 10% of the population in France, *trans* to a severe mutation in the *FECH* gene.[22] We have found a similar gene frequency for the low expression allele in the United Kingdom. This low expression allele, in which a C nucleotide is substituted for a T at position IVS3-48, modulates the use of an alternative splice acceptor site within the intron generating an aberrant mRNA that is rapidly degraded by the nonsense mediated mRNA decay mechanism. The severe *FECH* mutation, which is inherited in an autosomal dominant pattern in EPP families, and the low expression allele together decrease FECH activity to below about 35% of normal, the threshold below which sufficient protoporphyrin accumulates to produce photosensitivity. EPP might therefore be considered to be an autosomal recessive condition as the disease requires a genetic contribution from both alleles. However, in contrast to a classical recessive disorder, homozygosity for one of these alleles (IVS3-48C) does not cause disease, and it is perhaps more accurate to regard it as an autosomal dominant disorder in which penetrance is modulated by expression of wild-type FECH.[22] A bonus of these findings is that mutational analysis to determine the low expression allele status, which is relatively simple, has markedly improved preconceptual counselling. Investigation of the partner of an EPP patient for the presence of the low expression allele allows the risk of having a child with overt disease to be calculated. In addition, offspring can also be investigated soon after birth for the likelihood of developing photosensitivity. A minority of patients with EPP have true autosomal recessive disease with *FECH* mutations on both alleles.[23] The clinical significance of this mode of inheritance has not been fully established, although it may increase the risk of severe liver disease[23] as it has been suggested that this risk is inversely proportional to the level of residual enzyme activity.[24] In about 3% of EPP families, ferrochelatase enzyme activity and mutation analysis are normal. Recent investigation has shown that patients from most of these families have C-terminal deletions in the ALAS2 gene that lead to gain of function and produce a previously unrecognised form of porphyria, X-linked dominant protoporphyria.[3b] Clinically, this disorder is characterised by high clinical penetrance, markedly increased erythrocyte protoporphyrin and zinc-proto porphyrin levels, and an increased incidence of liver disease.[3b]

Familial PCT

About 25% of patients with PCT have the familial (type 2) form in which half-normal UROD activity is inherited in an autosomal dominant pattern and *UROD* mutations are present on one allele. Clinical penetrance is low, there is extensive allelic heterogeneity and mutations are found throughout the gene.[25,26] Most other patients have sporadic (type 1) PCT in which UROD deficiency is restricted to the liver and no mutations have been detected at the *UROD* locus.[24] In both types, accumulation of sufficient porphyrin to cause disease appears to require a decrease in hepatic UROD activity to about 25% of normal. A number of factors that increase the risk of appearance of symptoms in sporadic and familial PCT have been identified. These include alcohol abuse, infection with hepatotropic viruses, principally HCV and HIV, oestrogens, and mutations in the haemochromatosis (*HFE*) gene, particularly homozygosity for the C282Y mutation.[25,26] In addition, most patients have biochemical evidence of iron overload; clinically overt haemochromatosis is uncommon. Depletion of hepatic iron stores leads to clinical remission.[3] Clinical and experimental evidence indicates that the decrease in hepatic UROD activity from half-normal in familial PCT, and from normal in the sporadic form, results from an iron-dependent, reversible inactivation process.[25] Experiments in *UROD* knock-out mice and other animal models of PCT have identified iron, CYP1A2 and, possibly, other CYPs, and an unidentified inhibitor of UROD derived from ALA, probably by an iron-dependent oxidation, as key components in this process.[25-29]

Molecular investigation of a rare childhood-onset cutaneous porphyria, hepatoerythropoietic porphyria (HEP), with similar urinary and faecal porphyrin patterns to PCT has confirmed that this is the homozygous or compound heterozygous variant of PCT with some *UROD* mutations being common to both conditions (Table 2).[25]

Autosomal Recessive Porphyrias

Characterisation of the *UROS* gene that codes for the enzyme that is markedly deficient in CEP, has allowed some understanding of why this severe disorder demonstrates wide clinical heterogeneity. Affected individuals present with symptoms that start in utero (hydrops fetalis), during infancy, in childhood or later.[30] This clinical heterogeneity relates to the level of residual enzyme activity and by extension to the severity of the mutation as defined by in-vitro expression studies (summarised in ref. 30). Mutations of all types have been described in most exons, several introns and within the erythroid-specific promoter region.[30] The C73R mutation has been found in a third of reported allele mutations,[30] although we have found a lower percentage (2 of 28 alleles) in the UK population possibly due to increased representation from ethnic minorities. Mutational analysis has therefore proven helpful in evaluating the likely phenotype and for prenatal testing in families with an affected child.[30]

ADP, the rarest of the porphyrias, is an autosomal recessive acute porphyria that results from mutations in the *ALAD* gene.[3] Worldwide only seven patients have been described and all appear to be compound heterozygotes. Interestingly despite residual ALAD enzyme activities of less than 1%, only one patient presented in infancy.[31]

Mechanisms of Disease

Acute Neurovisceral Attacks

Acute neurovisceral attacks are due to neuronal damage in the central, peripheral and autonomic nervous system. These attacks, and the identical crises that occur in hereditary tyrosinaemia where ALAD becomes inhibited by succinyl acetone, are all associated with elevation of ALA in plasma and urine. Accompanying biochemical changes include induction of ALAS1, relative haem deficiency in the liver and, in VP and HCP, probably inhibition of HMBS by haem precursors (protoporphyrinogen or coproporphyrinogen).[32] At a cellular level there is axonal degeneration and chromatolysis affecting motor neurons, brain stem nuclei and autonomic ganglia. There may also be some changes in the brain cortex, although these tend to be less marked. The exact pathogenesis of the neurological manifestations of acute attacks remains elusive; hypotheses include a neurotoxic effect of ALA, haem deficiency in the liver with secondary effects on neurotransmitter metabolism, haem deficiency in the nervous system or a combination of two or more of these.[33] The generation of a porphobilinogen deaminase deficient mouse has led to some insights into pathogenesis.[34] Although the mouse model develops a chronic axonal neuropathy, that may be caused by haem deficiency,[35] it does not show the drug-induced acute neurovisceral attacks that characterise the human acute porphyrias.

Skin Lesions

There appears to be no fundamental difference in the underlying mechanism that results in porphyrin initiated cellular damage between bullous porphyrias and EPP. Absorption of light with wavelengths in the region of 400-410 nm results in excited state porphyrins which decay releasing energy either as a photon of light, or more damagingly, interacting with oxygen via energy transfer to form singlet oxygen, or via electron transfer to form superoxide anion.[25] These toxic products are capable of damaging proteins, lipids and DNA resulting in cellular damage, which releases other activating compounds and results in damage to cells of the dermis and areas of basement membrane in their immediate vicinity. Histopathology of the resulting lesions shows a characteristic accumulation of amorphous hyaline material in the vicinity of small blood vessels supplying upper dermis. Production and accumulation of the more water soluble porphyrins e.g., uroporphyrin and coproporphyrin results in bullous skin lesions which are a consequence of damage to the epidermis. In certain areas this results in a split in the basement membrane and the development of fluid filled bullae. The floor of the bullous has characteristic projections into the space, which give it a "festooned" appearance.

The symptoms of acute photosensitivity in EPP result from the elevation of free protoporphyrin which, unlike the zinc-chelated protoporphyrin that is increased in iron deficiency or lead poisoning, is lipophilic. This physicochemical property probably underlies the difference between the symptoms of EPP and the other cutaneous porphyrias, where the accumulated porphyrins are more hydrophilic and therefore distributed in a different pattern within the dermis. To explain the acute reaction to sunlight, it has been proposed that light induces the transfer of free protoporphyrin from plasma or RBC's to dermal capillary endothelial cells where it becomes a target for photo-activation by ultraviolet light (λ400-410 nm).[36]

Liver Disease

Liver disease is a potential complication of all the porphyrias except ADP and CEP. The risk of developing hepatocellular carcinoma (HCC) is thirty-six-fold higher in patients with acute porphyria than in the general population and such tumours occur in the absence of liver cirrhosis.[37] The carcinogenic mechanism has not been established but it is suggested that auto-oxidation of increased hepatic mitochondrial ALA generates free radicals which are mutagenic.[37]

In PCT, histological abnormalities are seen in hepatic tissue from virtually all patients and range from minor changes to frank cirrhosis.[25] Patients with PCT and cirrhosis are more likely to develop HCC than those with cirrhosis alone.

Protoporphyrin is hepatotoxic and its accumulation in the liver underlies the liver dysfunction that arises in a proportion of EPP patients.[9] It is not clear why about 2% of patients go on to develop liver failure.[9] The factors that predispose to this complication are poorly understood and research is currently focused on possible contribution by the patient's genotype with evidence that missense mutations in the *FECH* gene are under-represented in patients with severe liver disease.[24,38]

New Approaches to Management

Management of both acute and cutaneous porphyrias include symptomatic treatment as well as prevention of further episodes and or complications by suggesting relatively simple changes to the patient's lifestyle.[3,39] Novel approaches are being actively pursued by several research groups especially for patients suffering from the more severe forms of porphyria.

Established Therapy

Management of an acute neurovisceral attack includes symptom relief and general supportive measures in the form of analgesia, intravenous fluid therapy and treatment of complications as they arise (Table 3).[40] Specific treatment aimed at limiting the severity and shortening the attack should be instituted without waiting for the porphyria type to be established. Although the exact pathogenesis of acute attacks is not understood, intravenous haem preparations (either haem arginate or haematin) have proved successful in shortening the duration and limiting the severity of acute attacks when given early.[41] The probable mode of action is suppression of the rate limiting step of the haem synthetic pathway, ALA-synthase, thereby decreasing hepatic production of haem precursors ALA and PBG and/or correcting the relative haem deficiency in the liver. Treatment is by rapid infusion on four consecutive days and is associated with few side effects apart from irritation of peripheral veins.[40] Once the acute attack is over management shifts to prevention, both in the patient and in affected family members identified by family screening. Both symptomatic and presymptomatic patients are provided with patient information leaflets and counselled to avoid factors which increase the risk of an acute attack including, prescribed and illicit drugs, alcohol, smoking and calorie restriction. Our approach in Cardiff has been to provide patients and their clinicians with information on which drugs are likely to be safe and to encourage these to be used as first choice. With the ever-increasing numbers of prescribed drugs entering the market, this is an ongoing process and patients and their clinicians need to be advised where to obtain up to date information. This is best achieved by a regularly updated website, with information collated by a group of experts and this has recently been established in Europe (www.porphyria-europe.com). Where this is not possible, clinicians are encouraged to assess the risk versus the benefit of choosing a particular agent, and if necessary to monitor the patient carefully during treatment.

A minority of patients suffer recurrent acute attacks and require more complex treatment. For women who suffer cyclical attacks that can be clearly linked to a particular menstrual phase, ovulation suppression using gonadotrophin releasing hormone (GnRH) analogues has proved successful.[40] Alternatively regular prophylactic treatment with a haem preparation such as haem arginate, usually via an indwelling central venous catheter, has been successfully used to decrease the frequency and severity of acute attacks.[40] However this mode of treatment is not without risks and a significant number of patients treated in this way develop complications such as iron overload or bacterial infection of the venous catheter leading to septicaemia. In addition, the recurrent acute attacks together with complex treatment regimes result in a significant reduction in quality of life in this group of young adults.

Transplantation

Transplantation has been used as an effective treatment method for a variety of purposes in porphyric patients with the aim of correcting the enzyme deficiency or treating an associated complication such as liver failure or hepatoma.

Bone marrow transplantation is an established treatment for patients with moderate or severe CEP and has proved highly effective.[30] Dramatic improvement in photosensitivity coincides with large reductions in urine and plasma porphyrin levels. The main limitations are availability of histocompatibly matched stem cells, which may be from unaffected siblings or nonrelated donors, and the morbidity and mortality associated with this medical procedure. With improvements in immunosuppression these risks have diminished since the first procedure was undertaken in 1991. While bone marrow transplantation would almost certainly be effective in EPP the risks are considered to outweigh the benefits and it has therefore only been used to treat coexistent haematological malignancy.[42] However this may change if methods can be developed to identify those EPP patients at most risk of fulminant liver failure, thereby tipping the balance towards benefit rather than risk.

Orthotopic liver transplantation has been used in two circumstances: to correct the enzyme deficiency associated with acute hepatic porphyria and as a life-saving procedure in EPP patients who develop fulminant liver failure. The most radical so far has been liver transplantation in a young patient with AIP who suffered recurrent severe attacks.[43] This approach resulted in complete clinical and biochemical remission, and has confirmed the liver as being central to the disorder. Having demonstrated that correction of the hepatic enzyme deficiency is an effective treatment option various less radical methods may now be pursued.[44] A less successful outcome was seen in a patient with the very rare ADP, in whom a liver transplant did not reverse the biochemical abnormalities nor protect the patient from recurrent acute attacks.[31] However this autosomal recessive condition is associated with lower residual enzyme levels and the failure to correct the enzyme deficiency in all cells especially nerve tissue is likely to have been the cause of the continued excess production of ALA and poor clinical outcome.

In EPP, liver transplantation is a life-saving but not curative procedure that is used in patients with protoporphyrin-induced fulminant liver failure.[9] The operation is complicated by the potential for severe phototoxicity induced by irradiation of the surgical field by intense theatre lights. Strategies to limit this effect include exchange transfusion, plasmapheresis and light filters.[45]

Gene Therapy and Enzyme Replacement Therapy

As monogenic disorders the porphyrias, especially the more severe forms, are good candidates for gene therapy. This is especially true for CEP and EPP, where the molecular defect primarily affects a single organ, the bone marrow, and bone marrow transplantation has already been shown to be an effective treatment. In addition, X-linked severe combined immunodeficiency, an inherited haematological disorder has recently been successfully treated by ex vivo gene therapy although this carries a risk of causing leukaemia.[46] Transfer of the wild-type *UROS* gene into cultured human cells using a variety of viral vectors has been shown to result in stable expression of the transgene in a high proportion of dividing cells and correction of the metabolic defect.[47] A retroviral vector has also been used to correct the molecular defect in bone marrow from an EPP mouse model,[48] paving the way for a clinical trial in humans. The HEP genetic defect has also been corrected in vitro by retroviral gene transfer into cultured cells[49] but this disorder is likely to prove much more technically demanding as gene transfer into hepatic cells will be required.

The successful treatment of AIP by liver transplantation is likely to drive forward research into novel treatments for the acute hepatic porphyrias. It demonstrated that correction of the hepatic defect alone is sufficient to control the condition, and should lead to other methods of hepatic enzyme replacement, either by recombinant enzyme infusion or gene therapy. Enzyme replacement therapy for AIP is currently being developed, although the enzyme has not been modified to enhance hepatic uptake. In addition, a safe, reliable and successful method for gene transfer into in situ hepatic tissue is actively being pursued for several inherited metabolic disorders[50] and once established should allow rapid development of vectors for treatment of hepatic porphyrias.

Concluding Remarks

Although rare, the human porphyrias are important because of the significant morbidity and mortality associated with these conditions. In particular the autosomal dominant acute porphyrias have the propensity to cause a devastating illness in a minority of patients. The most important future developments in our understanding of these disorders is likely to come from research into the genetic and other factors which influence clinical expression. The options for permanent treatment are also likely to change once safe and reliable methods of gene delivery to the liver or bone marrow have been developed.

References

1. Warren MJ, Jay M, Hunt DM et al. The maddening business of King George III and porphyria. Trends Biochem Sci 1996; 21:229-234.
2. Loftus LS, Arnold WN. Vincent van Gogh's illness: Acute intermittent porphyria? BMJ 1991; 303:1589-1591.
3. Anderson KE, Sassa, S, Bishop DF et al. Disorders of heme biosynthesis: X-linked sideroblastic anemia and the porphyrias. In: Scriver CR, Beaudet AL, Sly WS et al, eds. The Metabolic and Molecular Basis of Inherited Disease. 8th ed. New York: McGraw-Hill, 2001:2961-3062.
3b. Whatley SD, Ducamp S, Gouya L et al. C-terminal deletions in the ALAS2 gene lead to gain of function and cause a previously undefined type of human porphyria, X-linked dominant protoporphyria, without anemia or iron overload. Am J Human Genet 2008; 83:408-414.
4. Ponka P. Cell biology of heme. Am J Med Sci 1999; 318:241-56.
5. Elder GH, Hift RJ, Meissner PJ. The acute porphyrias. Lancet 1997; 349:1613-161.
6. Millward LM, Kelly P, Deacon A et al. Self-rated psychosocial consequences and quality of life in the acute porphyrias. J Inherit Metab Dis 2001; 24:733-747.
7. Andant C, Puy H, Bogard C et al. Hepatocellular carcinoma in patients with acute hepatic porphyria: Frequency of occurrence and related factors. J Hepatol 2000; 32:933-939.
8. Andersson C, Wikberg A, Stegmayr B et al. Renal symptomatology in patients with acute intermittent porphyria. A population-based study. J Intern Med 2000; 248:319-325.
9. Meerman L. Erythropoietic protoporphyria. An overview with emphasis on the liver. Scand J Gastroenterol Suppl 2000; 232:79-85.
10. Mauzerall D, Granick S. The occurrence and determination of δ-aminolaevulinate and porphobilinogen in urine. J Biol Chem 1956; 219:435 446.
11. Deacon AC, Elder GH. Front line tests for the investigation of suspected porphyria. J Clin Pathol 2001; 54:500-507.
12. Whatley SD, Puy H, Morgan RR et al. Variegate porphyria in Western Europe: Identification of PPOX gene mutations in 104 families, extent of allelic heterogeneity, and absence of correlation between phenotype and type of mutation. Am J Hum Genet 1999; 65:984-994.
13. Lamoril J, Puy H, Whatley SD et al. Characterization of mutations in the CPO gene in British patients demonstrates absence of genotype-phenotype correlation and identifies relationship between hereditary coproporphyria and harderoporphyria. Am J Hum Genet 2001; 68:1130-1138.
14. Whatley SD, Woolf JR, Elder GH. Comparison of complementary and genomic DNA sequencing for the detection of mutations in the HMBS gene in British patients with acute intermittent porphyria: Identification of 25 novel mutations. Hum Genet 1999; 104:505-510.
15. Nordmann Y, Puy H, DaSilva V et al. Acute intermittent porphyria: Prevalence of mutations in the porphobilinogen deaminase gene in blood donors in France. J Int Med 1997; 242:213-7.
16. Kauppinen R, Mustajoki P. Prognosis of acute porphyria: Occurrence of acute attacks, precipitating factors, and associated diseases. Medicine 1992; 71:1-15.
17. Andersson C, Floderus Y, Wikberg A et al. The W198X and R173W mutations in the porphobilinogen deaminase gene in acute intermittent porphyria have higher clinical penetrance than R167W. A population based study. Scand. J Clin Lab Invest 2000; 60:643-648.
18. Meissner PN, Dailey TA, Hift RJ et al. A R59W mutation in human protoporphyrinogen oxidase results in decreased activity and is prevalent in South Africans with variegate porphyria. Nat Genet 1996; 13:95-97.
19. Lee JS, Anvret M. Identification of the most common mutation within the porphobilinogen deaminase gene in Swedish patients with acute intermittent porphyria. Proc Natl Acad Sci USA 1991; 88:10912-10915.
20. Elder GH. Hepatic porphyrias in children. J Inher Metab Dis 1997; 20:237-46.
21. Lamoril J, Puy H, Gouya L et al. Neonatal anaemia due to inherited harderoporphyria: Clinical characteristics and molecular basis. Blood 1998; 91:1453-7.
22. Gouya L, Puy H, Robreau AM et al. The penetrance of dominant erythropoietic protoporphyria is modulated by expression of wildtype FECH. Nat Genet 2002; 30:27-28.
23. Cox TM. Protoporphyria. In: Kadish KM, Smith KM, Guilard R, eds. The Porphyrin Handbook, vol. 14, Medical aspects of porphyrias. Amsterdam: Academic Press, 2003:121-150.

24. Chen FP, Risheg H, Liu Y et al. Ferrochelatase gene mutations in erythropoietic protoporphyria: Focus on liver disease. Cell Mol Biol 2002; 48:83-89.
25. Elder GH. Porphyria cutanea tarda and related disorders. In: Kadish KM, Smith KM, Guilard R, eds. The Porphyrin Handbook: Vol. 14 Medical Aspects of Porphyrins. San Diego: Academic Press, 2003:67-92.
26. Bulaj ZJ, Phillips JD, Ajioka RS et al. Hemochromatosis genes and other factors contributing to the pathogenesis of porphyria cutanea tarda. Blood 2000; 95:1565-1571.
27. Phillips JD, Jackson LK, Bunting M et al. A mouse model of familial porphyria cutanea tarda. Proc Natl Acad Sci USA 2001; 98:259-264.
28. Franklin MR, Phillips JD, Kushner JP. Uroporphyria in the uroporphyrinogen decarboxylase-deficient mouse: Interplay with siderosis and polychlorinated biphenyl exposure. Hepatology 2002; 36:805-811.
29. Sinclair PR, Gorman N, Trask HW et al. Uroporphyria caused by ethanol in Hfe(-/-) mice as amodel for porphyria cutanea tarda. Hepatology 2003; 37:351-8.
30. Desnick RJ, Astrin KH. Congenital erythropoietic porphyria: Advances in pathogenesis and treatment. Br J Haem 2002; 117:779-795.
31. Thunell S, Henrichson A, Floderus Y et al. Liver transplantation in a boy with acute porphyria due to aminolaevulinate dehydratase deficiency. Eur J Clin Chem Clin Biochem 1992; 30:599-606.
32. Meissner PN, Adams P, Kirsch R. Allosteric inhibition of human lymphoblasts and purified porpho-bilinogen deaminase by protoporphyrinogen and coproporphyrinogen: A possible mechanism for the acute attack of variegate porphyria. J Clin Invest 1993; 91:1436-1444.
33. Meyer UA, Schuurmans MM, Lindberg RLP. Acute porphyrias: Pathogenesis of neurological manifestations. Sem Liv Dis 1998; 18:43-52.
34. Lindberg RLP, Porcher C, Grandchamp B et al. Porphobilinogen deaminase deficiency in mice causes a neuropathy resembling that of human hepatic porphyria. Nat Genet 1996; 12:195-199.
35. Lindberg RLP, Martini R, Baumgartner M et al. Motor neuropathy in porphobilinogen deaminase-defient mice imitates the peripheral neuropathy of human acute porphyria. J Clin Invest 1999; 103:1127-1134.
36. Sandberg S, Brun A. Light-induced protoporphyrin release from erythrocytes in erythropoietic protoporphyria. J Clin Invest 1982; 70:693-698.
37. Andant C, Puy H, Bogard C et al. Hepatocellular carcinoma in patients with acute hepatic porphyria: Frequency of occurrence and related factors. J Hepatol 2000; 32:933-9.
38. Minder EI, Gouya L, Schneider-Yin X et al. A genotype-phenotype correlation between null-allele mutations in the ferrochelatase gene and liver complication in patients with erythropoietic protoporphyria. Cell Mol Biol (Noisy-le-grand) 2002; 48:91-96.
39. Badminton MN, Elder GH. Management of acute and cutaneous porphyrias. Int J Clin Pract 2002; 56:272-8.
40. Elder GH, Hift RJ. Treatment of acute porphyria. Hospital Medicine 2001; 62:422-425.
41. Mustajoki P, Nordmann Y. Early administration of haem arginate for acute porphyric attacks. Arch Intern Med 1993; 153:2004-2008.
42. Poh-Fitzpatrick MB, Wang X, Anderson KE et al. Erythropoietic protoporphyria: Altered phenotype after bone marrow transplantation for myelogenous leukemia in a patient heteroallelic for ferrochelatase gene mutations. J Am Acad Dermatol 2002; 46:861-6.
43. Soonawalla ZF, Orug T, Badminton MN et al. Liver Transplantation as a cure for acute intermittent porphyria. Lancet 2004; 363:705-6.
44. Wiesner RH, Ralela J, Ishitani MB et al. Recent advances in liver transplantation. Mayo Clin Proc 2003; 78:197-210.
45. Meerman L, Verwer R, Slooff MJ et al. Perioperative measures during liver transplantation for erythropoietic protoporphyria. Transplantation 1994; 88:541-549.
46. Kohn DB, Sadelain M, Glorioso JC. Occurrence of leukaemia following gene therapy of X-linked SCID. Nat Rev Cancer 2003; 3:477-88.
47. Geronimi F, Richard E, Lamrissi-Garcia I et al. Lentivirus-mediated gene transfer of uroporphyrinogen III synthase fully corrects the porphyric phenotype in human cells. J Mol Med 2003; 81:310-320.
48. Pawliuk R, Bachelot T, Wise RJ et al. Long-term cure of the photosensitivity of murine erythropoietic protoporphyria by preselective gene therapy. Nat Med 1999; 7:768-772.
49. Fontanellas A, Mazurier F, Moreau-Gaudry F et al. Correction of uroporphyrinogen decarboxylase deficiency (Hepatoerythropoietic porphyria) in Epstein-Barr virus-transformed B-cell lines by retrovirus-mediated gene transfer: Fluorescence-based selection of transduced cells. Blood 1999; 94:465-474.
50. Raper SE, Yudkoff M, Chirmule N et al. A pilot study of in vivo liver-directed gene transfer with an adenoviral vector in partial ornithine transcarbamylase deficiency. Hum Gene Ther 2002; 13:163-175.

CHAPTER 6

Heme Degradation:
Mechanistic and Physiological Implications

Angela Wilks*

Abstract

Heme oxidation catalyzed by heme oxygenase has evolved to carry out a number of important and diverse physiological processes including iron reutilization and cellular signaling in mammals, synthesis of light-harvesting pigments in cyanobacteria, light perception in plants, and the acquisition of iron in bacterial pathogens. In mammals the evolution of biliverdin IXα reductase in the conversion of biliverdin IXα to bilirubin IXα provides an important link between heme metabolism and antioxidant protection. The following review will address heme degradation in the context of the physiological and mechanistic aspects of heme oxidation and biliverdin reduction in animals and bacteria, and the possible clinical ramifications of modulation of heme oxygenase and biliverdin reductase activity.

Introduction

Heme oxygenase is the first step in the metabolism of heme, and is universally found in all heme-containing organisms. In animals, its major role is in the disassembly of red blood cells, allowing turnover of heme, and the utilization of the amino acids of hemoglobin. The reaction requires three molecules of oxygen and a total of seven electrons, from NADPH, for the conversion of one heme to biliverdin, CO and iron.[1] In mammals, the transfer of electrons to the heme oxygenases is mediated by cytochrome P450 reductase, the same redox partner that is responsible for electron transfer to the cytochrome P450 enzymes.[2] The mammalian heme oxygenases cleave heme in a regiospecific manner to yield biliverdin IXα as the sole enzymatic product (Fig. 1). The conversion of biliverdin IXα to bilirubin IXα is then catalyzed by biliverdin IXα reductase (BVR-A).[3] This chapter outlines our current understanding of heme degradation in animals and pathogenic bacteria. Heme metabolism in plants and cyanobacteria will be the focus of Chapter 12.

Evolution and Biological Function of Heme Oxygenase

In the three or so decades since Schmid and coworkers first described a heme oxygenase activity the role of this unique class of enzymes has extended well beyond the physiological degradation of hemoglobin.[4] The existence of two isoforms of mammalian heme oxygenase designated HO-1 and HO-2 respectively, have been known for sometime.[5] A third isoform HO-3 that appears to have little catalytic activity has recently been characterized and it has been suggested that its function may largely be regulatory.[6]

In mammals the heme oxygenase pathway is the only known physiologically relevant mechanism for the degradation of heme. The oxidative cleavage of heme not only removes a potentially toxic agent, but also provides a mechanism for the reutilization of iron. The critical role of heme oxygenase in iron-reutilization was further demonstrated with the genetic knockout mice of HO-1.[7] The HO-1

*Angela Wilks—Department of Pharmaceutical Sciences, School of Pharmacy, University of Maryland, 20 Penn Street, Baltimore, Maryland 21201, USA. Email: awilks@rx.umaryland.edu

Tetrapyrroles: Birth, Life and Death, edited by Martin J. Warren and Alison G. Smith.
©2009 Landes Bioscience and Springer Science+Business Media.

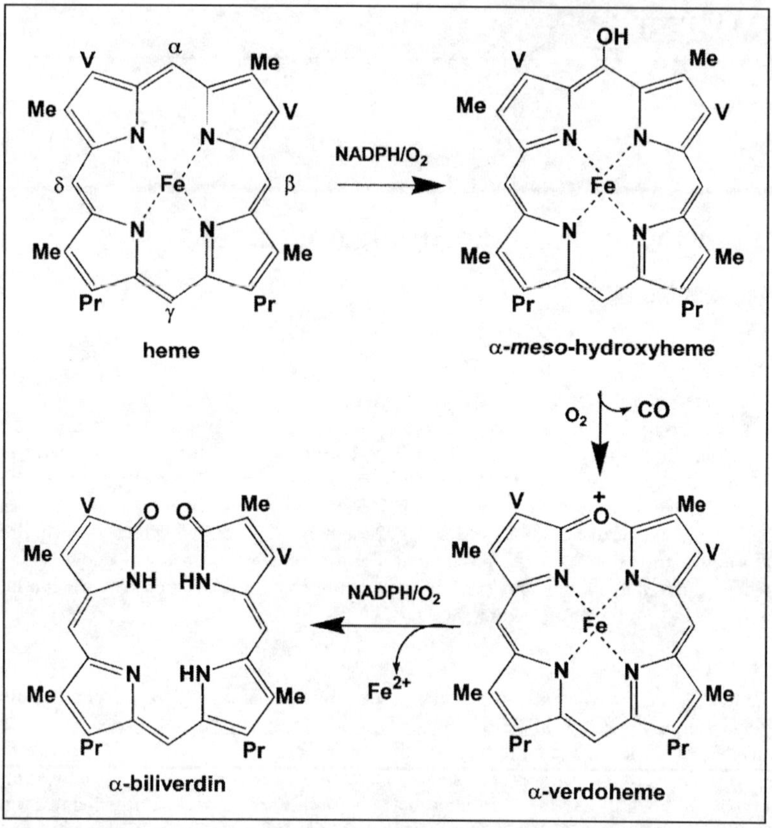

Figure 1. Heme oxygenase catalyzed conversion of heme to biliverdin. Abbreviations are as follows; M, methyl (-CH₃), V, vinyl (-CH=CH₂), Pr, propionate (-CH₂-CH₂-COOH). The *meso*-carbons are labeled α,β,δ, and γ.

deficient adult mice developed chronic anemia associated with low serum iron levels but an increased hepatic and renal iron content. The deposition of iron within the liver and kidneys promoted oxidative damage, chronic inflammation and tissue injury. The critical role of HO-1 in both antioxidant protection and iron homeostasis places HO-1 at a unique junction in research disciplines as diverse as cardiovascular physiology, the central nervous system, renal and hepatic function, inflammation and transplant surgery.

The heme oxygenase reaction produces biliverdin, which by the action of biliverdin IXα reductase (BVR-A) is converted to bilirubin. Bilirubin has long been known as a neurotoxic agent when it rises above physiological levels as is observed in neonatal jaundice or Crigler-Najjar syndrome.[8] However, bilirubin also acts as a lipid soluble antioxidant, and may have physiological relevance in protection against intracellular oxidative damage due to free-radical formation.[9] More recently it has been suggested that the oxidation of bilirubin to biliverdin, and its subsequent conversion back to bilirubin by the action of BVR-A, acts as an amplification cycle to protect cells from oxidative damage.[10] In addition to its role as an antioxidant bilirubin, as well as biliverdin, has also been implicated in the modulation of the immune response.[11] Therefore, the formation of bilirubin like many endogenous chemicals is beneficial at physiological concentrations, but becomes toxic when produced in excess.

The oxidative cleavage of heme by HO releases carbon monoxide (CO) as a product of the reaction. Carbon monoxide has been proposed to function as a neural messenger.[12] However, the implication of CO as a neural messenger in processes such as long-term memory potentiation, neuroendocrine regulation, and vasodilation is not without controversy as much of the experimental evidence has involved the use of metalloporphyrin inhibitors of heme oxygenase. Metalloporphyrins such as tin-protoporphyrin (Sn-PPIX) or zinc-protoporphyrin (Zn-PPIX) have inhibitory activity against other hemeproteins including guanylyl cyclase (sGC), a possible receptor target of CO. More recently, evidence has suggested that the anti-inflammatory response and inhibition of apoptosis by CO is independent of sGC and results from direct activation of the mitogen-activated protein (MAP kinase) pathway by CO.[13] While the receptors and direct sites of action of CO remain open to question and debate, enough evidence has now accumulated implicating CO as a cellular activator in many diverse regulatory functions.

Heme oxygenase also plays an important role in lower eukaryotic and prokaryotic organisms. In the cyanobacteria and higher plants heme oxygenase is essential for the biosynthesis of light-absorbing pigments (see Chapter 6). The formation of biliverdin by the action of heme oxygenase is required for the biosynthesis of phycobilins and phytochromobilins, the receptors involved in responses such as phototropism and regulation of chlorophyll biosynthesis. More recently bacteriophytochrome proteins have been observed in the nonphotosynthetic bacterium *Pseudomonas aeruginosa*, and unlike its photosynthetic counterparts, which utilize phytochromobilin, this organism directly incorporates biliverdin and functions as a photochromic kinase that is modulated by light.[14]

The utilization of heme as a source of iron by bacterial pathogens has highlighted yet another niche for which heme oxidation has an important function. Bacterial pathogens require iron for their survival and pathogenesis and as such have developed sophisticated mechanisms by which they can utilize the host heme and heme-containing proteins as a source of iron.[15] Cleavage of the porphyrin macrocycle facilitated by heme oxygenase is critical in the release of iron for further use by the pathogen, and in preventing toxicity associated with heme once it is internalized within the cell. The identification and characterization of iron-regulated heme oxygenase genes in *Corynebacterium diptheriae*, *Neisseriae meningitidis* and *Pseudomonas aeruginosa* provides another example of a conservation of mechanism in an essential physiological process.[16-18] In each of the physiological roles described above heme oxygenase has evolved to carry out a specific function essential to the survival of the organism, while essentially retaining the same structural and functional motif.

Sequence and Structural Conservation within the Heme Oxygenase Enzymes

The mammalian heme oxygenases are in all cases microsomal membrane bound proteins. The amino acid sequences of the human isoforms HO-1 (hHO-1) and HO-2 (hHO-2) are approximately 42% identical, significantly lower than the 80% identity between the human and rat HO-2 isoforms (Fig. 2). The recently identified HO-3 (hHO-3) has a much lower similarity to either HO-1 or HO-2 but appears to be more closely related to HO-2. Regions of high sequence identity between the three isoforms are found in the sequence corresponding to residues 125-150 and 11-40 in hHO-1. A histidine (His-25) in the latter sequence has been identified as the proximal histidine to the heme (see below).

Interestingly, the region of highest sequence identity, residues 125-150 in HO-1 and 144-169 in HO-2, corresponds to the sequence of the distal helix, which lies directly above the heme. The distal helix region was thought to be a fingerprint motif for the heme oxygenase proteins, however, we have recently characterized heme oxygenases from *Neisseria meningitides* (nm-HO)[17] and *Pseudomonas aeruginosa* (pa-HO)[18] that have a much lower degree of sequence identity within this region than the more well characterized heme oxygenases (Fig. 2). It is evident from the sequence alignments that there are no apparent critical active site residues that may be directly involved in catalysis.

Crystallographic Studies

The crystal structures of the truncated soluble forms of the human and rat HO-1 enzymes complexed with heme, have provided significant insight into both the mechanism and regioselectivity of the heme oxygenase reaction.[19,20] The structure of the human HO-1 lacking the terminal 55 amino acids, including a 23 C-terminal hydrophobic membrane anchor has been refined to 1.3 Å. Despite the truncation of the protein the activity and regioselectivity of the reaction were largely

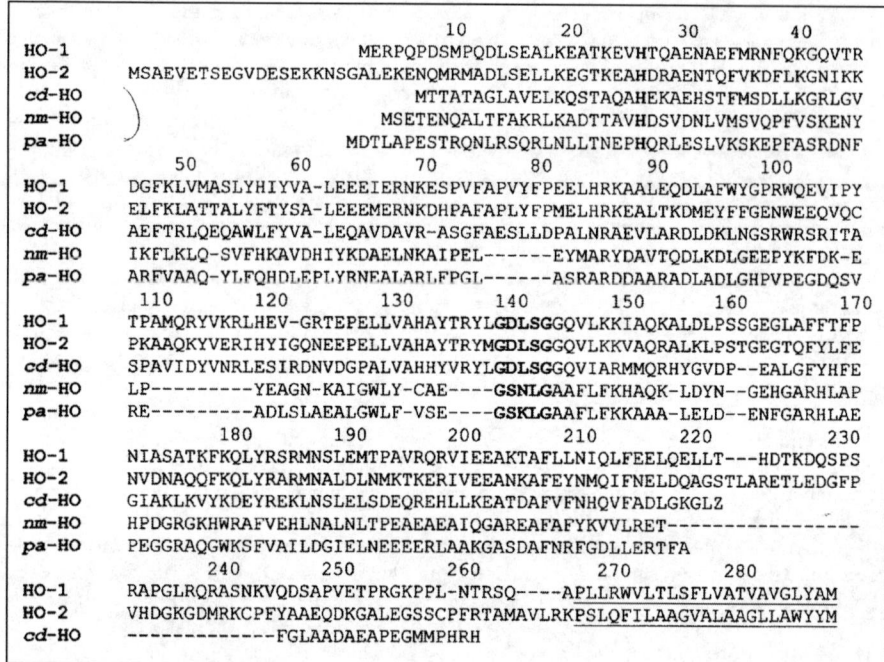

```
                                      10        20        30        40
HO-1                           MERPQPDSMPQDLSEALKEATKEVHTQAENAEFMRNFQKGQVTR
HO-2     MSAEVETSEGVDESEKKNSGALEKENQMRMADLSELLKEGTKEAHDRAENTQFVKDFLKGNIKK
cd-HO  \            MTTATAGLAVELKQSTAQAHEKAEHSTFMSDLLKGRLGV
nm-HO   )          MSETENQALTFAKRLKADTTAVHDSVDNLVMSVQPFVSKENY
pa-HO  /           MDTLAPESTRQNLRSQRLNLLTNEPHQRLESLVKSKEPFASRDNF
              50        60        70        80        90       100
HO-1     DGFKLVMASLYHIYVA-LEEEIERNKESPVFAPVYFPEELHRKAALEQDLAFWYGPRWQEVIPY
HO-2     ELFKLATTALYFTYSA-LEEEMERNKDHPAFAPLYFPMELHRKEALTKDMEYFFGENWEEQVQC
cd-HO    AEFTRLQEQAWLFYVA-LEQAVDAVR-ASGFAESLLDPALNRAEVLARDLDKLNGSRWRSRITA
nm-HO    IKFLKLQ-SVFHKAVDHIYKDAELNKAIPEL------EYMARYDAVTQDLKDLGEEPYKFDK-E
pa-HO    ARFVAAQ-YLFQHDLEPLYRNEALARLFPGL------ASRARDDAARADLADLGHPVPEGDQSV
              110       120       130       140       150       160       170
HO-1     TPAMQRYVKRLHEV-GRTEPELLVAHAYTRYLGDLSGGQVLKKIAQKALDLPSSGEGLAFFTFP
HO-2     PKAAQKYVERIHYIGQNEEPELLVAHAYTRYMGDLSGGQVLKKVAQRALKLPSTGEGTQFYLFE
cd-HO    SPAVIDYVNRLESIRDNVDGPALVAHHYVRYLGDLSGGQVIARMMQRHYGVDP--EALGFYHFE
nm-HO    LP---------YEAGN-KAIGWLY-CAE----GSNLGAAFLFKHAQK-LDYN--GEHGARHLAP
pa-HO    RE---------ADLSLAEALGWLF-VSE----GSKLGAAFLFKKAAA-LELD--ENFGARHLAE
              180       190       200       210       220       230
HO-1     NIASATKFKQLYRSRMNSLEMTPAVRQRVIEEAKTAFLLNIQLFEELQELLT---HDTKDQSPS
HO-2     NVDNAQQFKQLYRARMNALDLNMKTKERIVEEANKAFEYNMQIFNELDQAGSTLARETLEDGFP
cd-HO    GIAKLKVYKDEYREKLNSLELSDEQREHLLKEATDAFVFNHQVFADLGKGLZ
nm-HO    HPDGRGKHWRAFVEHLNALNLTPEAEAEAIQGAREAFAFYKVVLRET----------------
pa-HO    PEGGRAQGWKSFVAILDGIELNEEEERLAAKGASDAFNRFGDLLERTFA
              240       250       260       270       280
HO-1     RAPGLRQRASNKVQDSAPVETPRGKPPL-NTRSQ----APLLRWVLTLSFLVATVAVGLYAM
HO-2     VHDGKGDMRKCPFYAAEQDKGALEGSSCPFRTAMAVLRKPSLQFILAAGVALAAGLLAWYYM
cd-HO    ------------FGLAADAEAPEGMMPHRH
```

Figure 2. Sequence alignment of the human and bacterial heme oxygenases. HO-1 and HO-2, human isozymes, *cd*-HO, *Corynebacterium diphtheriae*, *nm*-HO, *Neisseriae meningitidis*, and *pa*-HO, *Pseudomonas aeruginosa*. The sequence numbering shown is that of the human HO-1 isozyme. The proximal His and the conserved glycine motif are shown in bold. The C-terminal membrane anchor of the human enzymes is underlined.

unaffected.[19] More recently we have solved the crystal structure of the soluble 24 kDa heme oxygenase from *Neisseriae meningitidis* (*nm*-HO) to 1.5 Å.[21] The first striking observation, given the limited sequence identity between hHO-1 and *nm*-HO, is the high degree of structural conservation. The heme oxygenase proteins have a novel fold which is primarily α-helical, with the δ-meso edge of the heme and the propionates exposed at the molecular surface of the protein (Fig. 3).

The heme is held between two helices, referred to as the proximal and distal helices. The proximal helix donates the histidine ligand, His-25 in hHO-1 and His-23 in *nm*-HO. Additional contact residues with the heme include Ala-28 and Glu-29 of hHO-1 and the corresponding Val-26 and Asp-27 in *nm*HO. Glu-29 in HO-1 and Asp-27 in *nm*-HO on the proximal side of the heme are both close enough to form a hydrogen bond with their respective proximal histidines.

The distal face of the heme where the oxygen ligand binds, and the reaction takes place, has a distal helix with a kink of approximately 50° directly over the heme (Fig. 3). This kink in the helix is provided by the glycines of the highly conserved sequence [139]Gly-Asp-Leu-Ser-Gly-Gly[144] in hHO-1 and [117]Gly-Ser-Asn-Leu-Gly-Ala[122] in *nm*-HO and is thought to provide the flexibility required for binding of the substrate (heme) and release of the product (biliverdin). The helix closely approaches the heme with direct backbone contacts from Gly-139 and Gly-143 in hHO-1. Recently, it has been shown that mutation of Gly-139 resulted in a loss of oxygenase activity and an increase in peroxidase activity.[22] The authors concluded that these residues are critical in maintaining an environment conducive to oxygenase activity, and that the key role of the heme oxygenase protein environment may be to suppress the formation of a ferryl species (see following sections).

In the human enzyme a cluster of polar residues near pyrrole B including Asn-210, Arg-136 and Asp-140 appear to form a hydrogen-bonding network of polar residues with a second tier of residues Tyr-58 and Tyr-114. In an effort to understand the role of this hydrogen-bonding network Ortiz de

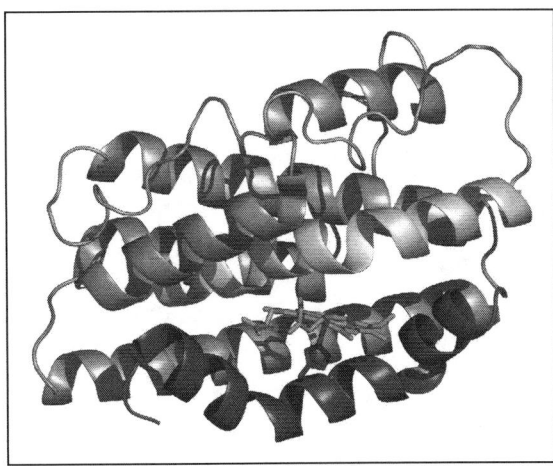

Figure 3. The overall structure of human HO-1. The proximal helix shown in blue donates the His-25 ligand to the heme iron. The distal helix in cyan lies across the heme shown in red. A color version of this figure is available online at www.eurekah.com.

Montellano and coworkers have replaced Asp-140 in the active site of hHO-1 with a number of amino acid residues, resulting in a decrease in the formation of biliverdin and an increase in peroxidase activity.[23] The authors concluded that replacement of Asp-140 disrupted oxygen binding resulting in a much smaller proportion of the Fe^{III}-O-OH intermediate being directed to the α-*meso* carbon, with the resulting heterolytic O-O bond cleavage yielding the ferryl species. The recent crystal structures of the ferric, ferrous and ferrous -NO forms of the wild type and Asp-140-Ala mutant of hHO-1 reveal a hydrogen-bonding network that is critical for protonation and activation of the oxygen.[24] In *nm* HO many of the conserved residues involved in formation of the hydrogen-bonding network have been replaced, but a similar hydrogen-bonding network is conserved which directly delivers the proton to the oxygen ligand (Fig. 4).[25]

With regard to the regioselectivity of the reaction it is clear that steric hindrance plays a role in determining which *meso*-carbon is hydroxylated in the initial reaction step. In both hHO-1 and *nm*-HO only the δ-*meso*-carbon is exposed at the surface of the protein with the other three *meso*-carbons being buried within the protein. Propionate charge interactions with conserved residues appear to be critical in binding and orienting the α-meso carbon of the heme for hydroxylation.[19] Residues in close proximity to the propionates in hHO-1 are Lys-18, Lys-22, Lys-179, and Arg-183, as well as Tyr-134. In the case of *nm*-HO Lys-16 and Tyr-112 corresponding to Lys-18 and Tyr-134 are absolutely conserved within the same orientation in the structure.

We have recently investigated the role of the heme propionate interactions with the protein in the bacterial heme oxygenases, and specifically *pa*-HO, which displays δ-regioselectivity, and is therefore unique in that all other heme oxygenases that have so far been characterized are α-selective.[18] Interestingly, on analysis of the sequence of *pa*-HO, which displays high sequence identity with *nm*-HO, Lys-18 and Tyr-112 in *nm*-HO are replaced by Asn-19 and Phe-117. Site-directed mutagenesis in which Asn-19 and Phe-117 of *pa*HO were replaced with Lys and Tyr, resulted in a protein with altered regioselectivity, resulting in 40% α-biliverdin and 60% δ-biliverdin on HPLC analysis of the final products of the reaction.[26] The altered regioselectivity was determined by ^{13}C-heme NMR studies to be the result of a dynamic equilibrium between two alternate heme seatings, one in keeping with the wild type *pa*-HO and one more closely related to the α-selective enzymes.[26]

One of the more interesting aspects of heme oxygenase reactivity is how the enzyme has evolved to discriminate between O_2 and CO. The O_2-affinity of ferrous heme-HO-1 and HO-2 are 30- to 90-fold higher than those of mammalian myoglobins.[27] Comparison of the dissociation constants indicates a much slower rate of dissociation from HO than the globins. In contrast the affinities for CO in HO are only six-fold higher than those for oxygen in contrast to the much higher affinities

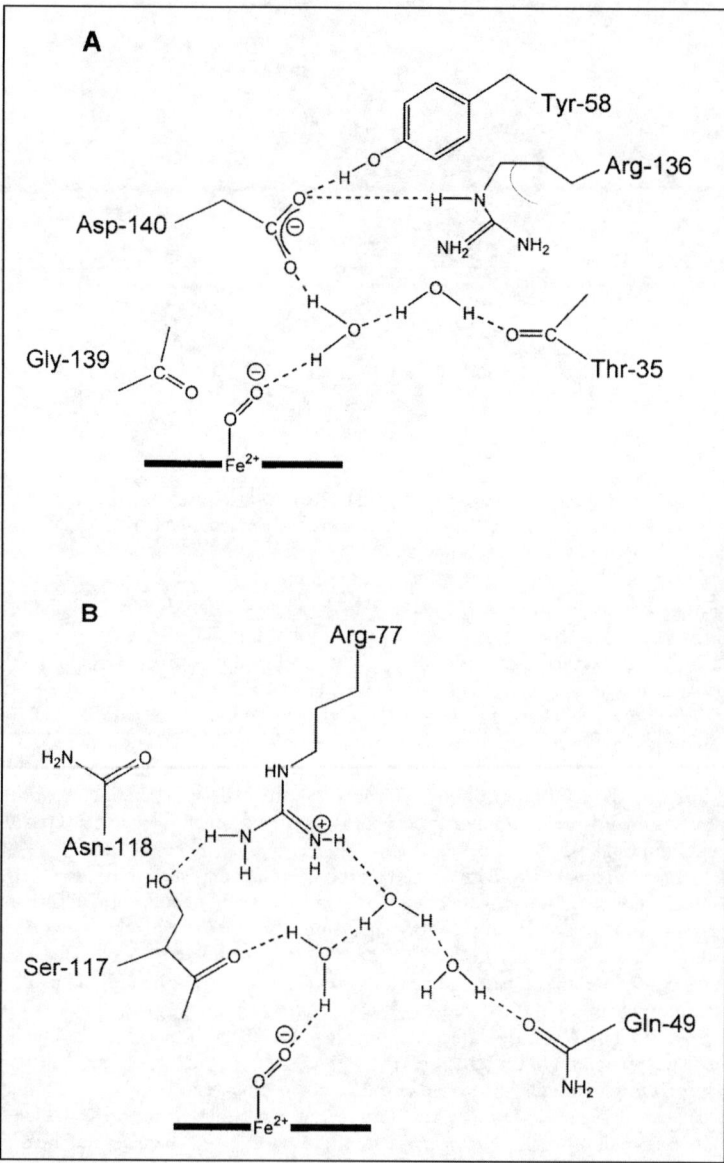

Figure 4. Schematic of the ordered hydrogen bonding network in hHO-1 (A) and nm-HO (B). The hydrogen-bonding network involved in stabilization of the bound dioxygen as proposed from the crystal structures of the NO-ferrous complexes of the heme-hHO-1 and heme-*nm*-HO. (Adapted from refs. 24, 25, respectively.)

for CO compared to O_2 in the globins. Thus the ratios of K_{CO}/K_{O2} are ~5.4 and ~25 for HO and the globins, respectively.[27] One possible explanation for the decrease in O_2 dissociation and CO affinity is suggested by the crystal structures of HO. Although there is no distal base such as histidine to coordinate directly to the bound oxygen, the ordered set of water molecules within the active site could function to stabilize the bound O_2, and participate in proton delivery to the bound oxygen. It is evident from the recent crystal structure of the ferrous-NO and CO complexes of *nm*-HO that the more polar O_2 ligand would form a more stable set of hydrogen-bonding interactions than CO perhaps accounting for the slower off rate of O_2 and hence the increased affinity for O_2 versus CO.[25]

Mechanism of Heme Oxygenase

Oxygen Activation and Meso-Hydroxylation

The first step in the catalytic turnover of heme is the reduction of the ferric Fe^{III}-heme-HO-1 complex to the ferrous state by NADPH-cytochrome P450 reductase. On reduction of the iron to the ferrous state, oxygen is bound to give the Fe^{II}-O_2 complex with a Soret maximum at 410 nm, and a striking resemblance to the Fe^{II}-O_2-myoglobin spectrum. Recent resonance Raman studies on the isotopically labelled Fe^{II}-O_2 complex have shown an isotope shift pattern that suggests the oxygen is highly bent, with the terminal oxygen closely approaching the α-*meso*-carbon.[28] The highly bent state of the bound dioxygen contrasts sharply with the more linear geometry of the Fe^{II}-O_2 myoglobin complex. The authors concluded, based on the similarity of the His-Fe stretching frequencies in both the Fe^{II}-O_2 myoglobin and HO complexes, that the effect was due to steric factors within the heme-binding pocket, a conclusion consistent with the recent crystal structures.[19-21] The hydrogen-bonding network in the reduced ferrous-NO and CO complexes of the *nm*-HO and hHO-1 crystal structures provide further evidence for the bent geometry, which facilitates in directing the bound dioxygen toward the α-meso-carbon.[24]

The dioxygen geometry has obvious implications for the hydroxylation of the α-*meso*-heme edge. Once the Fe^{II}-O_2 complex is formed a second electron transfer to the heme reduces the Fe^{II}-O_2 complex to the activated peroxide Fe^{III}-O-OH intermediate. This intermediate in heme oxygenase differs significantly from the more well characterized heme enzymes that undergo heterolytic cleavage of the O-O bond to yield the active ferryl complex (Fe^{IV} = O). The addition of H_2O_2 to the Fe^{III}-heme-HO-1 complex supports the hydroxylation of the α-*meso*-carbon to yield α-*meso*-hydroxyheme, which in the presence of oxygen is directly converted to verdoheme.[29] Therefore, H_2O_2 is equivalent to a molecule of oxygen and two electrons in the reaction of heme to verdoheme, but does not support the catalytic conversion of verdoheme to biliverdin.[29]

The nature of the peroxide species involved in hydroxylation of the heme was further studied with the acyl and alkyl peroxides.[29] Reaction of the Fe^{III}-heme-HO-1 complex with *meta*-chloroperbenzoic acid generated a ferryl complex that did not support hydroxylation. Similarly, the larger alkyl peroxides such as *tert*-butyl hydroperoxide and cumene hydroperoxide formed a ferryl species with no evidence of heme modification. The reaction with the much smaller ethyl hydroperoxide gave the most interesting and informative results.[30] Although ethyl hydroperoxide also gives rise to a ferryl species, reduction of the peroxo intermediate to prevent nonspecific degradation of the heme, followed by HPLC isolation identified a fraction of the heme that was modified. The modified heme on characterization by absorption, NMR, and mass spectroscopy was identified as the α-*meso*-ethoxyheme adduct.[30] The formation of α-*meso*-ethoxyheme paralleled that of α-*meso*-hydroxyheme, the first step in the enzymatic reaction. The α-*meso*-ethoxyheme is stable compared to α-*meso*-hydroxyheme, which is rapidly deprotonated as part of the normal reaction pathway to verdoheme. These experiments highlighted one important aspect of the heme oxygenase reaction and that is the nature of the activated peroxide intermediate. The reaction specifically rules out a mechanism by which the terminal oxygen of a peroxo anion (Fe^{III}-O-O$^-$) adds as a nucleophile to the α-meso carbon, because in the ethyl hydroperoxide reaction the terminal oxygen is blocked. These results therefore imply that the Fe^{III}-O-OH intermediate undergoes electrophilic addition at the α-meso-edge, in a mechanism that supports the data from both the studies with H_2O_2 and ethyl hydroperoxide (Fig. 5).

A series of crucial experiments utilizing ENDOR and EPR spectroscopy techniques were carried out on the cryogenically reduced Fe^{II}-O_2 HO-1 complex.[31] In these studies direct demonstration of a one-electron reduction of the Fe^{II}-O_2 HO-1 complex identified α-*meso*-hydroxyheme as a true intermediate in HO catalysis, and corroborated the hydroperoxide species as the activated oxygen in HO substrate hydroxylation. A series of EPR studies on model Fe^{III}-porphyrinates showed that the sum of the squares of the principal g-values (Σg), when less than 14, suggests a unique electronic configuration whereby the unpaired electron on the iron resides in the d_{xy} orbital. This perhaps has implications for reactivity of the heme oxygenase reaction. In such a system whereby the unpaired electron resides in the d_{xy} orbital, the resulting $(d_{xz}d_{yz})^4(d_{xy})^1$ electronic structure requires the macrocycle to ruffle to allow delocalization of the spin density of the d_{xy} orbital into the porphyrin π-system, resulting in an increase in the electron density at the α-*meso*-carbons. A recent reevaluation of the data in the context of a model complex of a low spin Fe^{III}-O-OH intermediate of heme oxygenase by EPR and NMR has been carried out.[32] The data suggest that a d_{xy} electron configuration is favored at physiological temperatures and as such may exist in the enzyme system. The summation of the g values (Σg) of 14.1 from the EPR data of the Fe^{III}-O-OH heme-HO-1 complex further suggests that the unpaired electron may reside in the d_{xy} orbital. The resulting electronic configuration, which places a significant electron spin density at the *meso*-carbons may help explain the readiness of the Fe^{III}-O-OH intermediate to attack the heme *meso*-carbon. In addition a flexible heme binding cavity would be essential to allow the resulting ruffling of the heme necessary for the orbital overlap in the $(d_{xz}d_{yz})^4(d_{xy})^1$ electronic configuration. While it is evident that reactivity of hemeproteins is both defined and modulated by the protein ligand to the heme, it is interesting to speculate that HO may also have evolved to prime the heme for hydroxylation by allowing a unique heme electronic structure that promotes the reaction. The combination of steric access being restricted to the α-meso-carbon, the bent geometry of the activated oxygen and an increased electron density together with the heme ruffling, which would cause the electron dense α-*meso*-carbon to be above the plane of the heme-iron, would clearly facilitate hydroxylation. We have recently examined the role of heme electronics in facilitating the initial hydroxylation step utilizing ^{13}C NMR spectroscopic studies with the hydroxide complex (Fe^{III}-OH) of heme oxygenase as a model of the hydroperoxide intermediate (Fe^{III}-OOH). The results strongly suggest that the coordination of hydroxide in the distal pocket induces the formation of at least three populations of Fe^{III}-OH complexes with distinct electronic configurations and nonplanar ring distortions.[33] These findings were interpreted to suggest that if the ligand field strength of the coordinated $^-$OOH ligand in HO is modulated by the distal pocket in a manner similar to that seen with $^-$OH, the resultant large spin density at the meso carbon and nonplanar deformations of the macrocycle prime the heme to participate actively in its own hydroxylation. Thus, it is likely that the regioselectivty of heme oxygenation is determined by a combination of both steric and electronic factors. In addition one could speculate that the flexibility of the distal helix and specifically the glycine motif may be critical in allowing the ruffling required to allow the reactive heme electronic structure. It would be pertinent to reexamine mutations within the flexible glycine motif in the context of heme electronics to determine if reduced flexibility of the helix contributes to a heme electronic structure that favors formation of a ferryl rather than heme hydroxylation. Furthermore, it is possible that hemeproteins such as the peroxidases and cytochromes P450, which undergo dioxygen bond cleavage to give the ferryl intermediate, have evolved a protein scaffold that essentially disallows access of the heme to an electronic configuration similar to that observed in HO, and hence shifts the reaction toward ferryl formation rather than heme hydroxylation.

Conversion of α-Mesohydroxyheme to Verdoheme

The conversion of α-*meso*hydroxyheme to verdoheme is an oxygen dependent process as evidenced from the reaction with H_2O_2 or on reconstitution with synthetic α-*meso*hydroxyheme where, in the absence of oxygen, α-*meso*hydroxyheme is a stable intermediate.[34] Addition of oxygen to the anaerobic α-*meso*hydroxyheme complex results in the appearance of an organic radical in the EPR spectrum at g = 2.008 with a rhombic signal at g = 6.07 and 5.71 characteristic of the Fe^{III} species.[34]

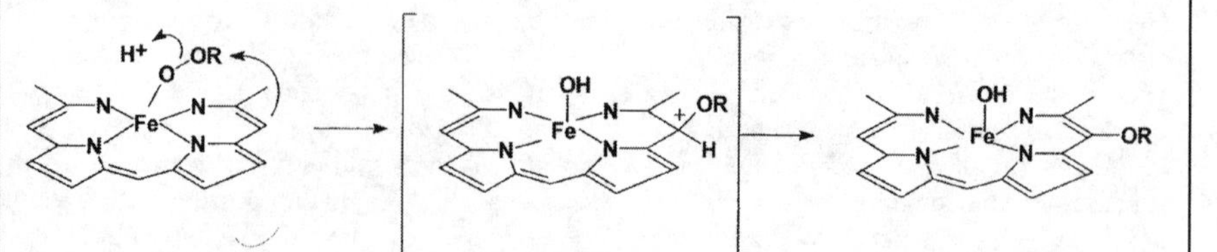

Figure 5. Electrophilic oxidation of the heme as catalyzed by HO. With H_2O_2 as the oxidant R=H and α-*meso*hydroxyheme is the product; with ethyl hydroperoxide as oxidant R= -CH_2CH_3 and α-*meso*ethoxyheme is the product.

Although it is largely agreed that the Fe^{III}-*meso*hydroxyheme is in equilibrium with the Fe^{II} radical form some questions remain on the requirement of reducing equivalents for the conversion of α-*meso*hydroxyheme to verdoheme. In studies with synthetically reconstituted apo-HO-1 and α-*meso*hydroxyheme it has been reported that the formation of verdoheme required input of a further reducing equivalent as well as O_2.[35] In contrast Liu et al, reported that only the addition of oxygen was required for the conversion of the H_2O_2 generated α-*meso*hydroxyheme to verdoheme.[34] There is no question that the conversion of Fe^{III}-verdoheme to Fe^{II}-verdoheme requires an electron. The observation that Fe^{III}-verdoheme accumulates in high yields in the H_2O_2 driven reaction in the presence of O_2 clearly indicates that an electron is not required at this step. Under physiological conditions in the presence of NADPH and cytochrome P450 reductase Fe^{III}-verdoheme does not accumulate and is rapidly converted to Fe^{II}-verdoheme. The step at which the electron is introduced is debatable but clearly it is not introduced prior to the formation of Fe^{III}-verdoheme, rather it is either following formation of Fe^{III}-verdoheme, after CO elimination, or prior to CO release directly in the formation of Fe^{II}-verdoheme.[34]

In the proposed mechanism, deprotonation of the α-*meso*hydroxyheme is required to yield the oxophlorin radical, which can react with oxygen to give a peroxy-radical intermediate, either through direct interaction with the iron or following internal electron transfer to give the Fe^{III}-hydroperoxy species. Either of these species would give rise to an unstable ferryl alkoxy radical with the consequent release of CO as a result of carbon-carbon bond cleavage. The resulting unstable carbon radical is then presumably internally oxidized to the cation by electron transfer to the ferryl species (Fig. 6).

While none of the intermediates beyond the addition of oxygen to the porphyrin radical species have been identified spectrally, the mechanism provides a reasonable explanation of the steps involved in CO release and formation of verdoheme.

Verdoheme to Biliverdin

The final stage of the HO catalyzed reaction, and the least well understood in mechanistic terms, is the conversion of verdoheme to biliverdin. Early $^{18}O_2$-labelling studies on the reconstituted HO-1 system provided an important foundation for the mechanism of oxygen insertion. Under a partial atmosphere of $^{16}O_2/^{18}O_2$ one atom each of ^{18}O and ^{16}O were incorporated into the final biliverdin product.[36] The incorporation of two atoms of oxygen, from separate oxygen molecules, indicates that the reaction solely by oxidation and not via a hydrolytic mechanism.

One important distinction with the utilization of H_2O_2 instead of NADPH-cytochrome P450 reductase is the inability of the reaction to proceed beyond verdoheme; in fact, the presence of excess H_2O_2 results in the nonspecific degradation of the heme to mono- and dipyrroles. This observation suggests that formation of Fe^{III}-verdoheme yields an intermediate in a higher oxidation state than that required for subsequent conversion to biliverdin. Presumably the formation of biliverdin from verdoheme requires a two-electron reduction precluding H_2O_2 from carrying out the final step in the reaction. A hypothetical mechanism proposed by Ortiz de Montellano to account for these observations is that the first step is reduction of the Fe^{III}-verdoheme to the Fe^{II} state, followed by binding of dioxygen and a second one electron reduction would then give an intermediate formally equivalent to that obtained on reaction with H_2O_2 (Steps 1 and 2 in Fig. 7).[37] Following dioxygen bond cleavage to give an alkoxy radical or anion, the subsequent ring opening and two-electron reduction would give rise to the final Fe^{III}-biliverdin product. An alternative pathway may involve the stepwise addition of molecular oxygen to the porphyrin edge (Step 3 in Fig. 7) and therefore the reaction with H_2O_2 would result in side reactions that give alternate products than that of molecular oxygen. At the present time it is not known at which step in the pathway the two electrons required to produce Fe^{III}-biliverdin are delivered.

The final step in the complete HO-dependent reaction is the conversion of Fe^{III}-biliverdin to Fe^{II}-biliverdin. This step in the pathway has important implications for the subsequent release of biliverdin and its conversion to bilirubin by the action of biliverdin reductase (BVR-A). Recently, Liu et al have confirmed that the rate-limiting step in the enzymatic degradation of heme to biliverdin, in the absence of BVR-A, is biliverdin release. In the presence of BVR-A, the rate-limiting step in the single turnover studies was the conversion of Fe^{II}-verdoheme to Fe^{III}-biliverdin. One possible factor that may contribute to biliverdin release is direct protein-protein interaction of biliverdin reductase with the biliverdin-HO-1 complex that somehow facilitates allosteric release of the product. While the kinetic studies have implicated certain steps in the reaction as being rate-limiting it must be remembered that these conditions are not physiological. Firstly, the rate of the overall reaction is somewhat sensitive to the ratio of NADPH-cytochrome P450 reductase to HO-1, and secondly under physiological conditions the oxidation state of the heme is not known, and this may have consequences for the overall rate of the reaction.

Biliverdin Reduction to Bilirubin

As outlined in the previous sections the oxidation of heme to biliverdin has physiological consequences far beyond that of disposal of a waste product, and as such the regioselectivity of the reaction has implications for many of the functions described in the previous section. The conversion of biliverdin, a relatively nontoxic soluble waste product, to bilirubin a substance that is neurotoxic and lipophilic would be puzzling if this was solely a function of metabolite excretion. However, as mentioned in previous sections, the role of bilirubin as a regeneratable anti-oxidant significantly alters this perspective and provides a rationale for the conversion of biliverdin to bilirubin in mammals. Therefore, while biliverdin in plants, prokaryotes and lower eukaryotic organisms has a number of varied physiological functions, it is bilirubin, the product of biliverdin reduction by biliverdin reductase that has physiological relevance in mammals.

Biliverdin IXα Reductase

Biliverdin IXα reductase (BVR-A) is a soluble cytoplasmic protein that catalyzes the regiospecific addition of hydrogen to the -HC(10) = C-N = group of biliverdin IXα (Fig. 8).[3] In conversion of the polar flexible biliverdin into a more conformationaly restricted lipophilic bilirubin, BVR-A is highly selective for the biliverdin IXα isomer. BVR-A can utilize either NADH or NADPH as cofactors, although NADPH appears to be the preferred cofactor at physiological pH.

The recent crystal structure of the apo-BVR-A revealed a characteristic dinucleotide binding fold, or Rossman fold, at the N-terminal domain, whereas the C-terminal section of the protein is largely composed of an anti-parallel six-stranded β-sheet.[38] Although in the apo-BVR-A structure neither the cofactor NADPH nor the substrate biliverdin were present, a mechanism for binding and activation was proposed. The dinucleotide-binding site is large enough to accommodate both NADPH, and biliverdin in a helical conformation, directly above the NADPH in a manner conducive to stereospecific reduction of the C10 bond. The basic residues Arg-171, Lys-218, Arg-224 and Arg-226 have been implicated in substrate binding through electrostatic interactions with the propionate side chains of biliverdin.

A more recent structure of the rat BVR-A -NADH complex reveals further insight into the stereospecific mechanism of this reaction.[39] The cofactor nicotinamide ring is buried within the BVR-A active site cleft packing against a hydrophobic surface. Invariant polar and charged residues within the vicinity of the C4 carbon of the nicotinamide ring that extends into the substrate-binding site include Ser-170, Tyr-97, and Arg-171. In modeling biliverdin IXα in the active site, the hydroxyl hydrogen of Ser-170 appears to be H-bonded to both the tyrosyl oxygen and the side chain nitrogen of Arg-171. This positions the phenolic oxygen of Tyr-97 4 Å from the pyrrole nitrogen and the central carbon of the γ-methene bridge and the nicotinamide C4 carbon, from which the hydride is transferred approximately 4 Å apart. This is consistent with a mechanism involving direct hydride transfer from C4 of the nicotinamide ring to the central methane, while the phenolic oxygen of Tyr-97 forms a hydrogen bond with the pyrrole (Fig. 8). The importance of Tyr-97 in catalysis is somewhat controversial given the observation that on mutation to Phe the enzyme retains 50% of the wild type activity.[39] Although Tyr-97 contributes to efficient catalysis it is not essential and may be required only to orient the substrate for efficient catalysis. The drive for catalysis may be directly due to hydride transfer from NADPH, a process that may be facilitated by the proximity of Glu-26, Glu-123 and Glu-126 in stabilizing NADP+. It is also feasible that distortion of the flexible substrate by the enzyme may facilitate catalysis, a possibility that awaits the successful crystallization of the substrate bound BVR-A complex.

The recent crystal structure of BVR-A also provides significant insight into the proposed interaction with HO and facilitation of biliverdin release. As outlined in the previous section the rate-limiting step of the HO reaction in the absence of BVR-A is the release of biliverdin, whereas

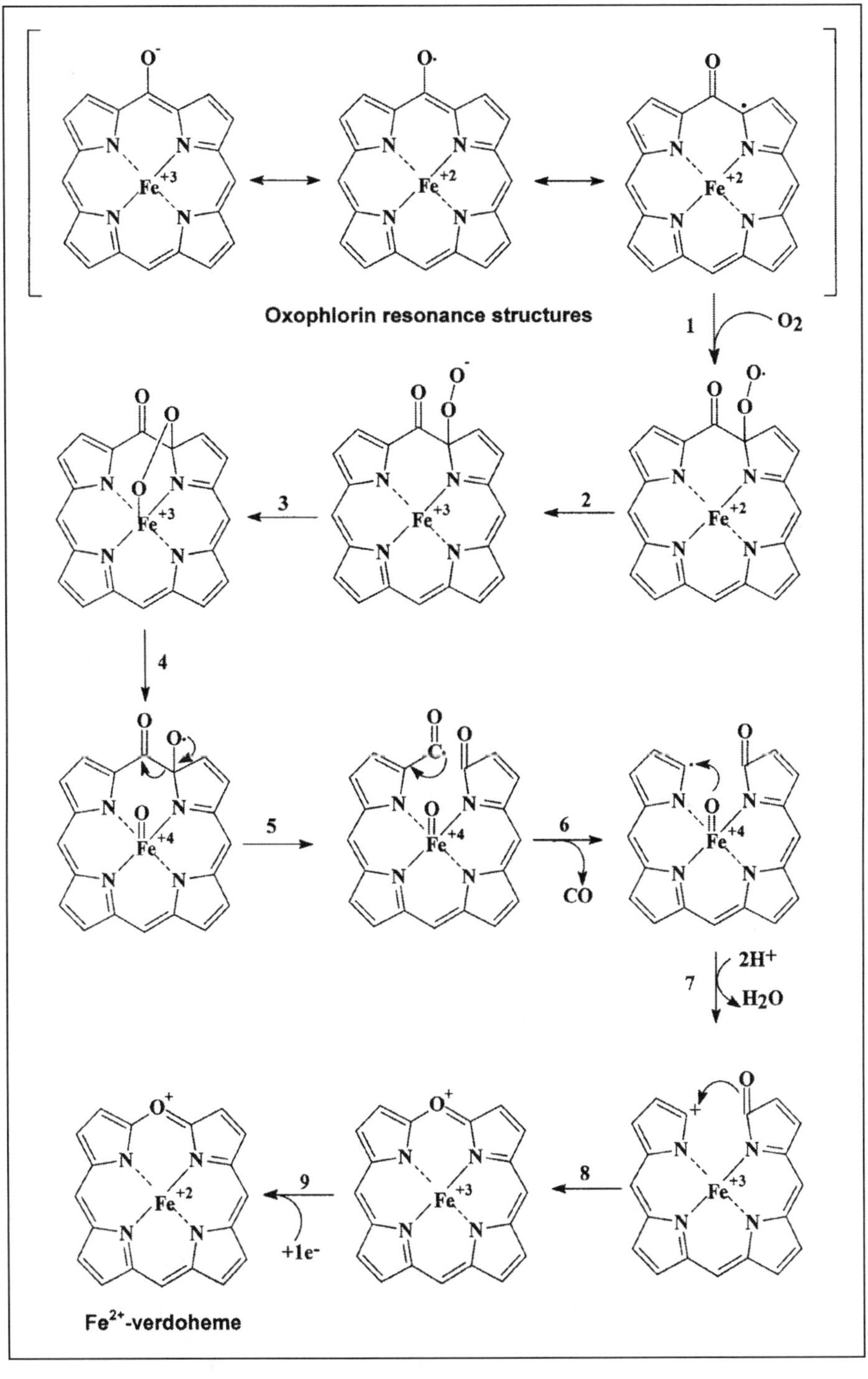

Figure 6. Proposed mechanism for the conversion of α-*meso*-hydroxyheme to Fe^{II}-verdoheme. The α-*meso*-hydroxyheme is shown deprotonated. (Adapted from ref. 36.)

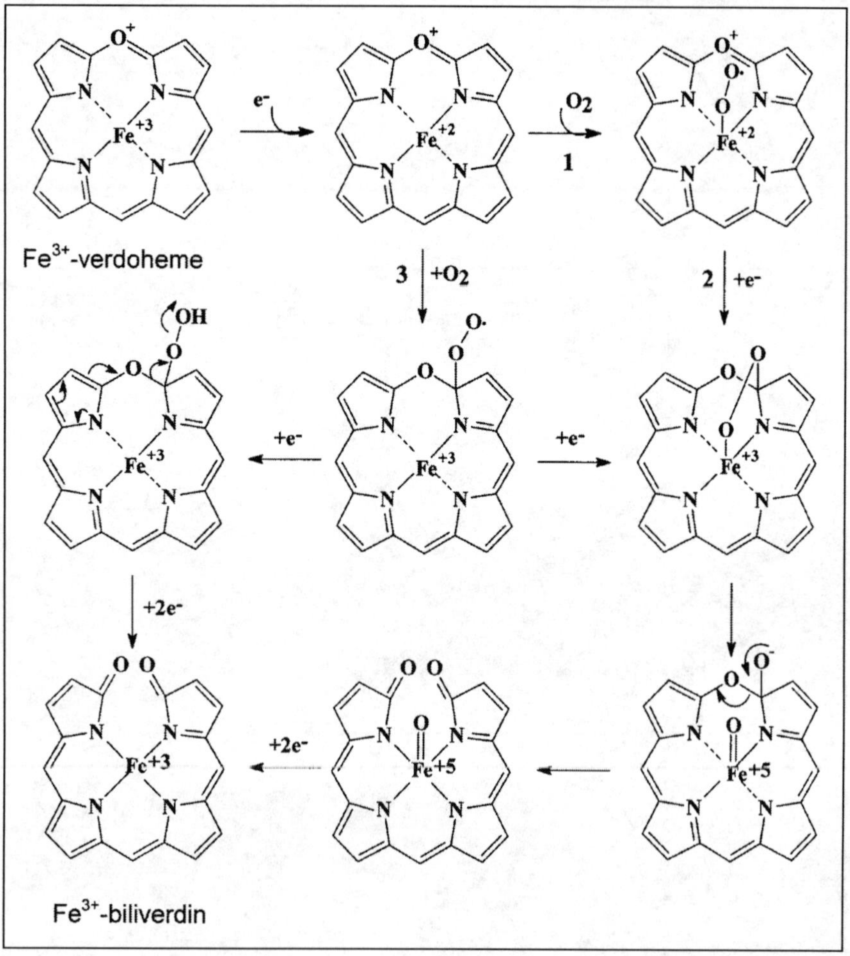

Figure 7. Proposed mechanism for the conversion of Fe^{II}-verdoheme to Fe^{III}-biliverdin. Two alternative pathways are shown that differ in the step at which the electrons are introduced into the system. (Adapted from ref. 36.)

in the presence of BVR-A the rate limiting step is the conversion of Fe^{II}-verdoheme to Fe^{III}-biliverdin. It is therefore feasible that a direct protein-protein interaction between BVR-A and the biliverdin-HO-1 complex is required for allosteric release of the product. Indeed, just as the propionate groups of the heme are important in binding and orienting the heme within HO-1, the presence of a number of Arg and Lys residues in BVR-A are ideally situated for interaction with the propionates of the resulting biliverdin. Interestingly, this exposed surface of the BVR-A protein is also somewhat electronegative, and may directly interact with the more electropositive face surrounding the exposed heme edge in the heme-HO-1 complex.[19,38,39] A more recent study utilizing fluorescence resonance energy transfer (FRET) technology has identified an overlapping binding site for both NADPH cytochrome P450 reductase and BVR-Aα on hHO-1 which includes residues Lys-18, Lys-22, Lys-179, Arg-183 and Arg-185, providing direct evidence for an interaction of BVR-A with hHO-1, which given the previously mentioned positive charged residues on the surface of hHO-1, most likely involves the electronegative surface of BVR-A.[40]

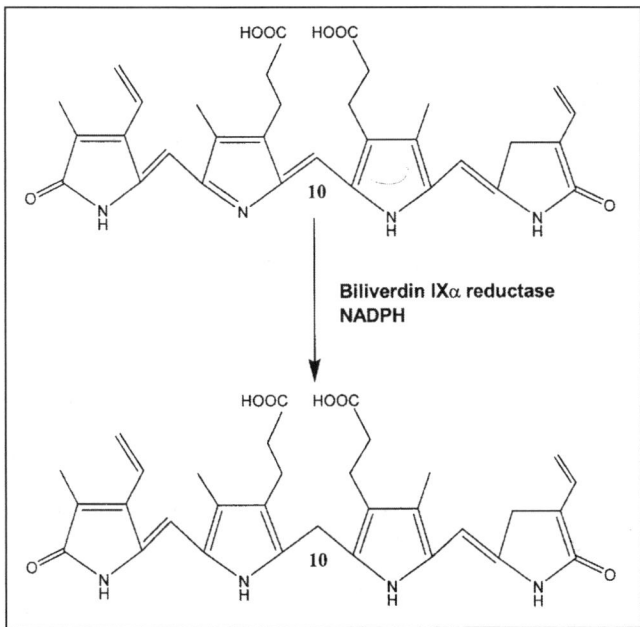

Figure 8. Conversion of biliverdin IXα to bilirubin IXα by the action of biliverdin IXα reductase. The C10 methene carbon is labeled in both biliverdin IXα and bilirubin IXα.

Biliverdin IXβ Reductase

An early hypothesis for the energetically demanding and potentially toxic conversion of biliverdin IXα to bilirubin IXα in mammals was the ability of bilirubin IXα, and not biliverdin IXα, to cross the placenta, and hence bilirubin was the means by which the fetus could transfer this waste product to the maternal circulation. However, the first heme catabolite identified in the fetus is biliverdin IXβ,[41] which unlike bilirubin IXα does not undergo internal H-bonding and can be excreted without glucuronidation. Clearly, as outlined in previous sections BVR-A has evolved in adults as a means by which to synthesize bilirubin IXα, a critical antioxidant molecule.

Heme cleavage at the β-*meso*-carbon produces biliverdin IXβ, which is then reduced by BVR-B to bilirubin IXβ.[42] The BVR-B enzyme is somewhat promiscuous in that it can also catalyze the reduction of the IXδ- and IXγ-biliverdin isomers as efficiently as the IXβ-isomer, and is identical in sequence to a flavin reductase isolated from adult erythrocytes.[42,43] The switch from a distinct heme degradative pathway in utero involving oxidation of the β-*meso*-carbon to the more well characterized α-*meso*-carbon oxidative cleavage in adults, may be coupled to the switch from fetal to adult hemoglobin. Flavin reductase is abundant in adult erythrocytes where its main purpose may be in the production of reduced flavins for the reduction of methemoglobin.[43]

The structure of BVR-B displays an α/β-dinucleotide binding fold which can accommodate NADPH and biliverdin IXβ, which binds in helical conformation in close proximity to the nucleotide.[44] The location of the propionates is critical for determining the substrate profile of BVR-B with substrates being aligned so that the C10 bridge is close to the labile C4 hydride on the nicotinamide ring. It is evident from the structure that biliverdin IXα would be bound in an orientation whereby the C10 methene bridge is unfavorably placed with respect to the NADPH.

It is apparent from the crystal structures of both BVR-A and BVR-B that despite significant sequence and structural differences the reduction of biliverdin to bilirubin is mechanistically conserved. These recent crystal structures of BVR-A and BVR-B have provided greater insight into the mechanism of biliverdin reduction to bilirubin, and provide a template for future studies related to therapeutic modulation of such activities.

Summary

In writing a review of heme degradation some twenty years ago one could not have imagined the surge of interest in heme catabolism and its physiological implications. Once thought to be solely a "housekeeping enzyme" in the degradation of hemoglobin from erythrocytes, heme oxygenase has evolved as a key protein in the cells response to oxidative stress, and as such has become a therapeutic target in areas of pathophysiology such as atherosclerosis, Alzheimer's and organ transplant rejection. Although still somewhat controversial, the unfolding story of CO and its role as a potential neurotransmitter promises to captivate the heme oxygenase community for the next decade. The rehabilitation of bilirubin IXα from a purely toxic molecule, to one which functions in antioxidant protection, has provided new insight into how nature has evolved to use potentially harmful by-products as important physiological molecules. Although bilirubin at physiologically relevant concentrations is beneficial, in excess it is extremely toxic and as such biliverdin reductase is a potentially attractive therapeutic target in the modulation of bilirubin levels in diseases such as hyperbilirubinemia and neonatal jaundice. Our increased understanding of the mechanism and regulation of these enzymes will aid in the development of new therapies for the aforementioned disease states.

In closing it is clear that the balance between a protective or toxic function is a fine one, and as such the therapeutic benefits of modulating heme oxygenase and biliverdin reductase activity, while potentially great are far from clear. It will be of great importance over the next decade to understand how these varying physiological functions interact as we begin to address the many unanswered questions regarding heme metabolism and its physiological and pathological functions.

References

1. Liu Y, Ortiz de Montellano PR. Reaction intermediates and single turnover rate constants for the oxidation of heme by human heme oxygenase-1. J Biol Chem 2000; 275(8):5297-307.
2. Yoshida T, Noguchi M, Kikuchi G. Oxygenated form of heme. Heme oxygenase complex and requirement for second electron to initiate heme degradation from the oxygenated complex. J Biol Chem 1980; 255(10):4418-20.
3. Noguchi M, Yoshida T, Kikuchi G. Purification and properties of biliverdin reductases from pig spleen and rat liver. J Biochem (Tokyo) 1979; 86(4):833-48.
4. Tenhunen R, Marver HS, Schmid R. Microsomal heme oxygenase. Characterization of the enzyme. J Biol Chem 1969; 244(23):6388-94.
5. Maines MD, Trakshel GM, Kutty RK. Characterization of two constitutive forms of rat liver microsomal heme oxygenase. Only one molecular species of the enzyme is inducible. J Biol Chem 1986; 261(1):411-9.
6. McCoubrey WK, Huang TJ, Maines MD. Isolation and characterization of a cDNA from the rat brain that encodes hemoprotein heme oxygenase-3. Eur J Biochem 1997; 247(2):725-32.
7. Poss KD, Tonegawa S. Heme oxygenase 1 is required for mammalian iron reutilization. Proc Natl Acad Sci USA 1997; 94(20):10919-24.
8. Maines MD. Heme Oxygenase: Clinical Applications and Functions. Boca Raton: CRC Press, 1992.
9. Stocker R, Yamamoto Y, McDonagh AF et al. Bilirubin is an antioxidant of possible physiological importance. Science 1987; 235(4792):1043-6.
10. Baranano DE, Rao M, Ferris CD et al. Biliverdin reductase: A major physiologic cytoprotectant. Proc Natl Acad Sci USA 2002; 99(25):16093-8.
11. Willis D, Moore AR, Willoughby DA. Heme oxygenase isoform expression in cellular and antibody-mediated models of acute inflammation in the rat. J Pathol 2000; 190(5):627-34.
12. Verma A, Hirsch DJ, Glatt CE et al. Carbon monoxide: A putative neural messenger [see comments] [published erratum appears in Science 1994 Jan 7;263(5143):15]. Science 1993; 259(5093):381-4.
13. Otterbein LE, Bach FH, Alam J et al. Carbon monoxide has anti-inflammatory effects involving the mitogen-activated protein kinase pathway. Nat Med 2000; 6(4):422-8.
14. Bhoo SH, Davis SJ, Walker J et al. Bacteriophytochromes are photochromic histidine kinases using a biliverdin chromophore. Nature 2001; 414(6865):776-9.
15. Wandersman C, Stojiljkovic I. Bacterial heme sources: The role of heme, hemoprotein receptors and hemophores. Curr Opin Microbiol 2000; 3(2):215-220.
16. Wilks A, Schmitt MP. Expression and characterization of a heme oxygenase (Hmu O) from Corynebacterium diphtheriae. Iron acquisition requires oxidative cleavage of the heme macrocycle. J Biol Chem 1998; 273(2):837-41.
17. Zhu W, Wilks A, Stojiljkovic I. Degradation of heme in gram-negative bacteria: The product of the hemO gene of neisseriae is a heme oxygenase. J Bacteriol 2000; 182(23):6783-6790.

18. Ratliff M, Zhu W, Deshmukh R et al. Homologues of neisserial heme oxygenase in gram-negative bacteria: Degradation of heme by the product of the pigA gene of pseudomonas aeruginosa. J Bacteriol 2001; 183(21):6394-403.
19. Schuller DJ, Wilks A, Ortiz de Montellano PR et al. Crystal structure of human heme oxygenase-1. Nat Struct Biol 1999; 6(9):860-7.
20. Sugishima M, Omata Y, Kakuta Y et al. Crystal structure of rat heme oxygenase-1 in complex with heme. FEBS Lett 2000; 471(1):61-66.
21. Schuller DJ, Zhu W, Stojiljkovic I et al. Crystal structure of heme oxygenase from the gram-negative pathogen Neisseria meningitidis and a comparison with mammalian heme oxygenase-1. Biochemistry 2001; 40(38):11552-8.
22. Liu Y et al. Replacement of the distal glycine 139 transforms human heme oxygenase-1 into a peroxidase. J Biol Chem 2000; 275(44):34501-7.
23. Lightning LK et al. Disruption of an active site hydrogen bond converts human heme oxygenase-1 into a peroxidase. J Biol Chem 2000; 276(14):10612-9.
24. Lad L et al. Crystal structures of the ferric, ferrous, and ferrous-No forms of the Asp140Ala mutant of human heme oxygenase-1: Catalytic implications. J Mol Biol 2003; 330(3):527-38.
25. Friedman J et al. Crystal structures of the NO- and CO-bound heme oxygenase from neisseriae meningitidis: Implications for O_2 activation. J Biol Chem 2003; 278(36):34654-9.
26. Caignan GA et al. Oxidation of heme to beta- and delta-biliverdin by Pseudomonas aeruginosa heme oxygenase as a consequence of an unusual seating of the heme. J Am Chem Soc 2002; 124(50):14879-92.
27. Migita CT et al. The oxygen and carbon monoxide reactions of heme oxygenase. J Biol Chem 1998; 273(2):945-9.
28. Takahashi S et al. Oxygen-bound heme-heme oxygenase complex: Evidence for a highly bent structure of the coordinated oxygen. J Am Chem Soc 1995; 117:6002-6006.
29. Wilks A, Ortiz de Montellano PR. Rat liver heme oxygenase. High level expression of a truncated soluble form and nature of the meso-hydroxylating species. J Biol Chem 1993; 268(30):22357-62.
30. Wilks A, Torpey J, Ortiz de Montellano PR. Heme oxygenase (HO-1). Evidence for electrophilic oxygen addition to the porphyrin ring in the formation of alpha-meso-hydroxyheme. J Biol Chem 1994; 269(47):29553-6.
31. Davydov RM et al. Hydroperoxy-heme oxygenase generated by cryoreduction catalyzes teh frmation of a-mesohydroxyheme as detected by EPR and ENDOR. J Am Chem Soc 1999; 121:10656-10657.
32. Rivera M et al. Models of the low-spin iron(III) hydroperoxide intermediate of heme oxygenase: Magnetic resonance evidence for thermodynamic stabilization of the d(xy) electronic state at ambient temperatures. J Am Chem Soc 2002; 124(21):6077-89.
33. Caignan GA et al. The hydroxide complex of Pseudomonas aeruginosa heme oxygenase as a model of the low-spin iron(III) hydroperoxide intermediate in heme catabolism: 13C NMR spectroscopic studies suggest the active participation of the heme in macrocycle hydroxylation. J Am Chem Soc 2003; 125(39):11842-52.
34. Liu Y et al. Heme oxygenase-1, intermediates in verdoheme formation and the requirement for reduction equivalents. J Biol Chem 1997; 272(11):6909-17.
35. Matera KM et al. Oxygen and one reducing equivalent are both required for the conversion of alpha-hydroxyhemin to verdoheme in heme oxygenase. J Biol Chem 1996; 271(12):6618-24.
36. Docherty JC et al. Mechanism of action of heme oxygenase. A study of heme degradation to bile pigment by 18O labeling. J Biol Chem 1984; 259(21):13066-9.
37. Ortiz de Montellano PR, Wilks A. Heme oxygenase structure and mechanism. Adv Inorg Chem 2000; 51:359-402.
38. Kikuchi A et al. Crystal structure of rat biliverdin reductase. Nat Struct Biol 2001; 8(3):221-5.
39. Whitby FG et al. Crystal structure of a biliverdin IXalpha reductase enzyme-cofactor complex. J Mol Biol 2002; 319(5):1199-210.
40. Wang J, Ortiz De Montellano PR. The binding sites on human heme oxygenase-1 for cytochrome P450 reductase and biliverdin reductase. J Biol Chem 2003.
41. Yamaguchi T, Nakajima H. Changes in the composition of bilirubin-IX isomers during human prenatal development. Eur J Biochem 1995; 233(2):467-72.
42. Yamaguchi T, Komoda Y, Nakajima H. Biliverdin-IX alpha reductase and biliverdin-IX beta reductase from human liver. Purification and characterization. J Biol Chem 1994; 269(39):24343-8.
43. Shalloe F et al. Evidence that biliverdin-IX beta reductase and flavin reductase are identical. Biochem J 1996; 316(Pt 2):385-7.
44. Pereira PJ et al. Structure of human biliverdin IXbeta reductase, an early fetal bilirubin IXbeta producing enzyme. Nat Struct Biol 2001; 8(3):215-20.

CHAPTER 7

Regulation of Mammalian Heme Biosynthesis

Amy E. Medlock and Harry A. Dailey*

Abstract

Regulation of the heme biosynthetic pathway in mammals occurs via two distinct mechanisms. These mechanisms reflect the fact that while most cells need to closely regulate relatively low levels of intracellular heme, differentiating erythroid cells must produce massive amounts of heme during a short period to satisfy the needs of hemoglobinization. In erythroid precursor cells all pathway enzymes are induced via erythroid-specific promoter elements and the first enzyme, erythroid-specific 5-aminolevulinate synthase (ALAS-2) encoded by a gene on the X chromosome, is also subject to translational regulation due to the presence of an iron-responsive element located in the 5' end of the mRNA. In nonerythroid cells a house-keeping regulatory scheme exists where most regulation appears to be via transcriptional regulation of a housekeeping 5-aminolevulinate synthase (ALAS-1) that is encoded on human chromosome 3. While the proteins of the mature forms of ALAS-1 and ALAS-2 are highly similar, the regulatory elements that control their expression are distinctly different and only ALAS-2 mRNA possesses an iron-regulatory element. Additional regulatory features exist throughout the pathway, but the major regulation appears to occur at the level of ALAS.

Introduction

In 1964 Lascelles authored "Tetrapyrrole Biosynthesis and Its Regulation".[1] This short tome outlined the state of the field at that time and presented a model for regulation of the pathway that involved both end product inhibition and repression of ALA synthesis. These discussions were based largely upon some elegant biochemical studies carried out in the facultative photosynthetic bacterium *Rhodobacter sphaeriodes* two decades before molecular techniques became available. While the model was relatively simple, it was noted that in this and other organisms there are a wide variety of factors that profoundly influence heme synthesis in unknown ways. Interestingly, in the book epilog Lascelles concluded a recap of control mechanisms by stating "in seeking for control mechanisms in tetrapyrrole synthesis the possibility of entirely novel principles must be constantly kept in mind." This statement remains true four decades later.

Regulation of heme synthesis has become fertile ground for discussion. Despite the fact that the model of pathway regulation at a single point (5-aminolevulinate synthase; ALAS) by a single regulator (heme) was never intended to be comprehensive and did not adequately explain data that were in the literature, this model became dogma and has only been slowly modified in the face of overwhelming data. Hence the study of regulation of heme biosynthesis has been a relatively perplexing field. Such confusion can be attributed to two main factors. Firstly, investigators have employed a multitude of cell types and organisms as model systems and, secondly, there has been a desire to explain all data with a single model. The first breakthrough in the evolution of new regulatory models occurred when it was discovered that animals possess two ALAS genes that are subject to tissue-specific regulatory mechanisms.[2] This was followed by the identification of tissue specific regulators for other heme pathway enzymes and characterization

*Corresponding Author: Harry A. Dailey—Biomedical and Health Sciences Institute, Paul D. Coverdell Center, University of Georgia, Athens, Georgia 30602, USA. Email: hdailey@uga.edu

Tetrapyrroles: Birth, Life and Death, edited by Martin J. Warren and Alison G. Smith.
©2009 Landes Bioscience and Springer Science+Business Media.

of some translational regulation. The existence of two distinct regulatory mechanisms reflects the fact that erythroid cells require large amount of heme for incorporation into hemoglobin, while in other cells, such as hepatocytes, significantly less heme is required for incorporation into hemoproteins such as cytochrome P450s. In addition in nonerythroid cells the level of heme is closely regulated over the life of the cell in response to changing demand for hemoprotein production while heme synthesis in erythroid cells occurs in the relatively short hemoglobinization stage of erythroid precursors.

Here we review current thinking and data on regulatory mechanisms for heme synthesis in animal cells. Regulation of tetrapyrrole synthesis in plants is reviewed in Chapter 15, and the complex and diverse regulation of microbial tetrapyrrole synthesis has been reviewed recently[3] and is outside the realm of this review. Likewise the crucial role played by heme oxygenases in overall heme homeostasis is not a subject for this chapter.

Regulation of Heme Biosynthesis by ALA Synthase

The first committed step in heme synthesis is catalyzed by 5-aminolevulinate synthase (ALAS) whose properties are reviewed in Chapter 2. While the dogma referred to in the Introduction generally states that ALAS is the first and rate-limiting step, it is clear that it is not the sole regulatory point in the pathway. Nonetheless this has been the most studied of all potential sites of regulation and is the one for which the largest body of data exists.

In animal cells there are two ALAS genes. The enzyme induced during erythroid differentiation is referred to as ALAS-E (erythroid) or ALAS-2. The enzyme present in nonerythroid cells is referred to as ALAS-H (housekeeping), ALAS-N (nonerythroid) or ALAS-1. In humans the gene encoding ALAS-1 is located on chromosome 3[4,5] and that encoding ALAS-2 on the X chromosome.[4,6] The recognition of two such distinctly regulated genes by Riddle et al in 1989[2] was a turning point in studies of heme regulation. Their data provided the seminal clue to unraveling the diverse and at times apparently contradictory data that existed in the literature. The key was that studies done on reticulocytes or other erythroid precursor cells reflected the distinct erythroid regulatory elements and other studies on liver or other nonerythroid cells reflected the housekeeping regulation.

The ALAS-1 and ALAS-2 proteins are highly similar although they differ noticeably near their amino-terminal ends.[7] Most significant, however, is that the regulatory regions of the two genes and the untranslated regions of the respective mRNAs are distinctly different.[8] These differences reflect the tissue-specific expression of the two ALAS enzymes.

ALAS-2

The human ALAS-2 gene is composed of eleven exons spread over a 22 Kb length.[9] Translation is initiated in exon 2 with exon 1 containing a 5' iron responsive element (IRE).[10-12] The mitochondrial targeting sequence is coded in exon 2. In all forms of ALAS the catalytic core is encoded by exons 5 to 11.[9] Interestingly, splice variants have been found in human (exon 4)[9] and mouse (exon 3).[13] The biological role of these variants is not known.

Much of what is currently known about mammalian ALAS-2 gene regulation comes from the work of Mays' group on human ALAS-2 (reviewed recently ref. 14). Regulation of ALAS-2 is complex and not yet completely understood (Fig. 1). It is induced only during erythroid differentiation and this program is erythropoietin dependent.[15] Examination of erythroid differentiation in cultured murine erythroleukemia (MEL) cells[16] and mouse embryonic stem (ES) cells[17,18] gives an interesting overview of the process. ALAS-1 diminishes as ALAS-2 activity is induced. Indeed, if ALAS-2 activity is knocked out, cell death occurs following induction of erythroid differentiation since insufficient ALAS activity exists for synthesis of heme.[17] However, induction of an exogenously regulated ALAS-2 alone, in the absence of the normal erythroid differentiation program, does not result in erythroid differentiation. So it is clear that ALAS-2 induction is necessary, but not sufficient in itself to support normal erythroid differentiation.

Transcriptional Regulation

Regulation of human ALAS-2 occurs in part at the transcriptional level and a number of erythroid-specific transcriptional factor binding sites have been identified in the 5'-untranslated region (UTR) as well as in several introns. In the 5'-UTR, the usual variety of erythroid-specific

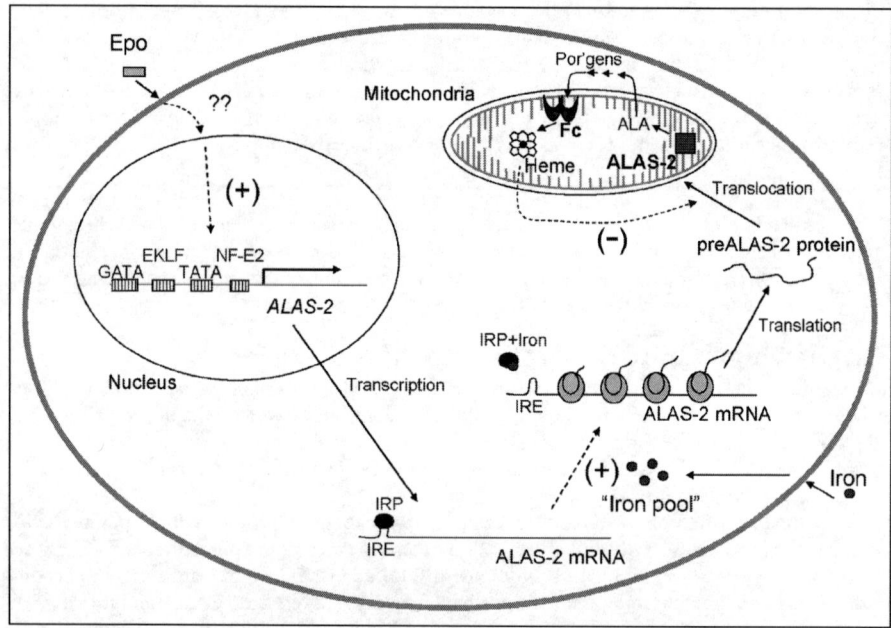

Figure 1. Regulation of ALAS-2. Transcription of *ALAS-2* is induced via erythropoietin (Epo). Translation of the ALAS-2 is inhibited in the absence of iron by iron regulatory protein (IRP) interacting with the 5' IRE. This translational inhibition is relieved when iron is present. The ALAS-2 protein is translocated to the mitochondria where it catalyzes the first step of the heme biosynthetic pathway. Heme synthesis then occurs through the modification of a series of precursors and porphyrinogens (por'gens). The end product of the pathway heme is involved in regulation by inhibiting the translocation of ALAS-2 into the mitochondria. Points of regulation are indicated as dashed lines.

transcription factors including binding sites of GATA-1, erythroid Krüppel-like factor (EKLF), and nuclear factor erythroid-derived 2 (NF-E2) are found.[14] In addition, the -300 bp region of the human ALAS-2 gene contains erythroid-specific elements that support maximal expression of ALAS-2 mRNA in erythroid cells.[19] In vitro mutagenesis studies of the promoter identified functional GATA-1 and GATA-1/TATA-binding protein sites, as well as a site for EKLF. However, the putative NF-E2 site did not appear to be functional.[19] Several erythroid-specific enhancers have been identified in introns 1 and 8.[20] Within intron 8 functional sites for GATA-1 and Sp1 were identified by transient expression assays. These intron 8 transcriptional elements are strong erythroid-specific enhancers and are orientation dependent. Introns 1 and 3 were noted to promote minor stimulatory and marginally inhibitory functions respectively, but these effects are yet to be fully characterized. Furthermore, a series of transactivation experiments clearly demonstrated that unidentified erythroid-specific factors, in addition to those outlined above, are necessary for activation to occur.

Of interest is that comparison of the human intron 8 sequence with canine and rodent show extensive sequence similarity in the set of putative transcriptional factor binding sites.[20] This, along with comparison of their promoter regions, indicates that the overall regulation in these mammals must be similar. This is in contrast to what has been found in *Drosophila melanogaster* where a single ALAS gene is subject to quite different regulatory mechanisms.[21] Even with the current data on transcriptional regulation it is still not possible to present a clear and complete model for ALAS-2 induction. Other elements, such as chromatin structure and possible locus control region (LCR) interactions, remain to be characterized.

Translational Regulation

A second site of ALAS-2 regulation occurs at the translational level by iron. ALAS-2 mRNA possesses a 5'-IRE that responds in the anticipated fashion, i.e., ALAS-2 protein is maximally synthesized only under iron-sufficient conditions and is poorly synthesized under iron-deficient conditions.[12,22,23] As much has been published and excellent reviews on the role of the IRE in ALAS-2 regulation exist,[24] this topic will not be reviewed in detail here. In the current context it is important to realize that only ALAS-2 and not ALAS-1 contains an IRE and is sensitive to translational regulation by iron. Additionally, regulation of ALAS-2 by iron must be viewed on a whole body basis (even though it is an erythroid precursor cell phenomenon) since iron homeostasis is complex and multifactorial involving uptake, transit across lumenal cells, transport throughout the body, uptake by erythroid precursor cells, intracellular trafficking, and delivery to ferrochelatase for insertion into heme.

A further point of translational regulation is that of translocation. ALAS-2 (like ALAS-1) is nuclear-encoded, synthesized in precursor form (preALAS protein) in the cytoplasm and translocated into the mitochondrial matrix in a step that involves proteolytic processing to remove the amino-terminal targeting sequence.[25] Interestingly, putative heme-binding motifs have been found in the amino-terminal region of both ALAS-1 and ALAS-2 as well as in *D. melanogaster* ALAS.[21] In vitro experiments demonstrated that two of these motifs in the mitochondrial targeting sequence of ALAS-2 respond to micromolar concentrations of heme by inhibiting translocation of ALAS-2 into isolated mitochondria.[26] Recently it has been found that all three putative heme binding motifs in ALAS-1 function in heme regulation of intracellular translocation of ALAS-1 in cultured Hepa 1-6 cells (Dailey TA, Dailey HA. unpublished observations). Clearly, further investigations will elucidate the in vivo role of heme and iron in regulation of ALAS-2 and ALAS-1.

ALAS-1

Since ALAS-2 is induced only during erythroid differentiation, ALAS-1 is essential to heme synthesis in all nonerythroid cells (Fig. 2). As mentioned above ALAS-1 and ALAS-2 are highly similar proteins with respect to their catalytic core. However, they differ significantly at their amino-terminal ends.[7] Unlike ALAS-2, ALAS-1 does not possess an IRE in its mRNA. However, since there have been no definitive experiments on regulation of ALAS-1 by iron, it remains to be determined if ALAS-1 activity is sensitive to cellular or body iron stores.

Transcriptional Regulation

Studies on regulation of ALAS-1 have focused largely upon transcriptional regulation of the gene by the pathway end product heme. It was a long held tenet that a regulatory "heme pool" existed and that when an apoprotein, such as apocytochrome P450, was induced by specific agents, then its depletion of this "heme pool" in the formation of the holohemoprotein would result in derepression of ALAS-1.[27] It is now apparent that such a system of secondary control is overly simplistic. Coordinate regulation of both the production of heme and apohemoproteins must clearly exist in at least some systems.[28] Coordinate regulation of two parallel synthetic assembly lines is not only intellectually satisfying, but could also ensure stoichiometric production of apoprotein and cofactor (heme). Clearly, regulation involving a "heme pool" (either hypothetically, or in reality) might then serve as a fine point of control that would respond to elements such as holoenzyme turnover rates and cellular heme oxygenase levels.

Recently Roberts and Elder[29] examined transcriptional regulation of ALAS-1 in human and rat. Interestingly, as was found with ALAS-2, human ALAS-1 is subject to alternate splicing, although for ALAS-1 these splice variants occur only in the untranslated region of the mRNA. No tissue specificity was found for these splice variants. It was noted that the human ALAS-1 gene possesses an additional untranslated exon that is lacking in the rat ALAS-1 gene. The core promoter regions of both human and rat ALAS-1 contain putative Sp1, nuclear factor kappa B (NFκB), nuclear respiratory factor-1 (NRF-1), and TATA sites.[29] Experiments on the rat ALAS-1 promoter established the functionality of two NRF-1 motifs, suggesting a role for ALAS-1 in overall cellular regulation of mitochondrial biogenesis.[30] In rat ALAS-1 there is also some evidence for the presence of an enhancer element in intron 1, but this element(s) was not identified.

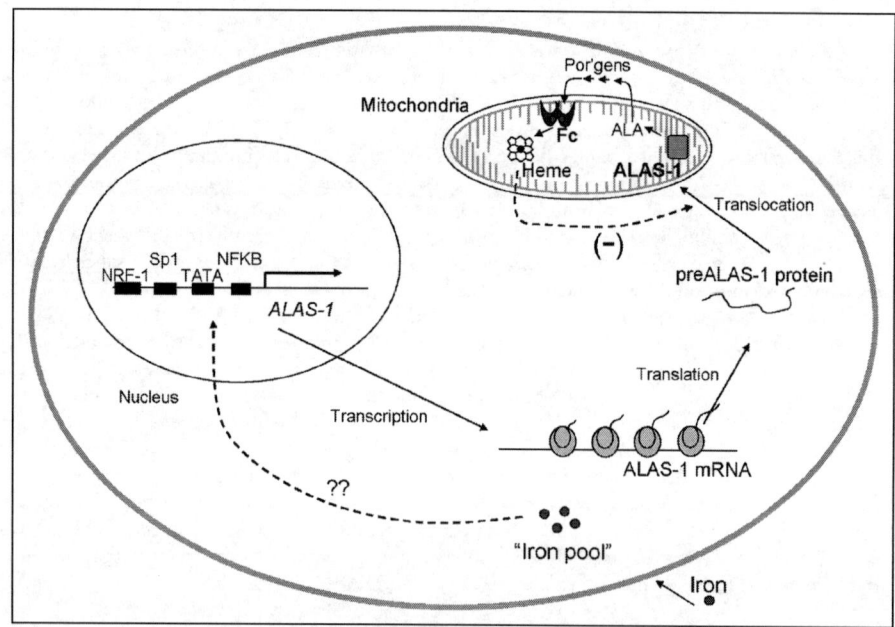

Figure 2. Regulation of ALAS-1. Transcription factor binding sites found in the promoter of ALAS-1 include NRF-1, Sp1, NFκB, and TATA box. Currently, no role for heme or iron has been demonstrated at the level of transcription. Heme is also thought to regulate translocation of the preALAS-1 protein into the mitochondria. Points of putative regulation are indicated as dashed lines.

Although there is no tissue specificity for the splice variants, it was found that tissue-specific differences exist for regulation of human ALAS-1.[29] In liver, initiation occurs downstream from a TATA box and regulation of ALAS-1 by cytochrome P450-stimulating agents (e.g., drugs) may be restricted to this region. In brain, however, initiation occurred from sites within and upstream from the TATA box. Based upon these findings it was proposed that even with the single promoter sequence of ALAS-1 in both liver and brain there are factors which result in tissue-specific expression. These currently unidentified factors may have an impact on the availability of the TATA box which appears necessary for the inducibility of this gene in some tissues such as liver and kidney. Unlike ALAS-2,[20] at the present time there are no studies to indicate that additional regulatory elements may exist in any internal introns other than the one report of an uncharacterized enhancer in rat intron 1.

ALAS Activity

It is important to note that the majority of studies on both ALAS-1 and ALAS-2 have been concerned with identification and characterization of transcriptional and translational elements along with studies on putative translocational features, but have not included actual enzyme activity studies. Since both ALAS proteins are synthesized as cytoplasmic precursor apoproteins, measurement of mRNA levels alone may not be informative about final ALAS activity if any posttranslational modifications and/or intracellular translocation and cofactor insertion are regulated. The relative lack of activity data reflects the difficulty of the ALAS assay and the need for significant quantities of material to conduct the procedure.

In one study where enzyme activity was reported during the induction of erythroid heme synthesis in MEL cells, data demonstrated a 14-fold increase in ALAS-2 mRNA in wild-type MEL cells, but no net increase in cellular ALAS activity.[31] In an ALAS-2 knockout cell line, induction resulted in a rapid and dramatic decrease in cellular ALAS activity which was proposed to reflect the repression of ALAS-1 synthesis in anticipation of the expected induction of ALAS-2.[17] These data were consistent with a previous study on MEL cells that specifically demonstrated a drop in ALAS-1 mRNA and immunodetectable ALAS-1 protein with a concomitant rise in ALAS-2 mRNA and protein following induction of erythroid differentiation.[32] Interestingly in this study there was a reported 15-fold increase in ALAS-2 mRNA but only a 2.5-fold increase in ALAS-2 protein compared with a 5-fold decrease in ALAS-1 protein. Actual ALAS enzyme activity was not reported in that study.

Regulation at Sites Other Than ALAS

While largely controlled by ALAS levels in all cell types, it has become clear that heme pathway regulation involves modulation of all heme biosynthetic enzymes. Abundant data exist to show upregulation of pathway enzymes during erythroid differentiation. All pathway enzyme genes possess both housekeeping and erythroid-specific promoter elements and a few exhibit erythroid-specific splice variants. However, no enzyme other than ALAS is represented by two distinct genes and no mRNA other than ALAS-2 has been found to possess an IRE for iron regulation.

Proximal Heme Synthetic Enzymes

Erythroid-specific promoter elements were first identified and characterized for porphobilinogen deaminase (PBGD) (also named hydroxymethylbilane synthase).[33] This gene contains erythroid-specific and housekeeping promoter elements embedded within a single promoter region (Fig. 3). The proteins encoded by these two promoters differ in their first exon via a promoter-specific splice variant.[34] For the housekeeping protein exon 1 is utilized, while exon 2 is skipped. This results in a mature protein that is 17 amino acid residues longer than the erythroid PBGD. For erythroid-specific expression exon 1 is skipped and the noncoding exon 2 is transcribed. The translational initiation site of the erythroid-specific form is in exon 3. The difference in the PBGD proteins thus expressed does not appear to have an effect on enzyme activity.[34] Recently it has been reported that there is a splice variant in human erythroid PBGD.[35] In this variant the intron between exons 2 and 3 is transcribed and has been found in a number of erythroid-related libraries. It is not known if this transcript gives rise to protein or if such a protein has altered activity or intracellular location.

The mouse and human PBGD erythroid-specific promoters have both similarities and differences.[33,36] They do not contain TATA or CAAT motifs, but do have CACCC motifs and a GF-1-binding site. Interestingly, neither promoters possess GATA-1-binding sites. Human possesses NF-E2 and CCAAT-binding site sequences which are absent in the mouse. In mouse the CACCC motif is duplicated while in human there exists only one. In in vitro reporter assays it was found that minimal erythroid-specific expression requires only the GF-1-binding site and duplicated CACCC motifs for mouse PBGD.[37] In human PBGD the GF-1-binding site along with the NF-E2-binding site are sufficient for erythroid-specific expression.[33]

The promoter region of mammalian housekeeping PBGD possesses two Sp1 sites but does not contain either TATA or CAAT boxes.[36] The housekeeping promoter is located upstream of the erythroid-specific promoter. There is evidence that another level of control exists during erythroid expression in which elements in the erythroid-specific region promote premature termination of housekeeping PBGD transcripts.

In vivo studies in differentiating MEL[16,31] and mouse ES cells[17,18] demonstrated that induction of PBGD mRNA occurs in both normal and ALAS-2 knockout cell lines although the level of induction in the knockout lines is somewhat less. PBGD enzyme activity in MEL cells increases

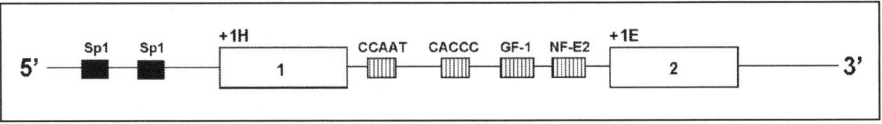

Figure 3. Promoter region of PBGD housekeeping and erythroid genes. Transcription factor binding sites for the erythroid specific promoter (striped squares) reside upstream of exon 2. The transcriptional start site is designated +1E for the erythroid specific form of the enzyme. The Sp1 sites (black squares) involved in regulation of transcription of the housekeeping form are located upstream of exon 1. The transcriptional start site for the housekeeping form is in exon 1 is designated as +1H.

approximately 4-fold in wild-type cells and 2-fold in ALAS-2 knockout cells.[31] Since PBGD is synthesized as an apoprotein that requires a source of PBG to make its own dipyrrole cofactor, it is not clear if the diminished PBGD activity reflects a decrease in apoprotein synthesis or a decreased ability to synthesize its cofactor due to limitations in ALA synthesis.

One unexplored possibility is that PBGD may be subject to additional regulation by direct feedback inhibition of the enzyme by protoporphyrinogen, the product of the pathway enzyme coproporphyrinogen oxidase. It has been shown that PBGD activity is inhibited in vitro by this pathway intermediate and a role for this interaction in the genetic disease variegate porphyria has been suggested.[38] However, no role for this inhibition in normal cellular metabolism has been demonstrated.

In addition to PBGD, two other pathway enzymes possess genes that give rise to alternate transcripts. 5-Aminolevulinate dehydratase (ALAD; also named PBG synthase) and uroporphyrinogen III synthase (UROS; also called uroporphyrinogen cosynthase) both have promoters that possess housekeeping and erythroid-specific elements although neither of these have been studied as extensively as PBGD.

The human ALAD gene contains two noncoding exons, 1A and 1B, along with 11 coding exons[39] (Fig. 4). Tissue-specific expression involves differential utilization of exons 1A and 1B, but since these are both untranslated the protein products for both housekeeping and erythroid-specific expression are identical. The noncoding exon 1A represents the start of the housekeeping form of ALAD mRNA while 1B is present only in the erythroid ALAD mRNA. A promoter region upstream of 1A is GC rich and possesses Sp1, AP1, and CCAAT-binding motifs but no TATA-binding site. Upstream from 1B is a promoter region with GATA-1 binding motifs and several CACCC boxes. Promoter reporter studies demonstrated that the 1B promoter was not effective in nonerythroid cells, but is active in erythroid cell types while the 1A promoter is active in all cell types. One interesting finding in mice is that it is not unusual for there to be multiple gene copies of ALAD. Indeed, 7 of 24 wild-caught mice from around the world had multiple copies of the ALAD gene.[40]

For UROS there are two physically distinct promoter regions[41,42] (Fig. 5). A single TATA-less housekeeping promoter is present upstream of exon 1 and was reported to have putative sites for Sp1, NF1, AP1, Oct1 and NRF2. This promoter drives the transcription of exons 1 (A/B), 2B-10. The initiation site is in exon 2. Interestingly exon 1 possesses an internal splice site that results in transcripts that contain either the entire exon 1 (1A + 1B) or just the first portion of exon 1 (1A). In both cases the exon 1 portion splices to a site internal to exon 2 (2B). An erythroid-specific promoter is located just upstream from exon 2 in the intronic region between exons 1 and 2. This promoter region also lacks a TATA box but does possess multiple GATA-1 sites. Reporter assay analysis of this and the housekeeping promoter region exhibited the expected tissue-specific expression of the reporter-fusion protein. The erythroid-specific transcript of UROS starts at exon 2. Since exon 1 lacks an initiation site both housekeeping and erythroid-specific UROS proteins are identical.

Terminal Heme Synthetic Enzymes

The remaining pathway enzymes, uroporphyrinogen decarboxylase (UROD), coproprophyrinogen oxidase (CPO), protoporphyrinogen oxidase (PPO), and ferrochelatase (Fc), all possess single promoter regions that contain both housekeeping and erythroid-specific elements (Table 1). None exhibit alternate splicing variants, but some evidence exists for variations in polyadenylation sites for mouse Fc and PPO that give rise to two transcript sizes.

The human UROD gene is composed of 10 exons and intervening introns spread over 3Kb.[43] Less work has been done to characterize the UROD promoter region than has been carried out with most pathway enzyme gene promoters. It has been shown that a single major mRNA species is found in all tissue types and a simple Sp1 and pseudo-TATA box promoter exists to drive housekeeping expression.

Regulation of CPO represents an interesting and, at present, incomplete story. While it would seem most reasonable that regulation of a pathway would not involve significant modulation or limitation of intermediate steps, the observation that in many organisms, including mammals, there is an overproduction, or at least relative accumulation, of coproporphyrin.[44] This may suggest that either a physical limitation to efficient intracellular substrate transfer, or a limitation in enzyme activity at CPO must exist. Studies in differentiating MEL cells demonstrated that during erythroid differentiation CPO activity at first drops before rising slightly.[45] Concomitantly coproporphyrin concentrations in the culture medium increase.[46] Interestingly the mRNA levels were shown to

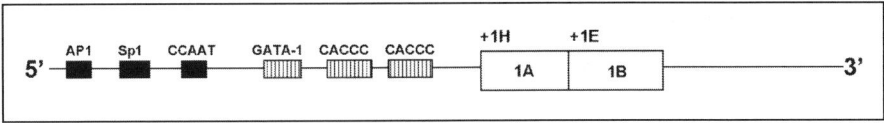

Figure 4. Promoter region of ALAD housekeeping and erythroid genes. The noncoding exon 1 is the start site of transcription of the housekeeping mRNA (+1H). Transcription factor binding sites found upstream of exon 1 A are shown as black squares. The start site for transcription of the erythroid form of the ALAD mRNA is the noncoding exon 1B (+1E). Transcription factor binding sites found upstream of exon 1 B are shown as striped squares.

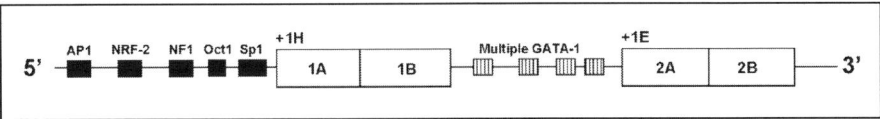

Figure 5. Promoter region of UROS housekeeping and erythroid genes. The promoter region for the housekeeping form resides upstream of exon 1 (1A/1B) and contains Sp1, NF1, and NRF-2 sites (shown as black squares). Transcription of the housekeeping form begins in exon 1 at +1H. The promoter region for the erythroid form of UROS resides between exons 1 and 2 (2A/2B) and contains multiple GATA-1 sites (striped squares). The transcriptional start site for the erythroid form is found in exon 2 and is designated by +1E.

increase without a drop, which suggests some possible posttranslational limitation or posttranslational regulation. In addition, expression of an exogeneously supplied and constitutively expressing mouse CPO in human K562 cells (an erythroleukemia cell line) resulted in cells that induced heme synthesis more rapidly and at a higher level than normal K562 cells.[47] These data along with the previous MEL cell data strongly suggest a role for CPO in regulation of heme synthesis. One additional feature of interest is that CPO possesses an unusually long leader sequence (120 amino acid residues), the function of which is not presently defined other than for mitochondrial targeting.

The gene for mouse CPO[48] has been examined and found to possess seven exons spanning approximately 11Kb, similar to that for human CPO.[49] Promoter analysis of the mouse CPO identified housekeeping and erythroid-specific elements within a single promoter region. An Sp1 site along with a GATA-1 binding site and a newly identified coproporphyrinogen oxidase [CPO] gene promoter regulatory element (CPRE) site were suggested to differentially regulate housekeeping vs. erythroid-specific expression of CPO through the action of as yet unidentified Factor X and Factor Y.[48]

The regulation of the penultimate and terminal enzymes of the pathway, PPO and Fc respectively, in differentiating MEL cells appear to be sensitive to regulation that is distinct from earlier pathway enzymes. In MEL cells lacking ALAS-2, neither of these enzyme activities increase following treatment of cells with DMSO to induce the erythroid program in contrast to all other pathway enzymes.[31] In an ALAS-2-deficient cell line that has had exogeneously supplied ALAS-2 and been induced with DMSO, the activities of both PPO and Fc are induced to wild-type levels. These data suggest that erythroid-specific induction of PPO and Fc require not only the erythroid-specific inducers that turn up synthesis of the earlier pathway genes, but also an additional factor, possibly heme itself.

The gene for both human[50] and mouse PPO[51] have been examined. Both contain 13 exons in a region approximately 4.2 Kb in size. Examination of the -600bp 5' flanking region of human PPO revealed putative Sp1, CCAAT and GATA-1 sites.[52] Promoter reporter assays of the -1.2 Kb 5' flanking region of mouse PPO identified a number of housekeeping and erythroid-specific elements.[51] The mouse PPO promoter lacks a TATA box and is not GC rich. The single promoter region contains two functional GATA-1 sites and an Sp1 site. Additional putative sites for NFκB, CAAT boxes, and drug and hormone elements exist. Erythroid-specific induction occurred when -1.2 Kb of the promoter region was present. However, inclusion of -5 Kb of 5' region resulted in decreased levels of reporter activity which suggested the possible presence of repressor elements in the -1 Kb to -5 Kb region.

Table 1. *Transcription factor binding sites found in the promoters of UROD, CPO, PPO, and ferrochleatase*

Enzyme	Transcription Factor Sites
UROD	Sp1
	pseudo TATA
CPO	GATA-1
	Sp1
	CPRE
PPO	Sp1
	CCAAT
	GATA-1
Ferrochelatase	CpG Island
	NF-E2
	GATA-1
	FKLF-2
	Sp1

The human ferrochelatase gene is composed of 11 exons covering a region of approximately 45 Kb.[53] Exon 1 contains both 5' untranslated sequence as well as the proposed first 22 amino-terminal residues. Exon 11 encodes the carboxyl-terminal domain as well as the entire 3'UTR. Evidence exists demonstrating the presence of alternative 3' polyadenylation sites in mouse ferrochelatase[54,55] and it appears that discrimination between these sites has something to do with erythroid vs nonerythroid cell expression.[54]

The single promoter region in Fc lacks both TATA and CAAT box motifs, but contains a CpG island in the region from -160 to +400 bp.[56] Transient expression studies demonstrate that an Sp1-driven promoter is sufficient for expression in nonerythroid cell lines and that a minimal promoter of 0.15 Kb, which lacks both the GATA-1 and NF-E2 elements, is sufficient to confer erythroid-specific expression. Interestingly, inclusion of the larger 4 Kb promoter fragment which contains a number of putative erythroid-specific elements failed to increase erythroid-specific expression over the level found with just the 0.15 Kb fragment. In transgenic mice, however, maximal erythroid-specific induction occurred with the 4.0 Kb region which may reflect organized chromatin structure.[57] In transgenic embryonic mouse cells, where a single copy of the reporter construct was inserted, it was found that developmentally-specific expression was obtained with the human ferrochelatase promoter fragment containing the Sp1, NF-E2 and GATA elements. In vivo erythroid specificity is presumably mediated via GATA-1 and NF-E2 elements present around -300 bp along with additional erythroid specific elements approximately -2 kb upstream from the transcriptional start site. FKLF-2, a novel Krüppel-like transcriptional factor, is also involved in activation of transcription of ferrochelatase.[59]

Studies with murine ferrochelatase in differentiating MEL cells suggests a putative erythroid-specific element in the -80 to -72 region.[60] A role for hepatocyte nuclear factor 1 alpha (HNF-1α) in regulation of mouse ferrochelatase has been suggested from studies with HNF-1α-null mice.[61]

One additional aspect of interest in considering the role of ferrochelatase in possible regulatory schemes relates to the presence of a [2Fe-2S] cluster in the mammalian enzyme.[62] The role of this cluster is not clear other than the fact that mutations which cause the loss of the cluster result in the loss of enzyme activity. Because assembly of the cluster is a necessary prerequisite for a functional holoenzyme, deficiency in cellular iron can result in diminished ferrochelatase activity.[63] Thus, while there is no iron responsive element or known gene regulation by iron, the [2Fe-2S] cluster becomes a defacto site of iron regulation of heme synthesis.

Conclusions

It is clear that while ALAS-1 and ALAS-2 are key regulators of heme synthesis via a variety of mechanisms, up regulation of the entire pathway during erythroid differentiation requires that all pathway genes possess both housekeeping and erythroid-specific promoter elements. Considering the key role that heme plays in cells as a cofactor in many redox reactions, along with the fact that iron is an essential component of heme, it is clear that regulation of heme synthesis must be viewed as an intricate global network rather than regulation at a single site by one regulatory element. Analysis of regulation of the pathway for individual cell types must therefore be considered in the context of the complete network. New players in the regulation of heme synthesis continue to arise and it is clear that we should still heed Lascelles' admonition to keep in mind "the possibility of entirely novel principles." Indeed, yet to be examined in detail are possible drug and hormonal responsive promoter elements whose existence is suggested by a variety of published observations, but for which no hard experimental data exist.

Acknowledgements

The authors wish to acknowledge P.N. Meissner at the University of Cape Town and T. Dailey at the University of Georgia for their helpful discussion and suggestions. Research in the laboratory of HAD was supported by NIH grant DK 32303.

References

1. Lascelles J. Tetrapyrrole Biosynthesis and Its Regulation. New York: WA Benjamin, 1964.
2. Riddle RD, Yamamoto M, Engel JD. Expression of delta-aminolevulinate synthase in avian cells: Separate genes encode erythroid-specific and nonspecific isozymes. Proc Natl Acad Sci USA 1989; 86:792-6.
3. O'Brian MR, Thony-Meyer L. Biochemistry, regulation and genomics of haem biosynthesis in prokaryotes. Adv Microb Physiol 2002; 46:257-318.
4. Bishop DF, Henderson AS, Astrin KH. Human delta-aminolevulinate synthase: Assignment of the housekeeping gene to 3p21 and the erythroid-specific gene to the X chromosome. Genomics 1990; 7:207-14.
5. Sutherland GR, Baker E, Callen DF et al. 5-Aminolevulinate synthase is at 3p21 and thus not the primary defect in X-linked sideroblastic anemia. Am J Hum Genet 1988; 43:331-5.
6. Cox TC, Bawden MJ, Abraham NG et al. Erythroid 5-aminolevulinate synthase is located on the X chromosome. Am J Hum Genet 1990; 46:107-11.
7. May BK, Bhasker CR, Bawden MJ et al. Molecular regulation of 5-aminolevulinate synthase. Diseases related to heme biosynthesis. Mol Biol Med 1990; 7:405-421.
8. May BK, Dogra SC, Sadlon TJ et al. Molecular regulation of heme biosynthesis in higher vertebrates. Prog Nucleic Acid Res Mol Biol 1995; 51:1-51.
9. Conboy JG, Cox TC, Bottomley SS et al. Human erythroid 5-aminolevulinate synthase. Gene structure and species-specific differences in alternative RNA splicing. J Biol Chem 1992; 267:18753-8.
10. Dandekar T, Stripecke R, Gray NK et al. Identification of a novel iron-responsive element in murine and human δ-aminolevulinic acid synthase mRNA. EMBO J 1991; 10:1903-9.
11. Cox TC, Bawden MJ, Martin A et al. Human erythroid 5-aminolevulinate synthase: Promoter analysis and identification of an iron-responsive element in the mRNA. EMBO J 1991; 10:1891-1902.
12. Dierks P. Molecular biology of eukaryotic 5-aminolevulinate synthase. Biosynthesis of Hemes and Chlorophylls. New York: McGraw Hill: 1990:201-233.
13. Schoenhaut DS, Curtis PJ. Structure of a mouse erythroid 5-aminolevulinate synthase gene and mapping of erythroid-specific DNAse I hypersensitive sites. Nucleic Acids Res 1989; 17:7013-28.
14. Sadlon TJ, Dell'Oso T, Surinya KH et al. Regulation of erythroid 5-aminolevulinate synthase expression during erythropoiesis. Int J Biochem Cell Biol 1999; 31:1153-67.
15. Zoller H, Decristoforo C, Weiss G. Erythroid 5-aminolevulinate synthase, ferrochelatase and DMT1 expression in erythroid progenitors: Differential pathways for erythropoietin and iron-dependent regulation. Brit J Haematol 2002; 118:619-26.
16. Fukuda Y, Fujita H, Garbaczewski L et al. Regulation of beta-globin mRNA accumulation by heme in dimethyl sulfoxide (DMSO)-sensitive and DMSO-resistant murine erythroleukemia cells. Blood 1994; 83:1662-7.

17. Yin X, Dailey HA. Erythroid 5-aminolevulinate synthase is required for erythroid differentiation in mouse embryonic stem cells. Blood Cells Mol Dis 1998; 24:41-53.
18. Harigae H, Suwabe N, Weinstock PH et al. Deficient heme and globin synthesis in embryonic stem cells lacking the erythroid-specific delta-aminolevulinate synthase gene. Blood 1998; 91:798-805.
19. Surinya KH, Cox TC, May BK. Transcriptional regulation of the human erythroid 5-aminolevulinate synthase gene. Identification of promoter elements and role of regulatory proteins. J Biol Chem 1997; 272:26585-94.
20. Surinya KH, Cox TC, May BK. Identification and characterization of a conserved erythroid-speciofic enhancer located in intron 8 of human 5-aminolevulinate synthase 2 gene. J Biol Chem 1998; 273:16798-809.
21. Ruiz de Mena I, Fernandez-Moreno MA, Bornstein B et al. Structure and regulated expression of the delta-aminolevulinate synthase gene from Drosophila melanogaster. J Biol Chem 1999; 274:37321-8.
22. Bhasker CR, Burgiel G, Neupert B et al. The putative iron-responsive element in the human erythroid 5-aminolevulinate synthase mRNA mediates translational control. J Biol Chem 1993; 268:12699-705.
23. Melefors O, Goossen B, Johansson HE et al. Translational control of 5-aminolevulinate synthase mRNA by iron-responsive elements in erythroid cells. J Biol Chem 1993; 268:5974-8.
24. Ponka P. Cell biology of heme. Am J Med Sci 1999; 318:241-56.
25. Ades IZ, Harpe KG. Biogenesis of mitochondrial proteins: Identification of the mature and precursor forms of the subunit of δ-aminolevulinic acid synthase from embryonic chick liver. J Biol Chem 1981; 250:9329-33.
26. Lathrop JT, Timko MP. Regulation by heme of mitochondrial protein transport through a conserved amino acid motif. Science 1993; 259:522-5.
27. Andrew TL, Riley PG, Dailey HA. Regulation of heme biosynthesis in higher animals. In: Dailey HA, ed. Biosynthesis of Heme and Chlorophylls. New York: McGraw-Hill, 1990:163-233.
28. Jover R, Hoffmann K, Meyer UA. Induction of 5-aminolevulinate synthase by drugs is independent of increased apocytochrome P450 synthesis. Biochem Biophys Res Commun 1996; 226:152-7.
29. Roberts AG, Elder GH. Alternative splicing and tissue-specific transcription of human and rodent ubiquitous 5-aminolevulinate synthase (ALAS1) genes. Biochim Biophys Acta 2001; 1518:95-105.
30. Li B, Holloszy JO, Semenkovich CF. Respiratory uncoupling induces delta-aminolevulinate synthase expression through a nuclear respiratory factor-1-dependent mechanism in HeLa cells. J Biol Chem 1999; 274:17534-40.
31. Lake-Bullock H, Dailey HA. Biphasic ordered induction of heme synthesis in differentiating murine erythroleukemia cells: Role of erythroid 5-aminolevulinate synthase. Mol Cell Biol 1993; 13:7122-32.
32. Fujita H, Yamamoto M, Yamajami T et al. Erythroleukemia differentiation. Distinctive responses of the erythroid-specific and the nonspecific 5-aminolevulinate synthase mRNA. J Biol Chem 1991; 266:17494-502.
33. Chretien S, Dubart A, Beaupain D et al. Alternative transcription and splicing of the human porphobilinogen deaminase gene result either in tissue-specific or in housekeeping expression. Proc Natl Acad Sci USA 1988; 85:6-10.
34. Grandchamp B, De Verneuil H, Beaumont C et al. Tissue-specific expression of porphobilinogen deaminase. Two isoenzymes from a single gene. Eur J Biochem 1987; 162:105-10.
35. Gubin AN, Miller JL. Human erythroid porphobilinogen deaminase exists in 2 splice variants. Blood 2001; 97:815-7.
36. Beaumont C, Porcher C, Picat C et al. The mouse porphobilinogen deaminase gene. Structural organization, sequence, and transcriptional analysis. J Biol Chem 1989; 264:14829-34.
37. Porcher C, Pitiot G, Plumb M et al. Characterization of hypersensitive sites, protein-binding motifs, and regulatory elements in both promoters of the mouse porphobilinogen deaminase gene. J Biol Chem 1991; 266:10562-69.
38. Meissner P, Adams P, Kirsch R. Allosteric inhibition of human lymphoblast and purified porphobilinogen deaminase by protoporphyrinogen and coproporphyrinogen. A possible mechanism for the acute attack of variegate porphyria. J Clin Invest 1993; 91:1436-44.
39. Kaya AH, Plewinska M, Wong DM et al. Human delta-aminolevulinate dehydratase (ALAD) gene: Structure and alternative splicing of the erythroid and housekeeping mRNAs. Genomics 1994; 19:242-8.
40. Bishop TR, Miller MW, Wang A et al. Multiple copies of the ALA-D gene are located at the Lv locus in Mus domesticus mice. Genomics 1998; 48:221-31.
41. Aizencang G, Solis C, Bishop DF et al. Human uroporphyrinogen-III synthase: Genomic organization, alternative promoters, and erythroid-specific expression. Genomics 2000; 70:223-31.

42. Aizencang GI, Bishop DF, Forrest D et al. Uroporphyrinogen III synthase. An alternative promoter controls erythroid-specific expression in the murine gene. J Biol Chem 2000; 275:2295-304.
43. Romana M, Dubart A, Beaupain D et al. Structure of the gene for human uroporphyrinogen decarboxylase. Nucleic Acids Res 1987; 15:7343-56.
44. Moore MR, McColl KEL, Rimington C et al. Disorders of Porphyrin Metabolism. New York: Plenum: 1987.
45. Conder L, Woodard SI, Dailey HA. Multiple mechanisms for the regulation of haem synthesis during erythroid cell differentiation. Possible role for coproporphyrinogen oxidase. Biochem J 1991; 275:321-6.
46. Woodard SI, Dailey HA. Multiple regulatory steps in erythroid heme biosynthesis. Arch Biochem Biophys 2000; 384:375-8.
47. Taketani S, Furukawa T, Furuyama K. Expression of coproporphyrinogen oxidase and synthesis of hemoglobin in human erythroleukemia K562 cells. Eur J Biochem 2001; 268:1705-11.
48. Takahashi S, Taketani S, Akasaka JE et al. Differential regulation of coproporphyrinogen oxidase gene between erythroid and nonerythroid cells. Blood 1998; 92:3436-44.
49. Delfau-Larue MH, Martasek P, Grandchamp B. Coproporphyrinogen oxidase: Gene organization and description of a mutation leading to exon 6 skipping. Hum Mol Genet 1994; 3:1325-30.
50. Roberts AG, Whatley SD, Daniels J et al. Partial characterization and assignment of the gene for protoporphyrinogen oxidase and variegate porphyria to human chromosome 1q23. Hum Mol Gen 1995; 4:2387-90.
51. Dailey TA, McManus JF, Dailey HA. Characterization of the mouse protoporphyrinogen oxidase gene. Cell Mol Biol 2002; 48:61-9.
52. Taketani S, Inazawa J, Abe T et al. The human protoporphyrinogen oxidase gene (PPOX): Organization and location to chromosome 1. Genomics 1995; 29:698-703.
53. Taketani S, Inazawa J, Nakahashi Y et al. Structure of the human ferrochelatase gene. Exon/intron gene organization and location of the gene to chromosome 18. Eur J Biochem 1992; 205:217-22.
54. Chan RY, Schulman HM, Ponka P. Expression of ferrochelatase mRNA in erythroid and nonerythroid cells. Biochem J 1993; 292:343-9.
55. Brenner DA, Frasier F. Cloning of murine ferrochelatase. Proc Natl Acad Sci USA 1991; 88:849-53.
56. Tugores A, Magness ST, Brenner DA. A single promoter directs both housekeeping and erythroid preferential expression of the human ferrochelatase gene. J Biol Chem 1994; 269:30789-97.
57. Magness ST, Tugores A, Diala ES et al. Analysis of the human ferrochelatase promoter in transgenic mice. Blood 1998; 92:320-8.
58. Magness ST, Tugores A, Brenner DA. Analysis of ferrochelatase expression during hematopoietic development of embryonic stem cells. Blood 2000; 95:3568-77.
59. Asano H, Li XS, Stamatoyannopoulos G. FKLF-2: A novel Krüppel-like transcriptional factor that activates globin and other erythroid lineage genes. Blood 2000; 95:3578-84.
60. Taketani S, Mohri T, Hioki K et al. Structure and regulation of the mouse ferrochelatase gene. Gene 1999; 227:117-124.
61. Muppala V, Lin CS, Lee YH. The role of HNF-1 alpha in controlling hepatic catalyase activity. Mol Pharm 2000; 57:93-100.
62. Dailey HA, Dailey TA, Wu CK et al. Ferrochelatase at the millennium: Structures, mechanisms and [2Fe-2S] clusters. Cell Mol Life Sci 2000; 57:1909-1926.
63. Taketani S, Adachi Y, Nakahashi Y. Regulation of the expression of human ferrochelatase by intracellular iron levels. Eur J Biochem 2000; 267:4685-4692.

CHAPTER 8

Tetrapyrroles in Photodynamic Therapy

David I. Vernon* and Ian Walker

Abstract

T he destructive photosensitising ability of porphyrins in patients suffering from porphyria has been well documented. Patients with those porphyrias that accumulate high levels of porphyrins in the skin suffer from a range of light-activated skin manifestations, the type and severity of which depends on the nature of the porphyrin that accumulates. In all cases the effect of solar irradiation of the porphyrin in the skin is to damage the tissue surface layers and its underlying structures. In most of the porphyrias this tissue damage can be severe in those areas of skin that are repeatedly exposed to sunlight and this is one of the major manifestations of these diseases. This inherent photosensitising ability of porphyrins can however be turned to good therapeutic use. The treatment of tumours and other lesions by the combined action of porphyrins, light and molecular oxygen has been developed into an effective technique called photodynamic therapy (PDT). The aim of this article is to give a brief overview of the development of tetrapyrroles in clinical photodynamic therapy and to report on those tetrapyrroles currently in clinical trials. There are several other classes of compounds being developed for PDT but this review will only highlight the tetrapyrrole based photosensitisers.

Brief History

Photodynamic therapy stemmed from studies in the early 1960s by Lipson and Baldes[1] who used a derivative of haematoporphyrin (Fig. 1) as a fluorescent stain for neoplastic tissue.[2] It was subsequently shown that this material not only had a potential use for the delineation and diagnosis of tumours but also had an inherent photosensitising ability that could be used in the eradication of the cancerous tissue. The first published clinical use of this technique was in 1976 when Kelly and Snell described the PDT of carcinoma of the bladder.[3] For reviews of the historical aspects of PDT see Bonnett,[4] Ackroyd,[5] Moan and Peng[6] and Dolmans.[7]

It was not until 1993 that the first photosensitiser gained regulatory approval. Photofrin®, an haematoporphyrin based photosensitiser preparation, was approved in Canada for the treatment of recurrent superficial bladder papilloma. It gained FDA approval in the USA in 1995 for treatment of obstructive oesophageal cancer and has subsequently been approved in several European countries. It was approved for treatment of advanced lung and oesophageal cancer in the United Kingdom in 1999.

The use of photodynamic therapy has not been restricted to oncology but has also been developed for a number of therapies including the treatment of noncancerous conditions such as Barrett's Oesophagus, psoriasis and age-related macular degeneration (AMD), one of the causes of blindness. Recently there has also been considerable interest in the photodynamic treatment of localised bacterial infection, due to the increase in the incidence of infections by antibiotic resistant bacteria.

*Corresponding Author: David I. Vernon—Institute of Molecular and Cell Biology, Department of Biological Sciences, University of Leeds, Leeds LS2 9JT, UK. Email: d.i.vernon@leeds.ac.uk

Tetrapyrroles: Birth, Life and Death, edited by Martin J. Warren and Alison G. Smith.
©2009 Landes Bioscience and Springer Science+Business Media.

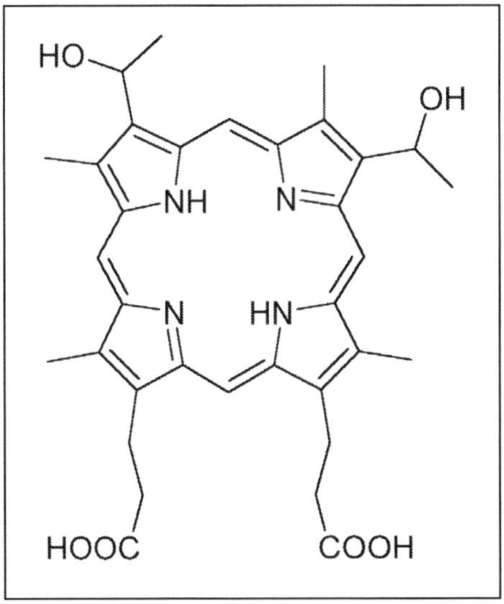

Figure 1. Structure of haematoporphyrin IX.

Singlet Oxygen: The Cytotoxic Agent

Photosensitising compounds in their lowest singlet energy state (S_o) absorb energy, in the form of photons, and enter a short-lived but excited singlet state (S_1*). In this state one of a pair of electrons is energised into a higher energy orbital. The extra energy from this state can be released again as photons of a lower energy (longer wavelength) giving rise to fluorescence, as demonstrated by the characteristic red fluorescence of the natural porphyrins. Alternatively the excited molecule can undergo an electron spin inversion (intersystem crossing) to give rise to a longer-lived triplet state (T_1). Normally electrons in each orbital are in pairs each possessing an opposite spin. According to the Pauli exclusion principle, electrons with the same spin cannot be in the same orbital. The inverted spin electrons in the triplet molecules are therefore trapped in separate orbitals. In general triplet state molecules can only interact with other molecules in a triplet state. There are however only a few molecules that are found in a triplet ground state. One such molecule is molecular oxygen therefore a photosensitiser in an excited triplet state can interact with it. The transfer of energy from the sensitiser to oxygen, in what is termed a type II process, results in an excited singlet form of oxygen that is highly reactive. The photosensitiser returns to the normal singlet ground state, which can be reexcited by further photons and go through the cycle again. The sensitiser can therefore be described as catalysing the conversion of molecular oxygen from a relatively unreactive triplet state ($^3O_2$) to a highly reactive singlet state ($^1O_2$). Oxygen itself is unable to absorb the photons from visible light but can acquire this energy from the porphyrin in its excited triplet state. The energy required to make this transformation from ground-state molecular oxygen to its singlet state is 94 kilojoules per mole of oxygen and porphyrins in their triplet state have sufficient excess energy to enable this transformation.

Although all the tetrapyrrole based photosensitisers used in photodynamic therapy are good singlet oxygen generators in vitro, it is only very recently that suitable technology has been available to show that singlet oxygen can be generated in vivo.[8]

The photosensitiser in its triplet state can in certain circumstances interact directly with bio-molecules, or solvent, by electron or hydrogen atom transfer. This generally results in substrate radical or radical ion formation with the photosensitiser acting as an oxidant. This mechanism is termed a Type I mechanism and although it has been shown to occur under certain circumstances in PDT,[9] it is not thought to be the major mechanism.

Singlet Oxygen Targets

Singlet oxygen is an extremely reactive molecule and will react with many bio-molecules. It has been demonstrated that its diffusion distance from the point of origin within a cell is approximately 0.02 μm.[10] This suggests that it would rarely escape the cell in which it was produced. Moan and coworkers however have suggested that singlet oxygen mediated damage of one cell could affect its nearest neighbour.[11]

This short diffusion distance is a result of the high reactivity of singlet oxygen with a large variety of cellular components.[12] It will oxidise several amino acids particularly cysteine, histidine, tryptophan, methionine and tyrosine. Damage to these critical residues within proteins can therefore result in alteration in cellular function and viability. Singlet oxygen will also oxidise unsaturated lipids initiating a lipid peroxidation cascade and ultimately causing membrane damage. Nucleosides, particularly guanosine, are also susceptible to singlet oxygen oxidation.[13]

Because of the high reactivity of singlet oxygen and therefore its short diffusion distance, any primary photodynamic action is restricted to those locations where the photosensitiser, light and oxygen are found. If any one of these components is missing, the photodynamic effect will not occur. This combination makes photodynamic therapy a local treatment, which is targeted by both the light delivery and the photosensitiser accumulation.

Light Delivery and Requirement

The activation of photosensitisers requires an efficient absorption of photons of a suitable wavelength. The absorption spectra of the porphyrins and chlorins have strong absorption bands around 400 nm (Soret band), but the penetration of light through tissue is strongly wavelength-dependent with red light (600-750 nm) being much more effective than shorter wavelengths. This is because of strong absorption by some tissue components such as haemoglobin or melanin at the lower wavelengths, as well as the fact that the scatter of photons decreases as wavelength increases.

The technological advances in lasers, particularly the recent development of small relatively inexpensive diode lasers, has enabled PDT to become much more widely available. In parallel the development of optical fibres of different geometries has allowed a variety of different organs in the body to be treated. Using endoscopy and optical fibre tips the delivery of light to various tissues can be achieved in a uniform and predetermined dose. Cylindrical diffusing fibre tips are used for treating areas such as the oesophagus and bronchus, whereas spherical, or geometrically modified fibre tips can be used for treating the bladder, uterus and brain. Straight cut fibres and microlens fibres are used to irradiate flat areas such as the skin.

The advantage of using lasers is that they can be used to deliver the required light dose down an optical fibre. However where large external areas are to be treated noncoherent polychromatic light sources may be used and several of these sources are now being developed.[14]

Photodynamic Damage

The overall effect of the photodynamic treatment of a lesion is the destruction of tumour tissue and complete healing of the area treated. In ideal circumstances this would result in the complete eradication of the tumour. The early processes involved in tumour photo-destruction are not completely understood even after almost 30 years of clinical use. There are known to be two main mechanisms by which tumours are initially injured during the photodynamic action, a direct tumour cell damaging effect and an indirect cell-killing due to vascular damage.[15] Both of these mechanisms can be demonstrated following most PDT protocols, but the mechanism that dominates varies with a number of factors such as the type of photosensitiser used, and the drug-to-light interval.

The blood vessels in tumours have been shown to shut down rapidly during photodynamic treatment[16,17] and this initial vasoconstriction may be followed by thrombosis and a longer term blockage (stasis) of the tumour vessels.[18,19] This stasis results in the tumour being starved of oxygen and nutrients, eventually resulting in its death. Initial rapid vasoconstriction however is thought to limit the photodynamic damage due to the hypoxia created in the tumour hence decreasing the production of further singlet oxygen. Recent studies on the effect of the irradiation regime has suggested that delivering the light dose at lower fluence rates can reduce this initial vasoconstriction whilst producing good tissue damage.[20] PDT at very short times after injection of a photosensitiser gives rise predominantly to a vascular effect and this has been exploited for the treatment of age-related macular degeneration (AMD), a condition involving abnormal blood vessel development in the eye.

The direct effect of PDT on the tumour cells results in cell death by two possible mechanisms, necrosis and programmed cell death (apoptosis). Necrosis generally occurs in response to external physical or chemical damage and typically results in cell swelling, metabolic collapse and loss of membrane integrity. Apoptosis is a set of events that are programmed to bring about cell destruction and their subsequent removal from the organism. This pathway is particularly active during the early development of an organism. It is characterised by blebbing of the plasma membrane giving rise to the formation of apoptotic bodies containing organelles, the condensation of nuclear material and fragmentation of DNA. The cellular fragments are engulfed by macrophages and other cells and further degraded by the phagocytic pathway. This method of cellular disposal prevents the initiation of an inflammatory response which is generally associated with necrotic cell death.[21]

There is considerable evidence that apoptosis plays a role in tumour eradication by PDT[22] Studies in cultured cells[23] and in tumours[24] using a number of photosensitisers have shown that the apoptotic pathways can be activated at several points. Release of cytochrome c from mitochondria is one of the primary initiators of apoptosis[25] and it is known that those photosensitisers that localise predominantly in the mitochondria are efficient initiators of apoptosis following light activation.[23,26] Localisation in the mitochondria however, is not a prerequisite for an apoptotic mechanism of PDT death as sensitisers that are predominantly lysosomal or show a general membrane localisation can also lead to cell death by apoptosis due to activation of the pathway at different points.[27] Those sensitisers that show a general membrane localisation are thought however to give rise predominantly to death of the cells by necrosis.[23,28]

Mechanisms of Tumour and Cellular Uptake

It is often stated in the literature that tumours accumulate photosensitisers to a greater extent than the normal surrounding tissue. This has proved to be the case in almost all animal tumour models where transplantable tumours are used. Where chemically induced tumours have been studied however, little tumour selectivity has been observed.[29] Unfortunately, there is very little patient data on the levels of photosensitiser present in tumours and the surrounding normal tissues of their origin. Ethical considerations have made it difficult to obtain both normal tissue as well as tumour tissue data. In a limited study of patients with lung or bronchial tumours, the levels of Photofrin® were found to be, at best, twice the levels of the normal surrounding tissue, with the magnitude and specificity of this being time-dependent.[30]

With the exception of brain tumours it has proven difficult in patients to get a significantly greater accumulation of photosensitiser in the tumour compared to the surrounding tissue, but the use of pro-drugs such as aminolaevulinic acid and its esters have dramatically overcome this problem.[31]

It has been shown that after intravenous administration, hydrophobic photosensitisers such as Photofrin® and Verteporfin are predominantly associated with the serum lipoproteins, particularly the Low (LDL) and High (HDL) density lipoproteins.[32,33] Association with LDL and subsequent uptake into tissues via receptor mediated endocytosis was initially thought to be the main mechanism of accumulation of these sensitisers in tissues.[34,35] Rapidly dividing tissues, including fast growing tumours, have higher numbers of LDL receptors than the normal surrounding tissues[34,36] and as a result could accumulate LDL bound sensitisers more rapidly. This is thought to be only part of the story, as several studies have now shown that the bio-distribution of the hydrophobic sensitisers does not necessarily reflect lipoprotein receptor levels.[37] Studies on the hydrophobic sensitiser, tin etiopurpurin, have demonstrated that a significant reduction in plasma lipoprotein levels does not alter the bio-distribution or tissue levels of the sensitiser,[38] suggesting that there are multiple pathways for the localisation of hydrophobic sensitisers. Hydrophilic sensitisers such as N-aspartylchlorin e_6 (Talaporfin) and haematoporphyrin bind predominantly to serum albumin and possibly other nonlipoprotein fractions with some binding to high density lipoprotein.[39,40] These sensitisers still show a good PDT response however. This may be due to the fact that sensitiser-protein complexes can be sequestered in the tumour as a consequence of poor lymphatic drainage and that the PDT effect is predominantly vascular. A similar situation may arise for hydrophobic sensitisers bound to the lipoproteins, in addition to sequestration by LDLs.

Tetrapyrroles in Photodynamic Therapy

There are many porphyrin and tetrapyrrole-based photosensitisers currently under investigation that have the correct physical and photochemical attributes required for a good photodynamic effect. Only a few of these however have suitable biological activity and there is not always a correlation between in vitro and in vivo efficacy. As a consequence only a few photosensitisers have progressed into clinical trials.

Haematoporphyrin Derivative (HpD) and Photofrin

The first photosensitiser to be used in clinical PDT research was called haematoporphyrin derivative. Haematoporphyrin itself (Fig. 1) is a powerful photosensitiser but in its purified form is not a good PDT agent due to its poor retention in tumours. Schwartz and coworkers developed a preparation of haematoporphyrin with improved tumour localisation.[41] This material was called haematoporphyrin derivative and was prepared in a two-step process. Initial acetylation of haematoporphyrin using 5% sulphuric acid in acetic acid followed by isolation of the product as a solid and subsequent alkaline hydrolysis and neutralisation, was used to prepare an injectable form of haematoporphyrin. This material was subsequently shown to be a complex mixture. The product of the acetylation step was identified as a mixture predominantly of di-acetylated haematoporphyrin but also contained mono-acetylated porphyrins and a number of dehydration products including protoporphyrin.[42] When dissolved in alkali at room temperature and then neutralised, an even more complex mixture was produced, containing not only a number of monomeric porphyrins (consisting predominantly of haematoporphyrin and its dehydration products) but also an oligomeric fraction.[43] This haematoporphyrin derivative (HpD) was the material used in many of the early clinical PDT studies. The discovery in 1982[44] that it was the oligomeric fraction that showed the greatest biological PDT activity led to the development of an HpD preparation that was enriched in this oligomeric fraction. This complex mixture which consists of a number of porphyrin molecules linked together via ester, ether and direct carbon-carbon bonds[45] (Fig. 2) was named Photofrin, and despite its complexity it became the first drug approved as a licensed PDT agent in Canada in 1993. Photofrin has subsequently been approved for a number of clinical indications including cancer of the bladder, oesophagus and lung,[46] and it has been used to treat many thousands of patients since it was first developed with undoubted success in many cases.[47] PDT using Photofrin has subsequently been used for treatment of other indications including nonsmall cell lung cancer with pleural spread,[48] cervical intraepithelial neoplasia[49] and pituitary adenoma.[50] It does however suffer from a number of drawbacks that have provided the impetus to develop new and more efficient photosensitisers.

Photofrin is normally photo-activated at 630 nm in order to maximise the light penetration through the tissue, but the specific absorption coefficient of this sensitiser at this wavelength is relatively low so the PDT efficiency is also relatively low. Another drawback of Photofrin is that it can cause prolonged skin photosensitivity. Pharmacokinetic studies in patients have shown that the active oligomeric component has a biological half-life of about 19 days.[51,52] This leads to prolonged skin photosensitivity, and patients may be photosensitive for several weeks after treatment, although good patient management can reduce this to an acceptable level.

The optical penetration depth of 630 nm light (the depth in tissue at which the light intensity is reduced to 37% of that at the surface, $1/e$) depends on the type of tissue but it is generally of the order of 1-4 mm.[53,54]

This depth increases as the wavelength of the light increases up to the point (about 850 nm) where light absorption by water becomes significant.[55] The increasing acceptability of PDT as a therapy brought about primarily by FDA approval of Photofrin with the less than ideal properties of Photofrin has led to a search for new and improved photosensitisers.

The Ideal Properties of a Photosensitiser

There are many properties of photosensitisers that need to be considered when developing them for clinical use. Not only are the chemical, photophysical and biological properties important to consider but there are also licensing, economic, and commercial aspects to keep in mind. The specifications required for new photosensitisers to reach the market place have changed as regulatory bodies, such as the Food and Drug Administration (FDA) in the USA, have become more stringent in their assessment of new drugs. The main considerations in photosensitiser design include:

- **Good singlet oxygen production.** It is generally accepted that singlet oxygen is the primary cytotoxic species produced during the photodynamic effect, although other reactive oxygen species (ROS) such as superoxide, hydroxyl radical and peroxides can be produced downstream. A good photosensitiser should therefore be an efficient producer of singlet oxygen with a good quantum yield.
- **Good spectral characteristics.** Photosensitisers with a high specific absorption coefficient at wavelengths longer than 630 nm, and preferably above 700 nm should be significantly better than Photofrin. The greater tissue penetration of the light needed to activate these sensitisers together with a higher absorbance at those wavelengths would facilitate the treatment of much larger tumours.

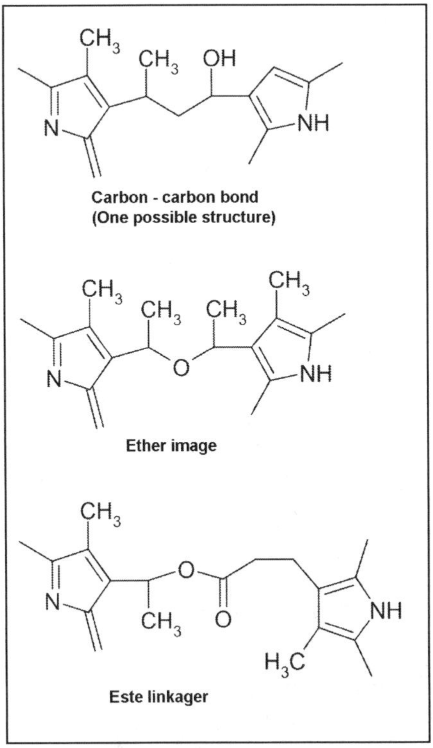

Carbon - carbon bond
(One possible structure)

Ether image

Este linkager

Figure 2. Examples of the types of bonds found in Photofrin. One pyrrole from each porphyrin monomer has been drawn.

In the bladder, for example, the optical penetration depth of light at 693 nm has been shown to be around 40% greater than at 633 nm.[54]

• **Suitable Pharmacokinetic Parameters.** The photosensitiser should show good tumour localisation and selectivity. Although selectivity of the PDT effect is also governed by the specificity of the light delivery, higher concentrations of the photosensitiser in the tumour compared to the normal surrounding tissue would minimise normal tissue damage. Sensitisers that have a low skin accumulation and a rapid elimination from the body would minimise the unwanted prolonged photosensitivity.

• **Purity and ease of synthesis.** There are many scientific and commercial difficulties encountered when developing complex drug mixtures for human administration. As legislation and the procedures for licensing of drugs becomes more rigorous it will be necessary to characterise all of the components of these mixtures for their pharmacological and toxic effects. Under the new regulations development of Photofrin would probably not have been considered as economical. The development of so called "second generation sensitisers" therefore will require the sensitisers to be single, pure and well characterised compounds. From a commercial point of view the compounds should be relatively simple to synthesise with little involvement of hazardous intermediates.

• **Low dark toxicity.** It is important that the photosensitiser shows no toxicity in the absence of the activating light. The advantage of PDT over chemotherapy is that in the absence of light the drug is harmless. The only side-effect encountered with Photofrin is a prolonged skin photosensitivity but this can generally be avoided by minimising exposure to bright sunshine and using sun blocks.

Second Generation Photosensitisers

There have been several hundred potential photosensitising drugs synthesised world-wide, all of which have suitable photo-physical properties but unfortunately many have no, or poor, biological activity. Those that show activity in cells in culture, often do not have suitable activity or tumour pharmacokinetics when tested in vivo. A large number of porphyrin-based photosensitisers have been considered for use in photodynamic therapy[56] but only few of these have gone into clinical trials. Those based on the tetrapyrrole nucleus are discussed below.

5,10,15,20 Tetrakis (meso-hydroxyphenyl) Chlorin (m-THPC, Foscan, temoporfin)

Berenbaum, Bonnett and coworkers developed a series of tetra hydroxyphenyl porphyrins with the hydoxyl group in either the para, ortho or meso position of the phenol ring. These showed good photosensitising properties[57] and subsequent biological testing was able to demonstrate that the meso isomer was the most active being almost 30 times more effective than HpD. The longest wavelength suitable for activation of this sensitiser in vivo is 648 nm and slightly red shifted compared to HpD (630 nm). The specific absorption coefficient was not significantly different.[58] Reduction of the porphyrin to the corresponding chlorin (Fig. 3) improves the photophysical properties considerably.[59] The lowest energy absorption peak is further red shifted to 652 nm and its specific absorption coefficient (22,000 $M^{-1}cm^{-1}$) is increased by a factor of about 20 compared to Photofrin.[60]

Preclinical studies with m-THPC have suggested that it is over 100 times more effective than Photofrin in terms of the photodynamic dose required i.e., drug dose x light dose.[61,62] The reasons for this effectiveness cannot be explained by the enhanced photophysical properties alone and must be due to as yet unidentified biological factors. m-THPC is a hydrophobic molecule and extensive studies have shown that it binds to the serum lipoproteins particularly HDL and LDL.[63] In human plasma it is thought to redistribute to the lipoproteins from an initial lipoprotein free complex containing highly aggregated sensitiser.[64]

Scotia Pharmaceuticals in the United Kingdom developed m-THPC for treatment of cancers of the head and neck, oesophagus and lung, particularly mesothelioma. Under the trade name Foscan®, it was approved in 2001 in the European Union, Norway and Iceland for the palliative treatment of advanced head and neck cancers. Foscan® is now being developed by Biolitec Pharma in Scotland for further clinical indications. The results of clinical trials have suggested that, for certain conditions, PDT with Foscan® could be considered as an alternative to surgery or radio-therapy.[65] Foscan has been used for the treatment of high grade dysplasia and early cancer in Barrett's oesophagus.[66] In a multi-centre study of 121 patients with oral squamous cell carcinoma m-THPC-PDT was shown to produce a complete response in 85% of the cases at twelve months and 77% at twenty four months. A major advantage of this treatment over surgery is the excellent cosmetic outcome.[67] Foscan has also been used for the treatment of head and neck tumours using interstitial PDT.[68]

Although the biological half-life of Foscan® is much shorter than that of Photofrin[69,70] it is still relatively long, and photosensitivity issues have delayed its development for use in other clinical areas.

5,10,15,20 Tetrakis (meso-hydroxyphenyl) Bacteriochlorin (m-THPBC)

Further reduction of the chlorin ring of m-THPC gives rise to the equivalent bacteriochlorin[59] (Fig. 3). Bacteriochlorins have the potential to be excellent PDT agents because of their high absorption in the near infra-red.[71-73] There is a considerable red shift in the longest wavelength Q band in converting the chlorin to the bacteriochlorin. m-THPBC shows strong absorption around 735 nm

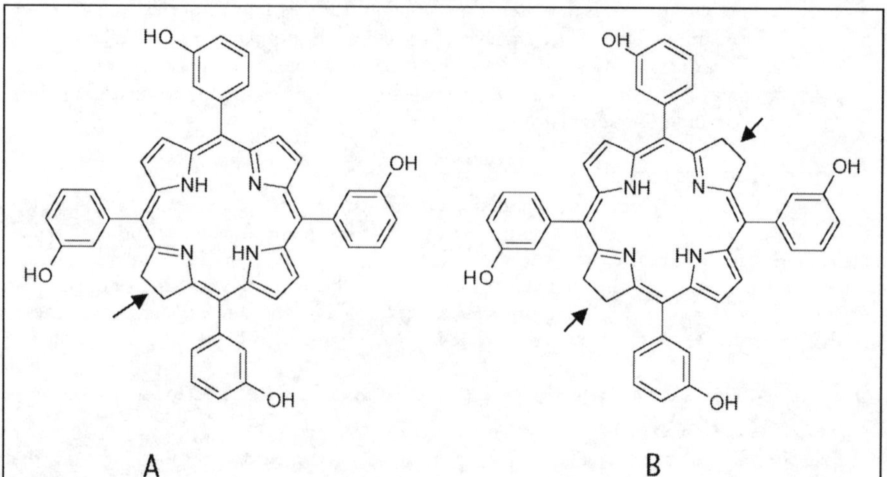

Figure 3. Structure of: A) m-tetra(hydroxyphenyl) chlorin; and B) m-tetra(hydroxyphenyl) bacteriochlorin (arrows indicate the reduced bonds).

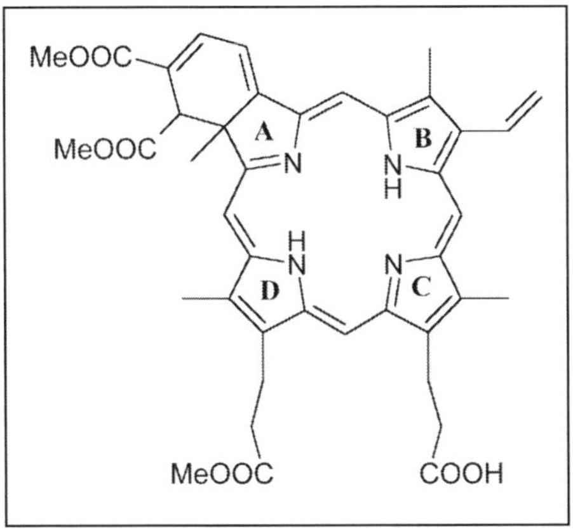

Figure 4. Structure of benzoporphyrin derivative.

as well as a six fold increase in the specific absorption coefficient from 22,400 $M^{-1}cm^{-1}$ to 136,000 $M^{-1}cm^{-1}$.[59]

A recent small Phase I clinical study using m-THPBC (SQ400) for the PDT treatment of liver metastases[74] showed that complete necrosis could be achieved within a radius of 1 cm of the optical fibre. The use of multiple optical fibres would therefore allow quite large tumours to be treated. The study also concluded that there were no significant side effects. The results from a larger trial of twenty four patients with colorectal liver metastases has supported these early findings.[75]

Benzoporphyrin Derivative (BPD, Verteporfin)

Benzoporphyrin derivative (Fig. 4) is a chlorin containing a cyclohexadiene ring fused to one of the pyrrole rings. It was first synthesised by David Dolphin and colleagues in the 1980s and was subsequently shown to be a good photosensitising agent.[76]

The original synthesis of BPD resulted in the formation of a di-acid derivative of the chlorin with the cyclohexadiene ring fused to ring B of the tetrapyrrole. This was formed by hydrolysis of the di-ester. Subsequent development gave rise to structural analogues containing only one free carboxylic acid and the cyclohexadiene ring fused to the ring A pyrrole.[33] This material was termed BPD mono-acid ring-A (BPD-MA) and it is this compound that has shown most promise as a PDT agent. As can be seen in Figure 4, there are two possible regio-isomers of this material differing only in the position of the free carboxyl group. It is this mixture of two isomers that has the generic name verteporfin.

Verteporfin has many of the characteristics of an ideal photosensitiser. It has a strong absorption peak at 686 nm with a specific absorption coefficient of around 34000 $M^{-1}cm^{-1}$.[77,78] The soret peak is also red shifted lowering the potential for UV induced skin photosensitivity. Like Foscan, verteporfin binds to serum lipoproteins and it has been shown in vitro that its uptake into cultured cells is almost exclusively via LDL receptor-mediated endocytosis.[79] In vivo however, other mechanisms may be important in tumour accumulation, particularly as it has been suggested that the main target of this sensitiser is the vasculature.[80] Another positive characteristic of verteporfin is that it is eliminated rapidly from the body, via a biliary route, and the retention in the skin is much lower than with Foscan.[29] The two regio-isomers have been shown to have slightly different tissue pharmacokinetic profiles.[81] Unfortunately the specificity of verteporfin for tumour tissue is not particularly good. Studied in animal models suggest that although significant levels of verteporfin can be achieved in tumour tissue the levels are not significantly greater than in the surrounding normal tissues.[29]

The results of clinical trials of the treatment of nonmelanoma skin cancers using BPD have been promising,[82] as have early clinical trials for the treatment of psoriasis and rheumatoid arthritis.

Although the selectivity of verteporfin for tumours is not ideal the observation that its mechanism of action is predominantly at the vascular level has led to one of the most significant developments in the treatment of age-related macular degeneration (AMD). This condition is the major cause of blindness in the over 50 age group. One form of this condition, the so-called wet form, is caused by new growth of abnormal blood vessels in the eye which are leaky and which eventually damage the macula, an area of the eye associated with central vision. There is no significantly effective routine treatment for this condition although several alternative therapies are currently being evaluated.

Verteporfin has been developed by QLT Photo Therapeutics Inc. and Novartis Ophthalmics as a liposomal preparation and given the name Visudyne™. This was launched in 2000 and is one of the most successful ophthalmology products. The treatment for AMD involves injection of the sensitiser then activation of it five minutes later using a precise dose of nonthermal light at 690 nm from a diode laser. Visudyne™ has now been used in several clinical trials, some of which are on-going but initial results are encouraging.[83] In a recent clinical trial verteporfin PDT has been used to treat choroidal

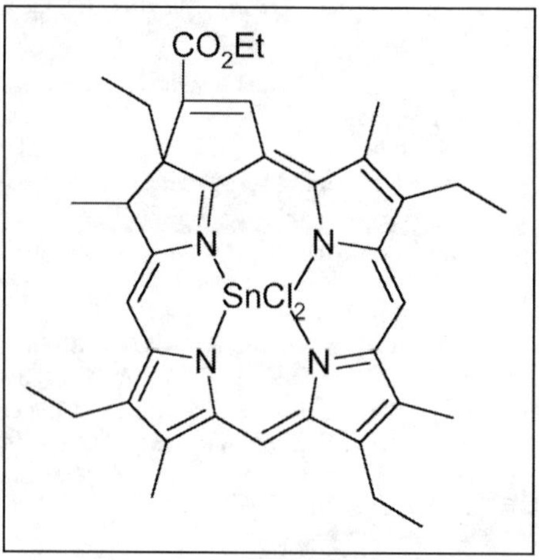

Figure 5. Structure of tin etiopurpurin.

hemangioma of the eye. The results from this trial demonstrate a good therapeutic response with some improvement in visual acuity.[84]

Tin Ethyl Etiopurpurin (SnEt₂, Purlytin, Rostaporfin)

The tin complex of ethyl etiopurpurin (Fig. 5) is a hydrophobic molecule that has several of the photophysical properties required of a good photosensitiser.[85] In methanol it has a strong absorption maximum at 656 nm with a specific absorption coefficient of 42,800 $M^{-1}cm^{-1}$. Because of its hydrophobic nature and its tendency to aggregate in aqueous solutions it has to be prepared as a Cremophor emulsion or similar formulation. It has been entered into a Phase II/III clinical trial for palliative treatment for patients presenting with recurrent metastatic breast cancer.[86] Patients were injected with tin ethyl etiopurpurin at a dose of 1.2 mg kg^{-1} and 24 hours later the tumours were irradiated with 660 ± 3 nm light at a total dose of 200 Joules/cm².

The results were encouraging with 92% of all lesions treated showing a complete initial response and 100% of lesions smaller than 0.5 cm showing a complete response. Although a long term response was not reported the data suggested that PDT using SnET₂ offers a good treatment for the local control of recurrent disease. SnET₂ has also been in clinical trials for PDT of skin cancer and AIDS-related Karposi's sarcoma.[87]

SnEt2 (rostaporfin) has been developed by Miravant Medical Technologies under the name Photrex for treatment of age-related macular degeneration (AMD) in competition with Verteporfin. It is in phase III development for the wet form of AMD, where it has shown a significant benefit to patients compared to placebo.

Mono-L-Aspartyl Chlorin e6 (Npe6, MACE, Talaporfin)

Another chlorin type photosensitiser is mono-L-aspartyl chlorin e6 (Fig. 6). This sensitiser has the advantage over some of the other chlorin sensitisers of being water soluble due to its four carboxylic acid groups. The parent tetrapyrrole, chlorin e6, is derivatised with aspartic acid and mono derivatisation results in three possible structural isomers. One of these (Fig. 6B) has been developed under the generic name talaporfin sodium. Like other chlorins Npe6 is a good singlet oxygen producer and has a good absorbance around 664 nm.[88] The specific absorption coefficient in methanol is 15,000 $M^{-1} cm^{-1}$ at 664 nm.

A small phase I clinical study of Npe6 for the treatment of cutaneous lesions including basal and squamous cell carcinomas as well as adenocarcinomas showed that the sensitiser was both safe and effective.[89] With a drug dose of around 3 mg kg^{-1} and a light dose of 100 Joules/cm², 66% of the tumour sites treated remained tumour free for the time of the study. One observation from this study suggested however that there was a lack of tumour selectivity at high sensitiser dose and some normal tissue damage was apparent. Although the follow up period for this study was short the data was sufficiently encouraging to warrant further evaluation. Another phase I clinical trial in patients with primary and secondary cancer of the skin and mucosal surfaces reports a 91% response rate. The photosensitiser produced little skin photosensitivity probably due to its rapid clearance from the circulation.[90]

A Phase II clinical study for early superficial squamous cell carcinoma in early stage lung cancer[91] confirmed that PDT with Npe6 was safe and gave excellent anti-tumour effects producing 85% complete tumour response in 35 patients. The authors concluded that use of Npe6 in combination with a suitable diode laser could become a standard modality for PDT of early superficial squamous cell carcinoma of the lung. More recently a multicentre phase I clinical study[92] has been carried out,

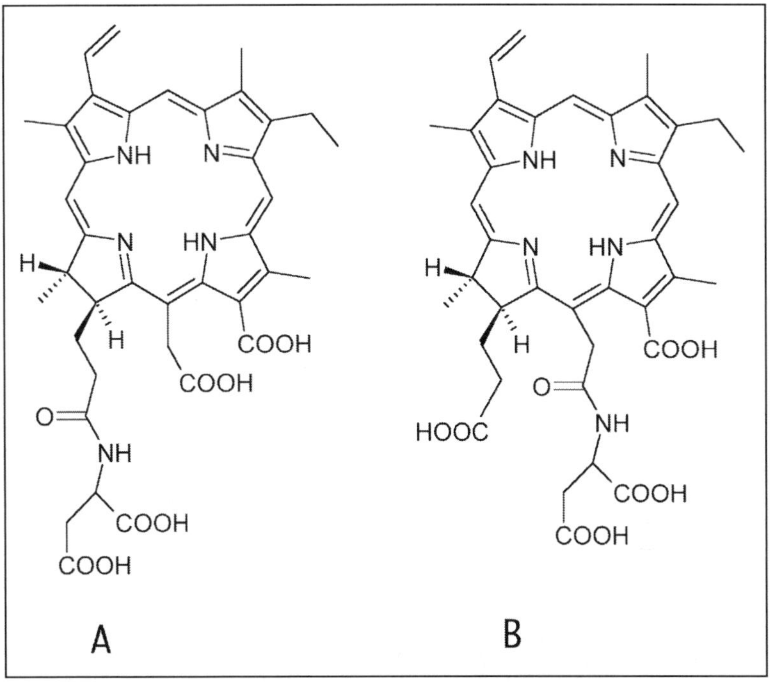

Figure 6. Structures of mono-L-aspartyl chlorin e6 (derivatisation on ring C is also possible).

using talaporfin sodium, to assess its safety for intratumoural PDT of a number of solid tumours. The study concluded that the treatment was safe and could offer patients with this type of resistant tumour another viable treatment option.

Palladium-Bacteriopheophorbide (TOOKAD, WST009)

Palladium-bacteriopheophorbide (Fig. 7) is a novel bacteriochlorophyll-a derived photosensitiser currently being developed by Steba Biotech and Negma-Lerands for the photodynamic treatment of prostate cancer. The first report of its photodynamic activity was from the group at the Weizmann Institute in Israel in 1999. Subsequent preclinical testing and development has demonstrated that it has many of the requirements of a good photosensitiser that is suitable for clinical use.

TOOKAD, as it is commonly known, has a very high extinction coefficient in the near infra-red and it has two main absorption maxima that can be utilised for photodynamic therapy. It is a hydrophobic sensitiser that is highly aggregated in aqueous media and therefore requires a disaggregating solvent formulation for efficient photodynamic activity. In the formulation developed for intravenous administration the photosensitiser (called WST09 in this solvent) has a major absorption peak at 762 nm with an extinction coefficient of 88,500 $M^{-1}cm^{-1}$ and a second suitable peak at 538 nm with an extinction coefficient of 23,100 $M^{-1}cm^{-1}$. It has been determined that the penetration depth (attenuation coefficient) of 762 nm light through canine prostate is about 2.5 times greater than that of 630 nm light used for Photofrin PDT. This therefore allows for much larger lesions to be treated.[93] Irradiation of the green absorption peak of TOOKAD (538 nm) could be utilised for superficial lesions where deep light penetration is not necessary. The quantum yield of singlet oxygen formed following Palladium-bacteriopheophorbide irradiation is very high but other reactive oxygen species such as hydroxyl radicals, superoxide and peroxides can be formed depending upon the cellular microenvironment.[94]

TOOKAD also has many of the biological characteristics required of a good photo-sensitising agent. A pharmacokinetic study in mice bearing the mammary adenocarcinoma EMT6 has shown

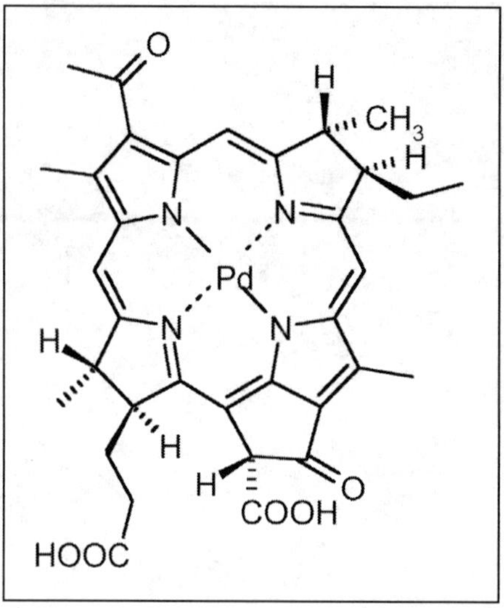

Figure 7. Structure of Pd-bacteriopheophorbide (TOOKAD).

that it is cleared from the body very quickly. The plasma levels of Pd-bacterio-pheophorbide (WST09) follow a biphasic decay with a very rapid initial half life of less than 2 minutes followed by a second short half-life of 1.3 hrs. The photo-sensitiser could not be observed in the tumour tissue at any time during this study suggesting that the photodynamic activity was more than likely due to an effect on the tumour vasculature and not as a result of direct tumour cell phototoxicity.[95] This hypothesis is supported by the observation that the optimum drug to light treatment interval for TOOKAD in a hamster cheek pouch model was as early as ten minutes post administration and that there was no response at four hours.[96] Similar photodynamic treatment of human prostatic small cell carcinoma xenographs using TOOKAD have shown good long term cures of relatively large tumours using a protocol whereby the light dose was administered immediately following drug administration. This further supports the idea that TOOKAD (WST009) acts by directly effecting the vasculature.[97]

Tookad is currently undergoing phaseI/II clinical trials for prostate cancer. The first reports of human clinical trials came from the groups in Canada and Israel in 2005 were dosimetry and initial efficacy were optimised.[98] TOOKAD was administered in a Cremophor-based formulation to patients with locally recurrent prostate cancer at doses up to 2 mg/kg. Light at 762 nm was delivered via interstitially placed optical fibres. Initial results have been promising and have demonstrated that TOOKAD-PDT could produce large volumes of necrosis in the prostate. In general the treatment was shown to be well tolerated with minimal systemic effects. Further clinical trials are currently underway.

2-[1-hexyloxyethyl]-2-Devinyl Pyropheophorbide-a (HPPH, Photochlor)

The chlorophyll-a derivative 2-[1-hexyloxyethyl]-2-devinyl pyropheophorbide-a (Fig. 8) is a hydrophobic chlorin-based photosensitiser derived from methyl pheophorbide-a extracted from *Spirulina* species.[99] The large specific absorption coefficient at 663 nm (~47,000 M^{-1} cm^{-1}) and the good singlet oxygen quantum yield (0.48)[100] makes it a good candidate for photodynamic therapy. Initial preclinical studies in a mouse model suggested that this sensitiser could produce significant tumour control. Mechanistic studies showed that the sensitiser was acting predominantly as a mediator of vascular photodamage and not causing substantial direct tumour cell phototoxicity.[101] The photosensitiser is now being developed at the Roswell Park Institute in Buffalo, New York as Photochlor and is being evaluated for PDT of basal cell carcinoma, obstructive oesophageal carcinoma and Barrett's oesophagus as well as early and late stage lung cancer. A recent pharmacokinetic and safety study in 25 patients[102] demonstrated that the treatment was well tolerated. Although sensitiser was detected in the plasma of these patients several months after administration there was no evidence of skin photosensitivity. Phase I/II clinical PDT studies using HPPH-PDT for the treatment of obstructive oesophageal cancer have been carried out as a palliative therapy. In these patients the sensitiser was administered at a dose of 0.15 mg kg^{-1} and the tumour irradiated endoscopically 48hours later with 665 nm light at a dose of 150 Joules cm^{-1}. The treatment was shown to improve the quality of life in six out of eight of these patients. Treatment of basal cell carcinomas resulted in most of the lesions giving a complete response at a drug to light interval of 24 hrs and initial data for the treatment of lung tumours are also encouraging.[102] A subsequent study of forty eight patients showed that Photochlor produced much lower skin photosensitisation than either Foscan or Photofrin. This was thought to be a result of rapid plasma clearance.[102,103]

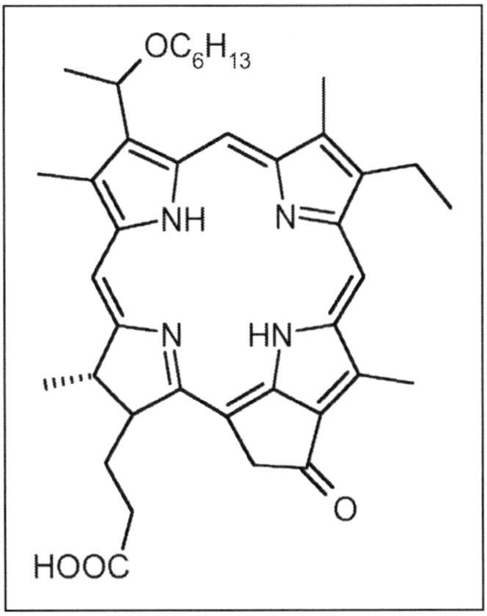

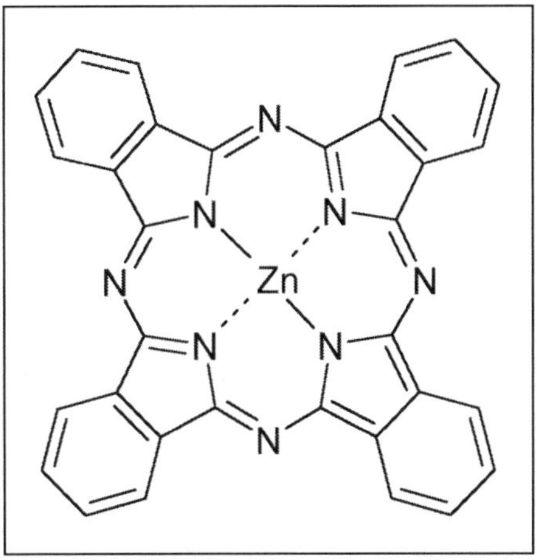

Figure 9. Structure of unsubstituted zinc phthalocyanine.

Figure 8. Structure of 2-[1-hexyloxyethyl]-2-devinyl pyropheophorbide-a.

Phthalocyanines

Phthalocyanines (Fig. 9) are synthetic tetrapyrrolic macrocyles that differ from porphyrins in that aza nitrogen atoms instead of methine bridges link their pyrrole rings. Phthalocyanines generally have high absorption in the 675-750 nm range of the spectrum due to the addition of benzene rings to the periphery of the macrocycle.[104] The macrocycle can chelate many elements and is easily modified.[105] It is known that a phthalocyanine chelated with a diamagnetic metal atom (such as aluminium or zinc) have extended triplet life-times and consequently a higher yield of singlet oxygen production. In contrast, phthalocyanines chelated with paramagnetic metals (such as iron and nickel) have very little photodynamic activity due to their shortened triplet lifetimes.[106] At present there is much interest in the phthalocyanine class of photosensitisers.

The Silicon-based phthalocyanine Pc4 (V.I Technologies), is currently being tested for the sterilisation of blood products[104] and also the treatment of neoplasms.[107] QLT in collaboration with Ciba-Geigy Ltd have developed a liposomal preparation of zinc phthalocyanine (CGP55847), which has shown promise in the treatment of squamous cell carcinomas.[107]

The Oncological Centre of the Russian Academy of Medical Sciences, in collaboration with the Moscow Medical Academy have used "Photosense", a mixture of aluminium sulphonated phthalocyanines (AlPcS$_n$) to treat many types of cancer including breast, skin, lip, lung and larynx. Their results as described elsewhere[107] have been encouraging.

Lutetium Texaphyrin (Lu-tex, Motexafin Lutetium)

Texaphyrins are not strictly tetrapyrroles but photosensitisers with an extended conjugation of the porphyrin ring system resulting in a strong absorbance peak at 732 nm making them suitable for the treatment of large or pigmented tumours. Lutetium texaphyrin, Lu-tex, (Fig. 10) has shown high selectivity for tumours, compared to normal skin, in subcutaneous melanoma lesions.[108] It is currently in phase I trials for recurrent prostate cancer and cervical intraepithelial neoplasia. Lu-tex is also currently in phase II clinical trials for treatment of breast cancer. Texaphyrin derivatives are also being developed by Pharmacyclics in the USA for use in the treatment of age-related macula degeneration (AMD) as the product motexafin lutetium (OPTRIN™).[104] A number of separate clinical trials are investigating the use of texaphyrin PDT for the prevention of restenosis and for the primary treatment of atherosclerotic lesions. Motexafin lutetium (as ANTRIN™) is also being developed for angioplasty of atherosclerotic cardiovascular disease. An open-label phase I trial in-

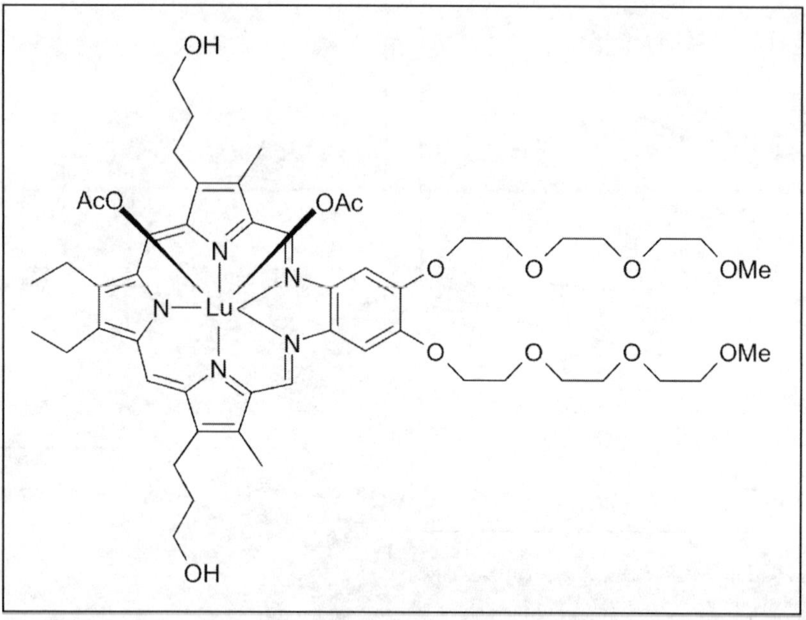

Figure 10. Structure of lutetium-texaphyrin, (Lu-tex).

volving eighty patients showed that the treatment of coronary atherosclerosis by motexafin lutetium-based PDT is safe and well tolerated.[109]

5-Aminolaevulinic Acid

The use of 5-aminolaevulinic acid (ALA) in photodynamic therapy is quite a different concept to that using the tetrapyrrole photosensitisers. ALA can be classed as a natural pro-drug as it is an early precursor of the natural porphyrins. Early human studies by Kennedy and Pottier[110] showed that topical administration of high concentrations of ALA overloads the haem biosynthetic pathway giving rise to a temporary accumulation of the intermediates of the haem biosynthetic pathway particularly protoporphyrin IX. The normal feedback inhibition of ALA synthase by haem is overcome by the addition of this excess ALA, and protoporphyrin accumulates. Furthermore, tumours and some rapidly dividing tissues are able to accumulate much more protoporphyrin than the surrounding tissue. The reasons for this are thought to be due to the differences in the activity of porphyrin biosynthetic pathway in tumours compared to normal tissues.[111] Measurement of the activity of the enzymes of the pathway in tumour tissue suggest that there is an increased activity of porphobilinogen deaminase together with a lower ferrochelatase activity. The ratio of the activity of these enzymes has been termed the power index and a high value for this increases the overall levels of protoporphyrin IX and decreases its metabolism to haem.[112] Another explanation is that the availability of ferrous iron is rate limiting preventing the efficient conversion of protoporphyrin to haem.[113] This accumulation of protoporphyrin is only transient and consequently the time between ALA administration and light treatment is critical. One of the advantages of this short window of opportunity is that there is no prolonged skin photosensitisation, a problem with many other photosensitisers. This favourable accumulation of protoporphyrin has been utilised and developed into a highly successful photodynamic treatment of certain superficial diseases. Many different treatment protocols have been tested but a typical ALA-PDT treatment for skin lesions involves the topical administration of 20% ALA in a oil-in-water emulsion. The ALA is left on the lesion for 3 to 4 hours during which time protoporphyrin accumulates to a therapeutic level. Irradiation of the area with a predetermined 630 nm light dose, usually lasting 10 to 20 minutes, is then carried out. During the light treatment some pain can be experienced, the severity of which depends not only on the site of treatment but also on the person being treated. In most cases pretreatment with some sort of analgesia overcomes this problem. The precise mechanism for the cause of this pain is not known but as it is only during the irradiation step of the treatment it is thought to be a photochemical effect probably due to accumulation of photosensitiser in specific nerve endings in the skin.

Clinical ALA-PDT

ALA-PDT has been used to treat a number of dermatological conditions.[114,115] Treatment of superficial epidermal premalignancies such as Bowen's disease (a squamous cell carcinoma) has typically achieved around 90% complete responses without recurrence and is now relatively routine.[116-119] A clinical trial comparing ALA-PDT to conventional cryotherapy[120] showed that PDT resulted in better clearance rates of the lesions than cryotherapy and was particularly better for treating larger areas. The cosmetic results observed following PDT were also much better due to excellent healing of the normal tissues. ALA-PDT has been used to treat actinic keratosis, a condition caused by chronic exposure of skin to sunlight and a condition that could lead to squamous cell carcinoma. DUSA Pharmaceuticals have developed Levulan® a stable preparation of ALA and this has now been approved by the FDA. A clinical Phase III trial of Levulan® PDT for actinic keratosis involving 243 patients and 1500 lesions showed around 90% complete response after two treatments. Clinical treatment of psoriasis using ALA-PDT has also been assessed [121-123] and the data obtained showed that although there was evidence of a therapeutic effect the overall conclusion was that the treatment was no better than other currently available therapies.

Basal cell carcinomas are small tumours that can be categorised into different types depending on whether they are nodular or invasive. ALA-PDT gives a good initial response with superficial BCCs after a single treatment[119,124] but long term follow-up studies suggest that this is not maintained and the recurrence rate can be as high as 50%.[117,125] In contrast, repeat treatments several days apart have improved the overall long term outcome.[126] The initial response of nodular BCCs to ALA-PDT however is much poorer. This failure is thought to be due to insufficient penetration of the ALA into the deeper regions of the lesion. Several methods have been used to try and increase the penetration of ALA. These have included the use of penetration enhancers such as dimethyl sulphoxide, tape stripping away the stratum corneum prior to ALA application and iontophoretic delivery.[127-129] The overall clinical effects of these have not been thoroughly investigated at present.

Other Applications of ALA-PDT

There are several nondermatological indications that have been treated with ALA-PDT. These include treatment of the uterus for the condition menhorragia,[130,131] treatment for cervical intraepithelial neoplasia (CIN), carcinoma in situ of the bladder and the metaplastic premalignant condition of the oesophagus called Barrett's Oesophagus. In the ALA-PDT treatment of Barrett's the ALA is given orally at a dose of around 30-60 mg kg^{-1}. The data from clinical studies using irradiation with either 635 nm red light[132] or 514 nm green light [133,134] are encouraging and have shown that ALA-PDT could become a suitable treatment for this condition.

Clinical data collected over the last few years has suggested that ALA-PDT has great potential for the treatment of some conditions but will probably be limited to superficial lesions. The treatment of larger tumours may be limited by the penetration of ALA through the tumour tissue. In order to try and overcome this problem ALA esters have been developed.[135,136]

Aminolaevulinic Acid Esters

Carboxylic acid esters of ALA can be thought of as preprodrugs. The increased lipophilicity of these compounds would enable them to cross cellular membranes more easily. Once inside the cell the esters are thought to be hydrolysed by nonspecific esterases releasing ALA, which then enters the haem biosynthetic pathway. In vitro studies on the formation of porphyrins in cells in culture have shown that the longer chain ALA esters, such as pentyl and hexyl esters, are far more efficient than ALA itself. The shorter alkyl esters such as the methyl, ethyl and propyl esters are no better than ALA alone.[135] The most efficient ester appears to be the hexyl ester which, in cultured cells, can produce the same amount of porphyrin at about one hundredth of the concentration. It has not been possible however, to demonstrate this dramatic improvement in skin models in vivo.[137]

ALA methyl ester (Metvix®) has been developed by a Norwegian company, Photocure ASA and a recent randomized clinical study[138] for treatment of actinic keratosis showed that the PDT treatment gave a 91% response rate whereas cryotherapy gave only a 68% response rate indicating that PDT using methyl aminolaevulinate is a good alternative particularly where large treatment areas are involved. Metvix® has also been used in the treatment of superficial and nodular basal cell carcinoma and under the protocol employed there was a 95% response rate. Clinical studies using ALA methyl ester for PDT treatment of basal cell carcinoma have also been encouraging.[139] It has recently been shown in a clinical trial involving two hundred and eleven patients that a single treatment with topical ALA-methyl ester-based PDT is an effective treatment for thin actinic keratosis lesions giving rise to a 93% complete response rate. Thicker lesions or nonresponding lesions were shown to benefit from repeat treatments, producing an 88% complete response rate.[140]

Photodetection of Tumours

The early work of Lipson and coworkers[2] used haematoporphyrin derivative as a fluorescent tumour marker that could be used to visualise tumours prior to surgery. HpD however did not have a particularly good specificity for the tumours and skin photosensitisation was also a problem. Because of this the photodiagnostic application of HpD was not studied with the same vigour as its photodynamic use.

The recent advances in photosensitiser development that have seen improvements in tumour selectivity and reductions in skin photosensitivity have brought about a renewed interest in photodiagnosis and photodetection. Foscan (m-THPC) has recently been used for the diagnosis and fluorescence guided resection of malignant brain tumours and initial patient data is encouraging.[141] The development of ALA and ALA esters for photodetection of tumours and lesions has continued over the last few years, and a number of centres and manufacturers are designing and producing specialised equipment for this purpose. The fact that highly fluorescent protoporphyrin is produced preferentially in tumours compared to the normal surrounding tissues when exogenous ALA is applied, has been utilised to develop techniques such as fluorescence guided surgery or biopsy acquisition.

Tumour photodetection and diagnosis using ALA has been assessed in a number of conditions including neoplasias of the bladder,[142] lung,[143] cervix,[144] and breast.[145] This fluorescence technique has been shown to be more specific and sensitive than conventional white light endoscopy.[146] More recently ALA esters have been employed for photo-detection and whereas ALA hexyl ester is not significantly better than ALA in dermatological PDT applications, it is considerably more efficient than ALA in the photo-detection of lesions in the bladder. A recent study[147] in 25 patients compared the ability of ALA and ALA hexyl ester to induce porphyrin fluorescence in neoplastic bladder tissue. The authors showed that, compared to ALA, the hexyl ester could produce a two-fold increase in protoporphyrin IX fluorescence at 5% of the dose . The hexyl ester is currently being developed by Photocure ASA as Hexvix® for photodetection in the bladder and is now approved in twenty six European countries. The benzyl ester (Benzvix®) is also being developed for similar procedures in gastroenterology.

Concluding Remarks

There are many photosensitisers being tested for use in photodynamic therapy but only a very few of these have progressed into clinical trials. Of these most are tetrapyrroles or closely related molecules. There are also several other tetrapyrrole-based sensitisers that show promise. These include Lemuteporfin (benzoporphyrin derivative 1,3-diene C,D-diethylene glycol ester A ring),[148] phthalocyanine PC4 (a silicon phthalocyanine),[149] BOPP (a boronated derivative of protoporphyrin)[150] and di-sulphonated aluminium phthalocyanine[151] to name only a few. As yet no clinical data has yet been published although some of these are currently undergoing clinical evaluation.

PDT is not a new therapy but has been developed over the last four decades to become a significant alternative approach to treatment of cancer and other diseases. Since the approval of Photofrin as the first photosensitiser and the subsequent accumulation of encouraging patient data there has been increased interest by a number of pharmaceutical companies in the development of new sensitisers and in the development of PDT as a therapy. It is true, I think, but not without bias that porphyrins and tetrapyrroles have paved the way for the acceptance of Photodynamic therapy as an alternative, and for some indications, the best treatment for certain cancers and other diseases.

Acknowledgements

We would like to thank Yorkshire Cancer Research for financial support.

Reference

1. Lipson RL, Baldes EJ. The photodynamic properties of a particular hematoporphyrin derivative. Arch Dermatol 1960; 82:508-516.
2. Lipson RL, Baldes EJ, Olsen AM. Hematoporphyrin derivative: A new aid for endoscopic detection of malignant disease. J Thorac Cardiovasc Surg 1961; 42:623-629.
3. Kelly JF, Snell ME. Hematoporphyrin derivative: A possible aid in the diagnosis and therapy of carcinoma of the bladder. J Urol 1976; 115:150-151.
4. Bonnett R. Photodynamic therapy in historical perspective. Rev Contemp Pharmacother 1999; 10:1-17.
5. Ackroyd R, Kelty C, Brown N et al. The history of photodetection and photodynamic therapy. Photochem Photobiol 2001; 74:656-669.
6. Moan J, Peng Q. An outline of the hundred-year history of PDT. Anticancer Res 2003; 23:3591-3600.
7. Dolmans DEJGJ, Fukumura D, Jain RK. Photodynamic therapy for cancer. Nat Rev Cancer 2003; 3:380-387.
8. Niedre M, Patterson MS, Wilson BC. Direct near-infrared luminescence detection of singlet oxygen generated by photodynamic therapy in cells in vitro and tissues in vivo. Photochem Photobiol 2002; 75:382-391.
9. Grossweiner LI, Patel AS, Grossweiner JB. Type I and type II mechanisms in the photosensitized lysis of phosphatidylcholine liposomes by hematoporphyrin. Photochem Photobiol 1982; 36:159-167.
10. Moan J, Berg K. The photodegradation of porphyrins in cells can be used to estimate the lifetime of singlet oxygen. Photochem Photobiol 1991; 53:549-553.
11. Dahle J, Angell-Petersen E, Steen HB et al. Bystander effects in cell death induced by photodynamic treatment UVA radiation and inhibitors of ATP synthesis. Photochem Photobiol 2001; 73:378-387.
12. Tuite EM, Kelly JM. Photochemical interactions of methylene blue and analogues with DNA and other biological substrates. J Photochem Photobiol B 1993; 21:103-124.
13. Dubbelman TMAR et al. Photodynamic therapy: Membrane and enzyme photobiology. Photodynamic Therapy: Basic Principles and Clinical Applications. New York: Marcel Dekker Inc., 1992:37-46.
14. Brancaleon L, Moseley H. Laser and nonlaser light sources for photodynamic therapy. Lasers Med Sci 2002; 17:173-186.
15. Henderson BW, Dougherty TJ. How does photodynamic therapy work? Photochem Photobiol 1992; 55:145-157.
16. Star WM, Marijnissen HP, van den Berg-Blok AE et al. Destruction of rat mammary tumor and normal tissue microcirculation by hematoporphyrin derivative photoradiation observed in vivo in sandwich observation chambers. Cancer Res 1986; 46:2532-2540.
17. Roberts DJH, Cairnduff F, Driver I et al. Tumor vascular shutdown following photodynamic therapy based on polyhematoporphyrin or 5-aminolevulinic acid. Int J Oncol 1994; 5:763-768.
18. Dolmans DE, Kadambi A, Hill JS et al. Vascular accumulation of a novel photosensitizer, MV6401, causes selective thrombosis in tumor vessels after photodynamic therapy. Cancer Res 2002; 62:2151-2156.
19. Krammer B. Vascular effects of photodynamic therapy. Anticancer Res 2001; 21:4271-4277.
20. Robinson DJ, de Bruijn HS, van d V et al. Protoporphyrin IX fluorescence photobleaching during ALA-mediated photodynamic therapy of UVB-induced tumors in hairless mouse skin. Photochem Photobiol 1999; 69:61-70.
21. Kanduc D, Mittelman A, Serpico R et al. Cell death: Apoptosis versus necrosis (review). Int J Oncol 2002; 21:165-170.
22. Luo Y, Chang CK, Kessel D. Rapid initiation of apoptosis by photodynamic therapy. Photochem Photobiol 1996; 63:528-534.
23. Kessel D, Luo Y, Deng Y et al. The role of subcellular localization in initiation of apoptosis by photodynamic therapy. Photochem Photobiol 1997; 65:422-426.
24. Zaidi SI, Oleinick NL, Zaim MT et al. Apoptosis during photodynamic therapy-induced ablation of RIF-1 tumors in C3H mice: Electron microscopic, histopathologic and biochemical evidence. Photochem Photobiol 1993; 58:771-776.
25. Petit PX, Susin SA, Zamzami N et al. Mitochondria and programmed cell death: Back to the future. FEBS Letters 1996; 396:7-13.
26. Kessel D, Luo Y. Mitochondrial photodamage and PDT-induced apoptosis. J Photochem Photobiol B 1998; 42:89-95.
27. Oleinick NL, Morris RL, Belichenko I. The role of apoptosis in response to photodynamic therapy: What, where, why, and how. Photochem Photobiol Sci 2002; 1:1-21.

28. Dellinger M. Apoptosis or necrosis following Photofrin photosensitization: Influence of the incubation protocol. Photochem Photobiol 1996; 64:182-187.

29. Blant SA, Ballini JP, van den Bergh H et al. Time-dependent biodistribution of tetra(m-hydroxyphenyl)chlorin and benzoporphyrin derivative monoacid ring A in the hamster model: Comparative fluorescence microscopy study. Photochem Photobiol 2000; 71:333-340.

30. Holroyd JA. The pharmacokinetics of the photosensitising drug polyhaematoporphyrin. PhD Thesis. University of Leeds, 1995.

31. Marti A, Jichlinski P, Lange N et al. Comparison of aminolevulinic acid and hexylester aminolevulinate induced protoporphyrin IX distribution in human bladder cancer. J Urol 2003; 170:428-432.

32. Kongshaug M. Distribution of tetrapyrrole photosensitizers among human plasma proteins. Int J Biochem 1992; 24:1239-1265.

33. Allison BA, Pritchard PH, Richter AM et al. The plasma distribution of benzoporphyrin derivative and the effects of plasma lipoproteins on its biodistribution. Photochem Photobiol 1990; 52:501-507.

34. Kessel D. Porphyrin-lipoprotein association as a factor in porphyrin localization. Cancer Lett 1986; 33:183-188.

35. Candide C, Morliere P, Maziere JC et al. In vitro interaction of the photoactive anticancer porphyrin derivative photofrin II with low density lipoprotein, and its delivery to cultured human fibroblasts. FEBS Lett 1986; 207:133-138.

36. Norata G, Canti G, Ricci L et al. In vivo assimilation of low density lipoproteins by a fibrosarcoma tumour line in mice. Cancer Lett 1984; 25:203-208.

37. Korbelik M. Low density lipoprotein receptor pathway in the delivery of Photofrin: How much is it relevant for selective accumulation of the photosensitizer in tumors? J Photochem Photobiol B 1992; 12:107-109.

38. Kessel D, Garbo GM, Hampton J. The role of lipoproteins in the distribution of tin etiopurpurin (SnET2) in the tumour-bearing rat. Photochem Photobiol 1993; 57:298-301.

39. Kessel D, Whitcomb KL, Schulz V. Lipoprotein-mediated distribution of N-aspartyl chlorin-E6 in the mouse. Photochem Photobiol 1992; 56:51-56.

40. Jori G. In vivo transport and pharmacokinetic behavior of tumour photosensitizers. Ciba Found Symp 1989; 146:78-86.

41. Lipson RL, Baldes EJ, Olsen AM. The use of a derivative of hematoporhyrin in tumor detection. J Natl Cancer Inst 1961; 26:1-11.

42. Bonnett R, Ridge RJ, Scourides PA et al. On the nature of hematoporphyrin derivative. J Chem Soc [Perkins 1] 1981; 3135-3140.

43. Kessel D, Cheng ML. Biological and biophysical properties of the tumor-localizing component of hematoporphyrin derivative. Cancer Res 1985; 45:3053-3057.

44. Berenbaum MC, Bonnett R, Scourides PA. In vivo biological activity of the components of haematoporphyrin derivative. Br J Cancer 1982; 45:571-581.

45. Byrne CJ, Marshallsay LV, Ward AD. The composition of Photofrin II. J Photochem Photobiol B 1990; 6:13-27.

46. Dougherty TJ. An update on photodynamic therapy applications. J Clin Laser Med Surg 2002; 20:3-7.

47. Dougherty TJ. Photodynamic therapy. Photochem Photobiol 1993; 58:895-900.

48. Friedberg JS, Mick R, Stevenson JP et al. Phase II trial of pleural photodynamic therapy and surgery for patients with nonsmall-cell lung cancer with pleural spread. J Clin Oncol 2004; 22:2192-2201.

49. Yamaguchi S, Tsuda H, Takemori M et al. Photodynamic therapy for cervical intraepithelial neoplasia. Oncology 2005; 69:110-116.

50. Marks PV, Belchetz PE, Saxena A et al. Effect of photodynamic therapy on recurrent pituitary adenomas: Clinical phase I/II trial—an early report. Br J Neurosurg 2000; 14:317-325.

51. Brown SB, Vernon DI, Holroyd JA et al. Pharmacokinetics of Photofrin in Man. Photodynamic Therapy and Biomedical Lasers. New York: Elsevier, 1992:475-479.

52. Bellnier DA, Dougherty TJ. A preliminary pharmacokinetic study of intravenous Photofrin in patients. J Clin Laser Med Surg 1996; 14:311-314.

53. Driver I, Lowdell CP, Ash DV. In vivo measurement of the optical interaction coefficients of human tumours at 630 nm. Phys Med Biol 1991; 36:805-813.

54. Shackley DC, Whitehurst C, Moore JV et al. Light penetration in bladder tissue: Implications for the intravesical photodynamic therapy of bladder tumours. BJU Int 2000; 86:638-643.

55. Wan S, Anderson RR, Parrish JA. Analytical modeling for the optical properties of the skin with in vitro and in vivo applications. Photochem Photobiol 1981; 34:493-499.

56. Sternberg ED, Dolphin D, Bruckner C. Porphyrin-based photosensitizers for use in photodynamic therapy. Tetrahedron 1998; 54:4151-4202.

57. Berenbaum MC, Akande SL, Bonnett R et al. meso-Tetra(hydroxyphenyl)porphyrins, a new class of potent tumour photosensitisers with favourable selectivity. Br J Cancer 1986; 54:717-725.

58. Berenbaum MC, Bonnett R. Tetra(hydroxyphenyl)porphyrins. Photodynamic Therapy of Neoplastic Disease. Vol. 2. Boca Raton: CRC Press Inc., 1990:169-177.
59. Bonnett R, White RD, Winfield UJ et al. Hydroporphyrins of the meso-tetra(hydroxyphenyl)porphyrin series as tumour photosensitizers. Biochem J 1989; 261:277-280.
60. Post JG, te Poele JA, Schuitmaker JJ et al. A comparison of functional bladder damage after intravesical photodynamic therapy with three different photosensitizers. Photochem Photobiol 1996; 63:314-321.
61. van Geel IP, Oppelaar H, Oussoren YG et al. Photosensitizing efficacy of MTHPC-PDT compared to photofrin-PDT in the RIF1 mouse tumour and normal skin. Int J Cancer 1995; 60:388-394.
62. Ball DJ, Vernon DI, Brown SB. The high photoactivity of m-THPC in photodynamic therapy. Unusually strong retention of m-THPC by RIF-1 cells in culture. Photochem Photobiol 1999; 69:360-363.
63. Michael Titus AT, Whelpton R, Yaqub Z. Binding of temoporfin to the lipoprotein fractions of human serum. Br J Clin Pharmacol 1995; 40:594-597.
64. Hopkinson HJ, Vernon DI, Brown SB. Identification and partial characterization of an unusual distribution of the photosensitizer meta-tetrahydroxyphenyl chlorin (temoporfin) in human plasma. Photochem Photobiol 1999; 69:482-488.
65. Copper MP, Tan IB, Oppelaar H et al. Meta-tetra(hydroxyphenyl)chlorin photodynamic therapy in early-stage squamous cell carcinoma of the head and neck. Arch Otolaryngol Head Neck Surg 2003; 129:709-711.
66. Lovat LB, Jamieson NF, Novelli MR et al. Photodynamic therapy with m-tetrahydroxyphenyl chlorin for high-grade dysplasia and early cancer in Barrett's columnar lined esophagus. Gastrointest Endosc 2005; 62:617-623.
67. Hopper C, Kubler A, Lewis H et al. mTHPC-mediated photodynamic therapy for early oral squamous cell carcinoma. Int J Cancer 2004; 111:138-146.
68. Lou PJ, Jager HR, Jones L et al. Interstitial photodynamic therapy as salvage treatment for recurrent head and neck cancer. Br J Cancer 2004; 91:441-446.
69. Jones HJ, Vernon DI, Brown SB. Photodynamic therapy effect of m-THPC (Foscan) in vivo: Correlation with pharmacokinetics. Br J Cancer 2003; 89:398-404.
70. Cramers P, Ruevekamp M, Oppelaar H et al. Foscan uptake and tissue distribution in relation to photodynamic efficacy. Br J Cancer 2003; 88:283-290.
71. Rovers JP, de Jode ML, Rezzoug H et al. In vivo photodynamic characteristics of the near-infrared photosensitizer 5,10,15,20-tetrakis(M-hydroxyphenyl) bacteriochlorin. Photochem Photobiol 2000; 72:358-364.
72. Bonnett R, Charlesworth P, Djelal BD et al. Photophysical properties of 5,10,15,20-tetrakis(m-hydroxyphenyl)porphyrin-(m-THPP), 5,10,15,20-tetrakis(m- hydroxyphenyl)chlorin (m-THPC) and 5,10,15,20-tetrakis(m-hydroxyphenyl)bacteriochlorin (m-THPBC): A comparative study. J Chem Soc, Perkin Transactions 2 1999; 325-328.
73. Rovers JP, de Jode ML, Grahn MF. Significantly increased lesion size by using the near-infrared photosensitizer 5,10,15,20-tetrakis (m-hydroxyphenyl)bacteriochlorin in interstitial photodynamic therapy of normal rat liver tissue. Lasers Surg Med 2000; 27:235-240.
74. Engelmann K, Mack MG, Eichler K et al. Interstitial photodynamic laser therapy for liver metastases: First results of a clinical phase I-study. Rofo-Fortschritte Auf dem Gebiet der Rontgenstrahlen und der Bildgebenden Verfahren 2003; 175:682-687.
75. van Duijnhoven FH, Rovers JP, Engelmann K et al. Photodynamic therapy with 5,10,15, 20-tetrakis(m-hydroxyphenyl) bacteriochlorin for colorectal liver metastases is safe and feasible: Results from a phase I study. Ann Surg Oncol 2005; 12:808-816.
76. Richter AM, Kelly B, Chow J et al. Preliminary studies on a more effective phototoxic agent than hematoporphyrin. J Natl Cancer Inst 1987; 79:1327-1332.
77. Aveline B, Hasan T, Redmond RW. Photophysical and photosensitizing properties of benzoporphyrin derivative monoacid ring A (BPD-MA). Photochem Photobiol 1994; 59:328-335.
78. Aveline BM, Hasan T, Redmond RW. The effects of aggregation, protein binding and cellular incorporation on the photophysical properties of benzoporphyrin derivative monoacid ring A (BPDMA). J Photochem Photobiol B 1995; 30:161-169.
79. Allison BA, Pritchard PH, Levy JG. Evidence for low-density lipoprotein receptor-mediated uptake of benzoporphyrin derivative. Br J Cancer 1994; 69:833-839.
80. Fingar VH, Kik PK, Haydon PS et al. Analysis of acute vascular damage after photodynamic therapy using benzoporphyrin derivative (BPD). Br J Cancer 1999; 79:1702-1708.
81. Richter AM, Jain AK, Canaan AJ et al. Photosensitizing efficiency of two regioisomers of the benzoporphyrin derivative monoacid ring A (BPD-MA). Biochem Pharmacol 1992; 43:2349-2358.

82. Lui H, Hobbs L, Tope WD et al. Photodynamic therapy of multiple nonmelanoma skin cancers with verteporfin and red light-emitting diodes: Two-year results evaluating tumor response and cosmetic outcomes. Arch Dermatol 2004; 140:26-32.

83. Brown SB, Mellish KJ. Verteporfin: A milestone in opthalmology and photodynamic therapy. Expert Opin Pharmacother 2001; 2:351-361.

84. Michels S, Michels R, Simader C et al. Verteporfin therapy for choroidal hemangioma: A long-term follow-up. Retina 2005; 25:697-703.

85. Pogue BW, Redmond RW, Trivedi N et al. Photophysical properties of tin ethyl etiopurpurin I (SnET2) and tin octaethylbenzochlorin (SnOEBC) in solution and bound to albumin. Photochem Photobiol 1998; 68:809-815.

86. Mang TS, Allison R, Hewson G et al. A phase II/III clinical study of tin ethyl etiopurpurin (Purlytin)-induced photodynamic therapy for the treatment of recurrent cutaneous metastatic breast cancer. Cancer J Sci Am 1998; 4:378-384.

87. Wilson BD, Bernstein Z, Sommer C et al. Photodynamic therapy for kaposis-sarcoma using photofrin and tin ethyl-etiopurpurin (Snet2). J Invest Dermatol 1995; 104:693.

88. Roberts WG, Shiau FY, Nelson JS et al. In vitro characterization of monoaspartyl chlorin e6 and diaspartyl chlorin e6 for photodynamic therapy. J Natl Cancer Inst 1988; 80:330-336.

89. Taber SW, Fingar VH, Coots CT et al. Photodynamic therapy using mono-L-aspartyl chlorin e6 (Npe6) for the treatment of cutaneous disease: A Phase I clinical study. Clin Cancer Res 1998; 4:2741-2746.

90. Chan AL, Juarez M, Allen R et al. Pharmacokinetics and clinical effects of mono-L-aspartyl chlorin e6 (NPe6) photodynamic therapy in adult patients with primary or secondary cancer of the skin and mucosal surfaces. Photodermatol Photoimmunol Photomed 2005; 21:72-78.

91. Kato H, Furukawa K, Sato M et al. Phase II clinical study of photodynamic therapy using mono-L-aspartyl chlorin e6 and diode laser for early superficial squamous cell carcinoma of the lung. Lung Cancer 2003; 42:103-111.

92. Lustig RA, Vogl TJ, Fromm D et al. A multicentre phase I safety study of intratumoral photoactivation of Talaporfin Sodium in patients with refractory solid tumors. Cancer 2003; 98:1767-1771.

93. Chen Q, Huang Z, Luck D et al. Preclinical studies in normal canine prostate of a novel palladium-bacteriopheophorbide (WST09) photosensitizer for photodynamic therapy of prostate cancers. Photochem Photobiol 2002; 76:438-445.

94. Vakrat-Haglili Y, Weiner L, Brumfeld V et al. The microenvironment effect on the generation of reactive oxygen species by Pd-bacteriopheophorbide. J Am Chem Soc 2005; 127:6487-6497.

95. Brun PH, DeGroot JL, Dickson EF et al. Determination of the in vivo pharmacokinetics of palladium-bacteriopheophorbide (WST09) in EMT6 tumour-bearing Balb/c mice using graphite furnace atomic absorption spectroscopy. Photochem Photobiol Sci 2004; 3:1006-1010.

96. Borle F, Radu A, Monnier P et al. Evaluation of the photosensitizer Tookad for photodynamic therapy on the Syrian golden hamster cheek pouch model: Light dose, drug dose and drug-light interval effects. Photochem Photobiol 2003; 78:377-383.

97. Koudinova NV, Pinthus JH, Brandis A et al. Photodynamic therapy with Pd-Bacteriopheophorbide (TOOKAD): Successful in vivo treatment of human prostatic small cell carcinoma xenografts. Int J Cancer 2003; 104:782-789.

98. Weersink RA, Bogaards A, Gertner M et al. Techniques for delivery and monitoring of TOOKAD (WST09)-mediated photodynamic therapy of the prostate: Clinical experience and practicalities. J Photochem Photobiol 2005; 79:211-222.

99. Pandey RK, Bellnier DA, Smith KM et al. Chlorin and porphyrin derivatives as potential photosensitizers in photodynamic therapy. Photochem Photobiol 1991; 53:65-72.

100. Pandey RK, Sumlin AB, Constantine S et al. Alkyl ether analogs of chlorophyll-a Derivatives: Part 1. Synthesis, photophysical properties and photodynamic efficacy. Photochem Photobiol 1996; 64:194-204.

101. Bellnier DA, Henderson BW, Pandey RK et al. Murine pharmacokinetics and antitumour efficacy of the photodynamic sensitizer 2-[1-hexyloxyethyl]-2-devinyl pyropheophorbide-a. J Photochem Photobiol B 1993; 20:55-61.

102. Bellnier DA, Greco WR, Loewen GM et al. Population pharmacokinetics of the photodynamic therapy agent 2-[1-hexyloxyethyl]-2-devinyl pyropheophorbide-a in cancer patients. Cancer Res 2003; 63:1806-1813.

103. Bellnier DA, Greco WR, Nava H et al. Mild skin photosensitivity in cancer patients following injection of Photochlor (2-[1-hexyloxyethyl]-2-devinyl pyropheophorbide-a; HPPH) for photodynamic therapy. Cancer Chemother Pharmacol 2005; 1-6.

104. Sharman WM, Allen CM, van Lier JE. Photodynamic therapeutics: Basic principles and clinical applications. Drug Discov Today 1999; 4:507-517.

105. Rosenthal I. Phthalocyanines as photodynamic sensitizers. Photochem Photobiol 1991; 53:859-870.
106. Chan WS, Marshall JF, Hart IR. Photodynamic therapy of a murine tumor following sensitisation with chloro aluminum sulfonated phthalocyanine. Photochem Photobiol 1987; 46:867-871.
107. Allen CM, Sharman WM, van Lier JE. Current status of phthalocyanines in the photodynamic therapy of cancer. Journal of Porphyrins and Phthalocyanines 2001; 5:161-169.
108. Dougherty TJ, Gomer CJ, Henderson BW et al. Photodynamic therapy. J Natl Cancer Inst 1998; 90:889-905.
109. Kereiakes DJ, Szyniszewski AM, Wahr D et al. Phase I drug and light dose-escalation trial of motexafin lutetium and far red light activation (phototherapy) in subjects with coronary artery disease undergoing percutaneous coronary intervention and stent deployment: Procedural and long-term results. Circulation 2003; 108:1310-1315.
110. Kennedy JC, Pottier RH. Endogenous protoporphyrin IX, a clinically useful photosensitizer for photodynamic therapy. J Photochem Photobiol B 1992; 14:275-292.
111. Batlle AM. Porphyrins, porphyrias, cancer and photodynamic therapy—a model for carcinogenesis. J Photochem Photobiol B 1993; 20:5-22.
112. Hinnen P, de Rooij FWM, Edixhoven A et al. Porphyrin biosynthesis in human Barrett's oesophagus and adenocarcinoma after ingestion of 5-aminolaevulinic acid. Br J Cancer 2000; 83:539-543.
113. Rittenhouse-Diakun K, Van Leengoed H, Morgan J et al. The role of transferrin receptor (CD71) in photodynamic therapy of activated and malignant lymphocytes using the heme precursor delta-aminolevulinic acid (ALA). Photochem Photobiol 1995; 61:523-528.
114. Van den Akker JTHM, Brown SB. Photodynamic Therapy based on 5-aminolevulinic acid: Applications in dermatology. Photobiology for the 21st Century. Overland Park: Valdenmar Publishing Company, 2002:165-181.
115. Kelty C, Brown NJ, Reed M et al. The use of 5-aminolaevulinic acid as a photosensitiser in photodynamic therapy and photodiagnosis. Photochem Photobiol Sci 2002; 1:158-168.
116. Kennedy JC, Pottier RH, Pross DC. Photodynamic therapy with endogenous protoporphyrin IX: Basic principles and present clinical experience. J Photochem Photobiol B 1990; 6:143-148.
117. Cairnduff F, Stringer MR, Hudson EJ et al. Superficial photodynamic therapy with topical 5-aminolaevulinic acid for superficial primary and secondary skin cancer. Br J Cancer 1994; 69:605-608.
118. Stables GI, Stringer MR, Robinson DJ et al. Large patches of Bowen's disease treated by topical aminolaevulinic acid photodynamic therapy. Br J Dermatol 1997; 136:957-960.
119. Svanberg K, Andersson T, Killander D et al. Photodynamic therapy of nonmelanoma malignant tumours of the skin using topical delta-amino levulinic acid sensitization and laser irradiation. Br J Dermatol 1994; 130:743-751.
120. Morton CA, Whitehurst C, Moseley H et al. Comparison of photodynamic therapy with cryotherapy in the treatment of Bowen's disease. Br J Dermatol 1996; 135:766-771.
121. Robinson DJ, Collins P, Stringer MR et al. Improved response of plaque psoriasis after multiple treatments with topical 5-aminolaevulinic acid photodynamic therapy. Acta Derm Venereol 1999; 79:451-455.
122. Collins P, Robinson DJ, Stringer MR et al. The variable response of plaque psoriasis after a single treatment with topical 5-aminolaevulinic acid photodynamic therapy. Br J Dermatol 1997; 137:743-749.
123. Schick E, Ruck A, Boehncke WH et al. Topical photodynamic therapy using methylene blue and 5-aminolaevulinic acid in psoriasis. J Dermatolog Treat 1997; 8:17-19.
124. Soler AM, Angell-Petersen E, Warloe T et al. Photodynamic therapy of superficial basal cell carcinoma with 5-aminolevulinic acid with dimethylsulfoxide and ethylendiaminetetraacetic acid: A comparison of two light sources. Photochem Photobiol 2000; 71:724-729.
125. Fink-Puches R, Soyer H, Hofer A et al. Long-term follow-up and histological changes of superficial nonmelanoma skin cancers treated with topical delta- aminolevulinic acid photodynamic therapy. Arch Dermatol 1998; 134:821-826.
126. Harth Y, Hirshowitz B, Kaplan B. Modified topical photodynamic therapy of superficial skin tumors, utilizing aminolevulinic acid, penetration enhancers, red light, and hyperthermia. Dermatol Surg 1998; 24:723-726.
127. Rhodes LE, Tsoukas MM, Anderson RR et al. Iontophoretic delivery of ALA provides a quantitative model for ALA pharmacokinetics and PpIX phototoxicity in human skin. J Invest Dermatol 1997; 108:87-91.
128. Peng QA, Warloe T, Moan J et al. Distribution of 5-aminolevulinic acid-induced porphyrins in noduloulcerative basal-cell carcinoma. Photochem Photobiol 1995; 62:906-913.
129. De Rosa FS, Marchetti JM, Thomazini JA et al. A vehicle for photodynamic therapy of skin cancer: Influence of dimethylsulphoxide on 5-aminolevulinic acid in vitro cutaneous permeation and in vivo protoporphyrin IX accumulation determined by confocal microscopy. J Control Release 2000; 65:359-366.

130. Gannon MJ, Johnson N, Roberts DJ et al. Photosensitization of the endometrium with topical 5-aminolevulinic acid. Am J Obstet Gynecol 1995; 173:1826-1828.
131. Wyss P, Caduff R, Tadir Y et al. Photodynamic endometrial ablation: Morphological study. Lasers Surg Med 2003; 32:305-309.
132. Gossner L, Stolte M, Sroka R et al. Photodynamic ablation of high-grade dysplasia and early cancer in Barrett's esophagus by means of 5-aminolevulinic acid. Gastroenterology 1998; 114:448-455.
133. Ackroyd R, Brown NJ, Davis MF et al. Photodynamic therapy for dysplastic Barrett's oesophagus: A prospective, double blind, randomised, placebo controlled trial. Gut 2000; 47:612-617.
134. Kelty CJ, Ackroyd R, Brown NJ et al. Photodynamic therapy for dysplastic Barrett's oesophagus: Long- term follow-up. Br J Surg 2001; 88:478.
135. Gaullier JM, Berg K, Peng Q et al. Use of 5-aminolevulinic acid esters to improve photodynamic therapy on cells in culture. Cancer Res 1997; 57:1481-1486.
136. Peng Q, Soler AM, Warloe T et al. Selective distribution of porphyrins in skin thick basal cell carcinoma after topical application of methyl 5-aminolevulinate. J Photochem Photobiol B 2001; 62:140-145.
137. Van den Akker JTHM, Holroyd JA, Vernon DI et al. Comparative in vitro percutaneous penetration of 5-aminolevulinic acid and two of its esters through excised hairless mouse skin. Lasers Surg Med 2003; 33:173-181.
138. Freeman M, Vinciullo C, Francis D et al. A comparison of photodynamic therapy using topical methyl aminolevulinate (Metvix) with single cycle cryotherapy in patients with actinic keratosis: A prospective, randomized study. J Dermatolog Treat 2003; 14:99-106.
139. Soler AM, Warloe T, Berner A et al. A follow-up study of recurrence and cosmesis in completely responding superficial and nodular basal cell carcinomas treated with methyl 5-aminolaevulinate-based photodynamic therapy alone and with prior curettage. Br J Dermatol 2001; 145:467-471.
140. Tarstedt M, Rosdahl I, Berne B et al. A randomized multicenter study to compare two treatment regimens of topical methyl aminolevulinate (Metvix)-PDT in actinic keratosis of the face and scalp. Acta Derm Venereol 2005; 85:424-428.
141. Zimmermann A, Ritsch-Marte M, Kostron H. mTHPC-mediated photodynamic diagnosis of malignant brain tumors. Photochem Photobiol 2001; 74:611-616.
142. Kriegmair M, Zaak D, Stepp H et al. Transurethral resection and surveillance of bladder cancer supported by 5-aminolevulinic acid-induced fluorescence endoscopy. Eur Urol 1999; 36:386-392.
143. Baumgartner R, Huber RM, Schulz H et al. Inhalation of 5-aminolevulinic acid: A new technique for fluorescence detection of early stage lung cancer. J Photochem Photobiol B 1996; 36:169-174.
144. Hillemanns P, Weingandt H, Baumgartner R et al. Photodetection of cervical intraepithelial neoplasia using 5-aminolevulinic acid-induced porphyrin fluorescence. Cancer 2000; 88:2275-2282.
145. Ladner DP, Steiner RA, Allemann J et al. Photodynamic diagnosis of breast tumours after oral application of aminolevulinic acid. Br J Cancer 2001; 84:33-37.
146. Jichlinski P, Leisinger HJ. Photodynamic therapy in superficial bladder cancer: Past, present and future. Urol Res 2001; 29:396-405.
147. Lange N, Jichlinski P, Zellweger M et al. Photodetection of early human bladder cancer based on the fluorescence of 5-aminolaevulinic acid hexylester-induced protoporphyrin IX: A pilot study. Br J Cancer 1999; 80:185-193.
148. Jiang H, Granville DJ, North JR et al. Selective action of the photosensitizer QLT0074 on activated human T lymphocytes. Photochem Photobiol 2002; 76:224-231.
149. Anderson CY, Freye K, Tubesing KA et al. A comparative analysis of silicon phthalocyanine photosensitizers for in vivo photodynamic therapy of RIF-1 tumors in C3H mice. Photochem Photobiol 1998; 67:332-336.
150. Hill JS, Kahl SB, Stylli SS et al. Selective tumor kill of cerebral glioma by photodynamic therapy using a boronated porphyrin photosensitizer. Proc Natl Acad Sci USA 1995; 92:12126-12130.
151. Chang SC, Buonaccorsi GA, MacRobert AJ et al. Interstitial photodynamic therapy in the canine prostate with disulfonated aluminum phthalocyanine and 5-aminolevulinic acid-induced protoporphyrin IX. Prostate 1997; 32:89-98.

CHAPTER 9

Heme Transport and Incorporation into Proteins

Linda Thöny-Meyer*

Abstract

Heme proteins are located in different compartments and organelles of pro- and eukaryotic cells. For their biosynthesis and assembly heme needs to be delivered to the polypeptides at the correct location. As heme is a hydrophobic molecule, it readily partitions into membranes and is bound to proteins, and the concentration of the free molecule in cells is extremely low. Heme is synthesized from the precursor 5-aminolevulinic acid, with the last step taking place in mitochondria in animals and fungi, and the plastid in plants. Some cells have the capability of taking heme up from the environment. The question therefore arises, how is heme transported across membranes and cellular compartments to its target proteins? The assembly of heme with these proteins is rather specific, and in many cases even stereospecific. For type c-type cytochromes, which bind heme covalently, specialized maturation systems have evolved that catalyze heme incorporation. It is likely that other heme proteins also need appropriate assembly factors for heme insertion. The current knowledge of intracellular heme transport and the biogenesis of heme-containing proteins is limited, and has only recently been recognized as an attractive field of research.

Introduction

Heme is the cofactor of a large number of cellular proteins and is important generally for aerobic life. Heme proteins are present in different subcellular compartments of bacterial and eukaryotic cells and in the blood serum. Heme is not water-soluble and readily aggregates in aqueous solutions at micromolar concentrations. It associates nonspecifically with lipids and proteins where it promotes peroxidations.[1] Due to this toxicity, the pool of free heme must be kept low. The last step of heme biosynthesis, the insertion of iron into protoporphyrin by the enzyme ferrochelatase occurs in the mitochondria, chloroplasts or in the bacterial cytoplasm. From these locations heme is distributed to the various cellular heme proteins by essentially unknown mechanisms. Importantly, it has to traverse biological membranes, to which it has a high affinity due to its amphiphilic nature. Predictions for heme transport range from postulating exclusive proteinaceous transport systems to autonomous diffusion across lipid membranes. The cellular interactions of heme with proteins and lipids are summarized in Figure 1. In the serum heme is bound by a number of specific binding proteins and transported to the liver, where it can be internalized. Pathogenic bacteria have developed strategies to capture heme for their own benefit as a source of iron. They require specific uptake systems for heme utilization. Our understanding of how heme is incorporated as a cofactor into its target proteins is very limited. The best investigated example is the posttranslational maturation of c-type cytochromes, where heme is attached covalently to the polypeptide.[2-8] Here, a specific heme delivery system is required. It is not known whether heme inserts spontaneously into proteins that

*Linda Thöny-Meyer—E. Blum & Co., Vorderberg 11, CH-80044, Zürich, Switzerland.
Email: lctm@bluewin.ch

Tetrapyrroles: Birth, Life and Death, edited by Martin J. Warren and Alison G. Smith.
©2009 Landes Bioscience and Springer Science+Business Media.

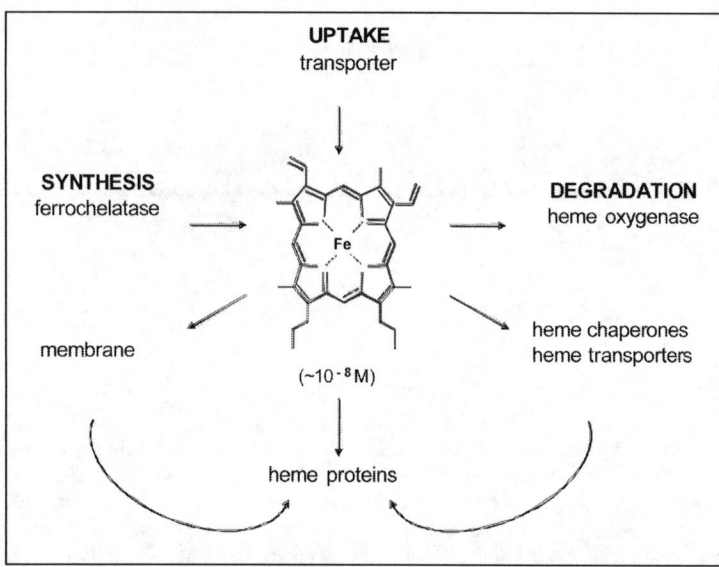

Figure 1. Cellular heme trafficking.

bind it in a noncovalent way in vivo, or if additional factors are needed to assist in this type of maturation. In this chapter, I summarize the current picture on heme traffic in the cell with emphasis on the recent progress that has been made in bacterial systems.

Localization of Heme in Prokaryotes

In bacteria, heme is synthesized in the cytoplasm; the last step is the incorporation of iron into protoporphyrin IX catalyzed by the enzyme ferrochelatase, which is a cytoplasmic enzyme[9] that has been shown to be membrane associated.[10] Some pathogens are able to capture heme from their environment as a source of iron. Heme formation or acquisition and its delivery to proteins within the bacterial cell is depicted in Figure 2. In the cytoplasm, heme is distributed to proteins such as bacterial hemoglobins, ferritin, catalase, or heme-containing sensor proteins involved in gene regulation. It is not known how heme is incorporated into these proteins. It is generally believed that the pool of free heme is extremely low under physiological conditions (in the range of 10^{-8} M).[11] Free heme can be defined as heme that is newly synthesized and has not been incorporated into heme proteins, or heme that has recently been released from heme proteins and is not immediately utilized or degraded. It seems unlikely that soluble heme proteins incorporate their prosthetic group randomly. Hence, the working hypothesis is that ferrochelatase delivers heme either directly to the hemoproteins or to intermediate cytoplasmic heme chaperones.

A number of respiratory and photosynthetic proteins known as the cytochromes contain heme as their prosthetic group for electron transport. Most of them are integral proteins of the cytoplasmic membrane. The question of how heme is incorporated into these polypeptides or polypeptide complexes has not been sufficiently addressed and our understanding of this problem is incomplete. Heme is found in protein domains of both, the inner and outer leaflet of the membrane. Many cytochromes also reside on the periplasmic side of the membrane or in the periplasm. This is particularly true for cytochromes of the c-type, where heme is bound covalently to the polypeptide by two thioether bonds formed between heme vinyls and cysteine thiols of the motif CXXCH that is a hallmark of this class of cytochromes.[12] In this case heme must be transported across the lipid bilayer of the cytoplasmic membrane. Type c cytochromes are ideal tools to study heme incorporation, as the covalent heme binding can be detected experimentally. Incorporation of heme into c-type cytochromes requires specific maturation systems of which three different types can be distinguished.[7] Two of them contain putative heme delivery systems that will be discussed

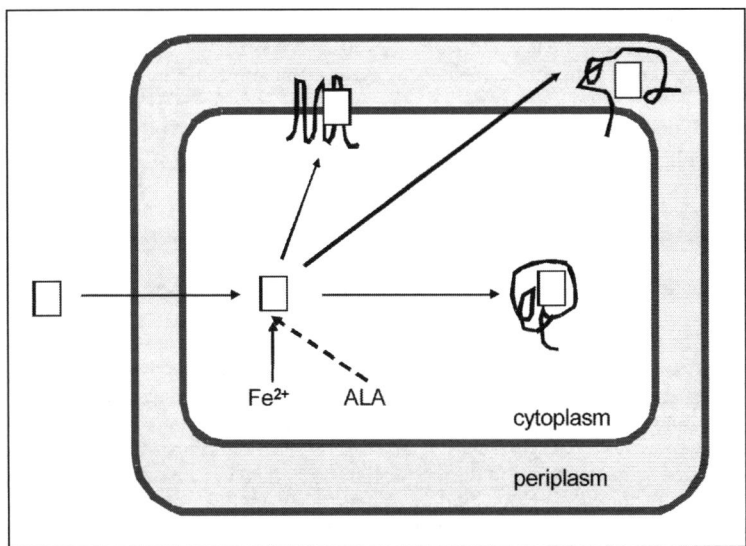

Figure 2. Bacterial heme routing. Heme (white squares) is either synthesized in the cytoplasm from its precursor ALA (δ-aminolevulinic acid), whereby iron is inserted in the last step. Alternatively, heme can be taken up by a specific transport system. The molecule is then distributed to cytoplasmic, membrane-integral or periplasmic proteins.

below. One of the cytochrome c maturation factors, the heme chaperone CcmE, is known to bind heme transiently in a covalent fashion, thus behaving like a c-type cytochrome.[13] Besides the c-type cytochromes, some periplasmic b-type cytochromes have been described, for example the *E. coli* cytochrome b_{562} whose function is unknown (for see ref. 8). Even though it has not been addressed how heme is inserted into these proteins, it is clear that they do not use the cytochrome c maturation (ccm) system,[14,15] because *ccm* deficient mutants can still incorporate heme into these cytochromes. In Gram-negative bacteria, certain heme binding proteins are also found in the outer membrane. These are c-type cytochromes[16,17] as well as heme receptors.[18-20]

Bacterial Heme Transport

Heme Uptake

The best example of heme transport is the uptake of heme by bacterial pathogens for iron utilization. Due to extremely low solubility of Fe (III) salts at physiological pH in the presence of oxygen (10^{-18} M) bacteria have evolved special strategies to capture heme from their host as a source of iron.[18-20] They contain specific receptors for heme or heme proteins in their outer membrane, which bind heme and import it across the outer membrane by a system that required the TonB/ExbBD proteins and an intact proton motive force across the cytoplasmic membrane.[21] Some Gram-negative bacteria even secrete small, so-called hemophores, which extract heme from a variety of heme containing proteins. The 19 kDa HasA of *Serratia marcescens* is excreted upon iron starvation and functions like a siderophore by binding circulating or hemoglobin-bound heme, returning to the cell surface and delivering heme to the outer membrane heme receptor HasR.[22] A similar function was described for the unrelated HxuA from *Haemophilus influenzae*.[23] Structural analysis of HasA indicates that heme can bind to this protein in two orientations relative to its α, γ-meso axis.[24,25] Heme is bound between two loops by the axial ligands His32 and Tyr75 and remains highly exposed to the solvent. Nevertheless it is well anchored to the protein by a network of hydrogen bonds around the heme ligands that involve the heme propionates, a situation referred to as "propionate zip". The formation and breaking of this propionate zip plays a key role in the mechanism of heme binding and release. HasA most likely captures heme diffusing from hemoglobin due to its higher affinity, protein-protein interactions between HasA and the HasR heme receptor have been reported, suggesting the formation of a (transient) complex.[22]

The mechanism of heme internalization by outer membrane heme receptors is not known. As heme receptors share homology with siderophore receptors, they probably have a beta barrel structure with an N-terminal cork-like domain that blocks the pore from the periplasmic side.[19,26] Once in the periplasm, heme is bound by a periplasmic heme binding protein (*Yersinia enterocolitica* HemT homologs) that associates with an ABC-type transporter. Members of this class of periplasmic binding proteins share a high degree of sequence identity (34-90% over the entire length) and are also homologous to the periplasmic vitamin B12 binding protein BtuF. Classical heme binding residues acting as heme iron ligands are not conserved strictly, with the exception of Tyr200 (*Y. enterocolitica* HemT numbering). The *hemT* homologous genes are coexpressed in operons also containing genes coding for the subunits of an ABC transporter that is needed for heme uptake through the inner membrane. The periplasmic heme binding proteins are believed to interact with the permease subunit of the ABC transporter. The ABC-type heme transporters are composed of a permease and an ATP subunit (e.g., *hemU* and *hemV*, respectively of *Y. enterocolitica*), which form a heterotetramer. The mechanism of heme transport by this ABC-type transporter is unknown, but based on proposed ideas of other ABC transporters it is likely that the periplasmic heme binding protein docks to the permease for heme delivery, and during the transport two ATPs are consumed, one before and one after substrate binding. A heme binding or heme interaction site within the permease has not been identified, nor is it known whether heme is released into the cytoplasm or into the membrane.

Certain Gram-positive bacteria such as *Corynebacterium diphteriae*[27] and *Staphylococcus aureus*[28] can use heme as an iron source. The *C. diphteriae* HmuT protein is a homolog of the periplasmic binding protein of heme uptake systems and is coexpressed with the genes *humU* and *hmuV* encoding the ABC transporter subunits. It was expressed in *E. coli* and shown to be a lipoprotein anchored to the cytoplasmic membrane. Affinity chromatography on hemin-and hemoglobin-agarose proved the heme binding capacity of this polypeptide, suggesting that it functions as a heme receptor at the cell surface.[27] Similar genes are present four times in the *S. aureus* genome. Moreover, characterization of the *isd* locus has identified additional genes, *isdA*, *isdB* and *isdC* encoding a novel set of cell wall associated heme relay factors. IsdA and IsdB bind heme and hemoglobin, respectively, at the cell surface. Heme is transferred to the cell-wall associated IsdC and then internalized by the membrane-bound IscD, IscE and IscF transporter.[29]

Cytoplasmic Heme Transport

Cytoplasmic heme exists as a result of either heme uptake or heme synthesis. How is cytoplasmic heme distributed to its target proteins or channeled into a degradation pathway? In *Shigella dysenteriae* the cytoplasmic protein ShuS binds heme with a Kd of 13 µM.[30] It has no sequence similarity to any proteins of known function, and its role in heme acquisition has not been determined. It forms soluble oligomers of ∼ 650 kDa that are composed of a single type of subunit with a molecular mass of 37 kDa and binds one heme per monomer. Since ShuS was also shown to nonspecifically bind double-stranded DNA, the protein may function both as a heme storage protein, during periods of active heme transport, and as a DNA binding protein to protect the DNA from any ensuing heme mediated oxidative damage. So far, this protein is the only candidate for a cytoplasmic heme chaperone.

Heme Export

An export of heme, i.e., the transport from the cytoplasm to the periplasm, must be postulated for biosynthesis of periplasmic *b*-type and for all *c*-type cytochromes. In fact, putative heme exporters are found in database entries and review articles because certain mutants in genes for typical ABC transporters have a defect in cytochrome *c* formation.[2,5,7,31-33] One class maps to *ccmABCD/helABCD* encoding components of a cytochrome *c* maturation system. Albeit the hypothesis that such a transporter is involved in heme export is highly attractive, no experimental proof for this has ever been presented. On the contrary, several lines of evidence point against this possibility,[34-36] most of all the fact that cytochrome *b* as well as the periplasmic heme chaperone CcmE can bind heme in the absence of CcmAB.[14,15,37] At present it is still not clear whether heme export is facilitated by an ABC transporter, by any other proteinaceous transporter, or if it can happen spontaneously by transmembrane diffusion at sufficient efficiency.

Heme Transport in Eukaryotes

Intracellular Heme Transport

In eukaryotes, the distribution of heme and its transport is more complex because heme is present in cytosolic, organellar and some nuclear proteins, but also in extracellular fluids such as the blood plasma (Fig. 3). Not much is known about the delivery of freshly synthesized heme in the eukaryotic cell. The last step of heme biosynthesis takes place in mitochondria in animals and fungi. Ferrochelatase is located in the inner mitochondrial membrane with its active side facing the mitochondrial matrix.[38] Newly synthesized heme is either incorporated into respiratory enzymes of the inner membrane and the intermembrane space, or must pass through both mitochondrial membranes before it can combine with cytosolic apoprotein such as hemoglobin type proteins (Hb Mb, Lb), or organellar heme proteins such as endoplasmic reticulum (ER)-localized P-450 oxygenases and cytochrome b, or peroxisomal catalase.[39] There is also evidence for heme in the nucleus,[40] where it is used for heme-responsive transcriptional regulators such as HAP1, a transcriptional activator in yeast.[41] In plants, plastids possess their own active ferrochelatase.[42]

It has not been elucidated how heme is exported from the mitochondrial matrix to the intermembrane space. A possible mitochondrial heme transporter is the erythroid ABC transporter ABC-me.[43] Overexpression of ABC-me was found to enhance hemoglobin biosynthesis, and the expression of the ABC-me transcript is induced during erythroid maturation. Alternatively, the transit of heme into the cytosol by diffusion cannot be excluded. In this case, heme has to penetrate across the inner and outer mitochondrial membrane, whereby a flipping from inner to outer leaflet is likely (Fig. 4). Uptake of heme from the mitochondrial outer membrane may then occur spontaneously due to higher affinity of cytosolic proteins to heme. A number of cytosolic proteins have been proposed to be involved in intracellular heme transport (Fig. 3), in particular the Z-protein,

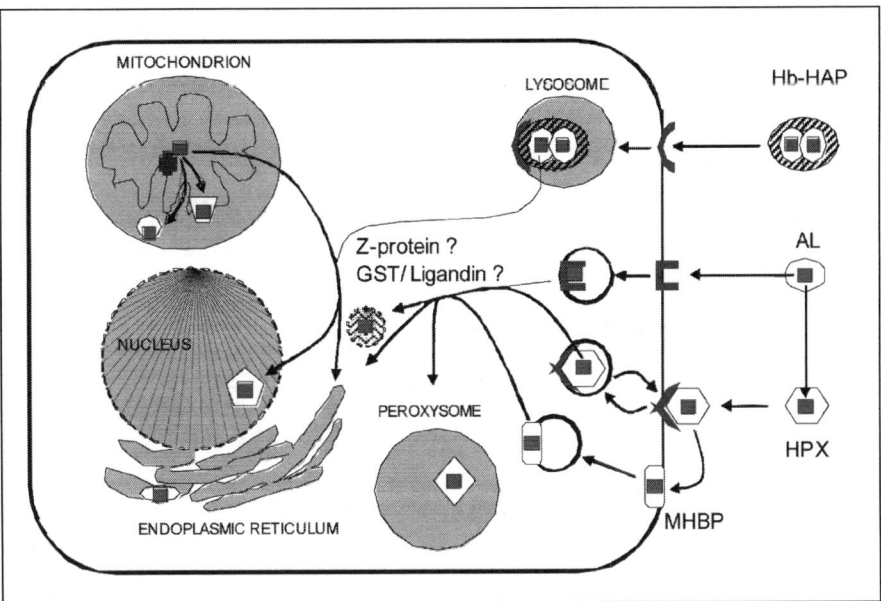

Figure 3. Heme uptake and distribution in the eukaryotic cell. Heme is represented by a gray square. Hb: hemoglobin; HAP: haptoglobin; AL: albumin; HPX: hemopexin; MHBP: membrane heme binding protein; GST: glutathione-S-transferase. The involvement of cytoplasmic heme carriers such as the Z-protein, or GST/ligandin in heme delivery to heme proteins or organelles is not clear. Whether or not cellular proteins are participating in the intracellular transfer steps indicated by arrows is an important issue of current and future research.

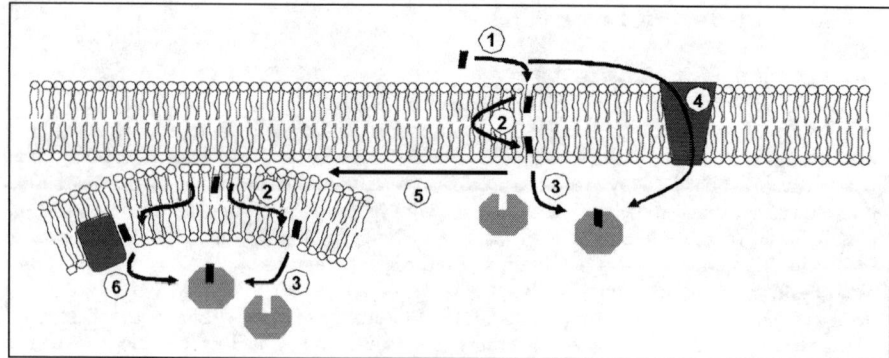

Figure 4. Heme translocation across membranes. Different possibilities are shown. 1) Spontaneous insertion of the hydrophobic heme molecule (black symbol) into a membrane leaflet. Ferrochelatase is membrane-associated and thus heme is close to the membrane at the last step of its synthesis. 2) Flipping of heme from one leaflet to the other. The flipping is more likely than a simple diffusion without reorientation, because the more hydrophilic propionate head groups of heme can be exposed to the solvent. 3) Extraction of heme from the membrane by a protein with high affinity. 4) Transport of heme across the membrane bilayer by a specific heme transporter. 5) Unassisted transfer of heme between different membranes: diffusion from one to another hydrophobic compartment. 6) Protein-assisted release of heme from the membrane by a heme chaperone for delivery to its target protein.

also known as the fatty acid binding protein, and glutathione transferases (GST) such as ligandin.[44] The former is also known as fatty acid binding protein or heme binding protein and has an affinity for heme in the range of 10^{-6} - 10^{-7} M, whereas the latter, besides its enzymatic function, can bind heme and other tetrapyrroles with similar affinities.

Extracellular Heme: Transport and Uptake

Extracellular heme occurs because it is released from senescent erythrocytes into the serum in the form of hemoglobin (Hb), which is bound as a dimer by haptoglobin (HAP), or in its free form, where essentially all is bound to albumin (AL) and hemopexin (HPX). These serum proteins have a high affinities for heme (Kd of 10^{-8} and <10^{-13} M, respectively), protect cells from oxidative damage, and preserve cellular iron by binding heme and transporting it to hepatocytes for receptor-mediated endocytosis.[11,44,45] Moreover, heme binding proteins are inhibitory to growth of microbial pathogens that depend on heme as a source of iron. Among the extracellular heme transport proteins albumin is the most abundant. It has a specific heme binding site and a K_d of 10^{-8} M and multiple secondary heme binding sites of lower affinity. Free heme in the blood stream is sequestered with great efficiency by HPX, one of the four most abundant proteins in the blood serum (5-20 mg/l). It is a glycoprotein that is synthesized by the liver and binds heme with the highest affinity known for a heme protein (K_d = 10^{-13} M).[44] The HPX crystal structure shows that heme is bound in a surface pocket formed at the interface between two structurally similar β-propeller domains, and embraced by an exposed flexible loop. The heme propionates are anchored into the protein domain-domain interface. The heme iron is bis-histidyl coordinated.[46] The ligand is released upon interaction with a specific heme receptor on the liver cells. HAP is a heterotetrameric glycoprotein circulating in the plasma that binds hemoglobin. It delivers Hb dimers to liver parenchymal cells through a receptor-mediated mechanism (Fig. 3).[11] Several other mechanisms have been postulated for heme uptake across the hepatocyte membrane (Fig. 3):[44,45] (i) interaction of heme-HPX with a HPX-receptor that would be transiently internalized by endocytosis and recycled after release of heme; (ii) delivery of heme from a HPX-receptor complex to a membrane heme binding protein (MHBP) that transports heme across the membrane and via vesicles to the ER; (iii) a plasma membrane heme receptor that binds heme delivered from albumin.

Unassisted Heme Transport

In vitro, heme can be transported into and within membranes without the assistance of special proteins, i.e., by passive diffusion (Fig. 4).[36,47-49] This is because it is an amphiphilic molecule which has no net charge at neutral pH in the reduced state. Heme can diffuse passively through model lipid bilayers. Its transmembrane movement from one leaflet to the other is a function of the chain length of the lipid[50] and perhaps also depends on the protonation state of the heme propionates.[18] However, more difficult is the exit of heme from the membrane. Heme has been demonstrated not to exit isolated intact mitochondria unless there is protein in the medium.[51] Hence, only proteins with a high affinity for heme are able to efficiently remove it from the lipid environment. Heme was shown to be extracted efficiently from single bilayer phospholipids and egg lecithin vesicles by albumin, hemopexin and apomyoglobin and apohemoglobin.[47,49] As transfer of heme between lecithin bilayers has also been observed, it has been postulated that migration of heme from the mitochondrial inner membrane to the ER and other subcellular compartments may occur spontaneously and rapidly without the help of protein heme carriers.[47] Figure 4 represents a scheme for both spontaneous as well as protein-assisted transmembrane movements of heme. How much unassisted heme diffusion contributes in vivo to intracellular heme trafficking is difficult to answer and may also depend on the efficiency of hemoprotein biosynthesis, in particular if heme-less apoproteins are unstable. If large amounts of heme proteins have to assemble in a short time, there may be a need for specific heme delivery mechanisms.

Heme-Protein Assembly

Noncovalent Assembly of Heme Proteins

The noncovalent association of heme and protein has been studied in detail for hemoglobin, which is a heterotetramer composed of two alpha and two beta chains, each carrying heme. Nevertheless, the precise pathway of subunit assembly and heme incorporation is still unclear. Heme can insert into apo-hemoglobin in two structurally distinct orientations, referred to as heme disorder.[52] In vitro heme assembly with apohemoglobins involves the formation of semihemoglobins, in which only one of the two chains is occupied with heme.[53] However, no specific chaperone type molecules are required for heme insertion. In vivo, an abundant erythroid protein, AHSP, was recently discovered to stabilize free alpha hemoglobin during assembly of hemoglobin that apparently prevents cytotoxic precipitation of free alpha subunits.[54]

For the noncovalent assembly of heme polypeptides in the outer leaflet of the membrane or in the periplasm such as in heme copper oxidases, the cytochrome bc_1 complex, the bd oxidase or the periplasmic b-type cytochromes (e.g., cytochrome b_{562}), heme must be transported across the lipid bilayer, but so far no heme-specific transporter has been identified. The *E. coli* cytochrome bd oxidase containing hemes b and d close to the periplasmic side of the membrane requires an ABC transporter for maturation.[55] Initially, it was believed that the CydDC transporter exports heme for assembly of this oxidase, but studies with radiolabelled heme for transport through membranes of a *cydD* mutant failed to prove such heme transport.[36] More recently, CydDC was reported to transport cysteine.[56] However, in an *E. coli* strain overexpressing the *cydDC* genes, a novel heme compound with an absorption maximum at 574 nm (thus called P-574) was discovered[57] and it was suggested that CydDC forms a complex with heme that establishes a link between heme transport and assembly with the oxidase.

As there is no other report in the literature on specific heme chaperones that direct heme to non$c$-type cytochromes, hemoglobin-like apoproteins, ferritin, catalases or heme regulatory proteins, it is possible that these proteins prefold and adopt a conformation which is suitable to acquire heme. Heme copper oxidases often contain more than one type of noncovalently-bound heme: for example, the *E. coli bo* oxidase carries one heme b and one heme o (having a hydroxyfarnesyl tail), and the heme o is found exclusively at the binuclear heme copper center. This indicates that there is specificity for the assembly of these hemes, either provided by the molecular structures of the hemes and the encompassing alpha helices, or by unknown assembly factors.

Heme Delivery during Cytochrome c Maturation

Much insight into heme delivery to proteins has come from studies on the biogenesis of *c*-type cytochromes, where the cofactor is bound covalently and irreversibly. Among a number of cytochrome *c* maturation factors, the heme chaperone CcmE found in *E. coli* and other Gram-negative bacteria[13] is of particular interest. A heme chaperone has a dual function of (i) transient storage of heme in a form which protects the cell from oxidative damage and (ii) the directed delivery of heme to the target protein. CcmE has three domains: an N-terminal membrane anchor, a periplasmic soluble beta barrel and a C-terminal, highly flexible tail of approximately 30 amino acids.[58,59] It binds heme transiently at a hydrophobic surface on the beta barrel and forms a covalent linkage to H130 that resides at the interface of the beta barrel and the flexible domain. The heme-binding form of CcmE is an essential intermediate of the cytochrome *c* maturation pathway. When heme is added to apo-CcmE in vitro, it initially binds noncovalently to the protein, which results in the spectral properties of a *b*-type cytochrome.[58] Upon reduction, heme is covalently crosslinked to CcmE, resulting in the typical shift of the α-band to a shorter wavelength, and from there the transfer of heme to apocytochrome *c* was achieved.[60] In vivo, heme binding to and release from CcmE requires two other cytochrome maturation factors, CcmC[61] and CcmF.[62] These are polytopic membrane proteins with a striking W-rich motif (WDXXWXWD) in a periplasmic domain and two or more conserved histidines, which are all essential for protein function and thus might participate in heme delivery. The current model is that heme reaches the periplasmic side of the membrane either by active transport or by diffusion, is bound and perhaps extracted from the membrane by the periplasmic domain of CcmC and then delivered to CcmE, which has been shown to form physical contacts with CcmC and to interact with hemin agarose.[61] CcmC is required for the formation of the covalent heme His130 bond. CcmE transiently buries heme, thereby preventing it from causing oxidative damage.[34,37] Finally, CcmE interacts with CcmF,[62] the putative cytochrome *c* heme lyase that is necessary for the formation of the thioether bonds. CcmF also interacts with CcmH, a periplasmic oxidoreductase that helps to keep the CXXCH heme binding site of apocytochrome *c* reduced in order to allow covalent heme attachment. Heme is transferred by CcmF to the reduced cysteines of apocytochrome *c*.[3] Heme insertion into *c*-type cytochromes occurs stereo-specifically in only one orientation. Two obvious questions regarding this type of cytochrome *c* maturation remain: (i) why does this process involve so many proteins, and (ii) what is the reason for using a covalent bond in the holo-CcmE intermediate that seems to have a primary role in heme transfer?

It has become clear in recent years that some Gram-negative bacteria, Gram-positive bacteria, cyanobacteria and plant chloroplasts use a cytochrome *c* maturation system (known as system II) that is simpler than that described above (system I) in that it does not involve a CcmE-type heme chaperone. Instead it uses a single polytopic membrane protein with a W-rich motif and flanking histidines for periplasmic heme delivery.[6,63,64] The two bacterial heme delivery and cytochrome *c* maturation systems are depicted in Figure 5.

Finally, most mitochondria (except those from plants and protests, which use system I) seem to use a simple cytochrome *c* heme lyase localized in the mitochondrial intermembrane space to attach heme to the imported apocytochrome. It is unknown whether the heme lyase can extract heme from the mitochondrial inner membrane (see Fig. 4) or whether there are additional as yet unidentified assembly factors in this organelle.

Outlook

Even though heme proteins have been known and heavily investigated for decades, we know little about how they are synthesized and matured in living cells. The hydrophobicity and soapiness of heme make studies of transport and assembly difficult as these usually are carried out in aqueous solutions. Moreover, free heme is almost absent in cellular fluids, and, when used in experiments, can lead to artifacts. Thus, studies of transport and assembly of heme bear a high risk of misinterpretation and must be carried out with great care. Important questions for the future are manifold: (i) does heme diffuse within membranes and subcellular compartments? (ii) Which are the proteins that enable or facilitate heme transport across membranes? Can the mechanisms of transport be defined? (iii) Are there intracellular and intra-organellar heme carriers/heme chaperones? (iv) Can specific assembly factors that interact with heme and/or apoprotein be identified? (v) What mechanisms direct the stereo-specific insertion of heme into a protein? Recent progress involving structure

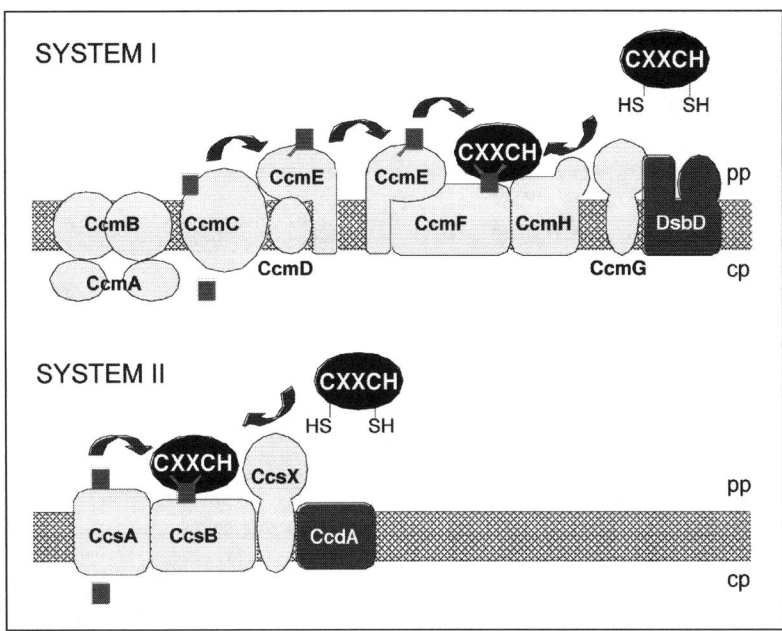

Figure 5. Bacterial systems for cytochrome *c* maturation. Specific factors are needed for transport and delivery of heme to apocytochrome *c*. Covalent attachment occurs at the SH groups in the heme binding site (CXXCH) of the apocytochrome that must be reduced. CcmGH and DsbD in system I and CcsX and CcdA in system II are thought to keep the heme binding site reduced. CcmC and CcmF in system I and CcsA in system II have a periplasmic W-rich motif that might interact with heme. System I but not system II involves a specific ABC transporter (CcmAB) and a heme chaperone (CcmE).

determination of heme binding and heme interacting proteins provides valuable information to design experiments addressing heme protein interactions. A combination of biochemical, biophysical, genetic and cell biological work will be necessary to give more insight in the biogenesis aspect of heme metabolism.

References

1. Vincent SH, Gradi RW, Shaklai N et al. The influence of heme-binding proteins in heme-catalyzed oxidations. Arch Biochem Biophys 1988; 265:539-550.
2. Thöny-Meyer L. Biogenesis of respiratory cytochromes in bacteria. Microbiol Mol Biol Rev 1997; 61:337-376.
3. Thöny-Meyer L. Haem-polypeptide interactions during cytochrome c maturation. Biochim Biophys Acta 2000; 1459:316-324.
4. Xie Z, Merchant S. A novel pathway for cytochromes c biogenesis in chloroplasts. Biochim Biophys Acta 1998; 1365:309-318.
5. Page MD, Sambongi Y, Ferguson SJ. Contrasting routes of c-type cytochrome assembly in mitochondria, chloroplasts and bacteria. Trends Biochem Sci 1998; 23:103-108.
6. Kranz RG, Beckett CS, Goldman BS. Genomic analyses of bacterial respiratory and cytochrome c assembly systems: Bordetella as a model for the system II cytochrome c biogenesis pathway. Res Microbiol 2002; 153:1-6.
7. Kranz R, Lill R, Goldman B et al. Molecular mechanisms of cytochrome c biogenesis: Three distinct systems. Mol Microbiol 1998; 29:383-396.
8. Allen JW, Daltrop O, Stevens JM et al. C-type cytochromes: Diverse structures and biogenesis systems pose evolutionary problems. Philos Trans R Soc Lond B Biol Sci 2003; 358(1429):255-266.
9. Hansson M, Hederstedt L. Purification and characterisation of a water-soluble ferrochelatase from Bacillus subtilis. Eur J Biochem 1994; 220:201-208.
10. O'Brian MR, Thöny-Meyer L. Biochemistry, regulation and genomics of heme biosynthesis in prokaryotes. Advances Microbial Physiol 2002; 46:257-318.

11. Sassa S, Kappas A. Disorders of heme production and catabolism. In: Handin RI, Lux SE, Stossel TP, eds. Blood: Principles and Practice of Hematology. Philadelphia: J.B. Lippincott Company, 1995:1473-1523.
12. Pettigrew GW, Moore GR. Cytochromes c. Biological Aspects. New York: Springer-Verlag, 1987.
13. Schulz H, Hennecke H, Thöny-Meyer L. Prototype of a heme chaperone essential for cytochrome c maturation. Science 1998; 281:1197-1200.
14. Goldman BS, Gabbert KK, Kranz RG. Use of heme reporters for studies of cytochrome biosynthesis and heme transport. J Bacteriol 1996; 178:6338-6347.
15. Throne-Holst M, Thöny-Meyer L, Hederstedt L. Escherichia coli ccm in-frame deletion mutants can produce periplasmic cytochrome b, but not cytochrome c. FEBS Lett 1997; 410:351-355.
16. Myers JM, Myers CR. Role of the tetraheme cytochrome CymA in anaerobic electron transport in cells of Shewanella putrefaciens MR-1 with normal levels of menaquinone. J Bacteriol 2000; 182(1):67-75.
17. Myers JM, Myers CR. Role for outer membrane cytochromes OmcA and OmcB of Shewanella putrefaciens MR-1 in reduction of manganese dioxide. Appl Environ Microbiol 2001; 67:260-269.
18. Genco CA, Dixon DW. Emerging strategies in microbial haem capture. Mol Microbiol 2001; 39:1-11.
19. Stojiljkovic I, Perkins-Balding D. Processing of heme and heme-containing proteins by bacteria. DNA Cell Biol 2002; 21:281-295.
20. Wandersman C, Stojiljkovic I. Bacterial heme sources: The role of heme, hemoprotein receptors and hemophores. Curr Opin Microbiol 2000; 3:215-220.
21. Braun V, Killmann H. Bacterial solutions to the iron-supply problem. Trends Biochem Sci 1999; 24(3):104-109.
22. Letoffe S, Nato F, Goldberg ME et al. Interactions of HasA, a bacterial haemophore with haemoglobin and with its outer membrane receptor HasR. Mol Microbiol 1999; 33:546-555.
23. Cope LD, Thomas SE, Hrkal Z et al. Binding of heme-hemopexin complexes by soluble HxuA protein allows utilization of this complexed heme by Haemophilus influenzae. Infect Immun 1998; 66:4511-4516.
24. Arnoux P, Haser R, Izadi N et al. The crystal structure of HasA, a hemophore secreted by Serratia marcescens. Nat Struct Biol 1999; 6:516-520.
25. Arnoux P, Haser R, Izadi-Pruneyre N et al. Functional aspects of the heme bound hemophore HasA by structural analysis of various crystal forms. Proteins 2000; 41:202-210.
26. Klebba PE, Newton SM. Mechanisms of solute transport through outer membrane porins: Burning down the house. Curr Opin Microbiol 1998; 1:238-247.
27. Drazek ES, Hammack CA, Schmitt MP. Corynebacterium diphtheriae genes required for acquisition of iron from haemin and haemoglobin are homologous to ABC haemin transporters. Mol Microbiol 2000; 36:68-84.
28. Mazmanian SK, Ton-That H, Su K et al. An iron-regulated sortase anchors a class of surface protein during Staphylococcus aureus pathogenesis. Proc Natl Acad Sci USA 2002; 99:2293-2298.
29. Mazmanian SK, Skaar EP, Gaspar AH et al. Passage of heme-iron across the envelope of Staphylococcus aureus. Science 2003; 299:906-909.
30. Wilks A. The ShuS protein of Shigella dysenteriae is a heme-sequestering protein that also binds DNA. Arch Biochem Biophys 2001; 387:137-142.
31. Beckman DL, Trawick DR, Kranz RG. Bacterial cytochromes c biogenesis. Genes Devel 1992; 6:268-283.
32. Ramseier TM, Winteler HV, Hennecke H. Discovery and sequence analysis of bacterial genes involved in the biogenesis of c-type cytochromes. J Biol Chem 1991; 266:7793-7803.
33. Goldman BS, Kranz RG. ABC transporters associated with cytochrome c biogenesis. Res Microbiol 2001; 152:323-329.
34. Schulz H, Pellicioli E, Thöny-Meyer L. New insights into the role of CcmC, CcmD, and CcmE in the haem delivery pathway during cytochrome c maturation by a complete mutational analysis of the conserved tryptophan-rich motif of CcmC. Mol Microbiol 2000; 37:1379-1388.
35. Page MD, Ferguson SJ. Mutational analysis of the Paracoccus denitrificans c-type cytochrome biosynthetic genes ccmABCDG: Disruption of ccmC has distinct effects suggesting a role for CcmC independent of CcmAB. Microbiology 1999; 145:3047-3057.
36. Cook GM, Poole RK. Oxidase and periplasmic cytochrome assembly in Escherichia coli K-12: CydDC and CcmAB are not required for haem-membrane association. Microbiology 2000; 146:527-536.
37. Schulz H, Fabianek RA, Pellicioli et al. Heme transfer to the heme chaperone CcmE during cytochrome c maturation requires the CcmC protein, which may function independently of the ABC-transporter CcmAB. Proc Natl Acad Sci USA 1999; 96:6462-6467.

38. Harbin BM, Dailey HA. Orientation of ferrochelatase in bovine liver mitochondria. Biochemistry 1985; 24(2):366-370.
39. Soga O, Kinoshita H, Ueda M et al. Evaluation of peroxisomal heme in yeast. J Biochem (Tokyo) 1997; 121:25-28.
40. Lo SC, Aft R, Mueller GC. Role of nonhemoglobin heme accumulation in the terminal differentiation of Friend erythroleukemia cells. Cancer Res 1981; 41:864-870.
41. Zhang L, Guarente L. Heme binds to a short sequence that serves a regulatory function in diverse proteins. EMBO J 1995; 14:313-320.
42. Cornah JE, Roper JM, Pal Singh D et al. Measurement of ferrochelatase activity using a novel assay suggests that plastids are the major site of haem biosynthesis in both photosynthetic and nonphotosynthetic cells of pea (Pisum sativum L.). Biochem J 2002; 362:423-432.
43. Shirihai OS, Gregory T, Yu C et al. ABC-me: A novel mitochondrial transporter induced by GATA-1 during erythroid differentiation. EMBO J 2000; 19:2492-2502.
44. Müller-Eberhard U, Nikkila H. Transport of tetrapyrroles by proteins. Semin Hematol 1989; 26:86-104.
45. Smith A. Transport of tetrapyrroles: Mechanisms and biological and regulatory consequences. In: Dailey HA, ed. Biosynthesis of heme and chlorophylls. New York: McGraw-Hill, 1990:435-490.
46. Paoli M, Anderson BF, Baker MH et al. Crystal structure of hemopexin reveals a novel high-affinity heme site formed between two beta-propeller domains. Nat Struct Biol 1999; 6:926-931.
47. Rose MY, Thompson RA, Light WR et al. Heme transfer between phospholipid membranes and uptake by apohemoglobin. J Biol Chem 1985; 260:6632-6640.
48. Tipping E, Ketterer B, Christodoulides L et al. The interactions of haem with ligandin and aminoazo-dye-binding protein A. Biochem J 1976; 157:461-467.
49. Cannon JB, Kuo FS, Pasternack RF et al. Kinetics of the interaction of hemin liposomes with heme binding proteins. Biochemistry 1984; 23:3715-3721.
50. Light IIIrd WR, Olson JS. Transmembrane movement of heme. J Biol Chem 1990; 265:15623-15631.
51. Yoda B, Israels LG. Transfer of heme from mitochondria in rat liver cells. Can J Biochem 1972; 50:633-637.
52. Mathews A, Brittain T. Haem disorder in recombinant- and reticulocyte-derived haemoglobins: Evidence for stereoselective haem insertion in eukaryotes. Biochem J 2001; 357:305-311.
53. Vasudevan G, McDonald MJ. Spectral demonstration of semihemoglobin formation during CN-hemin incorporation into human apohemoglobins. J Biol Chem 1997; 272:517-524.
54. Kihm AJ, Kong Y, Hong W et al. An abundant erythroid protein that stabilizes free alpha-haemoglobin. Nature 2002; 417:758-763.
55. Osborne JP, Gennis RB. Sequence analysis of cytochrome bd oxidase suggests a revised topology for subunit I. Biochim Biophys Acta 1999; 1410:32-50.
56. Pittman MS, Corker H, Wu G et al. Cysteine is exported from the Escherichia coli cytoplasm by CydDC, an ATP-binding cassette-type transporter required for cytochrome assembly. J Biol Chem 2002; 277:49841-49849.
57. Cook GM, Cruz-Ramos H, Moir AJ et al. A novel haem compound accumulated in Escherichia coli overexpressing the cydDC operon, encoding an ABC-type transporter required for cytochrome assembly. Arch Microbiol 2002; 178:358-369.
58. Enggist E, Thöny-Meyer L, Güntert P et al. NMR structure of the heme chaperone CcmE reveals a novel functional motif. Structure 2002; 10:1551-1557.
59. Arnesano F, Banci L, Barker PD et al. Solution structure and characterization of the heme chaperone CcmE. Biochemistry 2002; 41:13587-13594.
60. Daltrop O, Stevens JM, Higham CW et al. The CcmE protein of the c-type cytochrome biogenesis system: Unusual in vitro heme incorporation into apo-CcmE and transfer from holo-CcmE to apocytochrome. Proc Natl Acad Sci USA 2002; 99:9703-9708.
61. Ren Q, Thöny-Meyer L. Physical interaction of CcmC with heme and the heme chaperone CcmE during cytochrome c maturation. J Biol Chem 2001; 276:32591-32596.
62. Ren Q, Ahuja U, Thöny-Meyer L. A bacterial cytochrome c heme lyase: CcmF forms a complex with the heme chaperone CcmE and CcmH but not with apocytochrome c. J Biol Chem 2002; 277:7657-7663.
63. Beckett CS, Loughman JA, Karberg KA et al. Four genes are required for the system II cytochrome c biogenesis pathway in Bordetella pertussis, a unique bacterial model. Mol Microbiol 2000; 38:465-481.
64. Hamel PP, Dreyfuss W, Xie Z et al. Essential histidine and tryptophan residues in CcsA, a system II polytopic cytochrome c biogenesis protein. J Biol Chem 2003; 278:2593-2603.

CHAPTER 10

Heme and Hemoproteins

Andrew W. Munro,* Hazel M. Girvan, Kirsty J. McLean, Myles R. Cheesman
and David Leys

Introduction

Heme cofactors are found in a wide range of proteins, where they play various roles in e.g., steroid[1] and bioactive lipid synthesis,[2] energy transduction,[3] gene regulation,[4] cellular signaling,[5] oxygen transport[6] and antibiotic biosynthesis.[7] The diversity of physiological functions performed by hemoproteins means that heme is among the most versatile of protein cofactors.[8] Aside from electron transferase functions observed in respiratory cytochromes (e.g., mitochondrial cytochrome c ref. 9), several hemoprotein sensory or catalytic functions are recognized that involve the binding of gaseous ligands to the heme iron and/or the dissociation or switching of amino acid side chains as axial ligands to the iron.[4,10] Moreover, heme-dependent activation of iron-bound dioxygen (as seen in e.g., nitric oxide synthase and the cytochromes P450) enables a broad repertoire of reactions, including hydroxylation, epoxidation, demethylation and carbon-carbon bond cleavage.[11-13] This chapter reviews (i) the basic properties and synthesis of heme cofactors, (ii) the nature of their attachment to hemoproteins and the various types of protein scaffolds in which hemes are incorporated, (iii) exemplary functions of hemoproteins that demonstrate their broad range of biochemical functions, and (iv) analytical techniques that facilitate the understanding of the structural and redox characteristics of the protein-bound heme cofactor, and the mode of its ligation to the protein.

The Heme Synthetic Pathway

Hemes and chlorophylls are the most important examples of tetrapyrroles in nature. The tetrapyrroles are the result of four pyrrole molecules becoming linked via methine (-C=) bridges into a circular arrangement, with a system of conjugated double bonds giving rise to the UV-visible absorption characteristics of these compounds.[14] The later steps in the heme biosynthetic pathway are shown in (Fig. 1). The first common intermediate in heme synthesis across all life forms is 5-aminolevulinic acid (δ-aminolevulinic acid or δ-ALA). In mammals, birds and certain prokaryotes (e.g., *Rhodobacter*) the δ-ALA is formed from glycine and succinyl coenzyme A (succinyl CoA). In non-plant eukaryotes the reaction is catalysed in the mitochondrion by the enzyme 5-aminolevulinate synthase in a pyridoxal phosphate-dependent manner.[15] The intermediate 2-amino-3-oxoadipate is converted to δ-ALA with evolution of a molecule of CO_2, and the product is then transported into the cytoplasm. In plants and several bacteria, an alternative pathway uses glutamate as the starting material. The glutamate is attached to transfer RNA in an ATP-dependent manner by glutamate tRNA ligase. Following NAD(P)H-dependent reduction by glutamyl tRNA^{Glu} reductase, the intermediate (S)-glutamate-1-semialdehyde is transformed to δ-ALA by glutamate-1-semialdehyde 2,1-aminomutase with the movement of the amino group from the C-2- to the C-1 position.[14] In plants these processes are contained in the chloroplast, and the first two reactions are subject to feedback inhibition by heme.

*Corresponding Author: Andrew W. Munro—Faculty of Life Sciences, University of Manchester, Manchester Interdisciplinary Biocentre, 131 Princess Street, Manchester M1 7DN, UK. Email: andrew.munro@manchester.ac.uk

Tetrapyrroles: Birth, Life and Death, edited by Martin J. Warren and Alison G. Smith.
©2009 Landes Bioscience and Springer Science+Business Media.

Following convergence of the two different initial pathways at δ-ALA, the later steps (indicated in Fig. 1) involve (i) the condensation of two molecules of δ-ALA to form porphobilinogen (PBG) catalysed by PBG synthase (also known as δ-ALA dehydratase, AAD). There is species-specific metal ion dependence and negative regulation by heme. Condensation of four molecules of PBG results in first the formation of the linear tetrapyrrole hydroxymethylbilane, catalysed by hydroxymethylbilane synthase (also known as porphobilinogen deaminase, PBD), which contains an unusual dipyrromethane cofactor (formed from two molecules of PBG) linked to a cysteine residue. The hydroxymethylbilane would spontaneously cyclise into uroporphyrinogen I were it not for the action of uroporphyrinogen III synthase (UPS), which inverts the pyrrole D ring prior to cyclisation

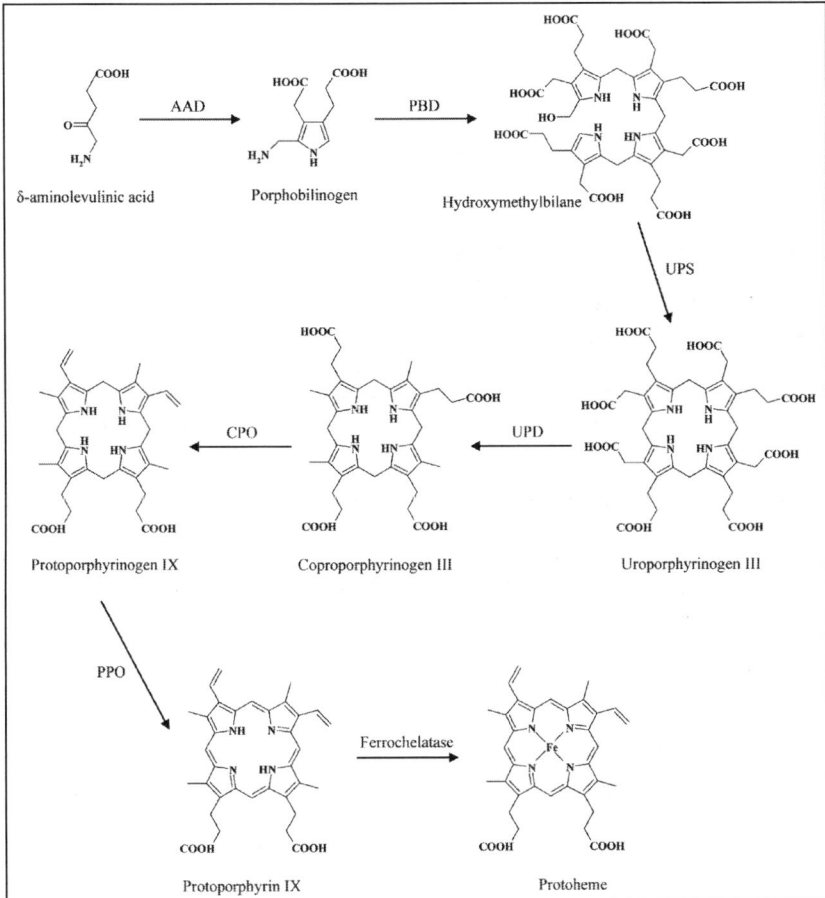

Figure 1. Pathway of heme *b* synthesis. As detailed in section 2, the first common intermediate in the pathways of heme synthesis in various organisms is δ-aminolevulinic acid (δ-ALA). Two molecules of δ-ALA are condensed to form porphobilinogen (PBG), catalysed by δ-ALA dehydratase (AAD), also known as PBG synthase. Condensation of four molecules of PBG leads to formation of hydroxymethylbilane, catalysed by porphobilinogen deaminase (PBD), also known as hydroxymethylbilane synthase. Uroporphyrinogen III synthase (UPS) inverts the pyrrole ring D and cyclisation of the molecule produces uroporphyrinogen III. Decarboxylation of the four acetate side chains by uroporphyrinogen decarboxylase (UPD) produces coproporphyrinogen III. Oxidative decarboxylation catalysed by coproporphyrinogen oxidase (CPO) results in conversion of two propionate side chains into vinyl groups and formation of protoporphyrinogen IX. Oxidation of methylene bridges connecting pyrroles (to form methenyl bridges) is catalysed by ferrochelatase (heme synthase)[14,16].

and leads to formation of uroporphyrinogen III. The next step is the decarboxylation of all four acetate side chains of uroporphyrinogen III by uroporphyrinogen decarboxylase (UPD, with evolution of four molecules of CO_2). In eukaryotes, the product (coproporphyrinogen III) is transferred back to mitochondria. Oxidative decarboxylation is then performed by the enzyme coproporphyrinogen oxidase (CPO), with resultant transformation of two propionate side chains into vinyl groups and formation of protoporphyrinogen IX along with two water molecules. The final step in the manufacture of the heme skeleton is the oxidation of methylene bridges connecting the pyrrole moieties into methenyl bridges, forming protoporphyrin IX. This reaction is catalysed by protoporphyrinogen oxidase (PPO) and also results in production of three water molecules.[14,16] Protoporphyrin is the branch point for synthesis of heme and chlorophyll. The insertion of ferrous iron by ferrochelatase (heme synthase) creates protoheme (heme *b* or ferriprotoporphyrin IX), the heme molecule found in proteins/enzymes such as myoglobin, hemoglobin and the cytochromes P450 (P450s).[17,18] Heme *b* is the precursor for several other types of hemes found in nature, as described in the section below (Fig. 2).

Structural Variations of the Heme Cofactor

A wide variety of different types of heme macrocycle are now recognized to occur naturally (Fig. 2). Heme *b* provides the starting point for synthesis of the other types of heme. Heme *a* is synthesized from heme *b* in two sequential reactions (with heme *o* as a stable intermediate) involving first the addition of a 17-hydroxyethylfarnesyl moiety at the 2-position on the heme (forming heme *o* and catalysed by heme *o* synthase), and then modification of the methyl group at the 8-position to a formyl group.[19] The latter reactions are catalysed by the enzyme heme *a* synthase, involving two successive hydroxylations at C8 (to heme *i*), with the resultant formylation and conversion to heme *a* likely a consequence of the breakdown of the unstable dihydroxylated intermediate with elimination of a water molecule. The *Bacillus subtilis* heme *a* synthase is a heme-containing integral membrane protein.[20] Heme *a* is the heme found exclusively in the terminal oxidases of aerobic respiratory chains - including cytochrome *c* oxidase (complex IV) of mitochondria and several bacteria.[21,22] As described, heme *o* also contains the 17-hydroxyethylfarnesyl moiety at the 2-position, but the 2-methyl group (as in heme *b*) is retained.[21] Heme *o* synthase performs the 17-hydroxyethylfarnesyl addition reaction to heme *b* at the vinyl group on the 2-position.[23,24] Recent studies indicate that the heterologously expressed heme *o* synthase and heme *a* synthase enzymes from both *B. subtilis* and *Rhodobacter sphaeroides* form complexes in vivo, helping to explain how the process of trafficking of heme *b* through to heme *a* occurs.[25]

In heme *s* (also referred to as chlorocruoroheme), a formyl group replaces the heme *b* vinyl group at the 2-position, and this form of heme is found in the hemoglobin of marine worms.[26] Heme *d* has a more profound alteration to the heme *b* core structure. The propionate group at position 6 of heme ring III is bound to the carbon both by the C-C bond seen in heme *b*, and by the carbonyl oxygen. This results in the formation of a lactone, and in a fifth ring for the heme *d* macrocycle.[27] Heme *d* is found in certain catalase enzymes and in many bacterial terminal respiratory oxidases, and is the site for oxygen reduction to water in the latter enzymes (e.g., cytochrome *bd* from *Escherichia coli*).[28] By contrast, heme d_1 is a variant ferric-dioxoisobacteriochlorin structure unique to the cytochrome cd_1 nitrite reductase enzyme class.[29,30]

The *c*-type heme is found in mitochondrial cytochrome *c* (which shuttles electrons between respiratory complexes III and IV), as well as in various other bacterial respiratory proteins.[31,32] It deviates from the *b*-type structure in that the vinyl groups of the heme are esterified to the apoprotein via two cysteine residues. The mechanisms by which these covalent linkages to the protein backbone occur are now becoming well understood. Heme attachment motifs in the protein are widely recognized in *c*-type cytochrome. The motif CXXCH is commonly conserved, where X represents any amino acid, the two cysteines are those involved in the covalent link to the heme and the histidine is usually an axial ligand to the heme iron.[33] Several multi-heme *c*-type cytochromes are now recognized in nature (e.g., a tetraheme flavocytochrome c_3 fumarate reductase from *Shewanella putrefaciens* and octaheme cytochrome c_3 from *Desulfovibrio desulfuricans*).[32,34] It is possible that covalent attachment of the macrocycle is important for heme retention and enables close packing of heme cofactors to facilitate efficient electron transport in such systems (vide infra).[35]

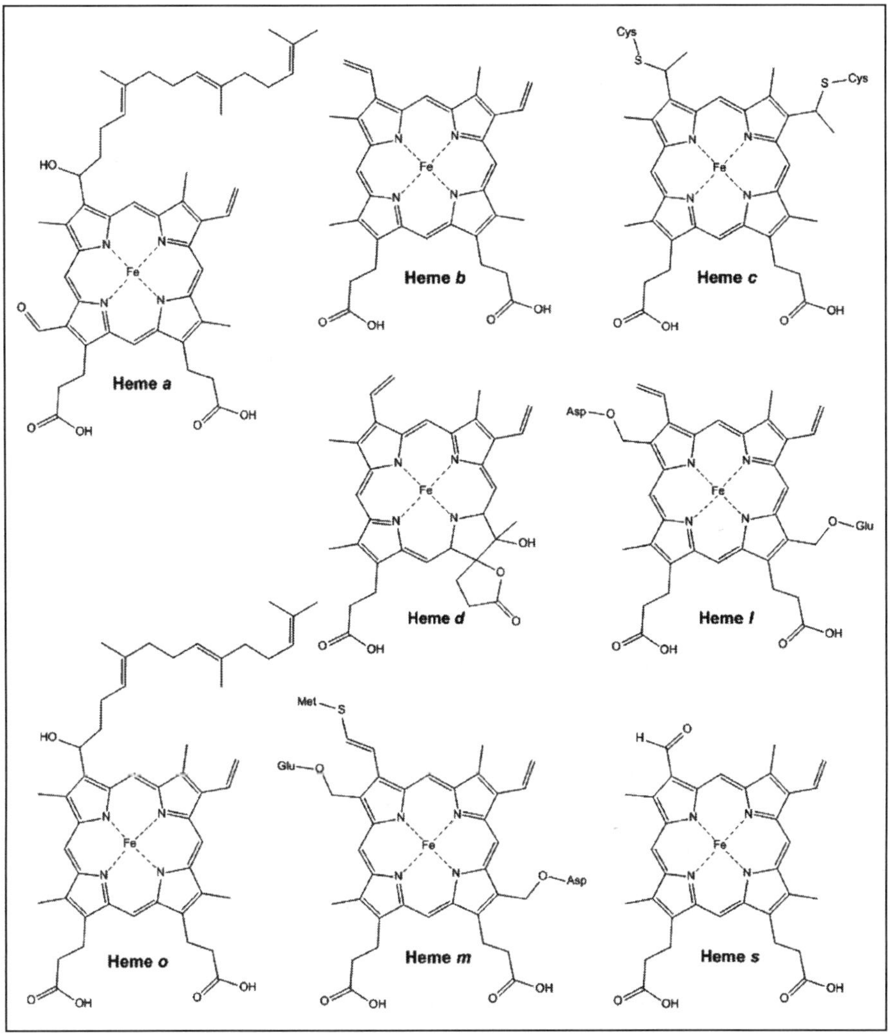

Figure 2. Stuctural variants of the heme cofactor in nature. As discussed in detail in this section, a series of structural variants of the heme macrocycle (i.e., heme *b*) are observed in nature. Heme *a* is found in terminal oxidases of aerobic respiratory chains (e.g., cytochrome *c* oxidase), and is derived from heme *b* via heme *o* as an intermediate.[21,22] The *c*-type hemes are widespread in nature and are covalently linked to their host proteins. They are derived from heme *b* through esterification of two heme vinyl groups to protein cysteine residues. Examples include mitochondrial respiratory cytochrome *c* and flavocytochrome *c* fumarate reductase from *Shewanella putrefaciens*.[30,31] Heme *s* (found in marine worm hemoglobins) is modified with a formyl group replacing the heme *b* 2-vinyl group, while heme *d* (found in certain catalases and bacterial terminal respiratory oxidases) contains an additional lactone structure.[26,27] Heme *l* (as found in the mammalian peroxidase lactoperoxidase) is covalently linked to the protein through esterification of the two heme methyl group substituents to glutamate/aspartate residues.[46,47] Heme *m* is observed in the mammalian neutrophil enzyme myeloperoxidase, and contains (in addition to lactoperoxidase-like covalent linkages between heme and protein) a sulfonium bond between the heme 2-vinyl group and a methionine residue.[48,49]

In cytochromes *c*, thioether bonds are formed between the heme vinyl groups and the two cysteines, with universally conserved stereochemistry and generally requiring the actions of other proteins.[36] Heme attachment systems can be as simple as a single heme lyase enzyme, as found in certain mitochondria. However, three different types of system have been characterized in detail. The system I or Ccm (cytochrome *c* maturation) apparatus involves up to nine different proteins and is found in most proteobacteria and in the mitochondria of several eukaryotes, including plants and algae.[37] System II occurs in other proteobacteria, Gram positive bacteria, cyanobacteria and in the chloroplasts of plants and algae. It appear to be composed of two essential proteins (ResB and ResC, sometimes as a single fusion protein) and two accessory proteins required for oxidation/reduction of disulfide bonds.[37,38] System III is the heme lyase found in fungal mitochondria and in the mitochondria of metazoans and some protozoa.[37,39] Biological diversity in these systems continues to be revealed. A fourth distinct system was proposed in trypanosomes, where a deviant heme binding motif with a single cysteine (XXXCH) is observed in certain cytochromes *c*, and in the genomes of which the genes for systems I-III are absent.[39] The Ccm system (system I) includes the CcmE membrane-anchored heme chaperone, which has been well studied in *E. coli*.[40,41] This protein appears to bind ferric heme via a conserved histidine residue and with linkage to the heme iron. In the ferrous form, other residues ligate the iron and the histidine (His130 in *E. coli*) is freed to enable a covalent linkage to a heme vinyl group. CcmE binds heme covalently in the periplasm prior to heme transfer to apocytochromes.[42] Again, diversity is observed in that motifs different from the typical CXXCH can be matured by the *E. coli* CcmE apparatus, and that an archaeal Ccm system lacks one of the major proteins (CcmH, a disulfide oxidoreductase) and has the conserved CcmE His residue (that ultimately covalently binds heme vinyl) replaced by a cysteine residue.[36]

While the above are among the most common types of heme observed naturally in biological systems, a number of other modified hemes are also recognized. In mammalian peroxidase enzymes (lactoperoxidase, eosinophil peroxidase and thyroid peroxidase), heme is generally covalently attached (esterified) to heme methyl groups autocatalytically.[44] Enzymes in this family oxidize a wide range of molecules (including phenols, catecholamines, steroid hormones and halides) in a peroxide-dependent manner and can generate a wide spectrum of products with antimicrobial activity.[45] Lactoperoxidase is synthesized in breast secretory epithelial cells and has important roles in host defence against bacterial infections.[46] For instance heme *l* (Fig. 2) is found covalently attached to lactoperoxidase. The heme *b* is covalently linked to the protein backbone in lactoperoxidase in a peroxide-dependent manner and via esterification of the two heme methyl substituents to glutamate 375 and aspartate 225 in the bovine enzyme.[47] Similar bonds form in the other mammalian-type peroxidases.[48]

Heme *m* is the name given to the form of heme *b* found covalently linked to the protein myeloperoxidase (Fig. 2). Myeloperoxidase is widely expressed in neutrophils and is considered to be a key component of the "respiratory burst" in host defence against infections. It may also be an important enzyme in inflammatory diseases.[49] Atomic structures have been solved for the human form of the enzyme.[50,51] In addition to the lactoperoxidase-like methyl linkages to glutamate and aspartate residues, there is an additional sulfonium ion linkage between a methionine side chain and the heme 2-vinyl group.[50] This gives this heme a green colour and the additional modification also allows the enzyme to oxidise chloride and bromide ions.[52]

Other heme modifications are recognized. For example, in another class of the peroxidase superfamily (class I, whose main members are cytochrome *c* peroxidase, ascorbate peroxidase and the bifunctional catalase-peroxidase [KatG] enzymes, which have a conserved distal histidine heme iron ligand), the linkage of a tryptophan residue to the heme macrocycle is observed in the case of the soybean soluble ascorbate peroxide (APX).[53] The relevant tryptophan becomes linked in a peroxide dependent manner, and in the absence of the substrate ascorbate. The reaction is proposed to occur via formation of the reactive ferryl-oxo intermediate of the enzyme (compound I), followed by oxidation and deprotonation of the relevant tryptophan residue (Trp41) and addition of the Trp41 radical across the heme 4-vinyl group.[53] The comparable tryptophan in *Mycobacterium tuberculosis* KatG (Trp107) does not interact with the heme, but instead becomes cross-linked to a tyrosine (Tyr229), which in turn is cross-linked to a methionine (Met255). The reaction again is peroxide dependent and proposed to involve KatG compound I.[54] A relatively recent discovery has been that the heme *b* bound to cytochrome P450 (P450 or CYP) enzymes is not exclusively non-covalently held in the protein matrix. In the case of various eukaryotic fatty acid oxidase P450s (CYP4 family members) it has been shown that enzyme turnover-dependent linkage of a heme methyl group to a conserved glutamate residue occurs.[55] The reaction is P450 turnover-dependent and is proposed to involve abstraction of a single electron from the glutamate carboxyl group by the P450 ferryl-oxo intermediate, giving a protein carboxyl radical. This radical then abstracts a hydrogen atom from the 5-methyl group of the heme to generate a benzylic-type radical, which is converted to a carbocation by electron transfer to the heme iron. Trapping of the carbocation by the carboxylate forms the heme-protein cross-link.[44] The oxidation (hydroxylation) of the 5-methyl to a hydroxymethyl group may occur if the glutamate carboxyl is not optimally positioned, and the carbocation is instead trapped by a water molecule. This reaction was observed in a mutant of CYP4F5, where the relevant glutamate is absent in wild-type enzyme. In the G330E mutant of CYP4F5, some glutamate-heme cross-linking is observed, along with a proportion of non-covalently bound 5-hydroxymethyl heme.[56] A proportion of heme was also shown to be autocatalytically cross-linked via the 5-methyl group in a G248E mutant of the well characterized *Pseudomonas putida* cytochrome P450 cam (CYP101A1) camphor hydroxylase, suggesting that rational protein engineering might provide a means of stabilizing heme binding by covalent cross-linking to these enzymes.[57]

Among other naturally occurring heme variants are siroheme and heme P460. In the recent atomic structure of the periplasmic cytochrome P460 protein from the ammonia oxidizing bacterium *Nitrosomonas europea*, a predominantly beta sheet structure was revealed and (in addition to the typical cytochrome *c* cysteine thioether links to heme vinyls) there is an additional covalent link between the side chain nitrogen of a lysine (Lys70) and the 13'-meso carbon of the heme. Loss of conjugation at this position in the porphyrin ring occurs as a result of sp^3 hybridization following cross-link formation.[58] In the earlier characterized hydroxylamine oxidoreductase (HAO) from the same bacterium, a similar Soret spectrum with maximum at 460 nm is observed as in the cytochrome P460. HAO catalyses the oxidation of hydroxylamine to nitrite, the second step in the ammonia oxidation process.[59] HAO is a 24-heme containing trimer, in which 21 of the hemes are "regular" *c*-type and participate in electron transfer from the three P460 hemes to acceptor cytochromes. Despite its similar spectral features, the heme in HAO is distinct from that of the cytochrome P460, with a tyrosine residue (Tyr467) cross-linked to the 5'-meso carbon directly opposite that which is lysine-linked in the cytochrome P460 in the heme macrocycle.[60] Siroheme is used as a cofactor (along with a 4Fe-4S iron sulfur cluster) in the six electron reduction reactions catalysed by bacterial nitrite (to ammonia) and sulfite (to sulfide) reductases.[61] It is distinct from "true" heme structures in that its structure originates from a branch point prior to formation of protoporpyrin IX (Fig. 1). Siroheme is an iron-containing isobacteriochlorin formed by successive methylation, oxidation and iron insertion into the tetrapyrrole uroporphyrinogen III.[62]

The modification of heme structure leads to perturbation of thermodynamic properties of the protein-bound hemes, and in many cases may enable them to participate in different types of redox reactions at different redox potentials. The covalent linkage of heme to the host protein backbone obviously stabilizes the binding of the cofactor, and again impacts on its thermodynamic features, and these may be the primary reasons underlying chemical modifications of this type in proteins such as cytochrome *c* and myeloperoxidase. Heme properties are obviously also modulated by the nature of coordination of the heme iron, as explained in the following section.

Heme Iron Coordination in Hemoproteins

A wide variety of heme axial ligands are observed in nature, and these are clearly critical to the modulation of heme iron reduction potential, and thus the ability of the relevant cytochrome to participate in the desired redox reactions. Ligands to heme iron may also be removed or replaced by other ligands (internal amino acid sidechains or external molecules) during catalytic cycles of certain hemoproteins to facilitate required reactivity. Variations in the nature of the protein environment, along with the changes in the types of axial ligands to the heme iron, can cause substantial variations in heme iron reduction potential. Examples of heme axial ligands observed in nature are shown in (Fig. 4).

Common amino acid axial ligands to heme iron include histidine, methionine and cysteine. In the cytochromes *c*, the heme is usually hexacoordinated with axial ligands typically provided by the side chain nitrogen atoms of two histidine residues (N^ε, *bis*-His ligation) or by a histidine and a methionine (S^δ, His-Met ligation). Examples of *bis*-His ligated proteins include hemes in cytochrome c_{554}, one of the proteins responsible for shuttling electrons from the HAO protein to the terminal oxidase in *Nitrosomonas europea*.[63] The protein's name derives from the spectral maximum for one of its major UV-visible absorption features (the α band) in the reduced form. In this tetraheme cytochrome, hemes 3 and 4 exemplify the typical *bis*-His coordination. However, in heme 1, one of the histidines (His 102) coordinates to the iron via the N^δ atom. The final heme (heme 2) is pentacoordinate, with a single histidine ligand.[64] The hexacoordinated hemes 1, 3 and 4 have a low-spin electronic configuration in the d-orbitals of the ferric heme iron, but the pentacoordinate heme 2 is high-spin.

Other bis-His coordinated c-type cytochromes include the tetraheme cytochromes c_3 from *Desulfovibrio gigas* and *Desulfuromonas acetoxidans*.[65,66] The mitochondrial respiratory cytochrome c oxidase (COX) enzyme (which catalyses the 4-electron reduction of oxygen to water) has an a-type heme with *bis*-His coordination (heme a). Electrons are transferred from cytochrome c to COX heme a through a copper centre (Cu_A, containing two copper atoms), and then on to a binuclear centre formed from another heme a (heme a_3) and a copper atom (Cu_B). The heme a_3 is the site for binding of oxygen, which occurs across the heme a_3 (in its ferrous state) and $Cu(I)_B$. This heme iron is pentacoordinate with a single histidine ligand in its resting (ferric) form, and with the vacant axial position ready for binding dioxygen on reduction of the heme.[67] *Bis*-His heme iron coordination is also seen in the cytochromes b_5, which are small (<20 kDa) electron transfer proteins widespread in eukaryotes, and also recognized in certain prokaryotes.[68,69] Among the many roles for eukaryotic cytochromes b_5 are participation in certain cytochrome P450 oxidation reactions, and reduction of methemoglobin in erythrocytes.[70,71]

His-Met heme coordination is also observed in certain c-type cytochromes. For example, the c-type heme component of mitochondrial respiratory complex III (cytochrome bc_1 or ubiquinol:ferricytochrome c oxidoreductase) has a His-Met coordinated iron.[72] This part of the bc_1 protein is responsible for shuttling electrons to the peripheral membrane protein cytochrome c, which then transfers them to COX. The cytochrome cd_1 nitrite reductase from *Paracoccus pantotrophus* is an important enzyme in bacterial denitrification, catalyzing the one electron reduction of nitrite to nitric oxide.[73] The four electron reduction of oxygen to water can also be catalysed by cytochrome cd_1. The enzyme binds both c- and d_1-type hemes, with the d_1 heme being the site of nitrite (and oxygen) reduction, and the c-heme being the electron conduit from partner proteins such as the copper protein azurin.[74] The c-type heme has *bis*-His coordination in the resting (ferric) state. However, it was observed in structural studies that, on reduction of a cd_1 crystal, one of the histidine ligands is exchanged for a methionine to give His-Met coordination in the c-heme.[73] Unusual heme ligation and ligation switches also occur in the non-covalently bound heme d_1. In the oxidised form the heme is coordinated by histidine and tyrosine ligands, with the tyrosine coordinating the heme iron (via the O^η atom) originating from the protein domain that binds the other (c-type) heme. On reduction, the tyrosine ligand is displaced to enable substrate binding to the heme d_1 iron.[73] The $E.$ *coli* periplasmic cytochrome b_{562} is another example of His-Met coordination.[75]

Several important enzymes are known to have cysteine (thiol or thiolate) coordinated heme iron. Examples include the cytochromes P450 (P450s), in which it is widely accepted that the proteinaceous ligand (proximal ligand) to the b-type heme is a cysteine thiolate.[76] The deprotonated form of the proximal ligand appears essential for the catalytic activity of the P450s, involving reductive activation of molecular oxygen bound in the final coordination position (as distal or 6th ligand) on the heme iron.[77] In the resting (ferric) state, the P450s generally have a weakly bound water (or hydroxide ion) as the 6th ligand. Binding of substrate molecules (endogenous substrates for eukaryotic P450s are typically lipophilic molecules such as steroids, fatty acids or hydrophobic xenobiotics) may lead to displacement of the 6th ligand and a shift in heme iron spin-state from low-spin towards high-spin, often accompanied by a large change in heme iron redox potential that enables electron transfer to the P450 from a redox partner enzyme.[78] Reduction of the P450 heme iron to its ferrous state allows oxygen binding, and a further single electron reduction and successive protonations of the oxy complexes ultimately leads to formation of the reactive ferryl-oxo compound I, which oxidises a substrate bound close to the heme.[79] The title P450 arises from the unusual and diagnostic position of the Soret maximum in the (inhibited) ferrous-carbon monoxy complex.[80] An inactive cysteine thiol-coordinated form of the enzyme has this Soret band at ~420 nm and is referred to as P420.[76,81] Other cystein(at)e-coordinated heme enzymes include chloroperoxidase and the nitric oxide synthases (NOS enzymes). The choroperoxidase enzyme from the fungus *Calderiomyces fumago* participates in the synthesis of the natural product caldariomycin, and has a range of activities including P450-, catalase- and peroxidase-like functions, and can catalyse halogenation reactions.[82] Heme coordination (as for the P450s and NOS enzymes) is via the cysteine S^γ atom. In the resting form, chloroperoxidase is pentacoordinate with a ferric, high-spin heme iron. The 6th coordination position is vacant for reaction with peroxide. The NOS enzymes catalyse the P450-like hydroxylation of L-arginine to form L-hydroxyarginine, and in a second oxidation step convert the first product to L-citrulline and nitric oxide (NO).[83] NO has important physiological roles in e.g., neurotransmission, immune response and vasodilation, and its binding to the protein guanlyate cyclase triggers signaling responses to give rise to these effects.[84,85] Electrons for NOS catalysis are derived from NADPH and delivered to the eukaryotic NOS heme via a fused diflavin reductase domain and likely via both a FMN cofactor in the reductase domain and a tetrahydropbiopterin cofactor bound close to the NOS heme in the oxygenase domain.[86] Known prokaryotic NOS enzymes are isolated heme-containing domains that communicate with separate cellular redox partners (e.g., flavodoxins).[87] In the resting state, the NOS heme iron is pentacoordinate with the 6th coordination position vacant for oxygen binding to the ferrous form. NOS is faced with the interesting problem of inhibitory affinity of the NO product for its heme iron, and the overall productive reaction rate in eukaryotic NOS enzymes is governed both by rates of heme reduction and the rates of dissociation of NO from the Fe(III)-NO adduct formed at the end of the NOS catalytic cycle.[86]

Recent studies indicate that cysteine is also a heme iron ligand in the heme-regulated eukaryotic initiation factor 2α (eIF2α) kinase (HRI) that catalyses the phosphorylation of the eIF2α subunit, which in turn inhibits protein synthesis under appropriate cellular stresses.[88] The HRI enzyme acts as a heme sensor, with heme dissociation activated by low heme levels in cells. The full-length HRI enzyme is likely to have a hexacoordinated heme iron with Cys/His ligands provided by distinct domains of the enzyme.[88] Other enzymes with Cys/His heme iron coordination include the triheme SoxAX enzyme complex from the bacterium *Rhodovulum sulfidophilum*, which is essential for photosynthetic thiosulfate and sulfide oxidation. There are two hemes in the SoxA protein, both of which are cysteinate-coordinated, and one of which has a histidine as the 6th axial ligand. The final heme in the SoxX component of the complex has His-Met coordination.[89]

Further distinct types of heme iron axial ligand coordination include Met-Met in bacterioferritin. The *E. coli* bacterioferritin is an oligomeric protein of 24 identical subunits that come together to form a hollow cube-like structure with rounded edges.[90] Iron is stored at the core in a hydrated oxide mineral form and the major physiological function of the protein appears to be iron uptake and storage. Each subunit contains a binuclear iron centre, and there are also 12 *b*-type hemes at the interfaces between pairs of subunits that are *bis*-Met coordinated by the same methionine residue in each of the two subunits.[90] Other relatively unusual types of heme ligation states include tyrosine (O^n) coordination in various catalases, where the resting state of the enzymes are 5-coordinate, enabling binding of peroxides in the vacant position on the heme iron.[91] Tyrosine coordination is also observed for the Tyr-His coordinated cytochrome *f*, a component of the chloroplast cytochrome b_6f complex. Cytochrome b_6f is an integral membrane protein that transfers electrons between reaction centre complexes in the photosynthetic membrane and participates in creation of an electrochemical proton gradient. Cytochrome *f* has one covalently-bound heme *c* and is the electron donor to the copper protein plastocyanin. Structural studies on this cytochrome indicate that the tyrosyl ligand is not derived from the amino acid side chain, but instead from the N^α that is the amino terminus of the polypeptide.[92] A further unusual ligation state is observed in the CooA class of transcriptional regulator proteins, which are carbon monoxide (CO) sensor proteins. The original CooA protein was from the photosynthetic bacterium *Rhodospirillum rubrum*, an organism able to derive energy from CO oxidation.[93] The *R. rubrum* CooA atomic structure has been resolved, revealing *b*-type ferrous heme iron coordination by histidine and by the nitrogen atom of the N-terminal proline residue in the ferrous state, and by cysteine and the proline side chain in the ferric state (Fig. 4).[94] The CooA proteins are homodimeric, and the proline of monomer A provides an axial ligand for the heme of monomer B, and vice versa. CooA is activated by CO binding, which displaces the proline as a ligand to the heme iron and induces structural reorientation of the DNA-binding domain of the protein to facilitate target DNA recognition.[95] Another unusual pattern of heme coordination comes in the siroheme-binding bacterial sulfite reductases (SiR's). The structure of the *E. coli* SiR hemoprotein domain reveals a pentacoordinate heme iron ligated by a cysteine thiolate, but where the thiolate sulfur is also bridged to one of the irons of a 4Fe-4S cluster.[96] His-Asn heme iron coordination is observed in the *c*-type cytochrome SHP from *Rhodobacter sphaeroides*. In this protein, the asparagine coordination is lost on heme reduction, enabling binding of nitric oxide.[97] Displacement of one of the two histidine axial ligands (and switch in heme iron spin-state from low- to high-spin) is also observed on reduction of the soluble cytochrome c" from *Methylophilus methylotrophus*.[98]

Diversity of Hemoprotein Form and Function

Heme is possibly the most functionally diverse of the protein-bound cofactors, participating in a wide range of chemical reactions.[8] These include electron transfer (e.g., mitochondrial cytochrome *c* and cytochrome *b₅*),[31,68] oxygen activation (e.g., P450 proteins and nitric oxide synthase)[77,83] and gene regulation (CooA).[4] Heme-containing proteins are essential for the physiology and viability of all living organisms and contribute to such diverse functions as respiration and oxygen carriage, steroid synthesis, drug metabolism, cellular signaling and apoptosis. As shown in (Fig. 3), there is no common protein fold for heme-binding proteins. Hemes are found in a diverse array of different protein scaffolds. However, the reactivity of the heme is linked to its protein environment, as discussed further in the Novel Aspects and Future Prospects section.

The roles of hemes in hemoproteins can be broken down into a number of general classes. These include (i) ligand-binding resulting in cellular transport/delivery or molecular response. Key examples of this function include association of gaseous molecules for storage, transport and/or delivery. These functions are exemplified by proteins such as globins (e.g., myoglobin and hemoglobin in their roles as oxygen carriers) and nitrophorins. Certain bloodsucking insects inject NO while feeding on the host in order to induce vasodilation and prevent coagulation of blood. The nitrophorins are the proteins used for NO storage and delivery. In the nitrophorin from the bedbug (*Cimex lectularius*), the *b*-type heme has a cysteine axial ligand with the other position on the iron free for NO binding. In this nitrophorin, NO can bind either as the 6th axial ligand to the heme iron or (at higher concentrations) as a S-NO conjugate of the cysteine, which dissociates from the heme iron following heme reduction.[99] Examples of histidine-coordinated nitrophorins also exist.[100] The gaseous ligand-binding function extends to sensing of diatomic gases. This role is exemplified by proteins such as the sensory kinases DevS (also known as DosS) and DosT from *Mycobacterium tuberculosis*. These two heme binding kinases control the activity of DevR (also known as DosR), and these systems are considered to be responsible for gene regulation that results in switching the bacterial pathogen from a replicating to a non-replicating and persistent infective state in response to NO and anaerobic conditions.[101] For both the DevS and DosT proteins, the kinase activity is coupled to the heme iron coordination state, such that carbon monoxide- (CO-) bound and NO-bound forms were active, but oxygen-bound forms were inactive as kinases.[102] The DevS and DosT proteins are histidine autokinases, and subsequently transfer the phosphate to an aspartate residue on DevR. In turn, activated (phosphorylated) DevR attains enhanced affinity for its DNA recognition sequences and activates the hypoxic response at the transcriptional level.[101] Other heme-dependent gas sensor regulatory proteins include CooA (CO and NO activated);[4] the NNR (nitrite reductase and nitric oxide regulator) from *Paracoccus denitrificans* that activates transcription in response to NO;[103] and the oxygen-responsive FixL sensor kinase from *Bradyrhizobium japonicum*.[104]

A related class of heme proteins (ii) are those involved in redox sensing. The *c*-type heme binding, chemotaxis signal transducing protein DcrA from *Desulfovibrio vulgaris* may act as both a redox sensor and an oxygen sensor. A reduction-dependent heme iron ligand switch appears to occur in DcrA, with a water ligand replaced by an amino acid in the reduced state. Oxygen and CO are both able to bind the ferrous heme iron in place of an endogenous ligand.[105] In *M. tuberculosis*, the DosS sensor kinase is a further heme-binding protein that seems responsive to redox potential, as opposed to hypoxia.[106] Cystathionine beta synthase (CBS) is a heme *b*-and pyridoxal phosphate- (PLP-) binding enzyme, with the heme iron hexacoordinated with histidine nitrogen and cysteine thiolate axial ligands in the ferric state.[107] CBS catalyses the condensation of homocysteine and serine to generate cystathionine, which may then be converted (via cysteine) to the antioxidant glutathione. Activity of CBS is decreased on heme reduction, suggesting that the heme functions as a redox sensor. However, CO and NO (that can displace endogenous amino acid ligands in the reduced state and inhibit activity) have also been suggested as potential physiological regulators of CBS activity.[107] A further redox sensor protein is the *E. coli* heme-regulated phosphodiesterase EcDOS, which senses cellular redox state through its heme. EcDOS has a N-terminal heme sensor domain

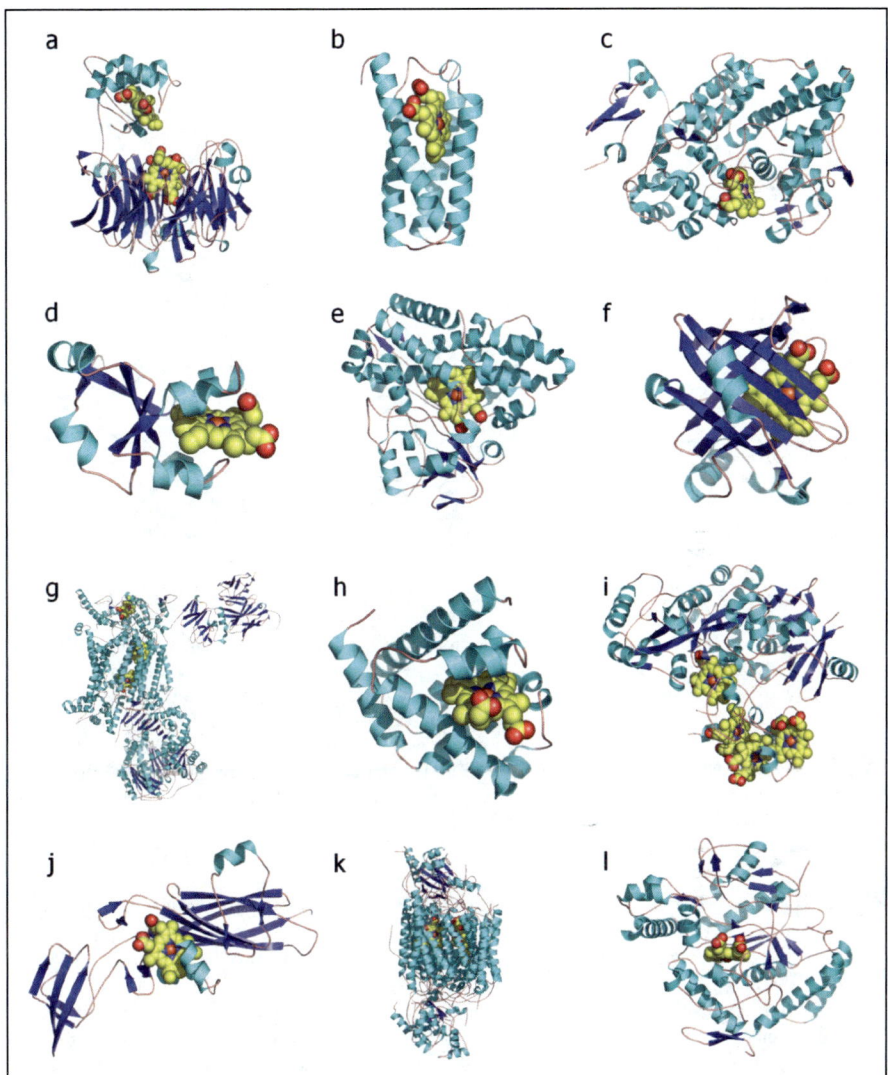

Figure 3. Biodiversity of hemoprotein structures. Heme is found in a huge number of different proteins and enzymes. There is no common hemoprotein fold, and examples are shown that illustrate the diversity of hemoprotein structures in nature. Heme cofactors are displayed in atom-coloured spacefill, with protein secondary structural elements displayed in cyan (alpha helices) and dark blue (beta sheets). Proteins shown are (a) cytochrome cd_1 from *Paracoccus pantotrophus* (PDB code 1DY7); (b) cytochrome *c'* from an *Alcaligenes* sp. (1CGO); (c) human cyclooxygenase (1IGZ); (d) bovine cytochrome b_5 (1CYO); (e) the *Mycobacterium tuberculosis* P450 CYP121 (1N4O); (f) *Rhodnius prolixis* nitrophorin (1D3S); (g) *Saccharomyces cerevisiae* cytochrome bc_1 (1P84); (h) sperm whale myoglobin (1AJG); (i) *Shewanella putrefaciens* (strain MR-1) flavocytochrome c_3 fumarate reductase (1D4D); (j) turnip (*Brassica rapa*) cytochrome *f* (1CTM); (k) bovine cytochrome *c* oxidase (2OCC); and (l) rat neuronal nitric oxide synthase (nNOS) oxygenase domain (1ZVI).

and a C-terminal effector domain (a common theme in this type of enzyme). Heme axial ligand switching occurs from His-water (or hydroxide) in the ferric enzyme to His-Met in the ferrous

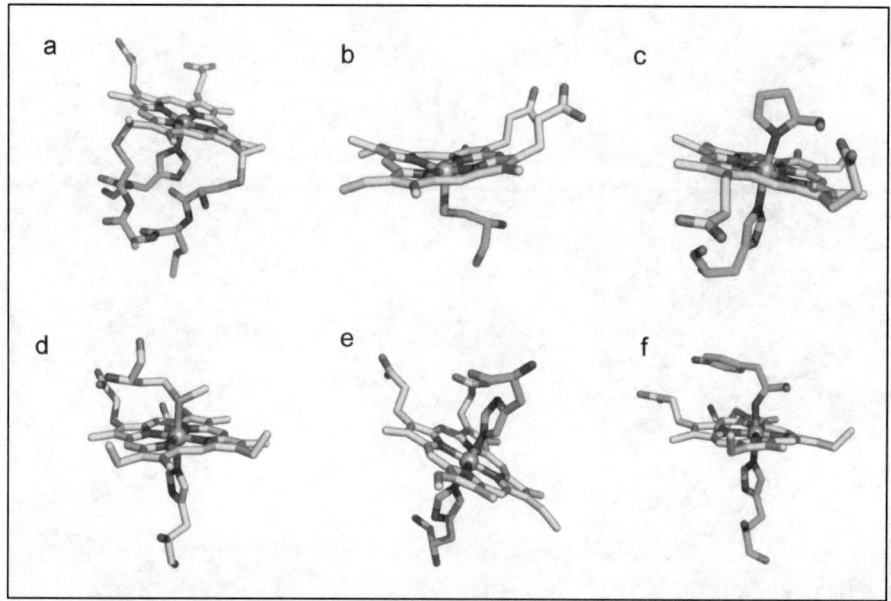

Figure 4. Axial coordination of heme iron in hemoproteins. As discussed in section 4, a variety of different types of heme iron axial coordination are observed in hemoproteins. Examples shown are (a) the pentacoordinate *c*-type heme in cytochrome *c'* from an *Alcaligenes* sp. (PDB code 1CGO); (b) the cysteinate-ligated pentacoordinate *b*-type heme in rat nNOS oxygenase domain (1ZVI); (c) the *b*-type heme in *Rhodospirillum rubrum* CooA in its ferrous His-Pro coordinated state (1FT9); (d) the *c*-type heme in *Paracoccus pantotrophus* cytochrome *cd*₁ in its ferrous His-Met coordinated form (1DY7); (e) the *b*-type heme in *Saccharomyces cerevisiae* cytochrome *bc*₁ with *bis*-His coordination (1P84); and (f) the hexacoordinated *c*-type heme in cytochrome *f* from turnip, with heme iron ligation from a histidine and from the N^{α} atom of the protein's N-terminal tyrosine residue.

enzyme. The ferrous enzyme is active as a cyclic AMP (cAMP) phosphodiesterase and responds to changes in environmental oxygen levels by modulating cellular cAMP concentration and altering transcription of target genes.[108] A wide range of different heme-based sensor proteins are now recognized, most having been characterized only within the last decade.[109]

A further functional class of hemoproteins (iii) are those involved purely in electron transfer reactions. These include mitochondrial and other cytochromes *c*, cytochromes *b*₅ and cytochrome *f*, as discussed earlier in the chapter.[8,31,68,92] However, electron transfer reactions through the heme iron can be readily linked to catalysis, and heme enzymes with true catalytic functions (iv) include the P450s, catalases and peroxidases, cytochrome *cd*₁ nitrite reductase, nitric oxide synthase and respiratory complexes.[3,12,45,74,86,91] The P450s are particularly versatile catalysts, and can couple their reductive activation of molecular oxygen to a variety of modifications of organic substrates bound close to the heme iron, including: aliphatic and aromatic carbon hydroxylation, dealkylation, epoxidation at carbon-carbon double bonds, reduction, carbon-carbon bond cleavage, dehydrogenation and deamination.[12] The much narrower range of oxidation reactions catalysed by the NOS enzymes are similar to those catalysed by several P450s and also involve a thiolate-coordinated heme *b*, but occur within a very different protein structural scaffold.[110] Soluble guanylate cyclase (sGC) links NO/CO sensing with the conversion of guanosine triphosphate (GTP) to the formation of the second messenger cyclic guanosine monophosphate (cGMP). The binding of NO to sGC leads to the formation of a 5-coordinate ferrous-NO heme *b* complex (via displacement of an endogenous histidine ligand) and to an active cyclase form.[5] Another class of hemoproteins (v) are those that appear to bind heme in a regulatory fashion and to possess a heme regulatory motif (HRM). The HRM is a cysteine-containing peptide sequence in a protein, and heme has been demonstrated to bind stoichiometrically to cysteine in synthetic HRM-containing peptides.[111] HRM's are found in proteins such as the mammalian 5-aminolevulinic acid synthases (ALAS 1 and 2) and erythroid-specific eIF2α kinase. Mutagenesis of HRM sequences showed that these were essential for heme-mediated inhibition of mitochondrial import of ALAS1.[112] A HRM sequence is also present in the iron regulatory protein 2 (IRP2), an important regulatory protein in mammalian iron homeostasis. IRP2 activity is modulated by ubiquitination and degradation, and IRP2 degradation may be effected by heme-mediated oxidation. The binding of heme to IRP2 was demonstrated, and binding of both ferric heme (to a cysteine in the HRM) and ferrous heme (to a HRM histidine) was suggested.[113] Heme also binds to a region containing four Cys-Pro dipeptide motifs in the the C-terminal region of the mammalian transcription factor Bach1. Heme inhibits DNA repressor activity of Bach1, but mutation of the Cys-Pro motifs abolished Bach1 interaction with heme. Thus, intracellular heme concentration modulates gene expression under control of Bach1.[114] A final class of hemoproteins (vi) would be those with uncertain physiological role. This class could include cytochromes c', which are periplasmic cytochromes widespread in protebacteria,[115] and the cytochrome c" from *Methylophilus methylotrophus*.[116] In the human DGCR8 protein involved in microRNA processing, heme binding is of uncertain catalytic function, but is important for dimerization of the protein and in enhancing primary microRNA cleavage activity.[117]

Spectroscopic Analysis of Hemoproteins

As discussed earlier in the chapter, there is a wealth of structural data available for hemoproteins in various redox states and ligand bound forms. The bulk of these data originate from X-ray crystallographic studies, although NMR structures are also available for some of the smaller cytochromes (e.g., cytochrome b_5).[118] However, in absence of crystallographic or NMR structural data there are a number of useful spectroscopic tools that can applied to define structural and mechanistic aspects of hemes in their protein environment. Some of the major methods are described briefly below. However, the list is not exhaustive and several other spectroscopic techniques (e.g., electron nuclear double resonance [ENDOR] and Mössbauer) have been applied successfully to analysis of hemoprotein structure and mechanism.[119,120]

The hemes are probably the most easily identifiable cofactors in proteins. Their conjugated porphyrin structures give rise to electronic transitions in the visible region and a strong red or red/brown colour according to redox state, the nature of the ligands to the heme iron, and the heme iron spin-state. This makes UV-visible absorption spectroscopy a very powerful tool in hemoprotein analysis. The major absorption feature of the protein-bound heme cofactor (the Soret band) is an excellent diagnostic feature and typically has a large extinction coefficient of around 100,000 M^{-1} cm^{-1} at its absorption maximum. Redox state change invariably leads to alteration of heme spectrum. For instance, mitochondrial respiratory cytochrome c undergoes substantial changes in optical spectrum on heme iron reduction, with large increases in absorption of the Soret band, and in the visible region in which the smaller heme α- and β-bands (or Q-bands) are found. Cytochrome c is often used as a substrate in reactions with enzymes such as cytochrome c oxidase and the diflavin reductase class of NADPH-dependent enzymes (e.g., NOS), and the large cytochrome c absorption increase at 550 nm on reduction (and vice versa on oxidation) is followed (*ca* 21-23,000 M^{-1} cm^{-1}).[121] Perturbation of a hemoprotein absorption spectrum on ligand association is common, particularly if the ligand replaces a heme iron axial ligand, or else binds at a vacant coordination position on the iron. Figure 5 panel A shows a typical situation in the case of a cytochrome P450 enzyme (CYP121 from *M. tuberculosis*). The resting (ferric) form of the b-type heme-containing enzyme has its Soret absorption maximum at ~416 nm. On binding the tightly-associating ligand econazole, the azole nitrogen coordinates to the ferric P450 heme iron *trans* to the protein cysteinate ligand, and the CYP121 Soret band shifts to 424 nm. On reduction of the P450 heme iron to its ferrous state followed by binding of CO, a split Soret feature is seen with maxima at ~419 nm and ~448 nm. These spectral features report on formation of hexacoordinate, ferrous Cys-Fe-CO complex in which the cysteine is in the thiol (419 nm) and thiolate (448 nm) forms, respectively.[122] The binding of substrate molecules to P450 enzymes also frequently effects the removal of a weakly bound water (or hydroxide) ligand from the 6th coordination position on the heme iron, and a concomitant switch in ferric heme iron spin-state from low-spin (S = 1/2) to high-spin (S = 5/2) as a result of electronic reorganization in the heme iron d-orbitals. A Soret spectral shift to ~390 nm is observed, and this absorption shift can provide the basis for spectroscopic titrations to enable the determination of K_d

for substrate binding.[123] Optical changes associated with transient reactive intermediate heme states (e.g., oxy complexes or ligand switchovers) are also accessible using low-temperature and/or stopped-flow absorption spectroscopy (e.g., ref. 124). In addition, the large optical changes associated with heme iron redox state change enable spectroelectrochemical approaches to be applied as a means of determining the midpoint reduction potential for various heme iron redox transitions relevant to enzyme function.[125] For example, Soret optical changes were titrated electrochemically in reductive and oxidative directions to determine the midpoint potential for the heme iron Fe^{3+}/Fe^{2+} transition in the P450 BM3 fatty acid oxidase system from *Bacillus megaterium*, and then to demonstrate that the binding of a lipid substrate led to the potential becoming more positive by ~130 mV.[78]

The heme chromophore is not fluorescent to any significant extent, meaning that fluorimetry of heme in generally not a viable approach to characterizing heme properties. However, other powerful spectroscopic approaches exist. Important techniques include magnetic circular dichroism (MCD) and electron paramagnetic resonance (EPR). MCD is a measure of the differential absorption of left and right circularly polarised light, as a function of wavelength, in the presence of a magnetic field orientated parallel to the direction of light propagation. Transitions observed in the absorption spectrum correspond to features in the MCD spectrum, but the signed nature of the latter provides far greater resolution and detail.[126,127] Furthermore, for paramagnetic chromophores (such as ferric heme iron and for selected higher oxidation states observed in certain heme enzyme reaction cycles), the MCD spectrum is also temperature dependent, allowing the extraction of magnetic parameters.[126,128] The electronic bands of heme in the UV-visible region are essentially π-π^* transitions located on the porphyrin, but these are significantly influenced by the properties of the iron. Variations in the form of the UV-visible MCD spectrum are thus diagnostic of spin and oxidation state. This correlation is well established for the common *b*- and *c*-type hemes and for others, for example heme *o*, which contain equivalent delocalised π-systems.[127,129] Comparable variations are observed in the MCD spectra of other modified hemes, such as heme d_1 or heme *a* which vary in the extent of the π-conjugation, but far fewer data are currently available with which to establish definitive benchmarks. Fe^{3+} *b*- and *c*-type hemes also give rise to additional porphyrin(π)-to-Fe^{III}(d) charge-transfer (CT) transitions at wavelengths longer than 600 nm and into the near-infrared (NIR) region at ~800-2500 nm.[127,130,131] Vibrational overtones arising from solvent, buffer and protein preclude the measurement of these bands using absorption spectroscopy.[126] They can, however, be located using MCD. It has been established that the energy of a single positively signed NIR-MCD CT transition varies systematically with the nature of the heme axial ligands, and this has been used to establish a diagnostic scale for the identification of these ligands.[127,132] This is illustrated in the lower part of (Fig. 5) panel C for four different heme iron axial ligand combinations. A similar correlation between the axial ligands and the position of two derivative-shaped MCD CT bands has now been established for high-spin Fe^{III} hemes, although fewer examples are available in these cases.[133] For example, (Fig. 5) panel C (upper) shows the blue shift in the position of these bands in alkaline myoglobin.

In EPR spectroscopy, transitions between ground state magnetic sub-levels (m_S components) are induced by fixed-frequency microwave radiation in a magnetic field swept experiment. Whereas

Figure 5, viewed on following page. Spectroscopic properties of hemoproteins. Panel a shows the UV-visible absorption spectra for the *M. tuberculosis* P450 enzyme CYP121 (*ca* 7 μM) in its oxidised (ferric) form (black spectrum); in complex with the heme-coordinating azole drug econazole (red); and in the reduced (ferrous) complex with CO (blue spectrum). The spectra detail the extensive changes feasible on alteration of heme iron oxidation-, spin- and ligand-bound- status. In oxidised form, the presence of species representing both low-spin (416 nm) and high-spin (~390 nm) forms are evident from the Soret absorption. In the Fe^{3+}-econazole complex, the absorption is shifted to 424 nm. In the Fe^{2+}-CO complex, peaks at both 448 nm and 419 nm reflect the presence of reduced heme iron coordinated by cysteine thiolate and thiol ligands, respectively. Panel b shows the X-band EPR spectrum of the decaheme protein OmcA from *Shewanella frigidimarina* (black line). The red and blue lines indicate the large g_{max} and rhombic EPR features arising from perpendicular and parallel histidine planes, respectively. The inset shows the single near-infra red (NIR) MCD charge-transfer (CT) feature at a wavelength characteristic of bis-histidine ligation. Panel c shows examples of MCD CT bands for high-spin (top) and low-spin (bottom) Fe^{III} hemes with various different axial heme iron ligand sets.

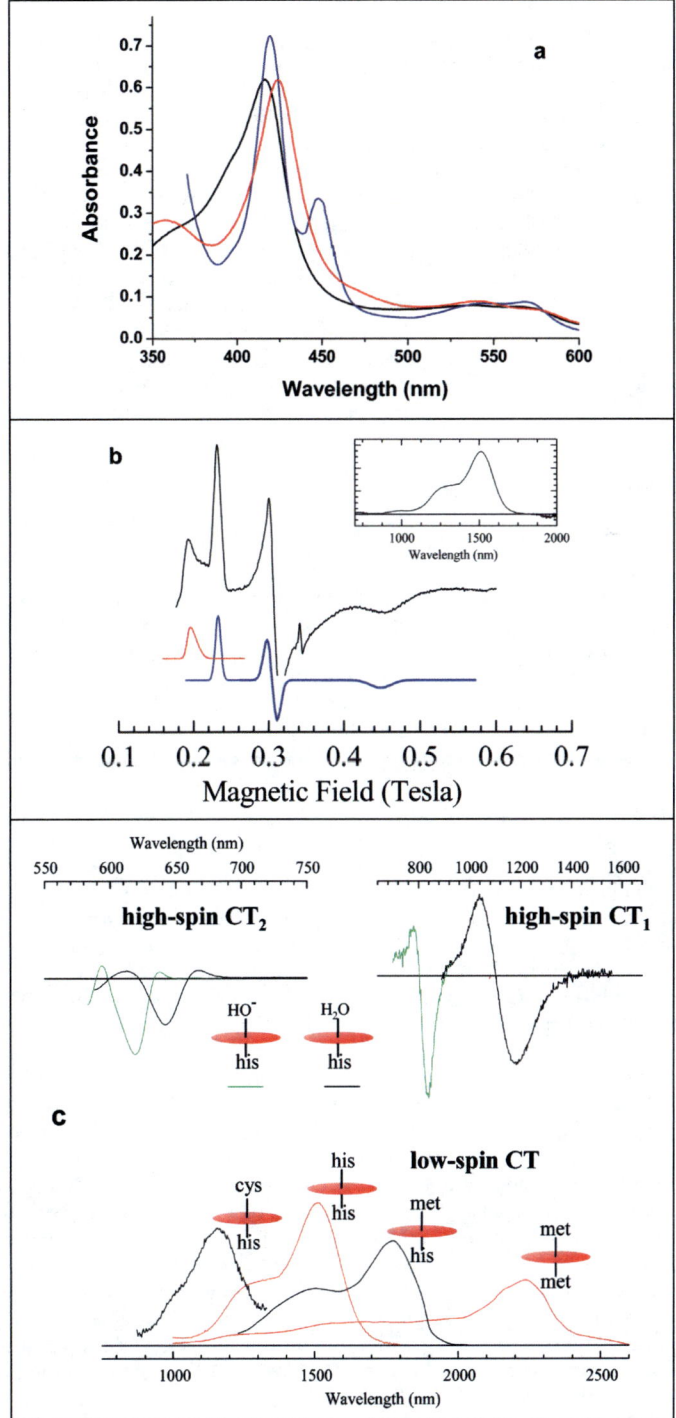

Figure 5, legend viewed on previous page.

detailed MCD spectra are obtained from hemes in any oxidation state, EPR spectra are, in general, detectable only for odd-electron (Kramers) systems; typically the Fe^{3+} state in the case of hemes. In both the high- and low-spin Fe^{3+} state, the EPR spectrum arises from transitions between the $m_S = \pm 1/2$ levels. While EPR can readily distinguish between the high- and low-spin states, it cannot unequivocally identify heme ligands. Although historically attempts have been made to do this for low-spin Fe^{3+} hemes,[134] the spectra are primarily determined by the relative orientation of the two ligands and not their identity. EPR spectra are therefore sensitive to more subtle changes in heme coordination. For example, *bis*-histidine ligation will invariably give rise to an MCD CT transition at 1500-1600 nm, but the EPR will vary according to the relative orientation of the two ligand planes. This gives rise to two limiting types of EPR spectrum, the "rhombic" for parallel planes and the "large g_{max}" for perpendicular.[135] This is illustrated in (Fig. 5) panel B for the case of OmcA, a decaheme protein from *Shewanella frigidimarina*.[136] All ten hemes have *bis*-histidine ligation and so the protein gives rise to a single MCD CT band in the NIR region at 1510 nm (inset). However, the ligand conformations are a mixture of perpendicular and parallel, resulting in both types of EPR spectrum. While EPR spectroscopy is perhaps a less useful technique than MCD with respect to identification of heme ligands, it is a powerful quantitative method, able to accurately define heme iron concentration. This is invaluable, since it frequently enables the definition of accurate optical extinction coefficients for individual heme proteins, through correlation of EPR signal intensity with UV-visible absorption properties.

Resonance Raman (RR) spectroscopy is another method frequently used to investigate structure and mechanism of heme cofactors. Raman spectroscopy involves inelastic scattering (or Raman scattering) of monochromatic light (usually delivered from a laser) by the irradiated sample, resulting in changes in energy of the laser photons that in turn report on the vibrational and rotational properties of the sample. However, Raman scattering signals are typically weak, and resonance Raman is a far more sensitive and informative method for interrogation of heme systems. In resonance Raman, the laser excitation is matched to a particular electronic transition of the sample, such that vibrational modes associated with the excited electronic state of the chromophore are substantially and selectively enhanced. RR often enables an increase in sensitivity of $\sim 10^6$ by comparison with ordinary Raman spectroscopy. Thus, despite a large hemoprotein having thousands of vibrational modes, RR spectroscopy enables one to focus specifically on the vibrational modes of the heme centre. In the case of heme-containing proteins, laser excitation in the Soret region is typically used, and a number of characteristic vibrational bands have been classified for heme and hemoproteins.[137]

RR spectroscopy, using Soret laser excitation, has been used successfully for several years in the study of numerous hemoproteins. A series of diagnostic heme vibrational features have been identified, and many have been correlated with specific properties of the heme iron and/or of the porphyrin macrocycle and its substituent groups.[137,138] Of particular importance is the high-frequency region (1000-1700 cm^{-1}) of the Raman spectrum, in which features reporting on e.g., the heme iron oxidation and coordination state are found. The position of the ν_4 vibrational mode (in the region from $\sim$1340-1380 cm^{-1}) is sensitive to electron density on the heme and is a good reporter of the heme oxidation state.[139] For instance, for the *B. megaterium* P450 BM3 enzyme it was shown that the ferric state of the P450 had its ν_4 feature at 1371 cm^{-1}, similar to that for the ferric *M. tuberculosis* CYP121 (1372 cm^{-1}).[122,140] The comparable feature for the ferric *Psuedomonas putida* P450 cam enzyme is at 1370 cm^{-1}, shifting to 1345 cm^{-1} on reduction of the heme iron.[141] The ν_3 vibrational band (in the region from 1475-1520 cm^{-1}) is sensitive to both coordination state and spin-state of the heme iron, while the ν_2 band (from 1560-1590 cm^{-1}) is sensitive to spin-state. For CYP121, the ν_3 feature is split into two components (at 1500 cm^{-1} and 1487 cm^{-1}), reflecting both low-spin and high-spin heme iron components, respectively.[122] For P450 BM3, there is a substrate (palmitate)-induced heme iron spin-state shift from low-spin towards high-spin, with concomitant shift of the ν_2 feature from 1585 cm^{-1} to 1575 cm^{-1}.[140] Other important features are located in the low frequency region of the heme RR spectrum ($\sim$200-800 cm^{-1}), including modes reporting on heme iron ligand bending and stretching, and heme out-of-plane modes.[139] For instance, in soluble guanylate cyclase, a mode at 204 cm^{-1} was assigned to a heme Fe^{2+}-histidine ligand stretch. On binding of NO, the histidine ligand was displaced in favour of NO (the ferrous heme remaining 5-coordinate) and new features at 1677 cm^{-1} and 525 cm^{-1} (both sensitive to isotopic labelling) were attributed to a N-O stretching vibration and a Fe-NO stretching vibration, respectively.[142] The heme iron-ligand stretch can be diagnostic of the nature of protein ligands, and in P450 BM3 a band at 350 cm^{-1} was assigned as the Fe^{3+}-cysteine thiolate stretch.[140] In P450 cam, a band associated with the stretching of NO in a Fe^{2+}-NO heme iron complex was shown to be resolved into two peaks (551 cm^{-1} and 561 cm^{-1}), and the binding of the redox partner (the 2Fe-2S protein putidaredoxin in its reduced form) was shown to increase intensity of the high frequency component, providing evidence for the perturbation of the electronic structure of the heme on redox partner binding.[143] In other RR studies with P450 BM3, it was shown that coordination of ferric heme iron by imidazole induced a more planar arrangement of the heme macrocycle and led to heme vinyl groups becoming coplanar with the heme ring.[138] Recent studies have highlighted the application of RR spectroscopy in detection of transient catalytic cycle intermediates, with the assignment of vibrational data to the highly transient ferric-hydroperoxo intermediate in the P450 cam catalytic cycle, and (again at cryogenic temperatures) for the short-lived ferric-hydroperoxo intermediate of myoglobin.[144,145]

Novel Aspects and Future Prospects

The field of heme and hemoproteins is vast, and this chapter is, by necessity, highly selective in terms of key details pertaining to aspects of structure, function and synthesis of these molecules. Structural studies have clearly defined that there is no "common fold" for hemoproteins and that an extremely diverse set of protein structural arrangements are compatible with the binding of heme macrocycles (see Fig. 3). It is clear that hydrogen bonding, ionic and other interactions between the non-covalently bound heme macrocycle and the protein scaffold are of at least equal importance to stabilizing heme binding as axial coordination by amino acid ligands.[146] However, much remains to be learned relating to why certain types of hemes are selected for different hemoprotein functions, and as to why heme is covalently linked to certain hemoproteins. In the former case, there is clear evidence that different heme types exhibit different thermodynamic properties. For instance, heme d_1 is more saturated than heme b and as a consequence the heme iron reduction potential is increased by ~200 mV.[147] The modifications made to heme a also make this heme more hydrophobic and more electron-withdrawing than heme b.[148] Redox potential of the heme centre can also be modified by the nature of the axial heme ligands. However, there are considerable heme iron reduction potential variations between proteins with the same ligand sets. For instance, bis-His coordinated hemes can vary in Fe^{3+}/Fe^{2+} midpoint potential by well over 700 mV in their various protein scaffolds, highlighting the importance of other protein-heme interactions and overall heme environment in controlling heme thermodynamic properties.[148] With respect to covalent heme binding (as in cytochromes c), it has been proposed that this mechanism has its origins in the prevention of heme loss from bacterial periplasmic cytochromes c (an environment in which de novo heme synthesis does not occur), and that the mitochondrial c-type cytochromes are evolutionary descendants of these periplasmic forms.[149] The requirement for covalent heme ligation (in this case involving an unusual histidine link to a heme b 2-vinyl group) to prevent ferrous heme dissociation was recently demonstrated in the case of a *Synechocystis* hemoglobin.[150] However, there are several arguments against the prevention of heme loss being the major reason for covalent heme ligation. These include the fact that yeasts do not possess any of the genes required for bacterial-type c-type heme biogenesis (i.e., distinct systems for heme c post-translational modification have evolved in prokaryotes and eukaryotes). More compelling explanations for covalent heme ligation may come from examination of the biological diversity of the c-type cytochromes, and could involve minimizing the ratio of amino acids to heme in a protein, or enabling different types of folded hemoprotein structures that would not be available were the heme(s) not covalently-linked.[35] Clearly much remains to be learned regarding the matching of heme type to protein function, and the mechanisms by which the hemoprotein scaffolds control the ligand-binding, thermodynamic and catalytic properties of their heme cofactors.

The heme cofactor exerts strong regulation over the pathway for its own synthesis, for instance at various steps in the pathway to δ-ALA in all organisms, and at the level of the PBG synthase.[14] It has been shown that regulation of mammalian 5-aminolevulinate synthase by heme occurs through heme-dependent destabilization of its mRNA.[151] At the translational level, it is well known that heme regulates translational control mediated by the mammalian heme-regulated inhibitor kinase. Effects are controlled by phosphorylation of the α subunit of the eukaryotic translational initiation factor 2 (eIF2α), which in turn inhibits global protein synthesis in erythroid cells and enables tight regulation over synthesis of alpha- and beta-globins; thus enabling tight control over hemoglobin formation and ensuring stoichiometric heme incorporation.[152] Aside from control of protein expression at the transcriptional and translational levels, heme is also now known to be a critical component of the DGCR8 protein, which is an essential protein for the preliminary step in microRNA processing.[117] Heme can also bind directly to protein to exert regulatory effects, as exemplified by the inhibition of DNA repressor activity in the mammalian Bach1 transcription factor on association of heme with Cys-Pro motifs in the protein.[114] The first enzyme in the pathway for heme degradation is also a hemoprotein. Heme oxygenase cleaves the heme to produce biliverdin, with release of both Fe^{2+} and CO, and this is the only known physiological pathway that generates CO.[153] CO is a heme toxin (i.e., inhibitory ligand to ferrous heme iron) and a respiratory inhibitor, but also a physiological effector (e.g., via action on sGC). CO is now well recognized to be involved in a number of physiological processes, such as vasodilation and apoptosis.[154] CO induces vasodilation through the activation of smooth muscle large-conductance Ca^{2+}-activated potassium channels. Recent studies indicate that heme itself is an inhibitor of these Ca^{2+} channels, but that CO binds to the heme to effect activation of the channel.[155] NO is also the product of a hemoprotein (NOS) and is similarly implicated in a plethora of physiological functions.[11] Thus, intricate regulatory networks between heme, its ligands and pathways of biological systems regulation at transcriptional/microRNA, translational and protein levels are obvious, and remain a major focus of scientific research.

With respect to hemoprotein catalysis, there is great research interest both in terms of identifying novel heme-containing catalysts and for adapting the activities of heme proteins and enzymes to make them useful for biotechnological exploitation, and to investigate structure and function relationships. Novel roles for heme and hemoproteins are regularly found. Recent examples include (i) the potentiation of the antibacterial activity of IgG following binding with heme, suggesting an inducible, innate-type defence mechanism against invading bacterial pathogens;[156] (ii) the identification of heme oxygenase-1 (also known as HSP32) as a survival factor and possible therapeutic target in neoplastic mast cells;[157] (iii) the demonstration that deoxymyoglobin is an efficient nitrite reductase (generating NO), with possible ramifications for the role of the protein in the cellular regeneration of NO;[158] (iv) a potential role for heme in Alzheimer's disease as a consequence of its formation of a complex with the amyloid beta peptide, and prevention of aggregation of the peptide. The amyloid beta peptide is strongly implicated in the pathophysiology of Alzheimer's disease;[159,160] and (v) the affinity of certain heme proteins (e.g., metmyoglobin and metcytochrome *c*) for double stranded DNA, and hydrogen peroxide driven endonuclease activity catalysed by these heme proteins.[161] As regards adaptation and evolution of existing hemoprotein activities, much attention has focused on P450 enzymes, with the aim being the generation of high value oxygenated molecules that may be difficult or inefficient to synthesise by organic chemistry routes. For instance, forced evolution approaches have been used to produce variants of the P450 BM3 enzyme with enhanced alkane hydroxylase activity, and chimeragenesis was employed to generate hybrid P450 enzymes with enhanced thermostability and some novel functions.[162,163] Rational mutagenesis approaches have also been used to generate novel P450 BM3 mutants with short chain fatty acid hydroxylase activity and, more recently, to generate a variety of novel heme iron ligation states by systematic variation of an amino acid residue at the distal face of the BM3 heme.[164-166] Heme ligand-switch mutagenesis enabled establishment of the spectroscopic and thermodynamic features of the novel heme iron ligand sets (Cys-Glu, Cys-His and Cys-Lys) in these cases. Much interest has also surrounded the manufacture of peptide "maquettes" that bind and coordinate heme in different ways, in order to provide simple and tractable models for the active sites of important heme enzymes and to understand factors (e.g., hydrophobicity) that govern the reactivity of the heme centre (e.g. ref. 167, 168). Protein engineering has also been used to create novel genetic fusions of heme proteins to physiological and non-physiological redox partners, usually producing functional electron transfer systems and sometimes enhancing their efficiency.[77,169] The field of hemoprotein biocatalysis and engineering is enormous and covers enzymes involved in functions as diverse as energy generation, oxidations of steroids and lipids, gene regulation, detoxification, immune response and cellular signaling.

In conclusion, the hemes are clearly among the most biologically diverse and versatile cofactors, and participate in an extremely broad range of physiologically important functions. Their natural repertoire of functions continues to expand as a result of identification of novel heme enzymes and novel activities associated with known hemoproteins. Rational and random mutagenesis approaches to evolving hemoproteins further broaden the spectrum of heme-associated activities. Collectively, the hemoprotein field embraces virtually all major aspects of physiology and fuels massive research efforts worldwide.

Acknowledgements

We are grateful to the UK Biotechnology and Biological Sciences Research Council and EC (network grant NM4TB) for support of our research.

References

1. Pikuleva IA. Cytochrome P450s and cholesterol homeostasis. Pharmacol Ther 2006; 112:761-763.
2. Sarkis A, Roman RJ. Role of cytochrome P450 metabolites of arachidonic acid in hypertension. Curr Drug Metab 2004; 5:245-256.
3. Yoshikawa S, Muramoto K, Shinzawa-Itoh K et al. Reaction mechanism of bovine heart cytochrome c oxidase. Biochim Biophys Acta 2006; 1757:395-400.
4. Roberts GP, Kerby RL, Youn H et al. CooA, a paradigm for gas sensing regulatory proteins. J Inorg Biochem 2005; 99:280-292.
5. Poulos TL. Soluble guanylate cyclase. Curr Opin Struct Biol 2006; 16:736-743.
6. Lukin JA, Ho C. The structure-function relationship of hemoglobin in solution at atomic resolution. Chem Rev 104:1219-1230.
7. Cupp-Vickery JR, Han O, Hutchinson CR et al. Substrate-assisted catalysis in cytochrome P450eryF. Nat Struct Biol 1996; 3:632-637.
8. Chapman SK, Daff S, Munro AW. Heme: The most versatile redox centre in biology? Structure and Bonding 1997; 88:39-70.
9. Michel H, Behr J, Harrenga A et al. Cytochrome c oxidase: Structure and spectroscopy. Annu Rev Biophys Biomol Struct 1998; 27:329-356.
10. Cutruzzola F, Rinaldo S, Centola F et al. NO production by Pseudomonas aeruginosa cd1 nitrite reductase. IUBMB Life 2003; 55:617-621.
11. Stuehr DJ. Mammalian nitric oxide synthases. Biochim Biophys Acta 1999; 1411:217-230.
12. Munro AW, Girvan HM, McLean KJ. Variations on a (t)heme-novel mechanisms, redox partners and catalytic functions in the cytochrome P450 superfamily. Nat Prod Rep 2007; 24:585-609.
13. Isin EM, Guengerich FP. Complex reactions catalyzed by cytochrome P450 enzymes. Biochim Biophys Acta 2007; 1770:314-329.
14. Michal G. Tetrapyrroles. In: Michal G, ed. Biochemical Pathways: An Atlas of Biochemistry and Molecular Biology. New York: John Wiley and Sons, 1999:68-74.
15. Ferreira GC, Zhang JS. Mechanism of 5-aminolevulinate synthase and the role of the protein environment in controlling the cofactor chemistry. Cell Mol Biol 2002; 48:827-833.
16. Warren MJ, Scott AI. Tetrapyrrole assembly and modification into the ligands of biologically functional cofactors. Trends Biochem Sci 1990; 15:486-491.
17. Perutz MF. Myoglobin and haemoglobin: Role of distal residues in reactions with haem ligands. Trends Biochem Sci 1989; 14:42-44.
18. Makris TM, Davydov R, Denisov IG et al. Mechanistic enzymology of oxygen activation by the cytochromes P450. Drug Metab Rev 2002; 34:691-708.
19. Mogi T, Saiki K, Anraku Y. Biosynthesis and functional role of haem O and haem A. Mol Microbiol 1994; 14:391-398.
20. Svensson B, Hederstedt L. Bacillus subtilis CtaA is a heme-containing membrane protein involved in heme A biosynthesis. J Bacteriol 1994; 176:6663-6671.
21. Lewin A, Hederstedt L. Compact archaeal variant of heme A synthase. FEBS Lett 2006; 580:5351-5356.
22. Michel H, Behr J, Harrenga A et al. Cytochrome c oxidase: Structure and spectroscopy. Annu Rev Biophys Biomol Struct 1998; 27:329-356.
23. Puustinen A, Wikström M. The heme groups of cytochrome o from Escherichia coli. Proc Natl Acad Sci USA 1991; 88:6122-6126.
24. Glerum DM, Tzagoloff A. Isolation of a human cDNA for heme a: Farnesyltransferase by functional complementation of a yeast COX10 mutant. Proc Natl Acad Sci USA 1994; 91:8452-8456.
25. Brown BM, Wang Z, Brown KR et al. Heme O synthase and heme A synthase from Bacillus subtilis and Rhodobacter sphaeroides interact in Escherichia coli. Biochemistry 2004; 43:13541-13548.
26. Lemberg R, Falk JE. Comparison of haem a, the dichroic haem of heart muscle, and of porphyrin a with compounds of known structure. Biochem J 1951; 49:674-683.
27. Timkovich R, Cork MS, Gennis RB et al. Proposed structure of heme d, a prosthetic group of bacterial terminal oxidases. J Am Chem Soc 1985; 107:6069-6075.
28. Vos MH, Borisov VB, Liebl U et al. Femtosecond resolution of ligand-heme interactions in the high-affinity quinol oxidase bd: A di-heme active site? Proc Natl Acad Sci USA 2000; 97:1554-1559.

29. Cheesman MR, Ferguson SJ, Moir JW et al. Two enzymes with a common function but different heme ligands in the forms as isolated: Optical and magnetic properties of the heme groups in the oxidised forms of nitrite reductase, cytochrome cd1, from Pseudomonas stutzeri and Thiosphaera pantotropha. Biochemistry 1997; 36:16267-16276.

30. Chang CK, Wu W. The porphinedione structure of heme d_1. Synthesis and spectral properties of model compounds of the prosthetic group of dissimilatory nitrite reductase. J Biol Chem 1986; 261:8593-8596.

31. Meyer TE. Evolution and classification of c-type cytocromes. Ch. 2. In: Scott RA, Mauk G, eds. Cytochrome c, A Multidisciplinary Approach. Sausalito: University Science Books, 1996:33-99.

32. Leys D, Tsapin AS, Nealson KH et al. Structure and mechanism of the flavocytochrome c fumarate reductase of Shewanella putrefaciens MR-1. Nat Struct Biol 1999; 6:1113-1117.

33. Stevens JM, Uchida T, Daltrop O et al. Covalent cofactor attachment to proteins: Cytochrome c biogenesis. Biochem Soc Trans 2005; 33:792-795.

34. Czjzek M, Guerlesquin F, Bruschi M et al. Crystal structure of a dimeric octaheme cytochrome c_3 (M_r 26,000) from Desulfovibrio desulfuricans Norway. Structure 1996; 4:395-404.

35. Barker PD, Ferguson SJ. Still a puzzle: Why is haem covalently attached in c-type cytochromes? Struct Fold Des 1999; 7:R281-R290.

36. Allen JW, Harvat EM, Stevens JM et al. A variant System I for cytochrome c biogenesis in archaea and some bacteria has a novel CcmE and no CcmH. FEBS Lett 2006; 580:4827-4834.

37. Feissner RE, Richard-Fogal CL, Frawley ER et al. Recombinant cytochromes c biogenesis systems I and II and analysis of haem delivery pathways in Escherichia coli. Mol Microbiol 2006; 60:563-577.

38. Stevens JM, Daltrop O, Allen JWA et al. C-type cytochrome formation: Chemical and biological enigmas. Acc Chem Res 2004; 37:999-1007.

39. Allen JW, Ginger ML, Ferguson SJ. Maturation of the unusual single-cysteine (XXXCH) mitochondrial c-type cytochromes found in trypanosomatids must occur through a novel biogenesis pathway. Biochem J 2004; 383:537-542.

40. Schulz H, Hennecke H, Thony-Meyer L. Prototype of a heme chaperone essential for cytochrome c maturation. Science 1998; 281:1197.

41. Schulz H, Fabianek RA, Pellicioli EC et al. Heme transfer to the heme chaperone CcmE during cytochrome c maturation requires the CcmC protein, which may function independently of the ABC-transporter CcmAB. Proc Natl Acad Sci USA 1999; 96:6462.

42. Allen JW, Ferguson SJ. The Escherichia coli cytochrome c maturation (Ccm) apparatus can mature cytochromes with an extra cysteine within or adjacent to the CXXCH motif. Biochem Soc Trans 2006; 34:91-93.

43. Allen JW, Harvat EM, Stevens JM et al. A variant System I for cytochrome c biogenesis in archaea and some bacteria has a novel CcmE and no CcmH. FEBS Lett 2006; 580:4827-4834.

44. Colas C, Ortiz de Montellano PR. Autocatalytic radical reactions in physiological prosthetic heme modification. Chem Rev 2003; 103:2305-2332.

45. Furtmüller PG, Zederbauer M, Jantscko W et al. Active site structure and catalytic mechanisms of human peroxidases. Arch Biochem Biophys 2006; 445:199-213.

46. Kussendrager KD, van Hooijdonk AC. Lactoperoxidase: Physico-chemical properties, occurrence, mechanism of action and applications. Br J Nutr 2000; 84:S19-S25.

47. Rae TD, Goff HM. The heme prosthetic group of lactoperoxidase: Structural characteristics of heme 1 and heme 1-peptides. J Biol Chem 1998; 43:27968-27977.

48. Oxvig C, Thomsen AR, Overgaard MT et al. Biochemical evidence for heme linkage through esters with Asp-93 and Glu-241 in human eosinophil peroxidise: The ester with Asp-93 is only partially formed in vivo. J Biol Chem 1999; 274:16953-16958.

49. Lau D, Baldus S. Myeloperoxidase and its contributory role in inflammatory vascular disease. Pharmacol Ther 2006; 111:16-26.

50. Fenna R, Zeng J, Davey C. Structure of green heme in myeloperoxidase. Arch Biochem Biophys 1995; 316:653-656.

51. Fiedler TJ, Davey CA, Fenna RE. X-ray crystal structure and characterization of halide-binding sites of human myeloperoxidase at 1.8 Å resolution. J Biol Chem 2000; 275:11964-11971.

52. Furtmüller PG, Burner U, Obinger C. Reaction of myeloperoxidase compound I with chloride, bromide, iodide, and thiocyanate. Biochemistry 1998; 37:17923-17930.

53. Pipirou Z, Bottrill AR, Metcalfe CM et al. Autocatalytic formation of a covalent link between tryptophan 41 and the heme in ascorbate peroxidase. Biochemistry 2007; 46:2174-2180.

54. Ghiladi RA, Knudsen GM, Medzihradszky KF et al. The Met-Tyr-Trp cross link in Mycobacterium tuberculosis catalase peroxidise. J Biol Chem 2005; 280:22651-22663.

55. Henne KR, Kunze KL, Zheng YM et al. Covalent linkage of prosthetic heme to CYP4 family P450 enzymes. Biochemistry 2001; 40:12925-12931.

56. LeBrun LA, Xu F, Kroetz DL et al. Covalent attachment of the heme prosthetic group in the CYP4F cytochrome P450 family. Biochemistry 2002; 41:5931-5937.
57. Limburg J, LeBrun LA, Ortiz de Montellano PR. The P450cam G248E mutant covalently binds its prosthetic heme group. Biochemistry 2005; 44:4091-4099.
58. Pearson AR, Elmore BO, Yang C et al. The crystal structure of cytochrome P460 of Nitrosomonas europaea reveals a novel cytochrome fold and heme-protein cross-link. Biochemistry 2007; 46:8340-8349.
59. Arciero DM, Hooper AB. Hydroxylamine oxidoreductase from Nitrosomonas europaea is a multimer of an octa-heme subunit. J Biol Chem 1993; 268:14645-14654.
60. Igarashi N, Moriyama H, Fujiwara T et al. The 2.8 Å structure of hydroxylamine oxidoreductase from a nitrifying chemoautotrophic bacterium, Nitrosomonas europaea. Nat Struct Biol 1997; 4:276-284.
61. Swamy U, Wang M, Tripathy JN et al. Structure of spinach nitrite reductase: Implications for multi-electron reactions by the iron-sulfur:siroheme cofactor. Biochemistry 2005; 44:16054-16063.
62. Kolko MM, Kapetanovich LA, Lawrence JG. Alternative pathways for siroheme synthesis in Klebsiella aerogenes. J Bacteriol 2002; 183:328-335.
63. Arciero DM, Collins MJ, Haladjian J et al. Resolution of the four hemes of cytochrome c_{554} from Nitrosomonas europaea by redox potentiometry and optical spectroscopy. Biochemistry 1991; 30:11459-11465.
64. Iverson TM, Arciero DM, Hsu BT et al. Heme packing motifs revealed by the crystal structure of the tetraheme cytochrome c554 from Nitrosomonas europaea. Nature Struct Biol 1998; 5:1005-1012.
65. Santos H, Moura JJG, Moura I et al. NMR studies of electron transfer mechanisms in a protein with interacting redox centres: Desulfovibrio gigas cytochrome c_3. Eur J Biochem 1984; 141:283-296.
66. Bruschi M, Woudstra M, Guigliarelli B et al. Biochemical and spectroscopic characterization of two new cytochromes isolated from Desulfuromonas acetoxidans. Biochemistry 1997; 36:10601-10608.
67. Tsukihara T, Aoyama H, Yamashita E et al. Structures of metal sites of oxidised bovine heart cytochrome c oxidase at 2.8 Å. Science 1995; 269:1069-074.
68. Schenkman JB, Jansson I. The many roles of cytochrome b_5. Pharmacol Ther 2003; 97:139-152.
69. Kostanjevecki V, Leys D, Van Driessche G et al. Structure and characterization of Ectothiorhodospira vacuolata cytochrome b_{558}, a prokaryotic homologue of cytochrome b_5. J Biol Chem 1999; 274:35614-35620.
70. Hildebrandt A, Estabrook RW. Evidence for the participation of cytochrome b_5 in hepatic microsomal mixed-function oxidation reactions. Arch Biochem Biophys 1971; 143:66-79.
71. Hegesh E, Hegesh J, Kaftory A. Congenital methemoglobinemia with a deficiency of cytochrome b_5. N Eng J Med 1986; 414:757-761.
72. Xia D, Yu CA, Kim H et al. Crystal structure of the cytochrome bc1 complex from bovine heart mitochondria. Science 1997; 277:60-66.
73. Williams PA, Fülöp V, Garman EF et al. Haem-ligand switching during catalysis in crystals of a nitrogen-cycle enzyme. Nature 1997; 389:416-412.
74. Fülöp V, Moir JWB, Ferguson SJ et al. The anatomy of a bifunctional enzyme: Structural basis for reduction of oxygen to water and synthesis of nitric oxide by cytochrome cd_1. Cell 1995; 81:369-377.
75. Hamada K, Bethge PH, Mathews FS. Refined structure of cytochrome b_{562} from Escherichia coli at 1.4 Å resolution. J Mol Biol 1995; 247:947-962.
76. Perera R, Sono M, Sigman JA et al. Neutral thiol as a proximal ligand to ferrous heme iron: Implications for heme proteins that lose cysteine thiolate ligation on reduction. Proc Natl Acad Sci USA 2003; 100:3641-3646.
77. Munro AW, Girvan HM, McLean KJ. Cytochrome P450-redox partner fusion enzymes. Biochim Biophys Acta 2007; 1770:345-359.
78. Daff SN, Chapman SK, Turner KL et al. Redox control of the catalytic cycle of flavocytochrome P-450 BM3. Biochemistry 2007; 36:13816-13827.
79. Denisov IG, Makris TM, Sligar SG et al. Structure and chemistry of cytochrome P450. Chem Rev 2005; 105:2253-2277.
80. Omura T, Sato R. The carbon monoxide binding pigment of liver microsomes. I. Evidence for its hemoprotein nature. J Biol Chem 1964; 239:2370-2378.
81. McLean KJ, Warman AJ, Seward HE et al. Biophysical characterization of the sterol demethylase P450 from Mycobacterium tuberculosis, its cognate ferredoxin, and their interactions. Biochemistry 2006; 45:8427-8443.
82. Sundaramoorthy M, Terner J, Poulos TL. The crystal structure of chloroperoxidase: A heme peroxidase-cytochrome P450 functional hybrid. Structure 1995; 3:137-1377.
83. Marletta MA, Hurshman AR, Ruschke KM. Catalysis by nitric oxide synthase. Curr Op Chem Biol 1998; 2:656-663.

84. Gorren AC, Mayer B. Nitric-oxide synthase: A cytochrome P450 family foster child. Biochim Biophys Acta 2007; 1770:432-435.
85. Stehle T, Schulz GE. Three-dimensional structure of the complex of guanylate kinase from yeast with its substrate GMP. J Mol Biol 1990; 211:249-254.
86. Stuehr DJ, Santolini J, Wang ZQ et al. Update on mechanism and catalytic regulation in the NO synthases. J Biol Chem 2004; 279:36167-36170.
87. Wang ZQ, Lawson RJ, Buddha MR et al. Bacterial flavodoxins support nitric oxide production by Bacillus subtilis nitric-oxide synthase. J Biol Chem 2007; 282:2196-2202.
88. Hirai K, Martinkova M, Igarashi J et al. Identification of Cys385 in the isolated kinase insertion domain of heme-regulated eIF2a kinase (HRI) as the heme axial ligand by site-directed mutagenesis and spectral characterization. J Inorg Biochem 2007; 101:1172-1179.
89. Cheesman MR, Little PJ, Berks BC. Novel heme ligation in a c-type cytochrome involved in thiosulfate oxidation: EPR and MCD of SoxAX from Rhodovulum sulfidophilum. Biochemistry 2001; 40:10562-10569.
90. Frolow G, Kalb (Gilboa) AJ, Yariv J. Structure of a unique twofold symmetric haem binding site. Nature Struct Biol 1994; 1:453-460.
91. Vainshtein BK, Melik-Adamyan WR, Barynin VV et al. Three dimensional structure of catalase from Penicillium vitale at 2.0 Å resolution. J Mol Biol 1986; 188:499-461.
92. Martinez SE, Huang D, Szczepaniak A et al. Crystal structure of chloroplast cytochrome f reveals a novel fold and unexpected heme ligation. Structure 1994; 2:95-105.
93. Yamashita T, Hoashi Y, Watanabe K et al. Roles of heme axial ligands in the regulation of CO binding to CooA. J Biol Chem 2004; 279:21394-21400.
94. Lanzilotta WN, Schuller DJ, Thorsteinsson MV et al. Structure of the CO sensing transcription activator CooA. Nat Struct Biol 2000; 7:876-880.
95. Borjigin M, Li H, Lanz ND et al. Structure-based hypothesis on the activation of the CO-sensing transcription factor CooA. Acta Crystallogr D 2007; 63:282-287.
96. Crane BR, Siegel LM, Getzoff ED. Sulfite reductase structure at 1.6 Å: Evolution and catalysis for reduction of inorganic anions. Science 1995; 270:59-67.
97. Leys D, Backers K, Meyer TE et al. Crystal structures of an oxygen-binding cytochrome c from Rhodobacter sphaeroides. J Biol Chem 2000; 275:16050-16056.
98. Klarskov K, Leys D, Backers K et al. Cytochrome c from the obligate methylotroph Methylophilus methylotrophus, an unexpected homolog of sphaeroides heme protein from the phototroph Rhodobacter sphaeroides. Biochim Biophys Acta 1999; 1412:47-55.
99. Weichsel A, Maes EM, Andersen JF et al. Heme-assisted S-nitrosation of a proximal thiolate in a nitric oxide transport protein. Proc Natl Acad Sci USA 2005; 102:594-599.
100. Berry RE, Shokhireva TK, Filippov I et al. Effect of the N-terminus on heme cavity structure, ligand equilibrium, rate constants, and reduction potentials of nitrophorin 2 from Rhodnius prolixus. Biochemistry 2007; 46:6830-6843.
101. Yukl ET, Ioanoviciu A, Ortiz de Montellano PR et al. Interdomain interactions within the two-component heme-based sensor DevS from Mycobacterium tuberculosis. Biochemistry 2007; 46:9728-9736.
102. Sousa EH, Tuckerman JR, Gonzalez G et al. DosT and DevS are oxygen-switched kinases in Mycobacterium tuberculosis. Protein Sci 2007; 16:1708-1719.
103. Lee YY, Shearer N, Spiro S. Transcription factor NNR from Paracoccus denitrificans is a sensor of both nitric oxide and oxygen: Isolation of nnr* alleles encoding effector-independent proteins and evidence for a haem-based sensing mechanism. Microbiology 2006; 152:1461-1470.
104. Sousa EH, Tuckerman JR, Gonzalez G et al. A memory of oxygen explains the dose response of the heme-based sensor FixL. Biochemistry 2007; 46:6249-257.
105. Yoshioka S, Kobayashi K, Yoshimura H et al. Biophysical properties of a c-type heme in chemot-axis signal transducer protein DcrA. Biochemistry 2005; 44:15406-15413.
106. Kumar A, Toledo JC, Patel RP et al. Mycobacterium tuberculosis DosS is a redox sensor and DosT is a hypoxia sensor. Proc Natl Acad Sci USA 2007; 104:11568-11573.
107. Singh S, Madzelan P, Banerjee R. Properties of an unusual heme cofactor in PLP-dependent cystathionine beta-synthase. Nat Prod Rep 2007; 24:631-639.
108. Sasakura Y, Yoshimura-Suzuki T, Kurokawa H et al. Structure-function relationships of EcDOS, a heme-regulated phosphodiesterase from Escherichia coli. Acc Chem Res 2006; 39:37-43.
109. Gilles-Gonzalez MA, Gonzalez G. Heme-based sensors: Defining characteristics, recent developments and regulatory hyptheses. J Inorg Biochem 2005; 99:1-22.
110. Li H, Raman CS, Glaser CB et al. Crystal structures of zinc-free and -bound heme domain of human inducible nitric-oxide synthase: Implications for dimer stability and comparison with endothelial nitric-oxide synthase. J Biol Chem 1999; 274:21276-21284.

111. Zhang L, Guarente L. Heme binds to a short sequence that serves a regulatory function in diverse proteins. EMBO J 1995; 14:313-320.
112. Munakata H, Sun JY, Yoshida K et al. Role of heme regulatory motif in the heme-mediated inhibition of mitochondrial import of 5-aminolevulinate synthase. J Biochem 2004; 136:233-238.
113. Ishikawa H, Kato M, Hori H et al. Involvement of heme regulatory motif in heme-mediated ubiquination and degradation of IRP2. Mol Cell 2005; 19:171-181.
114. Ogawa K, Sun J, Taketani S et al. Heme mediates derepression of Maf recognition element through direct binding to transcription repressor Bach1. EMBO J 2001; 20:2835-2843.
115. Huston WM, Andrew CR, Servid AE et al. Heterologous overexpression and purification of cytochrome c' from Rhodobacter capsulatus and a mutant (K42E) in the dimerization region. Mutation does not alter oligomerization but impacts the heme iron spin state and nitric oxide binding properties. Biochemistry 2006; 45:4388-4395.
116. Enguita FJ, Pohl E, Turner DL et al. Structural evidence for a proton transfer pathway coupled with haem reduction of cytochrome c" from Methylophilus methylotrophus. J Biol Inorg Chem 2006; 11:189-196.
117. Faller M, Matsunaga M, Yin S et al. Heme is involved in microRNA processing. Nat Struct Mol Biol 2007; 14:23-29.
118. Arnesano F, Banci L, Bertini I et al. The solution structure of rat microsomal cytochrome b_5. Biochemistry 1998; 37:173-184.
119. Kim SH, Perera R, Hager LP et al. Rapid freeze-quench ENDOR study of chloroperoxidase compound I: The site of the radical. J Am Chem Soc 2006; 128:5598-5599.
120. Garcia-Serres R, Davydov RM, Matsui T et al. Distinct reaction pathways followed upon reduction of oxy-heme oxygenase and oxy-myoglobin as characterized by Mossbauer spectroscopy. J Am Chem Soc 2007; 129:1402-1412.
121. Roitel O, Scrutton NS, Munro AW. Electron transfer in flavocytochrome BM3: Kinetics of flavin reduction and oxidation, the role of cysteine 999, and relationships with mammalian cytochrome P450 reductase. Biochemistry 2003; 42:10809-10821.
122. McLean KJ, Cheesman MR, Rivers SL et al. Expression, purification and spectroscopic characterization of the cytochrome P450 CYP121 from Mycobacterium tuberculosis. J Inorg Biochem 2002; 91:527-541.
123. Gustafsson MC, Roitel O, Marshall KR et al. Expression, purification, and characterization of Bacillus subtilis cytochromes P450 CYP102A2 and CYP102A3: Flavocytochrome homologues of P450 BM3 from Bacillus megaterium. Biochemistry 2004; 43:5474-5487.
124. Orii Y, Miki T. Oxidation process of bovine heart ubiquinol-cytochrome c reductase as studied by stopped-flow rapid-scan spectrophotometry and simulations based on the mechanistic Q cycle model. J Biol Chem 1997; 272:17594-17604.
125. Dutton PL. Redox potentiometry: Determination of midpoint potentials of oxidation-reduction components of biological electron-transfer systems. Methods Enzymol 1978; 54:411-435.
126. Thomson AJ, Cheesman MR, George SJ. Variable-temperature magnetic circular dichroism. Methods Enzymol 1993; 226:199-232.
127. Spinner F, Cheesman MR, Thomson AJ et al. The haem b_{558} component of the cytochrome bd quinol oxidase complex from Escherichia coli has histidine-methionine axial ligation. Biochem J 1995; 308:641-644.
128. Oganesyan VS, George SJ, Cheesman MR et al. A novel, general method of analyzing magnetic circular dichroism spectra and magnetization curves of high-spin metal ions: Application to the protein oxidised rubredoxin, Desulfovibrio gigas. J Chem Phys 1999; 110:762-777.
129. Cheesman MR, Watmough NJ, Pires CA et al. Cytochrome bo from Escherichia coli: Identification of haem ligands and reaction of the reduced enzyme with carbon monoxide. Biochem J 1993; 289:709-718.
130. Cheng JC, Osborne GA, Stephens PJ et al. Infrared magnetic circular dichroism in the study of metalloproteins. Nature 1973; 241:193-94.
131. McKnight J, Cheesman MR, Thomson AJ et al. Identification of charge-transfer transitions in the optical spectrum of low-spin ferric cytochrome P-450 Bacillus megaterium. Eur J Biochem 1993; 213:683-687.
132. Gadsby PMA, Thomson AJ. Assignment of the axial ligands of ferric ion in low-spin hemoproteins by near-infrared magnetic circular dichroism and electron paramagnetic resonance spectroscopy. J Am Chem Soc 1990; 112:5003-5011.
133. Dawson JH, Dooley DM. Magnetic circular dichroism spectroscopy of iron porphyrins and heme proteins. Ch. 1. Part 3. In: Lever ABP, Gray HB, eds. Iron Porphyrins. New York: VCH Publishers, 1989:1-135.

134. Blumberg WE, Peisach J. A unified theory for low spin forms of all ferric heme proteins as studied by EPR. In: Chance B, Yonetani T, Mildvan AS, eds. Probes of Structure and Function of Macromolecules and Membranes. Vol. 2. Probes of Enzymes and Hemoproteins. New York: Academic Press, 1971.

135. Walker FA. Magnetic spectroscopic (EPR, ESEEM, Mössbauer, MCD and NMR) studies of low-spin ferriheme centers and their corresponding heme proteins. Coord Chem Rev 1999; 185:471-534.

136. Field SJ, Dobbin PS, Cheesman MR et al. Purification and magneto-optical spectroscopic characterization of cytoplasmic membrane and outer membrane multiheme c-type cytochromes from Shewanella frigidimarina NCIMB400. J Biol Chem 2000; 275:8515-8522.

137. Abe M, Kitagawa T, Kyogoku T. Resonance Raman spectra of octaethylporphynato-Ni(II) and meso-deuterated and LSN substituted derivatives. II. A normal coordinate analysis. J Phys Chem 1978; 69:4526-4534.

138. Smith SJ, Munro AW, Smith WE. Resonance Raman scattering of cytochrome P450 BM3 and effect of imidazole inhibitors. Biopolymers 2003; 70:620-627.

139. Rousseau DL, Li D, Couture M et al. Ligand-protein interactions in nitric oxide synthase. J Inorg Biochem 2005; 99:306-323.

140. Miles JS, Munro AW, Rospendowski BN et al. Domains of the catalytically self-sufficient cytochrome P-450 BM-3: Genetic construction, overexpression, purification and spectroscopic characterization. Biochem J 1992; 288:503-509.

141. Yoshioka S, Tosha T, Takahashi S et al. Roles of the proximal hydrogen bonding network in cytochrome P450cam-catalysed oxygenation. J Am Chem Soc 2002; 124:14571-14579.

142. Deinum G, Stone JR, Babcock GT et al. Binding of nitric oxide and carbon monoxide to soluble guanylate cyclase as observed with Resonance Raman spectroscopy. Biochemistry 1996; 35:1540-1547.

143. Unno M, Christian JF, Sjodin T et al. Complex formation of cytochrome P450cam with Putidaredoxin. Evidence for protein-specific interactions involving the proximal thiolate ligand. J Biol Chem 2002; 277:2547-2553.

144. Mak PJ, Denisov IG, Victoria D et al. Resonance Raman detection of the hydroperoxo intermediate in the cytochrome P450 enzymatic cycle. J Am Chem Soc 2007; 129:6382-6383.

145. Ibrahim M, Denisov IG, Makris TM et al. Resonance Raman spectroscopic studies of hydroperoxo-myoglobin at cryogenic temperatures. J Am Chem Soc 2003; 125:13714-13718.

146. Schneider S, Marles-Wright J, Sharp KH et al. Diversity and conservation of interactions for binding heme in b-type heme proteins. Nat Prod Rep 2007; 24:621-630.

147. Barkigia KM, Chang CK, Fajer J et al. Models of heme d1: Molecular structure and NMR characterization of an iron(III) dioxoisobacteriochlorin (porphyrindione). J Am Chem Soc 1992; 114:1701-1707.

148. Reedy CJ, Gibney BR. Heme protein assemblies. Chem Rev 2004; 104:617-649.

149. Wood PM. Why do c-type cytochromes exist? Reprise Biochim Biophys Acta 1991; 1058:5-7.

150. Hoy JA, Smagghe BJ, Halder P et al. Covalent heme attachment in Synechocystis hemoglobin is required to prevent ferrous heme dissociation. Protein Sci 2007; 16:250-260.

151. Cable EE, Miller TG, Isom HC. Regulation of heme metabolism in rat hepatocytes and hepatocyte cell lines: Delta-aminolevulinic acid synthase and heme oxygenase are regulated by different heme-dependent mechanisms. Arch Biochem Biophys 2000; 384:280-295.

152. Chen JJ. Regulation of protein synthesis by the heme-regulated eIF2-alpha kinase: Relevance to anemias. Blood 2007; 109:2693- 2699.

153. Ryter SW, Alam J, Choi AM. Heme oxygenase-1/carbon monoxide: From basic science to therapeutic applications. Physiol Rev 2006; 86:583-650.

154. Ndisang JF, Tabien HE, Wang R. Carbon monoxide and hypertension. J Hypertens 2004; 22:1057-1074.

155. Jaggar JH, Li A, Parfenova H et al. Heme is a carbon monoxide receptor for large-conductance Ca^{2+}-activated K^+ channels. Circ Res 2005; 97:805-812.

156. Dimitrov JD, Roumenina LT, Doltchinkova VR et al. Antibodies use heme as a cofactor to extend their pathogen elimination activity and to acquire new effector functions. J Biol Chem 2007, (in press).

157. Kondo R, Gleixner KV, Mayerhofer M et al. Identification of heat shock protein 32 (Hsp32) as a novel survival factor and therapeutic target in neoplastic mast cells. Blood 2007; 110:661-669.

158. Shiva S, Huang Z, Grubina R et al. Deoxymyoglobin is a nitrite reductase that generates nitric oxide and regulates mitochondrial respiration. Circ Res 2007; 100:654-661.

159. Atamna H, Frey IInd WH. A role for heme in Alzheimer's disease: Heme binds amyloid beta and has altered metabolism. Proc Natl Acad Sci USA 2004; 101:11153-11158.

160. Atamna H. Heme binding to amyloid-beta peptide: Mechanistic role in Alzheimer's disease. J Alzheimers Dis 2006; 10:255-266.
161. Tan WB, Cheng W, Webber A et al. Endonuclease-like activity of heme proteins. J Biol Inorg Chem 2005; 10:790-799.
162. Peters MW, Meinhold P, Glieder A et al. Regio- and enantioselective alkane hydroxylation with engineered cytochromes P450 BM-3. J Am Chem Soc 2003; 125:13442-13450.
163. Li Y, Drummond DA, Sawayama AM et al. A diverse family of thermostable cytochrome P450s created by recombination of stabilizing fragments. Nat Biotechnol 2007; 25(9):1051-1059.
164. Ost TW, Miles CS, Murdoch J et al. Rational re-design of the substrate binding site of flavocytochrome P450 BM3. FEBS Lett 2000; 486:173-177.
165. Girvan HM, Seward HE, Toogood HS et al. Structural and spectroscopic characterization of P450 BM3 mutants with unprecedented P450 heme iron ligand sets: New heme ligation states influence conformational equilibria in P450 BM3. J Biol Chem 2007; 282:564-572.
166. Girvan HM, Marshall KR, Lawson RJ et al. Flavocytochrome P450 BM3 mutant A264E undergoes substrate-dependent formation of a novel heme iron ligand set. J Biol Chem 2004; 279:23274-23286.
167. Huang SS, Koder RL, Lewis M et al. The HP-1 maquette: From an apoprotein structure to a structured hemoprotein designed to promote redox-coupled proton exchange. Proc Natl Acad Sci USA 2004; 101:5536-5541.
168. Gibney BR, Huang SS, Skalicky JJ et al. Hydrophobic modulation of heme properties in heme protein maquettes. Biochemistry 2001; 40:10550-10561.
169. Sadeghi SJ, Meharenna YT, Fantuzzi A et al. Engineering artificial redox chains by molecular 'Lego'. Faraday Discuss 2000; 2000:135-153.

CHAPTER 11

Novel Heme-Protein Interactions— Some More Radical Than Others

Ann Smith*

Abstract

Heme (iron-protoporphyrin IX), as well as being an essential cofactor for cytochromes, oxidases and oxygen-binding proteins, is an important biological regulator acting via noncovalent, reversible interactions with a variety of proteins. These include DNA-binding transcription factors (e.g., Bach1, NPAS2 and Fur), and the cytoprotective, calcium-sensitive, K^+ Slo1 channel. The biological processes affected by heme may be as complex as the control of certain types of memory and the regulation of circadian rhythms. Heme contributes to the control of global iron homeostasis in mammals because heme catabolism generates iron, and heme and iron metabolism merge within the enterocyte after dietary heme uptake. Novel low-spin interactions of heme with proteins (e.g., hemopexin) protect cells from heme-mediated oxidative stress. Receptor-mediated uptake of heme from hemopexin regulates gene expression (e.g., heme oxygenase) and ligand binding to the hemopexin receptor activates a signaling network of three pathways. The acceptance of control of heme synthesis with degradation thus stringently regulating heme levels has been called into question with recent proposals that heme transporters (e.g., ABCG2 and FLVCR) function to limit intracellular heme toxicity in stem cells and that heme deficiency states develop that contribute to neurodegeneration. How, and whether, the balance of regulatory heme-protein interactions is controlled by moving heme via transporters through cell compartments as well as its uptake and export across the plasma membrane will soon be apparent.

Introduction

In this chapter we seek to present areas of cell biology and specific examples of novel heme-protein interactions in which heme is recognized as an important regulatory molecule. In many, if not all, cases heme binding drives the mechanism of these biochemical and regulatory effects. These functions of heme require its transport through the cell and, thus, across membranes. Therefore one focus here will be on the interaction of heme with carrier proteins that is noncovalent and reversible; a property needed for transporters and certain regulators. These heme-protein interactions are therefore distinct from the well known heme-proteins like the mitochondrial c-type cytochromes in the electron transport chain, where heme is covalently bound, and enzymes like catalase, a peroxidase, in which the heme group is responsible for catalysis. The degradation of hydrogen peroxide to oxygen and water is one of the most efficient enzymic reactions known (turnover number: 6 million molecules H_2O_2 per min) and, unusually, a tyrosine ligand of the heme-iron facilitates the oxidation of Fe(III) and removal of an electron from the heme ring, generating a heme radical.

Rapid progress has been made in understanding not only the mechanisms whereby heme-proteins act as regulators, but also how heme moves between subcellular compartments. Heme makes its way from the cell surface to the nucleus for gene regulation and, also, must travel

*Ann Smith—School of Biological Sciences, University of Missouri at Kansas City, Kansas City, Missouri 64110-2499, USA. Email: smithan@umkc.edu

Tetrapyrroles: Birth, Life and Death, edited by Martin J. Warren and Alison G. Smith.
©2009 Landes Bioscience and Springer Science+Business Media.

from its site of formation by mitochondrial ferrochelatase to other regions of the cell. A mitochondrial heme transporter has recently been characterized, e.g., ABCB6 (discussed below) and evidence provided that it moves heme from the cytosol across the outer mitochondrial membrane [Schuetz, J. et al, unpublished data]. This may contribute to the regulation of heme biosynthesis. Much remains to be discovered about the incorporation of heme into newly-synthesized apo-proteins in mammalian cells, a vital process for all cells that carry out aerobic metabolism. Insight is being gained from other systems including the work of Dr. L. Thöny-Meyer on the assembly of cytochrome c (Chapter 9).

The concept of heme as a regulator was first published by Shigeru Sassa in 1986[1] and, subsequently, by G. Padmanaban in 1989[2] who provided evidence for heme-mediated regulation of cytochrome P450.[3] It was an important concept for me when I investigated heme binding to the histidine-rich bacterial global regulator Fur[4,5] and for studies on the hemopexin system, which provided the first examples, respectively, of heme interacting with DNA binding proteins and of receptor-mediated heme uptake for the regulation of the heme oxygenase-1 (*ho1*) gene.[6] Interestingly, the scope of cellular processes in which heme is involved is intriguing and unexpected, including the regulation of the nucleo-cytoplasmic shuttling of transcription factors. The properties of heme are significantly altered depending upon the protein environment in which heme finds itself, yet little is established about how such redox potentials are controlled. Remarkably, the redox potential of the heme in heme-proteins of known structure spans values from -550 in a protein HasA, part of the heme acquisition system of *Serratia marcescens*[7] to +355 mV for heme f in cytochrome b_6f, the plastohydroquinone:plastocyanin oxidoreductase for photosynthesis with a similar role to mitochondrial bc_1 complexes.[8] Heme-protein assemblies have recently been reviewed (ref. 9). In addition, heme may drive the formation or dissolution of multimeric complexes that control transcription as shown for Hap1 in yeast.[10] New heme-protein-protein interactions will reveal more details of how this hydrophobic, amphipathic, redox active -molecule acts. Do heme chaperones exist? Are they needed to direct heme from a transporter to intracellular sites as shown for copper? How does heme arrive in the nucleus, from where in the cell, and how does it leave?

Current knowledge will be summarized here of how heme moves across the plasma membrane and through subcellular compartments in mammalian cells. These are two areas in which new molecular players, heme transporters, have recently been identified. Crystal structures of the heme transporter, hemopexin, have the heme binding site with a unique orientation of the heme. The heme propionates are anchored at the domain interface and the vinyl groups are exposed near the protein surface (see Fig. 2 below).

Proteins also sequester heme to minimize heme-mediated oxidative damage; proteins target heme to cells and transport heme across the plasma membrane; proteins move heme across the mitochondrial outer membrane; proteins export heme from cells across the plasma membrane; transcription factors bind heme and act to regulate gene expression positively or, alternatively, to relieve repression; heme binding regulates mitochondrial import of ALA-S1 enzyme and enzymic activity (see Chapter 7); and, heme-protein interactions regulate protein synthesis at the translational level: heme binds directly to eukaryotic transcription factor, eiF2α also termed heme regulated inhibitor [HRI].[11] Heme is used in both mammalian and prokaryotic cells to detect the gaseous signaling molecule nitric oxide (NO•) and its proposed "counterpart" neurotransmitter from heme catabolism, carbon monoxide (CO), as well as diatomic oxygen. Even complex biological processes like the maintenance of circadian rhythms, the "biological clock", are controlled in part by novel heme-protein interactions via the transcription factors, mPER2 and NPAS2 that are influenced by CO from heme catabolism (see ref. 12).

Intracellular heme levels and heme availability within a subcellular compartment will drive heme-protein interactions. The term "heme pool" to denote intracellular "free" heme, implying a discrete population of equivalent heme molecules, has its proponents and detractors. A concentration of 30nM - 1 μM has been calculated, assuming a steady state, from the known cellular content of heme-proteins and their established heme affinities.[13] For comparison, a concentration of ca. 3μM is equivalent to 1 million molecules of heme distributed within the volume of the mitochondria in a single cell (521 micron[3] in a cardiac myocyte http://www.nanomedicine.com/NMIIA/15.6.3.5.htm). Alternatively, "free heme" can be considered as any heme molecule moving from one protein binding site to another.

Heme levels within cells are tightly controlled by a balance between synthesis and degradation. The regulation of heme biosynthesis differs between erythroid and nonerythroid cells. The effects of heme, and hence heme-protein interactions, also differ in these cell types as described further below. Heme is known to regulate its synthesis at the level of feedback inhibition of the rate limiting step 5-aminolevulinate synthase (ALA-S). A housekeeping form ALA-S1 controls basal heme levels and ALA-S2 is responsible for generating heme in sufficient amounts for hemoglobin production in erythroid cells. If this balance is altered, heme transporters may play key roles in regulating intracellular heme levels in certain subcellular compartments, particularly in erythroid progenitor cells.

Heme deficiency states have been proposed to develop as a consequence of either decreased heme synthesis or elevated heme catabolism. Heme deficiency was first shown in cells from mice in which ALAS2 had been disrupted.[14] No hemoglobin was detected in embryos from these mice, which died by embryonic day 11. The differentiation of erythroid cells was arrested and iron accumulated in the cytoplasm of primitive cells. However, in adult mice made chimeric for ALAS2-null mutant cells, iron accumulated in the mitochondria of definitive erythroid cells. Thus, the mechanism of iron accumulation differs between primitive and definitive erythroid cells. Clearly, heme supply via ALAS-2 is necessary for the differentiation and iron metabolism of erythroid cells. Biochemical changes in the cells of patients with Alzheimer's disease (AD) are similar to those in cells treated with succinyl acetone or N-methyl protoporphyrin to inhibit, respectively, the second (ALA dehydratase) and last (ferrochelatase) steps in heme synthesis (see Chapters 3 and 4). Increased heme catabolism may contribute to the pathology in neurodegenerative conditions including AD,[15] although it is not known whether simultaneously there is a lack of compensatory heme synthesis. In acute intermittent porphyria, one type of genetic defect in heme biosynthesis (Chapter 5), a peripheral neuropathy develops in response to fasting and certain drugs by unknown mechanism(s), although a very recent publication provides a link with nutrition via the peroxisome proliferator-activated γ-coactivator 1α.[16] Generally, porphyrics have about 50% normal enzyme levels. Less than 30% of normal activity may be required to produce pathological symptoms that nevertheless do occur in this multifactorial disease.

Importantly, heme availability can contribute to yet another aspect of cellular homeostasis for processes that require functional heme-proteins. For example, in the control of surface events in mammalian cells such as the oxidative respiratory burst in neutrophils. When heme synthesis is decreased, synthesis of active holo cytochrome b_{558}, a component of the NADPH:O_2 oxido-reductase may be compromised. This may be a factor in chronic granulomatous disease.

Heme as a Sensor: Interactions of Heme with Proteins That Lead to Recognition of Gaseous Molecules: Oxygen, Carbon Monoxide and Nitric Oxide

In several bacterial and mammalian heme-proteins, heme acts as a sensor not only for oxygen (e.g., hemAT in *Bacillus subtilis*) but also for other gaseous signaling molecules including nitric oxide (NO•, e.g., soluble guanylyl cyclase) and carbon monoxide (CO, e.g., CooA in *Rhodospirullum rubrum*). Fifty such sensors have been identified, most within the past ten years and an excellent and comprehensive review has recently been published.[17] Briefly, these proteins are comprised of a domain or subunit that binds heme in a coordination that allows diatomic gases to bind to the second axial ligand position of the heme-iron. This heme binding controls the activity of another region or module of the protein that transmits signals to other nonheme proteins. These modulating domains include a histidine protein kinase, a cyclic-dinucleotide phosphodiesterase, a nucleotide cyclase, a chemotaxis receptor and DNA-binding transcription factors. The features of the heme-binding domains present novel heme binding sites that include modified globin domains, the Per-Arnt-Sim (PAS) domain and the cAMP-receptor-like domains.

Novel Mammalian Globins

The coordination of the heme iron that is most familiar to biologists is that for oxygen binding in the globins, hemoglobin (Hb) and myoglobin. The proximal ligand of the heme-iron is oxygen with a distal histidine and for such coordination the iron must be reduced (ferrous). The heme in Hb may also act as an NO• sensor by first binding NO• which is then transferred to a cysteine thiol. This reversible formation of S-nitrosylated hemoglobin (SNO-Hb) has been postulated to regulate blood pressure in humans.[18] It binds to Band 3 on red blood cells and upon deoxygenation the NO• is transferred to a cysteine thiol on Band 3. Although how this contributes to the vasodilation the mechanism is unknown. However, whether this is physiologically relevant has been disputed because of the known tight binding of oxygen to the ferrous heme iron,[19,20] which would compete with NO•. It may contribute to the vasocclusion in hypoxic vascular beds in sickle cell disease because HbS has a lower O_2 affinity.[21] Four unusual nonvertebrate globins have been identified that have unique structures and properties and their function has not been defined, but they have in common globin domains that bind heme.[22] In mammals, neuroglobin (Ngb), so named because it is expressed in neurons of brain and spinal cord, peripheral neurons, (but also in cells of endocrine glands), not only binds oxygen like the typical globins Hb and Mb but also binds NO•.[23] This is similar to several bacterial heme-proteins where heme is a sensor for diatomic gases including CO (e.g., CooA in *Rhodospirullum rubrum*), as well as oxygen. Ngb protein and mRNA are expressed in regions of the brain basal ganglia, cerebral cortex, hippocampus and cerebellum.[24] Ngb is induced by hypoxia and ischemia[25] which can be a cause of stroke. Interestingly, over-expression of Ngb allows cells to survive low levels of oxygen but also, paradoxically, oxidative stress.[26] Ngb levels decline with age in neurons of cerebellar neocortex and hippocampus of mouse brain.[24] These are the same areas of the brain that deteriorate in neurodegenerative conditions like Alzheimer's and Huntington's patients. Therefore, this globin is thought to have a protective function but no mechanism has been proposed.

NO• Binding to Heme-Proteins Leads to Activation of Signaling Events via Second Messengers

Nitric oxide synthase (NOS) is a dimeric heme-protein that catalyzes the conversion of L-arginine and O_2 to citrulline and NO•. NO• binds to a heme with one axial coordinated histidine in the N-terminus of the β1 subunit of soluble guanylate cyclase, an αβ heterodimer. It is this heme binding that activates the cyclase, which synthesizes the second messenger 3'5' -cyclic guanosine monophosphate (cGMP) from GTP. There are many downstream effector molecules established for this system that include protein kinases, ion channels and cyclic nucleotide phosphodiesterase that in turn control vascular smooth muscle tone improving blood flow and neurotransmission, among several other events. This important molecule has not been crystallized. However, a soluble ortholog of the NO• binding domain of human guanylyl cyclase was cloned from *Clostridium botulinum* which has a femtomolar affinity for NO• and an ortholog from the extremely thermophilic *Thermoanaerobacter tengongensis* was crystallized.[27] This protein, termed tSONO for sensor of NO•, has an αββαββ motif and a distal α-helical domain which is a novel fold. The heme-nitrosyl coordination is modified by a tyrosine residue. Furthermore, Raman and colleagues after comparing the domain architecture of tSONO with a SONO ortholog cloned from the eukaryotic unicellular green alga, *Chlamydomonas reinhardtii* propose that NO signaling via cGMP may have arisen before multicellular eukaryotes.

CO is one of three catabolites of heme generated by two heme oxygenase (HO) isozymes, HO-1 and 2. Both require oxygen and NADPH to cleave the substrate heme producing the linear tetrapyrrole, biliverdin (BV), and from the methene bridge carbon, CO. These two small molecules have protective roles in cells and evidence is accruing for additional roles for the protein itself. The iron is used for regulation inducing the storage protein ferritin and decreasing iron uptake via the transferrin receptor and increasing efflux via ferroportin. This can also be viewed as protective since the overall effect is to decrease intracellular iron levels. Unlike HO1, which is readily inducible, HO2 is constitutively expressed at high levels in the CNS and functions to protect neurons. Using a model of focal ischemia of vascular stress, mice with a targeted deletion of HO2 had far more extensive damage compared with control mice.[28] Intriguingly as now described, heme and its metabolites as well as HO2 interact with a membrane protein to regulate its transport functions.

Heme Binding to Ion Channels

Heme regulates, directly and indirectly, calcium sensitive, K^+ channels (i.e., Slo1 BK channels), expressed in neurons and residing in the inner mitochondrial membrane of human glioma cells.[29] When open they have a cytoprotective role after trauma, ischemia or hypoxia. Heme alters the opening and closing of the channel stabilizing the nonconducting conformation. This was shown by experimentally exposing heme to the intracellular side of the channel, which extensively reduces the amplitude of the K^+ current.[30] Expressed BK channels, comprising both hSlo1 and the auxiliary subunit β1, as well as native BK channels in single rat hippocampal neurons are inhibited by heme to the same extent. Calcium opens these BK channels and if sufficiently high (at least 5 μM) overcomes heme inhibition. EPR reveals that heme binds tightly (half maximal inhibition at 45-120 nM) to a 23 amino acid long peptide (hSloHBP23) containing CKACH (residues 612-616 of the channel protein) with a stoichiometry of 1:1 (heme: Slo-HBP23). Mutagenesis (either C615S or H616R) eliminated the inhibitory effect of heme. This heme site is therefore proposed to resemble that in a cytochrome c-derived undecapeptide CAQCH with binding to the histidine residue, but in c-type cytochromes the heme is covalently bound to an amino acid.

In the carotid body, BK channels are modulated by oxygen and recent evidence suggests that CO from heme catabolism is a physiological messenger affecting its activity. As oxygen levels decline, BK channels are acutely and reversibly inhibited leading to cell depolarization. Immunoprecipitation of rhBK channels stably expressed in human embryonic kidney cells revealed that γ-glutamyl transpeptidase and HO2 were physically associated with the channel. HO2 has been proposed to act as the oxygen sensor for the BK channel and to enhance channel activity in normoxia by generating CO from heme.[31] BV increased channel activity 4-fold and CO, from a chemical donor, activated the channel under normoxia (pO_2 ca. 150mmHg). Heme alone did not alter BK channel activity but low concentrations of heme (1 nM) with the coenzyme NADPH, under normoxia, increased the number of open channels. The CO is now derived from heme catabolism. However, even in the presence of substrates, when oxygen tension was decreased fewer channels were open. HO2 inhibition leads to excitation of the carotid body. The interesting and provocative model presented is that the channel senses changes in oxygen levels via the close proximity of HO2 which in normoxia produces CO and BV from heme, both of which open the channel. BV is rapidly reduced to bilirubin by cytosolic biliverdin reductase if NADPH is available. When oxygen levels are low, CO cannot be formed by HO2 and the channel remains closed.

Novel Low-Spin Heme-Protein Interactions

In proteins, the ligands for the heme iron in the fifth and sixth positions vary and the heme coordination by globin is a typical high spin heme with distinctive absorbance at 620nm. When the heme-iron is coordinated by two histidine residues (strong field ligands) a low spin heme-protein is generated. The spectral characteristics: α and β peaks in the region of about 530-550 nm without absorbance at 620 nm. The importance of low-spin bis-histidyl coordination of heme to dampen down the chemical reactivity of the heme as shown for hemopexin, has been reinforced by recent crystal structure data that reveal that a small protein α-hemoglobin-stabilizing protein (AHSP) binds to individual monomeric α-hemoglobin subunits. This leads to large conformational changes in the organization of helices producing an oxidation of the ferrous heme and a bis-histidyl coordination of the heme-iron.[32] αHb is inherently unstable releasing heme and generating reactive oxygen species, and both processes are prevented by AHSP.

Additional examples of low spin bis-histidyl coordination are the b-type cytochromes, e.g., cytochrome b_5 for fatty acid synthesis. Two forms of b_5 from different gene products exist: one in the endoplasmic reticulum and another in mitochondria[33] with a much lower redox potential.[34] Cytochrome b_5 together with P450 reductase increases the efficiency of cytochrome P450 mediated reactions by acting as another electron donor. The yeast, *Saccharomyces cerevisiae*, utilizes the damage resistance protein 1 (Dap1) to protect itself from DNA damage caused by exposure to the methylating agent, methyl methanesulfonate (MMS). Dap1 is predicted to resemble b_5 structurally with a homologous heme-domain. The rat and human homologs of Dap1 bind heme but their redox potentials have not yet been determined. Dap1 targets the p40 protein Erg11p/Cyp51 p in *S. cerevisiae* by utilizing heme to stabilize Erg11p which regulates both ergosterol synthesis and the resistance to MMS toxicity.[35] While many mechanistic details remain to be established, this regulatory role for the P450s is expected to be highly conserved because Dap homologs are present in genomes of human, porcine, eukaryotic and in fission yeast as well as flowering plants, *Drosophila* and *Caenorhabditis elegans*.

Heme-Binding Proteins That Are Protective, Preventing Heme-Mediated Oxidative Stress

Heme (iron-protoporphyrin IX) is vital to aerobic life. However, the heme-iron readily accepts and donates electrons (i.e., is redox active), and therefore heme is potentially toxic. The chemical reactions that occur in the presence of oxygen can destroy most biological molecules. Under normal circumstances the reactivity of heme is modified by its ligation to proteins. In a number of pathological circumstances, involving cell lysis and hemolysis, heme may be liberated from its conjugate protein.[36,37] In plasma there are several heme binding proteins and many of their counterparts are also found in fluids like CSF that bathe cells in barrier tissues like the brain and eye. Lipoprotein particles also bind heme[38] and heme readily causes the peroxidation of membrane lipids.

The Hemopexin System

Hemopexin has the highest affinity for heme (Kd less than pM) so even with a plasma concentration of 0.5-1.2 mg/ml (20 μM) it effectively competes with albumin (35-55 mg/ml; heme binding sites in HSA, Kds 1 and 10 μM). Hemopexin protects low density lipoproteins from heme-mediated oxidation.[39] Another low abundance protein, histidine proline rich glycoprotein (HPRG), also binds heme (Kd *ca.* 1 μM) with multiple sites due to its extensive histidine content[40,41] and heme bound to HPRG has been detected in patients after extensive hemolysis or heme-arginate treatment. However, HPRG induces the apoptosis of activated endothelial cells leading to potent anti-angiogenic effects.[42,43] The domain responsible is the histidine and proline rich domain that presumably would be influenced significantly during hemolysis by heme in the circulation.

Mechanisms whereby Heme Binding by Hemopexin Protects Cells against Redox-Active Heme

When hemoglobin is oxidized (methemoglobin), the ferri-heme is bound less tightly and is readily removed by hemopexin.[44] Hemopexin binds this heme and also that from other heme-proteins released by cell lysis not only with high affinity thus sequestering it from cells, but also in a manner that "dampens down" the chemical reactivity of heme. Hemopexin is therefore important in protecting cells against the toxicity of heme from hemoglobin especially in the absence of haptoglobin, which binds hemoglobin dimers.[45] Haptoglobin does not recycle but is degraded in the lysosomes after delivering hemoglobin to cells.[46] Protection by hemopexin is apparent when mice are exposed to extensive hemolysis. In *hpx* null mice, the kidney damage was far more extensive and life threatening in than in normal mice.[47] The *hpx* null mice have normal albumin levels demonstrating that the presence of albumin was not protective. Thus, hemopexin is considered the first line of defense against heme-mediated damage to cells during hemolysis, trauma and ischemia reperfusion injury.

The hemopexin system is a physiologically relevant model for heme transport and was the first system shown in mammalian cells to require specific hemopexin receptors.[48,49] Hemopexin was shown, both morphologically and biochemically, to colocalize with the iron transporter transferrin, the paradigm for recycling receptors. Apo-hemopexin recycles intact after heme delivery, and its endocytosis is modeled on that of transferrin.[50] The endocytosis of heme-hemopexin complexes provides a mechanism for the cellular accumulation of heme in a controlled and safe manner. Hemopexin receptors are expressed by several different cell types of the liver,[51] kidney (A. Agarwal and A. Smith, unpublished observations), and eye,[52] immune,[53] peripheral[54] and central nervous systems[55] [R.-C. Li, J. Lee, A. Smith and S. Doré, unpublished data]. When tissues are injured and red blood cells are destroyed, the ensuing inflammatory response leads to the generation of reactive oxygen species. These are the conditions in which the binding of heme by hemopexin plays a dual protective role in vivo. First, it acts as an extracellular anti-oxidant by sequestering heme from all cells, especially important for those that lack hemopexin receptors. Second, it targets heme and transports it to cells expressing hemopexin receptors for catabolism by heme oxygenase-1 (HO1).

A cell surface scavenger receptor protein, CD91/LRP (low density-like lipoprotein receptor) with more than 50 ligands has recently been shown to bind heme-hemopexin.[56] After uptake into CD91/LRP overexpressing COS-1 cells heme-^{125}I-hemopexin, labeled using chloramine T a strong oxidizing agent, was degraded within 1 hour. This contrasts with our published work, using different isolation and labeling procedures, where hemopexin has a long plasma half-life in intact rats.[57] It colocalizes with transferrin, as shown biochemically and morphologically, as well as recycles intact in cultured HepG2 cells[50] and is not rapidly degraded after injection as heme-^{125}I-hemopexin in intact rats.[48] The slow rate of catabolism of certain ligands of CD91/LRP has been ascribed to lack of consistent targeting to lysosomes[56] and to their resistance to protease digestion. More than 80% of the LRP1 ligand apolipoprotein E is recycled. While heme-hemopexin complexes are relatively resistant to plasmin and trypsin, asialo[125]I-hemopexin or heme-asialo ^{125}I-hemopexin is rapidly degraded in vivo[58,59] presumably after uptake via the galactose receptor. In contrast, a heme complex of the N-domain of hemopexin, but not intact apo-protein, was rapidly removed from the circulation and degraded.[57] Most of the ligands for CD91/LRP are no longer native but have been modified by cleavage and covalent modifications that occur during clotting or inflammation.[60] A sorting nexin 17 targets LRP1 to the recycling pathway rather than to lysosomes.[61] However, what determines whether hemopexin recycles intact after endocytosis or is degraded remains to be established. Significantly, we have shown that the binding of heme-hemopexin complexes to the hemopexin receptor leads not only to heme uptake but also to the activation of several signaling events. But these have not yet been shown to be due to binding to CD91/LRP. The consequent changes in cell behavior play a key role during oxidative stress in protecting cells from heme and its iron by inducing HO1, metallothioneins -1 and -2 (MTs) and ferritin. Elevated HO1 is anti-inflammatory and anti-apoptotic.[62] MTs protect against oxidative stress and redox-active copper, and ferritin stores iron in the cytosol. Receptor-mediated heme uptake from heme-hemopexinis associated with transcriptional activation of the *ho-1* gene.[6] Furthermore, signals from the hemopexin receptor, but not heme uptake, are associated with transcriptional activation of the *mt-1* gene,[63,64] activation of the stress activated protein kinase/ N-terminal cJun kinase (SAPK/JNK) cascade,[65] and stimulation of the nuclear translocation of the transcription factor MTF-1, which binds to metal responsive elements.[66] Evidence has been presented that these events are copper-dependent [see C below and refs 66,67]. In addition, after hemopexin-mediated heme uptake, the iron released from heme catabolism down-regulates expression of transferrin[68] and the transferrin receptor (TfR).[6] The release of iron from the breakdown of heme is generally viewed as dangerous for cells because the product of HO reaction in mammalian cells is Fe(II), a pro-oxidant. However, iron release can also be viewed as protective at least in certain cell types where iron regulates iron uptake systems (decreasing transferrin receptor) and iron exporters (increasing ferroportin), via iron responsive proteins and iron regulatory elements, thus lowering intracellular iron levels.

Novel Mode of Heme Binding by Hemopexin: A Paradigm for Heme Transporters?

The link between structure and function is a basic tenet of modern biology. Hemopexin is comprised of an edge-to-face association of two 4-bladed β-propeller "pexin" domains (Fig. 1, ref. 69), considered important for protein-protein and protein-ligand interactions. Hemopexin binds heme extremely tightly as required by an anti-oxidant molecule. Hemopexin also binds heme noncovalently and reversibly, as required by a transporter. The heme-binding site in hemopexin is unique (Fig. 1). Heme is anchored between the two domains by two histidine residues that coordinate with 5th and 6th axial ligand positions of the heme-iron (Fig. 1, center) and by ionic interactions between the propionate side chains and the basic R and K residues (Fig. 2, RHS).

This produces an unusual orientation of the heme in which the hydrophilic propionates are not exposed at the surface. Significantly, these propionates are anchored via ionic interactions with side chains of basic arginine residues (Fig. 2 RHS). This type of interaction may be a feature of transporters. It is an important means whereby the heme molecule is oriented for stereospecific cleavage of methene bridge carbons in the active site of heme oxygenase.[70] Perhaps the ligands of the heme-iron play lesser roles especially in transporters and proteins that bind both heme and porphyrins like ABCG2 (see below). Both ferric- and ferro-heme are bound by hemopexin and recent electrospectrochemistry studies support that ferro-heme is bound less tightly[143]. In hemopexin, the heme-iron may serve to help orientate the heme as well as to move it into or out of the binding site via docking at the wide end of the tunnel in the N-terminal pexin domain. The isolated N-, but not C-, terminal pexin domain binds heme.[57] The interactions at the domain interface lock the heme propionates into place and the linker peptide wraps around the heme, stabilizing the bis-histidyl coordination and the fit of heme within the binding pocket.

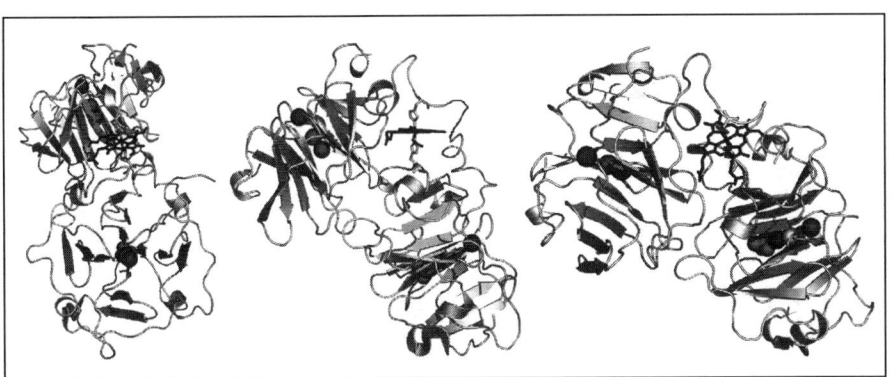

Figure 1. Crystal structure of heme-hemopexin (HPX) complexes. A novel high affinity heme-binding site is formed between two similar β-propeller domains [N-terminal (upper) and C-terminal (lower)] and bounded by the inter-domain linker peptide (see center). The heme-iron is coordinated by two histidine residues (center). Rotation of this structure around a vertical axis reveals the 4 blades of the propeller (C-domain, LHS) which is the smallest known domain with a folding topology. A third view (RHS) shwos the novel orientation of the heme between the two domains, the strand of the β-sheet blades and the ions that line the central tunnel of each domain. These figures were generated using pdb file 1QHU with PyMol.

Such tight binding must require a specific mechanism for release of the heme so that it can be transported into cells. We had proposed that interaction with the receptor contributes to this process, probably by a conformational change in hemopexin opening up the heme site. In addition, not only does reduction of the heme iron generate a weaker complex but reduction is facilitated as the pH is decreased, over the range pH 7.2 to pH 5.5[143]. A 100 mV increase in $E_{1/2}$ is brought about by a decrease of 1.8 pH units consistent with a hemopexin equilibrium reaction involving a single H^+ that stabilizes the reduced formed of the protein. Acidic pH is associated with the maturation of endosomes (ca. 6.5 in early endosomes) where heme is expected to be released from hemopexin. Models for heme uptake from hemopexin, characterized by ligand binding, include one process of binding and uptake shown to be of high affinity and low capacity we termed "specific", whereas another was of low affinity, high capacity termed "selective".[71] The latter was not due to nonspecific effects because heme-hemopexin complexes were required. Significantly, studies with several metabolic inhibitors distinguished between these two hemopexin-mediated processes, and between them and "uptake" of exogenous "free" heme. Currently, the role of CD91/LRP in the specific and selective processes is being evaluated.

Models of Iron Transport in Eukaryotic Cells Set Precedents for Heme Uptake into Mammalian Cells

Models for iron uptake in mammalian cells, particularly enterocytes and hepatocytes, are based on concepts developed from genetic studies in yeast. Certain parallels can be expected to exist as the mechanism of heme uptake is deciphered. Ferric iron uptake across the plasma membrane in yeast requires three proteins: a transporter (Ftr1p), a reductase (Fre1p) and a multi-copper oxidase (Fet3p). After reduction by Fre1p,[72,73] a flavo-hemoprotein of the cytochrome b-type NAD(P)H oxido-reductase class, the iron is reoxidized by Fet3p before transport across the plasma membrane by a high affinity, specific ferric transporter, Ftr1p. Fet3p resides with Ftr1p at the plasma membrane. Reduction generates a more soluble form of iron but the divalent transporters are not specific for iron whereas the ferric transporter is. Hence the proposed need for the "reoxidation" step although this seems like a futile cycle. Interestingly, Fet3p has a b-type heme and heme availability will affect the activity/amount of active holoenzyme. A lack of heme will decrease Fet3p function with lower Fe(III) uptake. The Fre1 system, which reduces both copper and iron, may be a multi-component electron transport chain.[74] Thus, heme metabolism is linked with iron and copper.

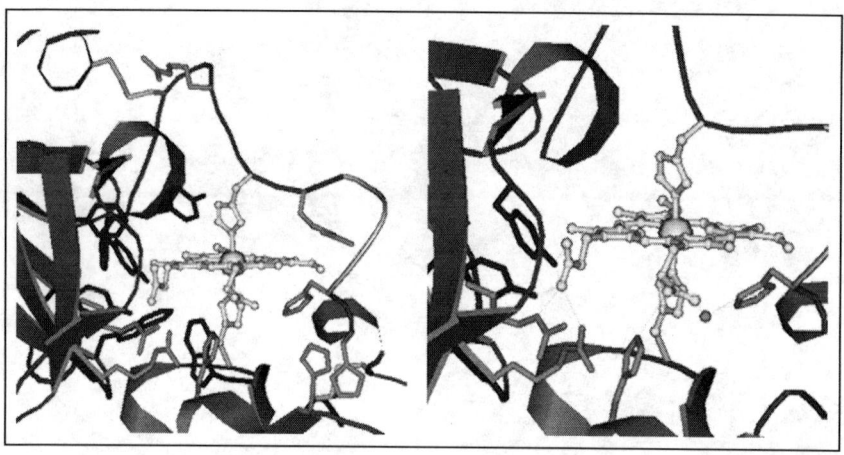

Figure 2. Heme binding site of hemopexin. LHS: a preponderance of aromatic and basic residues line the heme site, with side chain stacking of tryptophan 268 between histidine 266 and arginine 174 contribute to the tight binding. RHS: Propionates are anchored. A positive potential is derived from 4 arginine and 4 histidine residues and note the tyrosine and arginine residues around the pyrrole D ring.[69] These figures of the heme binding site of hemopexin were kindly generated by Dr. Max Paoli, (University of Nottingham, UK) using MOLSCRIPT program.

Iron in the gut exists predominantly as ferric iron. Therefore, before transport across the brush border membrane by the divalent metal transporter (DMT1; sometimes termed DCT1 for divalent cation transporter) iron is first reduced by a ferrireductase, duodenal cytochrome b (dCytb) also in the brush border.[75] Iron is transported out of intestinal cells via ferroportin (FPN1, also termed IREG1) for loading after diffusion across the extracellular matrix onto plasma transferrin (Tf). Tf binds ferric not ferrous iron, which is why a multi-copper oxidase is thought to be required for iron loading of transferrin. There are two Fet3p homologs: intracellularly, hephaestin, an unlikely candidate because it is found perinuclearly, or extracellularly, ceruloplasmin. In the "mobilferrin" system for ferric iron uptake by a cell surface β3-integrin, reduction of iron is proposed to take place in a cytosolic complex not at the cell surface.[76,77]

Iron in plasma is bound to Tf. Iron uptake by liver from diferric-Tf consists of two processes. Specific iron uptake occurs by the classical high affinity, low capacity clathrin-mediated endocytotic pathway mediated via transferrin receptor 1 (TfR1). A second pathway of low affinity and high capacity is mediated by a second Tf receptor, TfR2-α, via endocytosis.[78,79] The alternatively spliced form of TfR2, TfR2-β, may function as an iron binding protein in cytosol. TfR2-α is expressed predominantly in hepatocytes, binds iron with lower affinity than TfR1,[80] and is up-regulated by iron.[81] The Tf-bound ferric iron is reduced before export from early endosomes via DMT-1.[82-84] Iron uptake and efflux were impaired in cellular copper deficiency (Caco2 cell line from colon, one model of intestinal cells), implicating copper in iron transport in mammalian cells, as in yeast, but the mechanism remains to be defined.[75,85]

Heme Transporters

In mammalian cells, hemopexin is currently the only known heme transport protein with a specific cell surface receptor. Heme from [^{55}Fe]heme-hemopexin complexes is transported to the liver in intact rats,[48] and into isolated hepatocytes[71] and several distinct types of cultured cells including human HepG2 cells via endocytosis of heme-hemopexin complexes.[6,53,86-88]

Three trans-membrane heme transporters have recently been identified in mammalian cells (described further below). Heme carrier protein 1 (HCP1), at the brush border membrane of enterocytes of the duodenum for the uptake of dietary heme[89] and Caco2 cells take up ^{59}Fe-heme.[90] In erythroid and nonerythroid cells, heme is moved across intracellular membranes including the outer mitochondrial membrane and between subcellular compartments to HO1 in the smooth endoplasmic reticulum, and from its site of synthesis by ferrochelatase in the mitochondrion for holo-heme protein synthesis. Some heme transporters utilize energy from ATP hydrolysis e.g., ABCB6, a member of the family of ABC transporters. Another type, the feline leukemia virus receptor subgroup C, FLCVR (below), exports heme along a concentration gradient passively and is ubiquitously expressed.

Heme Transport across Enterocytes and Proof of Principle

Iron balance in the body is principally controlled by absorption of iron from the intestinal cells because humans are limited in their ability to excrete iron and there is no known active process.[91] Interestingly, studies with radioactive hemoglobin have established that heme iron is far more effectively absorbed than inorganic iron by cells of the gut. In humans eating an average normal diet, more than one third of the body's total daily iron requirement of 3 mg comes from dietary heme-iron [hemoglobin and myoglobin; refs. 92,93]. The duodenum and proximal jejunum are considered the main sites of absorption, not the stomach. There is indirect evidence that heme is absorbed intact from the gut lumen, since iron chelators do not inhibit absorption of iron.[94] Even in iron-replete rats, heme from [^{55}Fe]-heme-histidine is rapidly absorbed in vivo from the duodenum.[95] A putative heme receptor with a high affinity for heme (10^{-6}-10^{-7}M) was identified on the surface of pig intestinal cells[96] and on murine[97] and human erythroleukemic (K562) cells.[98] Most recent work on heme uptake utilizes in vitro studies with the line of Caco 2 cells.[90,99] Some morphological studies suggest that heme moves from the apical surface to within tubular structures and then to secondary lysosomes; but to date there is no clearly defined pathway. Thus, neither the chemical state of the heme transported nor the mechanism of heme absorption by duodenal enterocytes in vivo have yet been defined. Significant progress has been made in identifying the likely candidate for this and it is heme carrier protein 1 (HCP1) identified in a subtractive hybridization used to isolate several proteins involved in iron uptake by hypotransferrinemic mice.[89] HCP1 has homology to bacterial meta-tetracycline transporters with a conserved motif also present in the heme exporter FLVCR (see below). HCP1 is highly expressed in duodenum, not ileum; moves from the cytosol to the brush border in iron deficiency and, thus, would provide a clear mechanism for heme uptake from the gut lumen into duodenal enterocytes.

We had begun to characterize the route of heme and its iron across the enterocyte several years ago utilizing a monomeric form of heme (a *bis*-histidyl complex) that is soluble in aqueous solutions. This [^{55}Fe]-heme-histidine is rapidly absorbed in vivo from the duodenum of iron-replete adult rats.[95] The model under investigation is that a heme transporter is present in the brush border membrane of duodenal enterocytes for absorption of heme from the lumen of the gut. After uptake, intact heme may move to intracellular sites such as the nucleus to mediate gene regulation and in other cell types heme binds to a repressor, Bach1, allowing access of a robust transcription factor NFE-related factor, Nrf2, to the enhancer to activate the *hol* gene.[100,101] Heme is also expected to undergo catabolism by HO1 generating the linear tetrapyrrole biliverdin, CO and ferrous iron and heme is degraded by intestinal cells[102-104] presumably by HOs. Thus, heme-iron rather than intact heme is expected to be transported out of the cell across the basolateral membrane. Iron efflux via the iron transporter ferroportin[105] allows iron to eventually reach Tf in the plasma for circulation, although enterocytes are not bathed in plasma. This iron efflux from enterocytes requires a copper-protein with ferroxidase activity similar to hephestin or a homologous protein.

To address whether intact heme reaches the circulating plasma and is bound by hemopexin, or is degraded within the enterocyte and the heme-iron released reaches Tf in the circulation, blood was taken from the mesenteric vein that drains the duodenum, 30 minutes after the introduction of [^{55}Fe]-mesoheme-histidine solution (47 nmol heme per 100g body weight) into the lumen of the duodenum of an anesthetized rat.[95] Gel permeation of an aliquot of this plasma revealed a single symmetrical peak of radioactivity with an apparent Mr of 78 kDa (Fig. 3) close to Tf (Mol. wt. 79570). Immunoprecipitation of plasma samples with mono-specific, polyclonal anti-bodies to rat Tf and rat hemopexin showed that the [^{55}Fe]-radioactivity is associated with Tf, not hemopexin (Fig. 4). Thus, these studies demonstrate a proof of principle that radioactive heme presented to enterocytes in the lumen of the GI tract is absorbed and extensively catabolized by enterocytes and that, under physiological conditions, the pathway of heme-iron joins that of iron within the enterocytes and, furthermore, the iron of this heme is transferred to serum Tf (Fig. 5).

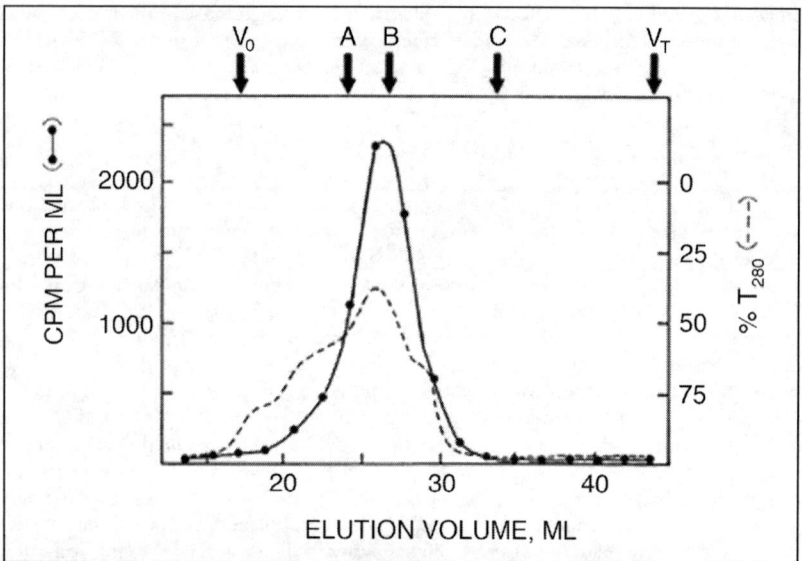

Figure 3. Detection of [^{55}Fe]-iron-transferrin, not [^{55}Fe]-heme-HPX, in plasma after administration of heme into the duodenal lumen of intact rats. An aliquot of serum was applied to a Sephacryl S-200 column (1.7 x 25 cm) and eluted (10.3 ml per h) with 10 mM sodium phosphate buffer, pH 7.4, containing 0.15M NaCl. Recovery of radioactivity was 88%. The arrows indicate the elution volume of standards used: aldolase (A), bovine serum albumin (B) and ribonuclease (C). Blue dextran was used to determine the void volume of the column (V_0) and L-tyrosine the total column volume (V_T). The radioactive iron bound was detected by liquid scintillation counting. Intact heme would bind to hemopexin and iron to transferrin.

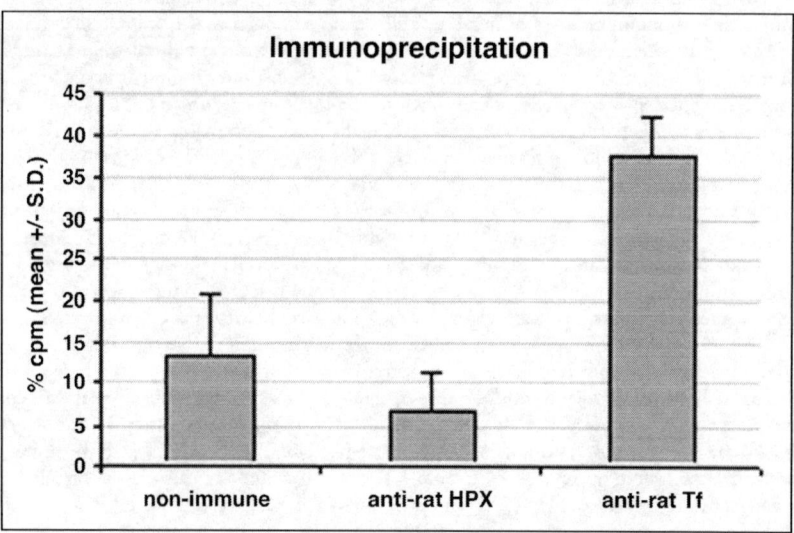

Figure 4. Identification of [^{55}Fe]-iron-transferrin (Tf), not [^{55}Fe]-heme-hemopexin (HPX), in plasma by immunoprecipitation of [^{55}Fe]radioactivity. These data represent the amount of radioactive [^{55}Fe]- cpm recovered after the proteins in aliquots of serum were precipitated by addition of mono-specific polyclonal antibodies raised against rat hemopexin or rat transferrin. Standard immunoprecipitation techniques using PEG 6000/8000 were employed and the data represents the mean +/- S.D. (n = 3).

One point for the coordinated regulation of uptake of dietary heme and iron is proposed to be at the basolateral membrane transporter, ferroportin (FPN1). Recently, FPN1 has been implicated in the global regulation of iron absorption,[106,107] it clearly represents a site where our data indicate that uptake of heme and iron would be coordinated. Circulating iron levels alter transferrin saturation[108] and regulate secretion of hepcidin by the liver which in turn affects FPN1 levels by binding physically to it and causing its internalization and degradation. Thus, iron release from enterocytes into plasma is decreased.[106] Nevertheless, a heme exporter, that is expressed on enterocytes (see below), the receptor for feline leukemia virus group C (FLVCR) has recently been identified. The studies in rats described above were designed to use doses of heme that represented normal and ten times normal levels of daily heme requirement for humans. The data demonstrate extensive catabolism of heme by enterocytes. However, excretion of low levels of heme might have been undetected. Ingestion of large amounts of heme, perhaps a "half-pounder" hamburger on an empty stomach, may generate sufficient heme within enterocytes to cause export across the basolateral membrane of intact heme into the systemic circulation.

Interactions of Heme with Transcription Factors

Heme participates in gene regulation in prokaryotes, yeast and also higher eukaryotes in part by binding directly to several functionally distinct DNA binding proteins. In bacteria, heme binds to the global regulator, ferric uptake regulator [Fur; refs. 4,5], in yeast, to the damage resistance protein 1 (Dap1p ref. 35) and HAP1,[10] and in mammalian cells, to the basic leucine zipper (bZip) protein Bach1.[109] One short motif for heme binding that has been identified in these proteins is the heme regulatory motif (HRM) that contains cysteine and proline residues often referred to as a "CP" motif. HRMs were first shown to bind heme in a functional manner in yeast to a multi-protein complex containing HAP1,[110,111] which often include heat shock proteins 70 and 90.[112,113] Hap1 binds to its own promoter and represses transcription in a heme-independent but Hsp70-dependent manner.[112] How heme binds to proteins containing HRMs has not been defined since no crystal structures are yet available. A role for heme-transcription factor interactions for HO1 regulation was first suggested after the discovery of the *E. coli* Fur-heme binding.[5] Here, we focus on heme-Bach1 interactions in mammalian cells for regulation of heme oxygenase1 gene (*ho-1*), an enzyme catalyzing the rate-limiting step in heme catabolism.

Bach 1, a basic leucine zipper (bZip) protein, is the specific target for Maf recognition element (MARE) binding proteins that regulate the expression of proteins including HO1 for protection

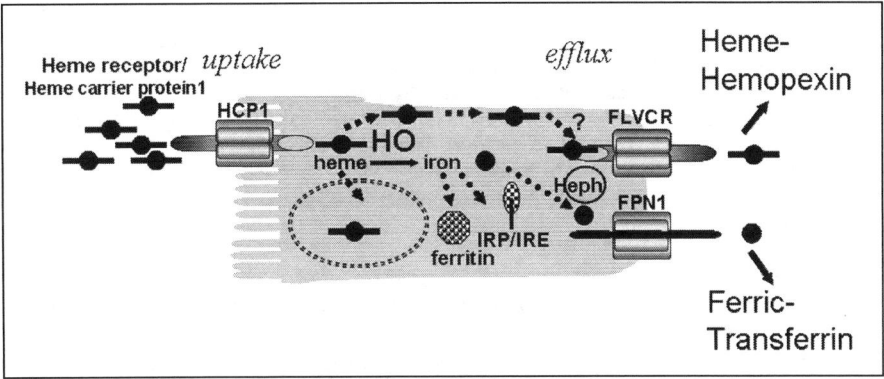

Figure 5. Model for heme uptake in enterocytes. Heme is transported from the gut lumen by HCP1 and then catabolized in the enterocyte, presumably by heme oxygenase-1 (HO). In experiments described here, no heme was detected on circulating hemopexin but radioactive iron from ^{55}Fe-heme was bound to transferrin in plasma. Thus, these data provide evidence that the pathway for heme-iron joins that of iron within the duodenal enterocyte. One point for the coordinated regulation of uptake of dietary heme and iron is, therefore, at the basolateral membrane transporter, FPN1. Additonal levels of regulation may occur via HCP1 whose mechanism of action is not yet established.

against oxidative stress and also enzymes for phase II metabolism, e.g., quinone oxidoreductase, GSH S-transferase and enzymes needed for glutathione synthesis. When heme levels rise in the nucleus, heme binds to the repressor, Bach1, which then dissociates from its DNA binding partner, a small Maf.[100,109] These heterodimers reside on two enhancers at binding sites termed stress response elements (StREs), which share some sequence homology with antioxidant response elements (AREs). Transcription is activated when the robust transcription factor Nrf2 also binds to a small Maf K on the enhancers of *ho1*.[101]

When over-expressed, Bach1 was serendipitously found to be colored brown.[114] Subsequently, binding studies with radioactive heme and absorbance spectroscopy (λ max in the Soret region of 421 nm) have revealed a stoichiometry of 1.3 mol heme bound per molecule of Bach1 (Kd 140 nM). In mammalian cells, spectral data in the Soret region indicate that heme binds directly to the transcription factor Bach 1. Unlike in Hap1, where a single peptide with a CP motif is sufficient, Bach 1 has six CP motifs, four of which are involved in heme binding as shown by deletion and mutation analyses (Cys to Ala). Thus, each CP is considered a potential heme binding ligand with a stoichiometry of binding of 1:1, and the Kd of 140 nM represents a high affinity interaction. These CPs surround the bZip domain and act independently of the BTB [BR-C, ttk and bab;[115]] domain needed for DNA binding. Heme binding is presumably noncovalent. Heme inhibits the DNA binding of Bach1/MafK in EMSAs but not by inhibiting the interaction of these transcription factors via their leucine zipper domains. The metal iron is important for the binding of heme to Bach1 since both proto- and meso-heme bind to Bach1, whereas the porphyrins, PPIX and mesoPPIX, do not. Importantly, Bach1 also represses the β-globin genes via a MARE and this repression is relieved by heme.[114,116] Additional Bach1- regulated genes will soon be identified and it will be interesting to see how regulation by this heme-responsive transcription factor differs in erythroid and nonerythroid cells. While repression by Bach1 is a common control, activation of the genes differ and globin is not activated by Nrf2 as is the *ho1* gene. Significantly, two CPs are needed for heme-mediated export of Bach1 that requires the nuclear export receptor Crm1.[117] Thus, heme contributes to the regulation of nucleo-cytoplasmic shuttling of DNA-binding proteins.[117]

HRMS Mediate Effects on Proteins That Are Needed for Heme Biosynthesis and the Regulation of Hemoglobin Synthesis

In the rate limiting enzyme for heme synthesis, ALA-S, two heme-binding motifs in the leader sequence, as well as one present in the N-terminus of the mature ALAS-1 function in vivo in the heme-regulated translocation of ALA-S1.[118] Heme also regulates ALA-S for feed back inhibition (for detailed review of ALA-S regulation see Chapter 7). The eukaryotic initiation factor 2α kinase (eIF α kinase) also needs an HRM for the heme-mediated regulation of hemoglobin synthesis.[11,119]

The ALAS-2 gene is specifically expressed in erythroid cells producing heme in the mitochondria for hemoglobin synthesis from the condensation of glycine and succinyl CoA. Heme regulates both the expression and localization of ALA-S2 and the housekeeping ALA-S1 that is expressed in all tissues. Mutation of the HRM (cysteine to serine) inhibited the mitochondrial import of ALAS-1 but not ALAS-2 expressed in quail QT6 fibroblasts.[120] Thus, heme acts differently in the regulation of these two isoenzymes. In erythroid cells, it appears that iron responsive elements play a more important role in regulation of ALA-S2 reflecting the predominant role of iron in heme synthesis in this cell type.

Heme-Protein Interactions and the Control of Circadian Rhythms

Every 24 hours we each experience a rhythmic fluctuation of our body temperatures, endocrine function, physical activity and metabolic rates - due to the well-known biological clock. This cycle is extremely well conserved in metazoans from fruit flies to mammals. It is synchronized (or entrained) by the dark-light cycle but persists even without environmental cues not only in cells of the brain in the suprachiasmiatic nucleus but also, and somewhat unexpectedly, in nonneuronal cells even in vitro. In the mouse, the rhythm requires a feedback loop at the level of transcription including cyclical changes in histone H3 acetylation synchronous with a cycle that controls steady state messenger RNA level of key molecular players.[121] In the light, two transcription factors Clock and BMAL1(brain and muscle aryl hydrocarbon receptor nuclear translocator-like protein 1) bind as a heterodimer to the promoter of 3 period (*Per*) and 2 cryptochrome (*Cry*) genes to activate transcription. However, the Per and Cry proteins do not accumulate in the cytosol due to rapid turnover mediated by phosphorylation before proteasomal degradation. Per:Cry heterodimers migrate to the nucleus where they positively and negatively regulate Clock:BMAL1 transcriptional activity - hence the oscillation. However, the regulatory interactions of Per and Cry with Clock:BMAL1 are complex, and it is not clear that the interactions modeled completely explain the circadian timing of Clock:BMAL1 activity. Much remains to be understood about the mechanism of negative and positive feedback (Fig. 6A).

From this perspective, an extremely exciting area in heme-protein interactions is the recently described link between heme metabolism and a Clock-related protein, neuronal PAS domain protein 2 (NPAS2). This is part of the clock in mammalian forebrain[122] and NPAS2 is similar to the heme-based PAS sensor proteins in bacteria. NPAS2 has a DNA-binding N-terminal basic helix-loop-helix region (bHLH) next to two PAS domains before the C-terminal 400 amino acids. Both PAS domains bind heme.[123] (It is not yet established whether Clock binds heme.) Food intake regulates and is regulated by the biological clock. NPAS2-deficient mice do not adapt to food restriction,[124] and entrainment of the cycle by food restriction may be linked to NPAS2:BMAL1 regulation of the promoter of lactate dehydrogenase (LDH).[125] This enzyme is needed to generate NAD^+ under anerobic conditions to maintain glycolysis. To bind to DNA in enhancers and regulate genes NPAS2, like Clock, needs BMAL1, also a bHLH PAS protein, as partner. Intriguingly, this DNA binding is increased in vitro as the ratio of NADH or NADPH relative to NAD^+ and $NADP^+$ increases at physiologically relevant concentrations (low mM) for normal cell metabolism.[125] The reduced coenzymes were proposed to increase the affinity of NPAS2:BMAL1 for DNA by acting as a molecular ligand rather than modifying the redox state of the transcription factors.[125] Currently, it is not clear what the redox state of the heme was or even whether it was bound to these proteins under the experimental conditions. A heme-mediated regulator of the NPAS2:BMAL1 heterodimer is likely to be CO. Heme catabolism by HOs release CO, which at concentrations >3 μM inhibits the binding of NPAS2:BMAL1 to DNA.[123] Generally, CO binds to ferro-heme. Kaasik and Lee[126] have demonstrated a reciprocal relationship between heme synthesis and clock gene expression. Their data provide a model linking heme synthesis and its degradation, with the generation of CO, to clock gene expression. The circadian regulators include another PAS domain transcription factor mPER2 as well as NPAS2. Heme synchronizes clock gene expression in NIH 3T3 cells. When intact mice housed in constant darkness are injected with heme (intraperitoneally at 30 mg/kg body weight, which is often employed route but neither the site of action or form of heme can be discerned) there is a decrease in mPER2 mRNA particularly during subjective night (during constant darkness, the daily interval in which lights were turned off in the previously imposed light/dark cycle). Wheel running activity, used to demonstrate regulation of circadian rhythms, was also inhibited by heme when injected during subjective night but not subjective day. Knock out mice lacking the circadian factors NPAS2, mPER2 and mPER1 ($Npas2^{m/m}$, $mPER1/2^{m/m}$) revealed that the effects of heme required NPAS2 and mPER2, but not mPER1. A weak circadian rhythm has also been observed for ALA-S1 mRNA, with peak expression late in the subjective day, which is altered in $Npas2^{m/m}$ mice. The expression of promoter constructs of fragments of the *Alas1* promoter and a luciferase reporter gene were stimulated 2-fold by the coexpression of both BMAL1 and NPAS, and up to 3.5-fold by NPAS2 and BMAL 1 together with mPER2. Thus, NPAS2 is proposed to be a primary regulator of *Alas1* in vivo, and mPER2 stimulates this positive regulation (Fig. 6C).

The role for heme seemed likely because both mPERs and NPAS2 contain the PAS motif, a known heme based PAS sensor protein domain in bacteria (see above). A direct interaction with heme is possible because mPER2 binds to a heme-Agarose affinity column but mPER2 with a mutated PAS domain does not. Reporter gene assays in culture suggest that mPER2 is a positive regulator of transcriptional activation by NPAS2:BMAL1. For the full expression of *Alas1*, the transactivators BMAL1 and NPAS2 are needed under the positive control of mPER2. The proposed model is that since ALA-S catalyzes the rate limiting step in heme synthesis, it controls intracellular heme levels. Although sufficient iron must be available for ferrochelatase to synthesize heme from protoporphyrin IX in the mitochondrial matrix, heme can be viewed as the prosthetic group of these two transcription factors, NPAS2 and mPER2. As heme levels rise a balance may be achieved by degradation of excess heme via heme oxygenase that releases CO (as long as oxygen levels and sufficient NADPH are available). With CO bound to the heme on BMAL1 and NPAS2, they no longer bind to DNA and transcription stops. Also, contributing to this decrease in transcription, there will

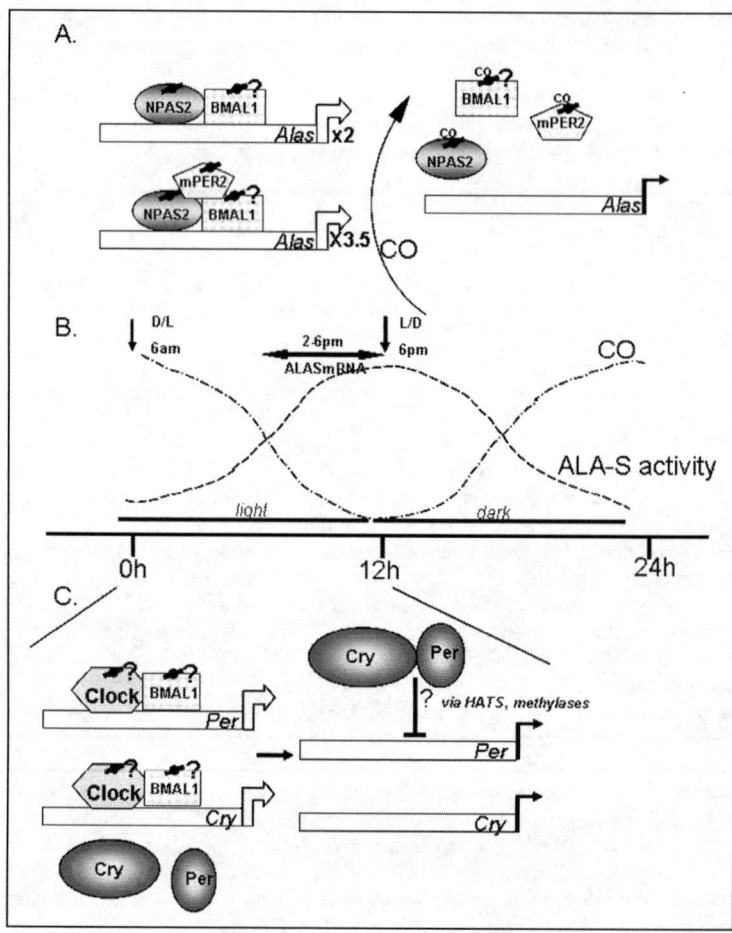

Figure 6. Relationships between endogenous CO production and regulation of ALA-S, HO and clock proteins. Panel A represents the increased transcription of murine ALA-S in response to activation by the heterodimer of PAS domain proteins NPAS2 and BMAL1, further augmented by mPER2. Carbon monoxide (CO) binds to NPAS2 and BMAL1 leading to their dissociation from the *Alas* promoter. Panel B shows two theoretical curves: one for CO production (dotted and dashed line) and the other for ALA-S activity levels (dashed line). Key questions here concern the phase lag between ALA-S activity increasing heme levels that are then degraded by HO generating CO; since ALA-S mRNA accumulates during the late day in murine liver,[126] there would have to be a phase lag in CO production to maintain high levels of CO during the dark phase, when NPAS2:BMAL1 transcriptional activity is low. In addition, whether local cellular or global levels of CO are most important in regulation is not known. There are diurnal variations in pulmonary levels of CO in humans.[140] In hamsters, HO increases in the suprachiasmic nucleus (SCN) at night in the dark phase.[141] However, in Drosophila, ALA-S mRNA increases when the lights are off and is regulated in a reciprocal fashion with HO-like Mrna.[142] Heme acts to regulate ALA-S protein directly by direct repression of its enzymic activity as well as via effects on its cellular location. Perhaps these occur initially in response to changes in heme levels followed by the transcriptional effects shown here. Panel C shows the links between the activation of Per and Cry genes by Clock/BMAL1 and the consequent chromatin modification by Cry/Per heterodimer that that turns off transcription of their own genes thus generating the cyclic protein modulations. The time course of Per and Cry nuclear localization (high during the period when lights are on and Clock/BMAL1 dependent transcription is also high) does not fit neatly with the postulated negative regulation of Clock:BMAL1 activity, suggesting a potentially important role for CO and heme oscillations in this negative regulation.

be less mPER2 in the cell. This provides a neat and efficient regulatory cycle that may generate part of the negative feedback on gene products essential for circadian rhythms. Nevertheless, several important questions remain to be answered especially with respect to the known phase lags between several key molecules (see Fig. 6).

Novel Heme-Protein Interactions for the Control of Intracellular Heme Levels

Recently, heme transporters have been identified in erythroid and progenitor cells.[127,128] To provide a basis for their function and regulation, these transporters have been proposed by Jonker and colleagues[129] and John Schuetz and colleagues[128] to protect hematopoietic cells from the toxic effects of heme. Mitochondrial function becomes impaired when heme and porphyrin levels rise.[130,131] Protection is important for maintaining genetic integrity. There are distinct differences in the way intracellular heme levels are controlled in erythroid cells specialized to synthesize hemoglobin. During differentiation, heme synthesis increases then globin synthesis. Hemoglobin synthesis is regulated via a heme-mediated decrease of Bach1 repression of the β-globin gene and, at the translational level, via a heme-mediated inhibition of a specific eukaryotic initiation factor 2(eIF2) α kinase (also termed HRI). This kinase, active when heme levels are low, phosphorylates eIF2α thus inactivating it.[11] When heme concentration increases, HRI activity decreases and eIF2α, no longer phosphorylated, acts in the initiation phase of protein synthesis to translate globin mRNA. Iron uptake is required and maintained by suitable levels of expression of the transferrin receptor.[132] An erythroid specific ALA-S2 gene regulates heme synthesis.[14] ALAS-2 (as described above) is also regulated by heme. In contrast in nonerythroid cells, raising intracellular heme levels rapidly induces activation of transcription of the *ho1* gene for heme degradation. Oxygen tension regulates heme levels via HO1 since oxygen is a substrate. Certainly, there is evidence for clear differences in the way iron responsive proteins regulate iron metabolism at physiological oxygen tension (2-6%) where IRP2 is the principal regulator and hyperoxia (20%) where IRP1 prevails.[133] Heme is involved directly in the regulation of IRP1 and IRP2[134,135] but perhaps only by providing iron to regulate IRP1 at higher oxygen concentrations.

Cellular heme levels respond to environmental oxygen and the heme content of cells increases in hypoxia.[136] Perhaps this is due in part to impaired heme catabolism because oxygen is required by HO1. Stem cells, unusually, thrive in hypoxia. Hematopoietic stem cells are found in large numbers in areas of the body where oxygen tension is low and as oxygen concentrations decline adapt by increasing surface glucose transporters, obtaining sufficient ATP from glycolysis. ATP hydrolysis provides the energy for the active transport of substrate molecules across the plasma membrane.[137] In humans, 49 ATP binding cassette (ABC) genes have been described with a variety of substrates, including heme. They are highly conserved proteins containing both ABC and transmembrane domains and are either a single polypeptide chain or two separate proteins (half transporters) that must dimerize for function. An ABC transporter subfamily G, member 2 (ABCG2) is also called the breast cancer resistance protein (BRCP), since several substrates for it are chemotherapeutic agents like mitoxantron. BRCP is needed for stem cells to survive hypoxia and bcrp null erythrocytes accumulate the fluorescent heme precursor, protoporphyrin IX [PPIX, refs. 129,138]. Patients with erythroprotoporphyria develop extreme photosensitivity due to elevated PPIX because heme synthesis is impaired due to abnormally low levels of mitochondrial ferrochelatase, the enzyme that incorporates iron into PPIX generating heme. Photosensitivity developed when bcrp null animals were fed a diet containing high levels of a molecule similar in structure to PPIX, the degradation product of chorophyll pheophorbide a.[129] BCRP expressing cells do not survive hypoxia when BCRP function is blocked. Jonker et al therefore concluded that BRCP transports PPIX and heme also. When heme synthesis is blocked with succinyl acetone at the ALA dehydratase step, bcrp null cells survive hypoxia, suggesting that at low oxygen tensions heme is toxic. Using membrane vesicles prepared from insect cells that expressed wild type human BRCP, heme increased the velocity of transport of another BRCP substrate, E3S. Such cooperativity is found with the multidrug resistance protein, another ABC transporter. Thus, BRCP is needed by hematopoietic cells to regulate intracellular porphyrin IX and/or heme levels in hypoxia.

Heme must be available for the continued development of erythroid progenitor cells. Burst forming units (BFU-Es) mature into colony forming units (CFU-Es) and then erythroid precursors cells take up iron and heme biosynthesis begins. Cats become severely anemic when infected with a nononcogenic retrovirus, the feline leukemia virus subgroup C (FeLV-C). As the infection develops, the levels of BFU-E cells are normal but there are very few CFU-E and erythroid precursors, suggesting that erythropoiesis is arrested before maturation to the CFU-E and proerythroblast stages. A cell surface protein that serves as a receptor for the feline leukemia virus subgroupC (FLVCR), is homologous to members of the major super facilitator group of proteins. These are transporters that move small solutes, including sugars and amino acids, across membranes in response to chemi-osmotic gradients. Thus, in contrast to ABCG2 (BRCP), ATP is not required.

Feline FLVCR is down regulated by the production of viral envelope components intracellularly with a specific loss of CFU-E, implicating FLVCR in CFU-E differentiation or survival. Heme synthesis is decreased at the CFU-E stage of differentiation or soon afterwards. Heme export, measured in kidney (NRK) cells expressing human FLVCR, is rapid complete within 30 min, but only when HO1 is inhibited by the heme analog ZnPPIX. Otherwise, heme levels decrease rapidly and independent of FLVCR expression. Inhibition of cell surface expression of FLVCR or FLVCR function impaired differentiation into erythroid cells and induced apoptosis. FLVCR expression is down regulated both at the protein and mRNA levels in mobilized PBCD34+ cells undergoing differentiation. Thus, heme excess is proposed to prematurely initiate globin translation producing apoptosis. FLVCR may prevent a rise in heme levels, interfering with normal hematopoeisis at the CFU-E stage of differentiation. This regulation by heme may or may not override the regulation of heme synthesis by iron availability. If in early stages of erythropoiesis heme levels fluctuate, FLVCR may provide a safety mechanism or "overflow".[127]

FLVCR orthologs are found in bacterial, plant and animal genomes and FLVCR is fairly ubiquitously expressed in tissues and in several different cell lines (hepatoma, neuroblastoma and in Caco2 cells derived from a primary human colonic adenocarcinoma). High levels of FLVCR protein were found on mobilized PB CD34+ cells and lines expressing the early erythroid phenotype (HEL-DR and K562) but was not detected in the mature erythroid cell line, HEL-R. Heme can potentially regulate FLVCR expression since a consensus sequence for an NF-E2 site has been identified in the region between -370 and -1030 nucleotides relative to the initiation site as well as GATA1, -2 and c-myb sites. It will be interesting to determine whether FLVCR responds to intracellular heme in nonerythroid cells including enterocytes. Schuetz and coauthors conclude that the estimated 100 μM intracellular heme levels in erythroid cells during globin synthesis is probably high enough to drive export by passive diffusion via FLVCR although its capacity is not yet established. Important biological processes often have backup systems, perhaps FLVCR is needed as well as ABCG 2 to maintain heme levels even if ATP is unavailable.

Another ABC transporter, ABCB6, has been linked to sensing both heme and porphyrin levels within erythroid cells and proposed to import heme into mitochondria [Krishnamurthy, P., Du, G., Sun, D., Fukuda, Y., Sampath, J., Mercer, K., Wang, J., Sosa-Pineda, B., Murti, G. and Schuetz, J.D., unpublished data]. Once synthesized, heme is conserved by transfer from cytosol into the mitochondrion rather than to HO for degradation or export. Heme, not iron, induces both ABCB6 mRNA and protein levels in murine erythroleukemia (MEL) cells but only when globin is being synthesized. ABCB6 levels and PPIX increase when MEL cells are incubated with the heme precursor, ALA, but not when heme synthesis is inhibited by succinyl acetone. Succinyl acetone alone did not alter ABCB6 levels revealing that while ABCB6 expression is linked with porphyrin levels bypassing the regulation normally imposed by rate- limiting ALA-S is required. Over expression of ABCB6, but not BRCP, blocks induction of globin expression by heme, but not by DMSO and prevented the decrease in phosphorylation of eIF2α implicating additional effects on eiF2α kinase (HRI).

Most ABC transporters are a single peptide with two transmembrane and two nucleotide binding regions. ABCB6 is a homodimer, forming a functional transporter from two half transporter peptides each with a single transmembrane and nucleotide site. An expressed C-terminal FLAG tagged ABCB6, although lacking a mitochondrial target sequence, was detected in the outer mitochondrial membrane with its hydrophilic C-terminus that hydrolyzes cytosolic ATP exposed to cytosol. Site directed mutagenesis and heme Agarose affinity chromatography revealed that a transmembrane domain binds heme, a region in ABC transporters known to bind their substrates. The C terminal of ABCB6 is predicted to bind heme as a homodimer before transport.

Over expression of wt ABCB6 in MEL and Saos2 cells, but not ABCB6 mutated in the Walker ATP binding domain, increased PPIX content; porphyrin synthesis (6-to 13-fold); expression of the transferrin receptor and ALA-S2 as well as increased activity of ALA dehydratase and coproporphyrinogen oxidase (CPO). It is well established that ALA-Ss contain HRMs that bind heme which not only acts as a feed back inhibitor but also blocks translocation of the enzyme into the mitochondrial matrix for condensation of glycine and succinyl CoA. Schuetz and colleagues speculate that, in non erythroid tissues such as liver and also in other proliferating cells, ABCB6 transports heme into the mitochondria where it contributes to the regulation of ALA-S by releasing heme into the inner membrane space.

Relationships between ATP Concentrations, Oxygen Tension and Heme Transporters, Many of Which Also Interact with Porphyrins

The availability of oxygen affects HO enzymic activity for heme degradation as well as two enzymes for heme synthesis, coproporphyrinogen oxidase (CPO) and protoporpyrinogen oxidases since all three require oxygen as substrate. CPO has been suggested as another rate limiting step in heme synthesis. In addition, transport of coproporphyrinogen III into mitochondria via an 18 KDa peripheral-type benzodiazepine receptor,[139] which may also direct import of protoporphyrin IX into mitochondria,[138] may be a regulatory step. Heme transporters like ABCG2 (BRCA) and HCP1[89] are regulated by hypoxia inducible factor, HIF.[128] Consequently, they are induced under conditions of low oxygen tension. HO is also regulated by HIF but will be enzymically inactive if oxygen is unavailable. Do heme transporters, heme biosynthesis and heme oxygenase complement each other by responding not only to heme concentrations but also to oxygen levels? A hierarchy of events presumably unfolds, in turn regulated by the availability of ATP. Heme transporters like ABCB6 and ABCG2 use the energy of ATP hydrolysis whereas FLVCR does not and heme flows down a concentration gradient. Under anaerobic conditions, given a supply of glucose and NAD$^+$ regeneration, glycolysis provides ATP. However, significantly less ATP is generated than from aerobic metabolism via the electron transport chain. With sufficient ATP, ABCB6 may bring heme into the mitochondria and decrease de novo heme synthesis by decreasing ALAS activity. FLVCR remains as a cytosolic safety valve for heme efflux. Overall, oxygen levels will influence not only the extent of heme synthesis and its catabolism but also heme movement by ABC and other transporters perhaps with consequences for gene regulation.

Concluding Remarks

These most recently described heme-protein interactions provide examples of a plethora of diverse environments for heme as it moves through the cell, allowing a wide range of biological processes to be influenced. Are all of these heme molecules "regulatory hemes"? Genetic model systems are currently being used to probe in detail the role of heme in complex biological processes and, importantly, how the homeostasis of heme also affects, or is affected by, several metals, in particular iron and copper. Further insights into vertebrate heme and iron metabolism are being revealed in zebrafish, and even fruit flies can potentially help contribute to our understanding of the biological roles of heme in mammals, for example in controlling the circadian rhythm. Unusually, heme together with at least one of its catabolites CO, often acts in synchrony to regulate gene expression, e.g., via PAS domains in neuronal cells. Thus, heme continues to intrigue its aficionados and to surprise its skeptics, and the research summarized here is providing avenues for enhanced knowledge of the biological roles of many metal-tetrapyrroles and tetrapyrroles.

Acknowledgements

This review is dedicated to the memory of Dr. Dennis Nicholson and Professor Charles Gray without whose support at King's College Hospital Medical School, London, UK, I would not have had a career in scientific research.

I am also extremely grateful to colleagues Dr. John Schuetz (Department of Pharmaceutical Sciences, St. Jude Children's Research Hospital, Memphis TN) and Dr. Andy McKie (Life Sciences, Kings College London, UK) who have shared un-published information with me and allowed its incorporation here. I would like to acknowledge the generosity of Dr. Max Paoli (University of

Nottingham, Nottingham, UK) who kindly provided the unpublished figures on the heme-binding site of hemopexin that are included here and for helpful discussions on the structure of hemopexin. I am indebted to Dr. Jeffrey Price (University of Missouri at Kansas City) for stimulating discussions about the regulation of clock genes in mice and Drosophila.

References

1. Galbraith RA, Sassa S, Kappas A. Induction of haem synthesis in Hep G2 human hepatoma cells by dimethyl sulphoxide. A transcriptionally activated event. Biochem J 1986; 237:597-600.
2. Padmanaban G, Venkateswar V, Rangarajan PN. Haem as a multifunctional regulator. Trends Biochem Sci 1989; 14:492-496.
3. Rangarajan PN, Padmanaban G. Regulation of cytochrome P-450b/e gene expression by a heme- and phenobarbitone-modulated transcription factor. Proc Natl Acad Sci USA 1989; 86:3963-3967.
4. Hooper N, Morgan WT, Neilands JB et al. Heme binding by the gene product of the fur (Ferric Uptake Regulation) gene. FASEB J 1992; 6:Abs. No.487.
5. Smith A, Hooper NI, Shipulina N et al. Heme binding by a bacterial repressor protein, the gene product of the ferric uptake regulation (fur) gene of Escherichia coli. J Protein Chem 1996; 15:575-583.
6. Alam J, Smith A. Receptor-mediated transport of heme by hemopexin regulates gene expression in mammalian cells. J Biol Chem 1989; 264:17637-17640.
7. Izadi N, Henry Y, Haladjian J et al. Purification and characterization of an extracellular heme-binding protein, HasA, involved in heme iron acquisition. Biochemistry 1997; 36:7050-7057.
8. Alric J, Pierre Y, Picot D et al. Spectral and redox characterization of the heme ci of the cyto-chrome b6f complex. Proc Natl Acad Sci USA 2005; 102:15860-15865.
9. Reedy CJ, Gibney BR. Heme protein assemblies. Chem Rev 2004; 104:617-649.
10. Zhang L, Hach A, Wang C. Molecular mechanism governing heme signaling in yeast: A higher-order complex mediates heme regulation of the transcriptional activator HAP1. Mol Cell Biol 1998; 18:3819-3828.
11. Chen JJ, London IM. Regulation of protein synthesis by heme-regulated eIF-2 alpha kinase. Trends Biochem Sci 1995; 20:105-108.
12. Boehning D, Snyder SH. Circadian rhythms: Carbon monoxide and clocks. Science 2002; 298:2339-2340.
13. Sassa S. Why heme needs to be degraded to iron, biliverdin IXalpha, and carbon monoxide? Antioxid Redox Signal 2004; 6:819-824.
14. Nakajima O, Takahashi S, Harigae H et al. Heme deficiency in erythroid lineage causes differentiation arrest and cytoplasmic iron overload. EMBO J 1999; 18:6282-6289.
15. Atamna H, Killilea DW, Killilea AN et al. Heme deficiency may be a factor in the mitochondrial and neuronal decay of aging. Proc Natl Acad Sci USA 2002; 99:14807-14812.
16. Handschin C, Lin J, Rhee J et al. Nutritional regulation of hepatic heme biosynthesis and porphyria through PGC-1alpha. Cell 2005; 122:505-515.
17. Gilles-Gonzalez MA, Gonzalez G. Heme-based sensors: Defining characteristics, recent developments, and regulatory hypotheses. J Inorg Biochem 2005; 99:1-22.
18. Cosby K, Partovi KS, Crawford JH et al. Nitrite reduction to nitric oxide by deoxyhemoglobin vasodilates the human circulation. Nat Med 2003; 9:1498-1505.
19. Xu X, Cho M, Spencer NY et al. Measurements of nitric oxide on the heme iron and beta-93 thiol of human hemoglobin during cycles of oxygenation and deoxygenation. Proc Natl Acad Sci USA 2003; 100:11303-11308.
20. Gladwin MT, Lancaster Jr JR, Freeman BA et al. Nitric oxide's reactions with hemoglobin: A view through the SNO-storm. Nat Med 2003; 9:496-500.
21. Pawloski JR, Hess DT, Stamler JS. Impaired vasodilation by red blood cells in sickle cell disease. Proc Natl Acad Sci USA 2005; 102:2531-2536.
22. Shikama K, Matsuoka A. Structure-function relationships in unusual nonvertebrate globins. Crit Rev Biochem Mol Biol 2004; 39:217-259.
23. Van Doorslaer S, Dewilde S, Kiger L et al. Nitric oxide binding properties of neuroglobin. A characterization by EPR and flash photolysis. J Biol Chem 2003; 278:4919-4925.
24. Sun Y, Jin K, Mao XO et al. Effect of aging on neuroglobin expression in rodent brain. Neurobiol Aging 2005; 26:275-278.
25. Sun Y, Jin K, Mao XO et al. Neuroglobin is up-regulated by and protects neurons from hypoxic-ischemic injury. Proc Natl Acad Sci USA 2001; 98:15306-15311.
26. Sun Y, Jin K, Peel A et al. Neuroglobin protects the brain from experimental stroke in vivo. Proc Natl Acad Sci USA 2003; 100:3497-3500.

27. Nioche P, Berka V, Vipond J et al. Femtomolar sensitivity of a NO sensor from Clostridium botulinum. Science 2004; 306:1550-1553.

28. Doré S, Sampei K, Goto S et al. Heme oxygenase-2 is neuroprotective in cerebral ischemia. Mol Med 1999; 5:656-663.

29. Siemen D, Loupatatzis C, Borecky J et al. Ca^{2+}-activated K channel of the BK-type in the inner mitochondrial membrane of a human glioma cell line. Biochem Biophys Res Commun 1999; 257:549-554.

30. Tang XD, Xu R, Reynolds MF et al. Haem can bind to and inhibit mammalian calcium-dependent Slo1 BK channels. Nature 2003; 425:531-535.

31. Williams SE, Wootton P, Mason HS et al. Hemoxygenase-2 is an oxygen sensor for a calcium-sensitive potassium channel. Science 2004; 306:2093-2097.

32. Feng L, Zhou S, Gu L et al. Structure of oxidized alpha-haemoglobin bound to AHSP reveals a protective mechanism for haem. Nature 2005; 435:697-701.

33. Cowley AB, Rivera M, Benson DR. Stabilizing roles of residual structure in the empty heme binding pockets and unfolded states of microsomal and mitochondrial apocytochrome b_5. Protein Sci 2004; 13:2316-2329.

34. Altuve A, Wang L, Benson DR et al. Mammalian mitochondrial and microsomal cytochromes b(5) exhibit divergent structural and biophysical characteristics. Biochem Biophys Res Commun 2004; 314:602-609.

35. Mallory JC, Crudden G, Johnson BL et al. Dap1p, a heme-binding protein that regulates the cytochrome P450 protein Erg11p/Cyp51p in Saccharomyces cerevisiae. Mol Cell Biol 2005; 25:1669-1679.

36. Gutteridge JM, Smith A. Antioxidant protection by haemopexin of haem-stimulated lipid peroxidation. Biochem J 1988; 256:861-865.

37. Vincent SH, Grady RW, Shaklai N et al. The influence of heme-binding proteins in heme-catalyzed oxidations. Arch Biochem Biophys 1988; 265:539-550.

38. Miller YI, Shaklai N. Kinetics of hemin distribution in plasma reveals its role in lipoprotein oxidation. Biochim Biophys Acta 1999; 1454:153-164.

39. Miller YI, Smith A, Morgan WT et al. Role of hemopexin in protection of low-density lipoprotein against hemoglobin-induced oxidation. Biochemistry 1996; 35:13112-13117.

40. Burch MK, Muhoberac BB, Morgan WT. Characterization of Cu^{2+} and Fe^{3+} -mesoporphyrin complexes with histidine-rich glycoprotein: Evidence for Cu^{2+} -Fe^{3+} mesoporphyrin interaction. J Inorg Biochem 1988; 34:135-148.

41. Burch MK, Morgan WT. Preferred heme binding sites of histidine-rich glycoprotein. Biochemistry 1985; 24:5919-5924.

42. Guan X, Juarez JC, Qi X et al. Histidine-proline rich glycoprotein (HPRG) binds and transduces anti-angiogenic signals through cell surface tropomyosin on endothelial cells. Thromb Haemost 2004; 92:403-412.

43. Juarez JC, Guan X, Shipulina NV et al. Histidine-proline-rich glycoprotein has potent antiangiogenic activity mediated through the histidine-proline-rich domain. Cancer Res 2002; 62:5344-5350.

44. Hrkal Z, Kalousek I, Vodrazka Z. Haeme binding to albumin and equilibria in the albumin-ferrihaemoglobin and albumin-haemopexin systems. Int J Biochem 1980; 12:619-624.

45. Lim SK, Kim H, Ali A et al. Increased susceptibility in Hp knockout mice during acute hemolysis. Blood 1998; 92:1870-1877.

46. Higa Y, Oshiro S, Kino K et al. Catabolism of globin-haptoglobin in liver cells after intravenous adminstration of hemoglobin-haptoglobin to rats. J Biol Chem 1981; 256(23):12322-12328.

47. Tolosano E, Hirsch E, Patrucco E et al. Defective recovery and severe renal damage after acute hemolysis in hemopexin-deficient mice. Blood 1999; 94:3906-3914.

48. Smith A, Morgan WT. Haem transport to the liver by haemopexin. Receptor-mediated uptake with recycling of the protein. Biochem J 1979; 182:47-54.

49. Smith A, Morgan WT. Transport of heme by hemopexin to the liver: Evidence for receptor-mediated uptake. Biochem Biophys Res Commun 1978; 84:151-157.

50. Smith A, Hunt RC. Hemopexin joins transferrin as representative members of a distinct class of receptor-mediated endocytic transport systems. Eur J Cell Biol 1990; 53:234-245.

51. Smith A, Morgan WT. Hemopexin-mediated heme uptake by liver. Characterization of the interaction of heme-hemopexin with isolated rabbit liver plasma membranes. J Biol Chem 1984; 259:12049-12053.

52. Hunt RC, Hunt DM, Gaur N et al. Hemopexin in the human retina: Protection of the retina against heme-mediated toxicity. J Cell Physiol 1996; 168:71-80.

53. Smith A, Eskew JD, Borza CM et al. Role of heme-hemopexin in human T-lymphocyte proliferation. Exp Cell Res 1997; 232:246-254.

54. Camborieux L, Julia V, Pipy B et al. Respective roles of inflammation and axonal breakdown in the regulation of peripheral nerve hemopexin: An analysis in rats and in C57BL/Wlds mice. J Neuroimmunol 2000; 107:29-41.
55. Morris CM, Candy JM, Edwardson JA et al. Evidence for the localization of haemopexin immunoreactivity in neurones in the human brain. Neurosci Lett 1993; 149:141-144.
56. Hvidberg V, Maniecki MB, Jacobsen C et al. Identification of the receptor scavenging hemopexin-heme complexes. Blood 2005; 106:2752-2579.
57. Morgan WT, Smith A. Domain structure of rabbit hemopexin. Isolation and characterization of a heme-binding glycopeptide. J Biol Chem 1984; 259:12001-12006.
58. Conway TP, Morgan WT, Liem HH et al. Catabolism of photo-oxidized and desialylated hemopexin in the rabbit. J Biol Chem 1975; 250:3067-3073.
59. Smith A. Intracellular distribution of haem after uptake by different receptors. Haem-haemopexin and haem-asialo-haemopexin. Biochem J 1985; 231:663-669.
60. Strickland DK, Gonias SL, Argraves WS. Diverse roles for the LDL receptor family. Trends Endocrinol Metab 2002; 13:66-74.
61. van Kerkhof P, Lee J, McCormick L et al. Sorting nexin 17 facilitates LRP recycling in the early endosome. EMBO J 2005; 24:2851-2861.
62. Morse D, Choi AM. Heme oxygenase-1: The "emerging molecule" has arrived. Am J Respir Cell Mol Biol 2002; 27:8-16.
63. Ren Y, Smith A. Mechanism of metallothionein gene regulation by heme-hemopexin. Roles of protein kinase C, reactive oxygen species, and cis-acting elements. J Biol Chem 1995; 270:23988-23995.
64. Alam J, Smith A. Heme-hemopexin-mediated induction of metallothionein gene expression. J Biol Chem 1992; 267:16379-16384.
65. Eskew JD, Vanacore RM, Sung L et al. Cellular protection mechanisms against extracellular heme: Heme-hemopexin, but not free heme, activates the N-terminal c-Jun kinase. J Biol Chem 1999; 274:638-648.
66. Vanacore R, Eskew J, Morales P et al. Role for copper in transient oxidation and nuclear translocation of MTF-1, but not of NFκB, by the hemopexin heme transport system. Antioxid Redox Signal 2000; 2:739-752.
67. Sung L, Womack M, Shipulina N et al. Cell surface events for metallothionein-1 and heme oxygenase-1 regulation by the hemopexin heme transport system. Antioxid Redox Signal 2000; 2:753-765.
68. Smith A. Role of redox-active metals in the regulation of the metallothionein and heme oxygenase gene by heme and hemopexin. In: Ferreira GC, Mourra JJG and Franco R, eds. Iron Metabolism. Weinheim: Wiley-VCH Publishing Co., 1999:65-93.
69. Paoli M, Anderson BF, Baker HM et al. Crystal structure of hemopexin reveals a novel high-affinity heme site formed between two beta-propeller domains. Nat Struct Biol 1999; 6:926-931.
70. Zeng Y, Deshmukh R, Caignan GA et al. Mixed regioselectivity in the Arg-177 mutants of Corynebacterium diphtheriae heme oxygenase as a consequence of in-plane heme disorder. Biochemistry 2004; 43:5222-5238.
71. Smith A, Morgan WT. Hemopexin-mediated transport of heme into isolated rat hepatocytes. J Biol Chem 1981; 256:10902-10909.
72. Askwith C, Eide D, Van Ho A et al. The FET3 gene of S. cerevisiae encodes a multicopper oxidase required for ferrous iron uptake. Cell 1994; 76:403-410.
73. De Silva DM, Askwith CC, Eide D et al. The FET3 gene product required for high affinity iron transport in yeast is a cell surface ferroxidase. J Biol Chem 1995; 270:1098-1101.
74. Lesuisse E, Casteras-Simon M, Labbe P. Evidence for the Saccharomyces cerevisiae ferrireductase system being a multicomponent electron transport chain. J Biol Chem 1996; 271:13578-13583.
75. Han O, Wessling-Resnick M. Copper repletion enhances apical iron uptake and transepithelial iron transport by Caco2 cells. Am J Physiol Gastrointest Liver Physiol 2002; 282:G527-533.
76. Simovich MJ, Conrad ME, Umbreit JN et al. Cellular location of proteins related to iron absorption and transport. Am J Hematol 2002; 69:164-170.
77. Conrad ME, Umbreit JN, Moore EG et al. Alternate iron transport pathway. Mobilferrin and integrin in K562 cells. J Biol Chem 1994; 269:7169-7173.
78. Kawabata H, Yang R, Hirama T et al. Molecular cloning of transferrin receptor 2. A new member of the transferrin receptor-like family. J Biol Chem 1999; 274:20826-20832.
79. Trinder D, Baker E. Transferrin receptor 2: A new molecule in iron metabolism. Int J Biochem Cell Biol 2003; 35:292-296.
80. Fleming RE, Migas MC, Holden CC et al. Transferrin receptor 2: Continued expression in mouse liver in the face of iron overload and in hereditary hemochromatosis. Proc Natl Acad Sci USA 2000; 97:2214-2219.

81. Deaglio S, Capobianco A, Cali A et al. Structural, functional, and tissue distribution analysis of human transferrin receptor-2 by murine monoclonal antibodies and a polyclonal antiserum. Blood 2002; 100:3782-3789.

82. Hodgson LL, Quail EA, Morgan EH. Receptor-independent uptake of transferrin-bound iron by reticulocytes. Arch Biochem Biophys 1994; 308:318-326.

83. Trinder D, Morgan E. Inhibition of uptake of transferrin-bound iron by human hepatoma cells by nontransferrin-bound iron. Hepatology 1997; 26:691-698.

84. Trinder D, Morgan E. Mechanisms of ferric citrate uptake by human hepatoma cells. Am J Physiol 1998; 275:G279-286.

85. Yu J, Wessling-Resnick M. Influence of copper depletion on iron uptake mediated by SFT, a stimulator of Fe transport. J Biol Chem 1998; 273:6909-6915.

86. Hunt RC, Hunt DM, Gaur NK et al. Protection of the retina against heme-mediated toxicity. J Cell Physiol 1996; 168:71-80.

87. Taketani S, Kohno H, Tokunaga R. Receptor-mediated heme uptake from hemopexin by human erythroleukemia K562 cells. Biochem Int 1986; 13:307-312.

88. Taketani S, Kohno H, Tokunaga R. Cell surface receptor for hemopexin in human leukemia HL60 cells Specific binding, affinity labeling, and fate of heme. J Biol Chem 1987; 262:4639-4643.

89. Shayeghi M, Latunde-Dada GO, Oakhill JS et al. Identification of an intestinal heme transporter. Cell 2005; 122:789-801.

90. Follett JR, Suzuki YA, Lonnerdal B. High specific activity heme-Fe and its application for studying heme-Fe metabolism in Caco2 cell monolayers. Am J Physiol Gastrointest Liver Physiol 2002; 283:G1125-1131.

91. McCance R, Widdowson E. Absorption and excretion of iron. Lancet 1937; ii:680-684.

92. Hallberg L, Solvell L. Absorption of hemoglobin iron in man. Acta Med Scand 1967; 181:335-354.

93. Bjorn-Rasmussen E, Hallberg L, Isaksson B et al. Food iron absorption in man. Applications of the two-pool extrinsic tag method to measure heme and nonheme iron absorption from the whole diet. J Clin Invest 1974; 53:247-255.

94. Conrad ME, Cortell S, Williams HL et al. Polymerization and intraluminal factors in the absorption of hemoglobin-iron. J Lab Clin Med 1966; 68:659-668.

95. Smith A. Transport of tetrapyrroles: Mechanisms and biological and regulatory consequences. In: Dailey HA, ed. Biosynthesis of Heme and Chlorophylls. New York: McGraw-Hill, 1990:435-490.

96. Grasbeck R, Majuri R, Kouvonen I et al. Spectral and other studies on the intestinal haem receptor of the pig. Biochim Biophys Acta 1982; 700:137-142.

97. Galbraith RA, Sassa S, Kappas A. Heme binding to murine erythroleukemia cells. Evidence for a heme receptor. J Biol Chem 1985; 260:12198-12202.

98. Majuri R. Heme-binding plasma membrane proteins of K562 erythroleukemia cells: Adsorption to heme-microbeads, isolation with affinity chromatography. Eur J Haematol 1989; 43:220-225.

99. Uc A, Stokes JB, Britigan BE. Heme transport exhibits polarity in Caco2 cells: Evidence for an active and membrane protein-mediated process. Am J Physiol Gastrointest Liver Physiol 2004; 287:G1150-1157.

100. Kitamuro T, Takahashi K, Ogawa K et al. Bach1 functions as a hypoxia-inducible repressor for the heme oxygenase-1 gene in human cells. J Biol Chem 2003; 278:9125-9133.

101. Sun J, Brand M, Zenke Y et al. Heme regulates the dynamic exchange of Bach1 and NF-E2-related factors in the Maf transcription factor network. Proc Natl Acad Sci USA 2004; 101:1461-1466.

102. Weintraub LR, Weinstein MB, Huser HJ et al. Absorption of hemoglobin iron: The role of a heme-splitting substance in the intestinal mucosa. J Clin Invest 1968; 47:531-539.

103. Raffin SB, Woo CH, Roost KT et al. Intestinal absorption of hemoglobin iron-heme cleavage by mucosal heme oxygenase. J Clin Invest 1974; 54:1344-1352.

104. Hartmann F, Bissell DM. Metabolism of heme and bilirubin in rat and human small intestinal mucosa. J Clin Invest 1982; 70:23-29.

105. McKie AT, Marciani P, Rolfs A et al. A novel duodenal iron-regulated transporter, IREG1, implicated in the basolateral transfer of iron to the circulation. Mol Cell 2000; 5:299-309.

106. Nemeth E, Tuttle MS, Powelson J et al. Hepcidin regulates cellular iron efflux by binding to ferroportin and inducing its internalization. Science 2004; 306:2090-2093.

107. Anderson GJ, Frazer DM, Wilkins SJ et al. Relationship between intestinal iron-transporter expression, hepatic hepcidin levels and the control of iron absorption. Biochem Soc Trans 2002; 30:724-726.

108. Frazer DM, Wilkins SJ, Becker EM et al. Hepcidin expression inversely correlates with the expression of duodenal iron transporters and iron absorption in rats. Gastroenterology 2002; 123:835-844.

109. Sun J, Hoshino H, Takaku K et al. Hemoprotein Bach1 regulates enhancer availability of heme oxygenase-1 gene. EMBO J 2002; 21:5216-5224.
110. Pfeifer K, Kim KS, Kogan S et al. Functional dissection and sequence of yeast HAP1 activator. Cell 1989; 56:291-301.
111. Zhang L, Guarente L. Heme binds to a short sequence that serves a regulatory function in diverse proteins. EMBO J 1995; 14:313-320.
112. Hon T, Lee HC, Hu Z et al. The heme activator protein Hap1 represses transcription by a heme-independent mechanism in Saccharomyces cerevisiae. Genetics 2005; 169:1343-1352.
113. Lee HC, Hon T, Lan C et al. Structural environment dictates the biological significance of heme-responsive motifs and the role of Hsp90 in the activation of the heme activator protein Hap1. Mol Cell Biol 2003; 23:5857-5866.
114. Ogawa K, Sun J, Taketani S et al. Heme mediates derepression of Maf recognition element through direct binding to transcription repressor Bach1. EMBO J 2001; 20:2835-2843.
115. Zollman S, Godt D, Prive GG et al. The BTB domain, found primarily in zinc finger proteins, defines an evolutionarily conserved family that includes several developmentally regulated genes in Drosophila. Proc Natl Acad Sci USA 1994; 91:10717-10721.
116. Tahara T, Sun J, Nakanishi K et al. Heme positively regulates the expression of beta-globin at the locus control region via the transcriptional factor Bach1 in erythroid cells. J Biol Chem 2004; 279:5480-5487.
117. Suzuki H, Tashiro S, Hira S et al. Heme regulates gene expression by triggering Crm1-dependent nuclear export of Bach1. EMBO J 2004; 23:2544-2553.
118. Dailey TA, Woodruff JH, Dailey HA. Examination of mitochondrial protein targeting of haem synthetic enzymes: In vivo identification of three functional haem-responsive motifs in 5-aminolaevulinate synthase. Biochem J 2005; 386:381-386.
119. McEwen E, Kedersha N, Song B et al. Heme-regulated inhibitor kinase-mediated phosphorylation of eukaryotic translation initiation factor 2 inhibits translation, induces stress granule formation, and mediates survival upon arsenite exposure. J Biol Chem 2005; 280:16925-16933.
120. Munakata H, Sun JY, Yoshida K et al. Role of the heme regulatory motif in the heme-mediated inhibition of mitochondrial import of 5-aminolevulinate synthase. J Biochem (Tokyo) 2004; 136:233-238.
121. Etchegaray JP, Lee C, Wade PA et al. Rhythmic histone acetylation underlies transcription in the mammalian circadian clock. Nature 2003; 421:177-182.
122. Reick M, Garcia JA, Dudley C et al. NPAS2: An analog of clock operative in the mammalian forebrain. Science 2001; 293:506-509.
123. Dioum EM, Rutter J, Tuckerman JR et al. NPAS2: A gas-responsive transcription factor. Science 2002; 298:2385-2387.
124. Dudley CA, Erbel-Sieler C, Estill SJ et al. Altered patterns of sleep and behavioral adaptability in NPAS2-deficient mice. Science 2003; 301:379-383.
125. Rutter J, Reick M, Wu LC et al. Regulation of clock and NPAS2 DNA binding by the redox state of NAD cofactors. Science 2001; 293:510-514.
126. Kaasik K, Lee CC. Reciprocal regulation of haem biosynthesis and the circadian clock in mammals. Nature 2004; 430:467-471.
127. Quigley JG, Yang Z, Worthington MT et al. Identification of a human heme exporter that is essential for erythropoiesis. Cell 2004; 118:757-766.
128. Krishnamurthy P, Ross DD, Nakanishi T et al. The stem cell marker Bcrp/ABCG2 enhances hypoxic cell survival through interactions with heme. J Biol Chem 2004; 279:24218-24225.
129. Jonker JW, Buitelaar M, Wagenaar E et al. The breast cancer resistance protein protects against a major chlorophyll-derived dietary phototoxin and protoporphyria. Proc Natl Acad Sci USA 2002; 99:15649-15654.
130. Antolin I, Uria H, Tolivia D et al. Porphyrin accumulation in the harderian glands of female Syrian hamster results in mitochondrial damage and cell death. Anat Rec 1994; 239:349-359.
131. Sandberg S, Romslo I. Porphyrin-sensitized photodynamic damage of isolated rat liver mitochondria. Biochim Biophys Acta 1980; 593:187-195.
132. Ponka P, Richardson DR. Can ferritin provide iron for hemoglobin synthesis? Blood 1997; 89:2611-2613.
133. Meyron-Holtz EG, Ghosh MC, Rouault TA. Mammalian tissue oxygen levels modulate iron-regulatory protein activities in vivo. Science 2004; 306:2087-2090.
134. Bourdon E, Kang DK, Ghosh MC et al. The role of endogenous heme synthesis and degradation domain cysteines in cellular iron-dependent degradation of IRP2. Blood Cells Mol Dis 2003; 31:247-255.

135. Ishikawa H, Kato M, Hori H et al. Involvement of heme regulatory motif in heme-mediated ubiquitination and degradation of IRP2. Mol Cell 2005; 19:171-181.
136. Hofer T, Wenger RH, Kramer MF et al. Hypoxic up-regulation of erythroid 5-aminolevulinate synthase. Blood 2003; 101:348-350.
137. Sarkadi B, Ozvegy-Laczka C, Nemet K et al. ABCG2 - A transporter for all seasons. FEBS Lett 2004; 567:116-120.
138. Abbott BL. ABCG2 (BCRP) expression in normal and malignant hematopoietic cells. Hematol Oncol 2003; 21:115-130.
139. Rebeiz N, Arkins S, Kelley KW et al. Enhancement of coproporphyrinogen III transport into isolated transformed leukocyte mitochondria by ATP. Arch Biochem Biophys 1996; 333:475-481.
140. Levitt MD, Ellis C, Levitt DG. Diurnal rhythm of heme turnover assessed by breath carbon monoxide concentration measurements. J Lab Clin Med 1994; 124:427-431.
141. Rubio MF, Agostino PV, Ferreyra GA et al. Circadian heme oxygenase activity in the hamster suprachiasmatic nuclei. Neurosci Lett 2003; 353:9-12.
142. Ceriani MF, Hogenesch JB, Yanovsky M et al. Genome-wide expression analysis in Drosophila reveals genes controlling circadian behavior. J Neurosci 2002; 22:9305-9319.
143. Flaherty M, Smith A, Crumbliss A. An Investigation of Hemopexin Redox Properties by Spectroelectrochemistry: Biological Relevance for Heme Uptake. Biometals 2007; [epub ahead of print].

CHAPTER 12

Synthesis and Role of Bilins in Photosynthetic Organisms

Nicole Frankenberg-Dinkel* and Matthew J. Terry

Abstract

Bilins are linear tetrapyrroles that function as the chromophores of the light-harvesting phycobiliproteins and light-signalling phytochromes in photosynthetic organisms. The biosynthesis of bilins proceeds through a ferredoxin-dependent heme oxygenase that oxidises heme to biliverdin IXα, the precursor of all functional bilins in these organisms. Biliverdin IXα is subsequently converted to one of three major bilin classes, phytochromobilin, phycocyanobilin and phycoerythrobilin, by a recently discovered family of related ferredoxin-dependent bilin reductases. Bilins are usually bound to apo-proteins through single or double covalent linkages and can be further modified during this process. This reaction is autocatalytic for the photoreceptor phytochrome but requires special lyases for phycobiliproteins. The binding to the latter results in a great diversity of bilin chromophores that completely span the visible light spectrum.

Introduction

Bilins are open-chain tetrapyrroles that are usually derived by cleavage of cyclic metalloporphyrins such as heme and chlorophyll. As the first bilins were isolated from animal bile, the term bilin was adopted. The basic structure of bilins is shown in Figure 1A and is characterised by a conjugated double bond system that results in very distinct colouration. The different colours of the individual bilins arise from the various number of double bonds in conjugation and often determine the pigmentation of the organisms in which they are found. Bilins occurring in photosynthetic organisms have two major roles. They act as sensors of light quality, intensity, duration and direction when associated with the protein moiety of the photoreceptor phytochrome (Fig. 1A).[1] When bound to phycobiliproteins, phycobilins, like phycocyanobilin (PCB), harvest visible light and efficiently transfer this energy to the photosynthetic reaction centres (Fig. 1C).[2] This enables certain photosynthetic bacteria to colonise light-limiting environments unsuitable for other photosynthetic organisms. This chapter focuses on the biochemical properties of the key enzyme families of bilin biosynthesis, the heme oxygenases and bilin reductases, and discusses the variety of roles played by bilins in photosynthetic organisms.

Structure and Spectral Properties of Protein-Bound Bilins

Most naturally-occurring bilins in photosynthetic organisms are covalently attached to proteins via one or two thioether bonds and examples of different bilin attachments are shown in Figures 1B,C and 2. Single thioether bond-attached bilins can be released from the protein by heating to reflux temperatures in methanol. The methanolysis of the thioether bond results in the generation of an ethylidine group that is conjugated with the chromophoric double bond system. Therefore, "free" bilins have slightly different absorption properties than the protein-bound form of the pigment.[3,4]

*Corresponding Author: Nicole Frankenberg-Dinkel—Physiology of Microorganisms,
Ruhr-University Bochum, Universitaetsstr. 150, 44780 Bochum, Germany.
Email: nicole.frankenberg@rub.de

Tetrapyrroles: Birth, Life and Death, edited by Martin J. Warren and Alison G. Smith.
©2009 Landes Bioscience and Springer Science+Business Media.

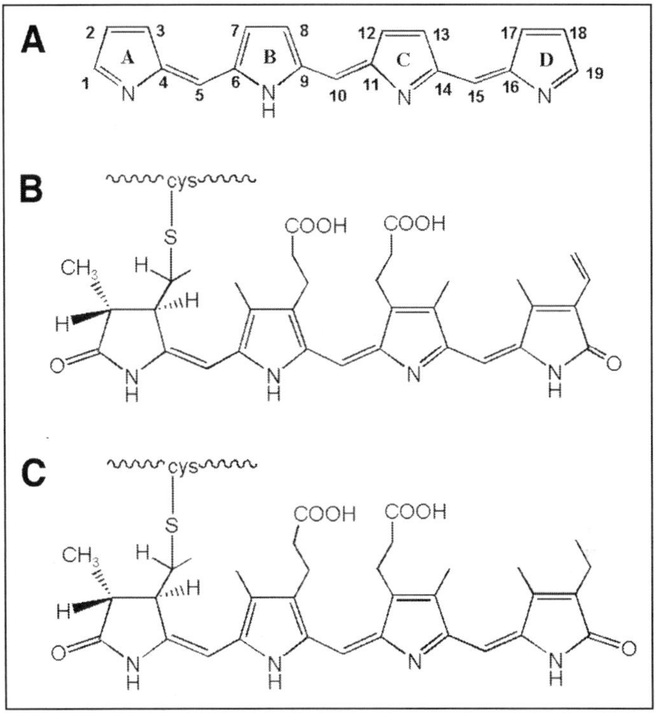

Figure 1. Structure of bilins. A) Basic bilin structure according to the IUPAC convention.[57] B) Phytochromobilin is bound to the plant photoreceptor phytochrome via a thioether linkage. C) Phycocyanobilin in its protein-linked conformation as it occurs in cyanobacterial phytochromes and phycobiliproteins.

Furthermore, binding to apoproteins can result in additional bilin isomers yielding a variety of phycobiliproteins whose absorption spectra completely span the visible light spectrum. These additional bilins do not occur in their free form and include phycoviolobilin (PVB), phycourobilin (PUB), mesobiliverdin (MBV), bilin 584 and bilin 618 (Fig. 2B). Based on their spectroscopic properties phycobiliproteins have been classified into four different spectral groups: 1. phycoerythrins (PE) with λ_{max} = 565-575 nm; 2. phycoerythrocyanin (PEC) with λ_{max} = 575 nm, 3. phycocyanins (PC) with λ_{max} = 615-640 nm and 4. allo-phycocyanin (APC) with λ_{max} = 650-655 nm. This remarkable spectral diversity is due to the chemically distinct chromophores with varying numbers of double bonds, different chromophore-protein linkages (single versus double linked; see Figs. 1A,B and 2), and distinct chromophore environments provided by the different polypeptide chains.[5] Furthermore, chromophore-chromophore interactions, as well as interactions with non-pigmented linker polypeptides, contribute to the spectral diversity of phycobiliproteins. Due to their role in light harvesting and energy transfer reactions phycobiliproteins exhibit not only strong absorption properties but are also highly fluorescent.[5] This property has provided a helpful cell biological tool. Phycobiliprotein-labeled detection reagents have been used extensively in flow cytometry to detect cell-specific expression of surface antigens.

In contrast to the wide variety of spectral forms of phycobiliproteins, the spectral properties of the phytochromes are relatively simple. Phytochromes can exist in two distinct spectral forms that are photointerconvertible. The P_r form of PΦB-bound oat phytochrome has a λ_{max} = 668 nm (with a second peak at λ = 378 nm) and can be converted into the P_{fr} form through absorption of red light. The resulting P_{fr} form has λ_{max} = 730 nm with the second peak shifted to λ = 402 nm.[6] For cyanobacterial phytochromes that utilise a PCB chromophore both the P_r and P_{fr} maxima are blue shifted. For Cph 1 from *Synechocystis* sp. PCC 6803 the λ_{max} = 656 nm and 704 nm respectively.[7,8]

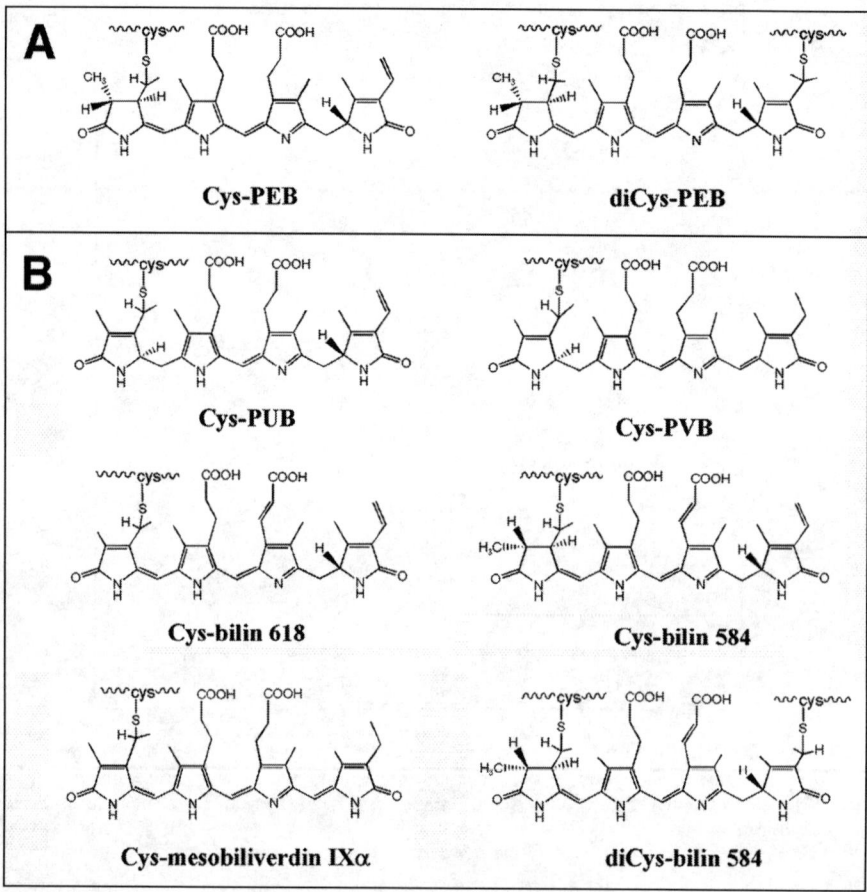

Figure 2. Representatives of protein-linked bilins. A) Cys-phycoerythrobilin (PEB) and di-Cys PEB as examples of the attachment of bilins to phycobiliproteins via either one or two thioether bonds (B) Cys-phycourobilin (PUB), Cys-phycoviolobilin (PVB), Cys-mesobiliverdin IXα, Cys-bilin 618, Cys-bilin 584 and diCys-bilin 584 are bilins that do not occur in their free form and are only found linked to phycobiliproteins.

Synthesis of Biliverdin IXα by Heme Oxygenases

Biliverdin (BV) IXα is the precursor of all functional bilins found in photosynthetic organisms and is derived by the action of heme oxygenase (HO; [E.C. 1.14.99.3]), which oxidises heme at the α-*meso* position exclusively to give the IXα isomer of BV (Fig. 3). There is evidence for at least one HO with a different regiospecificity in the opportunistic pathogenic bacterium *Pseudomonas aeruginosa*,[9] but no equivalent enzyme has been found in photosynthetic bacteria. The HO reaction requires molecular oxygen and electrons from, ultimately, NADPH and also results in the release of Fe^{2+} and CO. Extensive mechanistic studies on mostly mammalian HOs have led to a good understanding of the basic HO mechanism[10] (see Chapter 6) and current evidence suggests that this is broadly conserved in HOs from photosynthetic organisms.[11]

One property in which substantial differences have been observed between mammalian HOs and HOs from photosynthetic organisms is the direct source of electrons for the reaction. Mammalian HOs require cytochrome P450 reductase while HOs from photosynthetic organisms use ferredoxin. This was first shown by Cornejo and Beale[12] following partial purification of HO from the red alga *Cyanidium caldarium*, but has subsequently been demonstrated for the cyanobacterial[13] and plant[11] enzymes. In this latter case it was shown that ferredoxin-dependent HO activity could be

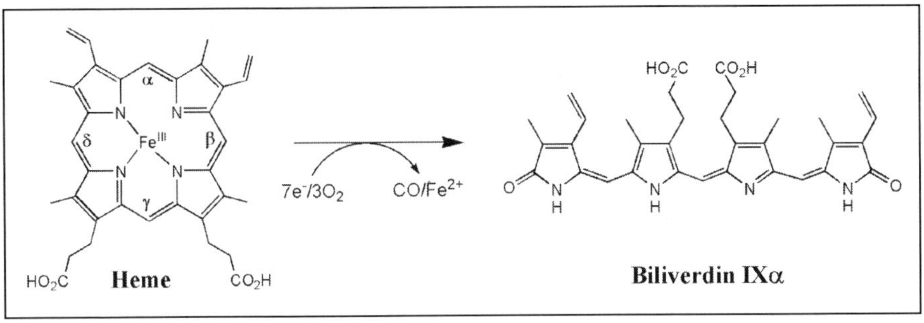

Figure 3. The heme oxygenase reaction. In most heme oxygenases the heme macrocycle is cleaved at the α-meso position to yield BV IXα. To date, only one other heme oxygenase has been described that has a different regiospecificity towards heme.[9]

driven in vitro by light in the presence of thylakoid membranes, although an NADPH:ferredoxin oxidoreductase can also be used.[11] It was recently observed that flavodoxin can partially substitute for ferredoxin in the cyanobacterial phycocyanobilin:ferredoxin oxidoreductase reaction[14] and it is possible that flavodoxin could also be used by HOs, although this has yet to be tested.

Two further features of the HO reaction in photosynthetic organisms differ from that previously observed for HOs from other organisms. Firstly, there appears to be a strong requirement for a second reductant. Ascorbate is the most effective reagent found to date and stimulates the algal,[15] cyanobacterial[16] and plant[11] enzymes. Indeed in the absence of any second reductant the activity of HO1 from *Arabidopsis thaliana* (AtHO1) was reduced by 90%.[11] The algal[15] and plant[11] enzymes also show a requirement for an iron chelator for full activity. This may reflect the fact that these plastid-localised HOs are usually operating under very high endogenous iron concentrations that are present in chloroplasts.

The progress in our understanding of the HO reaction in photosynthetic organisms is a reflection of recent developments in the cloning of HO-encoding genes. Two genes with homology to HOs were identified following the sequencing of the genome of the cyanobacterium *Synechocystis* sp. PCC 6803.[17] The *ho1* gene encodes a protein with a predicted molecular mass of 27 kDa, which was shown to be a functional HO by expression in *Escherichia coli*.[16] However, recombinantly expressed *ho2* did not show HO activity nor could *ho2* mRNA be detected in *Synechocystis* cell cultures.[16] It is therefore unclear whether *ho2* is a functional gene. Genes encoding putative HOs have also been found in the chloroplast genomes of red algae (e.g., *Rhodella violacea*; ref. 18), but it has yet to be demonstrated that they encode proteins with HO activity.

Plant *HO* genes were first cloned through the isolation of HO-deficient mutants. The *hy1* mutant of *Arabidopsis* was originally identified based on its phytochrome-deficient phenotype.[19] The *HY1* gene was cloned by chromosome walking[20] and a combination of mapping, genome sequencing and candidate gene complementation studies.[21] The HY1 protein, now called AtHO1, is comprised of 282 amino acids with a predicted mass of 32.6 kDa. This includes a chloroplast transit peptide of 55 amino acids, which is cleaved to give a mature protein of 26.6 kDa following chloroplast import.[20] Mature AtHO1 has also been demonstrated to have HO activity in vitro.[11,20] Both the plant and cyanobacterial enzymes contain the conserved His residue (His-86 [His-31 after transit peptide cleavage] and His-17 respectively refs. 11,16) that serves as the proximal heme ligand in the mammalian enzymes (His-25; ref. 10).

Remarkably, there are actually four putative *HO* genes in *Arabidopsis*, more than have been identified in any other organism to date.[22,23] These genes fall into two sub-families: the *HO1*-like genes that include *HO1*, *HO3* and *HO4*, and *HO2*. HO activity has yet to be demonstrated for any *Arabidopsis* HO other than AtHO1. There is some physiological evidence that AtHO2 can contribute to phytochromobilin (PΦB) synthesis implying some functional activity,[22] but given that AtHO2 lacks the conserved His residue required for heme binding, this observation will require further verification. Analysis of the expression of *AtHO1* and *AtHO2* indicate that both genes are expressed throughout the plant,[22] but in contrast to AtHO1 the cellular location of AtHO2 is not known. This is also true for AtHO3 and AtHO4 although all three also have putative chloroplast transit peptides.[22]

Comparative analysis of HO sequences across all organisms demonstrates that HOs from photosynthetic organisms all lack the C-terminal membrane anchor present in mammalian HOs.[10] However, while the biochemical properties of HOs from photosynthetic organisms are conserved, and indeed nuclear-encoded *Arabidopsis* proteins predicted to be of plastid origin have generally been shown to be similar to proteins from cyanobacteria, such a close relationship is not supported by current sequence alignments of the core catalytic domain (i.e., excluding N-terminal targeting sequences and C-terminal anchors; refs. 22,23). In these cases, plant HOs were equally divergent from cyanobacterial and mammalian *HO* genes with these latter two groups being more closely related. Whether these analyses adequately reflect the evolutionary relationship of these genes is unclear at present.

Biosynthesis of Bilins by Ferredoxin-Dependent Bilin Reductases

The major breakthrough in the study of bilin reductases was made through the identification of the genetic defect in the PΦB-deficient *hy2* mutant of *Arabidopsis thaliana*.[24] The cloning of the *HY2* gene was the basis for an extensive bioinformatics approach that led to the identification of a whole new class of enzymes.[24,25] As shown in Figure 4, ferredoxin-dependent bilin reductases catalyse the second step of the biosynthesis of the chromophore precursors of phytochromes and phycobiliproteins via three major independent pathways. Enzymes belonging to the ferredoxin-dependent bilin reductase family are involved in the two- or four-electron reduction of open-chain tetrapyrroles. Formally, these reactions resemble those catalysed by mammalian biliverdin reductases (BVRs; see also Chapter 6), but instead use ferredoxin as a reductant (Fig. 4). However, ferredoxin-dependent bilin reductases and BVRs are not evolutionarily related.[26] The following paragraphs will describe the distinct biochemical pathways leading to the three major bilins found in photosynthetic organisms: phytochromobilin (PΦB), phycoerythrobilin (PEB) and phycocyanobilin (PCB).

Biosynthesis of Phytochromobilin

The biosynthesis of the direct chromophore precursor of the plant photoreceptor phytochrome, PΦB is catalysed by phytochromobilin synthase (PΦB synthase; E.C. 1.3.7.4) through the two-electron reduction of BV IXα to 3Z-PΦB. The reaction requires reduced ferredoxin as an electron donor and targets the 2, 3, $3^1,3^2$-diene system of BV IXα and therefore formally represents a phytochromobilin: ferredoxin oxidoreductase.[26] In contrast to the multiple genes encoding HOs, the *Arabidopsis thaliana* PΦB synthase is encoded by a single gene.[24] Moreover, PΦB synthase is a very low abundance enzyme in plants with a high turnover rate.[27] The *HY2* gene encodes a protein with a predicted molecular mass of 38.1 kDa including a 45 amino acid chloroplast transit peptide.[24] Chloroplast localisation was confirmed both biochemically[27] and through the use of GFP fusion proteins.[24] The cloning of *HY2* enabled the direct comparison of the biochemical properties of the recombinant and native enzyme.[25,27] Both enzymes are monomeric in structure and require ferredoxin for catalytic activity. In contrast to the native PΦB synthase that exclusively produces the 3Z-isomer of PΦB, recombinant HY2 converts BV to a mixture of 3E-and 3Z-isomers. However, it is postulated that 3Z-PΦB is the sole product and formation of 3E-PΦB is triggered by glutathione in the chromatography buffer.[25] Interestingly, PΦB synthase activity has also been identified in the methylotrophic yeast *Pichia pastoris*.[28] This phenomenon was first identified during the expression of a cDNA clone encoding algal phytochrome, which led to the synthesis of holo-phytochrome without the addition of exogenous bilin. Although the gene encoding the *P. pastoris* PΦB synthase has not yet been cloned, it was shown that *P. pastoris* extracts harbour an activity that can convert BV to 3Z-PΦB.[28] To date, no *HY2* homologues have been identified in the genomic databases of nonphotosynthetic organisms.

Biosynthesis of Phycoerythrobilin

Until recently, all data on phycobilin biosynthesis were obtained from classical biochemical experiments on native enzyme preparations or extracts. Beale's pioneering work on phycobilin biosynthesis in the mesophilic red alga *Cyanidium caldarium* contributed much to our current knowledge about these enzymes.[3,29-32] These investigations revealed that the formation of the two major phycobilins, PCB and PEB is catalyzed by two ferredoxin-dependent bilin reductases as well as several double bond isomerases with the biosynthesis of PCB proceeding through the isomerisation of PEB. Starting from the common precursor of all phycobilins, BV IXα, the first bilin reductase catalyses the two-electron reduction at the C15 methine bridge of BV IXα yielding 15, 16-dihydrobiliverdin IXα (15, 16-DHBV). This semireduced intermediate is then further reduced to 3Z-PEB by a second bilin reductase.[13,32] Beale's original data could partly be confirmed by the cloning of genes encoding the bilin reductases 15,16 dihydrobiliverdin:ferredoxin oxidoreductase [E.C. 1.3.7.2] (*pebA*) and phycoerythrobilin:ferredoxin oxidoreductase [E.C. 1.3.7.3] (*pebB*) from various cyanobacteria and oxyphotobacteria.[25] These genes are usually organised in operons, in *Prochlorococcus* sp. even forming an operon together with the gene encoding HO (see Table 1). This is most intriguing since these organisms lack phycobilisomes. In *Synechococcus* sp. WH8020 the *pebA* and *pebB* genes are in close association with the phycobiliprotein gene cluster supporting their role in phycobilin biosynthesis.[33] Recombinant expression of *pebA* and *pebB* genes from various sources verified their involvement in ferredoxin-dependent reduction of BV and 15,16-DHBV, respectively.[25]

Biosynthesis of Phycocyanobilin

The biosynthesis of PCB seems to be rather more complex than for the other major bilins and to date three different biochemical pathways have been described (Fig. 4). The best characterised of these pathways is that mediated by the ferredoxin-dependent bilin reductase phycocyanobilin:ferredoxin oxidoreductase, PcyA [E.C. 1.3.7.5]. The *pcyA* gene was identified following the cloning of *HY2* and is found in all phycocyanin containing cyanobacteria as well as in oxyphotobacteria.[25] PcyA catalyses an atypical four-electron reduction of BV IXα to form 3Z-PCB (Fig. 4). In agreement with other bilin reductases, PcyA is a monomeric enzyme with a relative molecular mass of 30 kDa that forms a tight and stable stoichiometric complex with its substrate BV IXα.[14] The enzyme favours plant-type [2Fe-2S]-ferredoxins as reductants, however, flavodoxin can also serve as an electron donor. Interestingly, chloroplast-targeted PcyA can function in *Arabidopsis* to fully rescue the phytochrome-deficient *hy2* mutant.[34] Recent HPLC analyses have established that the catalysis proceeds via the semireduced intermediate $18^1,18^2$-DHBV, indicating that exo-vinyl reduction precedes A-ring (endo-vinyl) reduction (Fig. 4). Moreover, the use of an anaerobic assay for PcyA has established that both reductions proceed via radical intermediates.[35] Substrate specificity studies determined that the arrangement of the A- and D-ring substituents alters the positioning of the bilin substrate within the enzyme, therefore influencing the course of catalysis.[14]

Another pathway leading to PCB has been described in *Mesotaenium caldariorum*. In this green alga, 3Z-PΦB is a detectable intermediate in the conversion of BV to 3Z-PCB, which functions as the phytochrome chromophore in this organism.[36] This formal phytochromobilin reductase (designated PcyB by Frankenberg and Lagarias ref. 26) remains to be isolated and the corresponding genes to be cloned.

In contrast to the identified bilin reductase that catalyses a direct four electron conversion of BV to PCB, the biosynthesis of PCB in the red alga *C. caldarium* proceeds via 15, 16-DHBV and PEB.[13,30-32] This result was surprising since *C. caldarium* does not contain phycoerythrin or other PEB-containing phycobiliproteins.[3] A similar pathway was postulated for certain cyanobacteria.[13] Theoretically this reaction is likely to involve the sequential action of the two bilin reductases PebA and PebB (as described above) and the subsequent action of an as yet unidentified PEB:PCB isomerase (designated PcyC by Frankenberg and Lagarias ref. 26). Interestingly, the bilin reductase genes *pebA* and *pebB* are only found in phycoerythrin-containing cyanobacteria (Table 1). Exclusively phycocyanin containing cyanobacteria carry the single bilin reductase gene, *pcyA* (Table 1; ref. 25). Until the bilin reductases and isomerases are cloned from *C. caldarium* it remains unclear which enzymes/genes are involved in the formation of PCB in this and other related organisms.

Biosynthesis of Chlorophyll-Derived Bilins

The bioinformatics approach to identifying the HY2-family of bilin reductases revealed another interesting enzyme that had weak homologies to HY2.[25] Red chlorophyll catabolite reductase (RCCR) had been previously cloned and characterised from barley and is involved in chlorophyll degradation.[37] During chlorophyll degradation the tetrapyrrole macrocycle is dephytylated and the metal is removed which results in pheophorbide *a* (see Chapter 17). This macrocyclic intermediate is ring-opened by pheophorbide *a* oxygenase which yields a chlorophyll-derived bilin product, red chlorophyll catabolite (RCC). RCC is then reduced by the ferredoxin-dependent RCCR to produce the colourless primary fluorescent chlorophyll catabolite.[38] The reaction catalysed by RCCR is chemically similar to that catalysed by PebA; both reactions involve the reduction of a 15, 16-double bond. The chemically analogous pathways of heme and chlorophyll degradation therefore seem to utilise at

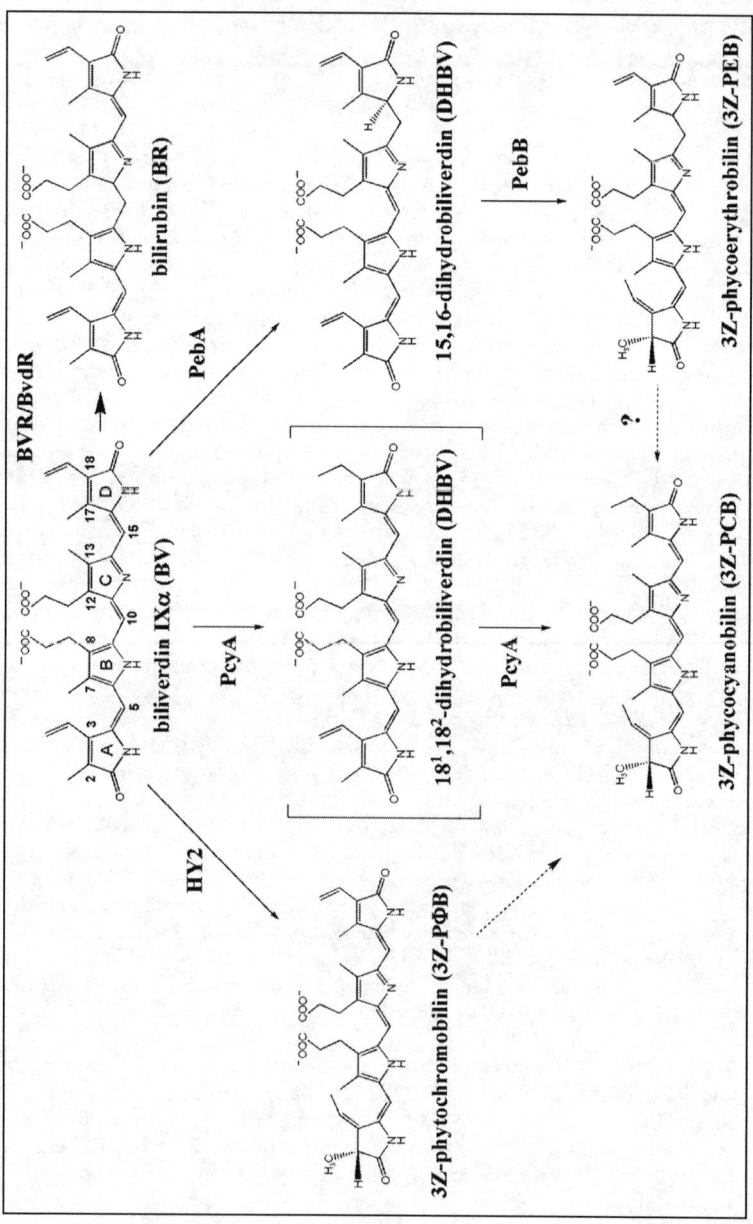

Figure 4. Biosynthesis of phycobilins in plants and cyanobacteria. The linear tetrapyrrole precursors of the phytochrome and phycobiliprotein chromophores of plants, algae, and cyanobacteria and the mammalian bile pigment bilirubin share the common intermediate biliverdin IXα. Enzyme abbreviations indicated on the arrows include: HY2, phytochromobilin synthase (3Z-phytochromobilin:ferredoxin oxidoreductase); PcyA, 3Z-phycocyanobilin:ferredoxin oxidoreductase; PebA, 15,16-dihydrobiliverdin:ferredoxin oxidoreductase; PebB, 3Z-phycoerythrobilin:ferredoxin oxidoreductase; BVR/BvdR, biliverdin IXα:NAD(P)H oxidoreductase. The dashed lines indicate activities that have been described biochemically, but neither the enzymes nor their corresponding genes have been identified (see text for details).

Synthesis and Role of Bilins in Photosynthetic Organisms

Table 1. Distribution and occurrence of bilin reductases, phytochromes and phycobiliproteins currently available in the genomic databases

Organism	PΦB		PebA	PebB	PcyA	Phytochrome	Phycobiliproteins
	Oxygenase	Synthase					
Arabidopsis thaliana	+	+	-	-	-	+	-
Oryza sativa L.	+	+	-	-	-	+	-
Lycopersicon sp.	+	+	-	-	-	+	-
Nostoc (Anabaena) sp. PCC 7120	+	-	-	-	+	+	APC[a], PC
Synechocystis sp. PCC 6803	+	-	-	-	+	+	APC, PC
Nostoc punctiforme	+	-	+	+	+	+	APC, PC, PE
Thermosynechococcus elongatus BP-1	+	-	-	-	+	(+)	APC, PC
Prochlorococcus marinus MED4	+	-	+	+	+	-	PE (?)
Prochlorococcus marinus SS120	+	-	+	+	+	-	PE
Prochlorococcus marinus MIT 9319	+	-	+	+	+	-	PE (?)
Synechococcus sp. WH 8020	+	-	+	+	+	+	APC, PC, PE
Synechococcus sp. PCC 7942	+	-	+	+	+	+	APC, PC, PE
Trichodesmium erythraeum	+	-	+	+	+	+	APC, PC, PE

a. Abbreviations used: APC, allophycocyanin; PC, phycocyanin; PE, phycoerythrin; ? indicates that function is not yet confirmed.

least one enzyme from the same enzyme family. Although the reactions are similar, RCCR is the most diverged member of the bilin reductase family, having a defined substrate specificity for chlorophyll-derived bilins. Therefore, it was not surprising that RCCR is unable to metabolise heme-derived bilins such as BV or PΦB.[37]

Cyanobacterial Biliverdin Reductase

A very interesting and unique enzyme was identified by Schluchter and Glazer[39] during a search for genes encoding enzymes that catalyse reactions in the phycobilin biosynthetic pathway. They identified an open reading frame (*bvdR*) in *Synechocystis* sp. PCC 6803 that had sequence similarity to mammalian BVR (see Chapter 6). Recombinant cyanobacterial BvdR catalyses the NADPH-dependent reduction of BV IXα at the C10 methine bridge to bilirubin (BR) IXα (Fig. 4). In contrast to mammalian BVR, the cyanobacterial enzyme has a pH-optimum of 5.8 (pH 8.8 in mammals) with a high substrate specificity towards BV. This suggests that BvdR does not metabolise phycobilins, the key precursors of the phycobiliproteins.[39] Interestingly, the conversion of BV to BR seems to interfere with the normal phycobiliprotein biosynthesis since a *bvdR* interposon mutant lacks phycocyanin.[39]

Assembly of Phycobiliproteins and Phytochromes

As discussed above, bilins are covalently linked to phycobiliprotein or phytochrome apo-proteins via thioether linkages and these linkages are much more varied in phycobiliproteins than in the phytochromes (Figs. 1B,C and 2). Consequently the assembly of the bilin chromophores is also highly divergent. Phytochrome apoproteins contain their own C-S lyase activity and bilin ligation is therefore autocatalytic in these proteins. This was first demonstrated for full-length recombinant phytochrome by Lagarias and Lagarias.[40] Subsequently, the bilin lyase domain has been further defined by a combination of sequence analysis and recombinant expression to a region of just 130-180 amino acids.[41] Interestingly, this domain is related to a broader class of small-molecule-ligand-binding motifs called GAF domains.[1,41] Since phytochromes in higher plants are synthesised in the cytosol and, as we have seen, PΦB is synthesised in plastids, the assembly of holophytochrome poses an interesting cell biological problem. Presumably, assembly also takes place in the cytosol, a result that is supported by the phytochrome-deficient phenotype of transgenic *Arabidopsis* plants expressing a cytosolically-localised BVR.[42] Whether PΦB is specifically transported or chaperoned prior to assembly is unknown.

The first studies on phycobiliprotein assembly studies also revealed that PCB and PEB spontaneously attach to recombinant subunits of apo-phycocyanin.[43-45] However, although this spontaneous assembly resulted in thioether adducts, in this case the structural and spectral properties were mainly unnatural and a requirement of a C-S lyase for correct bilin attachment to apo-phycobiliproteins was postulated. This was later confirmed by the identification of the *cpcE* and *cpcF* genes encoding subunits of a phycocyanin lyase.[46] The properties of this bilin lyase complex are well characterised and even have been shown to function in a heterologous host *E. coli*.[47] The CpcEF complex catalyses the attachment and removal of PCB to the α-subunit of phycocyanin.[46] Bilin lyases that attach the additional chromophores to the β-phycocyanin subunit remain to be identified. Additional bilin lyases that harbour an internal isomerisation activity have been identified in *Mastigocladus laminosus* (*Fischerella*).[48] The *pecE* and *pecF* genes encode a bilin lyase complex that catalyses the attachment of PCB to apo-phycoerythrocyanin and its subsequent isomerisation to PVB.[49,50] Other bilin lyases involved in attachment and/or isomerisation of bilins to apo-phycobiliproteins remain to be identified.

The Roles of Bilins in Photosynthetic Organisms

The two major roles for bilins in photosynthetic organisms are relatively well understood. In land plants PΦB functions as the chromophore of the phytochromes, a small family (three-five) of dimeric red/far-red reversible photoreceptors. These photoreceptors mediate a broad range of physiological responses to light (see Chapter 13). Phytochromes or phytochrome-related proteins are thought to be present in most photosynthetic organisms (Table 1), but interestingly their evolutionary history appears to be even older than that, as phytochromes have been found in nonphotosynthetic eubacteria and fungi (see ref. 1 for a discussion of phytochrome evolution). In contrast to the situation in land plants, cyanobacterial and algal phytochromes appear to utilise PCB as the chromophore. This has been established by comparison of purified phytochromes from endogenous and recombinant systems in both the green alga *Mesotaenium caldariorum*[36] and *Synechocystis* sp. PCC 6803.[8] Remarkably, phytochrome-like proteins in the nonphotosynthetic eubacteria *Deinococcus radiodurans* and *Pseudomonas aeruginosa* use BV IXα as a chromophore, which was first suggested to bind to a His residue rather than the conserved Cys present in other phytochromes.[51] BV IXα can also be utilised by the two phytochromes from *Agrobacterium tumefaciens*, it appears to be bound to a Cys outside the defined bilin lyase domain and seems to be the chromophore attachment site in most bacterial phytochromes.[52] Although a great diversity of phytochrome-like proteins have now been identified in bacteria their physiological roles are still poorly understood (see Chapter 13).

In contrast to phytochromes, phycobiliproteins are arranged into larger aggregates called phycobilisomes (PBS; ref. 2). PBS are peripheral membrane complexes that efficiently collect light and, through fluorescence resonance energy transfer, conveying it to a special pair of chlorophyll molecules located in the photosynthetic reaction centre. PBS can comprise approximately 30% of the cellular protein and can be divided into two structural domains, core and rods. Besides the chromophorylated phycobiliproteins, these domains also contain nonpigmented linker polypeptides. All PBS consist of the PCB-containing phycobiliproteins APC and PC. Depending on the organism, many PBS also contain PE and PEC (Table 1). In many cyanobacteria pigmentation changes with environmental light quality. In a process called complementary chromatic adaptation certain cyanobacteria can alter their protein-pigment composition to ensure maximal quantum efficiency for light harvesting (for a review see ref. 53). Interestingly, in the cyanobacterium *Fremyella diplosiphon* this process is regulated through a phytochrome-like sensor RcaE that perceives the light signal and starts a phosphorelay that controls the transcription of the phycobiliprotein operon.[54]

In addition to their roles in light sensing and harvesting, recent evidence is increasingly pointing to a signalling role for bilins. One dramatic example from the animal kingdom is the role of BV IXα in dorsal axis development in *Xenopus laevis* embryos,[55] but the phenotypes of plants with altered bilin metabolism also suggest such a possibility.[42] Another example is the regulatory phenotype of the *bvdR* mutant of *Synechocystis* sp. PCC 6803 mentioned earlier.[39] Certainly the importance and prevalence of bilins for photosynthetic light harvesting makes a putative signalling role for bilins in coordinating the synthesis and assembly of the light-harvesting apparatus a strong possibility and it has been hypothesised that progenitor phytochromes were bilin sensors.[1]

A slightly overlooked role for bilins and their biosynthetic enzymes in photosynthetic organisms is a catabolic one. This can be clearly seen for RCCR which is required for chlorophyll degradation, but it should be borne in mind that in plants HOs in particular may be just as important for heme degradation as PΦB synthesis. One important aspect of this process may be iron reutilisation as has been seen in animals (see Chapter 6). In this regard, it is interesting that while *Arabidopsis* contains only a single gene encoding PΦB synthase, it has four *HO* genes. AtHO1 clearly has a dominant role of PΦB synthesis for phytochrome, but it is possible that the other HOs are more important for heme degradation. The control of heme degradation is also likely to be significant because of the important regulatory role of heme in tetrapyrrole metabolism in plants.[56]

Note Added in Proof

During the proof stage of this chapter significant contributions have been made on the enzymology of ferredoxin-dependent bilin reductases. While PebA and PebB were shown to be involved in metabolic channeling of the intermediate 15, 16-DHBV from one enzyme to the other,[58] a new enzyme involved in PEB biosynthesis has been discovered in the cyanophage P-SSM2. This unusual enzyme, designated PEB synthase (PebS, E.C. 1.3.7.6) catalyzes the four-electron reduction of BV to PEB with the intermediacy of 15, 16-DHBV. Therefore, PebS represent a perfect form of metabolic channeling by combining two activities in one enzyme.[59] Furthermore, the crystal structures of this enzyme as well as those of PcyA from *Synechocystis* sp. PCC 6803 and *Nostoc* sp. PCC 7120 have been solved.[60-62] They all revealed the same overall fold comprising an a/b/a-sandwich. In the structures with bound substrate, BV adopts a porphyrin-like conformation with the propionate side chains facing the solvent side.

Acknowledgements

Both authors are grateful to Professor J. Clark Lagarias for his continued support. Work on bilin biosynthesis was supported by grants from the Deutsche Forschungsgemeinschaft to N. Frankenberg-Dinkel and the UK. Biotechnology and Biological Sciences Research Council to M.J. Terry.

References

1. Montgomery BL, Lagarias JC. Phytochrome ancestry: Sensors of bilins and light. Trends Plant Sci 2002; 7:357-366.
2. Grossman AR, Bhaya D, Apt KE, Kehoe DM. Light-harvesting complexes in oxygenic photosynthesis: Diversity, control, and evolution. Annu Rev Genet 1995; 29:231-288.
3. Beale SI. Biosynthesis of phycobilins. Chem Rev 1993; 93:785-802.
4. Terry MJ. Analysis of bilins. In: Smith AG, Witty M, eds. Heme, Chlorophyll and Bilins. Totowa: Humana Press, 2002:273-291.
5. Sidler W. Phycobilisomes and phycobiliprotein structure. In: Bryant DA, ed. The Molecular Biology of Cyanobacteria. Dordrecht: Kluwer Academic Publishers, 1994:139-216.
6. Litts JC, Kelly JM, Lagarias JC. Structurefunction studies on phytochrome. Preliminary characterization of highly purified phytochrome from Avena sativa enriched in the 124-kilodalton species. J Biol Chem 1983; 258:11025-11031.
7. Yeh KC, Wu SH, Murphy JT et al. A cyanobacterial phytochrome two-component light sensory system. Science 1997; 277:1505-1508.
8. Hübschmann T, Börner T, Hartmann E et al. Characterization of the Cph1 holo-phytochrome from Synchocystis sp. PCC 6803. Eur J Biochem 2001; 268:2055-2063.
9. Ratliff M, Zhu W, Deshmukh R et al. Homologues of Neisserial heme oxygenase in gram-negative bacteria: Degradation of heme by the product of the pigA gene of Pseudomonas aeruginosa. J Bacteriol 2001; 183:6394-6403.
10. Ortiz de Montellano PR, Wilks A. Adv Inorg Chem 2000; 51:359-407.
11. Muramoto T, Tsurui N, Terry MJ et al. Expression and biochemical properties of a ferredoxin-dependent heme oxygenase required for phytochrome chromophore synthesis. Plant Physiol 2002; 130:1958-1966.
12. Cornejo J, Beale SI. Algal heme oxygenase from Cyanidium caldarium. Partial purification and fractionation into three required protein components. J Biol Chem 1988; 263:11915-11921.
13. Cornejo J, Beale SI. Phycobilin biosynthetic reactions in extracts of cyanobacteria. Photosynth Res 1997; 51:223-230.
14. Frankenberg N, Lagarias JC. Phycocyanobilin: Ferredoxin oxidoreductase of Anabaena sp. PCC 7120. Biochemical and spectroscopic characterization. J Biol Chem 2003; 278:9219-9226.
15. Rhie GE, Beale SI. Phycobilin biosynthesis: Reductant requirements and product identification for heme oxygenase from Cyanidium caldarium. Arch Bioch Biophys 1995; 320:182-194.
16. Cornejo J, Willows RD, Beale SI. Phytobilin biosynthesis: Cloning and expression of a gene encoding soluble ferredoxin-dependent heme oxygenase from Synechocystis sp. PCC 6803. Plant J 1998; 15:99-107.
17. Kaneko T, Sato S, Kotani H et al. Sequence analysis of the genome of the unicellular cyanobacterium Synechocystis sp. strain PCC6803. II. Sequence determination of the entire genome and assignment of potential protein-coding regions. DNA Res 1996; 3:109-136.
18. Richaud C, Zabulon G. The heme oxygenase gene (pbsA) in the red alga Rhodella violacea is discontinuous and transcriptionally activated during iron limitation. Proc Natl Acad Sci USA 1997; 94:11736-11741.
19. Koornneef M, Rolff E, Spruit CJP. Genetic Control of light-inhibited hypocotyl elongation in Arabidopsis thaliana L. Heynh. Z Pflanzenphys 1980; 100:147-160.
20. Muramoto T, Kohchi T, Yokota A et al. The Arabidopsis photomorphogenic mutant hy1 is deficient in phytochrome chromophore biosynthesis as a result of a mutation in a plastid heme oxygenase. Plant Cell 1999; 11:335-347.
21. Davis SJ, Kurepa J, Vierstra R. The Arabidopsis thaliana HY1 locus, required for phytochrome-chromophore biosynthesis, encodes a protein related to heme oxygenases. Proc Natl Acad Sci USA 1999; 96:6541-6546.
22. Davis SJ, Bhoo SH, Durski AM et al. The heme-oxygenase family required for phytochrome chromophore biosynthesis is necessary for proper photomorphogenesis in higher plants. Plant Physiol 2001; 126:656-669.
23. Terry MJ, Linley PJ, Kohchi T. Making light of it: The role of plant haem oxygenases in phytochrome chromophore synthesis. Biochem Soc Trans 2002; 30:604-609.

24. Kohchi T, Mukougawa K, Frankenberg N et al. The Arabidopsis HY2 gene encodes phytochromobilin synthase, a ferredoxin-dependent biliverdin reductase. Plant Cell 2001; 13:425-436.
25. Frankenberg N, Mukougawa K, Kohchi T et al. Functional genomic analysis of the HY2 family of ferredoxin-dependent bilin reductases from oxygenic photosynthetic organisms. Plant Cell 2001; 13:965-978.
26. Frankenberg N, Lagarias JC. Biosynthesis and biological functions of bilins. In: Kadish KM, Smith KM, Guilard R, eds. The Porphyrin Handbook. Vol 13. New York: Elsevier Science, 2003:211-234.
27. McDowell MT, Lagarias JC. Purification and properties of phytochromobilin synthase from etiolated oat seedlings. Plant Physiol 2001; 126:1546-1554.
28. Wu SH, Lagarias JC. The methylotrophic yeast Pichia pastoris synthesizes a functionally active chromophore precursor of the plant photoreceptor phytochrome. Proc Natl Acad Sci USA 1996; 93:8989-8994.
29. Beale SI, Cornejo J. Enzymic transformation of biliverdin to phycocyanobilin by extracts of the unicellular red alga Cyanidium caldarium. Plant Physiol 1984; 76:7-15.
30. Beale SI, Cornejo J. Biosynthesis of phycobilins. Ferredoxin-mediated reduction of biliverdin catalyzed by extracts of Cyanidium caldarium. J Biol Chem 1991; 266:22328-22332.
31. Beale SI, Cornejo J. Biosynthesis of phycobilins. 3(Z)-Phycoerythrobilin and 3(Z)-phycocyanobilin are intermediates in the formation of 3(E)-phycocyanobilin from biliverdin IXα. J Biol Chem 1991; 266:22333-22340.
32. Beale SI, Cornejo J. Biosynthesis of phycobilins. 15,16-Dihydrobiliverdin IXα is a partially reduced intermediate in the formation of phycobilins from biliverdin IXα. J Biol Chem 1991; 266:22341-22345.
33. Wilbanks SM, Glazer AN. Rod structure of a phycoerythrin-II-containing phycobilisome 1. Organization and sequence of the gene cluster encoding the major phycobiliprotein rod components in the genome of marine Synechococcus sp.WH8020. J Biol Chem 1993; 268:1226-1235.
34. Kami C, Mukougawa K, Muramoto T et al. Complementation of phytochrome chromophoredeficient Arabidopsis by expression of phycocyanobilin: Ferredoxin oxidoreductase. Proc Natl Acad Sci USA 2004; 101:1099-1104.
35. Tu SL, Gunn A, Toney MD et al. Biliverdin reduction by cyanobacterial phycocyanobilin:ferredoxin oxidoreductase (PcyA) proceeds via linear tetrapyrrole radical intermediates. J Am Chem Soc 2004; 126:8682-8693.
36. Wu SH, McDowell MT, Lagarias JC. Phycocyanobilin is the natural precursor of the phytochrome chromophore in the green alga Mesotaenium caldariorum. J Biol Chem 1997; 272:25700-25705.
37. Wüthrich KL, Bovet L, Hunziker PE et al. Molecular cloning, functional expression and characterization of RCC reductase involved in chlorophyll catabolism. Plant J 2000; 21:189-198.
38. Matile P, Hörtensteiner S. Chlorophyll degradation. Annu Rev Plant Physiol 1999; 50:67-95.
39. Schluchter WM, Glazer AN. Characterization of cyanobacterial biliverdin reductase—Conversion of biliverdin to bilirubin is important for normal phycobiliprotein biosynthesis. J Biol Chem 1997; 272:13562-13569.
40. Lagarias JC, Lagarias DM. Self assembly of synthetic phytochrome holoprotein in vitro. Proc Natl Acad Sci USA 1989; 86:5778-5780.
41. Wu SH, Lagarias JC. Defining the bilin lyase domain: Lessons from the extended phytochrome superfamily. Biochemistry 2000; 39:13487-13495.
42. Montgomery BL, Yeh KC, Crepeau MW et al. Modification of distinct aspects of photomorphogenesis via targeted expression of mammalian biliverdin reductase in transgenic Arabidopsis plants. Plant Physiol 1999; 121:629-639.
43. Arciero DM, Bryant DA, Glazer AN. In vitro attachment of bilins to apophycocyanin. I. Specific covalent adduct formation at cysteinyl residues involved in phycocyanobilin binding in C-phycocyanin. J Biol Chem 1988; 263:18343-18349.
44. Arciero DM, Dallas JL, Glazer AN. In vitro attachment of bilins to allophycocyanin. II. Determination of the structures of tryptic bilin peptides derived from the phycocyanobilin adduct. J Biol Chem 1988; 263:18350-18357.
45. Arciero DM, Dallas JL, Glazer AN. In vitro attachment of bilins to allophycocyanin. III. Properties of the phycoerythrobilin adduct. J Biol Chem 1988; 263:18358-18363.
46. Fairchild CD, Zhao JD, Zhou JH et al. Phycocyanin alpha-subunit phycocyanobilin lyase. Proc Natl Acad Sci USA 1992; 89:7017-7021.
47. Tooley AJ, Cai YA, Glazer AN. Biosynthesis of a fluorescent cyanobacterial C-phycocyanin holo- α subunit in a heterologous host. Proc Natl Acad Sci USA 2001; 98:10560-10565.
48. Zhao KH, Deng MG, Zheng M et al. Novel activity of a phycobiliprotein lyase: Both the attachment of phycocyanobilin and the isomerization to phycoviolobilin are catalyzed by the proteins PecE and PecF encoded by the phycoerythrocyanin operon. FEBS Lett 2000; 469:9-13.

49. Storf M, Parbel A, Meyer M et al. Chromophore attachment to biliproteins: Specificity of PecE/ PecF, a lyase-isomerase for the photoactive 31-Cys-α84-phycoviolobilin chromophore of phycoerythrocyanin. Biochemistry 2001; 40:12444-12456.

50. Zhao KH, Wu D, Wang L et al. Characterization of phycoviolobilin phycoerythrocyanin-alpha 84-cystein-lyase-(isomerizing) from Mastigocladus laminosus. Eur J Biochem 2002; 269:4542-4550.

51. Bhoo SH, Davis SJ, Walker J et al. Bacteriophytochromes are photochromic histidine kinases using a biliverdin chromophore. Nature 2001; 414:776-779.

52. Lamparter T, Michael N, Mittmann F et al. Phytochrome from Agrobacterium tumefaciens has unusual spectral properties and reveals an N-terminal chromophore attachment site. Proc Natl Acad Sci USA 2002; 99:11628-11633.

53. Grossman AR, Bhaya D, He Q. Tracking the light environment by cyanobacteria and the dynamic nature of light harvesting. J Biol Chem 2001; 276:11449-11452.

54. Kehoe DM, Grossman AR. Similarity of a chromatic adaptation sensor to phytochrome and ethylene receptors. Science 1996; 273:1409-1412.

55. Falchuk KH, Contin JM, Dziedzic TS et al. A role for biliverdin IXα in dorsal axis development of Xenopus laevis embryos. Proc Natl Acad Sci USA 2002; 99:251-256.

56. Cornah JE, Terry MJ, Smith AG. Green or red: What stops the traffic in the tetrapyrrole pathway? Trends Plant Sci 2003; 8:224-230.

57. Moss GP. Nomenclature of tetrapyrroles. Recommendations 1986 IUPAC-IUB Joint Commission on Biochemical Nomenclature (JCBN). Eur J Biochem 1988; 178:277-328.

58. Dammeyer T, Frankenberg-Dinkel N. Insights into phycoerythrobilin biosynthesis point toward metabolic channeling. J Biol Chem 2006; 281:27081-27089.

59. Dammeyer T, Bagby SC, Sullivan MB et al. Efficient phage-mediated pigment biosynthesis in oceanic cyanobacteria. Curr Biol 2008; 18:442-448.

60. Dammeyer T, Hofmann E, Frankenberg-Dinkel N. Phycoerythrobilin synthase (PebS) of a marine virus: Crystal structure of the biliverdin-complex and the substrate free form. J Biol Chem 2008; 283:27547-27554.

61. Hagiwara Y, Sugishima M, Takahashi Y, Fukuyama K. Crystal structure of phycocyanobilin:ferredoxin oxidoreductase in complex with biliverdin IXalpha, a key enzyme in the biosynthesis of phycocyanobilin. Proc Natl Acad Sci USA 2006; 103:27-32.

62. Tu SL, Rockwell NC, Lagarias JC, Fisher AJ. Insight into the radical mechanism of phycocyanobilin-ferredoxin oxidoreductase (PcyA) revealed by X-ray crystallography and biochemical measurements. Biochemistry 2007; 46:1484-1494.

Phytochromes:
Bilin-Linked Photoreceptors in Bacteria and Plants

Matthew J. Terry* and Alex C. McCormac

Abstract

The phytochromes and related prokaryotic photoreceptors utilise a linear tetrapyrrole chromophore to monitor the surrounding light environment. They are found in most photosynthetic organisms, and also some nonphotosynthetic bacteria from which they most likely evolved. A key feature of the phytochromes is that they respond to light in a photoreversible manner. In plants, red light leads to the formation of Pfr, the physiologically-active phytochrome species, while far-red light reverses this process to give the Pr form. Prokaryotic phytochromes are also photoreversible, and are thought to be involved in regulating processes such as chromatic adaptation, phototaxis and pigment synthesis. In the model plant *Arabidopsis thaliana* there are five phytochromes that are involved in regulating all stages of plant development from germination to flowering. They also play a key role in chloroplast development and the regulation of the tetrapyrrole pathway.

Introduction

The phytochromes are large, soluble proteins that perceive light quantity (intensity and period length) and spectral quality (i.e., wavelength composition) to enable organisms to respond optimally to the changing environment. Such photoreceptors were thought to have evolved for plant adaptation to unfavourable light environments, but phytochromes and related proteins have now been identified in photosynthetic bacteria (cyanobacteria and purple bacteria), nonphotosynthetic eubacteria and fungi and are clearly much more ancient in origin.[1,2]

The existence of a red/far red (R/FR) light photoreceptor in plants was established during the last century through a series of pioneering experiments on the regulation of flowering and germination by light (see ref. 3 for a history of phytochrome research). The photoreceptor was termed phytochrome and purification of this protein from oat seedlings led to the proposal that phytochrome ultilised a linear tetrapyrrole-like chromophore. The structure and thio-ether linkage of the bound phytochromobilin (PΦB) chromophore was subsequently established by ^{1}H-NMR spectroscopy and is shown in Figure 1 (reviewed in ref. 4). The precursor of the bound phytochrome chromophore, PΦB, is synthesised in the plastid from the heme branch of the tetrapyrrole pathway[4] (see Chapter 12). Molecular cloning of phytochrome led to the identification of multiple phytochrome species in higher plants, with the model plant *Arabidopsis thaliana* L. (and many other dicotyledenous plants) containing five phytochromes (phyA-E), while monocotyledonous plants typically have three.[5]

Extensive, but unsuccessful, attempts to identify phytochrome in the model green alga *Chlamydomonas reinhardii* had led to the assumption that this photoreceptor was not widespread in nature. However, the sequencing of the *Synechocystis* sp. PCC 6803 genome opened the floodgates of

*Corresponding Author: Matthew J. Terry—School of Biological Sciences, University of Southampton, Bassett Crescent East, Southampton SO16 7PX, UK. Email: mjt@soton.ac.uk

Tetrapyrroles: Birth, Life and Death, edited by Martin J. Warren and Alison G. Smith. ©2009 Landes Bioscience and Springer Science+Business Media.

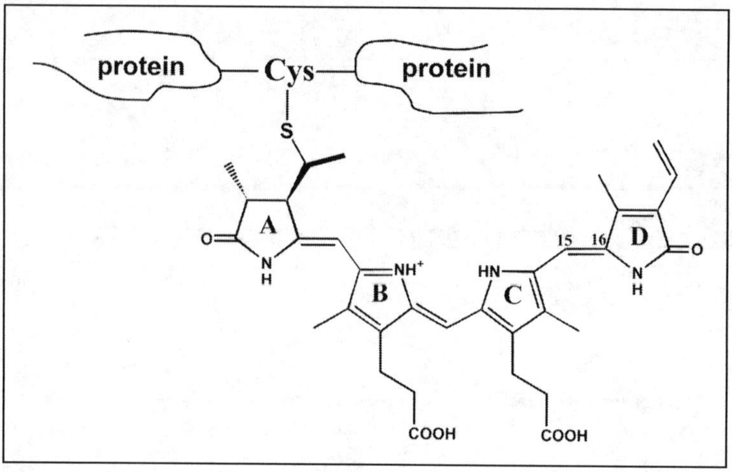

Figure 1. The structure of the bound phytochrome chromophore, PΦB. PΦB has only been identified in phytochromes from flowering plants. Algal and some cyanobacterial phytochromes use PCB while all other bacterial phytochromes use BV (see Chapter 12 for additional structures).

phytochrome diversity[6,7] and phytochromes have now been identified not only in plants and cyanobacteria, but also in purple bacteria, proteobacteria and fungi.[1,2] As discussed later the apoproteins of these divergent phytochromes have various combinations of different structural and functional motifs. They also use different bilin chromophores. Phycocyanobilin (PCB) is used by cyanobacterial and some algal phytochromes,[8,9] while bacterial phytochromes (BphPs) use biliverdin IXα (BV; see Chapter 12 for structures).[10] However, they all appear to share a common mechanism of bilin and light-dependent conformational changes resulting in altered biochemical and physiological activities.

The Phytochromes—A Diverse Family of Photoreversible Photoreceptors

As illustrated in Figure 2, the phytochromes show considerable variation in their molecular construction, and range in size from the 1465 amino acids of the phy3 protein of the fern *Adiantum capillus-veneris* to the cyanobacterial Cph1 proteins of approximately 750 amino acids. The smallest phytochrome-like protein described to date is the cyanobacterial photoreceptor RcaE at only 655 amino acid residues.[11] The plant phytochromes are all typically about 1150 residues (with molecular masses of 120-130 kDa). However, although phytochromes show great diversity at the molecular level they mostly conform to a simple model of an N-terminal photosensory or input domain containing the bound linear tetrapyrrole chromophore linked to a C-terminal regulatory domain (Fig. 2). Higher plant phytochromes are well conserved and contain a serine-threonine kinase regulatory domain[12] and a PAS-related domain (PRD) containing two PAS repeats[13] that may be involved in dimerization and/or allosteric regulation.[1] However, it is important not to assume that the regulatory domain is the primary mediator of signal propagation. It was recently shown in *Arabidopsis* that the N-terminal domain of phyB can complement a *phyB* mutant when dimerized and correctly targeted to the nucleus[14] indicating that this region of the protein alone can fulfil most of the signalling functions of phytochrome. It is possible that the kinase domain may have a role in regulating the extent of phytochrome signalling through autophosphorylation. The C-terminus is also clearly important for mediating light-dependent nuclear localization[15] and protein-protein interactions. Indeed, various potential signalling partners interacting with the C-terminal domain of plant phytochromes have now been identified.[16] In addition, as mentioned above, regions within both the PRD and kinase (HKRD) domains have been implicated in phytochrome dimerization in plants.[17]

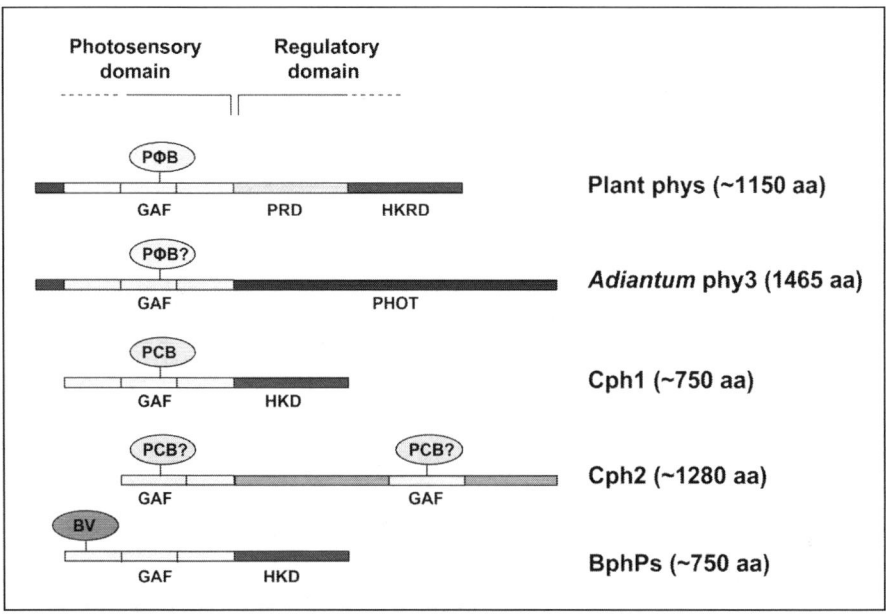

Figure 2. Domain structures of phytochromes and phytochrome-like proteins illustrating the diversity of this class of photoreceptor. All phytochromes possess an N-terminal photosensory domain including a bilin covalently bound to a cysteine of a small-molecule-ligand-binding (GAF) domain[1] and a C-terminal regulatory domain that is much more variable. Examples of regulatory domains include the histidine kinase domains (HKD) of Cph1 and the BphPs and the histidine kinase related domain (HKRD) of almost all plant phytochromes. One exception is the phototropin-like domain (PHOT) of Adiantum capillus-veneris phy3. This phytochrome-like protein is found only in a subclass of ferns.[19] Plant phytochromes also possess a PRD domain containing two PAS domains. The binding of the chromophores PphiB, PCB and BV are indicated.

Kinase domains, in various forms, are a common feature of the phytochrome regulatory domain. One unusual example is the phy3 protein from *Adiantum* (see Fig. 2). In this case a higher plant-like photosensory region is linked to a nearly full-length homologue of the blue-light photoreceptor PHOT1 containing a serine-threonine kinase domain.[18] Like its PHOT1 counterpart, *Adiantum* phy3 is required for chloroplast relocation and phototropism, but in this case to R light.[19] Another surprising example is the Ppr protein from the purple photosynthetic bacterium *Rhodospirillum centenum*.[20] This photoreceptor kinase has sequence similarity with both photoactive yellow protein and the phytochromes, but although it shows similarity throughout the phytochrome sequence it is more likely to use the blue-light absorbing *p*-hydroxycinnamic acid chromophore than a linear tetrapyrrole.[20] Most cyanobacterial phytochromes also contain kinase domains. The first of these to be characterized, Cph1 from *Synechocystis*, was shown to exhibit histidine kinase activity and appears to function as part of a two-component regulatory system, phosphorylating the response regulator Rcp1, in a light-dependent manner.[7] Cph1 histidine kinase activity is specific to the R-absorbing Pr form of the protein. However, although the serine-threonine kinase domain of eukaryotic phytochromes is evolutionarily related to the prokaryotic histidine kinase domain, this activity is greater in the Pfr form, the physiologically active form of phytochrome in plants.[12] Most prokaryotic phytochrome-like proteins characterised to date exhibit histidine kinase activities (e.g., refs. 21,22) with the notable exception of Cph2 from *Synechocystis*[23] and it is likely that this is the primary signal transduction mechanism for phytochromes in bacteria.

Phytochrome Photosensory Domains

One property considered to be characteristic of the phytochromes is that they exist in two photoreversible conformational forms that are relatively stable in vitro. Indeed the absorption spectra of phytochromes across the whole extended family are remarkably similar. Higher plant phytochromes are synthesised in the Pr form with a λ_{max} = 668 nm (for oat phyA).[24] Irradiation with R light results in conformational changes in the protein and an altered, Pfr absorption spectrum (λ_{max} = 730 nm). These conformational changes can be reversed with FR light. However, since both Pr and Pfr also absorb FR and R light respectively, even monochromatic light treatments result in a photoequilibrium between these two states. This is illustrated in Figure 3, which shows spectra of CphA from the cyanobacterium *Calothrix* sp. PCC 7601 expressed and purified using an *E. coli* expression system. For CphA the absorption maxima are blue-shifted compared with those of oat phyA primarily because CphA uses PCB as the chromophore (Table 1). PCB has a D-ring ethyl group rather than the vinyl group present on PΦB and thus a shorter π-electron system that shifts the absorption maxima to shorter wavelengths. The primary photoconversion event is the isomerization of the C15-C16 double bond (see Fig. 1) from *Z* to *E* such that the extended *ZZZ* configuration of the chromophore in Pr is changed to *ZZE*.[25,26] The chromophore is thought to remain protonated throughout photoconversion.[27] The photochemistry of the photoconversion process has been extensively studied and although some areas of the photoreaction mechanism are still controversial there is substantial information on the intermediates present during both the conversion from Pr to Pfr and the reverse Pfr to Pr reaction.[28]

The spectral properties of the phytochromes are dependent on both the chromophore used and its protein environment. As discussed above, the first phytochrome chromophore to be characterised was PΦB (Fig. 1) and it is likely that all higher plant phytochromes use this chromophore. Although it had long been known that purified apophytochromes could assemble with other bilin chromophores[29] the first phytochrome shown to use an alternative to PΦB was from the green alga *Mesotaenium caldarorium*, which uses PCB.[8] We now know that many prokaryotic phytochromes also use PCB,[7,9] which is already produced in abundance in these photosynthetic species. The bilin chromophores, PΦB and PCB, are both synthesized via the precursor BV. Both of these bilins contain an A-ring ethylidene group (see Chapter 12 for structures and biosynthetic details) thought to be essential for covalent attachment and it was therefore a surprise to find that BV, which lacks this group, appears to be the chromophore for a broad range of BphPs that have been identified through evermore extensive bioinfomatic analyses in both photosynthetic and nonphotosynthetic bacteria (see Fig. 2). They were originally identified in the eubacteria *Deinococcus radiothurans* and *Pseudomonas aeruginosa*,[10,30] but have been more extensively characterized from *Agrobacterium tumefaciens*, a soil-borne plant pathogen.[21,31-33] The BV chromophore is also covalently bound resulting in an absorption spectrum that is red-shifted compared with PΦB due to an extended π-electron system (see Table 1). Interestingly, the range of bilin chromophores utilised in nature may yet be extended. For example, the RcaE photoreceptor can bind bilins covalently and is required for R and green responsiveness,[11] while PixJ1 from *Synechocystis* exhibits a blue (λ_{max} = 435 nm)-green (λ_{max} = 535 nm) difference spectrum.[34]

In eukaryotic phytochromes the chromophore is covalently-bound via a thio-ether linkage to a unique cysteine residue that resides within a defined bilin lyase domain of about 130-180 amino acids.[23] This bilin lyase domain is related to GAF domains that have been identified as binding small molecules and ligands in a wide variety of contexts[35] and it has been proposed that phytochromes evolved from bilin sensors.[1] It is a feature of all phytochromes that assembly is autocatalytic and in this way differs considerably from the assembly of bilins into phycobiliproteins that requires additional bilin lyases.[36] Studies on the kinetics of holophytochrome assembly have resolved two major steps: a rapid noncovalent association followed by a slower phase during which the covalent bond is formed.[37,38] Interestingly, not all phytochromes contain the conserved cysteine within the GAF domain, most notably in the BphPs. The search for additional covalent attachment sites for their BV chromophores led to some controversy in the field and two sites have been proposed. The original suggestion was that the chromophore was attached to a conserved histidine residue immediately C-terminal to the cysteine via a Schiff base-like linkage.[30,39] However, a more likely hypothesis has been put forward in which BV binds covalently to an alternative cysteine (Cys20 in Agp1) that

Table 1. Spectral properties of the phytochromes

Organism	Protein	Chromophore[1]	Pr Difference λ_{max} (nm)	Pfr Difference λ_{max}(nm)	Reference
Avena sativa L.	purified phyA	PΦB	668	732	24
Arabidopsis thaliana L.	recombinant phyB	PΦB	665	728	42
Arabidopsis thaliana L.	recombinant phyC	PΦB	661	725	85
Arabidopsis thaliana L.	recombinant phyE	PΦB	670	724	85
Mesotaenium caldariorum	purified and recombinant	PCB	646	720	8
Synechocystis sp. PCC 6803	purified Cph1[2]	PCB	656	704	9
Synechocystis sp. PCC 6803	recombinant Cph1	PCB	656	706	9
Synechocystis sp. PCC 6803	recombinant Cph2	PCB	643	690	86
Calothrix sp. PCC7601	recombinant CphA	PCB	664	708	39
Calothrix sp. PCC7601	recombinant CphB	BV	702	754	39
Agrobacterium tumefaciens	recombinant Agp1	BV	701	750	31
Deinococcus radiodurans	recombinant *Dr* BphP	BV	698	750	10

[1]Abbreviations: BV, biliverdin IXα; PCB, phycocyanobilin; PΦB, phytochromobilin.
[2]A His-6 tagged Cph1 was expressed in Synechocystis sp. PCC 6803 under the control of the psbA2 promoter.
The difference maxima and minima for phytochromes characterised from higher plants, a green alga, cyanobacteria and non-photosynthetic eubacteria are shown.

resides N-terminal to the conventional GAF domain.[31-33] Covalent attachment at this site has an absolute requirement for an A-ring vinyl side chain[32] in contrast to the A-ring ethylidene group that is a prerequisite for holophytochrome assembly within the GAF domain.[40] It has been proposed that this N-terminal cysteine is the original site of chromophore ligation in the phytochromes and that the site subsequently switched to the cysteine within the GAF domain.[2] One reason for the original confusion over chromophore binding sites was that mutation of the conserved histidine residue in the GAF domain abolishes covalent attachment to that cysteine residue.[30,41] In the absence of detailed structural information on the phytochromes little progress has been made in identifying additional residues critical for chromophore binding and spectral integrity with the exception of a conserved glutamate N-terminal to the GAF-domain cysteine.[23] In contrast, experiments using chromophore analogues have identified the D-ring as having a key role in photoreversibility while the propionic acid side chains of the B- and C-rings appear crucial for correct chromophore positioning.[42] Interestingly, it has also been shown that the D-ring vinyl side chain of PΦB is essential for the biological activity of phytochrome A with PCB unable to mediate responses to continuous FR light.[43]

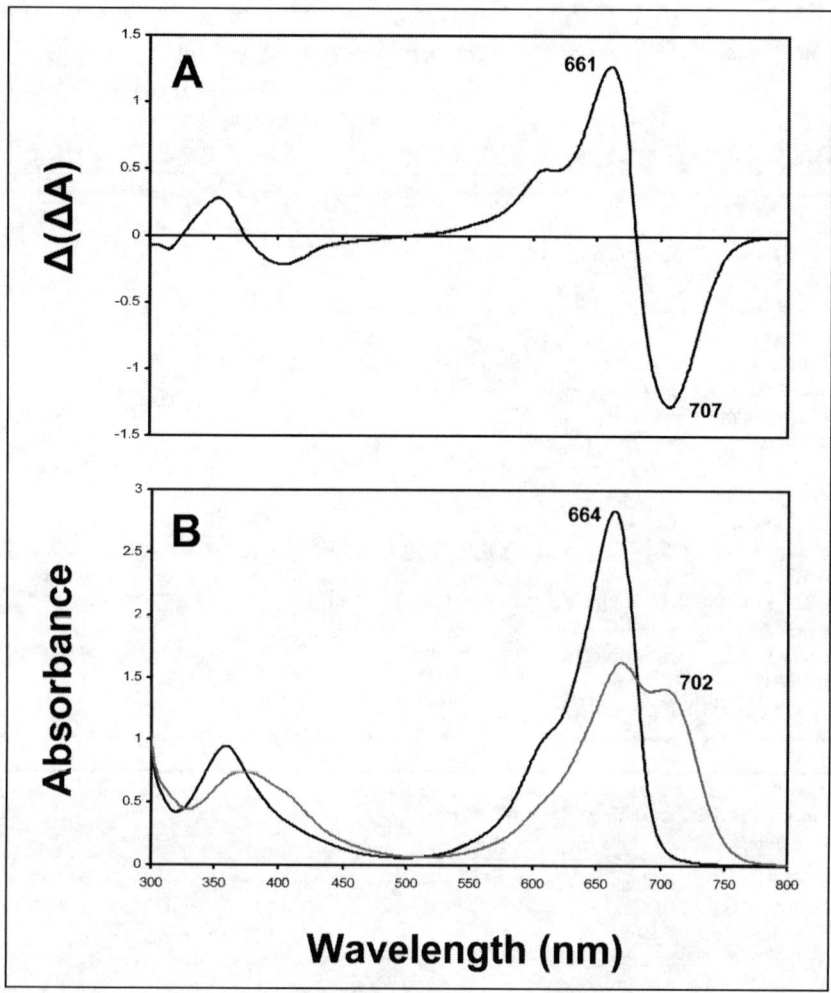

Figure 3. UV/visible absorption spectra for CphA from *Calothrix* sp. PCC 7601 reconstituted in vitro with PCB. A) Difference spectrum created by the subtraction of the Pfr from Pr spectra shown in B. B) UV/visible absorption spectra for the red-light (Pr; dark (blue) line) and far-red (Pfr; light (turquoise) line) absorbing forms of CphA. Absorption maxima and minima are indicated. (Adapted from ref. 87). A color version of this figure is available online at www.eurekah.com.

Physiological Roles of Phytochrome-Like Proteins in Prokaryotes

The proliferation of information about phytochrome-like sequences in prokaryotes has not been matched by information about their physiological roles. Indeed, for phytochromes identified through bioinformatic approaches this information is only just beginning to appear. The only reported phenotype for a *Synechocystis* mutant deficient in Cph1, the best characterised of the prokaryotic phytochromes at the molecular level, is a reduced growth rate in both high light (165 μmolm^{-2}s^{-1}) and FR light.[44] No effect was seen on the acclimation of the photosynthetic apparatus. Similarly a *cph2* mutant grew more slowly in both high and R light.[44] Cph2 also appears to have a role in inhibiting phototaxis towards blue light.[45] More defined physiological roles have been determined for phytochrome-related proteins identified through genetic screens for impaired photoresponses in cyanobacteria. This approach has led to the identification of PixJ1 required for positive phototaxis in *Synechocystis*,[34] CikA, a BphP required for light input to the circadian clock in *Synechococcus*[22] and RcaE that is involved in regulating complementary chromatic adaptation in *Fremyella diplosiphon*.[11] Another interesting example comes from the symbiotic photosynthetic *Bradyrhizobium* ORS278 strain. In this strain a gene encoding a BphP was located downstream of a photosynthetic gene cluster and expression of the complete photosynthetic apparatus was dependent on this BphP.[46] All of these examples might be interpreted as providing information about the light environment to optimise photosynthesis: a primary role rather similar to their higher plant counterparts (see later). But what of BphPs in nonphotosynthetic bacteria? One example is from the heterotrophic eubacterium *Deinococcus radiodurans* in which a BphP is required for the light-induced synthesis of protective carotenoid pigments.[30]

Phytochrome Function in Flowering Plants

From a physiological perspective the phytochromes from flowering plants are the best understood of this photoreceptor class. This reflects both the long history of phytochrome research by plant physiologists and a broad range of agricultural benefits from understanding and manipulating plant growth and development. The ratio of R and FR light provides spectral cues that are particularly useful for signalling to the plant information about the surrounding vegetative environment. This has important implications for avoidance and adaptive strategies in light-seeking and -harvesting. The phytochrome-mediated control of growth and development occurs throughout the plant's life cycle from seed germination, through sculpturing of vegetational architecture, to floral induction and senescence. In fact, the various phytochrome isoforms have extremely pleiotropic and overlapping roles encompassing transcriptional regulation, chloroplast development, growth inhibition, leaflet/organelle movement and daylength perception in floral initiation (for comprehensive reviews see refs. 3,47-51).

The most dramatic, and therefore most studied, of the phytochrome-regulated developmental steps is the progression from the etiolated form of young seedlings growing in complete darkness, to the light-grown plant that is competent for photosynthesis. In lower plants, phytochrome photoperception has a primary role in the detection of the direction of light.[52] In the evolutionary development to angiosperms a new developmental strategy in darkness evolved, termed skotomorphogenesis. Whereas sporophytes of algae, mosses and ferns follow almost identical developmental patterns in light or dark, in angiosperms the developmental pathways leading to a photoautotrophic plant are suppressed in the dark. The transition from skotomorphogenesis (where seedlings are existing on seed-reserves) to photomorphogenesis (where the development of functional chloroplasts allows photosynthesis to occur) requires a phytochrome-mediated light switch. This transition is characterised by a distinct change in seedling morphology. Dark-grown seedlings have elongated hypocotyls, closed cotyledons, an apical hook and undifferentiated plastid structures whereas, the autotrophic seedling has inhibited hypocotyl growth and expanded cotyledons with functional chloroplasts.[53,54] Gymnosperms represent an intermediate state where the skotomorphogenic lifestyle is characterised by some typically photomorphogenic features such as chlorophyll synthesis and the initiation of chloroplast development even in complete darkness.

The second photomorphogenic switch that has attracted substantial attention is the shade avoidance syndrome of light-grown plants.[48,55] This is a composite of growth responses that are initiated in plants threatened by shading from neighbouring plants. Light that is reflected from, or transmitted through, vegetation is depleted in the R region relative to the FR region as a consequence of absorption by chlorophyll. This reduced ratio of R:FR wavelengths is perceived by phytochrome. Many plant species, in an attempt to avoid shading, respond by increasing elongation of internodes and/or petioles, reducing leaf growth, increasing apical dominance and accelerating flowering. In dictating this response, the relative amount of R and FR light represents one of the most important aspects of photomorphogenesis in the green plant.

Specific Roles for Specific Phytochromes

As mentioned previously, in angiosperms, phytochrome apoproteins are encoded by a small gene family of 3-5 genes[5] with *Arabidopsis* possessing 5 *PHY* genes encoding phytochromes A to E (phyA-phyE). All angiosperm phytochromes isolated to date conform to the basic plant phytochrome structure shown (Fig. 2), but there is significant diversity in the amino acid sequence between the family members within a given plant species. *Arabidopsis* phyA and phyB show 52% identity in the amino acid sequence, whereas *Arabidopsis* phyB and phyD polypeptides are approximately 80% identical and more related to phyE than to either phyA or phyC.[5] Phylogenetic analysis indicates a duplication event in early seed-plant evolution giving rise to 2 initial *PHY* gene lineages, *PHYA/PHYC* and *PHYB/PHYD/PHYE*, which underwent later duplication in angiosperms to generate the divergent *PHYA, PHYC, PHYE* and *PHYB/PHYD* lineages. This divergence is reflected in the fact that the various phytochrome forms have both different functions and modes of response.

The most revealing approaches to the identification of the distinct photosensory roles of the individual photoreceptors have been those involving photomorphogenetic analysis under various light conditions of mutants that are deficient for the product of one or more of the *PHY* genes. Additional insights have also come from the study of transgenic plants that over-express specific photoreceptor apoprotein genes. The vast majority of this work has been carried out in the model plant *Arabidopsis*, but where parallel studies have been conducted in other species there is good agreement, making it possible to draw general conclusions about the roles of the various phytochrome members. Null mutations in the various *PHY* genes (*phy*) have revealed that the individual phytochromes have unique and redundant roles in light signalling.[49,50,56] In some cases family members monitor essentially the same light signals, but regulate different physiological responses. Conversely, different members of the family monitor different light parameters, but control essentially the same physiological responses.

Phytochrome B-deficient (*phyB*) *Arabidopsis* mutants were one of the first photoreceptor mutants to be isolated and this was achieved on the basis of their long hypocotyl phenotype following growth under white light.[57,58] Etiolated *phyB*-nulls are virtually insensitive to R light in almost all aspects of de-etiolation including inhibition of hypocotyl elongation and cotyledon expansion. However, they respond normally to FR and blue light. This indicates that phytochrome B is the principal photoreceptor for de-etiolation processes under R. Subsequent to de-etiolation, phyB plays a predominant role in controlling the shade avoidance syndrome as displayed by both young green seedlings and mature plants. Null mutants of *phyB* have a greatly attenuated (but not completely absent) shade avoidance syndrome, demonstrating that it is the major photoreceptor in this response. The phyB photoreceptor also plays a principal role in the photorequirement for *Arabidopsis* seed germination,[58,59] with germination being promoted markedly by a single brief R pulse (Rp), which can be nullified by a subsequent FRp. This R/FR reversible response is the classic hallmark of phytochrome action working in the low fluence response (LFR) mode. Seed can be cycled through sequential R/FR treatments and final germination levels reflect the wavelength of the last delivered light pulse only.

phyA shares many of the photomorphogenetic roles of phyB in the de-etiolation response, but does so predominately in response to FR rather than R light. Mutant *phyA* nulls are completely insensitive to continuous FR in all the morphogenetic aspects of the de-etiolation response and look essentially like dark-grown seedlings with long hypocotyls and closed cotyledons.[60,61] Therefore, phyA is the sole photoreceptor species mediating the de-etiolation response to this waveband and does so via the FR-high irradiance response (HIR). This response mode is characterised by a requirement for prolonged irradiation with continuous (or very rapidly pulsing[62]) FR of a reasonably high fluence rate ($0.1 - 50$ $\mu mol.m^{-2}.s^{-1}$). Hence, phyA perception of FR is strikingly different from that of phyB. Another phyA-specific response to FR is the very low fluence response (VLFR) of etiolated seedlings, but by contrast this can be initiated by a single pulse providing a total fluence, as the name suggests, as low as 10^{-9} $mol.m^{-2}$ and is readily saturated. The distinctions of the phyA and phyB modes of light perception are well described by Casal et al.[63] Although pure FR light is not encountered in the natural environment, a role for these FR-specific responses has been indicated by the demonstration that seedlings emerging under a dense canopy of vegetative shade were far less likely to survive if lacking these phyA responses. The phyA species does not function in the shade avoidance syndrome, but does have a specific role in the developed light-grown plant in sensing daylength for the initiation of flowering.[49,63]

The phyD and phyE proteins are most closely related in sequence to phyB but in single mutant lines of *phyD* or *phyE* there was no discernible effect on photomorphogenesis. However, when a *phyD* mutation was introgressed into a *phyB* mutant background a role for phyD was revealed that was similar to phyB in responding to R to inhibit hypocotyl growth and promote cotyledon expansion.[64] phyD is also involved in initiating the shade avoidance syndrome.[65] The early flowering and longer petioles resulting from *phyD* were clearly observed only in a double *phyBphyD* mutant, indicating a strong hierarchy of importance between these photoreceptors. However, this also demonstrates a certain level of redundancy within the phytochrome family in the perception R:FR ratio and initiation of shade avoidance. This is reinforced by the demonstration that there is also a role for phyE in this response. The *phyE* mutant in Arabidopsis was isolated in the background of a *phyAphyB* double mutant, which still exhibited internode elongation and early flowering in response to low R:FR.[66] In a mutagenised population, plants deficient in these remaining shade avoidance responses mapped to *PHYE*. Interestingly, phyE was also required for seed germination under continuous FR, a response that was thought to be mediated solely by phyA;[67] another good illustration of redundancy in the phy family.

The role of the phyC species had, until very recently, proved more elusive, but with the recent identification of *phyC* mutants these roles can be pieced together.[68,69] These studies showed that phyC is a weak R sensor that is involved throughout plant development. It also has an important role in modulating the effects of other phytochromes, most importantly acting redundantly with phyA to modulate phyB-mediated hypocotyl inhibition under R. This points to a highly complex interaction of the various phytochrome family members in achieving the precise morphological response to any given set of light conditions.

Phytochrome Mode of Action

Studies on phytochrome expression in plants suggest that in general all phytochromes are expressed in all plant tissues.[70] However, they are not equally abundant. In the dark phyA is the predominant species, comprising about 85% of the total phytochrome molecules, but this is reduced by up to 500-fold in continuous white light as the Pfr form of phyA is rapidly degraded.[70] In contrast, phyB-E are less abundant in the dark, but more stable in light. For this reason they have often been referred to as type II phytochromes with phyA behaving as the sole example of type I. In the light phyB is the most abundant phytochrome species consistent with the strong phenotype of the *phyB* mutant.[58] However, phyB and phyC are reduced by 4-5 fold in red or white light so the term stable is somewhat misleading.[70] In general, all phytochromes are found in all plant tissues although in light-grown plants phyA is only detectable in roots.[70]

The mode of phytochrome action has been the subject of intense research for many years, however it is still true to say that we don't know the primary mechanism by which the signal is transduced. Evidence from functional complementation experiments has established that both dimerization and nuclear localization are essential for full phytochrome activity at least with respect to the regulation of hypocotyl elongation.[14] Interestingly although phyA only exists as homodimers, the type II phytochromes can interact extensively with each other, thus complicating the interpretation of the mutant phenotypes described above.[71] In the dark, the phytochromes are localised in the cytoplasm, but rapidly relocate to the nucleus after light treatments.[15] Since nuclear localization is required for phytochrome complementation of a *phyB* phenotype it seems reasonable to assume that the primary mode of phytochrome action is in the nucleus.[14] However, as discussed by Nagy and Schäfer,[51] extensive evidence from other studies suggests that there are aspects of phytochrome signalling that depend on interactions within the cytoplasm. Until recently the search for the signalling mechanism tended to focus on the C-terminal domain of the phytochrome protein (see Fig. 2) and indeed numerous phytochrome-interacting factors have been identified using predominantly the C-terminus as bait.[16] The discovery that the N-terminal domain alone has full signalling activity has led to a reevaluation of the primary mechanism of phytochrome signalling.[14] As this fragment shows hyperactivity it appears that the C-terminus functions to attenuate signalling responses in addition to its roles in nuclear targeting and protein-protein interaction. It is possible that this attenuation is mediated through autophosphorylation events. While some of the primary events remain elusive, incredible progress has been made in recent years in the identification of components of phytochrome signalling pathways. This is a fertile and fast-changing field and the reader is directed to the following reviews for a more comprehensive summary of the current status.[16,54,72]

Phytochrome Regulation of Tetrapyrrole Synthesis

The regulation of chloroplast development is a key process in the successful transition from an etiolated, heterotrophic seedling to a photosynthetically-competent photoautotrophic seedling. The importance of phytochrome in mediating this process is illustrated clearly by the pale phenotype of phytochrome-deficient mutants such as *phyB* or the chromophore synthesis mutants (lacking all phytochromes) *hy1* and *hy2*.[58,73] Indeed, recent transcriptomic studies have revealed that essentially all photosynthetic genes and in particular those encoding components of the light-harvesting photosystems are light activated (e.g., ref. 74). The role of the phytochromes in regulating *Lhc* genes encoding the chlorophyll *a/b*-binding proteins has been extensively studied and is well established.[75,76] It is important to coordinate the synthesis of these apoproteins with their pigment molecules in order to stabilise the photosynthetic complexes and it might therefore be expected that genes encoding the enzymes of the chlorophyll biosynthesis pathway would be light regulated. This is indeed the case, with most showing a diurnal regulation.[77] However, the majority of the vast number of studies looking at light induction of chlorophyll synthesis enzymes have simply compared dark- and light-grown seedlings or at best the effect of white light. Relatively few have established a specific role for phytochrome in these responses. Early studies focussed on the role of phytochrome in the potentiation of greening response in which an anticipating light treatment results in an increased rate of chlorophyll synthesis on transfer to white light.[78] The ability of continuous FR or a R pulse to mediate this response established that it was under phytochrome control and is the direct result of a phytochrome-mediated increase in 5-aminolevulinic acid (ALA) synthesizing capacity,[79] the rate-limiting step in the tetrapyrrole pathway (see Chapter 15). The response to R pulses is a classic R/FR reversible low fluence response mediated by both phyA and phyB[58] while the response to FR is phyA-mediated.[80] The molecular basis of this response has now been elucidated. Of the 3 enzymes of the C_5 ALA synthesis pathway (Chapter 2), glutamyl-tRNA reductase (GluTR) holds a strategic position in catalysing the first committed step of the tetrapyrrole pathway. In *Arabidopsis*, this enzyme is encoded by a small nuclear gene family with *AtHEMA1* being the predominant family member in chloroplast-containing organs. In an etiolated seedling exposed to light *AtHEMA1* mRNA upregulation is strong (x10) and rapid (detectable by 2h) with a time course of transcript accumulation that is compatible with the kinetics of greening rates, i.e., a lag phase of around 2h with increasing rates of chlorophyll accumulation in the first 12h.[81] Also, the response kinetics of *AtHEMA1* following a single Rp mirror the potentiating effect of an Rp on greening. Hence, *AtHEMA1* is most likely the primary gene target of the phytochrome-transcriptional cascade during the early stages of chloroplast biogenesis. Further analysis of these phytochrome signalling pathways using light-signalling mutants demonstrate that, under continuous R, induction of *AtHEMA1* is under the control of phyA, phyB and an, as yet unidentified, additional phytochrome.[77] In contrast to the strong effect of specific phytochrome signalling on *AtHEMA1*, the next enzyme in the pathway, glutamate-1-semialdehyde-2-1 aminomutase (GSA), is only moderately light regulated with *GSA1* showing weak dependence on photoreceptor signalling.[76] It was hypothesized that the small increases in expression in the light seen for *GSA1* and indeed the genes encoding most pathway enzymes[77] could be mediated by retrograde signalling[82] from the developing (developed) chloroplast.[76] Indeed, strong dependence on chloroplast integrity has been shown for all enzymes required specifically for chlorophyll synthesis (McCormac and Terry, unpublished results).

Although the direct contribution of phytochrome in the induction of tetrapyrrole genes has not been established for any other gene in the pathway, the genes encoding the H subunit of magnesium chelatase (*CHLH*), a subunit of Mg-protoporphyrin IX monomethyl ester cyclase (*CRD1*) and chlorophyll *a* oxygenase (*CAO*) have all been identified as circadian regulated (as opposed to simple diurnal regulation).[77] Since phytochrome has a crucial role in setting the circadian clock in plants,[83] this suggests at least indirect regulation of these genes by phytochrome. In contrast to most tetrapyrrole genes it appears that the protochlorophyllide reductase genes *PORA* and to a lesser extent *PORB* are downregulated by light at least in Arabidopsis.[77] This finding is in agreement with earlier studies.[78,84] It remains to be seen whether other reports of phytochrome effects on the pathway such as phytochrome regulation of chlorophyll *b* synthesis from chlorophyll *a*[78] are similarly supported.

In conclusion, it is clear that the last few years has thrown up many surprises in phytochrome research both in the mechanism of phytochrome signalling and the diversity of organisms utilising phytochrome-like photoreceptors for a broad range of biological functions. This is an exciting time for this research field and future developments are eagerly awaited.

Note Added in Proof

Recently, there have been a number of major developments related to phytochrome research. Most significantly, the first crystal structure of the chromophore-binding domain was determined from Deinococcus phytochrome assembled with BV[88] and this has now been refined to 1.45Å.[89] For a summary of our current understanding of phytochrome structure see Rockwell et al.[90] Our understanding of the bacterial phytochromes is also developing fast with ever expanding families of phytochrome-related proteins. Progress in this area is summarised in recent reviews on bacteriophytochromes[91] and cyanobacterial phytochromes.[92]

References

1. Montgomery BL, Lagarias JC. Phytochrome ancestry: Sensors of bilins and light. Trends Plant Sci 2002; 7:357-366.
2. Lamparter T. Evolution of cyanobacterial and plant phytochromes. FEBS Lett 2004; 573:1-5.
3. Sage LC. Pigment of the imagination. A History of Phytochrome Research. New York: Academic Press Inc, Harcourt Brace Jovanovich Publishers, 1992.
4. Terry MJ, Wahleithner JA, Lagarias JC. Biosynthesis of the plant photoreceptor phytochrome. Arch Biochem Biophys 1993; 306:1-15.
5. Mathews S, Sharrock RA. Phytochrome gene diversity. Plant Cell Environ 1997; 20:666-671.
6. Hughes J, Lamparter T, Mittmann F et al. A prokaryotic phytochrome. Nature 1997; 386:663-663.
7. Yeh K-C, Wu S-H, Murphy JT et al. A cyanobacterial phytochrome two-component light sensory system. Science 1997; 277:1505-1508.
8. Wu S-H, McDowell MT, Lagarias JC. Phycocyanobilin is the natural precursor of the phytochrome chromophore in the green alga Mesotaenium caldariorum. J Biol Chem 1997; 272:25700-25705.
9. Hübschmann T, Börner T, Hartmann E et al. Characterization of the Cph1 holo-phytochrome from Synechocystis sp. PCC 6803. Eur J Biochem 2001; 268:2055-2063.
10. Bhoo S-H, Davis SJ, Walker J et al. Bacteriophytochromes are photochromic histidine kinases using a biliverdin chromophore. Nature 2001; 414:776-779.
11. Terauchi K, Montgomery BL, Grossman AR et al. RcaE is a complementary chromatic adaptation photoreceptor required for green and red light responsiveness. Mol Microbiol 2004; 51:567-577.
12. Yeh KC, Lagarias JC. Eukaryotic phytochromes: Light-regulated serine/threonine protein kinases with histidine kinase ancestry. Proc Natl Acad Sci USA 1998; 95:13976-13981.
13. Taylor BL, Zhulin IB. PAS domains: Internal sensors of oxygen, redox potential, and light. Microbiol Mol Biol Rev 1999; 63:479-506.
14. Matsushita T, Mochizuki N, Nagatani A. Dimers of the N-terminal domain of phytochrome B are functional in the nucleus. Nature 2003; 424:571-574.
15. Nagatani A. Light-regulated nuclear localization of phytochromes. Curr Opin Plant Biol 2004; 7:708-711.
16. Møller SG, Ingles PJ, Whitelam GC. The cell biology of phytochrome signalling. New Phytol 2002; 154:553-590.
17. Quail PH. An emerging molecular map of the phytochromes. Plant Cell Environ 1997; 20:657-665.
18. Nozue K, Kanegae T, Imaizumi T et al. A phytochrome from the fern Adiantum with features of the putative photoreceptor NPH1. Proc Natl Acad Sci USA 1998; 95:15826-15830.
19. Kawai H, Kanegae T, Christensen S et al. Responses of ferns to red light are mediated by an unconventional photoreceptor. Nature 2003; 421:287-290.
20. Jiang ZY, Swem LR, Rushing BG et al. Bacterial photoreceptor with similarity to photoactive yellow protein and plant phytochromes. Science 1999; 285:406-409.
21. Karniol B, Vierstra RD. The pair of bacteriophytochromes from Agrobacterium tumefaciens are histidine kinases with opposing photobiological properties. Proc Natl Acad Sci USA 2003; 100:2807-2812.
22. Mutsuda M, Michel K-P, Zhang X et al. Biochemical properties of CikA, an unusual phytochrome-like histidine protein kinase that resets the circadian clock in Synechococcus elongatus PCC 7942. J Biol Chem 2003; 278:19102-19110.
23. Wu S-H, Lagarias JC. Defining the bilin lyase domain: Lessons from the extended phytochrome superfamily. Biochem 2000; 39:13487-13495.
24. Lagarias JC, Kelly JM, Cyr KL et al. Comparative photochemical analysis of highly purified 124 kilodalton oat and rye phytochromes in vitro. Photochem Photobiol 1987; 46:5-13.
25. Kneip C, Hildebrandt P, Schlamann W et al. Protonation state and structural changes of the tetrapyrrole chromophore during the $P_r \rightarrow P_{fr}$ phototransformation of phytochrome: A resonance raman spectroscopic study. Biochem 1999; 38:15185-15192.

26. Andel IIIrd F, Murphy JT, Haas JA et al. Probing the photoreaction mechanism of phytochrome through analysis of resonance raman vibrational spectra of recombinant analogues. Biochem 2000; 39:2667-2676.
27. Foerstendorf H, Benda C, Gärtner W et al. FTIR studies of phytochrome photoreactions reveal the C=O bands of the chromophore: Consequences for its protonation states, conformation, and protein interaction. Biochem 2001; 40:14952-14959.
28. Braslavsky SE, Gärtner W, Schaffner K. Phytochrome photoconversion. Plant Cell Environ 1997; 20:700-706.
29. Elich TD, Lagarias JC. Formation of a photoreversible phycocyanobilin-apophytochrome adduct in vitro. J Biol Chem 1989; 264:12902-12908.
30. Davis SJ, Vener AV, Vierstra RD. Bacteriophytochromes: Phytochrome-like photoreceptors from nonphotosynthetic eubacteria. Science 1999; 286:2517-2520.
31. Lamparter T, Michael N, Mittmann F et al. Phytochrome from Agrobacterium tumefaciens has unusual spectral properties and reveals an N-terminal chromophore attachment site. Proc Natl Acad Sci USA 2002; 99:11628-11633.
32. Lamparter T, Michael N, Caspani O et al. Biliverdin binds covalently to Agrobacterium phytochrome Agp1 via its ring A vinyl side chain. J Biol Chem 2003; 278:33786-33792.
33. Lamparter T, Carrascal M, Michael N et al. The biliverdin chromophore binds covalently to a conserved cysteine residue in the N-terminus of Agrobacterium phytochrome Agp1. Biochem 2004; 43:3659-3669.
34. Yoshihara S, Ikeuchi M. Photatactic motility in the unicellular cyanobacterium Synechocystis sp. PCC 6803. Photochem Photobiol Sci 2004; 3:512-518.
35. Aravind L, Ponting CP. The GAF domain: An evolutionary link between diverse phototransducing proteins. Trends Biochem Sci 1997; 22:458-459.
36. Lagarias JC, Lagarias DM. Self-assembly of synthetic phytochrome holoprotein in vitro. Proc Natl Acad Sci USA 1989; 86:5778-5780.
37. Li L, Murphy JT, Lagarias JC. Continuous florescence assay of phytochrome assembly in vitro. Biochem 1995; 34:7923-7930.
38. Borucki B, Otto H, Rottwinkel G et al. Mechanism of Cph1 phytochrome assembly from stopped-flow kinetics and circular dichroism. Biochem 2003; 42:13684-13697.
39. Quest B, Gärtner W. Chromophore selectivity in bacterial phytochromes. Dissecting the process of chromophore attachment. Eur J Biochem 2004; 271:1117-1126.
40. Li L, Lagarias JC. Phytochrome assembly. Defining chromophore structural requirements for covalent attachment and photoreversibility. J Biol Chem 1992; 267:19204-19210.
41. Bhoo S-H, Hirano T, Jeong H-Y et al. Phytochrome photochromism probed by site-directed mutations and chromophore esterification. J Am Chem Soc 1997; 119:11717-11718.
42. Hanzawa H, Inomata K, Kinoshita H et al. In vitro assembly of phytochrome B apoprotein with synthetic analogs of the phytochrome chromophore. Proc Natl Acad Sci USA 2001; 98:3612-3617.
43. Hanzawa H, Shinomura T, Inomata K et al. Structural requirement of bilin chromophore for the photosensory specificity of phytochrome A and B. Proc Natl Acad Sci USA 2002; 99:4725-4729.
44. Fiedler B, Broc D, Schubert H et al. Involvement of cyanobacterial phytochromes in growth under different light qualities and quantities. Photochem Photobiol 2004; 79:551-555.
45. Wilde A, Fielder B, Börner T. The cyanobacterial photochrome Cph2 inhibits phototaxis towards blue light. Mol Microbiol 2002; 44:981-988.
46. Giraud E, Fardoux J, Fourrier N et al. Bacteriophytochrome controls photosystem synthesis in anoxygenic bacteria. Nature 2002; 417:202-205.
47. Kendrick RE, Kronenberg GHM, eds. Photomorphogenesis in Plants. 2nd ed. Dordrecht: Kluwer Academic Publishers, 1994.
48. Smith H. Physiological and ecological function within the phytochrome family. Ann Rev Plant Physiol Plant Mol Biol 1995; 46:289-315.
49. Whitelam GC, Patel S, Devlin PF. Phytochromes and photomorphogenesis in Arabidopsis. Phil Trans R Soc Lond B 1998; 353:1445-1453.
50. Smith H. Phytochromes and light signal perception by plants—An emerging synthesis. Nature 2000; 407:585-591.
51. Nagy F, Schäfer E. Phytochromes control photomorphogeneis by differentially regulated, interacting signalling pathways in higher plants. Ann Rev Plant Biol 2002; 53:329-355.
52. Wada M, Kadota A. Photomorphogenesis in lower green plants. Ann Rev Plant Physiol Plant Mol Biol 1989; 40:169-191.
53. von Arnim A, Deng X-W. Light control of seedling development. Ann Rev Plant Physiol 1996; 47:215-43.
54. Quail PH. Phytochrome photosensory signalling networks. Nature Rev Mol Cell Biol 2002; 3:85-93.

55. Morelli G, Ruberti I. Light and shade in the photocontrol of Arabidopsis growth. Trends Plant Sci 2002; 7:399-404.
56. Casal JJ. Phytochromes, cryptochromes, phototropin: Photoreceptor interactions in plants. Photochem Photobiol 2000; 71:1-11.
57. Koornneef M, Rolff E, Spruit CJP. Genetic control of light-inhibited hypocotyl elongation in Arabidopsis thaliana (L.) Heynh. Z Pflanzenphysiol 1980; 100:147-160.
58. Reed JW, Nagatani A, Elich TD et al. Phytochrome A and phytochrome B have overlapping but distinct functions in Arabidopsis development. Plant Physiol 1994; 104:1139-1149.
59. Shinomura T, Nagatani A, Hanzawa H et al. Action spectra for phytochrome A- and B-specific photoinduction of seed germination in Arabidopsis thaliana. Proc Natl Acad Sci USA 1996; 93:8129-8133.
60. Nagatani A, Reed JW, Chory J. Isolation and initial characterization of Arabidopsis mutants that are deficient in phytochrome A. Plant Physiol 1993; 102:269-277.
61. Whitelam GC, Johnson E, Peng J et al. Phytochrome A null mutants of Arabidopsis display a wild-type phenotype in white light. Plant Cell 1993; 5:757-768.
62. Shinomura T, Uchida K, Furuya M. Elementary processes of photoperception by phytochrome A for high-irradiance response of hypocotyl elongation in Arabidopsis. Plant Physiol 2000; 122:147-156.
63. Casal JJ, Sanchez RA, Yanovsky MJ. The function of phytochrome A. Plant Cell Environ 1997; 20:813-819.
64. Aukerman MJ, Hirschfeld M, Wester L et al. A deletion in the PHYD gene of the Arabidopsis Wassilewskija ecotype defines a role for phytochrome D in red/far-red light sensing. Plant Cell 1997; 9:1317-1326.
65. Devlin PF, Robson PRH, Patel SR et al. Phytochrome D acts in the shade-avoidance syndrome in Arabidopsis by controlling elongation growth and flowering time. Plant Physiol 1999; 119:909-915.
66. Devlin PF, Patel SR, Whitelam GC. Phytochrome E influences internode elongation and flowering time in Arabidopsis. Plant Cell 1998; 10:1479-1487.
67. Hennig L, Stoddart WM, Dieterle M et al. Phytochrome E controls light-induced germination of Arabidopsis. Plant Physiol 2002; 128:194-200.
68. Franklin KA, Davis SJ, Stoddart WM et al. Mutant analyses define multiple roles for phytochrome C in Arabidopsis photomorphogenesis. Plant Cell 2003; 15:1981-1989.
69. Monte E, Alonso JM, Ecker JR et al. Isolation and characterization of phyC mutants in Arabidopsis reveals complex crosstalk between phytochrome signaling pathways. Plant Cell 2003; 15:1962-1980.
70. Sharrock RA, Clack T. Patterns of expression and normalised levels of the five Arabidopsis phytochromes. Plant Physiol 2002; 130:442-456.
71. Sharrock RA, Clack T. Heterodimerization of type II phytochromes in Arabidopsis. Proc Natl Acad Sci USA 2004; 101:11500-11505.
72. Gyula P, Schäfer E, Nagy F. Light perception and signalling in higher plants. Curr Opin Plant Biol 2003; 6:446-452.
73. Terry MJ. Phytochrome chromophore-deficient mutants. Plant Cell Environ 1997; 20:740-745.
74. Tepperman JM, Hudson ME, Khanna R et al. Expression profiling of phyB mutant demonstrates substantial contribution of other phytochromes to red-light-regulated gene expression during seedling de-etiolation. Plant J 2004; 38:725-739.
75. Terzaghi WB, Cashmore AR. Light-regulated transcription. Ann Rev Plant Physiol Plant Mol Biol 1995; 46:445-474.
76. McCormac AC, Terry MJ. Light-signalling pathways leading to the coordinated expression of HEMA1 and Lhcb during chloroplast development in Arabidopsis thaliana. Plant J 2002; 32:549-559.
77. Matsumoto F, Obayashi T, Sasaki-Sekimoto Y et al. Gene expression profiling of the tetrapyrrole metabolic pathway in Arabidopsis with a mini-array system. Plant Physiol 2004; 135:2379-2391.
78. Kasemir H. Light control of chlorophyll accumulation in higher plants. In: Shropshire Jr W, Mohr H, eds. Encyclopedia of Plant Physiology, New Series, Vol 16B. Berlin: Springer-Verlag, 1983:662-686.
79. Huang L, Bonner BA, Castelfranco PA. Regulation of 5-aminolevulinic acid (ALA) synthesis in developing chloroplasts II. Regulation of ALA-synthesising capacity by phytochrome. Plant Physiol 1989; 90:1003-1008.
80. McCormac AC, Terry MJ. Loss of nuclear gene expression during the phytochrome A-mediated far-red block of greening response. Plant Physiol 2002; 130:402-414.
81. McCormac AC, Fischer A, Kumar AM et al. Regulation of HEMA1 expression by phytochrome and a plastid signal during de-etiolation in Arabidopsis thaliana. Plant J 2001; 25:549-561.
82. Gray JC, Sullivan JA, Wang J-H et al. Coordination of plastid and nuclear gene expression. Phil Trans R Soc Lond B 2003; 358:135-145.

83. Somers DE, Devlin PF, Kay SA. Phytochromes and cryptochromes in the entrainment of the Arabidopsis circadian clock. Science 1998; 282:1488-1490.
84. Apel K. The protochlorophyllide holochrome of barley (Hordeum vulgare L.). Phytochrome-induced decrease of translatable mRNA coding for the NADPH: protochlorophyllide oxidoreductase. Eur J Biochem 1981; 120:89-93.
85. Eichenberg K, Bäurle I, Paulo N et al. Arabidopsis phytochromes C and E have different spectral characteristics from those of phytochromes A and B. FEBS Lett 2000; 470:107-112.
86. Park C-M, Kim J-II, Yang S-S et al. A second photochromic bacteriophytochrome from Synechocystis sp. PCC 6803: Spectral analysis and down-regulation by light. Biochem 2000; 39:10840-10847.
87. Milford MI. Biosynthesis, properties and structure of phytochrome photoreceptors from cyanobacteria. Ph. D. Thesis. University of Southampton, 2001.
88. Wagner JR, Brunzelle JS, Forest KT, Vierstra RD. A light-sensing knot revealed by the structure of the chromophore-binding domain of phytochrome. Nature 2005; 438:325-331.
89. Wagner JR, Zhang JR, Brunzelle JS et al. High resolution structure of Deinococcus bacteriophytochrome yields new insights into phytochrome architecture and evolution. J Biol Chem 2007; 16:12298-12309.
90. Rockwell NC, Su YS, Lagarias JC. Phytochrome structure and signalling mechanisms. Ann Rev Plant Biol 2006; 57:837-858.
91. Giraud E, Verméglio A. Bacteriophytochromes in anoxygenic photosynthetic bacteria. Photosyn Res 2008; 97:141-153.
92. Ikeuchi M, Ishizuka T. Cyanobacteriochromes: a new superfamily of tetrapyrrole-binding photoreceptors in cyanobacteria. Photochem Photobiol Sci 2008; DOI: 10.1039/b802660m.

CHAPTER 14

Biosynthesis of Chlorophyll and Bacteriochlorophyll

Derren J. Heyes and C. Neil Hunter*

Abstract

The (bacterio)chlorophyll biosynthetic pathway is of profound importance to the biosphere. During the past 20 years, there have been major advances in the under standing of the genes involved in the pathway and, more recently, in the enzymes that they encode. Chlorophyll biosynthesis can be considered to start with protoporphyrin IX, which lies at the branchpoint with haem synthesis. Therefore, this chapter will summarise the steps in the pathway from protoporphyrin IX through to (bacterio)chlorophyll. The discussion focuses on the current understanding of the bacterial, algal, and plant enzymes with particular emphasis on their protein composition and structure, required cofactors, physical and catalytic properties, and protein-protein interactions.

Introduction

Chlorophyll is essential for life as the cofactor for the photosynthetic proteins that harvest sunlight and convert it to photochemical energy for the cell. It is the most abundant pigment on Earth, with at least 10^9 tonnes synthesised per annum; consequently, chlorophyll biosynthesis is one of the few biochemical processes that can be observed from outer space.[1] Apart from its obvious functional role in light-harvesting and photochemistry, chlorophyll biosynthesis also plays an important role in signal transduction. There are strong indications that signalling between the chloroplast and the nucleus involve intermediates (or their enzymes) in the chlorophyll biosynthesis pathway (see Chapter 15). Given the global significance of this pathway and its importance to plant development, the synthesis of chlorophyll and its bacterial equivalent, bacteriochlorophyll, has been an extensively studied area for many years.

Photosynthetic bacteria have often been used as a model system to study aspects of photosynthesis, due to their facultative growth, simple photosystems and ease of genetic manipulation. Amongst these bacteria, the purple nonsulphur bacteria *Rhodobacter capsulatus* and *Rhodobacter sphaeroides* have widely been used for molecular genetic analysis of bacteriochlorophyll biosynthesis as the photosynthesis gene cluster from each of these organisms has been sequenced.[2,3] These studies have provided the first detailed understanding of the genes involved in bacteriochlorophyll biosynthesis. Subsequently, most of the chlorophyll biosynthesis genes have been identified by virtue of their ability to complement bacteriochlorophyll biosynthesis mutants as well as by sequence similarity comparisons. As with many other areas of biochemistry the recent use of heterologous overexpression systems has been decisive, by providing abundant sources of many of the (bacterio)chlororophyll biosynthesis enzymes. Progress in the analysis of these enzymes has benefited from the fact that several of the intermediates in the (bacterio)chlorophyll biosynthesis pathway are rich in optical signals. Many of the steps in the pathway

*Corresponding Author: C. Neil Hunter—University of Sheffield, Western Bank, Sheffield S10 2TN, UK. Email: c.n.hunter@sheffield.ac.uk

Tetrapyrroles: Birth, Life and Death, edited by Martin J. Warren and Alison G. Smith.
©2009 Landes Bioscience and Springer Science+Business Media.

can therefore be studied by a variety of spectroscopic and kinetic techniques and these approaches are starting to provide the first detailed, quantitative information about the mechanisms of these enzymes.

The synthesis of (bacterio)chlorophyll shares a common set of steps with haem synthesis up to the branchpoint at protoporphyrin IX (see Chapters 2-4). The subsequent steps from protoporphyrin IX to (bacterio)chlorophyll, which are illustrated in Figure 1, comprise a diverse and chemically interesting set of reactions that will now be discussed individually (for reviews see refs. 1,2,4).

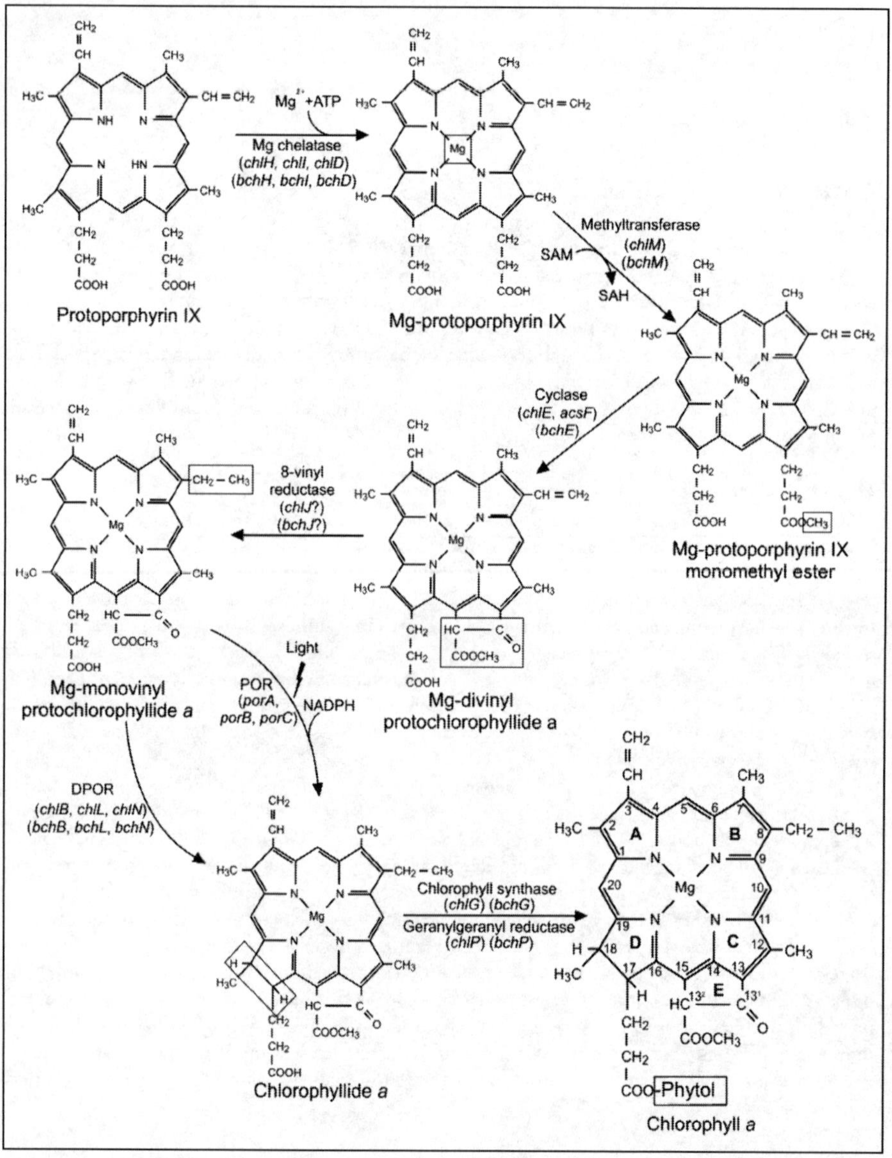

Figure 1. Outline of the chlorophyll biosynthetic pathway from protoporphyrin IX to chlorophyll *a*. The enzymes responsible for each step are shown together with the genes that have been identified to date. The boxed regions indicate the groups modified at each stage. The IUPAC numbering system of the carbon skeleton is indicated for the structure of chlorophyll *a*.

The Insertion of the Central Magnesium Ion

The first committed step of (bacterio)chlorophyll biosynthesis is the chelation of Mg^{2+} into the protoporphyrin macrocycle to form magnesium protoporphyrin IX (MgP). This reaction, which is catalysed by the enzyme Mg-protoporphyrin chelatase (Mg chelatase), lies at the branchpoint between haem and chlorophyll biosynthesis and has been the subject of much study in both bacteria and plants (for review see ref. 5). The insertion of Mg^{2+} into the porphyrin ring is energetically unfavourable as a result of removing H_2O molecules coordinated to the Mg^{2+} and many of the studies on Mg chelatase have revealed that ATP hydrolysis is essential for activity.[5] This is in contrast to the analogous reaction on the haem synthesis branch, where ferrochelatase catalyses the energetically favourable chelation of iron into protoporphyrin IX, without the involvement of ATP (see Chapter 4).

Analysis of *R. capsulatus* and *R. sphaeroides* mutants identified three loci, designated *bchH*, *bchI* and *bchD*, which were required for Mg chelation.[2,3] Insertional mutagenesis studies revealed that disruption to any of these three genes resulted in the accumulation of protoporphyrin IX and cells were no longer able to synthesise Mg-containing bacteriochlorophyll precursors.[6] Overexpression of the *bchH*, *bchI* and *bchD* genes from *R. sphaeroides* in *E. coli* resulted in the reconstitution of the ATP-dependent Mg chelatase in vitro and demonstrated that BchH, BchD and BchI, were the necessary components of this enzyme.[7] Homologous genes from *Synechocystis*, termed *chlH*, *chlI* and *chlD*, were subsequently identified, overexpressed in *E. coli* and, when all three gene products were combined, gave a functional Mg chelatase in vitro.[8] This provided the first direct evidence that Mg insertion in a chlorophyll-producing organism requires three different protein subunits, and suggested that the Mg chelatase enzyme is conserved between purple bacteria and cyanobacteria. Plant homologues of the Mg chelatase genes have been identified in a range of organisms although tobacco is so far the only plant species for which all three genes have been cloned and expressed to give a functional chelatase enzyme.[9] The three subunits of Mg chelatase, Bch/ChlD, Bch/ChlH and Bch/ChlI, have predicted molecular masses of 60-87 kDa, 120-155k Da and 37-46 kDa, respectively. The degree of identity between the deduced amino acid sequences of homologous proteins from various sources is 28–58% for the D subunit, 38–86% for the H subunit and and 50–90% for the I subunit.[5] Interestingly, the N-terminal half of ChlD shows approximately 40% identity to BchI and ChlI, suggesting that a duplication event of the I gene occurred, followed perhaps by gene fusion with the existing D gene, accompanying the evolutionary divergence of chlorophyll biosynthesis (and thus oxygenic photosynthesis) from bacteriochlorophyll biosynthesis.[8] The H subunit shares homology with CobN, the largest (140 kDa) subunit of cobaltochelatase, which is also a three-subunit enzyme requiring ATP for catalysis in the vitamin B_{12} biosynthetic pathway. Furthermore, all three Mg chelatase subunits are similar to the putative subunits of Ni chelatase.[5]

Although the mechanism of Mg chelatase is not fully understood, significant progress has been made in recent years to determine the functions of each subunit. It has been revealed that the I and D subunits catalyse the ATPase reaction that is presumably required to power the chelation reaction whereas the H subunit binds the porphyrin substrate. Detailed biochemical and structural work on the Mg chelatase enzyme has only become feasible with the use of heterologous expression systems to obtain an abundant source of purified proteins. The Mg chelatases from *R. sphaeroides*, *R. capsulatus* and *Synechocystis* have been studied in greatest detail and show an absolute requirement for ATP, Mg^{2+} and a suitable porphyrin (e.g., protoporphyrin IX). In all cases there is a lag phase in the reaction that can be reduced by preincubation with ATP and Mg^{2+}, suggesting that Mg chelation is a two-stage process, whereby Mg insertion follows initial activation of the components by ATP.[5,7,8]

In *Synechocystis* Mg chelatase, ChlH and ChlD are likely to be monomeric whereas ChlI can aggregate in an ATP-dependent manner from a dimeric structure to form high molecular weight aggregates (~6 to 8 subunits).[10] It has been shown using single particle analysis that these aggregates have a seven fold rotational symmetry and represent a heptamer.[11] The ATPase properties of wild-type and mutant *R. capsulatus* BchI subunits (D207N, R289K and L111F; *R. capsulatus* numbering) have been examined, and these residues appear to have a role in the interface between BchI molecules within the intact magnesium chelatase.[12] The first direct evidence for the formation of an I-D complex came when non-tagged BchI co-eluted with preimmobilised His-tagged BchD from a Ni^{2+}-chelating column equilibrated with Mg^{2+} and ATP.[13] Subsequently, it was revealed that only preincubation of the I and D subunits with Mg^{2+} and ATP was necessary to reduce the lag phase of Mg chelation.[10,11] The ATPase activity of the *Synechocystis* Mg chelatase coincides with that of Mg chelation and is 7-fold higher than that of the I-D complex alone.[14] Furthermore, the ChlI subunit had a significantly higher rate of ATP hydrolysis than the ChlI-ChlD complex. It has been demonstrated that the overall chelation reaction requires the hydrolysis of approximately 15 MgATPs.[15] A non-hydrolysable ATP analogue could replace ATP to maintain an I-D complex and remove the lag phase but not for the actual catalysis of Mg chelation.[14] It has been shown that ChlI from *Synechocystis* is a Mg^{2+} binding subunit that only hydrolyses ATP when an additional Mg^{2+} is bound. Upon binding of an ATP analogue the ChlI subunit underwent an isomerisation, which caused changes in Mg^{2+} affinity.[11] Consequently, the magnesium-bound form of the enzyme has been found to be a more effective catalyst of the chelation reaction, which suggests that there is a cooperative response with regards to magnesium.[15]

The crystal structure of the BchI subunit from *R. capsulatus* has been solved to 2.1Å resolution.[16] This protein belongs to the "ATPase associated with a variety of cellular activities" (AAA) family of ATPases, and contains a C-terminal helical domain and an N-terminal AAA module and nucleotide-binding site. The N-terminal helical domain is positioned roughly behind the nucleotide-binding domain, adjacent to the p-loop. The surface topology of this loop has revealed that it forms an exposed positively charged groove, suggested to be a candidate for a docking site for other magnesium chelatase subunits.[16] It has also been shown that the C-terminal domain of BchD shows homology to integrin I domains, which has led to the production of a modelled structure for this subunit. The metal ion binding sites found in integrin I domains were conserved, which suggested that the C terminal region of the D subunit may contain a Mg^{2+} binding site.[16]

Incubation of ChlH with protoporphyrin IX before starting a chelatase assay significantly reduced the length of the lag in product formation.[10] Consequently, the H subunit has been found to be the porphyrin-binding subunit and so probably contains the chelatase active site. Upon binding of deuteroporphyrin IX, a water-soluble analogue of protoporphyrin IX, to ChlH/BchH, red shifts in porphyrin fluorescence excitation and emission spectra were observed. This is consistent with nonplanar distortion of the macrocycle, and is likely to be accompanied by a conformational change in the H subunit.[17] It is likely that the importance of the H subunit extends beyond chlorophyll synthesis per se; recent studies of the *gun5* mutant of *Arabidopsis thaliana*, which has a lesion in the *chlH* gene, led to the conclusion that the Mg chelatase H subunit is involved in signal transduction between the plastid and the nucleus (see Chapter 15).

Based on all of these studies a provisional model for the Mg chelatase reaction has been proposed (Fig. 2). The I and D subunits are thought to interact with MgATP to form a complex. The H subunit with bound protoporphyrin IX reacts with I-D-MgATP complex to form a short-lived complex consisting of all three subunits and at least two of the substrates. Mg^{2+} is then inserted into protoporphyrin IX with concomitant hydrolysis of ATP, resulting in another short-lived complex consisting of the three subunits, MgADP and MgP. This complex dissociates into I-D-MgADP and H-MgP, which can then be recharged with MgATP and protoporphyrin IX, respectively.[10]

Methylation of Ring C

The enzyme, S-adenosyl-L-methionine MgP methyltransferase catalyses the methyl transfer step using S-adenosyl-L-methionine (SAM) as the methyl donor. The methyl group is translocated to the propionate side chain on ring C of MgP, forming magnesium protoporphyrin IX monomethylester (MgPME) and S-adenosyl-L-homocysteine (SAH). The reaction is thought to be essential to stop the spontaneous decarboxylation of the propionate group during isocyclic ring formation.[4] The methyltransferase enzyme is a soluble, single subunit enzyme that has a molecular weight of 25-27 kDa.[4,18]

The bacterial enzyme is encoded by the *bchM* gene and methyltransferase activity was confirmed by overexpression of the *R. capsulatus* gene in *E. coli*. The presence of a SAM-derived methyl group in the reaction product was confirmed by use of ^{14}C-methyl-labelled SAM as substrate.[19] The *bchM* gene of *R. sphaeroides* was similarly overexpressed in *E. coli*, resulting in a methyltransferase enzyme with a molecular weight of approximately 27.5 kDa.[20] A cyanobacterial homologue of *bchM* was isolated by functional complementation of a *R. capsulatus bchM* mutant with a cosmid library constructed from *Synechocystis* genomic DNA.[21] Expression of this gene, *chlM*, via the strong *R. capsulatus*

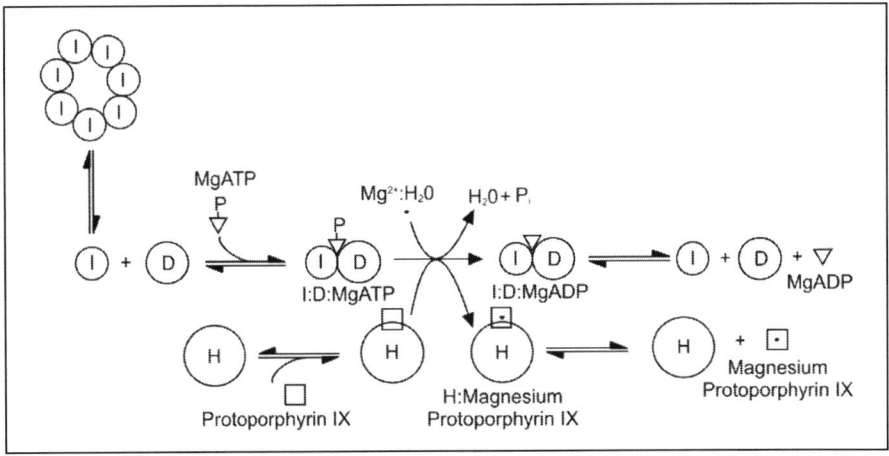

Figure 2. The proposed model for the Mg chelatase reaction. The model is based on the results of all recent mechanistic and structural studies on Mg-chelatase. The energy produced from the hydrolysis of ATP is used to remove the hydration shell of Mg^{2+}.

puc operon promoter supported nearly wild-type levels of bacteriochlorophyll biosynthesis. The deduced amino acid sequence of ChlM displayed relatively low sequence similarity (approximately 29% identity) with BchM. *R. sphaeroides* BchM, *R. capsulatus* BchM and *Synechocystis* ChlM each contain a putative SAM-binding motif.[19-21] It has been shown that the expression of *Synechocystis* *chlM* can be coupled to that of the Mg chelatase genes in *E. coli*, resulting in the conversion of protoporphyrin IX to MgPME.[18]

SAM-dependent methyltransferases act on a wide variety of target molecules, including DNA, RNA, proteins, polysaccharides and a variety of small molecules. Although the crystal structure of MgP methyltransferase has yet to be determined the enzyme from *Synechocystis* has been assigned to the small molecule class of methyltransferases, using structure based sequence searches.[22] All small molecule methyltransferases possess a catalytic domain containing both a SAM binding site, and a binding site for the other methyl acceptor. Many small molecule SAM-dependent methyltransferases have a core catalytic domain consisting of a mixed seven-stranded β-sheet. The methyl group in SAM is bound to a charged sulphur atom, which renders the relatively inert methyl moiety highly reactive towards nucleophiles and carbanions. Generally, the methyl group replaces a proton on the methyl acceptor, and bridging of this proton to an amino acid functional group can lead to transition state stabilisation.[22]

Little kinetic work has been performed on pure ChlM, possibly because the porphyrin substrate and product have similar spectral properties. It was not until recently when heterologous expression systems using *E. coli* were used to overexpress ChlM from *Synechocystis*, was it possible to characterise the kinetic properties of this enzyme. A more water-soluble substrate analogue of MgP, Mg deuteroporphyrin IX, was used to perform the steady-state kinetic assays and revealed that the reaction proceeds via a ternary complex. Product inhibition patterns using *S*-adenosyl-L-homocysteine (SAH) indicated a random binding mechanism, whereby SAH may bind productively to either free enzyme or a ChlM:MgD complex.[22]

The Missing Link in Chlorophyll Biosynthesis: The Formation of the Isocyclic Ring E of Protochlorophyllide

This step of the chlorophyll biosynthetic pathway has remained an enigma for over 50 years. The cyclase catalyses the conversion of MgPME to protochlorophyllide (Pchlide) and is responsible for converting the tetrapyrrole molecule from a red to a green colour. The methyl propionate side-chain of MgPME is modified to form an isocyclic ring (ring E). However, the cyclisation reaction is currently the most poorly understood step in the pathway as the cyclase has never been purified and studied.

Early studies on the reaction suggested that the formation of the isocyclic ring (ring E) might proceed in a similar way to the β-oxidation of fatty acids, via acrylate, β-hydroxy and β-keto intermediates.[4] However, organisms that form chlorophyll aerobically follow a different route, which does not involve the acrylate intermediate. All known functional chlorophylls have a keto group at position 13^1 arising from the introduction of an oxygen atom into the 13-methyl propionic acid side-chain during isocyclic-ring formation. The use of $^{18}O_2$-labelling has shown that this oxygen atom comes from atmospheric O_2 in higher plants.[23] This explains the ability of intact chloroplasts to cyclise synthetic β-hydroxy and β-keto derivatives, but not synthetic acrylate. However, in the photosynthetic bacterium *R. sphaeroides*, the origin of the 13^1-keto group was found to be H_2O, implying that cyclisation was an anaerobic process involving a hydratase.[24] The proposed reaction scheme involved a dehydrogenase removing two hydrogen atoms to produce the acrylate intermediate, which then undergoes hydration to the β-hydroxy intermediate.

Cyclase activity in cucumber chloroplasts has been resolved into membrane-bound and soluble fractions, suggesting that at least two protein components are required for the reaction.[4] The soluble component had a molecular mass greater than 30kDa and may bind porphyrins, while in vitro inhibition studies suggested that the membrane fraction may contain an essential heavy element.[25] The cyclase activity of *Synechocystis* was also found to consist of a membrane fraction and a soluble fraction, but that of *C. reinhardtii* appears to be exclusively membrane-associated. The membrane-associated component of the cyanobacterial cyclase has been solubilised in an active form. Oxidative cyclases have been found to have a general requirement for NADPH and Fe^{2+}, and so may incorporate a haem-containing monooxygenase.[26]

Although several genes are likely to be required for the cyclase reaction, only two have so far been identified, the *bchE* gene from *R. capsulatus*[6] and *R. sphaeroides*,[3] and the *acsF* (for aerobic cyclisation system Fe-containing subunit) gene from *Rubrivivax gelatinosus*.[27] Mutations in either gene resulted in the blockage of bacteriochlorophyll synthesis at the cyclisation step and the accumulation of MgPME. The putative gene product of *bchE* has a predicted molecular mass of 66kDa and has been shown to require a cobalamin (vitamin B_{12}) cofactor, which has led to a proposed mechanism for the anaerobic reaction (Fig. 3).[28] Adenosylcobalamin is thought to form an adenosyl radical, which in turn leads to the formation of the 13^1-radical of MgPME and the withdrawal of an electron to give rise to the 13^1-cation. This is subsequently attacked by a hydroxyl ion to give the 13^1-hydroxy intermediate of MgPME and the withdrawal of three hydrogen atoms leads to eventual cyclisation to form Pchlide. The AcsF protein contains putative binuclear-iron-cluster binding motifs and is thought to have a role in the aerobic cyclisation reaction.[27] However, this is the current limit of our biochemical understanding of this complex reaction in the pathway.

Reduction of the 8-Vinyl Group

Although the reduction of the 8-vinyl group of ring B to an ethyl group is an essential step in the biosynthesis of (bacterio)chlorophyll confusion still remains as to the exact stage in the pathway that the reaction occurs. Many monovinyl (MV) and divinyl (DV) chlorophyll precursors have been detected in plants and algae, which cannot be accounted for by the stepwise operation of a linear biosynthetic pathway.[1,4] In some cases, the ratios of accumulated MV and DV intermediates have been shown to be altered by physiological and environmental factors, such as age and light conditions, in addition to the intrinsic variation between species.[4] It has been proposed that MV and DV pools of different intermediates represent separate routes for chlorophyll biosynthesis and hence, a number of different 8-vinyl reductases may be responsible for reducing different DV intermediates. Furthermore, two sets of some of the other chlorophyll biosynthesis enzymes may exist in order to catalyse the independent conversion of MV and DV tetrapyrroles. Alternative models propose that there is a single 8-vinyl reductase enzyme with unique, intermediate-specific components, or one enzyme with broad substrate specificity. It has also been shown that the reduction of the 8-vinyl group may require the presence of NADPH as a reductant.[1,4]

Although no genes have yet been cloned that are essential for 8-vinyl reduction, mutants in the *bchJ* gene of *R. capsulatus* affected the ratio of MV and DV Pchlide and may help to shed some light on the reaction. A *bchJ*-disrupted strain was still able to synthesise bacteriochlorophyll but also accumulated DV Pchlide, whereas in a *bchJ/bchL* double mutant, the ratio of DV Pchlide to MV Pchlide was significantly heightened. Therefore it is likely that *bchJ* encodes an 8-vinyl reductase that is specific for

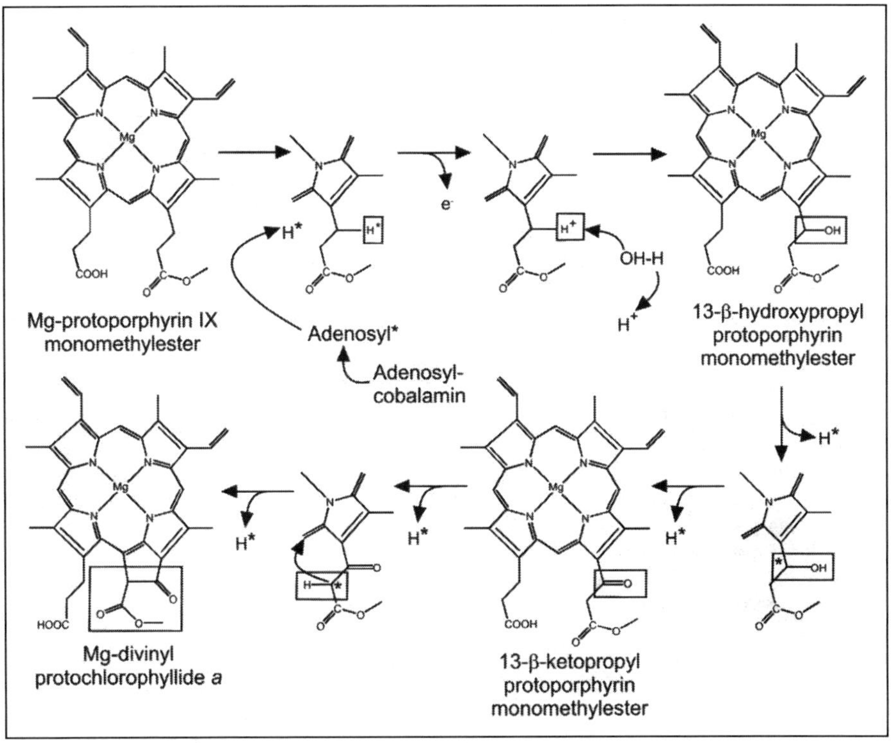

Figure 3. The proposed radical reaction mechanism for the adenosylcobalamin-dependent anaerobic Mg-protoporphyrin monomethylester cyclase.[27] The reaction proceeds via the β-hydroxy and β-keto intermediates, while the monovinyl species can also act as a substrate for the enzyme. Radicals are denoted by *.

DV Pchlide, although this remains to be proven.[29] No genes have yet been assigned to 8-vinyl reduction in chlorophyll-producing organisms, nor have any of the enzymes been purified.

Two Routes for the Reduction of Pchlide

The addition of hydrogen across the C17-C18 double bond of ring D of Pchlide leads to the formation of chlorophyllide (Chlide). This reaction can be catalysed by two completely different enzymes, one that requires light and another that does not. The light-dependent enzyme, NADPH-Pchlide oxidoreductase (EC 1.3.1.33, POR), appears to be found in all chlorophyll-containing organisms whereas the light-independent enzyme, Pchlide reductase (DPOR), is found in all except the angiosperms (flowering plants). Bacteriochlorophyll-containing photosynthetic bacteria appear to possess only the light-independent enzyme.[30]

POR: A Light-Driven Enzyme

When angiosperms are grown in the dark, large amounts of Pchlide accumulate and upon exposure to light, this is rapidly converted to Chlide. The enzyme responsible for this light-driven step in the pathway, POR, is one of only two enzymes known to require light for catalysis; the other is DNA photolyase. As a result of this requirement for light the reaction is an important regulatory step and is involved in the subsequent assembly of the photosynthetic apparatus (for a review see ref. 31). The enzyme requires NADPH as a cofactor and in the dark it is found in a ternary complex with the two substrates.[32] During the reaction, the Pchlide molecule performs the function of the photoreceptor and upon illumination, there is a very rapid transfer of hydride from the NADPH to Pchlide followed by Chlide release.[31]

In plants and algae, POR is encoded by the nuclear genome and contains a transit peptide at the N-terminus for import into plastids. Three different isoforms of POR have so far been identified in angiosperms, all of which display dramatically different patterns of light and developmental regulation. The physiological significance of the different isozymes, which are 75% similar, can only be speculated upon but it has been proposed that they are required at different stages of greening.[31] Two of the isoforms, PORA and PORB, have been proposed to form high molecular weight light-harvesting complexes in the prolamellar body of barley etioplasts. Such complexes, which consist of PORA and PORB in a ratio of 5:1, are thought to enable chlorophyll synthesis to occur rapidly even under low light intensities.[33] However, some organisms that can form chlorophyll in the dark, such as *Chlamydomonas reinhardtii* and *Synechocystis*, contain only a single POR-encoding gene.[31]

In etiolated plants POR is mainly found to exist as a ternary complex in highly organised networks of tubular membranes termed prolamellar bodies. Indeed, the formation of the prolamellar bodies is thought to be due to POR aggregation, thus explaining their regular, three-dimensional network. Several spectral forms of Pchlide have been identified in these membranes and the differences in the spectral properties have been attributed to the interactions of the pigment with the membranes and with POR, as well as pigment aggregation and structural arrangements. After illumination and subsequent formation of Chlide, there is a short wavelength shift in the absorption and fluorescence maxima of Chlide, known as the Shibata shift. This shift is thought to be caused by the disintegration of the POR-pigment aggregates, resulting in the dispersion of the prolamellar bodies. POR has also been found to be membrane-associated in cyanobacteria and in the chloroplasts of light-adapted plants and green algae.[31]

A comparison of the amino acid sequence of POR with other sequences in the protein database revealed that POR is a member of the 'RED' superfamily of enzymes (Reductases, Epimerases, Dehydrogenases).[34] The enzymes in this large protein family are all single domain NAD(P)/H-binding oxidoreductases and are generally dimers or tetramers. The crystal structures of several members of the family have been determined and have subsequently been used as a template to produce a homology model of POR from *Synechocystis*.[35] The structure consists of a central parallel β-sheet comprised of 7 β-strands, surrounded by 9 α-helices. The unique feature of POR within the family is the presence of a 33-residue insertion between the fifth and sixth β-sheets. The exact function of this large loop region is still unclear although roles in Pchlide binding, membrane association and protein-protein interactions have all been proposed. Binding of the cofactor is in the N-terminal part of the enzyme, where a common GXXXGXG nucleotide-binding motif occurs, which forms a tight βαβ fold, termed the Rossmann fold. Furthermore, a Tyr and a Lys residue, which are absolutely conserved throughout all members of the RED family, are located within the catalytic site and are essential for activity.[34]

Etioplast membranes have extensively been used to study the reaction catalysed by POR and have provided many clues about the catalytic mechanism.[31] Pchlide analogues with different side chains at the C17 position were not accepted as substrates by POR from oat etioplasts, suggesting that the free carboxylic acid group (C17) has an important role in the binding of the pigment to the enzyme. Similarly, the central Mg atom and the structure of the isocyclic ring of the Pchlide molecule have also been shown to be crucial for activity. However, the enzyme does not discriminate between vinyl and ethyl groups at the C8 position of ring B as both MV and DV forms of Pchlide can be converted to Chlide.[36] It has been shown that NADPH is essential for the photoreduction[32] and upon binding to the enzyme it protects one or more of the three conserved Cys residues in the protein from chemical modification.[37] The use of $4R$ and $4S$ ^{3}H-radiolabelled isomers of NADPH has revealed that the *pro-S* hydride of the nicotinamide ring is transferred to the C17 position of Pchlide.[38] When the two highly conserved residues, Tyr189 and Lys193 (numbering in *Synechocystis*), were mutated POR activity was abolished. A model was proposed whereby the hydride transferred to the C17 position of Pchlide is derived from NADPH and the proton at the C18 position is derived from the Tyr. The close proximity of the Lys residue was thought to be necessary to lower the pK_a of the phenolic group of the Tyr, thus facilitating deprotonation (Fig. 4).[34] After the reduction is completed, it has been suggested that the NADP+ product is exchanged for an NADPH molecule and then Chlide is released from the enzyme.[37]

The fact that POR is a light-driven enzyme means that catalysis can be initiated by illuminating samples at low temperatures.[39,40] In this way, certain steps in the photoreduction can be frozen out, allowing intermediates in the reaction pathway to be identified. A convenient approach to analyse these intermediates is to use low temperature fluorescence and absorbance spectroscopy. As a result many different spectroscopic Pchlide and Chlide species have been identified during the reaction in etioplasts, at temperatures ranging from 298 K to 4 K. However, the data are difficult to interpret as several processes, such as POR aggregation and prolamellar body formation can complicate the analysis of Pchlide photoreduction in such preparations.[31] The recent advent of heterologous expression systems eliminates such problems as the reaction mechanism can be studied in greater detail using purified POR proteins. For the *Synechocystis* enzyme a stable POR-NADPH-Pchlide ternary complex could be formed in the dark, which produced significantly red-shifted fluorescence and absorbance maxima compared to free Pchlide.[39] Three distinct steps were identified in the reaction pathway, an initial light-driven step, followed by two 'dark' reactions. The initial photochemical step, which can occur below 200 K, involves the formation of a nonfluorescent intermediate with a broad absorbance band at 696 nm that is suggested to represent an ion radical complex.[39] This species is then converted into a new intermediate that has an absorbance maximum at 681 nm and during the second 'dark' step this state gradually blue-shifts to yield the product, Chlide. Although the exact molecular nature of these intermediates is currently unknown the two 'dark' steps can only occur close to or above the 'glass transition' temperature of proteins, implying a role for domain movements and/or reorganization of the protein for these stages of the catalytic mechanism.[40] A thermophilic form of the enzyme has been used to identify two additional 'dark' steps, which were shown to represent a series of ordered product release and cofactor binding events. Firstly, NADP$^+$ is released from the enzyme and then replaced by NADPH, before release of the Chlide product and subsequent binding of Pchlide has taken place.[41] It has also been shown that the reaction, and hence the protein motions associated with it, can proceed on an ultrafast timescale after initiating catalysis with a 50 fs laser pulse. The study revealed that the reaction appears to be complete within 400 ps.[42]

DPOR: A Multi-Subunit Enzyme

Primitive anoxygenic (nonoxygen evolving) photosynthetic bacteria, cyanobacteria, algae, and all nonflowering plants including gymnosperms contain a light-independent (dark) Pchlide reductase (DPOR) that is absent from angiosperms. The presence of DPOR enables these organisms to synthesise bacteriochlorophylls and chlorophylls in the dark, whereas the lack of DPOR in angiosperms means light is essential for chlorophyll synthesis. While POR has been studied extensively, very little information is currently available about DPOR.[43]

Genetic studies of the purple nonsulphur bacterium *R. capsulatus* demonstrated that three genes, *bchL*, *bchN* and *bchB*, are essential for the light-independent Pchlide reduction during the biosynthesis of bacteriochlorophyll. Homologous genes, termed *chlB*, *chlL* and *chlN*, have been identified

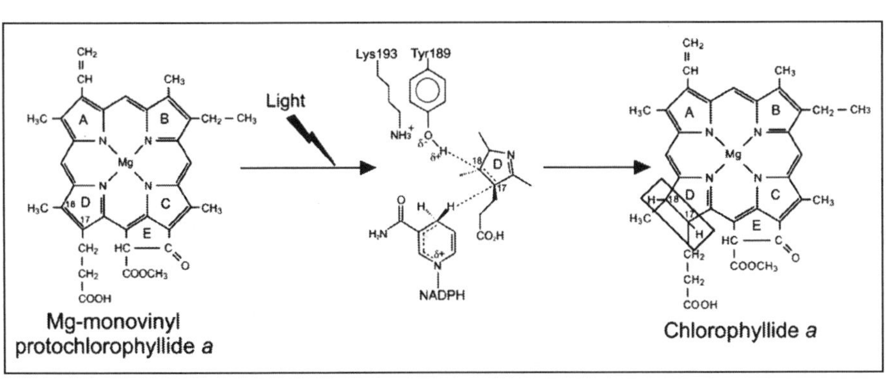

Figure 4. Model of the proposed mechanism of catalysis by POR.[33] The proton at the C18 position of Pchlide is derived from Tyr-189 (numbering in *Synechocystis* POR) and the hydride transferred to the C17 position is derived from the *pro-S* face of NADPH.

in red algae, the green alga *C. reinhardtii*, liverwort, gymnosperms and cyanobacteria and are required for Pchlide reduction during chlorophyll biosynthesis.[43] The amino acid sequences of the three DPOR subunits show significant similarities to the NifH, NifD and NifK subunits of nitrogenase.[44] Nitrogenase is an oxygen-sensitive, multi-subunit enzyme, which consists of two separable components, the Fe-protein (also called dinitrogenase reductase) and the MoFe protein complex, and catalyses the reduction of dinitrogen (N_2) to form ammonia ($2NH_3$). The Fe-protein, which is a dimer of NifH proteins and contains a 4Fe:4S redox cluster, transfers electrons from ferredoxin to the MoFe protein concomitant with Mg-ATP hydrolysis. The MoFe protein, which serves as the catalytic site for dinitrogen reduction, is an $\alpha_2\beta_2$ tetramer of the NifD and NifK proteins and contains 30 Fe atoms and 2 Mo atoms that are organised into two pairs of complex metalloclusters.[44]

The structural similarity between DPOR and nitrogenase is most apparent between the BchL/ChlL and NifH (dinitrogenase reductase) subunits where there is 33% overall identity and 50% similarity.[43] The ATP-binding motif of NifH and the four Cys residues involved in the chelation of the Fe-S cluster are completely conserved among BchL/ChlL proteins. This indicates that the BchL/ChlL proteins might catalyse ATP-dependent transfer of electrons from a reductant, such as ferredoxin, to a catalytic protein complex via the Fe:S centre. The amino acid sequences of the BchN/ChlN and BchB/ChlB subunits also exhibit similarity to NifD and NifK, respectively. However, there is no conservation of the residues that are involved in formation of the FeMo cofactor in nitrogenase, which indicates that the site where Pchlide is reduced is highly diverged from the catalytic site in nitrogenase.[43]

Despite the interesting structural similarity between DPOR and the well-characterised nitrogenase, biochemical analysis of DPOR has proved difficult. There have been a few studies of DPOR activity in crude cell-free extracts but purification of the DPOR enzyme has only recently been reported.[45] *R. capsulatus* was used to overexpress and purify two of the three DPOR subunits (BchL and BchN) whilst the third subunit (BchB) was copurified with the BchN protein, indicating that BchN and BchB form a tight complex. DPOR activity was shown to be dependent on the presence of all three subunits, ATP, and the reductant dithionite, which is supportive of strong "nitrogenase-like" behaviour.[45]

It remains unclear why some organisms possess both DPOR and POR for Pchlide reduction. The two systems may be differentially regulated in response to environmental signals, such as light intensity and nutritional signals, and in response to developmental stages, as is the case with PORA and PORB in angiosperms. In the cyanobacterium *Plectonema boryanum*, both POR and DPOR contribute to Pchlide reduction in the light but the extent of the contribution by POR increases at higher light intensities. In green algae POR and DPOR are both required to achieve maximal chlorophyll accumulation and it may be the case that POR has a regulatory role in these organisms.[30,31,43]

The Steps Unique to Bacteriochlorophyll Biosynthesis

Bacteriochlorophyll differs from chlorophyll in the substituents of rings A and B, and the steps from Chlide to bacteriochlorophyllide (Bchlide), which are unique to purple and green eubacteria, are responsible for these differences (Fig. 5). These latter stages in the bacteriochlorophyll synthesis pathway were deduced by identifying bacteriochlorophyll intermediates excreted by mutants of *R. sphaeroides*. Initially, 3-hydroxyethyl-Chlide, 3-hydroxyethyl-Bchlide and Bchlide were identified and it was proposed that hydration of the 3-vinyl group of ring A preceded the reduction of the C7-C8 double bond of ring B. However, it was later concluded that the order of these two steps was interchangeable, following the isolation of a *R. sphaeroides* mutant that accumulated 3-vinyl-Bchlide.[46]

The enzyme chlorin reductase is responsible for the reduction of ring B, a reaction that is chemically similar to the reduction of the C17-C18 double bond of ring D of Pchlide to form Chlide. Three gene products, *bchX*, *bchY* and *bchZ* have been shown to code for chlorin reductase in both *R. sphaeroides* and *R. capsulatus*.[2] It is interesting to note that the *bchXYZ* gene products show sequence similarity to the *bchLNB* gene products of DPOR and are likely to have arisen by gene duplication. The strongest conservation is between BchX and BchL, which also share significant sequence identity with the NifH subunit of nitrogenase. Hence it is likely that the mechanisms of both DPOR and chlorin reductase are homologous to the reaction catalysed by nitrogenase.[43]

Using ^{18}O-labelling it was shown that the carbonyl oxygen of the acetyl group at position 3 of ring A was found to be derived from water suggesting a hydratase mechanism rather than an oxygenase mechanism at this stage in the pathway.[24] The *bchF* gene product has been shown to be

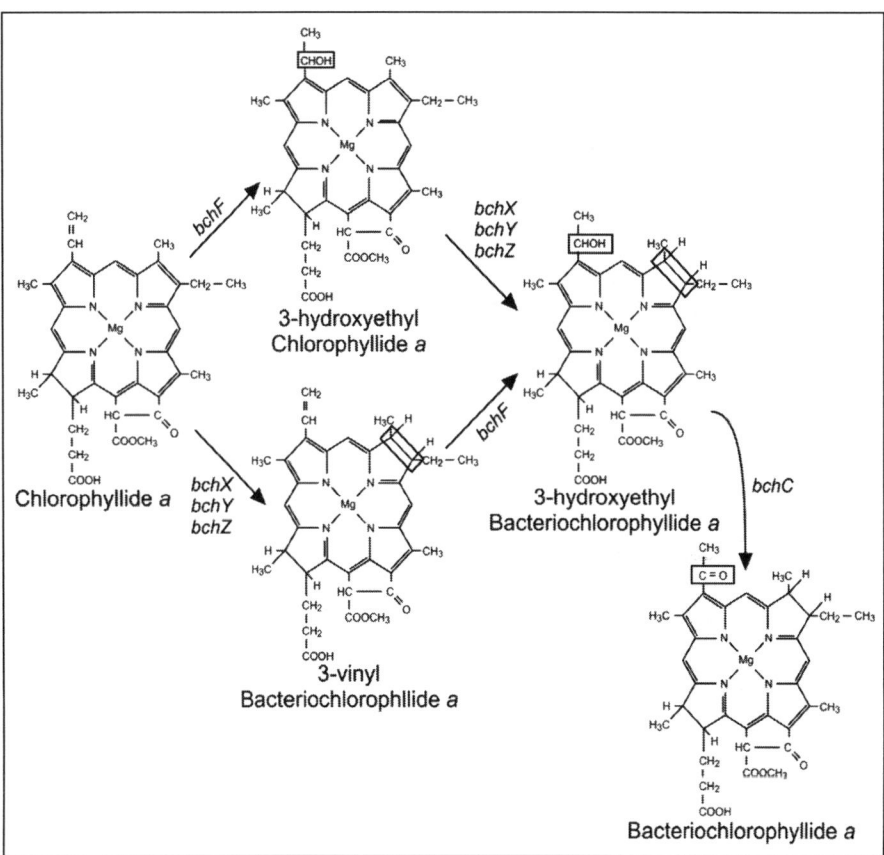

Figure 5. The steps in the bacteriochlorophyll biosynthesis pathway from chlorophyllide to bacteriochlorophyllide. The reactions catalysed by the *bchF* and *bchXYZ* gene products are interchangeable. The gene assigned to each step is shown in italics and boxed regions indicate the groups modified at each stage.

required for the hydration of the vinyl group in *R. capsulatus*.[6] However, the construction of a double mutant (*bchF/bchZ*) was required in order to obtain accumulation of Chlide, confirming the proposal that these steps are interchangeable.[6]

The next step is the oxidation of the hydroxyethyl group on ring A of 3-hydroxyethyl-Bchlide to form the acetyl group of Bchlide. This is catalysed by a dehydrogenase encoded by the *bchC* gene, which has been sequenced in both *R. sphaeroides* and *R. capsulatus*.[2] However, as yet, none of the enzymes of these latter stages of bacteriochlorophyll biosynthesis has been purified and consequently little is known about their reaction mechanisms.

The Final Steps: Addition and Reduction of the Phytol Tail

Chlorophylls and bacteriochlorophylls are hydrophobic molecules due to the incorporation of a long-chain alcohol residue, which also constitutes approximately 30% of their molecular weight. The esterification at the propionate side chain of ring D with phytol, a C_{20} isoprenoid alcohol, is generally the last stage of (bacterio)chlorophyll biosynthesis, although further modifications of the tetrapyrrole macrocycle may occur to generate, for example, chlorophyll *b*. It appears that the addition of the alcohol moiety is a crucial determinant of the assembly of the entire photosynthetic apparatus (see Chapter 15). The tails also play a major role in controlling the orientation of the transition dipoles of the tetrapyrrole rings, a function that is crucial for fast energy transfer.[47]

The enzyme (bacterio)chlorophyll synthase, which is found in the thylakoid membranes, cataly-ses the esterification reaction and is able to use either geranylgeranyl-pyrophosphate (PP) or phytyl-PP as a substrate, depending on the organism.[4] The enzyme is encoded by the *bchG/chlG* genes and heterologous expression in *E. coli* of the respective genes from *Synechocystis* and *R. capsulatus* was shown to be responsible for chlorophyll synthase and bacteriochlorophyll synthase activity, respec-tively. The overexpressed enzyme from *Synechocystis* can use Chlide as the pigment substrate but not Bchlide whereas the overexpressed enzyme from *R. capsulatus* uses Bchlide and not Chlide.[48] Chlo-rophyll synthase has also been shown to use both the DV and MV forms of Chlide but not the metal-free derivative, pheophorbide.[4]

It has become apparent that phytol is often not the initial esterifying alcohol and instead the tetrapyrrole is esterified with an activated form of the biosynthetic precursor of phytol, geranylgeranyl-PP. Shortly after esterification, the pigment-bound geranylgeranyl undergoes suc-cessive reductions of three of its four C-C double bonds to generate the final pigment in a process that requires NADPH and ATP (Fig. 6).[4] A direct phytylation route, involving the attachment of phytyl-PP to Chlide, has also been observed and the different pathways may predominate at different stages of development or with different substrate availabilities. A geranylgeranyl-PP re-ductase, which catalyses the conversion of free geranylgeranyl-PP to phytyl-PP has been located

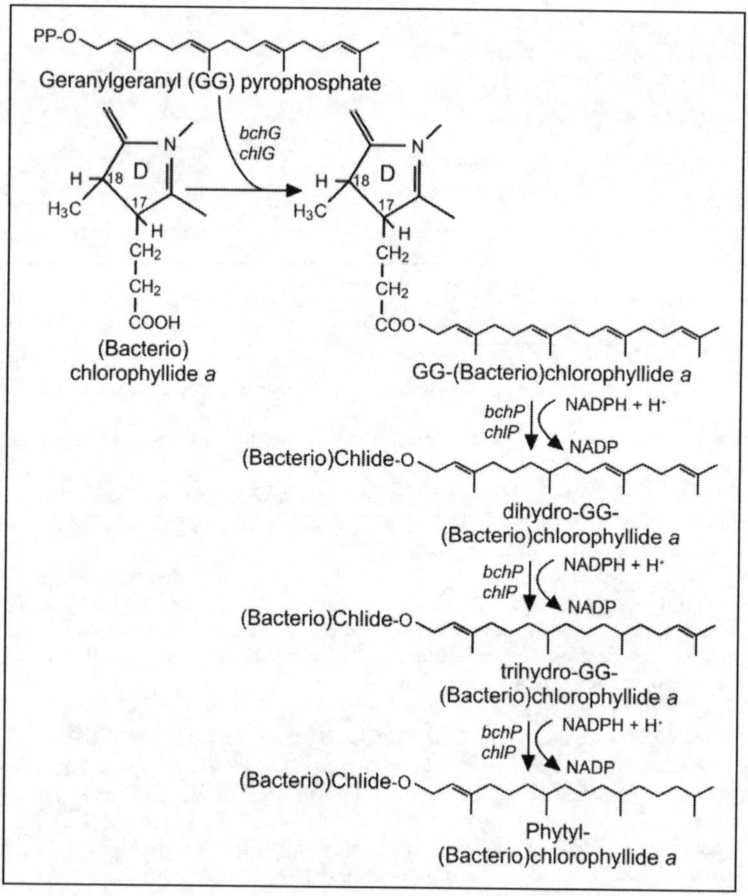

Figure 6. The proposed pathway for the phytylation of (bacterio)chlorophyllide. The genes for the esteri-fication and hydrogenation steps are shown in italics.

in the chloroplast envelope whereas the hydrogenase that catalyses the stepwise reduction of the geranylgeranyl side chain of chlorophyll to phytol was found in the thylakoid membrane.[4]

The product of a single gene is responsible for the three necessary reduction steps of the geranylgeranyl group to the phytyl group. A *bchP* mutant of *R. sphaeroides* was blocked in the terminal stages of bacteriochlorophyll biosynthesis and only contained bacteriochlorophyll that is esterified with geranylgeraniol. Upon heterologous expression of the *Synechocystis bchP* homologue, *chlP*, hydrogenated forms of esterified bacteriochlorophyll were detectable. It was concluded that *chlP* encodes the enzyme catalysing the stepwise hydrogenation of the geranylgeranyl group, which may occur either at the activated, pyrophosphate level or following chlorophyllide esterification.[49] The first direct demonstration of the activity of the product of the *bchP* gene, which encodes the enzyme geranylgeraniol-bacteriochlorophyll reductase, has since been reported.[50]

Concluding Remarks

Our understanding of the (bacterio)chlorophyll pathway has clearly been enhanced by the recent advances in genetic and molecular biological techniques. Sequence information for most of the enzymes in the pathway has now been obtained and many of the reactions have been analysed using purified recombinant protein. As a result structural and mechanistic information has been obtained on the various enzymes, which has provided valuable insights into the reactions that they catalyse. However, despite this many aspects of the pathway remain poorly understood. One of the primary objectives is to identify all of the missing components of the pathway, to allow the remaining steps to be characterised by quantitative biochemical approaches. It is also evident that for the majority of the enzymes there is a lack of a three-dimensional structure, which would further increase our understanding of the catalytic mechanisms. Ultimately, it will be fascinating to understand the way in which these enzymes interact and how the whole pathway delivers chlorophyll to nascent proteins of the photosynthetic apparatus.

References

1. Rudiger W. Chlorophyll metabolism: From outer space down to the molecular level. Phytochemistry 1997; 46:1151-1167.
2. Suzuki JY, Bollivar DW, Bauer CE. Genetic analysis of chlorophyll biosynthesis. Ann Rev Gen 1997; 31:61-89.
3. Naylor GW, Addlesee HA, Gibson LCD et al. The photosynthesis gene cluster of Rhodobacter sphaeroides. Photosynth Res 1999; 62:121-139.
4. Beale SI. Enzymes of chlorophyll biosynthesis. Photosynth Res 1999; 60:43-73.
5. Walker CJ, Willows RD. Mechanism and regulation of Mg-chelatase. Biochem J 1997; 327:321-333.
6. Bollivar DW, Suzuki JY, Beatty JT et al. Directed mutational analysis of bacteriochlorophyll a biosynthesis in Rhodobacter capsulatus. J Mol Biol 1994; 237:622-640.
7. Gibson LCD, Willows RD, Kannangara CG et al. Magnesium-protoporphyrin chelatase of Rhodobacter sphaeroides: Reconstitiution of activity by combining the products of the bchH, -I and -D genes expressed in Escherichia coli. Proc Natl Acad Sci USA 1995; 92:1941-1944.
8. Jensen PE, Gibson LCD, Henningsen KW et al. Expression of the chlI, chlD, and chlH genes from the cyanobacterium Synechocystis PCC6803 in Escherichia coli and demonstration that the three cognate proteins are required for magnesium-protoporphyrin chelatase activity. J Biol Chem 1996; 271:16662-16667.
9. Papenbrock J, Grafe S, Kruse E et al. Mg-chelatase of tobacco: Identification of a chl D cDNA sequence encoding a third subunit, analysis of the interaction of the three subunits with the yeast two-hybrid system, and reconstitution of the enzyme activity by coexpression of recombinant CHL D, CHL H and CHL I. Plant J 1997; 12:981-990.
10. Jensen PE, Gibson LCD, Hunter CN. Determinants of catalytic activity with the use of purified I, D and H subunits of the magnesium protoporphyrin IX chelatase from Synechocystis sp. PCC6803. Biochem J 1998; 334:335-344.
11. Reid JD, Siebert CA, Bullough PA et al. The ATPase activity of the ChlI subunit of magnesium chelatase and formation of a heptameric AAA$^+$ ring. Biochemistry 2003; 42:6912-6920.
12. Hansson A, Willows RD, Roberts TH et al. Three semidominant barley mutants with single amino acid substitutions in the smallest magnesium chelatase subunit form defective AAA$^+$ hexamers. Proc Natl Acad Sci USA 2002; 99:13944-13949.

13. Gibson LCD, Jensen PE, Hunter CN. Magnesium chelatase from Rhodobacter sphaeroides: Initial characterization of the enzyme using purified subunits and evidence for a BchI-BchD complex. Biochem J 1999; 337:243-251.

14. Jensen PE, Gibson LCD, Hunter CN. ATPase activity associated with the magnesium-protoporphyrin IX chelatase enzyme of Synechocystis sp. PCC6803: Evidence for ATP hydrolysis during Mg^{2+} insertion, and the MgATP-dependent interaction of the ChlI and ChlD subunits. Biochem J 1999; 339:127-134.

15. Reid JD, Hunter CN. Magnesium-dependent ATPase activity and cooperativity of magnesium chelatase from Synechocystis sp. PCC6803. J Biol Chem 2004; 279(26):26893-9.

16. Fodje MN, Hansson A, Hansson M et al. Interplay between an AAA module and an integrin I domain may regulate the function of magnesium chelatase. J Mol Biol 2001; 311:111-122.

17. Karger GA, Reid JD, Hunter CN. Characterization of the binding of deuteroporphyrin IX to the magnesium chelatase H subunit and spectroscopic properties of the complex. Biochemistry 2001; 40:9291-9299.

18. Jensen PE, Gibson LCD, Shephard F et al. Introduction of a new branchpoint in tetrapyrrole biosynthesis in Escherichia coli by coexpression of genes encoding the chlorophyll-specific enzymes magnesium chelatase and magnesium protoporphyrin methyltransferase. FEBS Letts 1999; 455:349-354.

19. Bollivar DW, Jiang ZY, Bauer CE et al. Heterologous expression of the bchM gene product from Rhodobacter capsulatus and demonstration that it encodes S-adenosyl-L-methionine: Mg-protoporphyrin IX methyltransferase. J Bacteriol 1994; 176:5290-5296.

20. Gibson LCD, Hunter CN. The bacteriochlorophyll biosynthesis gene, bchM, of Rhodobacter sphaeroides encodes S-adenosyl-L-methionine: Mg protoporphyrin methyltransferase. FEBS Letts 1994; 352(2):127-130.

21. Smith CA, Suzuki JY, Bauer CE. Cloning and characterization of the chlorophyll biosynthesis gene chlM from Synechocystis PCC 6803 by complementation of a bacteriochlorophyll biosynthesis mutant of Rhodobacter capsulatus. Plant Mol Biol 1996; 30:1307-1314.

22. Shepherd M, Reid JD, Hunter CN. Purification and kinetic characterisation of the magnesium protoporphyrin IX methyltransferase from Synechocystis PCC6803. Biochem J 2003; 371:351-360.

23. Walker CJ, Mansfield KE, Smith KM et al. Incorporation of atmospheric oxygen into the carbonyl functionality of the protochlorophyllide isocyclic ring. Biochem J 1989; 257:599-602.

24. Porra RJ, Schafer W, Gad'on N et al. Origin of the two carbonyl oxygens of bacteriochlorophyll a: Demonstration of two different pathways for the formation of ring E in Rhodobacter sphaeroides and Roseobacter denitrificans, and a common hydratase mechanism for 3-acetyl group formation. European J Biochem 1996; 239:85-92.

25. Walker CJ, Castelfranco PA, Whyte BJ. Synthesis of divinyl protochlorophyllide. enzymological properties of the Mg-protoporphyrin IX monomethyl ester oxidative cyclase system. Biochem J 1991; 276:691-697.

26. Bollivar DW, Beale SI. The chlorophyll biosynthetic enzyme Mg-protoporphyrin IX monomethyl ester (oxidative) cyclase: Characterization and partial purification from Chlamydomonas reinhardtii and Synechocystis sp. PCC 6803. Plant Physiol 1996; 112:105-114.

27. Pinta V, Picaud M, Reiss-Husson F et al. Rubrivivax gelatinosus acsF (previously ORF358) codes for a conserved, putative binuclear-iron-cluster-containing protein involved in aerobic oxidative cyclization of Mg-protoporphyrin IX monomethylester. J Bacteriol 2002; 184:746-753.

28. Gough SP, Petersen BO, Duus JO. Anaerobic chlorophyll isocyclic ring formation in Rhodobacter capsulatus requires a cobalamin cofactor. Proc Natl Acad Sci USA 2000; 97:6908-6913.

29. Suzuki JY, Bauer CE. Altered monovinyl and divinyl protochlorophyllide pools in bchJ mutants of Rhodobacter capsulatus—Possible monovinyl substrate discrimination of light-independent protochlorophyllide reductase. J Biol Chem 1995; 270:3732-3740.

30. Suzuki JY, Bauer CE. A prokaryotic origin for light-dependent chlorophyll biosynthesis of plants. Proc Natl Acad Sci USA 1995; 92:3749-3753.

31. Lebedev N, Timko MP. Protochlorophyllide photoreduction. Photosynth Res 1998; 58:5-23.

32. Griffiths WT. Reconstruction of chlorophyllide formation by isolated etioplast membranes. Biochem J 1978; 174:681-692.

33. Reinbothe C, Lebedev N, Reinbothe S. A protochlorophyllide light-harvesting complex involved in de-etiolation of higher plants. Nature 1999; 397:80-84.

34. Wilks HM, Timko MP. A light-dependent complementation system for analysis of NADPH:protochlorophyllide oxidoreductase. Identification and mutagenesis of two conserved residues that are essential for enzyme activity. Proc Natl Acad Sci USA 1995; 92:724-728.

35. Townley HE, Sessions RB, Clarke AR et al. Protochlorophyllide oxidoreductase: A homology model examined by site-directed mutagenesis. Proteins 2001; 44:329-335.

36. Klement H, Helfrich M, Oster U et al. Pigment-free NADPH:protochlorophyllide oxidoreductase from Avena sativa L. Eur J Biochem 1999; 265:862-874.
37. Oliver RP, Griffiths WT. Covalent labeling of the NADPH: Protochlorophyllide oxidoreductase from etioplast membranes with [^{3}H]N-phenylmaleimide. Biochem J 1981; 195:93-101.
38. Begley TP, Young H. Protochlorophyllide reductase. 1. Determination of the regiochemistry and the stereochemistry of the reduction of protochlorophyllide to chlorophyllide. J Am Chem Soc 1989; 111:3095-3096.
39. Heyes DJ, Ruban AV, Wilks HM et al. Enzymology below 200 K: The kinetics and thermodynamics of the photochemistry catalyzed by protochlorophyllide oxidoreductase. Proc Natl Acad Sci USA 2002; 99:11145-11150.
40. Heyes DJ, Ruban AV, Hunter CN. Protochlorophyllide oxidoreductase: "Dark" reactions of a light-driven enzyme. Biochemistry 2003; 42:523-528.
41. Heyes DJ, Hunter CN. Identification and characterization of the product release steps within the catalytic cycle of protochlorophyllide oxidoreductase, (in press).
42. Heyes DJ, van Stokkum IHM, Hunter CN et al. Ultrafast enzymatic reaction dynamics in protochlorophyllide oxidoreductase. Nature Struct Biol 2003; 10:491-492.
43. Fujita Y. Protochlorophyllide reduction: A key step in the greening of plants. Plant Cell Physiol 1996; 37:411-421.
44. Dean DR, Bolin JT, Zheng LM. Nitrogenase metalloclusters—structures, organization, and synthesis. J Bacteriol 1993; 175:6737-6744.
45. Fujita Y, Bauer CE. Reconstitution of light-independent protochlorophyllide reductase from purified BchL and BchN-BchB subunits—In vitro confirmation of nitrogenase-like features of a bacteriochlorophyll biosynthesis enzyme. J Biol Chem 2000; 275:23583-23588.
46. Pudek MR, Richards WR. A possible alternative pathway of bacteriochlorophyll biosynthesis in a mutant of Rhodopseudomonas sphaeroides. Biochemistry 1975; 14:3132-3137.
47. Freer AA, Prince S, Sauer K et al. Pigment-pigment interactions and energy transfer in the antenna complex of the photosynthetic bacterium Rhodopseudomonas acidophila. Structure 1996; 4:449-462.
48. Oster U, Bauer CE, Rüdiger W. Characterization of chlorophyll a and bacteriochlorophyll a synthases by heterologous expression in Escherichia coli. J Biol Chem 1997; 272:9671-9676.
49. Addlesee HA, Gibson LCD, Jensen PE et al. Cloning, sequencing and functional assignment of the chlorophyll biosynthesis gene, chlP, of Synechocystis sp. PCC 6803. FEBS Letts 1996; 389:126-130.
50. Addlesee HA, Hunter CN. Physical mapping and functional assignment of the geranylgeranyl- bacteriochlorophyll reductase gene, bchP, of Rhodobacter sphaeroides. J Bacteriol 1999; 181:7248-7255.

CHAPTER 15

Regulation of Tetrapyrrole Synthesis in Higher Plants

Matthew J. Terry and Alison G. Smith*

Abstract

Regulation of the tetrapyrrole pathway in plants is particularly crucial, since it is required for efficient synthesis of the photosynthetic apparatus, protection from the harmful phototoxicity of the pathway intermediates, and because of the proposed role played by some of the intermediates in signaling. The four major products, chlorophyll, haem, sirohaem and phytochromobilin, all need to be assembled with their respective apoproteins and the production of both components needs to be carefully coordinated. This is especially true as many tetrapyrroles such as the chlorophylls and their precursors are extremely phototoxic as free compounds. The major control points of the pathway are: the formation of the initial precursor, 5-aminolaevulinic acid; and the metal-ion insertion steps at the branchpoint between haem and Mg-protoporphyrin, and the formation of sirohaem. Because of the necessary complexity of tetrapyrrole regulation a wide range of regulatory mechanisms are employed, including transcriptional regulation of key genes in response to both environmental and internal cues, and internal pathway regulation by dedicated regulatory proteins and pathway intermediates. Moreover, genetic and microarray evidence indicates a link between the flux through the pathway and expression of genes encoding chlorophyll apoproteins. Here we discuss our current understanding of how these mechanisms are coordinated to control flux through the pathway to meet the requirements of the cell under different conditions.

Introduction

The tetrapyrrole pathway in higher plants is responsible for the synthesis of four major endproducts: chlorophyll, haem, sirohaem and phytochromobilin, via a common branched pathway[1-3] (Fig. 1). Tetrapyrrole synthesis takes place in the chloroplast, or plastids in nonphotosynthetic cells. In addition, the last two steps of haem synthesis, protoporphyrinogen IX oxidase (PPO) and ferrochelatase, are also found in mitochondria, suggesting that haem may be synthesized in these organelles in situ[4] (Chapter 4). Once synthesized, chlorophyll and sirohaem are confined to the chloroplast, but phytochromobilin is found in the cytosol and nucleus (Chapter 13) and haem is present in all cellular compartments.[4] All endproducts are bound to apoproteins, and it is likely that levels of "free" tetrapyrroles within the cell are extremely low. As for all metabolic pathways, regulation is crucial, but there are several special features of the plant tetrapyrrole pathway that cause its regulation to be extremely complex, involving not just control of the flux through the different branches, but also coordination with gene expression, both in the chloroplast and the nucleus. Thus in addition to feedback inhibition of enzyme activity and induction of key enzymes by external signals, there is also tight

*Corresponding Author: Alison G. Smith—Department of Plant Sciences, University of Cambridge, Downing Street, Cambridge, CB2 3EA, UK. Email: as25@cam.ac.uk

Tetrapyrroles: Birth, Life and Death, edited by Martin J. Warren and Alison G. Smith.
©2009 Landes Bioscience and Springer Science+Business Media.

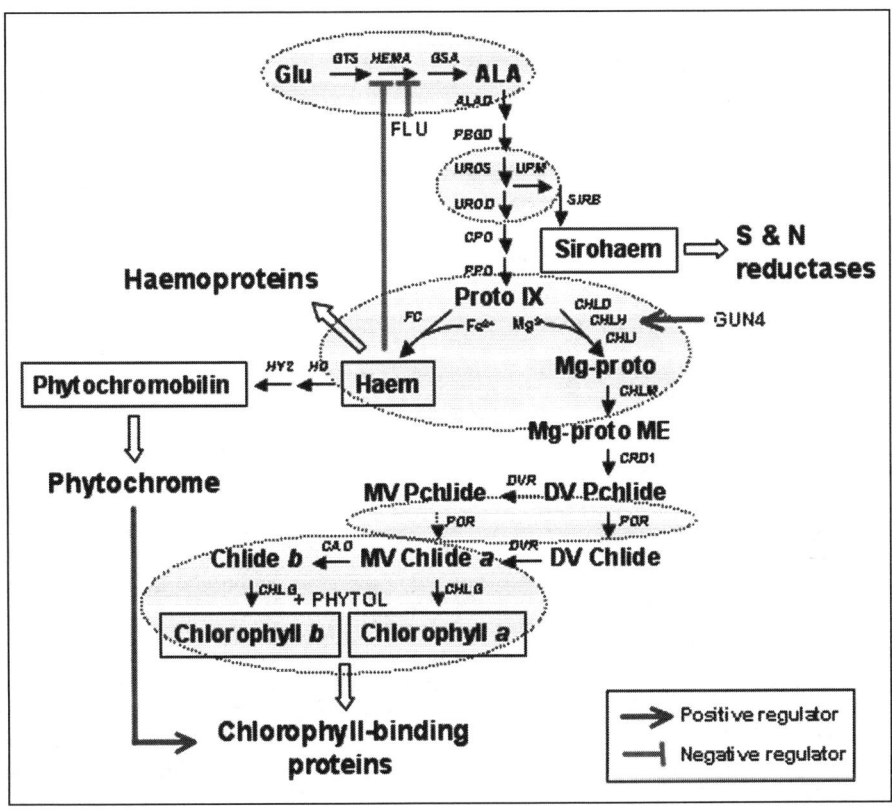

Figure 1. The tetrapyrrole synthesis pathway in higher plants, showing the four endproducts (boxed), chlorophyll, haem, phytochromobilin, the chromophore of the red-far red photoreceptors of higher plants, and sirohaem, one of the cofactors for sulphite and nitrite reductases (Si and Ni reductases). The key enzymes are included, together with ovals representing the five different major regulatory regions of the pathway: (i) synthesis of the initial precursor 5-aminolaevulinic acid (ALA); (ii) the production of sirohaem versus protoporphyrin IX; (iii) the branchpoint between haem and chlorophyll synthesis; (iv) and (v) regulation of the chlorophyll branch. The latter two are dealt with in detail in Chapter 16.

Haem is proposed to regulate ALA production via feedback inhibition of GluTR (glutamyl tRNA reductase). FLU also acts to inhibit GluTR in a haem-independent fashion. GUN4 activates Mg-chelatase (MgCh). Of the other *genomes uncoupled* mutants GUN5 corresponds to CHLH, one of the three MgCh subunits, and GUN2 and GUN3 are haem oxygenase (HO) and phytochromobilin synthase (PΦS) respectively (see text for details).

Other abbreviations: CPO, coproporphyrinogen III oxidase; CS, chlorophyll synthase; FC, (protoporphyrin IX) ferrochelatase; PChlide, protochlorophyllide; POR, protochlorophyllide oxidoreductase; PPO, protoporphyrinogen IX oxidase; SirB, sirohydrochlorin ferrochelatase; UPM, urogen III methyltransferase; UroD, urogen III decarboxylase; Urogen III, uroporphyrinogen III.

coordination with apoprotein synthesis. In this review we outline the reasons for the need to regulate the pathway so carefully, and describe recent advances in our understanding of the major control points at the synthesis of the initial precursor, 5-aminolaevulinate (ALA) and the branchpoints to sirohaem and between chlorophyll and haem. The regulation of the later stages of the chlorophyll pathway is described in Chapter 16.

Regulation of the Plant Tetrapyrrole Pathway—At the Heart of Plant Metabolism?

The essential role played by tetrapyrroles in plant metabolism—particularly chlorophyll—means that it is important for the plant to regulate their production. There are three main reasons for this regulation:

Relative Levels of Tetrapyrroles

A major consideration for the tight regulation of the tetrapyrrole biosynthesis pathway is that the relative levels of the end products are quite different, and that they vary according to tissue type and developmental stage. For instance mature tobacco leaves contain an estimated 1150 nmol/g fresh weight (g FW) chlorophyll, compared with 40 nmol/g FW haem,[5,6] whereas in nonphotosynthetic cells there is a complete absence of chlorophyll. Even within a single tissue, levels of tetrapyrroles can change dramatically. During the initial stages of greening of etiolated barley seedlings, chlorophyll levels increase from almost zero to 500 nmol/g FW in the first 12 h, whereas haem levels remain constant at 5 nmol/g FW.[7] In contrast, during root nodulation of legumes haem levels rise dramatically to 200 nmol/g FW to provide the cofactor for leghaemoglobin.[8]

Assembly with Apoproteins

All tetrapyrroles are assembled onto apoproteins, although the details of the assembly are still very poorly understood. Many of the tetrapyrrole intermediates, and the endproducts themselves, are readily excited by light and if allowed to accumulate, would lead to the formation of highly toxic radicals and reactive oxygen species.[3] It would thus be analogous to the situation in some porphyrias of humans and animals (see Chapter 5). Accordingly, the synthesis of tetrapyrroles is coordinated with expression of cognate apoproteins, to avoid the accumulation of excess intermediates. Genes for the light-harvesting chlorophyll *a/b* binding proteins (*Lhc*) and glutamyl-tRNA reductase, the rate-limiting enzyme of ALA and chlorophyll synthesis (see below), are both upregulated in the light by the phytochrome (responding to red/far-red light) and cryptochrome (blue-light) families of photoreceptors[9,10] (see also Chapter 13). Such coordination is also necessary because many apoproteins, including the chlorophyll *a/b* binding proteins, are unstable in the absence of their bound tetrapyrroles.[11] Induction of tetrapyrrole synthesis is therefore essential for the efficient assembly of the photosynthetic apparatus.

Tetrapyrroles as Signaling Molecules

There is now evidence that as well as a direct effect of phytochromobilin (the phytochrome chromophore) are the tetrapyrrole pathway itself can influence nuclear gene expression, following the characterisation of signaling pathways that indicate the physiological status of the plastid.[12-14] Mutant plants defective in this signaling pathway are called *genomes uncoupled* or *gun* mutants, and retain expression of *Lhcb* (encoding the major light-harvesting chlorophyll proteins) following transfer to light after treatment with the photobleaching herbicide norflurazon; in wild-type plants *Lhcb* expression is essentially abolished by this treatment. One group of *gun* mutants is in genes involved in tetrapyrrole synthesis: *GUN5* is *ChlH*, encoding one of the subunits of Mg-chelatase, *GUN2* is haem oxygenase, and *GUN3* is phytochromobilin synthase.[12] A fourth mutant from *Arabidopsis*, *gun4*, lacks a protein required for regulation of Mg-chelatase activity.[15] Similarly, a mutant of porphobilinogen deaminase and the *lin2* mutant, defective in coproporphyrinogen oxidase (CPO), also express *Lhcb* in the light in the presence of norflurazon,[14] as do lines overexpressing protochlorophyllide reductase (POR).[16] These observations strongly suggest that perturbation of flux through the tetrapyrrole pathway influences nuclear gene expression, and one model proposes that an intermediate of the pathway may act as the signal. Indeed, analysis of wild-type *Arabidopsis* plants treated with norflurazon reported increased levels of Mg-protoporphyrin IX relative to the untreated controls.[14] In the *gun* mutants, Mg-protoporphyrin accumulation was much less, as would be predicted from the site of the lesions (Fig. 1). Earlier work correlated Mg-protoporphyrin IX with *Lhcb* expression,[13,17] and in the green alga *Chlamydomonas reinhardtii*, nuclear gene expression was altered by feeding of cells with Mg-protoporphyrin IX, but not protoporphyrin IX.[18] However, there were discrepancies with this model, most notably the fact that mutants of another subunit of Mg-chelatase, CHLI, which would be predicted to have the same effect on the ability of the plant to accumulate Mg-protoporphyrin IX, do not exhibit a *gun* phenotype.[12] In a detailed mutant study, Mochizuki et al (2008)[18a] found no correlation between levels of chlorophyll synthesis intermediates and expression of *Lhcb*. At the same time, using a novel LC-MS method that can identify these intermediates unambiguously, Moulin et al (2008)[18b] showed that norflurazon severely represses the production of the chlorophyll branch. Nevertheless, whatever the nature of the signal, it is clear that careful regulation of flux through the tetrapyrrole pathway is essential to ensure correct retrograde signaling from the plastid to the nucleus.

Turning on the Tap—Regulation of the Synthesis of the Initial Precursor, ALA

Early Work

As in all other organisms studied to date, ALA synthesis is the major control point in the plant tetrapyrrole pathway (see Fig. 2), and determines the total flux through the pathway. Early work established ALA production as the rate-limiting step for chlorophyll synthesis during greening of etiolated seedlings (reviewed in refs. 19,20), and provided evidence that this involves feedback-inhibition by haem (Fig. 2A). This led to the proposal that in the dark, where there is no activity of the light-dependent protochlorophyllide reductase (Chapter 14), the block in the chlorophyll branch allows haem to accumulate, which then feedback inhibits ALA production. In the light where there is rapid chlorophyll accumulation, protohaem accumulates to a much lesser degree, thus releasing the block on the enzymes that synthesise ALA.[3] In support of this model, haem has been shown in many systems, from cyanobacteria through green algae to higher plants, to inhibit the synthesis of ALA from glutamate in vitro (e.g., refs. 21,22, reviewed in refs. 19,20), whereas other intermediates including Mg-protoporphyrin or protochlorophyllide had much less effect.[22] In addition, artificial depletion of the haem pool in intact plastids using apoperoxidase resulted in a 32% stimulation of ALA synthesis *in organello*.[23]

Evidence from Mutants

The role of haem as a key internal modulator of the pathway is further supported by the phenotype of the *aurea* and *yellow-green-2* mutants of tomato that are defective in phytochromobilin synthase and haem oxygenase activities, respectively, such that haem breakdown is likely to be reduced. These plants are pale with reduced chlorophyll, even though there is no block in the pathway of chlorophyll synthesis.[24] Dark-grown mutant seedlings have reduced amounts of protochlorophyllide resulting from an inhibition of ALA synthesis[25] and accumulate much less Mg-protoporphyrin after 2,2'-dipyridyl treatment than do wild-type plants treated with the same inhibitor.[24] This demonstrates that the mutants are deficient in their ability to degrade a haem pool that is inhibitory to ALA synthesis. Although phytochrome itself has a key role in the regulation of tetrapyrrole synthesis genes, in adult mutant plants phytochrome responses appear normal and yet the plants still display the reduced-chlorophyll phenotype.[24]

Similar conclusions can be drawn from study of other plants with altered levels of tetrapyrrole enzymes. Spontaneous lesion-forming mutants of maize (*les22*) and Arabidopsis (*lin2*) are defective in uroporphyrinogen decarboxylase (UroD) and CPO respectively.[26,27] The necrotic lesions result from the accumulation of the enzyme substrate. As these plants also have impaired haem synthesis, there is less ability to regulate ALA synthesis, leading to uncontrolled flux through the pathway that contributes further to the accumulation of the phototoxic tetrapyrrole intermediates. This is also observed in tobacco plants expressing antisense constructs for UroD,[28] CPO,[29] or ferrochelatase,[30] but no such accumulation is seen in tobacco plants antisense for any of the three subunits of Mg-chelatase, although they have less chlorophyll.[5,6] A plausible explanation for this is that in the antisense Mg-chelatase plants, the regulatory haem branch of the pathway remains intact, so that ferrochelatase could effectively 'mop up' any excess protoporphyrin, resulting in increased haem, and a new steady-state of ALA synthesis activity.[3] This model also explains the situation in different tissues of wild-type plants. In leaves of light-grown plants there is a high requirement for chlorophyll, so competition for protoporphyrin IX by the Mg-branch would reduce haem synthesis resulting in a transient decrease in a small, regulatory haem pool. When tetrapyrrole synthesis is in excess of requirements, protoporphyrin IX would rise, resulting in greater, transient production of haem, and hence a repression of ALA synthesis. This is effectively the situation in dark-grown seedlings in which protochlorophyllide reduction is prevented. Following a light flash there is an immediate increase in ALA synthesis activity [see ref. 19 for review]. This can be explained by the instant demand to replenish protochlorophyllide pools and thus the rapid removal of the inhibitory haem pool. Similarly, in nonphotosynthetic cells, this model can explain changes in flux, for example in root nodules, in which there is an increased requirement for haem for leghaemoglobin entirely independent of chlorophyll synthesis.[8] Clearly the absence of Mg-chelatase in these cells means that this branch cannot be responsible for the necessary increase in ALA synthesis.

ALA Synthesizing Enzymes

The site of down-regulation by haem appears to be at glutamyl-tRNA reductase (GluTR; Fig. 1; Fig. 2A). This is the first step in the pathway that is unique to tetrapyrrole synthesis because

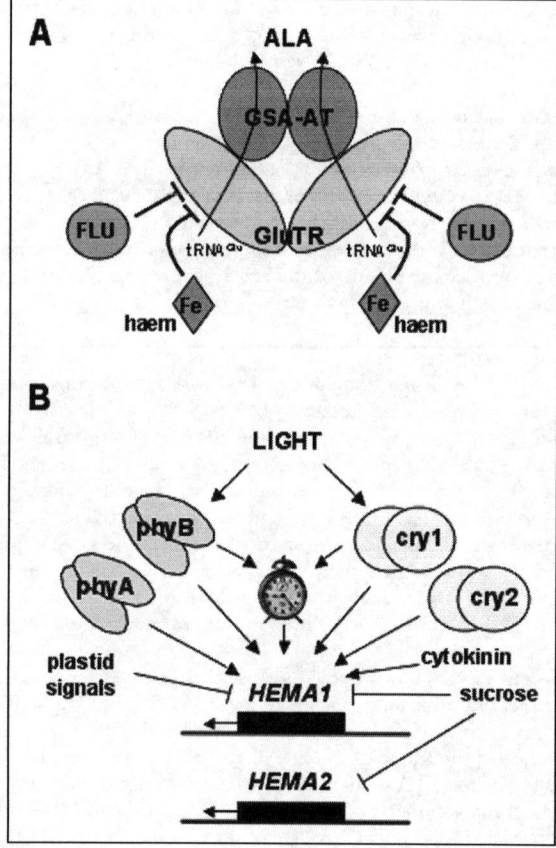

Figure 2. Regulation of ALA synthesis. A) Schematic showing a dimer of GluTR interacting with glutamate semialdehyde aminotransferase (GSA-AT) as proposed by Moser et al.[67] GluTR is directly inhibited by haem and the regulatory protein FLU. B) Two *HEMA* genes encode GluTR in *Arabidopsis*. *HEMA1* is regulated by multiple signals, including light via the phytochrome and cryptochrome families of photoreceptors. The light signals are closely associated with inputs from both the circadian clock and plastid signaling. In contrast, *HEMA2* shows little transcriptional regulation but, like *HEMA1*, is repressed by sucrose. See text for further details. Abbreviations for enzymes as in Figure 1.

glutamyl-tRNA synthetase (GluTS) also provides substrate for chloroplast protein synthesis. Exogenous haem can inhibit recombinant GluTR,[31] a process that involves the N-terminal 30 amino acids of the enzyme, since deletion of this region abolishes the response.[32] The idea that GluTR is the central controller of tetrapyrrole flux is supported by the observation that expression of the *HEMA1* gene, encoding one isoform of GluTR, is modulated by a wide range of regulatory signals (Fig. 2B).[33] Plant species commonly contain at least two *HEMA* genes. In *Arabidopsis*, *HEMA2* is expressed predominantly in roots and flowers and is not regulated by light or plastid signals.[34] This has led to the hypothesis that the primary role of *HEMA2* is to provide tetrapyrroles constitutively for normal cellular processes. In contrast *HEMA1* although expressed in dark-grown seedlings is induced up to 30-fold in the light,[33] suggesting a role more tailored to providing tetrapyrroles (both chlorophyll and haem) for biogenesis of the photosynthetic apparatus. Experiments using narrow waveband light sources showed that *HEMA1* expression was under the control of both phytochrome and cryptochrome photoreceptors with the proximal 451 bp of the promoter sufficient for full light responsiveness.[33] Further studies using photoreceptor-deficient *Arabidopsis* mutants have established that phytochrome A, phytochrome B, a third as yet unidentified phytochrome and both cryptochromes 1 and 2 all play a role in light regulation.[10] This study also identified a number of components, such as FHY1, FHY3, HY5 and COP1, involved in the phytochrome signaling pathways. More recently studies on a phytochrome-interacting protein PIF1 indicate that this may also play a key regulatory role in regulating ALA synthesis.[35] Although the effect of light is the best characterized *HEMA* regulatory response it is just one of a number of factors that regulate expression. *HEMA* expression has been shown to be regulated by the circadian clock in both barley[36] and tobacco.[37] This is consistent with the strong role for photoreceptors in entraining the clock.[38] *HEMA* genes have also been shown to be strongly regulated by plastid signals,[33] hormones, such as cytokinin,[16,39] and sugars.[34] The direct influence of these signals on GluTR gene expression therefore enables an integrated regulation of flux through the tetrapyrrole pathway, in which internal control mechanisms can modulate external inputs. By contrast, *GSA* genes encoding glutamate-1-semialdehyde aminotransferase (GSA-AT), the third enzyme of ALA synthesis, are rather weakly responsive to these signals.[10] Indeed, GSA-AT has a similar moderate induction under white light seen for the majority of genes encoding tetrapyrrole enzymes,[40] and it has been hypothesized that this may be the result of increased chloroplast development and division under these conditions resulting in increased plastid signaling input.[41]

Feedback from the Mg-Branch

Recently, a more prominent role has been proposed for the Mg-branch in regulating flux through the tetrapyrrole pathway. This proposal has come from two lines of evidence. Firstly, transgenic tobacco plants containing antisense constructs for CHLI and CHLH have both reduced Mg-chelatase activity and a concomitant reduction is ALA synthesis activity.[5,6] Where measured, the decrease in ALA synthesis correlated with a reduction in *HEMA* expression.[6] Since there was no significant increase in total extractable haem, which was actually reduced in some lines, it was concluded that flux through the Mg-branch of the pathway was the key determinant of ALA synthesis activity. The signal for down-regulation might be the enzyme itself, or an intermediate.[2,6] The second line of evidence comes from the isolation of the *flu* mutant of Arabidopsis.[42] FLU is a nuclear-encoded protein that is localized to chloroplast membranes and functions as a negative regulator of chlorophyll synthesis. Consistent with the key role of GluTR in integrating regulatory signals, FLU interacts with GluTR (but not GSA-AT) in yeast two-hybrid assays.[43] In etiolated *flu* seedlings, there is an over-accumulation of Pchlide, but no accumulation of 'free' haem and it was therefore proposed that FLU operates independently of haem.[42] This is further supported by the observation that FLU interacts with GluTR at a different position than that attributed to haem inhibition and that a haem oxygenase-deficient mutant *hy1* partly rescues the *flu* phenotype.[44] However, while it is clear that FLU has a crucial role to play in regulating the tetrapyrrole pathway independently of haem, there is no current evidence that it functions in response to the levels of Mg-porphyrins.

The Mg-porphyrins themselves are likely to have damaging photosensitizing effects on the chloroplast and mechanisms must be employed to prevent this. Such a pathway would not be specific to tetrapyrrole synthesis because genes for many chloroplast proteins are responsive to plastid signalling.

Nevertheless *HEMA1* expression is highly regulated by plastid signals[33] and thus changes in *HEMA1* expression could be modulated effectively through this pathway. Similarly, ferrochelatase-2 genes are responsive to light[45] and plastid signaling.[46] Interestingly, there are second genes for both ferrochelatase and GluTR, neither of which are responsive to plastid signals.[45,46,34] This would therefore provide an effective mechanism for regulating GluTR and ferrochelatase activity dedicated to the synthesis of the photosynthetic apparatus while allowing constitutive activity to remain independent of such control.

Decision Time at the Branchpoints

Sirohaem Branch

The first branchpoint in the pathway is between sirohaem and the other major endproducts (Fig. 3A), where uroporphyrinogen III methyltransferase must compete with UroD for the substrate uroporphyrinogen III.[47] How the regulation is achieved is completely unknown, either in higher plants, or indeed in other organisms such as yeast and bacteria, which also make sirohaem. The last enzyme of this branch, sirohydrochlorin ferrochelatase, catalyses the insertion of ferrous iron into sirohydrochlorin to make sirohaem. It has recently been identified in higher plants by sequence similarity to the *Bacillus megaterium* enzyme SirB.[48] Characterisation of the *Arabidopsis* homologue, AtSirB, revealed two distinct differences to the enzyme from bacteria. Firstly, it is only half the size of the bacterial enzyme, and in fact aligns with both the N- and C-terminal halves of *B. megaterium* SirB, demonstrating that the latter is likely to have arisen by a gene duplication and fusion event of an ancestral protein. A similar situation is found for CbiX, a sirohydrochlorin cobaltochelatase involved in vitamin B_{12} biosynthesis (a tetrapyrrole not made by higher plants), where there are two classes of enzyme, one of which (CbiX^L) appears to be a duplication of the other (CbiX^S).[49] By analogy, AtSirB is therefore referred to as AtSirB^S.

The second difference between the plant and bacterial enzymes is that AtSirB^S has an 2Fe.2S centre of unknown function at the C-terminus, a feature it shares in common with ferrochelatase (the terminal enzyme of haem synthesis) from animals and several microbes[50,51] but which, intriguingly is absent from plant protoporphyrin IX ferrochelatases (Fig. 3B). It is thus tempting to speculate that the Fe.S cluster is involved in the competition for Fe^{2+} between the two ferrochelatases within the plastid (Fig. 3A). There is currently no experimental evidence to indicate the role of this moiety, either in AtSirB or indeed animal protoporphyrin ferrochelatases, although its disruption inactivates the human ferrochelatase (see Chapter 4).[50]

Chlorophyll versus Haem Branch

Much more is known about the regulation of the major branchpoint in plant tetrapyrrole synthesis, namely between haem and chlorophyll. In this case two chelatases with different metal ion-specificities compete for the same porphyrin substrate, protoporphyrin IX. Given the great variation in flux through the chlorophyll and haem branches during development and in different cell types, it is likely that the chelatase enzymes play important regulatory roles. One of the clues to what the regulatory mechanisms might be comes from the structure and properties of the two chelatases. Ferrochelatase is a monomer or homodimer, with no cofactor requirement, and has structural (although little sequence) similarity to the cobatochelatase, CbiX.[49] In contrast, Mg-chelatase comprises a heterotrimer of three different subunits, CHLI (~40 kDa), CHLD (~70 kDa) and CHLH (~140 ka), and requires ATP for activity (see Chapter 14). It is analogous to another form of cobaltochelatase found in the aerobic pathway of vitamin B_{12} biosynthesis (Chapter 18), which comprises three subunits, CobS, CobT and CobN, although only CobN shares any significant sequence similarity with a Mg-chelatase subunit (CHLH). Nonetheless, the resemblance between the two classes of chelatase has led to the proposal that the cobaltochelatases represent the ancestral origin of these enzymes,[52] and the marked differences between Mg-chelatase and ferrochelatase could then be exploited to facilitate branchpoint regulation.

Mg-chelatase has a much lower K_m for protoporphyrin IX than ferrochelatase, suggesting that Mg-chelatase would compete effectively with ferrochelatase for this substrate. Furthermore, Mg-chelatase activity requires ATP whereas ATP inhibits ferrochelatase activity,[53] and addition of ATP to isolated chloroplasts results in increased Mg-chelation and decreased haem synthesis.[54] ATP also activates the heterotrimeric Mg-chelatase, by facilitating the formation of a complex between

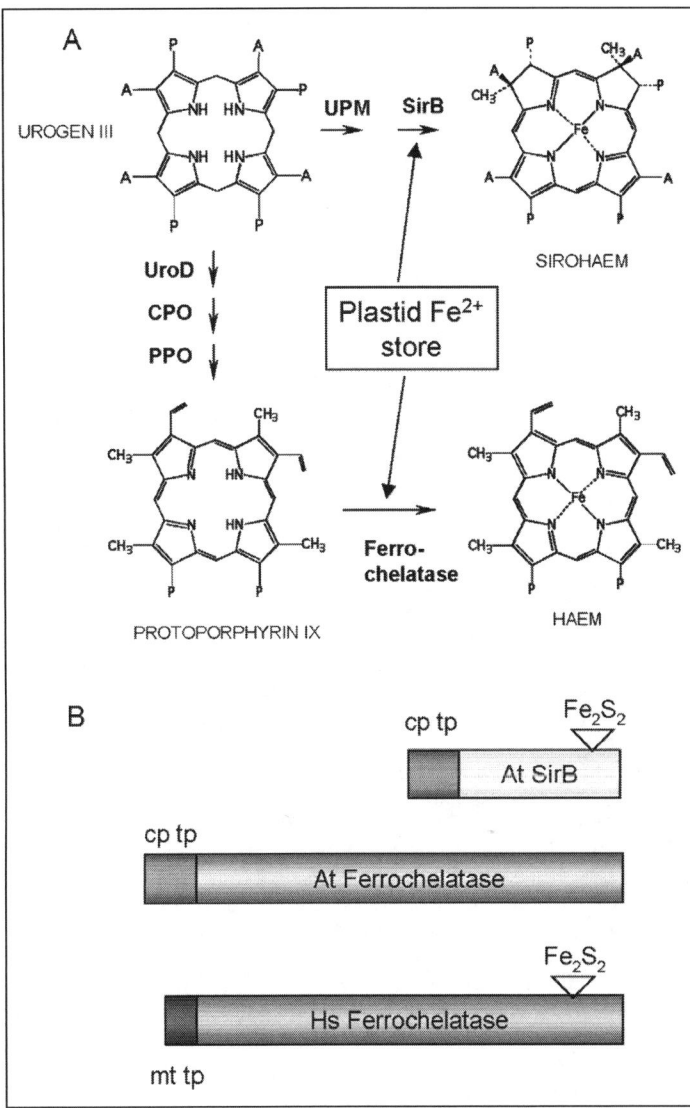

Figure 3. The branchpoint to sirohaem. A) Sirohaem is synthesised from the tetrapyrrole primogenitor urogen III by methylation (catalysed by UPM), followed by oxidation and then insertion of a ferrous iron by sirohydrochlorin ferrochelatase (SirB). Alternatively, urogen III is oxidatively decarboxylated by three enzymes to form protoporphyrin IX, then insertion of a ferrous iron by (protoporphyrin IX) ferrochelatase makes haem. A = acetate group; P = propionate group Abbreviations for enzymes as in Figure 1. B) Schematic of ferrochelatase precursor proteins. *Arabidopsis* SirB (AtSirB) has a C-terminal Fe_2S_2 cluster, as does human (Hs) ferrochelatase, whereas *Arabidopsis* ferrochelatase does not. Both *Arabidopsis* enzymes are synthesised as precursor proteins with N-terminal chloroplast transit peptides (cp tp), while the Hs ferrochelatase has a mitochondrial targeting sequence (mt tp).

CHLI and CHLD.[55,56] The third subunit, CHLH, which binds protoporphyrin IX, then associates with the other two subunits, and metal chelation can occur. Thus in the light, when ATP levels are high, the Mg-branch of the pathway would be favoured, and conversely, in the dark, Mg-chelation would be reduced. Another component that might play an important role in this process is the recently identified GUN4 protein.[15] This protein can bind both substrate and product of the Mg-chelatase reaction in vitro and copurifies with CHLH in a chloroplast protein complex. Purified recombinant GUN4 protein from the cyanobacterium *Synechocystis* PCC6803 was found to stimulate Mg-chelatase activity in a reconstituted reaction.[15] A *Synechocystis gun4* mutant in which the *GUN4* gene had been inactivated also had reduced Mg-chelatase activity, and lower chlorophyll content.[57] Perhaps more surprisingly ferrochelatase activity was also severely reduced in the *gun4* mutant, and there was an accumulation of protoporphyrin IX. The authors speculated that the protein may be involved in optimal distribution of the protoporphyrin IX substrate to the two chelatases, and/or stabilizing the enzymes.

Another mechanism for branchpoint regulation that has been proposed invokes substrate channeling such that different pools of protoporphyrin IX are available for each of the chelatases. Support for this comes from studies of tobacco plants in which tetrapyrroles enzymes have been inactivated by antisense technology (see earlier). In plants antisense for ferrochelatase-2, there is an uncontrolled accumulation of protoporphyrin IX,[30] but such an increase is not seen in plants with altered Mg-chelatase activity.[5,6] It therefore seems that a pool of protoporphyrin IX, once committed to the haem branch of the pathway, is no longer available as a substrate for Mg-chelatase.

Lastly, at least in mature tissues, control might be exerted through differential expression of the genes for the chelatases. In tobacco leaves grown in light/dark cycles, Mg-chelatase activity peaks at the beginning of the light phase and ferrochelatase activity is greatest at the end of the light phase,[37] indicating a diurnal rhythm. One basis for this pattern of activity appears to be the relative gene expression of the two chelatases. *HEMH2*, encoding ferrochelatase-2, gradually increases during the light phase whereas *CHLH* is highest at the beginning of this period[37] or at the end of the preceding dark period.[58] Interestingly, the expression levels of genes encoding other Mg-chelatase subunits did not change significantly under these conditions,[37] consistent with general observations that CHLH is the principal target for transcriptional regulation of Mg-chelatase.[58] Indeed, detailed analysis of gene expression profiles in *Arabidopsis* for all enzymes in the pathway has identified *CHLH* as one of only six genes that are critically responsive to light and circadian regulation;[40] the other genes being *HEMA1* (see earlier), *PORA, PORB* (which were both down-regulated by light) and two genes, *CRD1* and *CAO*, encoding enzymes of the chlorophyll branch of the pathway (discussed later). Another possibility for regulation at the level of gene transcription can be found in the expression patterns of the two ferrochelatase genes. Ferrochelatase-2 is characterised by the presence of the so-called generic LHC motif at the C-terminus of the protein,[45,59] found also in light-harvesting chlorophyll proteins.[60] In *Arabidopsis*, ferrochelatase-2 is expressed only in photosynthetic tissue[46] and, as discussed earlier, is severely repressed in the light on norflurazon. It is therefore likely to be involved in making haem for photosynthetic cytochromes.[46] Ferrochelatase-1, which does not contain the LHC motif, is ubiquitously expressed throughout the plant[46,59] and is implicated in synthesis of haem for respiratory cytochromes and haemoproteins involved in the defence response.[46] The marked upregulation of this gene in leaves by environmental stress such as wounding or viral infection might provide the means to channel tetrapyrrole intermediates away from chlorophyll and into haem production. The role of GUN4 should also be considered. Preliminary experiments have indicated that *GUN4* expression is strongly regulated by both light and plastid signaling (McCormac and Terry, unpublished results) thus providing an ideal mechanism for stimulating the chlorophyll branch of the pathway by activating a positive regulator of Mg-chelatase activity.

Regulation of the Chlorophyll Branch

A number of steps in the chlorophyll branch of the pathway may be important for regulation. The best characterised of these is the reduction of Pchlide to chlorophyllide (Chlide) by POR. As discussed in detail in Chapter 14, this enzyme requires light for activity and Pchlide accumulates in dark-grown seedlings. Once in the light, Pchlide is rapidly transformed to Chlide by POR, which is degraded resulting in major changes in plastid structure and initiating the transition from etioplasts and proplastids to chloroplasts (for a review of POR see ref. 61). As discussed earlier, photoactivation of POR results in a rapid increase in ALA synthesis. The mechanism for this is not known but it is likely that it results from reduced levels of free haem once the restriction of the flux through the Mg-branch of the pathway is lifted in the light by a rapid increase in chlorophyll synthesis which competes effectively with ferrochelatase. In *Arabidopsis*, POR is encoded by three genes *PORA-C*.[61] Both *PORA* and *PORB* are expressed in the dark, but are rapidly downregulated in the light, although *PORB* expression is retained in light-grown seedlings.[40,61] In contrast *PORC* is only expressed in light-grown tissues,[61] but is not critically responsive to light signals, being induced in parallel with most tetrapyrrole enzymes, about 2-3-fold.[40] However, it is not at all clear that such a complex regulatory pattern of *POR* expression is common to all plants and a number of species, including pea and cucumber appear to have a single light-induced *POR* gene.[61]

The comprehensive analysis of expression profiles for genes encoding the complete tetrapyrrole pathway in *Arabidopsis* have revealed that both *CRD1* encoding a subunit of Mg-protoporphyrin IX monomethylester cyclase and *CAO* encoding chlorophyll *a* oxygenase are both highly regulated, showing strong induction in light and circadian regulation.[40] That CAO levels are highly regulated is perhaps unsurprising given that this enzyme is the first committed step in chlorophyll *b* synthesis. Indeed, detailed analyses of CAO expression at the transcript and protein levels showed a strong correlation to chlorophyll *a/b* ratio under a variety of circumstances indicating a role in regulating the antenna size of light-harvesting complexes in during light acclimation responses.[62] The fact that *CRD1* expression is also highly regulated suggests that this step is also important for overall regulation of the pathway. An explanation for this is not currently apparent, but it is possible that regulation of this step impacts on the regulation of nuclear gene expression.

Conclusion

Clearly regulation of the tetrapyrrole pathway in higher plants takes place at many levels. There is internal feedback by haem and FLU, and extensive transcriptional regulation of some key genes encoding enzymes in the pathway, which acts as a means to integrate external signals with the internal regulatory mechanism. Expression of chloroplast-encoded genes is also intimately related to the tetrapyrrole pathway since Glu-tRNA is a substrate both for ALA synthesis and for plastid protein synthesis. There is increasing evidence for post-translational regulation too. Tobacco plants overexpressing rat biliverdin reductase that was targeted to the chloroplast exhibited a photobleached phenotype and poor correlation between gene expression and protein or enzyme activity levels.[63] Most notable was the turnover of CHLI in these plants and it is interesting that CHLI has been identified as a target of thioredoxin, which has been implicated in chloroplast protein degradation in response to light stress.[64] A recent study in barley of the chlorophyll-deficient *albostrians* mutant observed similar inconsistencies,[65] as did a comprehensive comparison of gene expression levels with the chloroplast proteome.[66] A regulatory pathway controlling the stability of key tetrapyrrole enzymes might therefore account for some of the anomalies in the current literature and is a major area for future work. Another key question for the future concerns the spatial relationships between the key pathway enzymes and regulators. In particular, branchpoint regulation is likely to be strongly affected by the spatial organization of enzymes and intermediates within the plastid, and by transport of intermediates and endproducts across membranes. Finally there is the association of the tetrapyrrole cofactors with their cognate apoproteins, an area of research in its infancy, with an adequate understanding only of haem binding to c-type cytochromes (Chapter 9).

Acknowledgements

Research funding in our laboratories has been received from BBSRC, the Royal Society, the Leverhulme Trust EU FP5, and the Gatsby Charitable Foundation.

References

1. Beale SI. Enzymes of chlorophyll biosynthesis. Photosynth Res 1999; 60:43-73.
2. Papenbrock J, Grimm B. Regulatory network of tetrapyrrole biosynthesis—Studies of intracellular signalling involved in metabolic and developmental control of plastids. Planta 2001; 213:667-681.
3. Cornah JE, Terry MJ, Smith AG. Green or red: What stops the traffic in the tetrapyrrole pathway? Trends Plant Sci 2003; 8:224-230.
4. Smith AG, Cornah JE, Roper JM et al. Compartmentation of tetrapyrrole metabolism in higher plants. In: Bryant JA, Burrell MM, Kruger NJ, eds. Plant Carbohydrate Metabolism. Oxford: BIOS Scientific Publishers, 1999:281-294.

5. Papenbrock J, Pfundel E, Mock HP et al. Decreased and increased expression of the subunit CHL I diminishes Mg chelatase activity and reduces chlorophyll synthesis in transgenic tobacco plants. Plant J 2000; 22:155-164.

6. Papenbrock J, Mock HP, Tanaka R et al. Role of magnesium chelatase activity in the early steps of the tetrapyrrole biosynthetic pathway. Plant Physiol 2000; 122:1161-1169.

7. Castelfranco PA, Jones OTG. Protoheme turnover and chlorophyll synthesis in greening barley tissue. Plant Physiol 1975; 55:485-490.

8. Santana MA, Pihakaski-Maunsbach K, Sandal N et al. Evidence that the plant host synthesizes the heme moiety of leghemoglobin in root nodules. Plant Physiol 1998; 116:1259-1269.

9. Hamazato F, Shinomura T, Hanzawa H et al. Fluence and wavelength requirements for Arabidopsis CAB gene induction by different phytochromes. Plant Physiol 1997; 115:1533-1540.

10. McCormac AC, Terry MJ. Light-signalling pathways leading to the coordinated expression of HEMA1 and Lhcb during chloroplast development in Arabidopsis thaliana. Plant J 2002; 32:549-559.

11. Murray DL, Kohorn BD. Chloroplasts of Arabidopsis thaliana homozygous for the ch-1 locus lack chlorophyll b, lack stable LHCPII and have stacked thylakoids. Plant Mol Biol 1991; 16:71-79.

12. Mochizuki N, Brusslan JA, Larkin R et al. Arabidopsis genomes uncoupled 5 (GUN5) mutant reveals the involvement of Mg-chelatase H subunit in plastid-to-nucleus signal transduction. Proc Natl Acad Sci USA 2001; 98:2053-2058.

13. La Rocca N, Rascio N, Oster U et al. Amitrole treatment of etiolated barley seedlings leads to deregulation of tetrapyrrole synthesis and to reduced expression of Lhc and RbcS genes. Planta 2001; 213:101-108.

14. Strand A, Asami T, Alonso J et al. Chloroplast to nucleus communication triggered by accumulation of Mg-protoporphyrin IX. Nature 2003; 421:79-83.

15. Larkin RM, Alonso JM, Ecker JR et al. GUN4, a regulator of chlorophyll synthesis and intracellular signaling. Science 2003; 299:902-906.

16. McCormac AC, Terry MJ. Loss of nuclear gene expression during the phytochrome A-mediated far-red block of greening response. Plant Physiol 2002; 130:402-414.

17. Oster U, HB, Rüdiger W. The greening process in cress seedlings. V. Possible interference of chlorophyll precursors accumulated after thujaplicin treatment with light-regulated expression of Lhc genes. J Photochem Photobiol 1996; 36:255-261.

18. Kropat J, Oster U, Rudiger W et al. Chlorophyll precursors are signals of chloroplast origin involved in light induction of nuclear heat-shock genes. Proc Natl Acad Sci USA 1997; 94:14168-14172.

18a. Mochizuki N, Tanaka R, Tanaka A et al. The steady state level of Mg-protoporphyrin IX is not a determinant of plastid-to-nucleus signaling in Arabidopsis. Proc Natl Acad Sci USA, 2008; in press.

18b. Moulin M, McCormac AC, Terry MJ, Smith AG. Tetrapyrrole profiling in Arabidopsis seedlings reveals that retrograde plastid nuclear signalling is not due to Mg-protoporphyrin IX accumulation. Proc Natl Acad Sci USA 2008; in press.

19. Beale SI, Weinstein JD. Biochemistry and regulation of photosynthetic pigment formation in plants and algae. In: Jordan PM, ed. Biosynthesis of Tetrapyrroles. Amsterdam: Elsevier, 1991:155-235.

20. Gough SP, Westergren T, Hansson M. Chlorophyll biosynthesis in higher plants. Regulatory aspects of 5-aminolevulinate formation. J Plant Biol 2003; 46:135-160.

21. Castelfranco PA, Zeng XH. Regulation of 5-aminolevulinic acid synthesis in developing chloroplasts.4. An endogenous inhibitor from the thylakoid membranes. Plant Physiol 1991; 97:1-6.

22. Weinstein JD, Beale SI. Enzymatic conversion of glutamate to delta-aminolevulinate in soluble extracts of the unicellular green-alga, Chlorella-vulgaris. Arch Biochem Biophys 1985; 237:454-464.

23. Thomas J, Weinstein JD. Free heme in isolated chloroplasts—An improved method of assay and its physiological importance. Plant Physiol Biochem 1992; 30:285-292.

24. Terry MJ, Kendrick RE. Feedback inhibition of chlorophyll synthesis in the phytochrome chromophore-deficient aurea and yellow-green-2 mutants of tomato. Plant Physiol 1999; 119:143-152.

25. Ryberg M, Terry MJ. Analysis of protochlorophyllide reaccumulation in the phytochrome chromophore-deficient aurea and yg-2 mutants of tomato by in vivo fluorescence spectroscopy. Photosynth Res 2002; 74:195-203.

26. Hu GS, Yalpani N, Briggs SP et al. A porphyrin pathway impairment is responsible for the phenotype of a dominant disease lesion mimic mutant of maize. Plant Cell 1998; 10:1095-1105.

27. Ishikawa A, Okamoto H, Iwasaki Y et al. A deficiency of coproporphyrinogen III oxidase causes lesion formation in Arabidopsis. Plant J 2001; 27:89-99.

28. Mock HP, Grimm B. Reduction of uroporphyrinogen decarboxylase by antisense RNA expression affects activities of other enzymes involved in tetrapyrrole biosynthesis and leads to light-dependent necrosis. Plant Physiol 1997; 113:1101-1112.

29. Kruse E, Mock HP, Grimm B. Reduction of coproporphyrinogen oxidase level by antisense RNA-synthesis leads to deregulated gene-expression of plastid proteins and affects the oxidative defense system. EMBO J 1995; 14:3712-3720.
30. Papenbrock J, Mishra S, Mock HP et al. Impaired expression of the plastidic ferrochelatase by antisense RNA synthesis leads to a necrotic phenotype of transformed tobacco plants. Plant J 2001; 28:41-50.
31. Pontoppidan B, Kannangara CG. Purification and partial characterization of barley glutamyl-tRNA(Glu) reductase, the enzyme that directs glutamate to chlorophyll biosynthesis. Eur J Biochem 1994; 225:529-537.
32. Vothknecht UC, Kannangara CG, von Wettstein D. Barley glutamyl tRNA(Glu) reductase: Mutations affecting haem inhibition and enzyme activity. Phytochem 1998; 47:513-519.
33. McCormac AC, AF, Kumar AM et al. Regulation of HEMA1 expression by phytochrome and a plastid signal during de-etiolation in Arabidopsis thaliana. Plant J 2001; 25:549-561.
34. Ujwal ML, McCormac AC, Goulding A et al. Divergent regulation of the HEMA gene family encoding glutamyl-tRNA reductase in Arabidopsis thaliana: Expression of HEMA2 is regulated by sugars, but is independent of light and plastid signalling. Plant Mol Biol 2002; 50:81-89.
35. Huq E, Al-Sady B, Hudson M et al. Phytochrome-interacting factor 1 is a critical bHLH regulator of chlorophyll biosynthesis. Science 2004; 305:1937-1941.
36. Kruse E, Grimm B, Beator J et al. Developmental and circadian control of the capacity for delta-aminolevulinic acid synthesis in green barley. Planta 1997; 202:235-241.
37. Papenbrock J, Mock HP, Kruse E et al. Expression studies in tetrapyrrole biosynthesis: Inverse maxima of magnesium chelatase and ferrochelatase activity during cyclic photoperiods. Planta 1999; 208:264-273.
38. Somers DE, Devlin PF, Kay SA. Phytochromes and cryptochromes in the entrainment of the Arabidopsis circadian clock. Science 1998; 282:1488-1490.
39. Masuda T, Ohta H, Shioi Y et al. Stimulation of glutamyl-tRNA reductase activity by benzyladenine in greening cucumber cotyledons. Plant Cell Physiol 1995; 36:1237-1243.
40. Matsumoto F, Obayashi T, Sasaki-Sekimoto Y et al. Gene expression profiling of the tetrapyrrole metabolic pathway in Arabidopsis with a mini-array system. Plant Physiol 2004; 135:2379-2391.
41. McCormac AC, Terry MJ. The nuclear genes Lhcb and HEMA1 are differentially sensitive to plastid signals and suggest distinct roles for the GUN1 and GUN5 plastid-signalling pathways during de-etiolation. Plant J 2004; 40:672-685.
42. Meskauskiene R, Nater M, Goslings D et al. FLU: A negative regulator of chlorophyll biosynthesis in Arabidopsis thaliana. Proc Natl Acad Sci USA 2001; 98:12826-12831.
43. Meskauskiene R, Apel K. Interaction of FLU, a negative regulator of tetrapyrrole biosynthesis, with the glutamyl-tRNA reductase requires the tetratricopeptide repeat domain of FLU. FEBS Lett 2002; 532:27-30.
44. Goslings D, Meskauskiene R, Kim C et al. Concurrent interactions of heme and FLU with Glu tRNA reductase (HEMA1), the target of metabolic feedback inhibition of tetrapyrrole biosynthesis, in dark- and light-grown Arabidopsis plants. Plant J 2004 2004; 40:957-967.
45. Suzuki T, Masuda T, Singh DP et al. Two types of ferrochelatase in photosynthetic and nonphotosynthetic tissues of cucumber—Their difference in phylogeny, gene expression, and localization. J Biol Chem 2002; 277:4731-4737.
46. Singh DP, Cornah JE, Hadingham S et al. Expression analysis of the two ferrochelatase genes in Arabidopsis in different tissues and under stress conditions reveals their different roles in haem biosynthesis. Plant Mol Biol 2002; 50:773-788.
47. Leustek T, Smith M, Murillo M et al. Siroheme biosynthesis in higher plants—Analysis of an S-adenosyl-L-methionine-dependent uroporphyrinogen III methyltransferase from Arabidopsis thaliana. J Biol Chem 1997; 272:2744-2752.
48. Raux-Deery E, Leech HK, Nakrieko KA et al. Identification and characterization of the terminal enzyme of siroheme biosynthesis from Arabidopsis thaliana—A plastid-located sirohydrochlorin ferrochelatase containing a 2Fe-2S center. J Biol Chem 2005; 280:4713-4721.
49. Brindley AA, Raux E, Leech HK et al. A story of chelatase evolution—Identification and characterization of a small 13-15-kda "ancestral" cobaltochelatase (CbiX(s)) in the archaea. J Biol Chem 2003; 278:22388-22395.
50. Wu CK, Dailey HA, Rose JP et al. The 2.0 A structure of human ferrochelatase, the terminal enzyme of heme biosynthesis. Nat Struct Biol 2001; 8:156-160.
51. Dailey TA, Dailey HA. Identification of [2Fe-2S] clusters in microbial ferrochelatases. J Bacteriol 2002; 184:2460-2464.
52. Schubert HL, Raux E, Wilson KS et al. Common chelatase design in the branched tetrapyrrole pathways of heme and anaerobic cobalamin synthesis. Biochemistry 1999; 38:10660-10669.

53. Cornah JE, Roper JM, Singh DP et al. Measurement of ferrochelatase activity using a novel assay suggests that plastids are the major site of haem biosynthesis in both photosynthetic and nonphotosynthetic cells of pea (Pisum sativum L.). Biochem J 2002; 362:423-432.
54. Walker CJ, Yu GH, Weinstein JD. Comparative study of heme and Mg-protoporphyrin (monomethyl ester) biosynthesis in isolated pea chloroplasts: Effects of ATP and metal ions. Plant Physiol Biochem 1997; 35:213-221.
55. Walker CJ, Willows RD. Mechanism and regulation of Mg-chelatase. Biochem J 1997; 327:321-333.
56. Jensen PE, Reid JD, Hunter CN. Modification of cysteine residues in the ChlI and ChlH subunits of magnesium chelatase results in enzyme inactivation. Biochem J 2000; 352:435-441.
57. Wilde A, Mikolajczyk S, Alawady A et al. The gun4 gene is essential for cyanobacterial porphyrin metabolism. FEBS Lett 2004; 571:119-123.
58. Gibson LC, Marrison JL, Leech RM et al. A putative Mg chelatase subunit from Arabidopsis thaliana cv C24. Sequence and transcript analysis of the gene, import of the protein into chloroplasts, and in situ localization of the transcript and protein. Plant Physiol 1996; 111:61-71.
59. Chow KS, Singh DP, Amanda RW et al. Two different genes encode ferrochelatase in Arabidopsis: Mapping, expression and subcellular targeting of the precursor proteins. Plant J 1998; 15:531-541.
60. Janssen S. A guide to the Lhc genes and their relatives in Arabidopsis. Trend Plant Sci 1999; 4:236-240.
61. Masuda T, Takamiya K. Novel insights into the enzymology, regulation and physiological functions of light-dependent protochlorophyllide oxidoreductase in angiosperms. Photosyn Res 2004; 81:1-29.
62. Harper AL, von Gesjen SE, Linford AS et al. Chlorophyllide a oxygenase mRNA and protein levels correlate with the chlorophyll a/b ratio in Arabidopsis thaliana. Photosyn Res 2004; 79:149-159.
63. Franklin KA, Linley PJ, Montgomery BL et al. Misregulation of tetrapyrrole biosynthesis in transgenic tobacco seedlings expressing mammalian biliverdin reductase. Plant J 2003; 35:717-728.
64. Balmer Y, Koller A, del Val G et al. Proteomics gives insight into the regulatory function of chloroplast thioredoxins. Proc Natl Acad Sci USA 2003; 100:370-375.
65. Yaronskaya E, Ziemann V, Walter G et al. Metabolic control of the tetrapyrrole biosynthetic pathway for porphyrin distribution in the barley mutant albostrians. Plant J 2003; 35:512-522.
66. Kleffmann T, Russenberger D, von Zychlinski A et al. The Arabidopsis thaliana chloroplast proteome reveals pathway abundance and novel protein functions. Curr Biol 2004; 14:354-362.
67. Moser J, Schubert WD, Beier V et al. V-shaped structure of glutamyl-tRNA reductase, the first enzyme of tRNA dependent tetrapyrrole biosynthesis. EMBO J 2001; 20:6583-6590.

Regulation of the Late Steps of Chlorophyll Biosynthesis

Wolfhart Rüdiger*

Abstract

The regulatory network that controls formation of the various components of the photosynthetic machinery becomes evident when late steps of chlorophyll biosynthesis are investigated by deregulation. A major regulatory point is revealed by a dark-to-light shift revealing the interplay between the light dependent reduction of protochlorophyllide a to chlorophyllide a with phase transitions of plastid membranes and stable accumulation of chlorophyll a-binding proteins. The second part deals with chlorophyll b formation, details of which are controversially disputed in the literature. This is connected with the formation of nuclear-encoded proteins of light-harvesting complexes; expression of their genes, in turn, responds to plastid signals one of which is a chlorophyll precursor. Finally, a hypothetical role for carotenoids to maintain a well-regulated tetrapyrrole pathway will be discussed.

Introduction

Chlorophyll (Chl) biosynthesis from simple precursors requires many steps; 16 enzymes of this pathway have been identified. For consideration of regulation, the pathway can be subdivided into five main parts (Fig. 1): (1) formation of 5-aminolaevulinate (ALA) from glutamate, (2) formation of protoporphyrin IX (Proto) from 8 molecules of ALA, (3) formation of protochlorophyllide a (Pchlide a) from Proto, (4) formation of Chl a from Pchlide a, and (5) formation of Chl b. A second branch from Proto ("iron branch") leads to haem and plant bilins, parallel to part 3 of the above scheme that belongs to the "magnesium branch". Up- and down-regulation of one branch influences the other, since they share a common precursor: a high substrate flux into one branch can partly deplete the other branch for substrate and vice versa. Therefore, it can be expected that regulation of Chl synthesis influences not only the flux within the magnesium branch but also the distribution of precursors between both the magnesium and the iron branch.

Chl has a dual function in photosynthesis: Chl is highly reactive in the reaction centres to mediate light-driven charge separation, and Chl is biochemically inert in the antennae to harvest light energy and transfer it to the reaction centres. The environment of specific proteins and cofactors such as carotenoids causes these diverse properties. All Chl molecules in green plants are embedded in specific proteins, and it will be hypothesized in this chapter that the same applies to most Chl precursors, too. That plants do not accumulate free Chl, is due to its chemical properties: a solution of pure Chl in an organic solvent bleaches within minutes or hours when exposed to sunlight. This photo-destruction is not restricted to Chl itself: the precursors with a porphyrin structure are photo-sensitizers, which in the light catalyse peroxidation of lipids and other macromolecules via production of singlet oxygen. This property is the reason for a strict coregulation of the tetrapyrrole pathway with formation of protecting proteins and cofactors.

*Wolfhart Rüdiger—Department Biologie I, Botanik, Universität München, Menzinger Str. 67, D-80638, München, Germany. Email: ruediger@lrz.uni-muenchen.de

Tetrapyrroles: Birth, Life and Death, edited by Martin J. Warren and Alison G. Smith.
©2009 Landes Bioscience and Springer Science+Business Media.

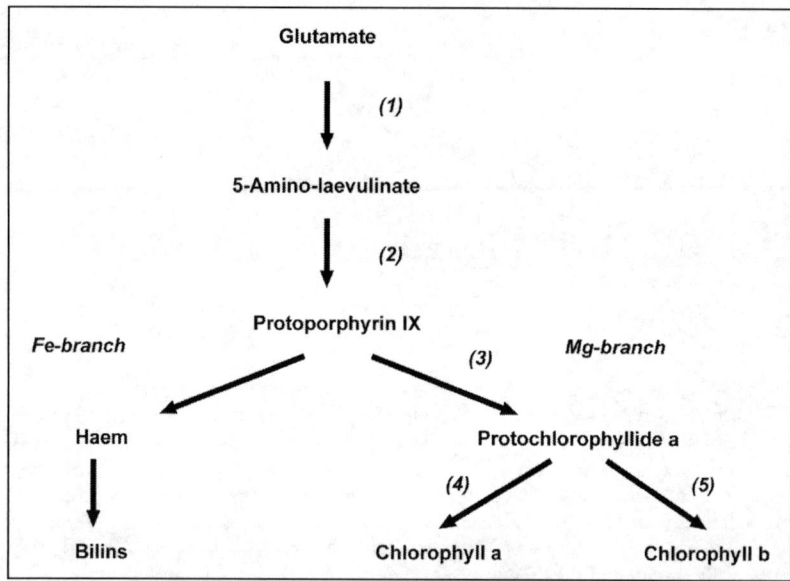

Figure 1. Scheme of chlorophyll biosynthesis. Parts 4 and 5 are discussed in this chapter.

Regulation of Chl biosynthesis follows the principle of other biosynthetic pathways: key steps that are rate-limiting for the overall metabolic flux are regulated at several levels and by several factors, starting from hormonal and light-regulated expression of the genes encoding the respective enzymes to feed-back control of the enzyme activity by end products or suitable intermediates. Such key enzymes catalyse the first specific step of the pathway or critical steps at branching points; these steps of Chl biosynthesis are dealt with in Chapters 14 and 15 of this book. This chapter will concentrate on parts 4 and 5 of Chl biosynthesis (see Fig. 1); emphasis will be on principles of regulation that are specific for Chl biosynthesis. Since Chl formation is embedded in a regulatory network that involves other compounds of the photosynthetic machinery, the significance of single regulatory steps can often be recognized by specific de-regulation. Further, the properties of single enzymes can sometimes give clues as to the mode of regulation. Both approaches will be described in this chapter.

Light Regulation via NADPH: Protochlorophyllide Oxidoreductase (POR)

Two different enzyme systems have evolved for reduction of Pchlide to Chlide, one operating in the dark and another one that catalyses a photoreaction. The "dark enzyme", consisting of 3 protein subunits, is present in phototrophic bacteria, cyanobacteria, algae, mosses, ferns, and gymnosperms; these organisms can form Chl in darkness. The "light enzyme", consisting of a single protein (POR), occurs in all organisms that carry out oxygenic photosynthesis, from cyanobacteria onward; in organisms containing both enzymes, POR can be assumed to play primarily the role of a light regulator. The most dramatic effect of light regulation can be seen in angiosperms that lack the "dark enzyme": when such seedlings are grown in absolute darkness, these "etiolated" plants do not synthesize any Chl but accumulate the precursor Pchlide a instead, bound together with the cofactor NADPH to the enzyme NADPH:protochlorophyllide oxidoreductase (POR, Fig. 2). Irradiation of this ternary substrate/cosubstrate/enzyme complex causes the immediate formation of NADP$^+$ and Chlide, and release of these products enables the enzyme to bind substrate and cosubstrate for a new photoreaction. The mechanism of the photoreaction is described in Chapter 14; emphasis in this chapter lies on the consequence of the photoreaction for regulation of part 4 of Chl biosynthesis, mainly deduced from investigation of flash-irradiated etiolated barley seedlings[1,2] (Fig. 2). A critical step that is normally

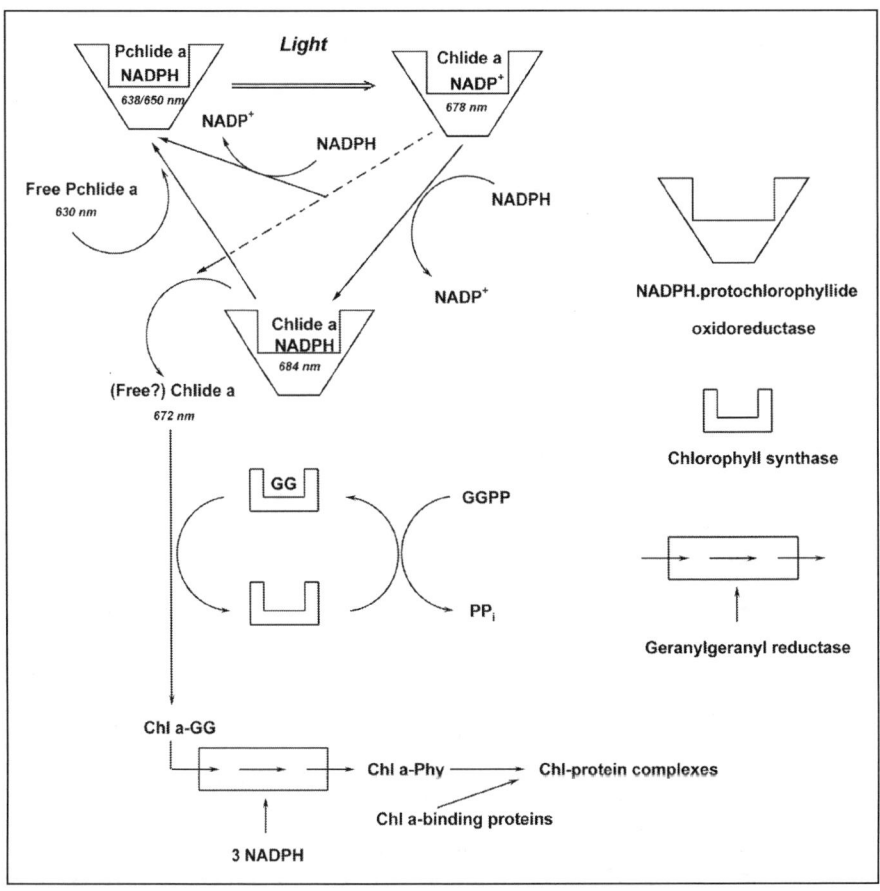

Figure 2. Light regulation of part 4 of chlorophyll biosynthesis in etiolated angiosperm seedlings. The absorption maxima of the pigment-protein complexes and their composition according to Oliver and Griffiths[4] are indicated. The dashed arrow signifies the rapid phase of chlorophyll synthesis, which typically makes up 1/7 of the accumulated Pchlide. Chl *a*-binding proteins are the six plastid-DNA encoded proteins P700A, P700B, D1, D2, CP43, CP47. Pchlide: protochlorophyllide; Chlide: chlorophyllide; GG: geranylgeraniol; Phy: phytol.

rate-limiting is the release of Chlide from POR; we hypothesize that the product can either be pulled out by binding to the next enzyme, Chl synthase, presuming that free binding sites are available, or pushed out by an excess of free Pchlide.

The hypothesis of a pulling mechanism came from the observation that about 1/7 of total Chlide formed by saturating irradiation was esterified within seconds accompanied by an absorption shift from 678 directly to 672 nm (dashed arrow in Fig. 2); the rapid reaction rate is compatible with the assumption of a 7: 1 complex between POR and Chl synthase that is preloaded with GGPP.[2] According to its "ping-pong" mechanism, Chl synthase must be preloaded with geranylgeraniol (GG) before it can bind Chlide,[3] and reloading after the first round of catalysis takes some time. Meanwhile, the residual 6/7 of total Chlide remain bound to POR. NADPH exchanges for NADP+ in the complex within few minutes; according to Oliver and Griffiths,[4] this is the molecular basis for the shift from 678 to 684 nm, observed in many laboratories. The subsequent shift from 684 to 672 nm ("Shibata shift") and esterification of residual Chlide require 30-60 min for completion, depending on the temperature.

The hypothesis of a pushing mechanism, i.e., exchange of Chlide by Pchlide, is based on the observation that the Shibata shift is accelerated when an excess of Pchlide accumulates after incubation with 5-aminolaevulinate,[4] whereas the kinetics of esterification of Chlide remain unchanged under these conditions.[2] Thus, the Shibata shift and the esterification of Chlide, which run in parallel under a variety of conditions,[55] can experimentally be separated from each other. Under such conditions, it is not the release of Chlide from POR but the reloading of Chl synthase with geranylgeranyl diphosphate (GGPP) for esterification of Chlide that limits the reaction rate of Chl formation.

Geranylgeranyl reductase is a multifunctional enzyme that catalyses both reduction of Chl-GG to Chl-Phy and of GGPP to phytyl diphosphate (PhyPP).[5] In etiolated plants, its activity is low but increases immediately after irradiation, probably by new synthesis of the enzyme because the increase is inhibited by cycloheximide;[55] its mRNA level is higher in green than in etiolated plants.[5]

Chl synthase accepts both GGPP and PhyPP;[6] since the level of PhyPP is at least 8 times lower than that of GGPP in etiolated plants,[7] esterification of Chlide with PhyPP plays only a marginal role in this tissue and is not included in Figure 1. It may be more important in green plants where only formation of Chl-Phy has been observed although a very rapid reduction of Chl-GG, detected in leaves of dark-grown seedlings containing proplastids,[8] is also compatible with the experimental results. A study with transgenic tobacco plants expressing antisense RNA for geranylgeranyl reductase revealed reduced amounts of total Chl; up to 58% of the remaining Chl were esterified with GG.[9] Thus, Chl-GG can be incorporated into pigment-protein complexes but causes reduced stability. In the same study, the authors found reduced levels of tocopherol, which requires PhyPP (and not GGPP) for its biosynthesis in tobacco. Light regulation of geranylgeranyl reductase affects not only Chl synthesis but also formation of tocopherol and presumably phylloquinone.[5]

Saturating irradiation of etiolated oat seedlings transiently depleted the pool of GGPP, probably by consumption by esterification of Chlide.[10] Since GGPP is involved as a precursor in several pathways, e.g., carotenoid biosynthesis, regulation of the esterification of Chlide can also influence the substrate flux into other pathways using GGPP.

The Chl *a*-binding proteins are included in Figure 2 because of their tight connection with Chl synthesis. It had been known for a long time that these proteins accumulate together with Chl when etiolated leaves are irradiated, but fail to accumulate when isolated, intact plastids are irradiated. With the knowledge that isolation of plastids leads to the loss of soluble GGPP and PhyPP,[11] Eichacker et al[12] demonstrated the accumulation of these proteins by synthesis of Chl in the dark from added Chlide and PhyPP. Addition of already esterified Chl did not lead to the accumulation of these proteins; thus, the interaction with the proteins requires either Chl "in statu nascendi" or the presence of the enzyme, Chl synthase. The effect of Chl synthesis turned out to be the stabilisation of the newly formed proteins that are proteolytically degraded in the absence of Chl.[13] The permanent synthesis and degradation of these proteins in darkness seems to be an oddly wasteful; it can be argued, however, that free Chl is such a potent sensitizer for membrane photo-oxidation that it is necessary to have Chl apoproteins always available in excess to be able to bind each newly-synthesized Chl molecule immediately.[13] There is no stabilisation with Chlide or PhyPP alone; however, artificial [Zn]-Chl, produced via the Chl synthase reaction, stabilises the proteins at low concentrations but leads to their destabilisation at higher concentrations.[14] Since free Chl is unstable, the results signify a mutual stabilisation of Chl and Chl-binding proteins, and the entire process, outlined in Figure 2, is light dependent, mediated by the photo-conversion of Pchlide to Chlide catalysed by POR.

The molecular basis for the spectral changes caused by irradiation of etioplasts is presented in Figure 2 according to the model of Oliver and Griffiths.[4] This model does not explain the two spectral forms of photoactive Pchlide in the ternary complex, absorbing at 638 and 650 nm. Both are connected to the inner membranes of etioplasts, the 638 nm form is enriched in prothylakoids and the 650 nm form in prolamellar bodies (PLBs). Circular dichroism measurements, which indicate different degrees of pigment-pigment interaction, have been interpreted such that the forms are aggregated Pchlide-POR complexes of different sizes.[15,16] Irradiation should therefore yield Chlide-POR complexes of similar heterogeneous sizes. The Shibata shift can then be understood as disaggregation from larger to smaller size and transfer from PLBs to prothylakoids,[17] perhaps involving conformational changes of the POR protein.[18]

I wish to draw the attention of the reader to still another aspect: the prothylakoid membrane has a lamellar structure while a cubic membrane phase has been attributed to PLBs, consisting of a regular network of tubules that can store large amounts of Pchlide, POR, and membrane lipids in a small volume.[19] Since POR has no membrane-spanning sequence it must be located at the membrane surface. A tubular structure has two different surfaces with opposite curvature, and it is reasonable to assume that POR complexes are more tightly packed at the inner surface than at the outer surface of the tubules. This should result in a higher degree of pigment-pigment interaction at the inner than at the outer side of the tubules, and it could easily explain the observed two spectral forms, the 650 nm form at the inner and the 638 nm form at the outer surface, while classical aggregation could lead to a continuous spectral shift depending on the size of the aggregates. It is interesting in this connection that Chl synthase activity is latent in isolated, intact PLBs, i.e., the enzyme is present but catalyses the esterification of Chlide with added GGPP or PhyPP only after dissociation of PLBs.[20] It is plausible to assume that Chlide esterification in intact etiolated leaves also requires dispersal of PLBs, and this emphasizes the significance of coregulation of metabolism with the ultra structure of the cell. The cubic structure is stable as long as the ternary POR complex contains Pchlide; after photo-conversion to Chlide, it becomes unstable and finally transforms into a lamellar membrane system.

A different model for PLBs with the assumption of large amounts of Pchlide *b* has been proposed.[21,22] This will be critically discussed in the next section on Chl *b*.

Regulation of Chlorophyll b Biosynthesis

In addition to Chl *a*, Chl *b* is a major pigment in LHCI and LHCII, the outer antenna complexes connected to photosystems I and II, respectively. Transfer of etiolated plants from darkness to light leads to formation of Chl *a*-containing reaction centres and inner antennae before the Chl *b*-containing outer antennae LHCI and LHCII appear. Together with the structural similarity of Chls *a* and *b*, the sequence of appearance led to the speculation of Chl *b* biosynthesis from Chl *a*. The reaction step was substantiated by isolation of the *CAO* genes from *Chlamydomonas reinhardtii*[23] and *Arabidopsis thaliana*,[24] and investigation of the recombinant enzyme revealed that the reaction takes place with Chlide *a* instead of Chl *a*, and is a two-step oxygenation.[25] A number of investigations had already shown that the back reaction, reduction of Chl *b* to Chl *a*, occurs under physiological conditions (for review see ref. 26); forward and backward reactions can be compiled to a cycle ("chlorophyll cycle", Fig. 3). It is reasonable to assume that the single reactions of the cycle are regulated according to the requirement of more or less Chl *a* and *b*, respectively; however, we are far from understanding the underlying mechanisms since single, scattered observations do not yet allow complete picture to be drawn.

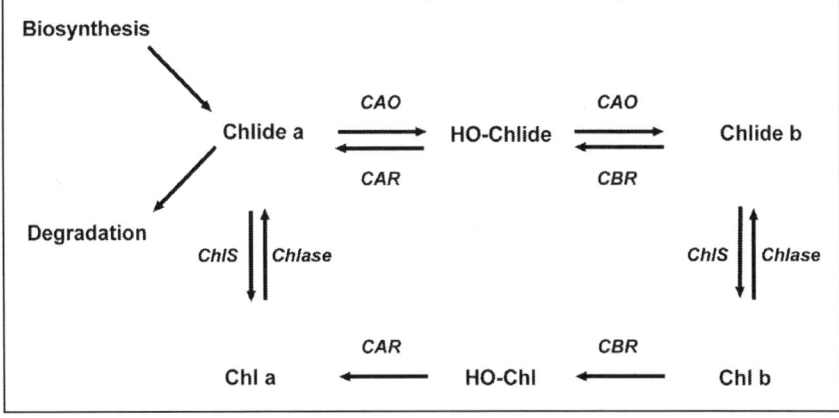

Figure 3. Scheme of chlorophyll b biosynthesis and the chlorophyll cycle. HO-Chl and HO-Chlide: 7^1-hydroxy-chlorophyll a and 7^1-hydroxy-chlorophyllide a, respectively. *CAO*: chlorophyllide a-oxygenase, *CAR*: 7^1-hydroxy-chlorophyll(ide) a-reductase, *CBR*: chlorophyll(ide) b-reductase, *ChlS*: chlorophyll synthase, *Chlase*: chlorophyllase.

In etiolated oat seedlings, the level of CAO mRNA was very small and increased after transfer of the plants to light (K. Nikulina, U. Oster and W. Rüdiger, unpublished results). Using appropriate *Arabidopsis* mutants and transfer from dim to medium light and back, Espineda et al[24] found an increase of the CAO mRNA level with increasing Chl *a/b* ratio while *Lhcb1* mRNA levels were less responsive. There must be some cross-talk, however: overexpression of CAO in *A. thaliana* yielded not only more Chl *b*, measured as decreased Chl *a/b* ratio, but also an enlargement of the antenna size, measured as ratio of LHCP to the core protein CP47 of photosystem II.[27] A dramatic effect of an opposite cross-talk was found in a photosystem I-deficient strain of the cyanobacterium *Synechocystis* sp. PCC 6803 that naturally does not contain Chl *b*: whereas introduction of the *Arabidopsis* CAO gene resulted in the production of very little Chl *b* (<10% of total Chl),[28] Chl *b* was the main Chl (60% of total) after the same transformation of a strain that contained the pea *Lhcb* gene.[29] Interestingly, Chl *b* had displaced Chl *a* in the photosystem II core complex without disturbing its function; no immuno-detectable amounts of LHCII were present and pulse labelling revealed only a transient accumulation of traces of Lhcb protein. The authors concluded that LHCII either activates CAO or serves as initial binding site for Chl *b*, which is released on degradation of LHCII and then incorporated into newly synthesized core complexes, the binding sites of which apparently do not differentiate between Chl *a* and *b*. Transfer of the control strain from light to heterotrophic growth in the dark led to very slow degradation of Chl *a* and steady accumulation of Pchlide *a* whereas the *lhcb*[+]/*cao*[+] strain exhibited equally fast degradation of Chls *a* and *b*, fast and transient accumulation of Pchlide *a* and transient accumulation of Pchlide *b*, trailing that of Pchlide *a* by several days; the maximum amount of Pchlide *b* was 10-15% that of Pchlide *a*.[30] Thus Pchlide *a* seems to be an inefficient substrate for CAO in this system. Further, the authors found that the presence of b-type Chl precursors lead to inhibition of ALA formation, a second block of the synthesis chain between Mg-Proto and Pchlide, and an alteration of the distribution of protoporphyrin between the iron and the magnesium branch of tetrapyrrole synthesis; they speculated that interaction of the precursor with regulatory proteins, e.g., the small cab-like proteins (SCPs), disturbs the regulation of this pathway. It remains to be shown whether the conclusions apply also to higher plants.

Coregulated synthesis of the nuclear-encoded LHCP with synthesis of Chl, in particular of Chl *b*, is more complex than coregulation of accumulation of plastid-encoded Chl *a*-binding proteins with Chl *a* synthesis. The basic principle is simple: expression of the *Lhcb* genes is phytochrome-regulated, and the proteins accumulate only when Chl *b* is synthesized. However, expression of *Lhcb* genes additionally requires the presence of intact plastids, sensed by a "plastid factor" (for review see ref. 31). At first, it was considered to be a positive signal that was not produced when the plastids were photo-oxidized until Strand et al[32] proposed that this "plastid factor" is Mg-Proto, which accumulates in excess when plastids are photo-oxidized and which inhibits expression of the *Lhcb* gene. Reduced *Lhcb* gene expression under conditions of increased Mg-Proto accumulation had been observed before,[33,34] but it had been difficult to separate the specific effect from possible destructive effects caused by photosensitization. Mg-Proto has been shown to mediate light-induced expression of the nuclear-encoded chaperone HSP70 in *Chlamydomonas reinhardtii*, which protects photosystem II against damage by high light, and thus qualifies as a "positive plastid signal" in this case.[35-37] The rapid and transient increase in Mg-Proto levels on dark-to-light shifts in green tobacco and barley seedlings[38] suggests that Mg-Proto mediates a light signal also in higher plants. Genetic evidence in *Arabidopsis* revealed the participation of Mg-Proto in up- and down-regulation of a large number of nuclear genes, mainly encoding proteins associated with photosynthetic reactions.[32]

Besides the Mg-Proto signalling pathway, there are other plastid-to-nucleus signalling pathways, mostly not yet characterized in detail. One signal seems to be connected with plastid translation,[39] another one that has been correlated with *Lhcb* gene expression, is the redox state of the plastoquinone pool. Experimental evidence came e.g., from light or temperature induced changes in the redox state,[40] studies with inhibitors of the photosynthetic electron transport,[41] investigation of a mutant with reduced plastoquinone synthesis caused by mutation in the phosphoenolpyruvate/phosphate translocator.[42] This pathway may be most important for acclimation of the photosynthesis apparatus to high and low light; however, it operates also during early development and affects many other gene products besides Chl *a/b*-binding proteins.[43] By stabilisation of Chl *b* through bonding to LHCP, it participates in the regulation of Chl accumulation.

As depicted in Figure 3, Chlide *a* is the compound entering the Chl cycle when Chls *a* and *b* are synthesized and likewise the compound for exit from the cycle when Chls are degraded in higher plants. The latter hypothesis is based on the observation that Chl degradation products are structurally derived from Chl *a* even when Chl *b* is degraded, and that Pheide *b* is a competitive inhibitor of Pheide oxygenase, the key enzyme for Chl degradation[44,45] (see also Chapter 17). The situation is different in algae: open-chained tetrapyrroles, structurally derived from both Chl *a* and Chl *b*, have been found on Chl degradation in *Chlorella pyrodenoides*.[46] In accordance with this result, no Chl *b* reductase activity was detected in *Chlamydomonas reinhardtii* (V. Scheumann and W. Rüdiger, unpublished result).

The hypothesis that Chlide *a* enters the cycle rather than compounds that are earlier or later in the biosynthetic chain came from the substrate specificity of recombinant CAO determined in vitro: within the reaction time of 3 h, neither Pchlide *a* nor esterified Chl *a* were oxygenated to the respective *b*-compound.[25] As mentioned before, it took several days after formation of Pchlide *a* until some Pchlide *b* appeared in a transgenic cyanobacterium indicating that Pchlide *a* is a very poor substrate for CAO also in this system.[30] Thus it can be expected that Pchlide *b*, if it is formed at all in higher plants, should be a minor compound besides the main intermediate Pchlide *a*. By contrast, Reinbothe et al[21] speculated that PLBs of barley etioplasts should mainly consist of 5 parts Pchlide *b* bound to PORA complexed with 1 part Pchlide *a* bound to PORB; the speculation was based on model experiments with [Zn]-Pchlides *a* and *b* and in vitro-translation products of PORA and PORB. Scheumann et al[47] and Armstrong et al[48] questioned whether this is the situation in vivo, given that the model was not compatible with data from several other laboratories. However, Reinbothe et al published two further papers in support of the above speculation without solving the discrepancies with other laboratories.[22,49] Three points of major dispute may be illustrated briefly: (1) Reinbothe et al[22] presented a highly unlikely explanation for the failure of other laboratories to detect any Pchlide *b* in etiolated tissue; they hypothesized the immediate enzymatic reduction of endogenous Pchlide *b* to Pchlide *a* during extraction of intact etioplasts or leaves with 80% acetone; the authors recommended 100% acetone containing 0.1% diethyl pyrocarbonate for detection of Pchlide *b*. However, all attempts of other laboratories to find Pchlide *b* in etiolated plants under the recommended conditions failed: the use of 100% acetone for extraction of fresh[50] or frozen leaves,[51] including the addition of diethyl pyrocarbonate[52] yielded only Pchlide *a* without traces of Pchlide *b*. Reinbothe et al stated that their etioplast preparations did not reduce exogenous [Zn]-Pchlide *b* while Scheumann et al[47] described reduction by etioplasts of exogenous Pchlide *b* to the 7¹-hydroxy-derivative but not further to Pchlide *a*; in fact, the pigment pattern of Pchlide *b* as the main compound with some 7¹-hydroxy-derivative and even less Pchlide *a*, shown by Reinbothe et al after extraction of etioplasts with 100% acetone, resembles the pattern obtained by Scheumann et al after incubation of etioplasts with an excess of Pchlide *b*. (2) A key statement by Reinbothe et al[21,49] is the lack of photo-conversion of Pchlide *b* bound to PORA; given the postulated 1 : 5 stoichiometry of Pchlide *a : b*, this should mean that only one part of total Pchlide should be photoactive and five parts should be nonphotoactive. This is in contrast to the findings in other laboratories that the majority of Pchlide in etiolated plants is photo-converted to Chlide *a* by flash illumination,[48] e.g., about 90% photo-conversion in etiolated barley leaves by a single saturating flash of 2 ms duration,[2] and Pchlide *b*, added to etioplast preparations, is readily photoconverted to Chlide *b*.[47] Poor photoconversion of Pchlides is normally considered as beginning denaturation of etioplast preparations. (3) Reinbothe et al ascribe a light-harvesting and photo-protective function specifically to the putative PORA-Pchlide b complex. There is no doubt that PLBs, in spite of containing only Pchlide *a*, play a photo-protective role: plants lacking PLBs caused by reduced formation of the POR protein, either by anti-sense transformants or by phytochrome-mediated down-regulation of POR-mRNA, suffer severe damage on transfer from darkness to light; this defect can be cured by overexpression of either PORA or PORB (reviewed in ref. 48). The exact mechanism of this protection is still unknown.

A New Role for Carotenoids?

There are manifold correlations between Chl and carotenoid formation and accumulation. Both pigments share GGPP as a precursor in their respective biosynthetic pathways, and a high flux of GGPP into one pathway can reduce the synthesis rate in the other pathway. The role of carotenoids for photo-protection of Chl is most obvious when there is no protection: in the absence of carotenoids, caused either by mutation or inhibitors of their biosynthesis, Chls are completely bleached in the light. Furthermore, carotenoids are believed to have a structural role for stabilisation of Chl-protein complexes, clearly demonstrated in reconstitution experiments of light-harvesting complexes.[53]

La Rocca et al[34] deduced a new role for carotenoids from the observation that some inhibitors of carotenoid biosynthesis caused deregulation of Chl biosynthesis. Such results were obtained after treatment of etiolated seedlings with amitrole, leading to accumulation of lycopene apparently through inhibition of the cyclase reaction, and more recently with 2-(4-chlorophenylthio)triethylamine (CPTA, ref. 56), a specific inhibitor of lycopene cyclase.[54] The deregulation manifested itself by accumulation of ALA, Mg-Proto and its methyl ester, and Pchlide *a*. Interestingly, the regular structure of PLBs changed to a mass of irregularly aggregated membranes, and most Pchlide was nonphototransformable with fluorescence emission at 632 nm instead of 657 nm in the controls. A detailed analysis that included incubation with ALA showed at least two regulatory sites that were affected by the amitrole treatment, one for ALA formation and a second one downstream of ALA. The analogous treatment with norflurazon, an inhibitor of phytoene desaturase, yielded plants in which these parameters did not differ from those of the water controls: they showed regular PLBs and normal levels of all Chl precursors. Only after incubation with ALA, did the Pchlide level increase over that of the water controls. These results were obtained when the plants were kept in absolute darkness. The situation changed when norflurazon-treated plants were transferred from darkness to light and suffered from photo-oxidation of plastid membranes: under these conditions, elevated levels of Mg-Proto were detected.[32] La Rocca et al discussed metabolic channelling for all steps of Chl biosynthesis, which is interrupted by amitrole treatment and not (or to a lesser extent) by norflurazon. To discuss this proposal in more detail, the chemical structures of the main carotenoids accumulated under the various treatments are shown (Fig. 4). The main carotenoid in control plants is lutein, and the minor carotenoids are also cyclic. It can be speculated that they bind to the enzymes of the Chl biosynthetic pathway for photo-protection of the intermediates. Furthermore, the cyclic carotenoids can play a

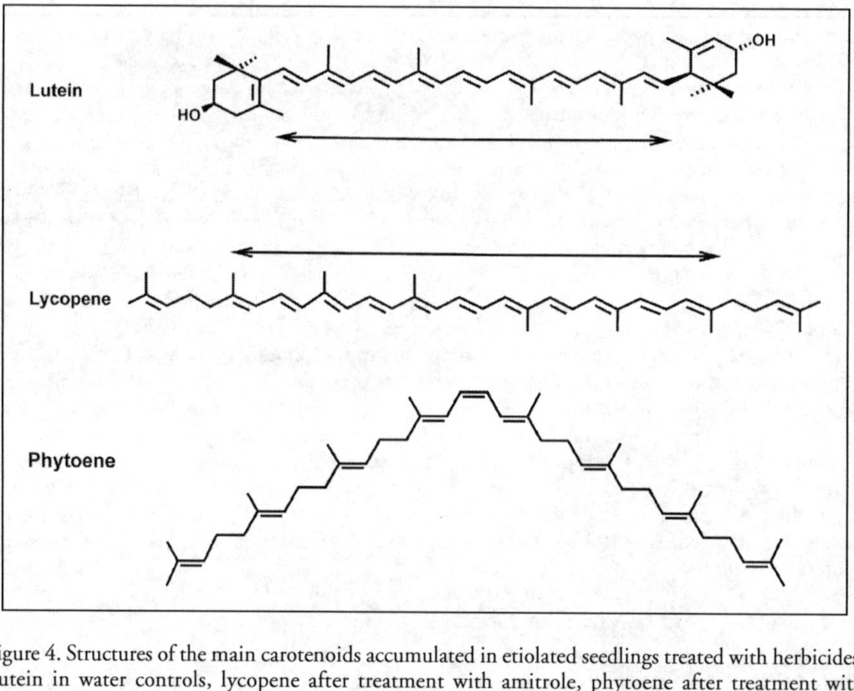

Figure 4. Structures of the main carotenoids accumulated in etiolated seedlings treated with herbicides. Lutein in water controls, lycopene after treatment with amitrole, phytoene after treatment with norflurazone. The bars indicate the length of the rigid part of the molecules.

structural role in enzyme-enzyme interactions that should facilitate metabolic channelling. Norflurazon treatment leads to accumulation of phytoene, which is noncyclic but flexible because of its many single bonds; the observation of normal levels of intermediates, interpreted as "normal channelling", indicates that phytoene can be accommodated in the carotenoid binding sites without disturbing the conformation of enzymes or enzyme complexes. The lack of photo-protection has no effect as long as the plants are kept in darkness; upon photo-oxidation, however, channelling is disturbed and abnormal levels of Mg-Proto accumulate.[32] Lycopene is a noncyclic carotenoid that has more conjugated double bonds and hence a longer rigid structure than cyclic carotenoids; binding to the carotenoid-specific sites is predicted to disturb the conformation of the enzyme proteins and/or of the interaction between several enzymes. This corresponds to the observed deregulation of the pathway and to the disturbance of the regular cubic structure of PLBs. Even if this explanation must remain hypothetical at present, it is plausible and valuable for planning of new experimental approaches.

Note Added in Proof

Recent results raised doubts about a possible role of Mg-Proto as signaling compound in higher plants; in particular, Moulin et al[57] and Mochizuki et al[58] showed that plastid-to-nucleus signaling in *Arabidopsis* as investigated by Strand et al[32] is not due to Mg-Proto accumulation.

Acknowledgement

The cited work of the author was supported by the Deutsche Forschungsgemeinschaft, Bonn, and the Fonds der Chemischen Industrie, Frankfurt.

References

1. Domanskii V, Rüdiger W. On the nature of the two pathways in chlorophyll formation from protochlorophyllide//Photosynthesis Research. Photosynth Res 2001; 68:131-139.
2. Domanskii V, Rassadina V, Gus-Mayer S et al. Characterization of two phases of chlorophyll formation during greening of etiolated barley leaves. Planta 2003; 216:475-483.
3. Schmid HC, Rassadina V, Oster U et al. Preloading of chlorophyll synthase with tetraprenyl diphosphate is an obligatory step in chlorophyll biosynthesis. Biol Chem 2002; 383:1769-1776.
4. Oliver RP, Griffiths T. Pigment-protein complexes of illuminated etiolated leaves. Plant Physiol 1982; 70:1019-1025.
5. Keller Y, Bouvier FD, Harlingue A. Metabolic compartmentation of plastid prenyllipid biosynthesis - Evidence for the involvement of a multifunctional geranylgeranyl reductase. Eur J Biochem 1998; 251:413-417.
6. Schmid HC, Oster U, Kögel J et al. Cloning and characterisation of chlorophyll synthase from Avena sativa. Biol Chem 2001; 382:903-911.
7. Benz J, Fischer I, Rüdiger W. Determination of phythyl diphosphate and geranylgeranyldiphosphate in etiolated oat seedlings. Phytochemistry 1983; 22:2801-2804.
8. Schoefs B, Bertrand M. The formation of chlorophyll from chlorophyllide in leaves containing proplastids is a four-step process. FEBS Lett 2000; 486:243-246.
9. Tanaka, R, Oster U, Kruse E et al. Reduced activity of geranylgeranyl reductase leads to loss of chlorophyll and tocopherol and to partially geranylgeranylated Chlorophyll in transgenic tobacco plants expressing antisense RNA for geranylgeranyl reductase. Plant Physiol 1999; 120:695-704.
10. Benz J, Haser A, Rüdiger W. Changes in the endogenous pools of tetraprenyl diphosphates in etiolated oat seedlings after irradiation. Z Pflanzenphysiol 1983; 111:349-356.
11. Benz J, Hampp R, Rüdiger W. Chlorophyll biosynthesis by Mesophyll protoplasts and plastids from etiolated oat (Avena sativa L.) leaves. Planta 1981; 152:54-58.
12. Eichacker LA, Soll J, Lauterbach P et al. In vitro synthesis of Chlorophyll a in the dark triggers accumulation of Chlorophyll a apoproteins in barley etioplasts. J Biol Chem 1990; 265:13566-13571
13. Kim J, Eichacker LA, Rüdiger W et al. Chlorophyll regulates accumulation of the plastid-encoded chlorophyll proteins P700 and D1 by increasing apoprotein stability. Plant Physiol 1994; 104:907-916.
14. Eichacker LA, Helfrich M, Rüdiger W et al. Stabilization of chlorophyll a-binding apoproteins P700, CP47, CP43, D2, and D1 by chlorophyll a or Zn-pheophytin a. J Biol Chem 1996; 271:32174-32179.
15. Böddi B, Lindsten A, Ryberg M et al. On the aggregational states of protochlorophyllide and its protein complexes in wheat etioplasts. Physiol. Plant 1989; 76:135-143.

16. Böddi B, Lindsten A, Ryberg M et al. Phototransformation of aggregated forms of protochlorophyllide in isolated etioplast inner membranes. Photochem Photobiol 1990; 52:83-87.

17. Ryberg M, Dehesh K. Localization of NADPH- protochlorophyllide oxidoreductase in dark-grown wheat (Triticum aestivum) by immuno-electron microscopy before and after transformation of the prolamellar bodies. Physiol Plant 1986; 66:616-624.

18. Zhong LB, Wiktorsson B, Ryberg M et al. The Shibata shift: Effects of in vitro conditions on the spectral blue shift of chlorophyllide in irradiated isolated prolamellar bodies. J Photochem Photobiol B Biol 1996; 36:263-270.

19. Selstam E, Widell Wigge A. Chloroplast lipids and the assembly of membranes. In: Sundqvist C, Ryberg M, eds. Pigment-protein complexes in plastids, synthesis and assembly. San Diego: Academic Press, 1993:241-277.

20. Lindsten A, Welch CJ, Schoch S et al. Chlorophyll synthetase is latent in well preserved prolamellar bodies of etiolated wheat. Physiol Plant 1990; 80:277-285.

21. Reinbothe C, Lebedev N, Reinbothe S. A protochlorophyllide light-harvesting complex involved in detiolation of higher plants. Nature 1999; 397:80-84.

22. Reinbothe S, Pollmann S, Reinbothe C. In situ conversion of protochlorophyllide b to protochlorophyllide a in barley. J Biol Chem 2003; 278:800-806.

23. Tanaka A, Ito H, Tanaka R et al. Chlorophyll a oxygenase (CAO) is involved on chlorophyll b formation from chlorophyll a. Proc. Natl Acad Sci USA 1998; 95:12719-12723.

24. Espineda CE, Linford AS, Devine D et al. The AtCAO gene, encoding chlorophyll a oxygenase, is required for chlorophyll b synthesis in Arabidopsis thaliana. Proc Natl Acad Sci USA 1999; 96:10507-10511.

25. Oster U, Tanaka R, Tanaka A et al. Cloning and functional expression of the gene encoding the key enzyme for chlorophyll b biosynthesis (CAO) from Arabidopsis thaliana. Plant J 2000; 21:305-310.

26. Rüdiger W. Biosynthesis of chlorophyll b and the chlorophyll cycle. Photosynth. Res 2002; 74:187-193.

27. Tanaka R, Koshino Y, Sawa S et al. Overexpression of chlorophyllide a oxygenase (CAO) enlarges the antenna size of photosystem II in Arabidopsis thaliana. Plant J 2001; 26:365-373.

28. Satoh S, Ikeuchi M, Mimuro M et al. Chlorophyll b expressed in cyanobacteria functions as a light-harvesting antenna in photosystem I through flexibility of the proteins. J Biol Chem 2001; 276:4293-4297.

29. Xu H, Vavilin D, Vermaas W. Chlorophyll b can serve as the major pigment in functional photosystem II complexes of cyanobacteria. Proc Natl Acad Sci USA 2001; 98:14168-14173.

30. Xu H, Vavilin D, Vermaas W. The presence of chlorophyllb in Synechocystis sp. PCC. J Biol Chem 2002; 277:42726-42732.

31. Rodermel S. Pathways of plastid-to-nucleus signalling. Trends Plant Sci 2001; 6:471-478.

32. Strand A, Asami T, Alonso J et al. Chloroplast to nucleus communication triggered by accumulation of Mg-protoporphyrinIX. Nature 2003; 421:79-83.

33. Kittsteiner U, Brunner H, Rüdiger W. The greening process in cress seedlings. II. Complexing agents and 5-aminolevulinate inhibit accumulation of cab-mRNA coding for the light-harvesting chlorophyll a/b protein. Physiol Plant 1991; 81:190-196.

34. La Rocca N, Rascio N. Oster U et al. Amitrole treatment of etiolated barley seedlings leads to deregulation of tetrapyrrole synthesis and to reduced expression of Lhc and RbcS genes. Planta 2001; 213:101-108.

35. Kropat J, Oster U, Rüdiger W et al. Chlorophyll precursors are signals of chloroplast origin involved in light induction of nuclear heat-shock genes. Proc Natl Acad Sci USA 1997; 94:14168-14172.

36. Kropat J, OsterU, RüdigerW et al. Chloroplast signalling in the light induction of nuclear HSP70 genes requires the accumulation of chlorophyll precursors and their accessibility to cytoplasm/nucleus. Plant J 2000; 24:523-531.

37. Schroda M, Kropat J, Oster U et al. Possible role for molecular chaperones in assembly and repair of photosystem II. Biochem. Soc Trans 2001; 29:413-418.

38. Pöpperl G, Oster U, Rüdiger W. Light-dependent increase in chlorophyll precursors during the day-night cycle in tobacco and barley seedlings. J Plant Physiol 1998; 153:40-45.

39. Sullivan JA, Gray JC. Plastid translation is required for the expression of nuclear photosynthesis genes in the dark and in roots of the pea lip 1 mutant. Plant Cell 1999; 11:901-910.

40. Maxwell DP, Laudenbach DE, Huner NPA. Redox regulation of light-harvesting complex II and cab mRNA abundance in Dunaliella salina. Plant Physiol 1995; 109:787-795.

41. Escoubas JM, Lomas M, La Roche J et al. Light intensity regulation of cab gene transcription is signaled by the redox state of the plastoquinone pool. Proc Natl Acad Sci USA 1995; 92:10237-10241.
42. Streatfield SJ, Weber A, Konsman EA et al. The phosphoenolpyruvate/phosphate translocator is required for phenolic metabolism, palisade cell development and plastid-dependent nuclear gene expression. Plant Cell 1999; 11:1609-1622.
43. Karpinski S, Reynolds H, Karpinska B et al. Systemic signalling and acclimation in response to excess excitation energy in Arabidopsis. Science 1999; 284:654-657.
44. Hörtensteiner S, Vicentini F, Matile P. Chlorophyll breakdown in senescent cotyledons of rape, Brassica napus L: Enzymatic cleavage of phaeophorbide a in vitro. New Phytol 1995; 129:237-246.
45 Kräutler B, Matile P. Solving the riddle of chlorophyll breakdown. Acc Chem Res 1999; 32:35-43.
46. Gossauer A, Engel N. New trends in photobiology: Chlorophyll catabolism—structures, mechanisms, conversions. J Photochem Photobiol B 1996; 32:141-151.
47. Scheumann V, Klement H, Helfrich M et al. Protochlorophyllide b does not occur in barley etioplasts. FEBS Lett 1999; 445:445-448.
48. Armstrong GA, Apel K, Rüdiger W. Does a light-harvesting protochlorophyllide a/b-binding protein complex exist? Trends Plant Sci 2000; 5:40-44.
49. Reinbothe C, Buhr F, Pollmann S et al. In vitro reconstitution of light-harvesting POR-protochlorophyllide complex with protochlorophyllides a and b. J Biol Chem 2003; 278:807-815.
50. Schoch S, Helfrich M, Wiktorsson B et al. Photoreduction of Zinc-protopheophorbide b with NADPH-protochlorophyllide oxidoreductase from etiolated wheat (Triticum aestivum L.). Eur J Biochem 1995; 229:291-298.
51. Helfrich M, Schoch S, Schäfer W et al. Absolute configuration of protochlorophyllide alpha and substrate specificity of NADPH-protochlorophyllide oxidoreductase. J Am Chem Soc 1996; 118:2606-2611.
52. Kolossov VL, Rebeiz CA. Chloroplast biogenesis 88. Protochlorophyllide b occurs in green but not in etiolated plants. J Biol Chem 2003; 278:49675-49678.
53. Paulsen H, Schmid VHR. Analysis and reconstitution of chlorophyll proteins. In: Witty M, Smith AG, eds. Analytical Methods in Heme, Chlorophyll, and Related Molecules. Natick: Eaton Publishing, 2001:235-254.
54. Böger P. Mode of action of herbicides affecting carotenogenesis. J Pesticide Sci 1996; 21:473-478.
55. Rassadina V, Domanskii V, Averina NG et al. Correlation between chlorophyllide esterification, Shibata shift and regeneration of protochlorophyllide650 in flash-irradiated etiolated barley leaves. Physiol Plant 2004; 121:556-567.
56. Rocca NL, Rascio N, Oster U et al. Inhibition of lycopene cylase results in accumulation of chlorophyll precursors. Planta 2007; 225:1019-1029.
57. Moulin M, McCormac AC, Terry MJ et al. Tetrapyrrole profiling in Arabidopsis seedlings reveals that retrograde plastid nuclear signaling is not due to Mg-protoporphyrin IX accumulation. Proc Natl Acad Sci USA 2008; 105:15178-15183.
58. Mochizuki N, Tanaka R, Tanaka A et al. The steady-state level of Mg-protoporphyrin IX is not a determinant of plastid-to-nucleus signaling in Arabidopsis. Proc Natl Acad Sci USA 2008; 105:15184-15189.

CHAPTER 17

Chlorophyll Breakdown

Bernhard Kräutler*

Abstract

Chlorophyll metabolism is probably the most visible manifestation of life. In spite of its obvious ecological importance, chlorophyll catabolism has remained an enigma until about twelve years ago. Contrary to all expectations, chlorophyll breakdown in vascular plants rapidly leads to colorless degradation products. It only fleetingly involves colored intermediates, which result from an oxidative opening of the chlorophyll macrocycle. This stage is rapidly followed by a reduction to shortly existent fluorescent catabolites, which isomerize rapidly to colorless and nonfluorescent tetrapyrrolic catabolites. These latter colorless bilanones accumulate in the vacuoles of the degreened plant material and may represent the final products of controlled chlorophyll breakdown in higher plants. This chapter delineates important structural features of chlorophyll catabolites from natural sources and some of the biochemistry of chlorophyll breakdown in higher plants.

Introduction

The emergence of autumnal colors in the foliage of deciduous trees belongs to the most fascinating natural phenomena. The associated disappearance of the green plant pigments by breakdown of the chlorophylls has remained a mystery until about twelve years ago.[1-4] In this chapter the current knowledge on the structures of the chlorophyll catabolites from vascular plants and from other sources is outlined. In addition, it recapitulates briefly the present day knowledge on the biochemical pathways of their formation from the plant chlorophylls.[3-6]

The chlorophylls hold a key position among the pigments of life, due to their unique roles in the capture and transformation of sun light.[7] Indeed, the seasonal appearance and disappearance of the green pigments is probably the most visual sign of life on earth, observable even from outer space.[4] It is estimated that more than 10^9 tons of chlorophyll are biosynthesized and degraded every year on the earth.[1]

Chlorophyll Breakdown in Higher Plants

The earlier, unsuccessful search for chlorophyll catabolites from vascular plants was directed at the finding of colored compounds.[1] Today, the major chlorophyll catabolites from senescent higher plants are known to be colorless.[2-6] Matile and coworkers provided first good evidence for the presence of colorless chlorophyll catabolites in senescent leaves (in *Festuca pratensis*[8,9] and in barley[10-12]). Surprisingly these catabolites, which easily decomposed into pink and rust-colored compounds, were found in the vacuoles, rather than in the degreened chloroplasts.[11,13] The main catabolite from barley, *Hv*-NCC-1 (2, $3^1,3^2,8^2$-trihydroxy- 1,4,5,10,15,20-($22H,24H$)-octahydro-13^2-[methoxycarbonyl]-4,5-dioxo-4,5-seco-phytoporphyrinate), was the first degreened chlorophyll catabolite from higher plants to be identified (see Scheme 1).[2,14] Its structure revealed the nongreen chlorophyll catabolite 2 to be derived from chlorophyll *a* (1a) and gave first-hand clues as to the major structural changes occurring in the degradation of chlorophyll during plant senescence.[2-6] It

*Bernhard Kräutler—Institute of Organic Chemistry, University of Innsbruck, A-6020 Innsbruck, Austria. Email: bernhard.kraeutler@uibk.ac.at

Tetrapyrroles: Birth, Life and Death, edited by Martin J. Warren and Alison G. Smith.
©2009 Landes Bioscience and Springer Science+Business Media.

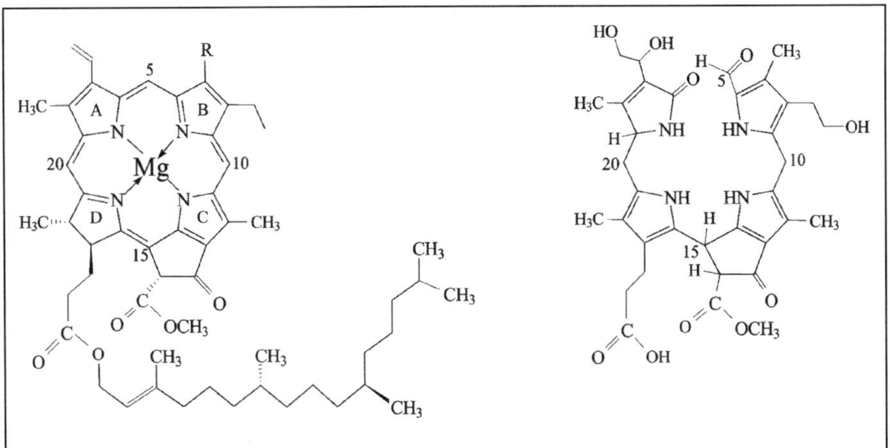

1a: chlorophyll *a* (R = CH$_3$)
1b: chlorophyll *b* (R = HC=O)

2: *Hv*-NCC-1

Scheme 1. Structural formulae of plant chlorophylls (left) and of *Hv*-NCC-1 (2) (right), the first nongreen chlorophyll catabolite to be characterized structurally.[2]

confirmed the suspected catabolic relevance of an oxygenolytic cleavage of the chlorin macrocycle, but indicated it to occur at the "northern" meso-position, contrary to all earlier speculations.[1,2]

Early Steps

The structure of *Hv*-NCC-1 (2) was consistent with the loss of the phytol side chain from chlorophyll *a* (1a) as one of the first events of chlorophyll breakdown. Hydrolysis of both the chlorophylls, of chlorophyll *a* (1a) and of chlorophyll *b* (1b) to phytol and to chlorophyllide *a* (3a) and *b* (3b), resp. (see Scheme 2), is catalyzed by chlorophyllase[15] and sets the stage for the rapid parallel degradation of the chlorophylls and of the chlorophyll binding proteins.[4,16,17]

All the chlorophyll catabolites detected in extracts from vascular plants were found to be derived from chlorophyll *a* (1a) and none of them from the minor chlorophyll component in plants, chlorophyll *b* (1b), the 7-formyl analogue of 1a.[2-6] The puzzling fate of the *b*-type chlorophylls in the course of chlorophyll breakdown was clarified by the recent discovery of a biochemical pathway from chlorophyllide *b* (3b) to chlorophyllide *a* (3a).[18-21] The established oxidative biochemical transformation of the *a*-type into the *b*-type chlorophylls[22] has thus obtained a reductive counterpart in a "chlorophyll *a* / chlorophyll *b* cycle".[16,23]

These findings were in line with the discovery that pheophorbide *a* oxygenase, the crucial and senescence specifically expressed oxygenase that cleaves the chlorin macrocycle, accepts pheophorbide *a* (4a), whereas it is inhibited by pheophorbide *b* (4b).[24] The reductive transformation of chlorophyllide *b* (3b) therefore is required, so that all the plant chlorophylls are made available for the degradative "pheophorbide *a*" pathway.[25,26] Dephytylation and reductive conversion of *b*-chlorophyll(ide)s to *a*-type analogues is indicated to precede the loss of the magnesium ion.[3-6,25-28] The magnesium dechelating enzyme was analyzed in senescent leaves of *Chenopodium album* and was found to require the assistance of a heat stable "magnesium dechelating substance".[27] Enzyme controlled removal of the magnesium ion from chlorophyllide a (3a) provided pheophorbide *a* (4a), the substrate of the crucial oxygenase. Pyropheophorbide *a* (5, see Scheme 2) has also been observed in *Chenopodium album* and has been considered likewise to be an "early" catabolite of chlorophyll degradation in this green plant.[29,30] So far, however, a nongreen tetrapyrrolic chlorophyll catabolite having a 13^2-methylene group (as in the pyropheophorbides) has not been isolated from senescent higher plants.[2-6,14,31-36]

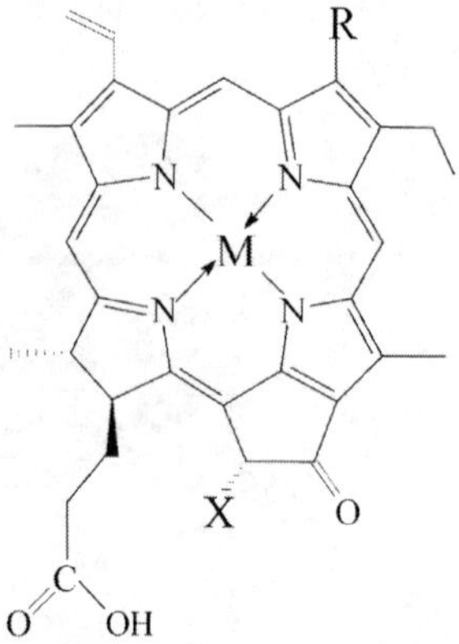

$M = Mg$, $X = C(O)OCH_3$
3a: chlorophyllide *a* ($R = CH_3$)
3b: chlorophyllide *b* ($R = HC=O$)

$M = 2H$, $X = C(O)OCH_3$
4a: pheophorbide *a* ($R = CH_3$)
4b: pheophorbide *b* ($R = HC=O$)

$M = 2H$, $X = H$, $R = CH_3$
5: pyropheophorbide *a*

Scheme 2. Structural formulae of some chlorins, isolated from senescent plants.

Cleavage of the Chlorophyll Macroring

All studies concerning chlorophyll breakdown in senescent vascular plants have documented the broad observation of colorless and nonfluorescent chlorophyll catabolites.[2,6,14,31-36] The discovery, that 4a accumulated in the absence of molecular oxygen in the higher plant *Festuca pratensis*,[9] suggested the involvement of both, O_2 and 4a, as common substrates in a key oxidative enzymatic step during chlorophyll breakdown. When the apparent rates of chlorophyll breakdown in senescent cotyledons of oil seed rape ("*Brassica napus*") were high, tiny amounts of fluorescent compounds (provisionally named "fluorescent chlorophyll catabolites") could be observed fleetingly.[37-39] None of these fluorescent compounds accumulated in vivo and they were considered to represent early products of cleavage of the porphinoid macrocycle of pheophorbide *a* (4a). Such a fluorescent chlorophyll catabolite (named *Bn*-FCC-2) became available by the use of a preparative enzymatic in-vitro system obtained from senescent oilseed rape (*Brassica napus*).[40] The structure elucidation of the fluorescent chlorophyll catabolite *Bn*-FCC-2 (6) indicated it to be a $3^1,3^2$-didehydro-1,4,5,10,17,18,20,($22H$)-octahydro-13^2-(methoxycarbonyl)-4,5-dioxo-4,5-seco-phytoporphyrin (see Scheme 3) and established the crucial oxygenolytic cleavage of the porphinoid macroring of pheophorbide *a* (4a) as a rather early step in chlorophyll breakdown.[24,40,41] This hypothetic oxygenolytic opening of the macroring would have to occur at the "northern" meso-position. The activity of the putative oxygenase (a nonheme iron-dependent enzyme[42]) was shown to depend upon the presence of reduced ferredoxin,[43] and to be remarkably specific for 4a, while 4b competitively inhibited it.[24]

The structure of 6 suggested the hypothetical oxygenolytic ring cleavage to be followed by a reduction step, involving the saturation of the "western" δ-meso position and directly generating the fluorescent chlorophyll catabolite (6).[40] According to this hypothesis, the direct product of the enzymatic oxygenolysis would most likely be the red tetrapyrrole 7, an elusive "red chlorophyll catabolite", that was thus considered also as the putative precursor of 6 in chlorophyll breakdown.[40] The red tetrapyrrole 7 would have the same chromophore structure as some of the red bilinones which had been found to be excreted as final degradation products of the chlorophylls in the green alga *Chlorella protothecoides*.[44,45] Authentic 7 ($3^1,3^2$-didehydro-4,5,10,17,18,($22H$)-hexahydro-13^2-(methoxycarbonyl)-4,5-dioxo-4,5-seco-phyto-porphyrin) could be prepared by chemical degradation of pheophorbide *a* (4a).[46] Tracing experiments indeed revealed the authentic "red chlorophyll catabolite" (7) to be identical with a red compound, of which minute amounts could be found in extracts from senescent plant material.[47]

The oxygenolytic transformation of 4a to (an enzyme bound form of) 7 is achieved by a single enzyme, a mono-oxygenase, termed pheophorbide a oxygenase (PaO).[48,49] Activity of PaO is detectable only in senescent leaves,[26] in contrast to chlorophyllase, as well as several other enzymes contributing to

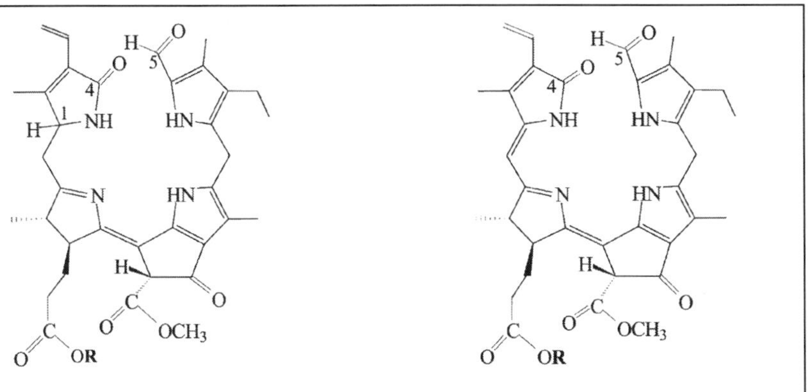

6 & 6-*epi*: R = H; "primary fluorescent chlorophyll catabolites" (FCCs)

Me-6: R = CH₃; FCC methyl ester

7: R = H; "red chlorophyll catabolite" (RCC)

Me-7: R = CH₃; RCC methyl ester

Scheme 3. Structural formulae of chlorophyll catabolites from plants: the "early" ring cleavage products "red chlorophyll catabolite" (RCC, 7) and the "primary fluorescent chlorophyll catabolites" (pFCCs) 6 and 6-*epi*.

chlorophyll breakdown. PaO is located in the chloroplast envelope and may be considered to represent the key enzyme of chlorophyll breakdown.[48]

The red chlorophyll catDabolite RCC (7) is bound strongly to pheophorbide *a* oxygenase (PaO), and inhibits it. Therefore only trace amounts of 7 may be found in incomplete in-vitro catabolic experiments.[47,49] However, a reductase, named red chlorophyll catabolite reductase (RCC-reductase), directly reduces the bound red catabolite RCC (7) and sets it free.[50] The central steps of chlorophyll breakdown in higher plants thus depend on the intimate cooperation of the membrane bound PaO with RCC-reductase, which is located in the stroma of the plastids and whose action is associated with the release of the reduction product 6 or 6-*epi*.[50,51]

The product of the reductase from oilseed rape, the fluorescent chlorophyll catabolite FCC (6,)3¹,3²-didehydro-1,4,5,10,17,18,20-(22H)-octahydro-13²-(methoxycarbonyl)-4,5-dioxo-4,5-seco-phytoporphyrin)[40] has been called a primary FCC (or pFCC), as it is an FCC devoid of further peripheral refunctionalization. While the reductase depends on reduced ferredoxin as electron donor, it is remarkable that (other) cofactors appear not to be involved in the task of reducing (bound) 7 to the pFCC 6 (see Scheme 4).[50]

A second fluorescent chlorophyll catabolite (*Ca*-FCC-2) was isolated from sweet pepper (*Capsicum annuum*) and could be shown to be 6-*epi*,[52] a stereoisomer of 6.[40] The two FCCs (6 and 6-*epi*) are the direct reduction products of RCC-reductase and differ only in their absolute configuration at C(1). The epimeric nature of 6 and 6-*epi* provides evidence for a remarkable, species dependent stereo-dichotomy of the RCC-reductases and for a functional irrelevance of the absolute configuration at the newly generated chiral center C(1).[51,52] The FCCs 6 and 6-*epi* are both devoid of further peripheral refunctionalization and are considered as the "primary" fluorescent chlorophyll catabolites (pFCCs) of the two plant species.[40,52]

The ferredoxin-driven reduction of RCC (7) apparently does not dependent on cofactors. This puzzling observation may be rationalized by the outcome of an electrochemical reduction of the methyl ester Me-7 of the red chlorophyll catabolite (available from chemical synthesis,[46] see Scheme 3), which produced the strongly luminescent tetrapyrroles Me-6 and Me-6-*epi* directly (but with little stereoselectivity and modest regioselectivity),[53] the methyl esters of the two epimeric pFCCs (6 and 6-*epi*). These electrochemical experiments suggested that i) RCC (7) might be inherently sufficiently redox-active to undergo ferredoxin-driven and enzyme mediated reduction to 6 or 6-*epi* and ii) the reduction of RCC by RCC-reductase may come about in enzyme controlled steps involving single electron reductions, followed by stereo- and regio-controlled protonation.[53]

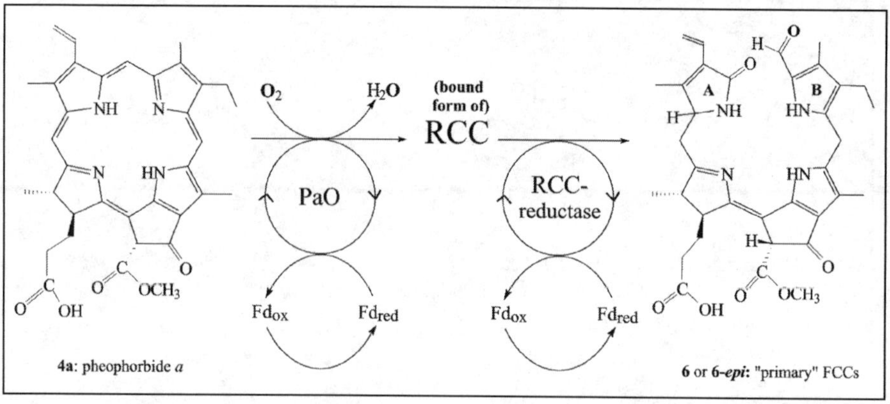

Scheme 4. The two enzymes, pheophorbide *a* oxygenase (PaO) and RCC-reductase (RCC-R), achieve the key catabolic ring cleaving transformation of pheophorbide *a* (4a) to the "primary" FCCs (6 or 6-*epi*) in higher plants.

The Arrival at Colorless and Nonfluorescent Chlorophyll Breakdown Products

The chemically rather labile, fluorescent chlorophyll catabolites, such as the pFCCs (6 and 6-*epi*), which do not accumulate during chlorophyll breakdown in vascular plants, are transformed further to the colorless and nonfluorescent chlorophyll catabolites (NCCs, such as *Hv*-NCC-1,(2, $3^1,3^2,8^2$-trihydroxy-1,4,5,10,15,20-($22H,24H$)-octahydro-13^2- [methoxycarbonyl]-4,5-dioxo-4,5-seco-phytoporphyrinate).[2-6,14,31-36] Complete de-conjugation of the four pyrrolic unit is characteristic of the chromophore of the tetrapyrrolic NCCs (see Schemes 1 and 5). This deconjugation may directly result from a tautomerization reaction of the FCCs into that of the NCCs. Such a transformation would be indicated to be thermodynamically rather favorable,[2,4-6] in analogy to the results of studies on the tautomerization chemistry of a range of hydro-porphinoids,[54] and may possibly be achieved even under rather mild conditions by nonenzymatic reactions.[5,55]

The constitution of *Hv*-NCC-1 (2)—the first tetrapyrrolic chlorophyll derivative to be identified and structurally characterized as a degreened chlorophyll catabolite from a senescent plant[2,14]—indicated (among other important information on chlorophyll breakdown) the relevance of several peripheral refunctionalization reactions, which were tentatively assumed to occur in the later stages of the breakdown pathway.[2-6,14] Three colorless and nonfluorescent chlorophyll catabolites (NCCs) were discovered in naturally degreened senescent cotyledons of the dicot canola (*Brassica napus*),[5,31,32] whose senesce occurs under natural growth conditions. The three nonfluorescent chlorophyll catabolites, termed *Bn*-NCCs (*Bn*-NCC-1 (8a), *Bn*-NCC-2 (8b), *Bn*-NCC-3 (8c)), were found to account for practically all of the chlorophyll broken down in the senescent cotyledons of oilseed rape (see Scheme 5).[31,32] In a variety of other senescent higher plants, such as the autumn leaves of sweet gum (*Liquidambar styraciflua*)[33] and of the tree *Cercidiphyllum japonicum* (the *Cj*-NCCs),[34,55] as well as in naturally degreened leaves of spinach (the *So*-NCCs)[35,36] several further NCCs were isolated. Most notably, the basic structure of all the known NCCs turned out to be the same as the one of *Hv*-NCC-1 (2) from barley and to be derived from chlorophyll a (1a) by an oxygenolytic ring opening at the α-meso position.

The indicated peripheral refunctionalizations are likely to arise in enzyme catalyzed processes, such as the remarkable and apparently characteristic terminal hydroxylation at the terminal position of the ethyl side chain at ring B.[2-6,31-36] This refunctionalization by a polar hydroxy group appears to serve the purpose of increasing the polarity of the catabolites and of providing an anchor point for further, secondary refunctionalization with hydrophilic groups. Reanalysis of extracts of senescent leaves of the tree *Cercidiphyllum japonicum* confirmed the dominant presence of the known nonfluorescent chlorophyll catabolite (*Cj*-NCC-1, 9a),[34] but led to the discovery of another, less

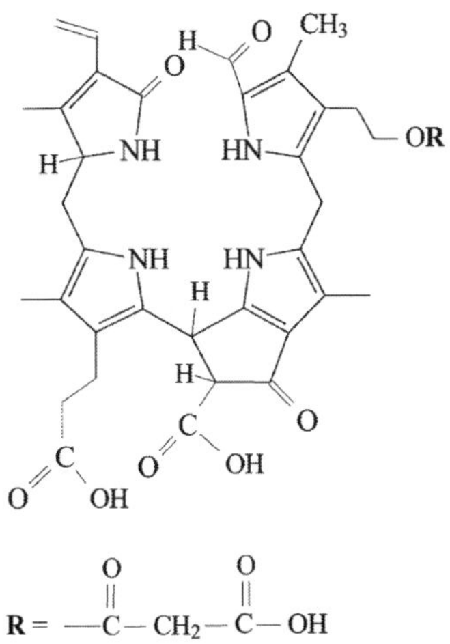

Scheme 5. Structural formulae of nonfluorescent chlorophyll catabolites from oil seed rape (*Bn*-NCC-1 (8a), *Bn*-NCC-2 (8b), *Bn*-NCC-3 (8c).

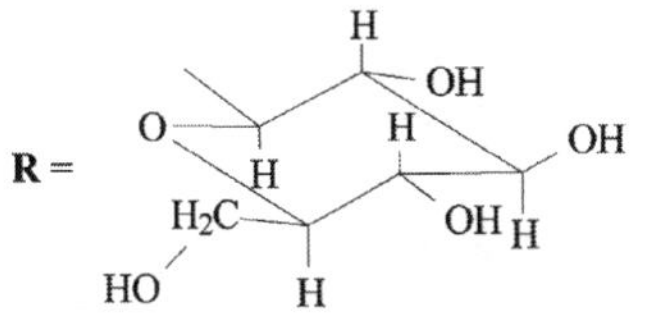

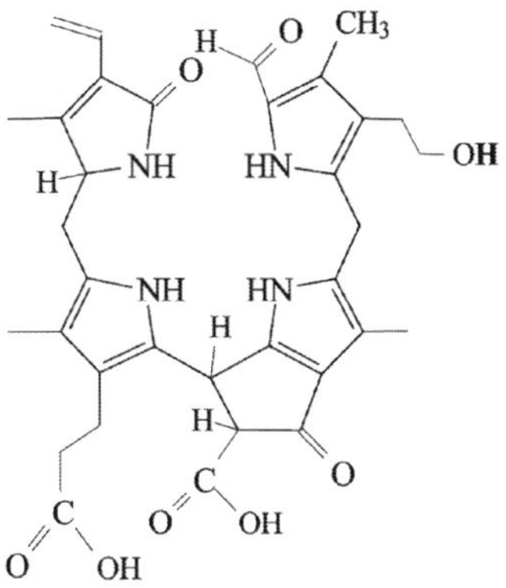

polar NCC (*Cj*-NCC-2, 9b).[55] The molecular formula of 9b was indicated to be the same as the one of the pFCCs (6 and 6-*epi*)[40,55] and spectroscopic analysis revealed it to be a $3^1,3^2$-didehydro-1,4,5,10,15,20-(*22H,24H*)- octahydro-13^2-(methoxycarbonyl)-4,5-dioxo-4,5-seco-phytoporphyrinate, i.e., to have the same structure as *Cj*-NCC-1 (9a) except for the crucial hydroxy-function at ring B.

The appearance of the NCCs in the vacuoles of senescent leaves, at first, was considered puzzling and revealed a new facette of the hypothetical intracellular transport processes involved in chlorophyll breakdown (see Fig. 1).[25,26,37] This finding, as well as the timing of the isomerization of FCCs into the corresponding NCCs during chlorophyll breakdown, are a matter of renewed current interest.[3-6] Indeed, considering the low pH values typical of the vacuoles, these "storage organs of the plant cell" now appear not only to be the final storage vessel for the NCCs, but they are also the likely sites for the final isomerization of FCCs to NCCs.[3,13,55-57] Indeed, exploratory experiments with an authentic FCC (*Ca*-FCC-2, 6-*epi*) showed the FCCs to be "programmed" to undergo a nonenzymatic acid-induced isomerization to the corresponding NCCs (see Scheme 6):[55] The pFCC 6-*epi* isomerized at pH 4.9 at room temperature to give the NCC 10 in a highly stereoselective nonenzymatic reaction. The NCC 10 was identified with *Cj*-NCC-2 (9b), which is to be considered as the structurally simplest NCC (of *Cercidiphyllum japonicum*).[55] The corresponding FCC methyl ester (Me-6) and its C-1 epimer (Me-6-*epi*), which were available from synthetic work,[53] were less ready to undergo similar isomerization reaction (Oberhuber M, Kräutler B, publication in preparation), indicating the propionic acid function of the FCC 6-*epi*, to play a relevant role in the isomerization to the NCC 9b.[55]

Breakdown Beyond the Stage of Colorless Tetrapyrrolic Catabolites

Endogenous breakdown of chlorophyll in senescent plant tissue beyond the stage of the nonfluorescent chlorophyll catabolites (NCCs) has not been well established and may not follow a specific pathway. Indeed, the NCCs are accumulated in the vacuoles of senescent leaves of higher plants.[11,13,58,59] The amount of *Bn*-NCCs present in degreened cotyledons from oilseed rape, corresponded roughly to the calculated amount of chlorophylls (*a* and *b*) present initially in the green leaf.[32] The total content of NCCs in degreened leaves of barley and of French beans appeared not to decrease strongly over a time of several days.[10,41,56,60] Accordingly, the NCCs were suggested to represent the final chlorophyll breakdown products in senescent vascular plants.[3-5,25,26]

However, evidence of tetrapyrrolic products of further degradation of NCCs was provided by the recent identification of colorless urobilinogenoidic linear tetrapyrroles, described as the two stereoisomers 11 and 11-*epi* (see Scheme 7)[61] in extracts of degreened primary leaves of barley. The tetrapyrroles 11 and 11-*epi* were associated with further degradation of *Hv*-NCC-1 (2), from which their constitution differs on account of the absence of the formyl group, which, in turn, corresponds

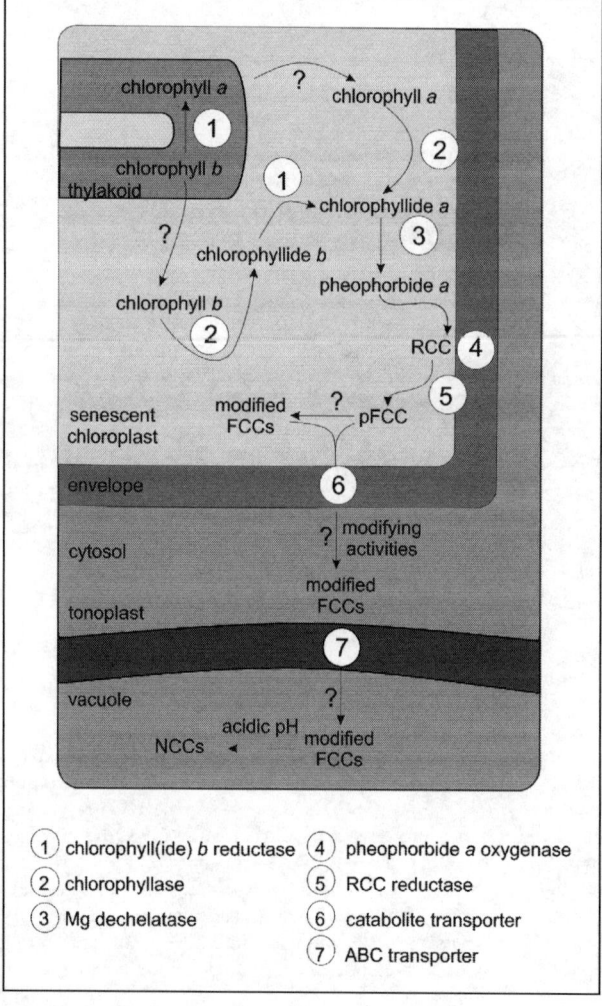

Figure 1. Topographical model of chlorophyll breakdown in senescent plants.[4,26]

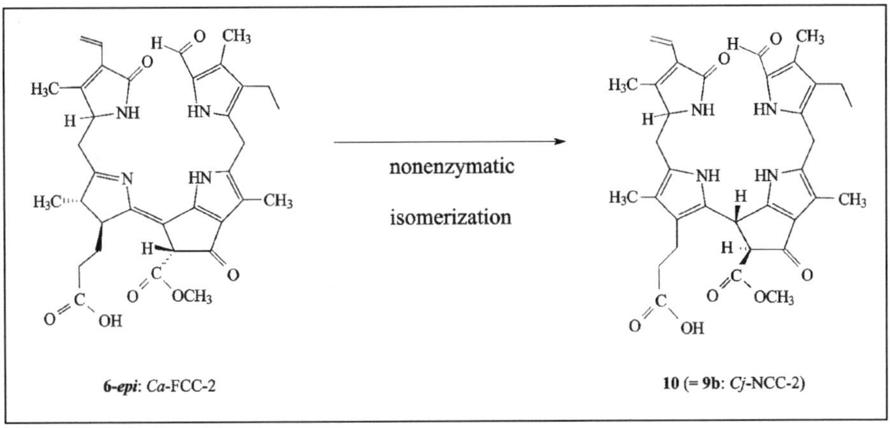

Scheme 6. A nonenzymatic isomerization transforms FCCs (such as the pFCC, *Ca*-FCC-2, *6-epi*) into NCCs (such as the *Cj*-NCC-2, 9b).

to the α-meso position of pheophorbide *a*. Accordingly, 11 and 11-*epi* were suggested to represent a new type of chlorophyll breakdown product from the degreened primary leaves of barley.[61] These newly found tetrapyrroles possibly arise from further endogenous, yet nonenzymatic, transformation of the NCCs in the tissue of the senescent barley leaves. Oxidative loss of the formyl group from related linear tetrapyrroles has been noted.[61] Indeed, the original characterization for *Hv*-NCC-1 (2) as a "rusty" pigment pointed to the readiness of these reduced linear tetrapyrroles to undergo further transformations in the presence of air (and weak acids), which become manifest by the appearance of the rust color.[2-4,14] Clearly, the later stages of further degradation of the chlorophyll catabolites will crucially depend also on the eventual fate of the degreened leaves or other plant tissues, as well as on further use and consumption by heterotrophic organisms. Fungal pathogens exploit the senescence processes in the plants and may exert control over it.[62]

The development of gerontoplasts plays a very prominent role in the recycling of nutrients, such as reductively fixed nitrogen and magnesium ions from senescent leaves to other parts of the plant.[60]

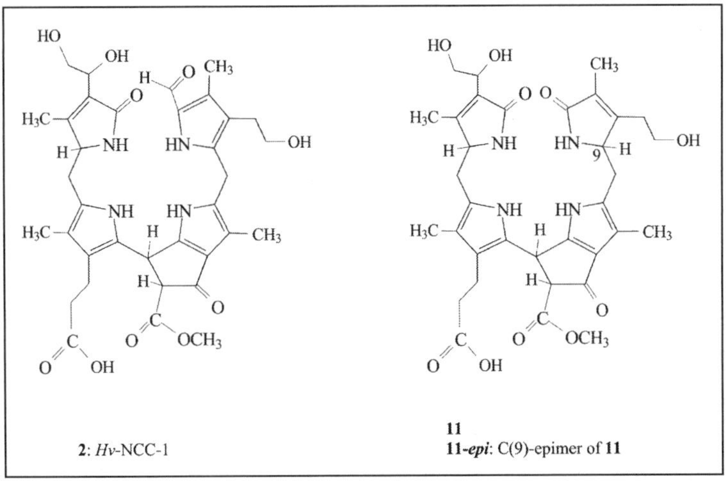

Scheme 7. Urobilinogenoidic linear tetrapyrroles 11 and 11-*epi*[61] from further oxidative degradation of the NCC *Hv*-NCC-1 (2)

There is some evidence of further (slower) breakdown of chlorophyll beyond the stage of tetrapyrroles and with formation of monopyrrolic products.[63,64] However, chlorophyll breakdown appears to be aimed primarily at the dismantling of chlorophyll protein complexes as a prerequisite of protein degradation,[26,65] rather than at reusing the four nitrogen atoms of the chlorin macrocycle (which represents only a few percent of total leaf nitrogen).[3-6] Dismantling of the chlorophyll protein complexes should be considered problematic for the living plant cell and the sophisticated machinery of chlorophyll catabolism must be interpreted as serving a vitally important detoxification process.[5,26]

Chlorophyll Catabolites from Other Sources

In the last fifteen years, significant steps have been accomplished towards the elucidation of chlorophyll breakdown in barley, oilseed rape and other higher plants.[2-6,26] Important parallel work has revealed the structures of chlorophyll catabolites from a green alga (*Chlorella protothecoides*)[44,45,66] and from marine organisms (the dinoflagellate *Pyrocystis lunula* and krill).[67,68] The pigments from *C. protothecoides* were determined to be linear tetrapyrroles with the same basic skeleton as found in the colorless chlorophyll catabolite *Hv*-NCC-1 (2), i.e., also to be formed from chlorophylls by an oxygenolytic cleavage of the macroring at the "northern" meso-position catalyzed by a monooxygenase.[69] However, different to the plant systems, the red catabolites from the alga were found to be derived from chlorophyll *a* (1a) as well as from chlorophyll *b* (1b).[44,45,70] In more general contrast, the tetrapyrrolic breakdown products from the marine systems were seen to be related to the chlorophylls by an oxidative opening at the western meso-position of the porphinoid macroring.[67,68,71]

Conclusion and Outlook

Gradually chlorophyll catabolism has revealed some of the major aspects of its "enigmatic" nature.[2-6,26] A well-controlled and sequentially operating enzymatic and chemical machinery appears as being characteristic of the degradation of the chlorophylls in higher plants (see Scheme 8).[71] A surprising (albeit still distant) parallel between heme catabolism in photosynthetic organisms[72] and

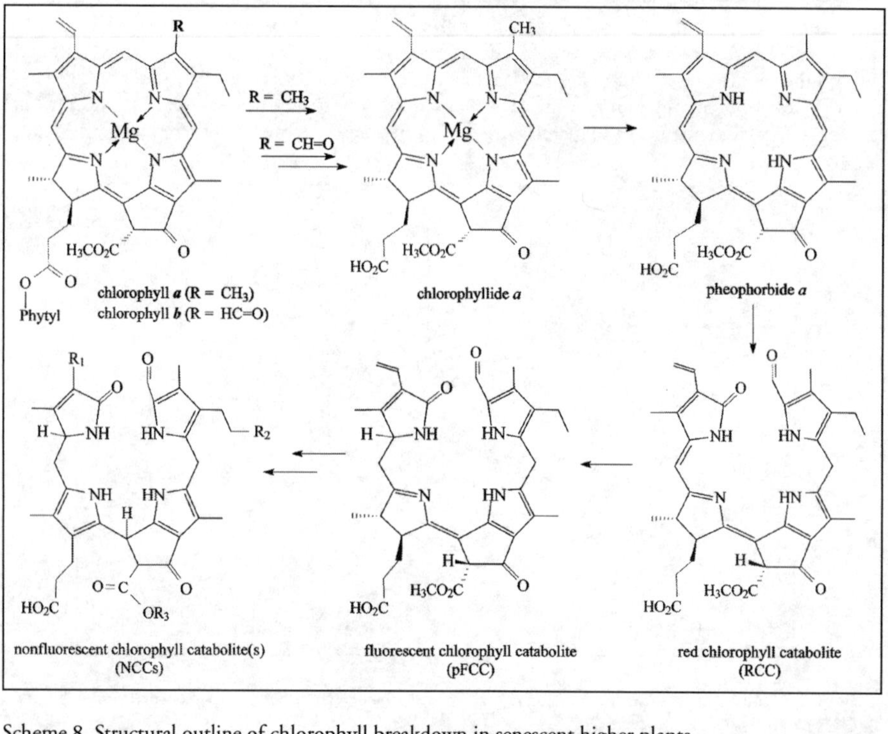

Scheme 8. Structural outline of chlorophyll breakdown in senescent higher plants.

chlorophyll breakdown in higher plants[71] has become apparent: The key step in both degradative pathways is implied to be an oxygenolytic cleavage of the porphinoid macrocycle involving the α-meso position (rather than the δ-meso position, as formerly assumed to occur in chlorophyll breakdown[1]).[2,4] The subsequent reductive enzymatic transformations in both degradative pathways also depend upon remarkably homologous reductases.[50,51,72]

In chlorophyll breakdown in higher plants, the key oxygenolytic cleavage occurs with pheophorbide *a* as an intriguingly specific substrate for a membrane-bound and specifically senescence-induced monooxygenase. In turn, this specificity requires reductive conversion of the *b*-type to *a*-type chlorophylls.[23] Clearly, the result of this primary ring cleavage reaction and of the subsequent (coupled) step(s) rapidly convert the chromophore of a photoactive and intensely colored chlorin into that of a colorless tetrapyrrole with de-conjugated heterocyclic rings.[3-6] Chlorophyll breakdown in senescent leaves may be classified, first of all, as a detoxification process.[3,4,26] A major goal will be to find and characterize the crucial ring-opening oxygenase, an elusive enzyme that has left its marks in early genetics already, in the investigations by Mendel.[73] The questions also continue to be of considerable interest, whether (i) colorless and nonfluorescent tetrapyrrolic chlorophyll catabolites do constitute the final products of controlled chlorophyll breakdown in senescent higher plants, and (ii) whether such colorless tetrapyrrolic remnants of the chlorophylls in the senescent leaves are to be considered as mere "waste" or whether they have a further physiological role in the plant. Another area of considerable interest will be the catabolism of chlorophylls in the marine environment, an important reservoir of photosynthetic activity with porphinoid pigments. Chlorophyll breakdown clearly will continue to be a fascinating natural phenomenon.

Acknowledgements

I would like to thank Philippe Matile and Stefan Hörtensteiner for their experimentally and intellectually fruitful collaboration, and Walter Mühlecker, Michael Oberhuber, Joachim Berghold and Kathrin Breuker, former and present members of my group for their important contributions to this work. I am grateful for financial support by the Austrian National Science Foundation (FWF, projects No. P13503-MOB and P16097-B07).

References

1. Brown SB, Houghton JD, Hendry GAF. Chlorophyll breakdown. In: Scheer H, ed. Chlorophylls. Boca Raton: CRC-Press USA, 1991:465-489.
2. Kräutler B, Jaun B, Bortlik K et al. On the enigma of chlorophyll degradation: The constitution of a secoporphinoid catabolite. Angew Chem Int Ed Engl 1991; 30:1315-1318.
3. Matile P, Hörtensteiner S, Thomas H et al. Chlorophyll breakdown in senescent leaves. Plant Physiol 1996; 112:1403-1409.
4. Kräutler B, Matile P. Solving the riddle of chlorophyll breakdown. Acc Chem Res 1999; 32:35-43.
5. Hörtensteiner S, Kräutler B. Chlorophyll breakdown in oilseed rape. Photosynth Res 2000; 64:137-146.
6. Kräutler B. Unravelling chlorophyll catabolism in higher plants. Biochem Soc Trans 2002; 30:625-630.
7. Scheer H, ed. Chlorophylls. Boca Raton: CRC-Press, 1991.
8. Matile P, Ginsburg S, Schellenberg M et al. Catabolites of chlorophyll in senescent leaves. J Plant Physiol 1987; 129:219-228.
9. Thomas H, Bortlik K, Rentsch D et al. Catabolism of chlorophyll in vivo: Significance of polar chlorophyll catabolites in a nonyellowing senescence mutant of Festuca pratensis Huds. New Phytol 1989; 111:3-8.
10. Bortlik K, Peisker C, Matile P. A novel type of chlorophyll catabolite in senescent barley leaves. J Plant Physiol 1990; 136:161-165.
11. Matile P, Ginsburg S, Schellenberg M et al. Catabolites of chlorophyll in senescing barley leaves are localized in the vacuoles of mesophyll cells. Proc Natl Acad Sci USA 1988; 85:9529-9532.
12. Peisker C, Thomas H, Keller F et al. Radiolabelling of chlorophyll for studies on catabolism. J Plant Physiol 1990; 136:544-549.
13. Matile P. The vacuole and cell senescence. Adv Bot Res 1997; 25:87-112.
14. Kräutler B, Jaun B, Amrein W et al. Breakdown of chlorophyll: Constitution of a secoporphinoid chlorophyll catabolite isolated from senescent barley leaves. Plant Physiol Biochem 1992; 30:333-346.
15. Willstätter R, Stoll A. Investigations on Chlorophyll. Lancaster: Science Printing Press, 1928.
16. Rüdiger W. Chlorophyll metabolism: From outer space down to the molecular level. Phytochemistry 1997; 46:1151-1167.

17. Thomas H, Hilditch P. Metabolism of thylakoid membrane proteins during foliar senescence. In: Thomas WW, Nothnagel EA, Huftakter RC, eds. Plant Senescence: Its Biochemistry and Physiology. Rockville: Am Soc Plant Physiologists, 1987:114-122.
18. Ito H, Tanaka Y, Tsuji H et al. Conversion of chlorophyll b to chlorophyll a by isolated cucumber etioplasts. Arch Biochem Biophys 1993; 306:148-151.
19. Ito H, Tanaka A. Determination of the activity of chlorophyll b to chlorophyll a conversion during greening of etiolated cucumber cotyledons by using pyrochlorophyllide b. Plant Physiol Biochem 1996; 34:35-40.
20. Scheumann V, Ito H, Tanaka A et al. Substrate specifity of chlorophyll(ide) b reductase in etioplasts of barley (Hordeum vulgare L). Eur J Biochem 1996; 242:163-170.
21. Scheumann V, Schoch S, Rüdiger W. Chlorophyll b reduction during senescence of barley seedlings. Planta 1999; 209:364-370.
22. Tanaka A, Ito H, Tanaka R et al. Chlorophyll a oxygenase (CAO) is involved in chlorophyll b formation from chlorophyll a. Proc Natl Acad Sci USA 1998; 95:12719-12723.
23. Rüdiger W. The last step of chlorophyll synthesis. In: Kadish KM, Smith KM, Guilard R, eds. The Porphyrin Handbook. Vol 13. New York: Academic Press, 2003:71-108.
24. Hörtensteiner S, Vicentini F, Matile P. Chlorophyll breakdown in senescent cotyledons of rape, Brassica napus L: Enzymatic cleavage of pheophorbide a in vitro. New Phytol 1995; 129:237-246.
25. Hörtensteiner S. Chlorophyll breakdown in higher plants and algae. Cell Mol Life Sci 1999, 56:330-347.
26. Matile P, Hörtensteiner S, Thomas H. Chlorophyll degradation. Annu Rev Plant Physiol Plant Mol Biol 1999; 50:67-95.
27. Shioi Y, Tomita N, Tsuchiya T et al. Conversion of chlorophyllide to pheophorbide by Mg-dechelating substance in extracts of Chenopodium album. Plant Physiol Biochem 1996; 34:41-47.
28. Langmeier M, Ginsburg S, Matile P. Chlorophyll breakdown in senescent leaves: Demonstration of Mg-dechelatase activity. Physiol Plant 1993; 89:347-353.
29. Shioi Y, Tatsumi Y, Shimokawa K. Enzymatic degradation of chlorophyll in Chenopodium album. Plant Cell Physiol 1991; 32:87-93.
30. Shioi Y, Watanabe K, Takamiya K. Enzymatic conversion of pheophorbide a to the precursor of pyropheophorbide a in leaves of Chenopodium album. Plant Cell Physiol 1996; 37:1143-1149.
31. Mühlecker W, Kräutler B, Ginsburg S et al. Breakdown of chlorophyll: A tetrapyrrolic chlorophyll catabolite from senescent rape leaves. Helv Chim Acta 1993; 76:2976-2980.
32. Mühlecker W, Kräutler B. Breakdown of chlorophyll: Constitution of nonfluorescing chlorophyll catabolites from senescent cotyledons of the dicot rape. Plant Physiol Biochem 1996; 34:61-75.
33. Iturraspe J, Moyano N, Frydman B. A new 5-formlybilinone as the major chlorophyll a catabolite in tree senescent leaves. J Org Chem 1995; 60:6664-6665.
34. Curty C, Engel N. Detection, isolation and structure elucidation of a chlorophyll a catabolite from autumnal senescent leaves of Cercidiphyllum japonicum. Phytochemistry 1996; 42:1531-1536.
35. Oberhuber M, Berghold J, Mühlecker W et al. Chlorophyll breakdown—On a nonfluorescent chlorophyll catabolite from spinach. Helv Chim Acta 2001; 84:2615-2627.
36. Berghold J, Breuker K, Oberhuber M et al. Chlorophyll breakdown in spinach: On the structure of five nonfluorescent chlorophyll catabolites. Photosynth Res 2002, 74:109-119.
37. Matile P, Schellenberg M, Peisker C. Production and release of a chlorophyll catabolite in isolated senescent chloroplasts. Planta 1992; 187:230-235.
38. Bachmann A, Fernández-López J, Ginsburg S et al. Stay-green genotypes of Phaseolus vulgaris L: Chloroplast proteins and chlorophyll catabolites during foliar senescence. New Phytol 1994; 126:593-600.
39. Ginsburg S, Schellenberg M, Matile P. Cleavage of chlorophyll-porphyrin. Requirement for reduced ferredoxin and oxygen. Plant Physiol 1994; 105:545-554.
40. Mühlecker W, Ongania KH, Kräutler B et al. Tracking down chlorophyll breakdown in plants: Elucidation of the constitution of a "fluorescent" chlorophyll catabolite. Angew Chem Int Ed Engl 1997; 36:401-404.
41. Ginsburg S, Matile P. Identification of catabolites of chlorophyll-porphyrin in senescent rape cotyledons. Plant Physiol 1993; 102:521-527.
42. Schellenberg M, Matile P, Thomas H. Breakdown of chlorophyll in chloroplasts of senescent barley leaves depends on ATP. J Plant Physiol 1990; 136:564-568.
43. Schellenberg M, Matile P, Thomas H. Production of a presumptive chlorophyll catabolite in vitro: Requirement for reduced ferredoxin. Planta 1993; 191:417-420.
44. Engel N, Curty C, Gossauer A. Chlorophyll catabolism in Chlorella protothecoides. Part 8: Facts and artifacts. Plant Physiol Biochem 1996; 34:77-83.

45. Gossauer A, Engel N. Chlorophyll catabolism—structures, mechanisms, conversions. J Photochem Photobiol 1996; 32:141-151.
46. Kräutler B, Mühlecker W, Anderl M et al. Breakdown of chlorophyll: Partial synthesis of a putative intermediary catabolite. Helv Chim Acta 1997; 80:1355-1362.
47. Rodoni S, Vicentini F, Schellenberg M et al. Partial purification and characterization of red chlorophyll catabolite reductase, a stroma protein involved in chlorophyll breakdown. Plant Physiol 1997; 115:677-682.
48. Hörtensteiner S, Wüthrich K, Matile P et al. The key step in chlorophyll breakdown in higher plants: Cleavage of pheophorbide a macrocycle by a monooxygenase. J Biol Chem 1998; 273:15335-15339.
49. Rodoni S, Mühlecker W, Anderl M et al. Chlorophyll breakdown in senescent chloroplasts. Cleavage of pheophorbide a in two enzymatic steps. Plant Physiol 1997; 115:669-676.
50. Wüthrich KL, Bovet L, Hunziker PE et al. Molecular cloning, functional expression and characterization of RCC reductase involved in chlorophyll catabolism. Plant J 2000; 21:189-198.
51. Hörtensteiner S, Rodoni S, Schellenberg M et al. Evolution of chlorophyll degradation: The significance of RCC reductase. Plant Biol 2000; 2:63-67.
52. Mühlecker W, Kräutler B, Matile P et al. Breakdown of chlorophyll: A fluorescent chlorophyll catabolite from sweet pepper (Capsicum annuum). Helv Chim Acta 2000; 83:278-286.
53. Oberhuber M, Kräutler B. Breakdown of chlorophyll: Electrochemical bilin reduction provides synthetic access to fluorescent chlorophyll catabolites. Chem Bio Chem 2002; 3:104-107.
54. Eschenmoser A. Vitamin B$_{12}$. Experimental work on the question of the origin of its molecular structure. Angew Chem 1988; 27:5-40.
55. Oberhuber M, Berghold J, Breuker K et al. Breakdown of chlorophyll: A nonenzymatic reaction accounts for the formation of the colorless "nonfluorescent" chlorophyll catabolites. Proc Natl Acad Sci USA 2003; 100:6910-6915.
56. Matile P. Chloroplast senescence. In: Baker NR, Thomas H, eds. Crop Photosynthesis: Special and Temporal Determinants. Amsterdam: Elsevier Science Publisher, 1992:413-440.
57. Staehelin AL, Newcomb EH. In: Buchanan BB, Gruissem W, Jones RL, eds. Biochemistry and Molecular Biology of Plants. Rockville: Am Soc Plant Physiologists, 2001:25-27.
58. Matile P, Düggelin T, Schellenberg M et al. How and why is chlorophyll broken down in senescent leaves. Plant Physiol Biochem 1989; 27:595-604.
59. Hinder B, Schellenberg M, Rodoni S et al. How plants dispose of chlorophyll catabolites. Directly energized uptake of tetrapyrrolic breakdown products into isolated vacuoles. J Biol Chem 1996; 271:27233-27236.
60. Matile P. Senescence in plants and its significance for nitrogen economy. Chimia 1987; 41:376-381.
61. Losey FG, Engel N. Isolation and characterization of a urobilinogenoidic chlorophyll catabolite from Hordeum vulgare. J Biol Chem 2001; 276:8643-8647.
62. Hammond-Kosack K, Jones JDG. Responses to plant pathogens. In: Buchanan BB, Gruissem W, Jones RL, eds. Biochemistry and Molecular Biology of Plants. Rockville: Am Soc Plant Physiologists, 2001:1102-1156.
63. Suzuki Y, Shioi Y. Detection of chlorophyll breakdown products in the senescent leaves of higher plants. Plant Cell Physiol 1999; 40:909-915.
64. Llewellyn CA, Fauzi R, Mantoura C et al. Products of chlorophyll photodegradation-2. Structural identification. Photochem Photobiol 1990; 52:1043-1047.
65. Thomas H. Chlorophyll: A symptom and a regulator of plastid development. New Phytol 1997; 136:163-181.
66. Engel N, Jenny TA, Mooser V et al. Chlorophyll catabolism in Chlorella protothecoides—Isolation and structure elucidation of a red bilin derivative. FEBS Lett 1991; 293:131-133.
67. Nakamura H, Musicki B, Kishi Y. Structure of the light emitter in krill bioluminescence. J Am Chem Soc 1988; 110:2683-2685.
68. Nakamura H, Kishi Y, Shimomura O et al. Structure of dinoflagellate Luciferin and its enzymatic and nonenzymatic air-oxidation products. J Am Chem Soc 1989; 110:7607-7611.
69. Curty C, Engel N, Gossauer A. Evidence for a monooxygenase-catalyzed primary process in the catabolism of chlorophyll. FEBS Lett 1995; 364:41-44.
70. Iturraspe J, Engel N, Gossauer A. Chlorophyll catabolism. Isolation and structure elucidation of chlorophyll b catabolites in Chlorella protothecoides. Phytochem 1994; 35:1387-1390.
71. Kräutler B. Chlorophyll breakdown and chlorophyll catabolites. In: Kadish KM, Smith KM, Guilard R, eds. The Porphyrin Handbook. Vol 13. New York: Academic Press, 2003:183-209.
72. Frankenberg N, Lagarias JC. Biosynthesis and biological functions of Bilins. In Kadish KM, Smith KM, Guilard R, eds. The Porphyrin Handbook. Vol 13. New York: Academic Press, 2003:211-235.
73. Thomas H, Schellenberg M, Vicentini F et al. Gregor Mendel's green and yellow pea seeds. Bot Acta 1996; 109:3-4.

CHAPTER 18

Vitamin B$_{12}$: Biosynthesis of the Corrin Ring

Ross M. Graham, Evelyne Deery and Martin J. Warren*

Abstract

Vitamin B$_{12}$ is a cobalt-containing modified tetrapyrrole, whose structural complexity and beguiling chemistry has fascinated scientists for over 80 years. As with all modified tetrapyrroles, its structure is derived from uroporphyrinogen III. This transformation requires a large number of enzyme-mediated steps that result in peripheral methylation, cobalt chelation, ring contraction, decarboxylation, amidation and adenosylation. There are two related though genetically distinct routes for cobalamin biosynthesis, which are referred to as the aerobic and anaerobic pathways. In this chapter the biosynthesis of the corrin ring component of vitamin B$_{12}$ along these two routes is described.

Introduction

Out of all the modified tetrapyrroles, perhaps the most aesthetic is vitamin B$_{12}$ (cobalamin), where the combination of its sheer utter complexity and the mesmerizing chemistry that it facilitates makes it a natural curiosity. Vitamin B$_{12}$ has the architecture of a molecular gyroscope, consisting of a corrin ring to which is bound a lower nucleotide and an upper ligand constituted by either a methyl or adenosyl group (Fig. 1). It is a vitamin, of course, because it is required by humans as an essential dietary supplement.[1] Deficiency can result in pernicious anaemia, neurological disorders and diseases associated with disorders of methionine metabolism. These medical conditions relate to the fact that B$_{12}$ is required for two key metabolic enzymes, methionine synthase and methylmalonyl CoA mutase. What makes vitamin B$_{12}$ unique in comparison to other vitamins is that it is made only by certain bacteria - it is not made de novo by any eukaryote.[2]

At its simplest, cobalamin is a modified tetrapyrrole, belonging to the same family as, *inter alia*, haem and chlorophyll.[3] For the synthesis of cobalamin, the basic tetrapyrrole primogenitor, uroporphyrinogen III (Fig. 1), is modified by the peripheral addition of methyl groups, amido groups, a nucleosyl side chain and an atom of cobalt, as well as by the loss of one of the integral macrocyclic framework carbon atoms in a ring contraction step.[4-8] Such modifications are reflected in a highly complex biosynthetic pathway, requiring a total of at least 19 enzyme-mediated reactions. In this chapter we shall deal with the biosynthesis of the corrin ring component of vitamin B$_{12}$. The synthesis and attachment of the nucleotide loop is dealt with in the subsequent chapter.

Cobalamin is synthesised in vivo by at least two independent pathways (Fig. 2). On the surface, these pathways differ in their requirement for oxygen. However, this requirement is not as simple as the substitution of a single oxygen-dependent reaction. The aerobic and anaerobic pathways exhibit different chemical approaches, using different sets of enzymes, to arrive at the same end product. However, some generalisations can be made about these pathways (Figs. 2, 3). They initiate from the dipyrrocorphin precorrin-2, from which the routes diverge into aerobic and anaerobic pathways before converging with the synthesis of the intermediate adenosylcobyrinic acid *a,c*-diamide (Fig. 2). The final reactions concern the synthesis and attachment of the lower axial ligand, a modified purine nucleotide. This part is covered in detail in the subsequent chapter. The

*Contributing Author: Martin J. Warren—Protein Science Group, Department of Bioscience, University of Kent, Canterbury, Kent CT2 7NJ, UK. Email: m.j.warren@kent.ac.uk

Tetrapyrroles: Birth, Life and Death, edited by Martin J. Warren and Alison G. Smith.
©2009 Landes Bioscience and Springer Science+Business Media.

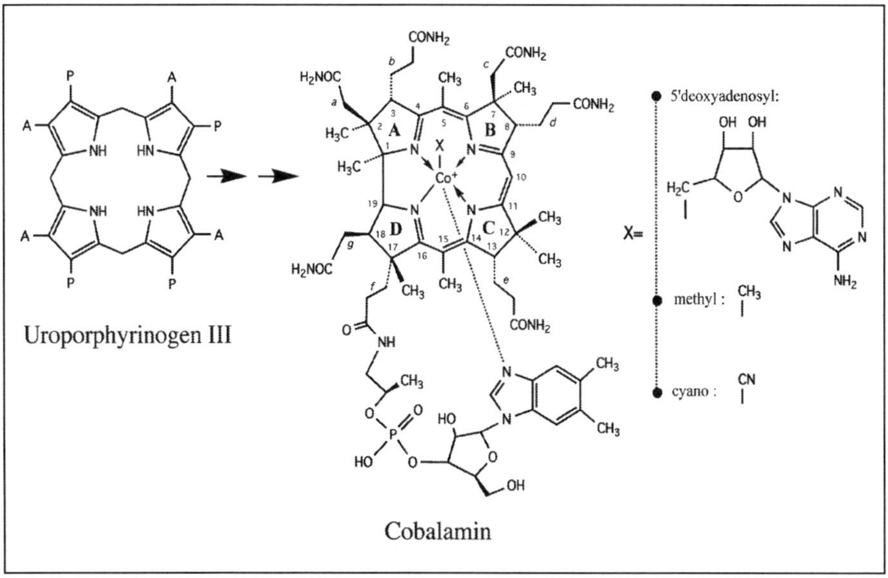

Figure 1. Uroporphyrinogen III and B_{12}. The transformation of uroporphyrinogen III into vitamin B_{12} requires a large number of enzyme-mediated steps to account for the peripheral methylation, ring contraction, decarboxylation, cobalt insertion, amidations, lower nucleotide loop assembly and attachment and upper ligand attachment. The numbering and lettering associated with cobalamin is also shown.

genetic requirements and an outline of the two pathways are shown in Figure 3. In general, the genes encoding the enzyme of the anaerobic pathway are given the prefix *cbi*, whereas the genes of the aerobic pathway are termed *cob*.

The First Common Step: Production of Precorrin-2

As shown in Figure 2, the first reaction, common to both cobalamin biosynthetic pathways, involves the methylation of uroporphyrinogen III to produce precorrin-2. The nomenclature of the intermediates ("precorrin-*n*") has been standardised to reflect the number (*n*) of methyl groups attached to the basic ring structure of the tetrapyrrole.[9] The first reaction is catalysed by *S*-adenosyl-L-methionine uroporphyrinogen III methyltransferase (SUMT, CobA).[10] This homodimer catalyses the ordered transfer of two methyl groups, each donated by *S*-adenosylmethionine (SAM), to the C2 and C7 positions of uroporphyrinogen III, respectively. The reaction is strongly and competitively inhibited by *S*-adenosyl-L-homocysteine (SAH), the breakdown product of SAM.[10] Interestingly, a second form of inhibition also acts on this reaction: uroporphyrinogen III, the substrate, inhibits the reaction at concentrations in excess of 0.5 to 2.0 μM.[10,11] Substrate inhibition of this type is likely to constitute a regulatory mechanism for the reaction and, hence, the pathways in which it is involved. However, this is not true of the reaction in all systems. For instance, the *cobA* orthologue from *Methanobacterium ivanovii*, a methanogenic member of the archeae, does not exhibit substrate inhibition.[12] Methanogenic organisms have a high requirement for coenzyme F_{430}, also a product derived from uroporphyrinogen III (Fig. 2). Thus, the difference in inhibition may reflect differences in both the quantity and type of modified tetrapyrrole required by each cell type.

The crystal structure of SUMT has recently been solved.[13] The enzyme is similar in structure to the methylase domain of CysG[14] and to a number of other methyltransferases involved in cobalamin biosynthesis, including CobI, CobJ, CobM, CobF and CobL.[15-18] This structural similarity is likely to reflect the evolution of these enzymes from a common ancestral enzyme. Mechanistically, the enzyme is likely to employ the inherent chemistry of the two substrates to promote catalysis by

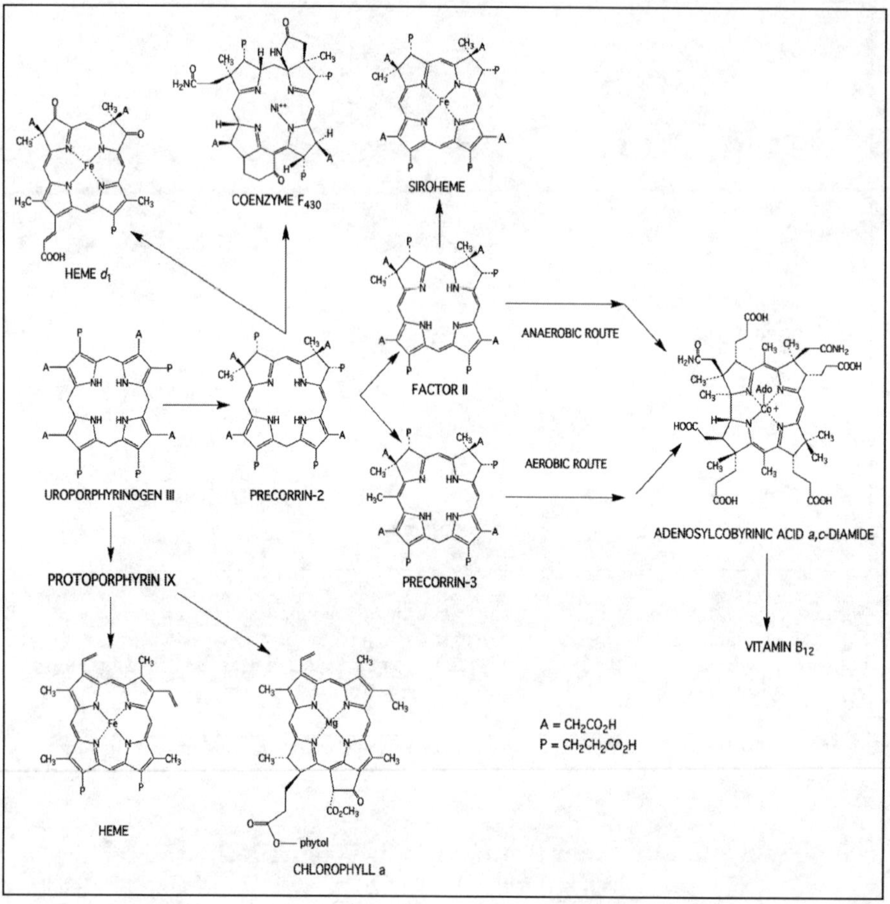

Figure 2. The central role of uroporphrinogen III in modified tetrapyrrole synthesis is shown. The aerobic and anaerobic routes for vitamin B_{12} synthesis are highlighted, showing that the two pathways diverge at the level of precorrin-2 and rejoin at cobyrinic acid *a,c*-diamide.

proximity, where the nucleophilicity of uroporphyrinogen III and electrophilicity of SAM are brought together by the enzyme to ensure the appropriate regiospecific methylation event takes place. Thus, initially, the enzyme would bind SAM in close proximity to the C-2 position of uroporphyrinogen III and subsequently, after dissociation of SAH and reorientation of precorrin-1 (Fig. 4), a new SAM molecule would be bound in close proximity to C-7 to permit synthesis of precorrin-2.

The Aerobic Pathway

An overview of the aerobic pathway is shown in Figure 5. The first reaction following divergence of the pathways along the aerobic route is the SAM-dependent methylation of precorrin-2 at the C-20 position to form precorrin-3A (Fig. 5).[19,20] This reaction is catalysed by the enzyme, SAM precorrin-2 methyltransferase (CobI)[19] a homodimer which has been reportedly isolated only from the obligate aerobe, *Pseudomonas denitrificans*. Like SUMT, it is strongly inhibited by its product SAH.[19,20] There is little mechanistic or structural information available on CobI.

The conversion of precorrin-3A to precorrin-3B (Fig. 5) is the obligate aerobic step in the aerobic cobalamin synthetic pathway. The reaction is catalysed by precorrin-3B synthase (CobG),[18]

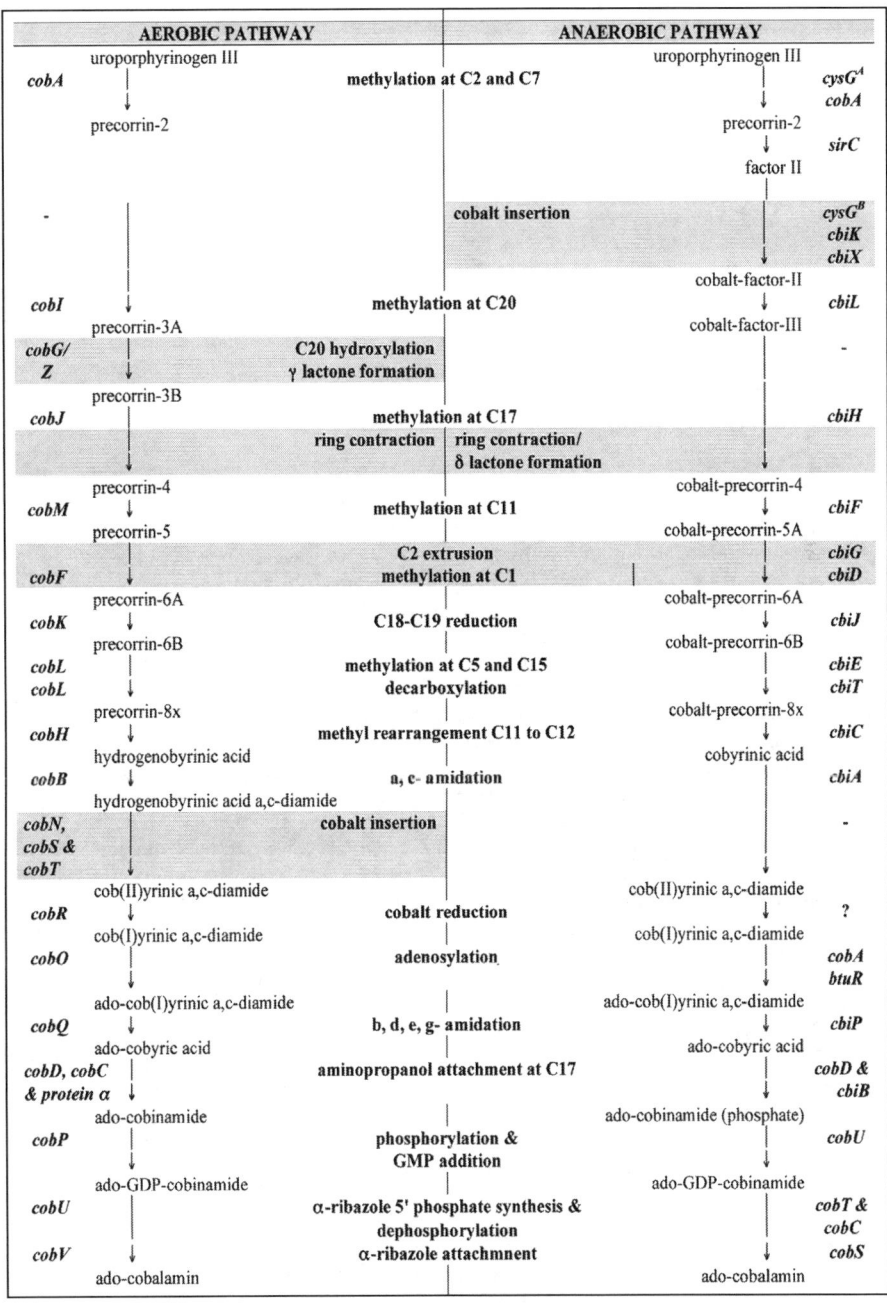

Figure 3. Pathway comparisons. The aerobic and anaerobic pathways for cobalamin biosynthesis are compared. The biochemical differences between the two pathway requires distinct repertoires of genes and enzymes, which are listed in temporal order.

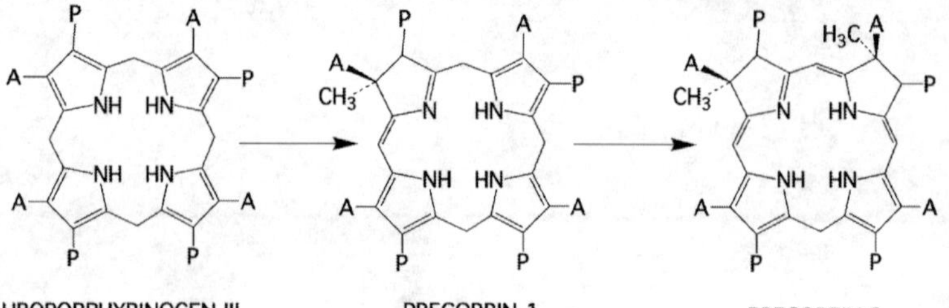

UROPORPHYRINOGEN III   **PRECORRIN-1**   **PRECORRIN-2**

Figure 4. Transformation of uroporpyrinogen III into precorrin-2. This *bis*-methylation is catalysed by the enzyme S-adenosyl-L-methionine uroporphyrinogen III methyltransferase (SUMT). The reaction proceeds in an ordered fashion, where the monomethylated intermediate, precorrin-1, is transiently released during the catalytic cycle of the enzyme.

a monooxygenase that is thought to contain a redox centre, either a single 4Fe-4S or two 2Fe-2S centres.[18,21] CobG is known to catalyse the transformation of precorrin-3A into precorrin-3B by hydroxylation at C-20 (Fig. 5). Elegant labelling studies have shown that the oxygen at C-20 is derived from molecular oxygen.[22] Subsequently, the acetate side chain attached to C-1 forma a γ-lactone with the C-1 position.[22,23] There is very little mechanistic information available about the reaction catalysed by CobG. However, the enzyme displays some similarity to sulphite reductase but in essence CobG catalyses a two electron oxidation rather than the six electron reduction mediated by sulphite reductase. It is not known how CobG binds molecular oxygen although it has been suggested this may involve a non-haem iron.[5]

Not all organisms that harbour the aerobic pathway contain a *cobG*. For instance, in *R. capsulatus* *cobG* is substituted by *cobZ*.[24] Through a series of pathway reconstitution experiments it was shown that CobG and CobZ are isofunctional and catalyse the same reaction even though they are quite distinct proteins. CobZ was shown to consist of two functional domains—an N-terminal flavin region containing a non-covalently attached FAD and a C-terminal membrane-bound region that houses a *b*-type haem.[24] The N- and C-terminal domains are connected by a bridging region that contains two 4Fe-4S centres. On the basis of the characterisation of the recombinant protein and a study of the redox groups, a mechanism has been proposed for CobZ as shown in Figure 6. In this case the reduced flavin binds oxygen and the Fe-S centres and haem are used to reduce the flavin during the catalytic cycle.

The synthesis of precorrin-3B primes the molecule for the ring contraction process, which is completed by three further steps (Fig. 5). These reactions are all mediated by SAM-dependent methyltransferases, where a methyl group is added to the C-17 (CobJ), C-11 (CobM) and C-1 (CobF) positions respectively.[18,25] The first of these methylations also initiates the ring contraction process. It is catalysed by SAM precorrin-3B methyltransferase (CobJ), and results in methylation of C-17, triggering contraction of the tetrapyrrole-derived ring between C-19 and C1.[18,21] The extruded C-20 position is left attached to C-1[26] and the intermediate formed is precorrin-4 (Fig. 5). The second of the three reactions is catalysed by SAM precorrin-4 methyltransferase (CobM)[18,25] and catalyses the transfer of a methyl group to the C-11 position of the porphyrin ring to form precorrin-5 (Fig. 5). Initially, it was suggested that this enzyme may also catalyse the deacetylation.[23] However, it was soon shown that loss of this moiety occurs after SAM-dependent methylation at C-1,[18,27] in a reaction catalysed by the SAM precorrin-5 methyltransferase (CobF). This reaction results in production of precorrin-6A (Fig. 5).[28]

The production of precorrin-3B raises the oxidation state of the intermediate above that of the later intermediate, hydrogenobyrinic acid (Fig. 5).[18,23] Thus, a reduction step is necessary somewhere between precorrin-3B and hydrogenobyrinic acid. It has been shown that precorrin-6A is reduced to precorrin-6B (Fig. 5) by the action of an NADPH-dependent precorrin-6A reductase (CobK),[28,29] which catalyses the transfer of H_R from NADPH to the C19 position of precorrin-6A.[30,31]

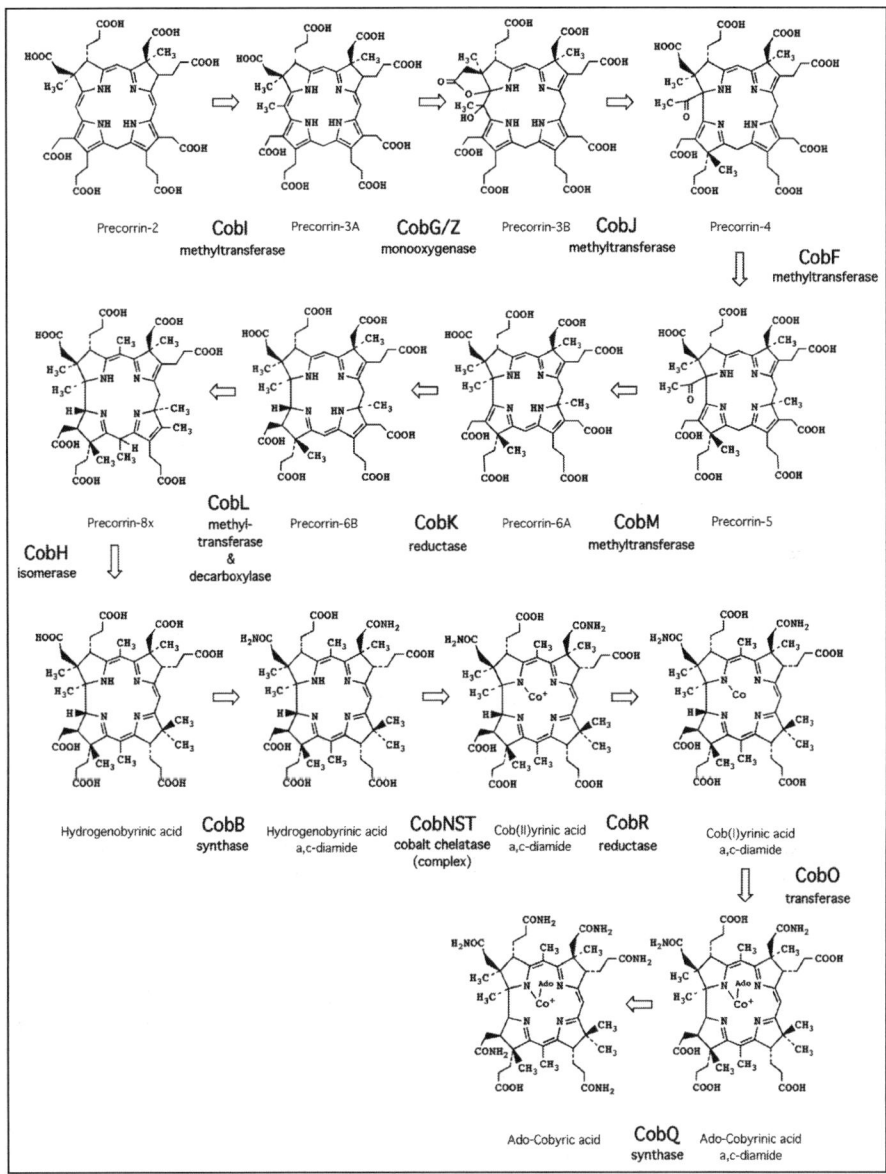

Figure 5. The aerobic pathway. This figure outlines the pathway intermediates that are formed along the aerobic cobalamin biosynthetic pathway between precorrin-2 and adenosylcobyric acid.

Production of Hydrogenobyrinic Acid

Precorrin-6B is converted into precorrin-8 by CobL (precorrin-8 synthase) (Fig. 5).[32] This enzyme catalyses not only the bismethylation of C-5 and C-15 but also the decarboxylation of the acetic acid side chain on C-12.[32] CobL appears to have arisen as the result of a gene fusion event[17] and it has been suggested, based on sequence similarity, that the N-terminal portion functions as the methyltransferase, while the C-terminal portion is involved in the decarboxylation.[32] More recently, the crystal structure of CbiT from *Methanobacterium thermoautotrophicum*, the anaerobic orthologue

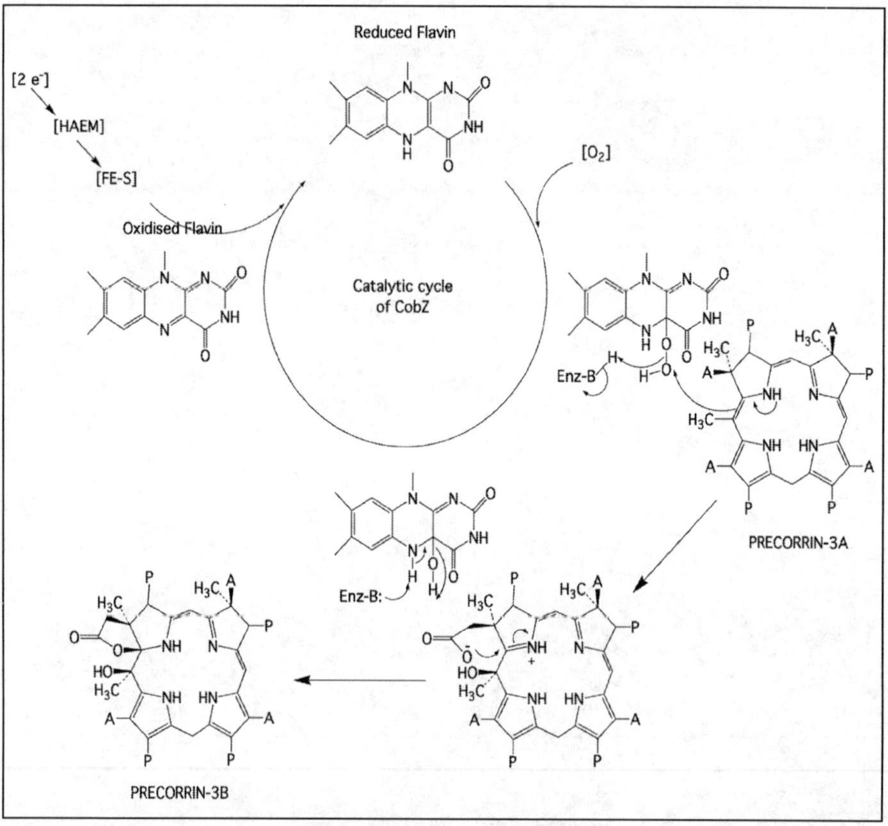

Figure 6. Catalytic cycle of CobZ in the synthesis of precorrin-3B. CobZ is isofunctional with CobG, both of which act as a monooxygenases in the hydroxylation of precorrin-3A. With CobZ, oxygen binds to a reduced flavin and hydroxylates precorrin-3A at the C-20 position. The flavin is then reduced back electrons fed via a heme and Fe-S centres to complete the catalytic cycle.

of the C-terminal portion of precorrin-8 synthase, has led to the proposal that the C-terminus of precorrin-8 synthase may catalyse methylation at C-5 and the N-terminus catalyses methylation at C-15. The C-15 methylation would then induce the decarboxylation reaction.[33] However, both scenarios, as well as the order of methylation, are yet to be tested experimentally.

The reaction catalysed by CobM saw the transfer of a methyl group to the C-11 position of the tetrapyrrole framework, but it is known that this methyl group is found attached to C-12 in vitamin B_{12}, indicating that the methyl group must rearrange during corrin ring synthesis. The rearrangement of the methyl group from C-11 to C-12 is catalysed by CobH (hydrogenobyrinic acid synthase) (Fig. 5).[34] Only one of five tautomers of precorrin-8 is a substrate for the reaction.[35] The structure of hydrogenobyrinic acid synthase indicates that selectivity is based on the size and shape of the active site being specific for the tautomer with an inflexible B-ring. Substrate binding invokes correct electrostatic interactions between the carboxylate groups of the *C* and *D* rings of precorrin-8 and the side chains of conserved residues in the enzyme.[36]

The final reactions in the aerobic pathway prior to reconvergence with the other synthetic pathways involve the transfer of six amido moieties, an atom of cobalt and an adenosyl residue to hydrogenobyrinic acid. The first of these reactions is catalysed by hydrogenobyrinic acid *a,c*-diamide synthase (CobB) (Fig. 5) which performs the energy-dependent (ATP) transfer of the amido group from each of two glutamine molecules to the acetyl side chains attached to the C7 (*c*-chain) and C2 (*a*-chain) atoms, respectively.[37]

The metal centre of cobalamin is cobalt, which along the aerobic pathway is inserted into hydrogenobyrinic acid *a,c*-diamide to generate cobyrinic acid *a,c*-diamide (Fig. 5). In *P. denitrificans*, the cobaltochelation reaction is catalysed by an oligomer consisting of the products of the *cobN*, *cobS* and *cobT* genes.[38-40] When isolated, one copy of CobN was present, and CobS and T were present as a 450 kDa component. CobN was shown to bind hydrogenobyrinic acid *a,c*-diamide and cobalt whilst the CobST component bound ATP.[40] Analysis of the primary structures suggest that CobS is an ATPase and that CobT contains an integrin I domain and a metal-ion dependent adhesion site (MIDAS).[41] Comparison with the magnesium chelatase active in the chlorophyll biosynthesis pathway of *R. capsulatus* and other photosynthetic organisms suggests that cobalt chelatase may act by forming, in the presence of ATP, a complex of CobS (possibly a homohexamer) which then binds a complex of CobT (also possibly a homohexamer). In the presence of cobalt and HBA-*a,c*-diamide, CobN binds to the CobST complex causing a change in its configuration which, in turn, results in ATP hydrolysis and insertion of the metal into the substrate.[41] If each molecule of CobS binds and hydrolyses one molecule of ATP, metal chelation by the corrin is an energetically expensive process, both in terms of the ATP hydrolysis directly involved in chelation and the energy expended in manufacture of the subunits comprising the chelatase. This is in contrast to metal insertion in the anaerobic pathway where cobalt insertion into factor II (sirohydrochlorin) is catalysed by a monomer without the involvement of ATP.[42] This is analogous to the contrast between Mg-chelation and ferrochelation at the chlorophyll/haem branchpoint. [see Chapters 4 and 14].

Immediately following the insertion of cobalt, the product, cob(II)yrinic acid-*a,c*-diamide is reduced to cob(I)yrinic acid-*a,c*-diamide by the action of a flavoprotein (Fig. 5),[43] an enzyme for which the gene has not yet been identified. Although it has been possible to purify the protein to homogeneity and to investigate some of the basic properties of the enzyme, N-terminal sequence analysis of the protein did not align with any known cobalamin biosynthetic gene. The cobalt in cobyrinic acid-*a,c*-diamide is reduced so as to allow it to act as a powerful nucleophile in the subsequent adenosylation reaction, which is catalysed by the ATP:co(I)rrinoid adenosyltransferase (CobO) (Fig. 5).[44] This reaction requires ATP as the donor of the adenosyl moiety and produces adenosylcob(III)yrinic acid-*a,c*-diamide.[4]

The final step in the biosynthesis of adenosylcobyric acid involves another series of amidation reactions, where the *b, d, e* and *g* carboxylic acid side chains are targeted (Fig. 5). This reaction is very similar to the amidation of the *a* and *c* side chains and, indeed, the enzyme that catalyses the *b, d, e* and *g* reactions, CobQ, displays some sequence similarity to CobB.[39] CobQ is a homodimeric enzyme (adenosylcobyric acid synthase), and amidates the side chains via an ordered, ATP-dependent transfer of the nitrogen groups from glutamine to the *b, d, e* and *g* side-chains of adenosylcobyrinic acid *a,c*-diamide.[45]

Proteins of Unknown Function

Two proteins, CobE and CobW, are consistently associated with the aerobic pathway but their functions have remained elusive. In *P. denitrificans*, CobE is a small protein of approximately 17 kDa.[46] It is not known whether CobE has enzymatic activity, or plays a structural or regulatory role. CobW is a 38 kDa protein which contains an ATP binding motif and may contain a NAD(H) binding site.[39] Also present in the C-terminal region is a multiple histidine-containing sequence. When present, the *cobW* gene is usually the open reading frame upstream of the *cobN* gene, for example in *P. denitrificans*, *P. aeruginosa* and *R. capsulatus*. Together, these observations suggest that CobW may be involved in cobalt chelation; however, this remains to be proven experimentally.[7]

Thus, the reactions specifically required for adaptation to an oxygen-containing environment consist of 19 modifications to precorrin-2, including transfer of many functional groups to the basic ring structure, changes to its oxidation state and expulsion of side chains. The remaining reactions, which are required to generate cobalamin, also involve some complex and interesting chemistry and are the subject of review in the subsequent chapter.

The Anaerobic Pathway

Obviously, an organism growing under anaerobic conditions will be unable to catalyse a reaction requiring molecular oxygen. The term "anaerobic pathway" suggests that the pathway operates only

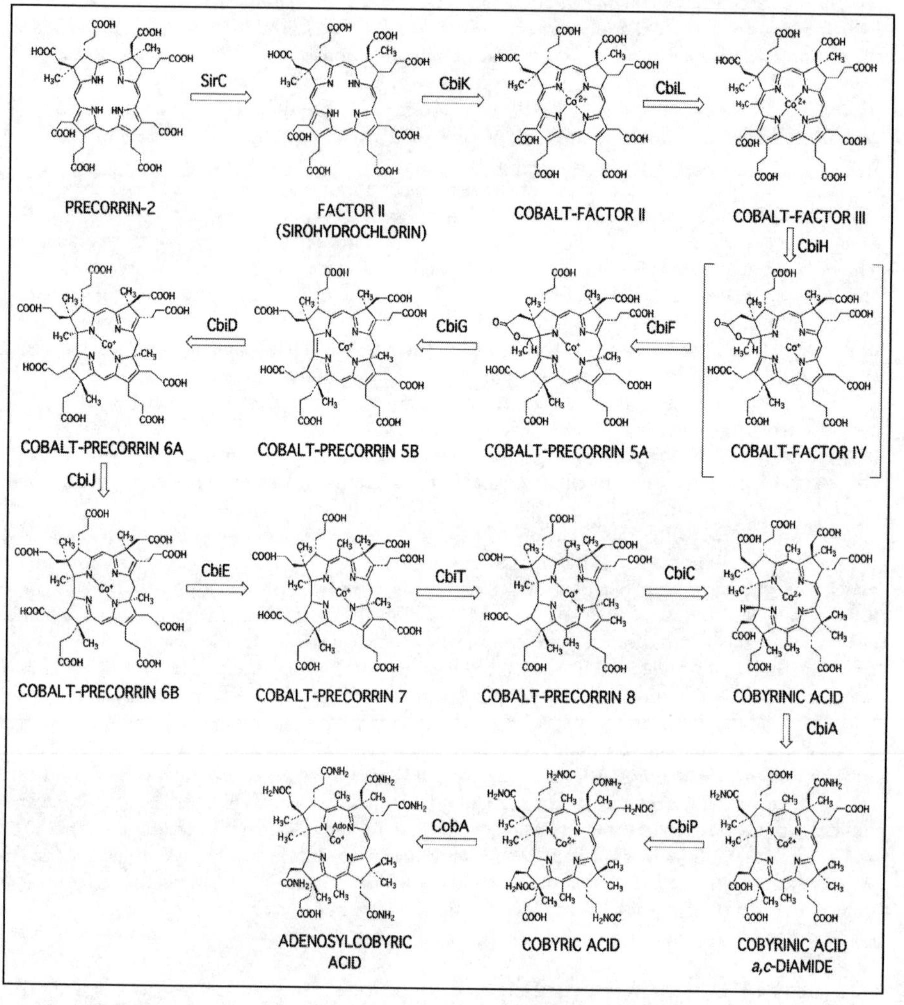

Figure 7. The anaerobic synthesis of the corrin component of B_{12}. The transformation of precorrin-2 into adenosylcobyric acid is shown.

in the absence of oxygen. In fact, by using the term "anaerobic pathway" we really mean that the pathway does not require molecular oxygen, since there are organisms that contain the anaerobic pathway but which are able to produce cobalamin under aerobic conditions. The anaerobic pathway has been largely studied in three organisms; *Salmonella enterica, Bacillus megaterium* and *Propionibacterium freudenreichii*.[2,5,47,48] Comparisons of the aerobic and anaerobic routes suggest that although the two routes are quite distinct there are also a significant number of similar enzymes employed on both routes.

An overview of the anaerobic pathway is shown in Figure 7. It had long been thought that the first committed step along the anaerobic cobalamin biosynthetic pathway was the insertion of cobalt into precorrin-2 to generate cobalt-precorrin-2. However, experimental evidence had accumulated over several years that cobalt was, in fact, inserted into the oxidised version of precorrin-2, factor-II or sirohydrochlorin (Fig. 7). More recently, specific precorrin-2 dehydrogenases have been characterised which are able to catalyse the oxidation of precorrin-2 in an NAD-dependent manner.[49] Moreover, it has also been shown that the cobaltochelatases of the anaerobic pathway have a greater specificity constant for factor II than precorrin-2.[42,50] All this evidence makes it clear that that the anaerobic pathway proceeds via factor II and cobalt-factor II (Fig. 7).

The cobaltochelatases associated with the anaerobic synthesis of vitamin B_{12} include CbiK and CbiX. The first of these enzymes to be discovered was the *S. enterica* CbiK,[51] whose structure has been solved and revealed that the enzyme has a very similar topology to the ferrochelatase of heme synthesis.[52] By analogy with porphyrin ferrochelatases,[53-55] the mechanism of cobalt insertion probably involves distortion of the bound tetrapyrrole-derived substrate into a puckered, or crown-like, shape. The cobaltochelatase has two active site histidine residues, which not only may assist in the binding of cobalt but may also play a role in the deprotonation of the macrocycle.[52] In organisms such as *B. megaterium*, CbiK is substituted by CbiX, an enzyme with a very low level of similarity to CbiK but is yet isofunctional.[42] The CbiX protein has recently been shown to contain an 4Fe-4S centre although the role played by this redox group has not yet been established.[42] In the archaebacteria, orthologues of CbiX are found that are approximately half the size of the CbiX present in *B. megaterium*. Moreover, these smaller CbiX proteins, termed CbiXˢ, align with both the N-terminal and C-terminal regions of the larger CbiX enzymes, indicating that the larger CbiX proteins probably evolved from a gene duplication and fusion event of two smaller *cbiXˢ* sequences.[50] Some CbiX proteins also contain a histidine-rich region, which could act as a metal store for the enzyme.[42]

After insertion of the metal ion, the aerobic pathway turns it attention to the peripheral methylations associated with corrin biosynthesis. Thus the next step involves methylation of C-20, in a reaction catalysed by CbiL, which generates cobalt-factor III (Fig. 7).[56] This enzyme is able to methylate both cobalt-precorrin-2 and cobalt-factor II,[57] although the preferred substrate would appear to be cobalt-factor II. The product of the reaction is cobalt-factor III. The recent crystallisation of the enzyme that belongs to the larger family of methyltransferases associated with cobalamin biosynthesis including CobA, CbiH, CbiF and CbiE,[16] has given a greater insight into the regiospecific methylase activity of the enzyme.[58,59]

Production of Cobalt-Precorrin-6A

Between cobalt-factor III and cobyrinic acid, the first intermediate to be characterised was cobalt-factor IV (Fig. 7).[60] This intermediate was isolated after incubation of factor II with a cell free extract of *P. freudenreichii*. A similar compound was isolated by incubating cobalt-precorrin-3 with CbiH,[61,62] the C-17 methyltransferase. In both cases the intermediate was isolated as cobalt-factor IV but it has been assumed that the true oxidation state of the intermediate is that of cobalt-precorrin-4. Cobalt-factor IV has already undergone ring contraction. It contains a δ-lactone structure, which includes the extruded methylated C-20 position (Fig. 7). The biointermediacy of the compound was shown by its conversion into cobyrinic acid by a cell free extract.[60] The identification of cobalt-factor IV represented a major achievement on the anaerobic pathway, not least since most of the intermediates appear to be highly unstable.[63] Its structure also has implications for the mechanism of the pathway, but its synthesis also raises some further questions. For instance, although it is known that cobalt-factor III is the product of the previous enzyme, CbiL, is this intermediate reduced to cobalt-precorrin-3 before it is acted upon by CbiH? Alternatively, is cobalt-factor III the true substrate for CbiH? Certainly, the low yield of cobalt-factor IV would suggest that perhaps some other enzyme is involved in the transformation.

By analogy with the aerobic pathway, it was assumed that cobalt-precorrin-4 is methylated at C-11 by CbiF to give cobalt-precorrin-5 (Fig. 3).[5,15] CbiF was first shown to be the C-11 methyltransferase when it was shown to methylate precorrin-3 in a multienzyme cocktail of cobalamin biosynthetic enzymes.[56] The 'out-of-turn' methylation was presumed due to the high concentration of recombinant enzyme present in the incubation. The structure of CbiF has also been solved,[15] revealing it also to be a member of the class III methyltransferases.[15] Experimental proof that CbiF does catalyse the synthesis of cobalt-precorrin 5 came from the isolation of the intermediate from an incubation containing cobalt-precorrin 3 together with CbiH, CbiF and SAM.[64] Significantly, though, the researchers found that addition of CbiG to this mixture resulted in the appearance of a derivative of cobalt-precorrin 5 in which the lactone ring had been opened and the "C_2" unit, corresponding to the methylated C20 position, had been extruded.[64] From these experiments it was possible to conclude that CbiF was responsible for the synthesis of cobalt-precorrin 5A whereas CbiG generated cobalt-precorrin 5B.

Previously, it had been established that the C_2 fragment is lost as acetaldehyde rather than acetic acid as occurs during the aerobic pathway.[59,65] The enzyme that catalyses the opening of the lactone ring, and the loss of acetaldehyde represents a significant discovery, since no equivalent step has been shown in the aerobic pathway. Interestingly, CbiG displays some sequence similarity to CobE, one of the proteins of the aerobic pathway for which no function has yet been ascribed.

The subsequent step in the aerobic pathway is the postulated synthesis of cobalt-precorrin 6A, which involves methylation at C1 of cobalt-precorrin5B.[63] The enzyme that performs this reaction is thought to be CbiD. The identification of this enzyme was revealed by an elegant multi-gene cloning approach, where omission of *cbiD* led to the biosynthesis of a 1-*des* methyl cobyrinic acid analogue.[66] CbiD would appear to an atypical methyltransferase since it does not align with any of the known (class III)[16] cobalamin biosynthetic enzymes. Nonetheless, CbiD can now be placed with greater confidence in the pathway for the synthesis of cobalt-precorrin-6A (Fig. 3).

Again, by analogy with the aerobic pathway, it is assumed that cobalt-precorrin-6A is reduced to cobalt-precorrin-6B by an NADPH-dependent enzyme called CbiJ (Fig. 3).[2,17] The subsequent actions of CbiE and T result in methylations of the northern (C-5) and southern (C-15) *meso* positions as well as decarboxylation of the acetate side chain attached to C-12 and generates cobalt-precorrin-8. Interestingly, CbiE and CbiT are found separate in some organisms but fused together in others.[67,68] As discussed above, in conjunction with the aerobic pathway, the structure of CbiT from *M. thermoautotrophicum* has been determined and shown to be methyltransferase albeit a non-canonical cobalamin biosynthetic methyltransferase. On this basis it has been suggested that CbiE catalyses methylation at C-5 and CbiT catalyses C-15 methylation followed by decarboxylation of the C-12 acetyl side chain.[33] Experimental evidence for the role of CbiT as a methylase and decarboxylase has come from the synthesis of some novel (non biological) corrinoids.[69] By incubating cobalt-precorrin 3 in the presence of CbiH, F, G and T, products were synthesized that were either methylated at C15 or were methylated and decarboxylated.[69] The absence of compounds that were decarboxylated but not methylated implies that CbiT first methylates at C15 and the decarboxylates the acetic acid side chain.[63,69]

Finally, cobalt-precorrin-8 undergoes isomerisation to form cobyrinic acid, the putative isomerase being the *cbiC* gene product,[5] followed by amidation of the *a* and *c* acetyl side chains, in a reaction catalysed by the *cbiA* gene product,[70] to form cobyrinic acid-*a,c*-diamide (Fig. 7). Recent studies on the *S. enterica* CbiA have shown that it is a monomer and is able to use either ammonia or glutamine as the amido donor. Analysis of the glutaminase partial reaction demonstrated that the hydrolysis of glutamine and the synthesis of the cobyrinic acid *a,c*-diamide product are uncoupled.[70] It would appear that CbiA catalyzes the sequential amidation of the *c*- and *a*-carboxylate groups of cobyrinic acid via the formation of a phosphorylated intermediate. The final amidations of the *b, d, e* and *g* side chains are catalysed by CbiP.[71] Recently, it has been reported that this enzyme is mechanistically similar to CbiA and amidates the corrin ring in the order *e, d, b, g*.[72] The product of the CbiP reaction is cobyric acid. However, it is likely that cobyric acid is synthesised as adenosylcobyric acid and that the adenosylation reaction occurs, by analogy with the aerobic pathway, at the level of cobyrinic acid *a,c*-diamide (Figs. 5, 7).[4] The enzymes responsible for reduction of the cobalt and the adenosylation process are discussed in the accompanying chapter.

In summary, the biosynthesis of the corrin ring component of vitamin B_{12} is mediated by complex pathways constituting aerobic and anaerobic routes.[4,5,7,73] In contrast to the complete elucidation of the aerobic pathway, there are many steps within the anaerobic synthesis that need clarification. Moreover, the pathway has to be managed to ensure that the synthesis is controlled to meet the demand for the coenzyme and the availability of the micronutrient cobalt. These considerations will undoubtedly be the subject of research in the coming years.

Acknowledgements

Financial support from the BBSRC and Wellcome Trust is gratefully acknowledged.

References

1. Stabler SP, Allen RH. Vitamin B$_{12}$ deficiency as a worldwide problem. Annu Rev Nutr 2004; 24: 299-326.
2. Roth JR, Lawrence JG, Bobik TA. Cobalamin (coenzyme B$_{12}$): synthesis and biological significance. Annu Rev Microbiol 1996; 50:137-81.
3. Warren MJ, Scott AI. Tetrapyrrole assembly and modification into the ligands of biologically functional cofactors. Trends Biochem Sci 1990; 15: 486-91.
4. Warren MJ, Raux E, Schubert HL et al. The biosynthesis of adenosylcobalamin (vitamin B$_{12}$). Nat Prod Rep 2002; 19: 390-412.
5. Roessner CA, Santander PJ, Scott AI. Multiple biosynthetic pathways for vitamin B$_{12}$: variations on a central theme. Vitamins and Hormones 2001; 61: 267-297.
6. Raux E, Schubert HL, Warren MJ. Biosynthesis of cobalamin (vitamin B$_{12}$): a bacterial conundrum. Cell Mol Life Sci 2000; 57: 1880-1893.
7. Blanche F, Cameron B, Crouzet J et al. Vitamin B12: How the problem of its biosynthesis was solved. Angew Chem Int Ed Engl 1995; 34(4):383-411.
8. Battersby A.R. How nature builds the pigments of life: the conquest of vitamin B12. Science 1994; 264(5165):1551-7.
9. Uzar HC, Battersby AR, Carpenter TA et al. Biosynthesis of porphyrins and related macrocycles. 28. Development of a pulse labeling method to determine the c-methylation sequence for vitamin-B12. Chem. Soc Perkin Trans I 1987; 1689-1696.
10. Blanche F, Debussche L, Thibaut D et al. Purification and characterization of S-adenosyl-L-methionine: uroporphyrinogen III methyltransferase from Pseudomonas denitrificans. J Bacteriol 1989; 171(8):4222-31.
11. Robin C, Blanche F, Cauchois L et al. Primary structure, expression in Escherichia coli, and properties of S-adenosyl-L-methionine:uroporphyrinogen III methyltransferase from Bacillus megaterium. J Bacteriol 1991; 173(15): 4893-6.
12. Blanche F, Robin C, Couder M et al. Purification, characterization, and molecular cloning of S-adenosyl-L-methionine: uroporphyrinogen III methyltransferase from Methanobacterium ivanovii. J Bacteriol 1991; 173(15):4637-45.
13. Vevodova J, Graham RM, Raux E et al. Structure/function studies on a S-adenosyl-L-methionine-dependent uroporphyrinogen III C methyltransferase (SUMT), a key regulatory enzyme of tetrapyrrole biosynthesis. J Mol Biol 2004; 344(2): 419-33.
14. Stroupe ME, Leech HK, Daniels DS et al. CysG structure reveals tetrapyrrole-binding features and novel regulation of siroheme biosynthesis. Nat Struct Biol 2003; 10(12):1064-73.
15. Schubert HL, Wilson KS, Raux E et al. The X-ray structure of a cobalamin biosynthetic enzyme, cobalt- precorrin-4 methyltransferase. Nat Struct Biol 1998; 5(7): 585-92.
16. Schubert HL, Blumenthal RM, Cheng X. Many paths to methyltransfer: a chronicle of convergence. Trends Biochem Sci 2003; 28(6):329-35.
17. Roth JR, Lawrence JG, Rubenfield M et al. Characterization of the cobalamin (vitamin B12) biosynthetic genes of Salmonella typhimurium. J Bacteriol 1993; 175(11): 3303-16.
18. Debussche L, Thibaut D, Cameron B et al. Biosynthesis of the corrin macrocycle of coenzyme B12 in Pseudomonas denitrificans. J Bacteriol 1993; 175(22): 7430-40.
19. Thibaut D, Couder M, Crouzet J et al. Assay and purification of S-adenosyl-L-methionine:precorrin-2 methyltransferase from Pseudomonas denitrificans. J Bacteriol 1990; 172(11): 6245-51.
20. Warren MJ, Roessner CA, Ozaki S et al. Enzymatic synthesis and structure of precorrin-3, a trimethyldipyrrocorphin intermediate in vitamin B12 biosynthesis. Biochemistry 1992; 31(2):603-9.
21. Scott AI, Roessner CA, Stolowich NJ et al. Biosynthesis of vitamin B12. Discovery of the enzymes for oxidative ring contraction and insertion of the fourth methyl group. Febs Letters 1993; 331(1-2):105-8.
22. Spencer JB, Stolowich NJ, Santander PJ et al. Mechanism of the ring contraction step in vitamin-B12 biosynthesis—the origin and subsequent fate of the oxygen functionalities in precorrin-3X. J Am Chem Soc 1994; 116:4991-4992.
23. Debussche L, Thibaut D, Danzer M et al. Biosynthesis of vitamin B12: Structure of precorrin-3B, the trimethylated substrate of the enzyme catalysing ring contraction. J Chem Soc Chem Commun 1993; 1100-1103.
24. McGoldrick HM, Roessner CA, Raux E et al. Identification and characterization of a novel vitamin B12 (cobalamin) biosynthetic enzyme (CobZ) from Rhodobacter capsulatus, containing flavin, heme, and Fe-S cofactors. J Biol Chem 2005; 280(2):1086-94.
25. Roessner CA, Spencer JB, Ozaki, S et al. Overexpression in Escherichia coli of 12 vitamin B12 biosynthetic enzymes. Protein Expression and Purification 1995; 6(2):155-63.

26. Thibaut D, Debussche L, Frechet D et al. Biosynthesis of vitamin-B12—the structure of factor-IV, the oxidized form of precorrin-4. J Chem Soc Chem Commun 1993; 513-515.

27. Min CH, Atshaves BP, Roessner CA et al. Isolation, structure, and genetically-engineered synthesis of precorrin-5, the pentamethylated intermediate of vitamin-B12 biosynthesis. J Am Chem Soc 1993; 115:10380-10381.

28. Thibaut D, Debussche L, Blanche F. Biosynthesis of vitamin B12: isolation of precorrin-6x, a metal-free precursor of the corrin macrocycle retaining five S-adenosylmethionine- derived peripheral methyl groups. Proc Natl Acad Sci USA 1990; 87(22):8795-9.

29. Thibaut D, Debussche L, Blanche F. Biosynthesis of vitamin B12: structure of precorrin-6x octamethyl ester. Proc Natl Acad Sci USA 1990; 87(22):8800-4.

30. Weaver GW et al. Biosynthesis of vitamin-B12—the site of reduction of precorrin-6X. J Chem Soc Chem Commun 1991; 976-979.

31. Kiuchi F, Thibaut D, Debussche L et al. Biosynthesis of vitamin-B12—stereochemistry of transfer of a hydride equivalent from NADPH by precorrin-6X reductase. J Chem Soc Chem Commun 1992; 306-308.

32. Blanche F, Famechon A, Thibaut D et al. Biosynthesis of vitamin B12 in Pseudomonas denitrificans: the biosynthetic sequence from precorrin-6y to precorrin-8x is catalyzed by the cobL gene product. J Bacteriol 1992; 174(3):1050-2.

33. Keller JP, Smith PM, Benach J et al. The crystal structure of MT0146/CbiT suggests that the putative precorrin-8w decarboxylase is a methyltransferase. Structure (Camb) 2002; 10(11): 1475-87.

34. Thibaut D, Couder M, Famechon A et al. The final step in the biosynthesis of hydrogenobyrinic acid is catalyzed by the cobH gene product with precorrin-8x as the substrate. J Bacteriol 1992; 174(3):1043-9.

35. Thibaut D, Kiuchi F, Debussche L et al. Biosynthesis of vitamin-B12—structural studies on precorrin-8X, an octamethylated intermediate and the structure of its stable tautomer. J Chem Soc Chem Commun 1992; 982-985.

36. Shipman LW, Li D, Roessner CA et al. Crystal Structure of Precorrin-8x Methyl Mutase. Structure with Folding & Design 2001; 9(7):587-596.

37. Debussche L, Thibaut D, Cameron B et al. Purification and characterization of cobyrinic acid a,c-diamide synthase from Pseudomonas denitrificans. J Bacteriol 1990; 172(11):6239-44.

38. Cameron B, Blanche F, Rouyez MC et al. Genetic analysis, nucleotide sequence, and products of two Pseudomonas denitrificans cob genes encoding nicotinate-nucleotide: dimethylbenzimidazole phosphoribosyltransferase and cobalamin (5'-phosphate) synthase. J Bacteriol 1991; 173(19):6066-73.

39. Crouzet J, Levyschil S, Cameron B et al. Nucleotide sequence and genetic analysis of a 13.1-kilobase-pair Pseudomonas denitrificans DNA fragment containing five cob genes and identification of structural genes encoding Cob(I)alamin adenosyltransferase, cobyric acid synthase, and bifunctional cobinamide kinase-cobinamide phosphate guanylyltransferase. J Bacteriol 1991; 173(19):6074-87.

40. Debussche L, Couder M, Thibaut D et al. Assay, purification, and characterization of cobaltochelatase, a unique complex enzyme catalyzing cobalt insertion in hydrogenobyrinic acid a,c- diamide during coenzyme B12 biosynthesis in Pseudomonas denitrificans. J Bacteriol 1992; 174(22):7445-51.

41. Fodje MN, Hansson A, Hansson M et al. Interplay between an AAA module and an integrin I domain may regulate the function of magnesium chelatase. J Mol Biol 2001; 311(1):111-22.

42. Leech HK, Raux E, McLean KJ et al. Characterization of the cobaltochelatase CbiXL: evidence for a 4Fe-4S center housed within an MXCXXC motif. J Biol Chem 2003; 278(43):41900-7.

43. Blanche F, Maton L, Debussche L et al. Purification and characterization of Cob(II)yrinic acid a,c-diamide reductase from Pseudomonas denitrificans. J Bacteriol 1992; 174(22):7452-4.

44. Debussche L, Couder M, Thibaut D et al. Purification and partial characterization of Cob(I)alamin adenosyltransferase from Pseudomonas denitrificans. J Bacteriol 1991; 173(19):6300-2.

45. Blanche F, Couder M, Debussche L et al. Biosynthesis of vitamin B12: stepwise amidation of carboxyl groups b, d, e, and g of cobyrinic acid a,c-diamide is catalyzed by one enzyme in Pseudomonas denitrificans. J Bacteriol 1991; 173(19):6046-51.

46. Crouzet J, Cauchois L, Blanche F et al. Nucleotide sequence of a Pseudomonas denitrificans 5.4-kilobase DNA fragment containing five cob genes and identification of structural genes encoding S-adenosyl-L-methionine: uroporphyrinogen III methyltransferase and cobyrinic acid a,c-diamide synthase. J Bacteriol 1990; 172(10):5968-79.

47. Roessner CA, Huang KX, Warren MJ et al. Isolation and characterization of 14 additional genes specifying the anaerobic biosynthesis of cobalamin (vitamin B12) in Propionibacterium freudenreichii (P. shermanii). Microbiology 2002; 148(Pt 6):1845-53.

48. Raux E, Schubert HL, Roper JM et al. Vitamin B12; insights into biosynthesis's Mount improbable. (Review) Bioorganic Chem 1999; 27:100-118.

49. Raux E, Leech HK, Beck R et al. Identification and functional analysis of enzymes required for precorrin-2 dehydrogenation and metal ion insertion in the biosynthesis of sirohaem and cobalamin in Bacillus megaterium. Biochem J 2003; 370(Pt 2):505-16.

50. Brindley AA, Raux E, Leech HK et al A story of chelatase evolution: identification and characterization of a small 13-15-kDa "ancestral" cobaltochelatase (CbiXS) in the archaea. J Biol Chem 2003; 278(25):22388-95.

51. Raux E, Thermes C, Heathcote P et al. A role for Salmonella typhimurium cbiK in cobalamin (vitamin B12) and siroheme biosynthesis. J Bacteriol 1997; 179(10):3202-12.

52. Schubert HL, Raux E, Wilson KS et al. Common chelatase design in the branched tetrapyrrole pathways of heme and anaerobic cobalamin synthesis. Biochemistry 1999; 38(33):10660-9.

53. Al-Karadaghi S, Hansson M, Nikonov S et al. Crystal structure of ferrochelatase: the terminal enzyme in heme biosynthesis. Structure 1997; 5(11):1501-10.

54. Blackwood ME, Rush TS, Romesberg F et al. Alternative modes of substrate distortion in enzyme and antibody catalyzed ferrochelation reactions. Biochemistry 1998; 37(3):779-82.

55. Lecerof D, Fodje M, Hansson A et al. Structural and mechanistic basis of porphyrin metallation by ferrochelatase. J Mol Biol 2000; 297(1):221-32.

56. Roessner CA, Warren MJ, Santander PJ et al. Expression of 9 Salmonella typhimurium enzymes for cobinamide synthesis. Identification of the 11-methyl and 20-methyl transferases of corrin biosynthesis. FEBS Lett 1992; 301(1):73-8.

57. Spencer P, Stolowich NJ, Sumner LW et al. Definition of the redox states of cobalt-precorrinoids: investigation of the substrate and redox specificity of CbiL from Salmonella typhimurium. Biochemistry 1998; 37(42):14917-27.

58. Frank S, Deery E, Brindley AA et al. Elucidation of substrate specificity in the cobalamin (vitamin B12) biosynthetic methyltransferases; structure and function of the C20 methyltransferase (CbiL) from Methanothermobacter thermautotrophicus. J Biol Chem 2007; 282(33):23957-69

59. Wang J, Stolowich NJ, Santander PJ et al. Biosynthesis of vitamin B12: concerning the identity of the two-carbon fragment eliminated during anaerobic formation of cobyrinic acid. Proc Natl Acad Sci USA 1996; 93(25):14320-2.

60. Scott AI, Stolowich NJ, Wang J et al. Biosynthesis of vitamin B₁₂: Factor IV, a new intermediate in the anaerobic pathway. Proc Natl Acad Sci USA 1996; 93:14316-14319.

61. Santander PJ, Roessner CA, Stolowich NJ et al. How corrinoids are synthesized without oxygen: nature's first pathway to vitamin B12. Chem Biol 1997; 4(9):659-66.

62. Santander PJ, Stolowich NJ, Scott AI. Chemoenzymatic synthesis of an unnatural tetramethyl cobalt corphinoid. Bioorg Med Chem 1999; 7(5):789-794.

63. Roessner CA, Scott AI. Fine-tuning our knowledge of the anaerobic route to cobalamin (vitamin B12). J Bacteriol 2006; 188(21):7331-4.

64. Kajiwara Y, Santander PJ, Roessner CA et al. Genetically engineered synthesis and structural characterization of cobalt-precorrin 5A and -5B, two new intermediates on the anaerobic pathway to vitamin B12: definition of the roles of the CbiF and CbiG enzymes. J Am Chem Soc 2006; 128(30):9971-8.

65. Scott AI Discovering nature's diverse pathways to vitamin B12: A 35-year odyssey. J Org Chem 2003; 68(7): 2529-2539.

66. Roessner CA, Williams HJ, Scott AI. Genetically engineered production of 1-desmethylcobyrinic acid, 1-desmethylcobyrinic acid a,c-diamide, and cobyrinic acid a,c-diamide in Escherichia coli implies a role for CbiD in C-1 methylation in the anaerobic pathway to cobalamin. J Biol Chem 2005; 280(17):16748-53.

67. Raux E, Lanois A, Rambach A et al. Cobalamin (vitamin B12) biosynthesis: functional characterization of the Bacillus megaterium cbi genes required to convert uroporphyrinogen III into cobyrinic acid a,c-diamide. Biochem J 1998; 335(Pt 1):167-73.

68. Raux E, Lanois A, Warren MJ et al. Cobalamin (vitamin B12) biosynthesis: identification and characterization of a Bacillus megaterium cobI operon. Biochem J 1998; 335(Pt 1):159-66.

69. Santander PJ, Kajiwara Y, Williams HJ et al. Structural characterization of novel cobalt corrinoids synthesized by enzymes of the vitamin B12 anaerobic pathway. Bioorg Med Chem 2006; 14(3):724-31.

70. Fresquet V, Williams L, Raushel FM. Mechanism of cobyrinic acid a,c-diamide synthetase from Salmonella typhimurium LT2. Biochemistry 2004; 43(33):10619-27.

71. Raux E, Lanois A, Levillayer F et al. Salmonella typhimurium cobalamin (vitamin B12) biosynthetic genes: functional studies in S. typhimurium and Escherichia coli. J Bacteriol 1996; 178(3): 753-67.

72. Williams L, Fresquet V, Santander PJ et al. The Multiple Amidation Reactions Catalyzed by Cobyric Acid Synthetase from Salmonella typhimurium Are Sequential and Dissociative. J Am Chem Soc 2007; 129(2):294-5.

73. Blanche F, Thibaut D, Debussche L et al. Parallels and decisive differences in vitamin B₁₂ biosyntheses. Angew Chem Int Ed Engl 1993; 32(11):1651-1653.

CHAPTER 19

Conversion of Cobinamide into Coenzyme B₁₂

Jorge C. Escalante-Semerena,* Jesse D. Woodson, Nicole R. Buan
and Carmen L. Zayas

Introduction

Cobamides are unique cyclic tetrapyrroles because their structures include an upper (Coβ) and a lower (Coα) axial ligand. The Coα and Coβ ligands are important for the interaction of the cobamide with enzymes, and for the chemistry of the reaction catalyzed by the enzyme. B₁₂ (5,6-dimethylbenzimidazolylcobamide, cobalamin) is the best-known cobamide, which in its vitamin form has a cyano group as Coβ ligand, and in its coenzymic form it has a 5'-deoxyadenosine group as Coβ ligand (Fig. 1). In this chapter we review the current understanding of upper and lower ligand attachment to the ring macrocycle. Most of the knowledge reviewed here was obtained using bacterial systems; we add to the discussion recent work from our laboratory that uncovered variations in the conversion of cobinamide to B₁₂ in archaea.

While most of the work on cobamide biosynthesis has been performed in bacteria, knowledge of how archaea synthesize cobamides is very sparse.[1,2] It is known, however, that at least some archaea synthesize and require cobamides for survival. The best-studied example may be the methanogenic archaea, which require cobamides for methanogenesis from H₂ and CO₂, acetate, or methanol.[3,4] Genome sequence analysis suggests that some archaea may require cobamides for a cobamide-dependent ribonucleotide reductase required for DNA synthesis. This idea is supported by the purification of active cobamide-dependent ribonucleotide reductases from the archaea *Thermoplasma acidophilum* and *Pyrococcus furiosus*.[5,6] Genome sequence analysis of archaeal genomes has also provided insights into cobamide biosynthesis in archaea. To date, every sequenced genome appears to encode several orthologs to genes required for nucleotide loop assembly and several have orthologs to most of the genes required for the de novo pathway. Analyses of the cobalamin biosynthetic genes in archaea suggest these prokaryotes may synthesize cobamides via pathways more related to the bacterial anaerobic pathway than to the aerobic pathway.[7]

Attachment of 5-Deoxyadenosine, the Upper (Coβ) Ligand of Coenzyme B₁₂

In this section we discuss upper ligand attachment because in *S. enterica* and probably in other prokaryotes, both de novo corrin ring biosynthesis and salvaging of exogenous, incomplete corrinoids proceeds via adenosylated intermediates.[8] Formation of the Co-C bond between the corrin ring and the upper ligand requires the Co ion of the ring to be in its Co(I) oxidation state before the 5'-deoxyadenosyl moiety of ATP can be enzymatically transferred to it. The existence of the corrinoid adenosylation pathway was first described in *Propionibacterium freundenreichii* ssp *shermanii* and *Clostridium tetanomorphum*,[9,10] but the identity of the gene encoding the ATP:co(I)rrinoid adenosyltransferase enzyme was first established in *E. coli*, *S. enterica* and *Pseudomonas denitrificans*.[8,11-13] Evidence was reported for the existence of two separate systems in *Clostridium tetanomorphum* to

*Corresponding Author: Jorge C. Escalante-Semerena—Department of Bacteriology, University of Wisconsin, 6478 Microbial Sciences Building, 1550 Linden Drive, Madison, Wisconsin 53706-1521, USA. Email: escalante@bact.wisc.edu

Tetrapyrroles: Birth, Life and Death, edited by Martin J. Warren and Alison G. Smith. ©2009 Landes Bioscience and Springer Science+Business Media.

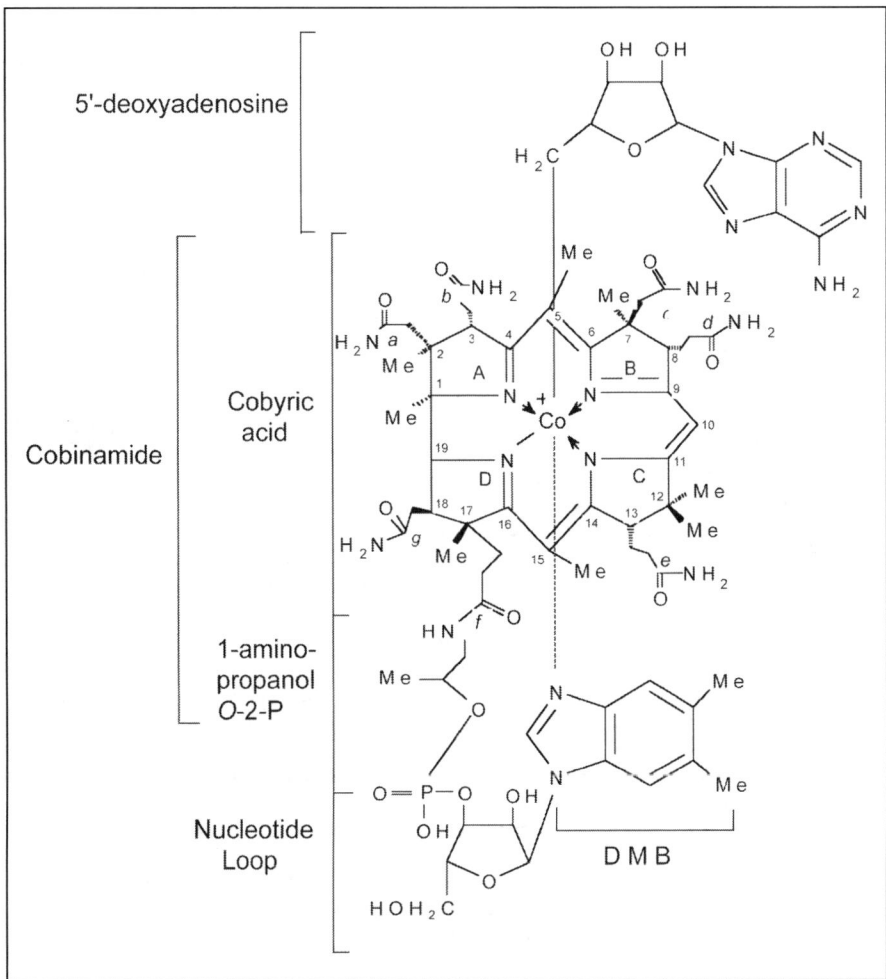

Figure 1. Chemical structure of coenzyme B_{12} (AdoB$_{12}$; adenosylcobalamin, AdoCbl). DMB, 5,6-dimethylbenzimidazole. Arrows and dashed line represent coordination bonds.

reduce Co(III) to Co(I), however, the genes encoding these functions were not identified. More recently, the cobalt reducing systems in *S. enterica* became known and are discussed below.[14,15]

Importance of the Corrinoid Adenosylation Pathway

Genetic evidence obtained in *S. enterica* showed that lack of ATP:co(I)rrinoid adenosyltransferase (CobA) enzyme blocks de novo synthesis of the corrin ring, indicating that this pathway proceeds via adenosylated intermediates.[8] Salvaging of incomplete corrinoids such as cobyric acid and cobinamide is also blocked in *S. enterica cobA* mutants, suggesting that corrinoid-binding enzymes involved in nucleotide loop assembly also require adenosylated substrates.[8] Adenosylation is also relevant to the expression of the gene encoding the transporter of corrinoids across the outer membrane of Gram-negative bacteria such as *E. coli* and *S. enterica*. In *E. coli*, inactivation of the *btuR* gene (encodes a CobA homolog) blocks the conversion of B_{12} into coenzyme B_{12}, resulting in the unregulated, constitutive expression of *btuB*.[16-19] The expression of the *btuB* gene is controlled by a riboswitch mechanism[18,20] discussed in detail in Chapter 20.

Reduction of Co^{3+} to Co^{1+}

Reduction of Co^{3+} to Co^{1+} requires consecutive one-electron reductive steps catalyzed by the cob(III)alamin reductase (EC 1.6.99.8) and the cob(II)alamin reductase (EC 1.6.99.9) enzymes. The reducing systems and the ATP:co(I)rrinoid adenosyltransferase (EC 2.5.1.17) enzymes responsible for the conversion of the vitamin to the coenzymic form of B_{12} were partially purified and characterized from *P. freundenreichii* and *C. tetanomorphum*.[21-25] These enzymes were recenty isolated to homogeneity from *S. enterica* and the genes encoding them were identified.[12,14,15,26] From the latter studies it was concluded that *S. enterica* probably lacks a dedicated cob(III)alamin reductase for the conversion of co(III)balamin to co(II)alamin. This conclusion was based on the spontaneous, enzyme-independent, dihydroflavin-dependent reduction of cob(III)alamin to cob(II)alamin.[14] On the other hand, the thermodynamically unfavorable reduction of cob(II)alamin to cob(I)alamin[27] was achieved in vitro using the reducing system comprised of ferrredoxin (flavodoxin):NADP[+] reductase (Fpr) and flavodoxin A (FldA) proteins to transfer electrons from NADPH + H[+] onto complete or incomplete co(III)rrinoids to generate co(I)rrinoids.[15] Although the Fpr/FldA system is the first enzymatic system to be coupled to the CobA adenosyltransferase enzyme to convert cobalamin to its coenzymic form, the involvement of this system in corrinoid reduction is not unprecedented, as shown by the studies of the cobalamin-dependent methionine synthase function in *E. coli*.[28-32] Our current working model for the corrinoid adenosylation pathway is shown in Figure 2. In *P. denitrificans*, an NADH-dependent flavoenzyme with cob(II)yrinic acid *a,c*-diamide reductase activity was isolated to homogeneity but the gene encoding this activity was not identified.[33] Use of the Fpr/FldA system for corrinoid reduction appears to be widely spread in nature since human cells also use it to maintain function of the methionine synthase enzyme.[34]

Interaction between Flavodoxin A (FldA) and CobA Proteins

Recent ^{1}H-NMR studies of the interactions between *E. coli* FldA and ferredoxin:flavoprotein reductase (Fpr) and methionine synthase (MetH) identified residues of FldA involved in docking to Fpr and MetH.[32] The docking surface of FldA is comprised of a hydrophobic patch housing the flavin cofactor that is surrounded by several acidic residues. Recent work from our laboratory identified the same residues of FldA as being involved in docking to CobA during cobalt ion reduction (Fig. 3).[35] In vivo and in vitro assessment of the functionality of site-directed variants of CobA identified residues Arg9 and Arg165 as critical for docking, but not for catalysis.

Redundancy in the Electron Transferases Used to Reduce Co^{2+} to Co^{1+}

A second flavodoxin was recently identified by the similarity of the nucleotide sequence of the putative ORF (referred to as *fldB*) to *fldA*.[36] Though the FldB protein has not been biochemically characterized as a flavodoxin, the predicted amino acid sequence of FldB is 45% identical to FldA. Unlike *fldA*, *fldB* function is not essential, and its expression is regulated by oxidative stress.[36] The

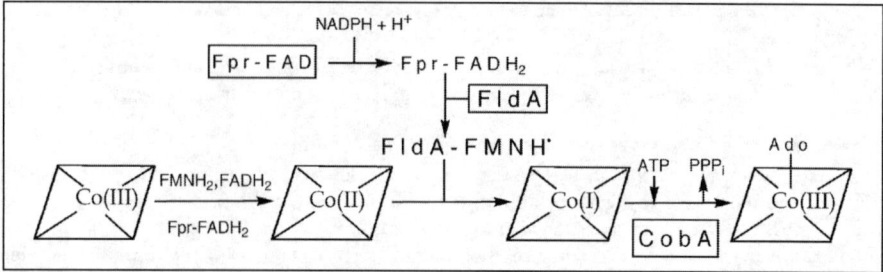

Figure 2. The corrinoid adenosylation pathway in *S. enterica*. In this model, reducing power for the reduction of Co^{3+}-corrinoids to Co^{1+}-corrinoids is derived from NADPH + H[+] by the action of the ferrredoxin (flavodoxin):NADP[+] reductase (Fpr) and flavodoxin A (FldA) proteins. Fpr-FAD, oxidized form of Fpr; Fpr-FADH$_2$, hydroquinone form of Fpr; FldA-FMN, oxidized form of FldA; FldA-FMN[•], semiquinone form of FldA; PPP$_i$, tripolyphosphate. Modified from reference 39, with permission from The American Society for Biochemistry and Molecular Biology.

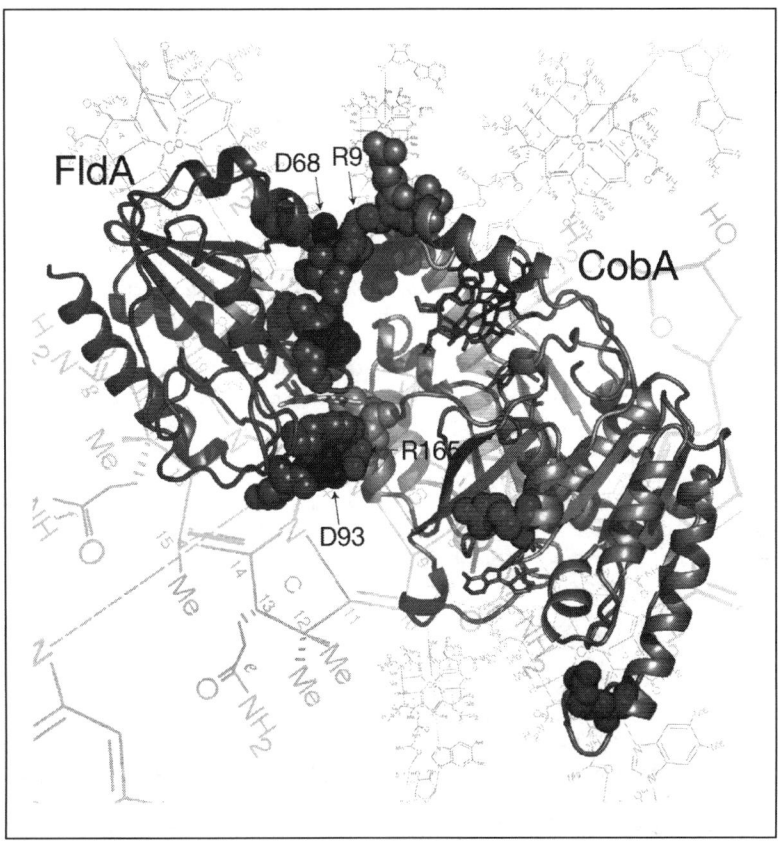

Figure 3. Docking surface of flavodoxin (FldA). Computer-assisted modeling was used to identify two putative arginyl residues of the cob(I)alamin adenosyl transferase enzyme (CobA, shown as a dimer), enzyme that are critical to docking with aspartyl residues of flavodoxin A (FldA).[35] The chemical structure of coenzyme B$_{12}$ is shown in the background.

physiological role of FldB is unknown, and unpublished work from our laboratory showed that reduced FldB cannot reduce cob(II)alamin to cob(I)alamin.

Formation of the Co-C Bond

Early studies of the conversion of B$_{12}$ to coenzyme B$_{12}$ showed that ATP was the donor of the 5'-deoxyadenosyl group.[9,37] A more detailed biochemical analysis of the reaction, was performed when the gene encoding the ATP:co(I)rrinoid adenosyltransferase (EC 2.5.1.17) enzyme responsible for catalyzing the reaction was identified in *P. denitrificans* (*cobO*) and *S. enterica* (*cobA*).[8,12,26,38] During the course of the reaction, the CobA enzyme releases tripolyphosphate (PPP$_i$) as by-product. ^{31}P-NMR spectroscopy data indicate that PPP$_i$ is not cleaved by CobA. Further support for the conclusion that PPP$_i$ is a by-product of the CobA reaction was obtained from inhibition studies.[39] CobA is only the second enzyme known to release PPP$_i$, the other such enzyme being SAM synthetase.[40]

Three-Dimensional Crystal Structure of the ATP:co(I)rrinoid Adenosyltransferase (CobA) Enzyme of S. enterica

The three-dimensional structure of an ATP:co(I)rrinoid adenosyltransferase enzyme is known. High-resolution crystal structures of the CobA enzyme of *S. enterica* in its apo form, complexed with MgATP, and complexed with Cbl and MgATP were solved (PDB entries 1G5R, 1G64).[41] The structures containing B_{12} revealed that CobA binds B_{12} differently than enzymes that use B_{12} as cofactor such as methionine synthase, methylmalonyl-CoA mutase, diol dehydratase, or glutamate mutase.[42-47] New protein/corrinoid interactions where the ring is held via a limited number of hydrophobic interactions were observed. This is likely the reason why CobA can bind several different corrinoids as substrates. The structure of the CobA complexed with its substrates also shows that ATP is bound to a unique P-loop (GNGKGKT, one residue shorter than the consensus GXXXXGKT/S), and that the nucleotide binds to the P-loop in opposite orientation to that of ATP hydrolases, i.e., in CobA the γ-phosphate of ATP binds to the location occupied by the α-phosphate in nucleoside triphosphate hydrolases. The structure of the ternary complex shows the cobalt ion of B_{12} located ~6.1Å away from the target C5' of the ribose, suggesting that the protein mv y undergo a conformational change to bring the Co(I) and C5' target to sufficient proximity for the nucleophilic attack to occur. The driving force for the proposed conformational change remains to be established.

Adenosylation during de Novo Corrin Ring Biosynthesis

The point of adenosylation during de novo corrin ring biosynthesis has been established in *P. denitrificans*, but not in any other prokaryote. Results from the analysis of corrinoid substrates that can be used by the *P.d.* CobO enzyme showed that cobyrinic acid is not a substrate for the enzyme, but cobyrinic acid *a,c*-diamide is, indicating a key role for the amide groups in substrate recognition.[38] The CobO (*P. denitrificans*) and CobA (*S. enterica*) enzymes can adenosylate cobyric acid, cobinamide and cobalamin.[26,38] The timing of adenosylation during de novo cobinamide synthesis may be different in *S. enterica*, because prokaryotes that use the anaerobic pathway of corrin ring biosynthesis (e.g., *S. enterica*, *P. freundenreichii*, *B. megaterium*) insert the cobalt ion into the ring early in the pathway,[48-52] and there are no data to suggest that conversion of cobalt-precorrin-2 ring to cobyrinic acid *a,c*,-diamide proceeds via nonadenoyslated intermediates.

Other Corrinoid Adenosyltransferases

Homologs of the *S. enterica* CobA enzyme can be found in a host of cobamide-producing prokaryotes. Of these, only the CobO enzyme from *P. denitrificans* and the CobA enzyme of the methanogenic archaeon *Methanosarcina mazei* Göl have been isolated to homogeneity and partially characterized.[38,53] Genes encoding additional adenosyltransferase activities in *S. enterica* are *pduO*,[54,55] and *eutT*.[56,57] Neither PduO nor EutT are homologous to CobA. The three-dimensional structures of PduO and EutT may reveal different folds that provide adenosyltransferase activity. Recently, the human and bovine ATP:adenolsyltransferases were identified by complementation of a *pduO* cobA mutant strain of *S. enterica* during growth on 1,2-propanediol.[58] Both, the human and bovine adenosyltransferases appear to be phylogentically related to PduO, but not to CobA. A high-resolution three-dimensional crystal structure of an archaeal homolog of the PduO-type human adenosyltransferase enzyme was recently reported in its apo form.[59]

The Nucleotide Loop Assembly (NLA) Pathway

The nucleotide loop is the structure within cobamides that tethers the lower ligand base to the ring macrocycle. The nucleotide loop features an α-N-glycosidic bond between the base and the ribosyl group, and a phosphodiester bond between the 3' hydroxyl group of the ribosyl moiety and the *(R)*-1-amino-2-propanol (AP) moiety of cobinamide (Cbi) (Fig. 1). From a biochemical standpoint the nucleotide assembly (NLA) pathway consists of three steps: (i) activation of the lower ligand base; (ii) activation of the precursor adenosylcobinamide-phosphate (AdoCbi-P); and (iii) joining of the activated precursors to yield adenosylcobamide (AdoCbi) (Fig. 4). The report in 1984 by Jeter et al that *S. enterica* synthesized B_{12} de novo, greatly accelerated the identification and analysis of genes, proteins and enzymes involved in the assembly of this complex coenzyme.[60] The genetic analysis of the region of the *S. enterica* genome containing the bulk of the *c*obalamin *b*iosynthetic (*cob*) genes[60-63] was consistent with the gene organization derived from the nucleotide sequence of the *cob* operon.[64] The location of the *S. enterica* genes whose functions are known to be required for the assembly of the nucleotide loop and the attachment of the upper ligand are shown in Figure 5. Subsequent work showed that the last three genes of this operon (i.e., *cobUST*) encode functions required for the assembly of the nucleotide loop,[61,62] with one additional function encoded outside of

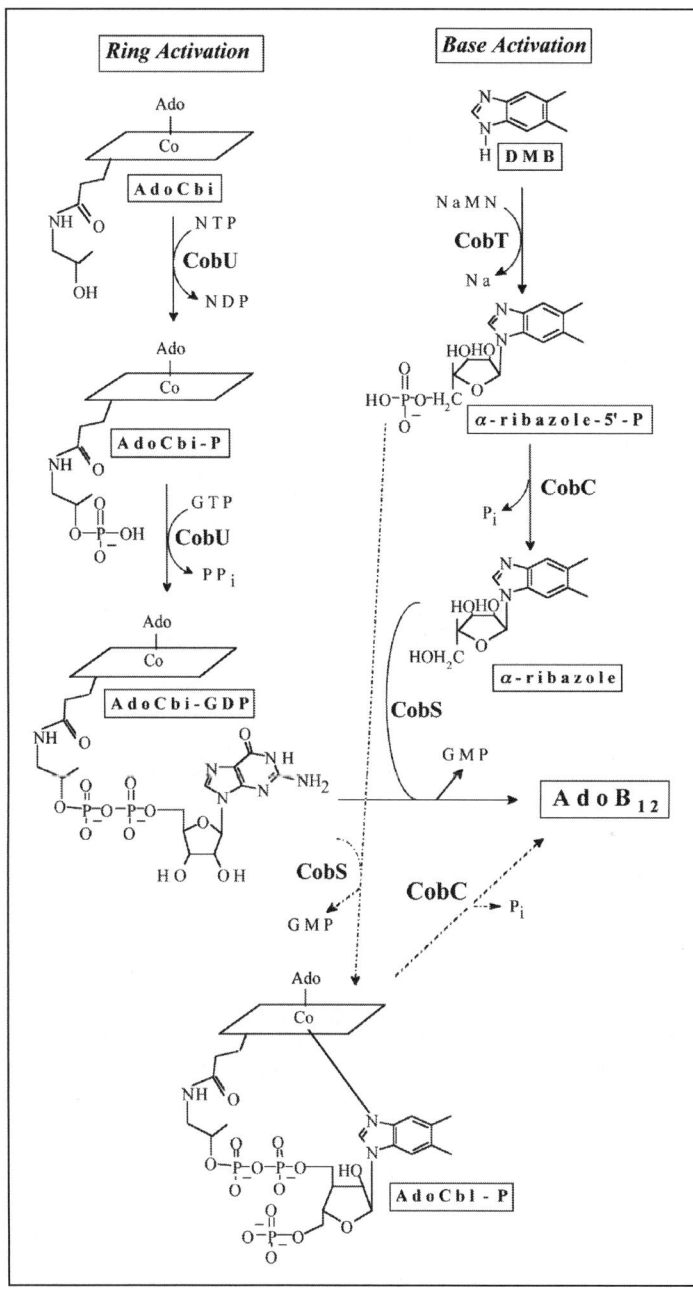

Figure 4. Nucleotide loop assembly pathway in *S. enterica*. Abbreviations used: GMP, guanylate; GTP, guanosine triphosphate; PPi, pyrophosphate; adoCbi, adenosylcobinamide; AdoCbi-P, adenosylcobinamide-phosphate; AdoCbi-GDP, adenosylcobinamide-guanosine diphosphate; NaMN, nicotinate mononucleotide, DMB, 5,6-dimethylbenzimidazole; CobU, AdoCbi kinase/guanylyltransferase; CobT, NaMN:DMB phosphoribosyltransferase; CobC, α-ribazole-5'-phosphate phosphatase; AdoB₁₂-P phosphatase; CobS, cobalamin (5'-phosphate) synthase. Modified from reference 102, with permission from *Microbiology*.

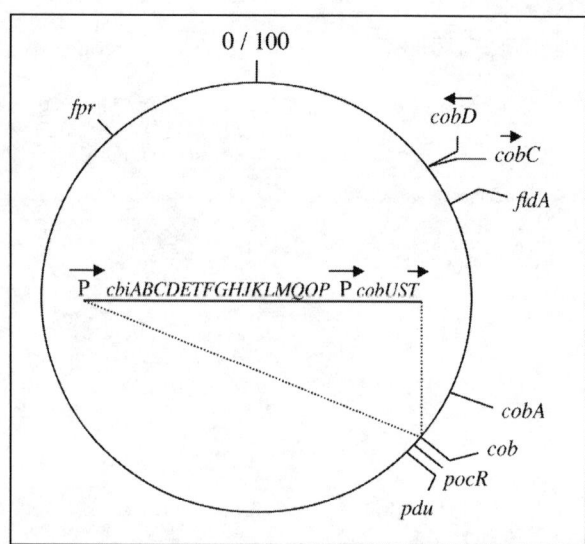

Figure 5. Organization and location of the B$_{12}$ biosynthetic genes in *S. enterica*. Location and organization of the cobalamin biosynthetic operon of *S. enterica*. The *pocR* gene encodes a positive regulatory protein that control expression of the *cob* operon as a function of 1,2-propanediol in the environment.[127] Abbreviations: *fldA*, flavodoxin A; *fpr*, flavodoxin reductase; *cobD*, L-threonine-*O*-3-phosphate decarboxylase; *cobC*, α-ribazole-5'-phosphate phosphatase; *pdu* propanediol utilization operon.

the *cob* operon.[65] In the last 10 years, our understanding of the biochemistry underpinning the nucleotide loop assembly and corrinoid adenosylation pathways was greatly advanced by the enzymological and structural analyses of the enzymes involved.

The Corrinoid Substrate

Although, discussion of the de novo synthesis of the corrinoid entering the NLA pathway is out of the scope of this review, it is important to cover some aspects of it. For several decades it was thought that the last step in de novo corrin ring biosynthesis consisted of the attachment of (*R*)-1-amino-2-propanol to the *f* propionic acid group of cobyric acid (corrin ring). This idea was reconsidered when we reported evidence supporting the hypotheses that (*R*)-1-amino-propanol-*O*-2-phosphate not (*R*)-1-amino-2-propanol was involved in the reaction.[66] Therefore, the end product of the reaction was AdoCbi-P, not AdoCbi. This finding modified our understanding of one of the branches in the nucleotide loop assembly pathway, identifying the kinase activity of CobU as a function exclusively dedicated to salvaging Cbi from the environment.[2]

Key to this discovery was the determination of the enzymatic activity of the CobD protein.[66] CobD is a novel pyridoxal 5'-phosphate (PLP)-dependent L-threonine-*O*-3-phosphate decarboxylase (EC 4.1.1.81) responsible for the synthesis of (*R*)-1-amino-propanol-*O*-2-phosphate (AP-P), the linker between the nucleotide loop and the corrin ring in cobalamin biosynthesis. The three-dimensional structure of CobD (PDB entry 1LKC), its apo state (PDB entry 1LC5), the apo state complexed with the substrate (PDB entry 1LC7), and its product external aldimine complex (PDB entry 1LC8) have been determined at 1.8, 1.46, 1.8, and 1.8Å resolution, respectively. In its native form, CobD is a dimer, with each monomer consisting of a small domain and a large domain. CobD has been classified as part of the α-family of PLP-dependent enzymes. In fact, CobD is remarkably similar to histidinol-phosphate aminotransferase (PDB entry 1GEW),[67] not only in its three-dimensional structure, but in the coordination of the pyridoxal 5'-phosphate and putative substrate binding site. Detailed analysis of CobD three-dimensional structures showed that decarboxylation occurs by placing the hydrogen of the α-carbon of L-threonine-phosphate opposite to the Lys[206] that forms the internal aldimine. An equivalent Lys residue is believed to

Halobacterium sp. strain NRC-1.[90] The initial biochemical characterization of Cbiz was performed with recombinant enzyme encoded by the *cbiZ* gene of the methanogenic archaeon *Methanosarcina mazei* strain Göl.[91]

Physiological Roles of the AdoCbi Kinase and Guanylyltransferase Enzyme in Bacteria

The existence of CobY helps us understand the physiological roles of the enzymatic activities of CobU in *S. enterica* and other bacteria with CobU-homologs. In contrast to the AdoCbi-P guanylyltransferase activity, the cobinamide kinase activity of CobU is not part of the de novo coenzyme B_{12} pathway, but it is essential to the conversion of exogenous, unphosphorylated cobinamide coenzyme B_{12}. These conclusions are consistent with previous findings, which suggested that the product of de novo corrin ring biosynthesis in *S. enterica* was AdoCbi-P, not AdoCbi as previously thought.[2,66]

Activation of the Lower (*Coα*) Ligand

The *Coα* ligand base of cobamides varies considerably,[92] and studies of lower ligand base activation have focused on 5,6-dimethylbenzimidazole (DMB), the lower ligand base of cobalamin.[63,93-97] The activation of DMB occurs in two steps catalyzed by a NaMN-dependent phosphoribosyltransferase (PRTase), and a phosphatase enzyme. The end result of the lower ligand activation branch of the nucleotide loop assembly pathway is the conversion of DMB to its nucleoside. Because the first enzyme of the pathway uses NaMN as substrate, the product of the reaction, N^1-(5-phospho-α-D-ribosyl)-DMB nucleotide (α-ribazole-5'-P), contains an α-*N*-glycosidic bond. This bond configuration is needed to make the 3'-OH of the ribosyl moiety available for the formation of the phosphodiester bond linking the lower ligand to the corrin ring. Because the 5'-OH is not phosphorylated in cobamides, the phosphate group in α-ribazole-5'-phosphate is removed by a phosphatase. At present, the PRTase reaction is better understood than the phosphatase reaction, both of which are discussed below.

General Features of the Phosphoribosyltransferase (PRTase) Enzyme

The nicotinate mononucleotide (NaMN):DMB phosphoribosyltransferase (PRTase) (EC 2.4.2.21) enzyme catalyzes the transfer of the phosphoribosyl moiety of NaMN onto DMB to yield α-ribazole-5'-P. The PRTase activity was studied in partially purified preparations of the enzyme isolated from *Propionibacterium freundenreichii* (formerly *P. shermanii*),[94] *Clostridium sticklandii*,[98] and *Propionibacterium arabinosum*,[99] and to homogeneity from *Pseudomonas denitrificans* (encoded by the *cobU* gene)[96] and *Salmonella enterica* (encoded by the *cobT* gene).[97] NaMN:DMB PRTase enzymes isolated from different sources appear to have different affinities for NaMN. For example, the affinity of the *P. denitrificans* CobU enzyme for NaMN was higher (K_m = 83 μM)[96] than the *S. enterica* CobT enzyme (K_m = 684 μM),[97] the reason for this drastic difference is unclear, but may be a result of structural differences between these homologous enzymes. The K_m for DMB has only been reported for the *P. denitrificans* CobU enzyme, and is considerably lower than the affinity for NaMN (K_m = 16 nM),[96] probably reflecting the low intracellular level of DMB in this bacterium.

Nicotinate Mononucleotide (NaMN) Is Not the Only Substrate for the PRTase Enzyme

Until recently, there was no reason to doubt that NaMN was the only substrate for the NaMN:DMB PRTase enzyme in vivo. Two pieces of information raised the possibility that NaMN may not be the only substrate (or perhaps in some prokaryotes it may not be the substrate) that the lower-ligand activating enzyme uses to activate DMB in vivo. In the case of the *S. enterica* CobT enzyme, the K_m of CobT for NaMN (0.51 mM) was unexpectedly high. This finding is important because the intracellular level of NaMN in wild-type *S. enterica* or *E. coli* is undetectable.[100,101] We now know that the CobT enzyme of *S. enterica* can use NAD⁺ as substrate to activate DMB, but the product of the reaction is not α-ribazole-5-P as in the case when NaMN is the substrate. Instead, the CobT enzyme generates α-DMB adenine dinucleotide (referred to as α-DAD).[102] Although the K_m of CobT for NAD⁺ (9 mM) is ca. 18-fold higher than that for NaMN (0.51 mM), the intracellular level of NAD⁺ (ca. 0.8 mM)[100,103] can compensate for the lack of affinity of the enzyme for NAD⁺. These results raised the

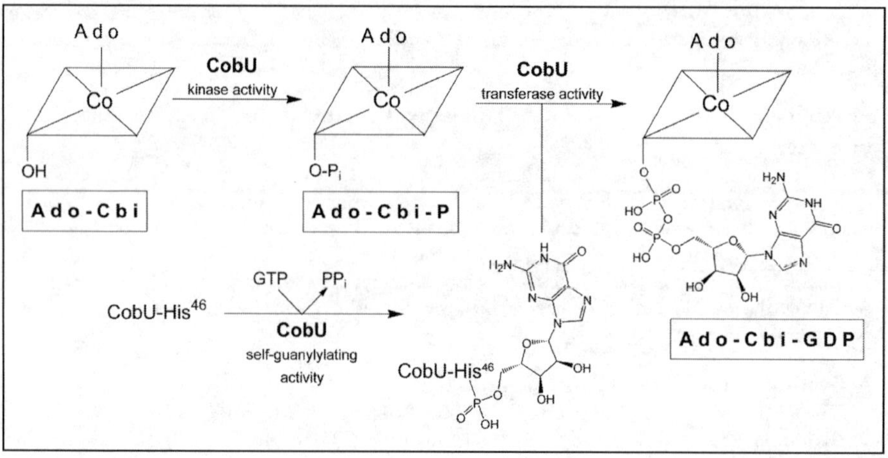

Figure 6. Reactions catalyzed by the CobU enzyme. The CobU enzyme is thought to be always guanylylated in vivo due to its auto-guanylylating activity. This activity is needed to induce the catalytically active conformation of the protein.[73] In its active conformation the enzyme uses either ATP or GTP (shown as NTP) to phosphorylate the hydroxyl group of the (*R*)-1-amino-2-propan-ol moiety of AdoCbi. The phosphate group of AdoCbi-P is retained in AdoCbi-GDP, which is generated by the transfer of the GMP moiety attached to CobU onto AdoCbi-P.

possibility that in *S. enterica,* the activation of DMB proceeds via α DAD. Support for this possibility was obtained in vitro when α-DAD-dependent synthesis of coenzyme B_{12} was demonstrated using cell-free extracts enriched for the enzymes required for the assembly of the nucleotide loop.[102] This new information may reflect ways in which the cell ensures synthesis of coenzyme B_{12} regardless of variations in the pools of phosphoribosyl donors. That is, under conditions where the level of NaMN increases, one would predict that the CobT enzyme would use NaMN over NAD^+. Under conditions where the level of NaMN is low, the cell could still use NAD^+ to make coenzyme B_{12}. Interestingly, the CobT homolog of *P. freundenreichii* was reported unable to use NAD^+ to activate DMB, suggesting that not all the CobT homologs may have NAD^+-dependent activity.[93]

Evidence that the NAD^+-dependent activity of CobT may be physiologically important was obtained from genetic and biochemical studies of a function present in *S. enterica* that compensates for the lack of CobT enzyme in *cobT* mutant strains under growth conditions that demanded coenzyme B_{12} biosynthesis. This alternative function for CobT is encoded by the *cobB* gene, which encodes a member of the SIR2 family of eukaryotic regulatory proteins (sirtuins).[104] Sirtuins are bifunctional enzymes with NAD^+-dependent ADPribosyltransferase and protein deacetylase activities with relevance to gene expression and metabolism.[105-113] Of relevance to coenzyme B_{12} biosynthesis is the NAD^+-dependent ADPribosyltransferase activity of the CobB sirtuin. In vitro CobB-dependent synthesis of α-ribazole-5'-P,[104] supports the hypothesis that the NAD^+-dependent ADPribosyltransferase activity of CobT is physiologically significant in *S. enterica*. Figure 7 incorporates α-DAD as a putative intermediate of the NLA pathway.

Are There Enzymes of the NLA Pathway Yet to Be Discovered?

If α-DAD were indeed an intermediate of the pathway, one would conclude that one or more coenzyme B_{12} biosynthetic enzymes have been overlooked. At present, it is unclear how α-DAD is incorporated into coenzyme B_{12}. If α-DAD is cleaved by a dinucleotide pyrophosphatase, the products of the reaction would be AMP and α-ribazole-5'-P, with the latter serving as substrate for the CobC phosphatase. Alternatively, if α-DAD were cleaved by a dinucleotide hydrolase, the products of the reaction would be ADP and α-ribazole, a reaction that would bypass the need for the CobC phosphatase (Fig. 7). Extensive mutant searches have not yielded B_{12} auxotrophs with lesions in genes other than the *cobCUST* genes, suggesting that the putative dinucleotide pyrophosphatase or

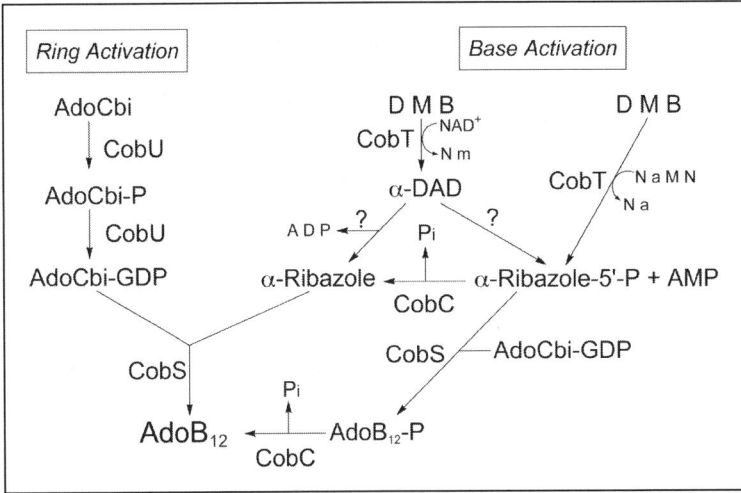

Figure 7. Current model for the assembly of the nucleotide loop of AdoB$_{12}$. Reactions indicated by dashed arrows have been shown to occur in vitro. Question marks indicate uncertainty about the cleavage of α-DAD. Modified from reference 102, with permission from *Microbiology*.

dinucleotide hydrolase either has other functions in the cell, there is redundancy of these functions in the cell, or both enzymes exist and inactivation of the gene encoding one of them is insufficient to generate a B$_{12}$ auxotrophy. Reverse genetic approaches to this problem are needed to identify the gene encoding the dinucleotide pyrophosphatase or the dinucleotide hydrolase.

Three-Dimensional Crystal Structure of the NaMN:5,6-Dimethylbenzimidazole (DMB) PRTase Enzyme

The X-ray structure of the apo form of the CobT enzyme from *S. enterica* was solved at 1.9Å resolution (PDB entry 1L4B).[114] The quaternary structure of native CobT enzyme is a dimer with 2-fold symmetry, and each subunit has two domains. A substantial section of the interface between the two subunits is contributed by the small domain and by the final helix of the large domain. The large domains consists of six-stranded β-sheets with connecting α-helices exhibiting Rossman fold topology. The small domain is comprised of components of the N- and C-termini of the polypeptide chain and contains a three-helix bundle. The active site of the enzyme is formed by the loops at the C-terminal end of the β strands and the small domain of the small subunit. The structure of the CobT enzyme suggests that residue Glu[317] is the catalytic base for the reaction. The fold of the CobT enzyme is very different than that of type I or type II phosphoribosylpyrophosphate (PRPP)-dependent PRTases, and the orientation of substrates and products is opposite to that expected for a Rossman fold.[114,115] The structural fold of CobT is not related to any other fold found in protein databases.

Specificity of the NaMN:DMB PRTase Enzyme for Its Base Substrate

The chemical nature of the base in the nucleotide of cobamides varies depending on their source. The list of base substrates includes DMB, 5-methylbenzimidazole, 5-methoxybenzimidazole, 5-hydroxybenzimidazole, 5-methoxy-6-methylbenzimidazole, adenine, 2-methylsulfinyladenine, 2-methylsulfonyladenine, *p*-cresol, and phenol.[61,116-122] Although the *S. enterica* CobT enzyme displays high specificity for NaMN, its specificity for the base substrate is very poor. This conclusion is based on results from in vitro studies performed with homogeneous *S. enterica* CobT enzyme, NaMN and benzimidazole, dimethylphenylenediamine, imidazole, histidine, adenine or guanine as substrates.[97] In all cases, the enzyme catalyzed the formation of the corresponding α-nucleotide, and in all cases the product of the reaction was incorporated into cobamides that were active in vivo.[97]

To understand the molecular basis for the lack of specificity of the CobT enzyme of *S. enterica* for its base substrates, crystal structures of the enzyme complexed with adenine, 5-methylbenzimidazole, 5-methoxybenzimidazole, imidazole, indole, 2-hydroxypurine, *p*-cresol or phenol were solved (PDB entries 1L4E, 1LRF, 1LRG 1L4H, 1L4K, 1L4M, 1L4N, 1L5F, 1L5K, 1L5L, 1L5M, 1L5N, 1L5O,1JHM, 1JHO, 1JHP, 1JHQ, 1JHR, 1JHU, 1JHV, 1JHX, 1JHY, 1DOS, 1DOV, 1JH8, 1JHA).[123] Adenine, 5-methylbenzimidazole, 5-methoxybenzimidazole, 2-hydroxypurine reacted with NaMN in the crystal lattice to form the corresponding α-nucleotide. The crystal structures of these complexes revealed that only minor conformational changes in the side chains that form the DMB binding site are needed to accommodate different base substrates. No product was formed in the crystal containing *p*-cresol or phenol even though *p*-cresol and phenol bound to CobT in approximately the same location as DMB. Analysis of the crystal structures showed that in both cases the active site of CobT was too large for the reaction to occur, suggesting that the CobT homolog in *Sporomusa ovata* (a strict anaerobic bacterium that synthesizes phenyl-cobamides) must have evolved an active site that brings the two substrates closer together than the *S. enterica* enzyme. From the mechanistic standpoint, it is of interest to learn how the *S. ovata* activates the substrate in the absence of an aromatic nitrogenous base.

Dephosphorylation of α-Ribazole-5'-Phosphate (α-Ribazole-5'-P)

Figure 4 shows the second step in the activation of DMB being catalyzed by the CobC phosphatase enzyme. In this reaction, α-ribazole-5'-P (the product of the NaMN:DMB PRTase enzyme) is converted to α-ribazole, or alternatively, AdoB$_{12}$-P is dephosphorylated to yield AdoB$_{12}$. This enzymatic activity was reported present in cell-free extracts of *P. freudenreichii*[94] and *P. denitrificans*,[96] but neither the enzyme nor the gene encoding it were identified in these bacteria. In *S. enterica*, the α-ribazole-5'-P phosphatase activity is encoded by the *cobC* gene. The CobC protein is homologous to phosphoglycerate mutases, acid phosphatases and to the biphosphatase domain of eukaryotic 6-phosphofructo-2-kinase/fructose-2,6-biphosphatase enzymes[65] with many of the residues needed for function conserved.[124] Mechanistic studies of the *S. enterica* CobC enzyme have not been reported. In *S. enterica*, the *cobC* gene is divergently transcribed from the *cobD* gene, a coenzyme B$_{12}$ biosynthetic gene encoding L-threonine O-3-phosphate decarboxylase.[66]

Phenotypes of a Phosphatase-Deficient Bacterial Strain

In *S. enterica*, the effect of the lack of CobC enzyme on coenzyme B$_{12}$ synthesis is most evident when the demand for coenzyme B$_{12}$ is high, e.g., during growth on ethanolamine as carbon and energy source.[65] However, when the demand for coenzyme B$_{12}$ is low (e.g., synthesis of methionine via the B$_{12}$-dependent methionine synthase), there is no detectable effect on B$_{12}$-dependent growth unless the strain also harbors a null allele of the *cobT* gene encoding the NaMN:DMB phosphoryboyltransferase enzyme. Explanations for these phenotypes are not obvious. It is possible that under conditions where low levels of coenzyme B$_{12}$ are sufficient to support growth, the activity of a nonspecific phosphatase generates α-ribazole bypassing the need for CobC activity. Alternatively, under these conditions the B$_{12}$-dependent enzymes in this bacterium may be able to use B$_{12}$-5'-P as coenzyme.[125] An intriguing, and yet unexplained phenotype of *cobC* mutant strains is the DMB auxotrophy observed when the strains are grown under high aeration.[65] How the lack of CobC function affects DMB synthesis when the level of environmental oxygen is elevated remains unclear. A physiological explanation for the observed DMB auxotrophy of *cobC* mutant strains will likely shed valuable insights into the pathway of DMB synthesis in *S. enterica*.

Timing of Phosphate Removal in Bacteria

In *S. enterica*, it is clear that the 5' phosphate group of α-ribazole-5'-P is absent in AdoB$_{12}$, the final product of the pathway.[61] It is unclear, however, whether α-ribazole-5'-P or coenzyme B$_{12}$-5'-P is the substrate for the CobC phosphatase enzyme in vivo. In vitro, however, CobC can use either α-ribazole-5'-P or coenzyme B$_{12}$-5'-P as substrate.[65,125] Insights into the timing of phosphate removal in *P. denitrificans* were obtained through the kinetic analysis of the CobS reaction when α-ribazole-5'-P or α-ribazole was used as substrate for the enzyme. The *P. denitrificans* enzyme showed a higher affinity for α-ribazole-5'-P ($K_m = 0.9\mu M$) relative to α-ribazole ($K_m = 7.8 \mu M$).[96] Although it is clear however, that the *S.e.* CobC enzyme can dephosphorylate either α-ribazole-5'-P or AdoB$_{12}$-P in vitro,[65,125] we have obtained in vivo and in vitro evidence to support the conclusion that α-ribazole α-ribazole-5'-P, is the preferred substrate for the *S. enterica* CobS enzyme (Zayas & Escalante-Semerena, unpublished results).

Archaea Do Not Have an Ortholog of the Phosphatase-Encoding Gene Present in Bacteria

Analysis of the archaeal genomes sequenced to date failed to reveal an archaeal ortholog to CobC. At this point is unclear if CobC has been replaced by a functionally similar nonorthologous enzyme (such as the case with CobU and CobY) or if the archaeal pathway differs in such a way that it does not require an α–ribazole-5'-P phosphatase. A better understanding of how archaea activate the lower ligand base and biochemical analysis of archaeal CobT orthologs will indicate if α–ribazole-5'-P phosphatase is an intermediate of the pathway and if an α–ribazole-5'-P phosphatase would be required. Another possibility would be that archaea produce phoshporylated cobamides, which are still biologically active and meet the requirements of the cell.

The Last Step of the Pathway: Joining AdoCbi-GDP and α-Ribazole

The enzyme that catalyzes the last step of the NLA pathway, and for that matter the last step of AdoCbl biosynthesis, is cobalamin (5'-phosphate) synthase. In *S. enterica*, this enzyme is encoded by the *cobS* gene, and in *P. denitrificans* it is encoded by the *cobV* gene.[13,64] The cobalamin (5'-phosphate) synthase enzyme has not been isolated to homogeneity, and in *P. denitrificans*, it appears to form complexes with a number of unidentified proteins.[96] Purified preparations of *S.e.* CobS enzyme were obtained to reconstitute the entire NLA pathway in vitro.[125] The in vitro data showed that the *S. enterica* CobS enzyme, like the *P. denitrificans* CobV enzyme, can use either α-ribazole or α-ribazole-5'-P as substrate. Recent work from our laboratory showed that the CobS enzyme in *S. enterica* and the methanogenic archaeon *Methanobacterium thermoautotrophicum* strain ΔH (and likely in other prokaryotes) is an integral membrane protein.[126] This finding begs the question of why the final step of cobalamin biosynthesis has been maintained through evolution as a membrane-associated function. Future research in this field will explore the molecular details of this very exciting question in an effort to uncover the selective pressures that both archaea and bacteria have been subjected to in order to maintain this intriguing association with the cell membrane.

Acknowledgements

This work was supported by grants GM40313 and GM62203 from the General Medical Sciences Institute of the National Institutes of Health, and by the Ira L. Baldwin Chair Professorship to J.C. E.-S.; N.R. Buan is a Howard Hughes Medical Institute Fellow; J. D. Woodson is the Department of Bacteriology's Ira Baldwin Fellow; C.L. Zayas is a NIH Predoctoral Fellow (F31 GM640009).

References

1. Blanche F, Robin C, Couder M et al. Purification, characterization, and molecular cloning of S-adenosyl-L-methionine: Uroporphyrinogen III methyltransferase from Methanobacterium ivanovii. J Bacteriol 1991; 173:4637-4645.
2. Thomas MG, Escalante-Semerena JC. Identification of an alternative nucleoside triphosphate: 5'- deoxyadenosylcobinamide phosphate nucleotidyltransferase in Methanobacterium thermoautotrophicum ΔH. J Bacteriol 2000; 182:4227-4233.
3. DiMarco AA, Bobik TA, Wolfe RS. Unusual coenzymes of methanogenesis. Ann Rev Biochem 1990; 59:355-394.
4. Thauer RK. Biochemistry of methanogenesis: A tribute to Marjory Stephenson. Microbiology 1998; 144:2377-2406.
5. Riera J, Robb FT, Weiss R et al. Ribonucleotide reductase in the archaeon Pyrococcus furiosus: A critical enzyme in the evolution of DNA genomes? Proc Natl Acad Sci USA 1997; 94:475-478.
6. Tauer A, Benner SA. The B12-dependent ribonucleotide reductase from the archaebacterium Thermoplasma acidophila: An evolutionary solution to the ribonucleotide reductase conundrum. Proc Natl Acad Sci USA 1997; 94:53-58.
7. Warren MJ, Raux E, Schubert HL et al. The biosynthesis of adenosylcobalamin (vitamin B₁₂). Nat Prod Rep 2002; 19:390-412.

8. Escalante-Semerena JC, Suh SJ, Roth JR. cobA function is required for both de novo cobalamin biosynthesis and assimilation of exogenous corrinoids in Salmonella typhimurium. J Bacteriol 1990; 172:273-280.

9. Peterkofsky A, Redfied B, Weissbach H. The role of ATP in the biosynthesis of coenzyme B_{12}. Biochem Biophys Res Comm 1961; 5:213-216.

10. Brady RO, Barker HA. Influence of oxygen on the photolysis of coenzyme B_{12}. Biochem Biophys Res Comm 1961; 4:373-378.

11. Lundrigan MD, Kadner RJ. Altered cobalamin metabolism in Escherichia coli btuR mutants affects btuB gene regulation. J Bacteriol 1989; 171:154-161.

12. Suh SJ, Escalante-Semerena JC. Cloning, sequencing, and overexpression of cobA which encodes ATP:corrinoid adenosyltransferase in Salmonella typhimurium. Gene 1993; 129:93-97.

13. Crouzet J, Levy-Schil S, Cameron B et al. Nucleotide sequence and genetic analysis of a 13.1-kilobase-pair Pseudomonas denitrificans DNA fragment containing five cob genes and identification of structural genes encoding Cob(I)alamin adenosyltransferase, cobyric acid synthase, and bifunctional cobinamide kinase-cobinamide phosphate guanylyltransferase. J Bacteriol 1991; 173:6074-6087.

14. Fonseca MV, Escalante-Semerena JC. Reduction of cob(III)alamin to cob(II)alamin in Salmonella enterica Serovar Typhimurium LT2. J Bacteriol 2000; 182:4304-4309.

15. Fonseca MV, Escalante-Semerena JC. An in vitro reducing system for the enzymic conversion of cobalamin to adenosylcobalamin. J Biol Chem 2001; 276:32101-32108.

16. Lundrigan MD, DeVeaux LC, Mann BJ et al. Separate regulatory systems for the repression of metE and btuB by vitamin B_{12} in Escherichia coli. Mol Gen Genet 1987; 206:401-407.

17. Franklund CV, Kadner RJ. Multiple transcribed elements control expression of the Escherichia coli btuB gene. J Bacteriol 1997; 179:4039-4042.

18. Nou X, Kadner RJ. Adenosylcobalamin inhibits ribosome binding to btuB RNA. Proc Natl Acad Sci USA 2000; 97:7190-7195.

19. Ravnum S, Andersson DI. Vitamin B_{12} repression of the btuB gene in Salmonella typhimurium is mediated via a translational control which requires leader and coding sequences. Mol Microbiol 1997; 23:35-42.

20. Nahvi A, Barrick JE, Breaker RR. Coenzyme B_{12} riboswitches are widespread genetic control elements in prokaryotes. Nucleic Acids Res 2004; 32:143-150.

21. Brady RO, Castanera EG, Barker HA. The enzymatic synthesis of cobamide coenzymes. J Biol Chem 1962; 223:2325-2332.

22. Walker GA, Murphy S, Huennekens FM. Enzymatic conversion of vitamin B_{12a} to adenosyl-B_{12}: Evidence for the existence of two separate reducing systems. Arch Biochem Biophys 1969; 134:95-102.

23. Vitols E, Walker GA, Huennekens FM. Enzymatic conversion of vitamin B_{12s} to a cobamide coenzyme, α-(5,6-dimethylbenzimidazolyl)deoxy-adenosylcobamide (adenosyl-B_{12}). J Biol Chem 1966; 241:1455-1461.

24. Fujii K, Huennekens FM. Activation of methionine synthetase by a reduced triphosphopyridine nucleotide-dependent flavoprotein system. J Biol Chem 1974; 249:6745-6753.

25. Fujii K, Galivan JH, Huennekens FM. Activation of methionine synthase: Further characterization of flavoprotein system. Arch Biochem Biophys 1977; 178:662-670.

26. Suh SJ, Escalante-Semerena JC. Purification and initial characterization of the ATP:corrinoid adenosyltransferase encoded by the cobA gene of Salmonella typhimurium. J Bacteriol 1995; 177:921-925.

27. Lexa D, Saveant JM. The electrochemistry of vitamin B_{12}. Acc Chem Res 1983; 16:235-243.

28. Hoover DM, Ludwig ML. A flavodoxin that is required for enzyme activation: The structure of oxidized flavodoxin from Escherichia coli at 1.8Å resolution. Protein Sci 1997; 6:2525-2537.

29. Hoover DM, Jarrett JT, Sands RH et al. Interactions of Escherichia coli cobalamin-dependent methionine synthase and its physiological partner flavodoxin: Binding of flavodoxin leads to axial ligand dissociation from the cobalamin cofactor. Biochemistry 1997; 36:127-138.

30. Jarrett JT, Hoover DM, Ludwig ML et al. The mechanism of adenosylmethionine-dependent activation of methionine synthase: A rapid kinetic analysis of intermediates in reductive methylation of cob(II)alamin enzyme. Biochemistry 1998; 37:12649-12658.

31. Hall DA, Jordan-Starck TC, Loo RO et al. Interaction of flavodoxin with cobalamin-dependent methionine synthase. Biochemistry 2000; 39:10711-10719.

32. Hall DA, Vander Kooi CW, Stasik CN et al. Mapping the interactions between flavodoxin and its physiological partners flavodoxin reductase and cobalamin-dependent methionine synthase. Proc Natl Acad Sci USA 2001; 98:9521-9526.

33. Blanche F, Maton L, Debussche L et al. Purification and characterization of cob(II)yrinic acid a,c diamide reductase from Pseudomonas denitrificans. J Bacteriol 1992; 174:7452-7454.

34. Olteanu H, Banerjee R. Human methionine synthase reductase, a soluble P-450 reductase-like dual flavoprotein, is sufficient for NADPH-dependent methionine synthase activation. J Biol Chem 2001; 276:35558-35563.

35. Buan NR, Escalante-Semerena JC. Computer-assisted docking of flavodoxin with the ATP:Co(I)rrinoid adenosyltransferase (CobA) enzyme reveals residues critical for protein-protein interactions but not for catalysis. J Biol Chem 2005; 280:40948-40956.

36. Gaudu P, Weiss B. Flavodoxin mutants of Escherichia coli K-12. J Bacteriol 2000; 182:1788-1793.

37. Peterkofsky A, Weissbach H. Release of inorganic triphosphate from adenosine triphosphate during vitamin B₁₂ coenzyme biosynthesis. J Biol Chem 1963; 238:1491-1497.

38. Debussche L, Couder M, Thibaut D et al. Purification and partial characterization of Cob(I)alamin adenosyltransferase from Pseudomonas denitrificans. J Bacteriol 1991; 173:6300-6302.

39. Fonseca MV, Buan NR, Horswill AR et al. The ATP:co(I)rrinoid adenosyltransferase (CobA) enzyme of Salmonella enterica requires the 2'-OH Group of ATP for function and yields inorganic triphosphate as its reaction byproduct. J Biol Chem 2002; 277:33127-33131.

40. Markham GD, Hafner EW, Tabor CW et al. S-Adenosylmethionine synthetase from Escherichia coli. J Biol Chem 1980; 255:9082-9092.

41. Bauer CB, Fonseca MV, Holden HM et al. Three-dimensional structure of ATP:corrinoid adenosyltransferase from Salmonella typhimurium in its free state, complexed with MgATP, or complexed with hydroxycobalamin and MgATP. Biochemistry 2001; 40:361-374.

42. Drennan CL, Huang S, Drummond JT et al. How a protein binds B₁₂: A 3.0A X-ray structure of B₁₂-binding domains of methionine synthase. Science 1994; 266:1660-1674.

43. Mancia F, Keep NH, Nakagawa A et al. How coenzyme B12 radicals are generated: The crystal structure of methylmalonyl-coenzyme A mutase at 2Å resolution. Structure 1996; 4:339-350.

44. Mancia F, Smith GA, Evans PR. Crystal structure of substrate complexes of methylmalonyl-CoA mutase. Biochemistry 1999; 38(25):7999-8005.

45. Shibata N, Masuda J, Tobimatsu T et al. A new mode of B12 binding and the direct participation of a potassium ion in enzyme catalysis: X-ray structure of diol dehydratase. Structure Fold Des 1999; 7:997-1008.

46. Hoffmann B, Konrat R, Bothe H et al. Structure and dynamics of the B12-binding subunit of glutamate mutase from Clostridium cochlearium. Eur J Biochem 1999; 263:178-188.

47. Reitzer R, Gruber K, Jogl G et al. Glutamate mutase from Clostridium cochlearium: The structure of a coenzyme B12-dependent enzyme provides new mechanistic insights. Structure Fold Des 1999; 7:891-902.

48. Müller G, Zipfel F, Hliney K et al. Timing of cobalt insertion in vitamin B₁₂ biosynthesis. J Amer Chem Soc 1991; 113:9893-9895.

49. Warren MJ, Bolt EL, Roessner CA et al. Gene dissection demonstrates that the Escherichia coli cysG gene encodes a multifunctional protein. Biochem J 1994; 302:837-844.

50. Schubert HL, Raux E, Wilson KS et al. Common chelatase design in the branched tetrapyrrole pathways of heme and anaerobic cobalamin synthesis. Biochemistry 1999; 38:10660-10669.

51. Fazzio TG, Roth JR. Evidence that the CysG protein catalyzes the first reaction specific to B₁₂ synthesis in Salmonella typhimurium, insertion of cobalt. J Bacteriol 1996; 178:6952-6959.

52. Santander PJ, Roessner CA, Stolowich N et al. How corrinoids are synthesized without oxygen: Nature's first pathway to vitamin B₁₂. Chem Biol 1997; 4:659-666.

53. Buan N, Rehfeld K, Escalante-Semerena JC. Studies of the CobA-type ATP:Co(I)rrinoid adenosyltransferase enzyme of Methanosarcina mazei strain Göl. J Bacteriol 2006; 188, (In press).

54. Walter D, Ailion M, Roth J. Genetic characterization of the pdu operon: Use of 1,2-propanediol in Salmonella typhimurium. J Bacteriol 1997; 179:1013-1022.

55. Johnson CL, Pechonick E, Park SD et al. Functional genomic, biochemical, and genetic characterization of the Salmonella pduO gene, an ATP:cob(I)alamin adenosyltransferase gene. J Bacteriol 2001; 183:1577-1584.

56. Buan NR, Suh SJ, Escalante-Semerena JC. The eutT gene of Salmonella enterica encodes an oxygen-labile, metal-containing ATP: Corrinoid adenosyltransferase enzyme. J Bacteriol 2004; 186:5708-5714.

57. Sheppard DE, Penrod JT, Bobik T et al. Evidence that a B12-adenosyl transferase is encoded within the ethanolamine operon of Salmonella enterica. J Bacteriol 2004; 186(22):7635-7644.

58. Leal NA, Park SD, Kima PE et al. Identification of the human and bovine ATP: Cob(I)alamin adenosyltransferase cDNAs based on complementation of a bacterial mutant. J Biol Chem 2003; 278:9227-9234.

59. Saridakis V, Yakunin A, Xu X et al. The structural basis for methylmalonic aciduria. The crystal structure of archaeal ATP:cobalamin adenosyltransferase. J Biol Chem 2004; 279:23646-23653.
60. Jeter RM, Olivera BM, Roth JR. Salmonella typhimurium synthesizes cobalamin (vitamin B_{12}) de novo under anaerobic growth conditions. J Bacteriol 1984; 159:206-213.
61. Johnson MG, Escalante-Semerena JC. Identification of 5,6-dimethylbenzimidazole as the Coα ligand of the cobamide synthesized by Salmonella typhimurium: Nutritional characterization of mutants defective in biosynthesis of the imidazole ring. J Biol Chem 1992; 267:13302-13305.
62. O'Toole GA, Rondon MR, Escalante-Semerena JC. Analysis of mutants of Salmonella typhimurium defective in the synthesis of the nucleotide loop of cobalamin. J Bacteriol 1993; 175:3317-3326.
63. Trzebiatowski JR, O'Toole GA, Escalante-Semerena JC. The cobT gene of Salmonella typhimurium encodes the NaMN: 5,6-dimethylbenzimidazole phosphoribosyltransferase responsible for the synthesis of N^1-(5-phospho-alpha-D-ribosyl)-5,6-dimethylbenzimidazole, an intermediate in the synthesis of the nucleotide loop of cobalamin. J Bacteriol 1994; 176:3568-3575.
64. Roth JR, Lawrence JG, Rubenfield M et al. Characterization of the cobalamin (vitamin B_{12}) biosynthetic genes of Salmonella typhimurium. J Bacteriol 1993; 175:3303-3316.
65. O'Toole GA, Trzebiatowski JR, Escalante-Semerena JC. The cobC gene of Salmonella typhimurium codes for a novel phosphatase involved in the assembly of the nucleotide loop of cobalamin. J Biol Chem 1994; 269:26503-26511.
66. Brushaber KR, O'Toole GA, Escalante-Semerena JC. CobD, a novel enzyme with L-threonine-O-3-phosphate decarboxylase activity, is responsible for the synthesis of (R)-1-amino-2-propanol O-2-phosphate, a proposed new intermediate in cobalamin biosynthesis in Salmonella typhimurium LT2. J Biol Chem 1998; 273:2684-2691.
67. Haruyama K, Nakai T, Miyahara I et al. Structures of Escherichia coli histidinol-phosphate aminotransferase and its complexes with histidinol-phosphate and N-(5'-phosphopyridoxyl)-L-glutamate: Double substrate recognition of the enzyme. Biochemistry 2001; 40:4633-46344.
68. Cheong CG, Escalante-Semerena JC, Rayment I. Structural studies of the L-threonine-O-3-phosphate decarboxylase (CobD) enzyme from Salmonella enterica: The apo, substrate, and product-aldimine complexes. Biochemistry 2002; 41:9079-9089.
69. Bernhauer K, Wagner F, Michna H et al. Zur chemie und biochemie der corrinoide. XXIX. Biogenesewege von der cobyrinsäure zur cobyrsäure und zum cobinamid bei Propionibacterium shermanii. Hoppe-Seyler's Z. Physiol Chem 1968; 349:1297-1309.
70. Ronzio RA, Barker HA. Enzymic synthesis of guanosine diphosphate cobinamide by extracts of propionic acid bacteria. Biochemistry 1967; 6:2344-2354.
71. Woodson JD, Peck RF, Krebs MP et al. The cobY gene of the archaeon Halobacterium sp. strain NRC-1 is required for de novo cobamide synthesis. J Bacteriol 2003; 185:311-316.
72. O'Toole GA, Escalante-Semerena JC. Purification and characterization of the bifunctional CobU enzyme of Salmonella typhimurium LT2. Evidence for a CobU-GMP intermediate. J Biol Chem 1995; 270:23560-23569.
73. Thompson T, Thomas MG, Escalante-Semerena JC et al. Three-dimensional structure of the guanylylated form of adenosycobinamide kinase/adenosylcobinamide-phosphate guanylyltransferase enzyme of Salmonella typhymurium at 2.3Å Resolution. Biochemistry 1999; 38:12995-13004.
74. Thomas MG, Thompson TB, Rayment I et al. Analysis of the adenosylcobinamide kinase / adenosylcobinamide phosphate guanylyltransferase (CobU) enzyme of Salmonella typhimurium LT2. Identification of residue H46 as the site of guanylylation. J Biol Chem 2000; 275:27376-27386.
75. Thompson TB, Thomas MG, Escalante-Semerena JC et al. Three-dimensional structure of adenosylcobinamide kinase/adenosylcobinamide phosphate guanylyltransferase from Salmonella typhimurium determined at 2.3Å resolution. Biochemistry 1998; 37:7686-7695.
76. Pal D, Chakrabarti P. Cis peptide Bonds in proteins: Residues involved, their conformations, interactions and locations. J Mol Biol 1999; 294:271-288.
77. Blanche F, Debussche L, Famechon A et al. A bifunctional protein from Pseudomonas denitrificans carries cobinamide kinase and cobinamide phosphate guanylyltransferase activities. J Bacteriol 1991; 173:6052-6057.
78. Martin SA, Moss B. Modification of RNA by mRNA guanylyltransferase and mRNA (guanine-7-)methyltransferase from vaccinia virions. J Biol Chem 1975; 250:9330-9335.
79. Shuman S, Hurwitz J. Mechanism of mRNA capping by vaccinia virus guanylyltransferase: Characterization of an enzyme-guanylate intermediate. Proc Natl Acad Sci USA 1981; 78:187-191.
80. Shuman S, Schwer B. RNA capping enzyme and DNA ligase: A superfamily of covalent nucleotidyl transferases. Molec Microbiol 1995; 17:405-410.
81. Hakansson K, Doherty AJ, Shuman S et al. X-ray crystallography reveals a large conformational change during guanyl transfer by mRNA capping enzymes. Cell 1997; 89:545-553.

82. Szumilo T, Drake RR, York JL et al. GDP-mannose pyrophosphorylase. Purification to homogeneity, properties, and utilization to prepare photoaffinity analogs. J Biol Chem 1993; 268:17943-17950.
83. Cartwright JL, McLennan AG. Formation of a covalent Nepsilon2-guanylylhistidyl reaction intermediate by the GTP:GTP guanylyltransferase from the brine shrimp Artemia. Arch Biochem Biophys 1999; 361:101-105.
84. Toyama R, Mizumoto K, Nakahara Y et al. Mechanism of the mRNA guanylyltransferase reaction: Isolation of N epsilon-phospholysine and GMP (5' leads to N epsilon) lysine from the guanylyl-enzyme intermediate. EMBO J 1983; 2:2195-2201.
85. Roth MJ, Hurwitz J. RNA capping by the vaccinia virus guanylyltransferase. Structure of enzyme-guanylate intermediate. J Biol Chem 1984; 259:13488-13494.
86. Liu JJ, McLennan AG. Purification and properties of GTP:GTP guanylyltransferase from encysted embryos of the brine shrimp Artemia J Biol Chem 1994; 269:11787-11794.
87. Frey PA, Richard JP, Ho HT et al. Stereochemistry of selected phosphotransferases and nucleotidyltransferases. Methods Enzymol 1982; 87:213-235.
88. Brown K, Pompeo F, Dixon S et al. Crystal structure of the bifunctional N-acetylglucosamine 1-phosphate uridyltransferase from Escherichia coli: A paradigm for the related pyrophosphorylase superfamily. EMBO J 1999; 18:4096-4107.
89. Stupperich E, Steiner I, Eisinger HJ. Substitution of Coα-(5-hydroxybenzimidazolyl)cobamide (factor III) by vitamin B₁₂ in Methanobacterium thermoautotrophicum. J Bacteriol 1987; 169:3076-3081.
90. Woodson JD, Zayas CL, Escalante-Semerena JC. A new pathway for salvaging the coenzyme B12 precursor cobinamide in archaea requires cobinamide-phosphate synthase (CbiB) enzyme activity. J Bacteriol 2003; 185:7193-7201.
91. Woodson JD, Escalante-Semerena JC. CbiZ, an amidohydrolase enzyme required for salvaging the coenzyme B₁₂ precursor cobinamide in archaea. Proc Natl Acad Sci USA 2004; 101:3591-3596.
92. Renz P. Biosynthesis of the 5,6-dimethylbenzimidazole moiety of cobalamin and of other bases found in natural corrinoids. In: Banerjee R, ed. Chemistry and Biochemistry of B₁₂. New York: John Wiley and Sons Inc., 1999:557-575.
93. Friedmann HC. Partial purification and properties of a single displacement trans-N-glycosidase. J Biol Chem 1965; 240:413-418.
94. Friedmann HC, Harris DL. The formation of alpha-glycosidic 5'-nucleotides by a single displacement trans-N-glycosidase. J Biol Chem 1965; 240:406-412.
95. Fyfe JA, Friedmann HC. Vitamin B₁₂ biosynthesis. Enzyme studies on the formation of the α-glycosidic nucleotide precursor. J Biol Chem 1969; 244:1659-1666.
96. Cameron B, Blanche F, Rouyez MC et al. Genetic analysis, nucleotide sequence, and products of two Pseudomonas denitrificans cob genes encoding nicotinate-nucleotide: Dimethylbenzimidazole phosphoribosyltransferase and cobalamin (5'-phosphate) synthase. J Bacteriol 1991; 173:6066-6073.
97. Trzebiatowski JR, Escalante-Semerena JC. Purification and characterization of CobT, the nicotinate mononucleotide:5,6-dimethylbenzimidazole phosphoribosyltransferase enzyme from Salmonella typhimurium LT2. J Biol Chem 1997; 272:17662-17667.
98. Friedmann HC, Fyfe JA. Pseudovitamin B₁₂ biosynthesis. Enzymatic formation of a new adenylic acid, 7-α-D-ribofuranosyladenine 5'-phosphate. J Biol Chem 1969; 244:1667-1671.
99. Schneider Z. Biosynthesis of vitamin B₁₂. In: Schneider Z, Stroiski A, eds. Comprehensive B₁₂. Berlin: Walter de Gruyter, 1987:93-110.
100. Lundquist R, Olivera BM. Pyridine nucleotide metabolism in Escherichia coli. I. Exponential growth. J Biol Chem 1971; 246:1107-1116.
101. Bochner BR, Ames BN. Complete analysis of cellular nucleotides by two-dimensional thin layer chromatography. J Biol Chem 1982; 257:9759-9769.
102. Maggio-Hall LA, Escalante-Semerena JC. Alpha-5,6-dimethylbenzimidazole adenine dinucleotide (alpha-DAD), a putative new intermediate of coenzyme B₁₂ biosynthesis in Salmonella typhimurium. Microbiology 2003; 149:983-990.
103. Lundquist R, Olivera BM. Pyridine nucleotide metabolism in Escherichia coli. II. Niacin starvation. J Biol Chem 1973; 248:5137-5143.
104. Tsang AW, Escalante-Semerena JC. CobB, a new member of the SIR2 family of eucaryotic regulatory proteins, is required to compensate for the lack of nicotinate mononucleotide: 5,6-dimethylbenzimidazole phosphoribosyltransferase activity in cobT mutants during cobalamin biosynthesis in Salmonella typhimurium LT2. J Biol Chem 1998; 273:31788-31794.
105. Starai VJ, Celic I, Cole RN et al. Sir2-dependent activation of acetyl-CoA synthetase by deacetylation of active lysine. Science 2002; 298:2390-2392.
106. Starai VJ, Takahashi H, Boeke JD et al. Short-chain fatty acid activation by acyl-coenzyme A synthetases requires SIR2 protein function in Salmonella enterica and Saccharomyces cerevisiae. Genetics 2003; 163:545-555.

107. Gulick AM, Starai VJ, Horswill AR et al. The 1.75Å crystal structure of acetyl-CoA synthetase bound to adenosine-5'-propylphosphate and coenzyme A. Biochemistry 2003; 42:2866-2873.
108. Bell SD, Botting CH, Wardleworth BN et al. The interaction of Alba, a conserved archaeal chromatin protein, with Sir2 and its regulation by acetylation. Science 2002; 296:148-151.
109. Smith JS, Brachmann CB, Celic I et al. A phylogenetically conserved NAD+-dependent protein deacetylase activity in the Sir2 protein family. Proc Natl Acad Sci USA 2000; 97:6658-6663.
110. Frye RA. Characterization of five human cDNAs with homology to the yeast SIR2 gene: Sir2-like proteins (sirtuins) metabolize NAD and may have protein ADP-ribosyltransferase activity. Biochem Biophys Res Commun 1999; 260:273-279.
111. Tanny JC, Moazed D. Coupling of histone deacetylation to NAD breakdown by the yeast silencing protein Sir2: Evidence for acetyl transfer from substrate to an NAD breakdown product. Proc Natl Acad Sci USA 2000; 98:415-420.
112. Tanny JC, Dowd GJ, Huang J et al. An enzymatic activity in the yeast Sir2 protein that is essential for gene silencing. Cell 1999; 99:735-745.
113. Imai S, Armstrong CM, Kaeberlein M et al. Transcriptional silencing and longevity protein Sir2 is an NAD-dependent histone deacetylase. Nature 2000; 403:795-800.
114. Cheong CG, Escalante-Semerena JC, Rayment I. The three-dimensional structures of nicotinate mononucleotide:5,6- dimethylbenzimidazole phosphoribosyltransferase (CobT) from Salmonella typhimurium complexed with 5,6-dimethybenzimidazole and its reaction products determined to 1.9Å resolution. Biochemistry 1999; 38:16125-16135.
115. Cheong CG, Escalante-Semerena JC, Rayment I. Capture of a labile substrate by expulsion of water molecules from the active site of nicotinate mononucleotide:5,6-dimethylbenzimidazole phosphoribosyltransferase (CobT) from Salmonella enterica. J Biol Chem 2002; 277:41120-41127.
116. Stupperich E, Kräutler B. Pseudo vitamin B$_{12}$ or 5-hydroxybenzimidazolyl-cobamide are the corrinoids found in methanogenic bacteria. Arch Microbiol 1988; 149:213-217.
117. Stupperich E, Eisinger HJ, Krautler B. Identification of phenolyl cobamide from the homoacetogenic bacterium Sporomusa ovata. Eur J Biochem 1989; 186:657-661.
118. Stupperich E, Eisinger HJ, Krautler B. Diversity of corrinoids in acetogenic bacteria. p-Cresolylcobamide from Sporomusa ovata, 5-methoxy-6-methylbenzimidazolylcobamide from Clostridium formicoaceticum and vitamin B$_{12}$ from Acetobacterium woodii. Eur J Biochem 1988; 172:459-464.
119. Kräutler B, Kohler HPE, Stupperich E. 5'-Methylbenzimidazolyl-cobamides are the corrinoids from some sulfate-reducing and sulfur-metabolizing bacteria. Eur J Biochem 1988; 176:461-469.
120. Renz P, Blickle S, Friedrich W. Two new vitamin B-12 factors from sewage sludge containing 2-methylsulfinyladenine or 2-methylsulfonyladenine as base component. Eur J Biochem 1987; 163:175-179.
121. Hodgkin DC, Pickworth J, Robertson JH et al. The crystal structure of the hexacarboxylic acid derived from B$_{12}$ and the molecular structure of the vitamin. Nature 1956; 176:325-328.
122. Hodgkin DC, Kamper J, Lindsey J et al. The structure of vitamin B$_{12}$. I. An outline of the crystallographic investigation of vitamin B$_{12}$. Proc Roy Soc 1957; A242:228-263.
123. Cheong CG, Escalante-Semerena JC, Rayment I. Structural investigation of the biosynthesis of alternative lower ligands for cobamides by nicotinate mononucleotide: 5,6-dimethylbenzimidazole phosphoribosyltransferase from Salmonella enterica. J Biol Chem 2001; 276:37612-37620.
124. Pilkis SJ, Claus IJ, Lange AJ. 6-Phosphofructose-2-kinase/fructose-2,6-biphosphatase: A metabolic signaling enzyme. Ann Rev Biochem 1995; 64:799-835.
125. Maggio-Hall LA, Escalante-Semerena JC. In vitro synthesis of the nucleotide loop of cobalamin by Salmonella typhimurium enzymes. Proc Natl Acad Sci USA 1999; 96:11798-11803.
126. Maggio-Hall LA, Claas KR, Escalante-Semerena JC. The last step in coenzyme B(12) synthesis is localized to the cell membrane in bacteria and archaea. Microbiology 2004; 150:1385-1395.
127. Rondon MR, Escalante-Semerena JC. The poc locus is required for 1,2-propanediol-dependent transcription of the cobalamin biosynthetic (cob) and propanediol utilization (pdu) genes of Salmonella typhimurium. J Bacteriol 1992; 174:2267-2272.

CHAPTER 20

The Regulation of Cobalamin Biosynthesis

Jeffrey G. Lawrence*

Introduction

Expositions on the regulation of biochemical pathways usually succeed in disappointing at least half of their potential audience. From a holistic standpoint, one could view gene regulation as the embodiment of the physiological significance of the encoded gene products. If one understood when, where and why genes were either active or inactive, one would gain insight into the selective forces retaining those genes within a genome. From a reductionist standpoint, gene regulation can be achieved in almost countless ways, each offering at worst insight into how a cell is controlling the dynamic expression of its inherently static genetic material and at best uncovering previously undiscovered mechanisms by which the activities of gene products are controlled. From an evolutionary standpoint, both views may differ between different organisms, allowing either fruitful comparative biology when the differences are recognized, or potentially misleading extrapolation when they are not.

The regulation of cobalamin (coenzyme B_{12}) biosynthesis offers several more twists. First, cobalamin biosynthesis is intimately coupled to the synthesis of other tetrapyrroles in the cells (notable examples include heme, siroheme and chlorophyll). Second, cobalamin is a cofactor that may be used by several different enzymes in any one cell. As a result, multiple signals may be integrated to control the rate of cobalamin biosynthesis; this integration will differ among organisms, since organisms differ greatly in which cobalamin-dependent enzymes are encoded in their genomes. Last, organisms may "synthesize" cobalamin when provided with complex intermediates, like cobinamide, and forego the synthesis of the tetrapyrrole backbone; some organisms can synthesize cobalamin only when provided with such an intermediate. The utilization of "imported" cobinamide adds yet another layer of complexity to the regulation of gene responsible for the biosynthesis of one of the largest, most complex and perhaps most poorly understood of bacterial cofactors.

This review will attempt to synthesize all of these viewpoints into a comprehensive treatment of how and why cobalamin is synthesized. By necessity, much of the discussion will focus on relatively few study organisms, those where cobalamin synthesis or use has been examined in detail. As I cannot discuss all of these topics in depth, I will discuss each briefly in context and direct the reader towards more comprehensive explorations of any one topic. What I hope the reader will extract is an appreciation for the vast number of strategies taken by microörganisms to synthesis this molecule at the appropriate time and place, as well as to harness the potential resource of corrinoid compounds present in the environment.

The Complexity of Cobalamin

Cobalamin is synthesized from uroporphyrinogen III (Uro-III, Fig. 1), the "basic" tetrapyrrole used to synthesize all tetrapyrrole derivatives, including chlorophyll, heme, siroheme and related molecules. Uro-III is one of three isomers of uroporphyrinogen, this one being asymmetric in

*Jeffrey G. Lawrence—Department of Biological Sciences, University of Pittsburgh, Pittsburgh, Pennsylvania 15260, USA. Email: jlawrenc@pitt.edu

Tetrapyrroles: Birth, Life and Death, edited by Martin J. Warren and Alison G. Smith.
©2009 Landes Bioscience and Springer Science+Business Media.

having acetyl- side chains attached to ring carbons 2 and 18; the juxtaposition of these acetyl-groups plays a critical role in aerobic cobalamin biosynthesis by mediating the elimination of carbon-20,[1] yielding the 19-member corrinoid ring found in cobalamin and other cobamides and distinguishing it from other tetrapyrroles. It is for this reason that some have speculated that cobalamin preceded heme and other tetrapyrroles in ancestral metabolism.[2,3] An alternative view is that Uro-III could have been the favored isomer in the soup of prebiotic molecules available,

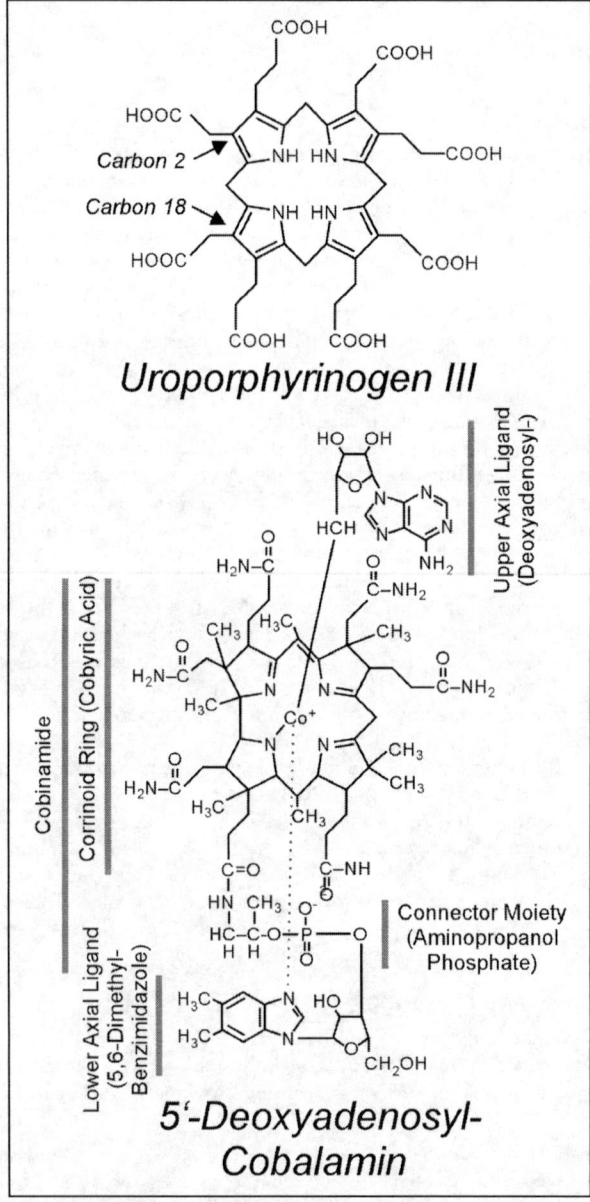

Figure 1. The structures of uroporphyrinogen-III, the tetrapyrrole precursor of corrinoids, and cobalamin. Individual portions of cobalamin referred to in the text are noted.

since experiments of the Miller/Urey variety yield this isomer in a 2:1 ratio over other forms.[4] Regardless of its origin, the asymmetry of the corrinoid ring provides cobalamin with some unique properties, due to the strain placed on the corrinoid ring.

The central cobalt atom has four planar ligands—all nitrogen atoms in the tetrapyrrole backbone, to one of which it is covalently bonded—and two axial ligands. The upper (Coβ) axial ligand is 5'-deoxyadenosine in coenzyme B_{12}; the strain of the corrinoid ring makes homolytic cleavage of this covalent carbon-cobalt bond quite facile, and allows the adenosyl- radical to mediate interesting chemical rearrangements. Other common upper axial ligand include a methyl- group (which is found which cobalamin participates in methylation reactions, like that catalyzed by the *E. coli* MetH methionine synthase and its human homologue), a cyano- group (found in commercially synthesized vitamin B_{12}) and water (which does not form a covalent bond).

The lower (Coα) axial ligand is 5,6-dimethylbenzimidazole in coenzyme B_{12}, where one of the nitrogen atoms acts as the electronegative donor group. Different cobamides have different lower axial ligands,[5] most of which contain electronegative nitrogen atoms that serve to coordinate the cobalt atom, although there are exceptions (like cresol derivatives found in *Sporomusa* spp.). When corrinoids are transported into the cell, these "alternative" forms of cobalamin are readily available, and the cell must be able to discriminate between cobamide forms to prevent the accumulation of potential inhibitory molecules. This idea will be discussed further below.

The lower axial ligand is connected to the corrinoid ring via an aminopropanol-phosphate group attached to the propyl-carboxamide chain at ring carbon 17.[6] When this group is attached during biosynthesis, cobinamide-phosphate is formed. Cobinamide-phosphate can be formed from the off-pathway intermediate cobinamide by kinasing the hydroxyl-group of the 1-amino, 2-propanol via the CobU kinase activity. This kinase activity—completely unnecessary during de novo biosynthesis—attests to the importance of cobinamide transport and utilization, and portends additional physiological significance (and therefore regulation) of the genes required for cobinamide utilization.

Cobalamin in Context: Regulating a Branch Point

There are at least two distinct pathways to synthesize cobalamin from Uro-III, functioning under either aerobic or anaerobic growth conditions;[7-10] it is possible that the timing of the insertion of the central cobalt atom is critical in allow for aerobic cobalamin biosynthesis, since the +1 oxidation state of this atom is difficult to maintain. In considering the regulation of cobalamin biosynthesis, we must assess its impact on the biosynthesis of other molecules derived from Uro-III. In many cases this is a moot point, since interwoven biosynthetic pathways often have separate regulatory apparati that allow for an increase in production of a critical intermediate if it is being utilized for another purpose. For example, a sudden increase in methionine biosynthesis would drain cysteine pools, since cysteine acts as the donor molecule for the sulfur atom in methionine.[11,12] An increase in sulfate reduction and cysteine biosynthesis follows indirectly in enteric bacteria like *Escherichia coli* and *Salmonella enterica*, since the depletion of cysteine pools results in a lack of metabolic inhibition of the CysE protein, allowing for more cysteine to be made. By analogy, one could postulate that activating the cobalamin biosynthetic pathway would drain Uro-III pools, and the cognate biosynthetic genes (often considered part of the heme regulon) would be up-regulated.

Yet the situation is more complex in *E. coli* and *S. enterica*, since a single enzyme controls the flow of tetrapyrroles into three molecules (Fig. 2). The CysG enzyme catalyzes both methylation reactions,[13] which divert Uro-III from the heme biosynthetic route, and chelation of either iron to form siroheme,[14] or cobalt to lead toward cobalamin formation.[15] (It should be noted that the role of other metal chelatases in cobalt and iron insertion into the methylated intermediate have been discussed.[16,17] Some perform both functions, and a putative "ancestral" form has been proposed.[18]) Therefore, CysG activity should respond to cellular needs for both cobalamin and siroheme; yet in these two organisms this does not appear to be the case. First, siroheme is used as a cofactor for two enzymes, sulfite reductase and nitrite reductase. Expression of the *cysG* gene in *E. coli* and *S. enterica* is induced by nitrite during anaerobic respiration, but not by sulfur starvation;[19] this pattern indicates that siroheme's role as a cofactor in the CysIJ sulfite reductase has not selected for regulation of *cysG* expression in response to sulfur starvation, and that baseline transcription is sufficient. Second, the intermediate produced by CysG after methylation (precorrin-2) is the precursor for cobalamin as well as siroheme.[14] CysG can balance the production of these tetrapyrroles by chelating either cobalt

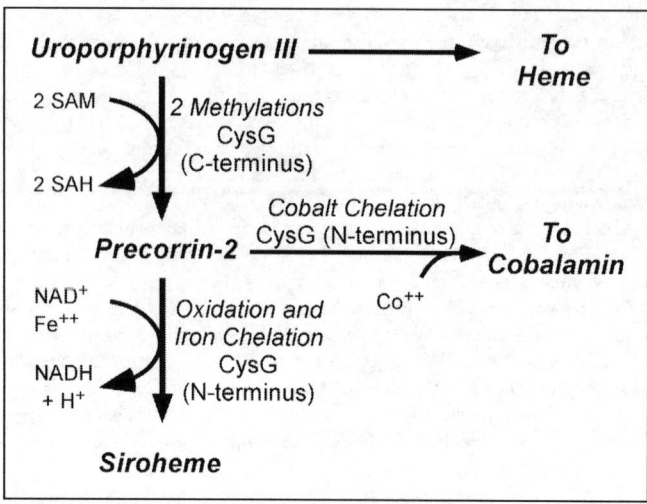

Figure 2. The flow of Uro-III to three different tetrapyrroles as controlled by the action of the CysG protein of enteric bacteria. SAM: S-adenosyl-methionine, SAH: S-adenosyl-homocysteine.

or iron. This activity does not appear to be regulated directly by the need for cobalamin in the cell. However, the amount of cobalamin produced may be controlled indirectly as a function of the amount of cobalt present in the cytoplasm; this may be achieved via an inducible cobalt transport system encoded by the *Salmonella cob* operon.[7]

While the CysG enzyme couples the methylation, oxidation and metal chelation activities of siroheme biosynthesis into a single enzyme, clearly intermediates are released which allow for cobalamin biosynthesis. Intermediate release is also more likely in organisms like *Bacillus megaterium*, where these activities are encoded by three separate polypeptides,[20,21] or Pseudomonas denitrificans, where the methylase activity is performed by a single-function protein, CobA.[22] In contrast, the flow of tetrapyrroles towards cobalamin is assisted by enzymes like the HemD-CobA hybrid protein of Selenomonas ruminantium, which combines the final step of Uro-III synthesis with the methylation steps, thus forming precorrin-2.[23] This arrangement is likely tolerated in this obligately fermentative organism since heme molecules are not required for respiratory cytochromes. This case also demonstrates the danger in naming genes by homology; here the "HemD" portion of this bifunctional gene plays no role in heme biosynthesis.

While the apportionment of tetrapyrroles between heme, siroheme and cobalamin appears to be primarily passive in *E. coli* and *S. enterica*—with more siroheme produced anaerobically in the presence of nitrate—the situation is more complex in *Klebsiella aerogenes*, which encodes a second siroheme synthase (CysF) homologous to CysG.[24] While *cysG* expression is regulated by nitrite anaerobically (as in *Salmonella*), the *cysFDNC* operon is controlled by the CysB activator in response to sulfur starvation. Yet this apparent redundancy reflects more than a selected ability to produce siroheme in response to sulfur starvation, an activity lost in *E. coli* and *Salmonella*. Unlike CysG, the CysF protein cannot produce an intermediate that is available for cobalamin biosynthesis; this constraint may be due to a difference in the order of reactions catalyzed by CysF, or the failure of the protein to release the intermediate during catalysis, making it unavailable to the cobalamin biosynthetic apparatus.

A question remains, then, as to why *Klebsiella* retains both the *cysF* and *cysG* genes, while the *cysF* gene was likely lost from the ancestor of *E. coli* and *S. enterica*. One model to explain the role of the *cysF* gene entails the production of cobalamin in *Klebsiella*, which (unlike in *Salmonella*) can occur during aerobic growth conditions. Since the *cysG* gene is not induced aerobically, flow of tetrapyrroles into the cobalamin pool would deplete the siroheme pool, interfering sulfate reduction; the action of the CysF protein would allow for siroheme production without any diversion of precorrin-2 into cobalamin. Since the ability to synthesize cobalamin de novo was lost from the ancestor of *E. coli* and *S. enterica*, the selection to retain the *cysF* gene would also have been lost (Fig. 3).

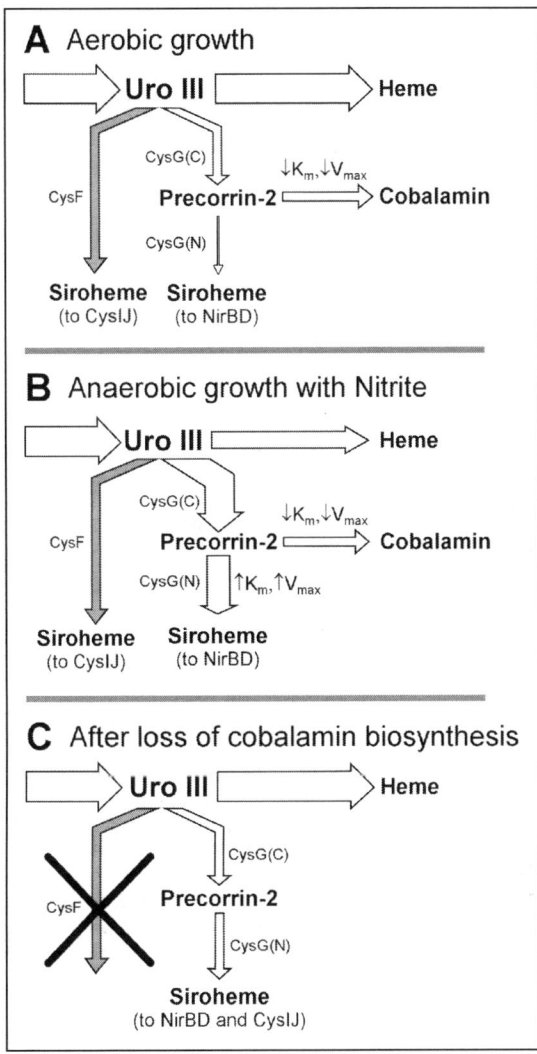

Figure 3. Regulatory scheme whereby loss of de novo cobalamin biosynthesis would lead to the loss of the *cysF* gene from the ancestor of *Escherichia coli* and *Salmonella enterica*. A) Under aerobic growth, the CysF protein provides siroheme for the CysIJ sulfite reductase. B) Under anaerobic growth, induction of the cysG gene produces siroheme for the NirBD nitrite reductase; its high Km and Vmax prevent excess synthesis of cobalamin due to this induction. C) After loss of cobalamin biosynthesis, basal expression of the *cysG* gene would provide sufficient amounts of siroheme to satisfy the requirements of the CysIJ sulfite reductase, leading to the loss of the unselected *cysF* gene.

Operon Induction and Physiological Significance

In *Salmonella enterica*, cobalamin is used as a cofactor for three different enzymes: the MetH methionine synthase,[11,25] the EutBC ethanolamine deaminase,[26-28] and the PduCDE propanediol dehydratase.[29,30] Its requirement for one or more of these functions could induce cobalamin biosynthesis. Careful analysis of the expression of the *Salmonella cob* operon has revealed that it is induced anaerobically in the presence of propanediol;[31] neither methionine starvation in *metE* mutants nor carbon/nitrogen starvation in the presence of ethanolamine resulted in greater production of cobalamin, suggesting that the selection retaining the *cob* genes in *Salmonella* lay in its role as a cofactor during propanediol utilization. This induction is mediated by the PocR regulatory protein (Fig. 4A), whose cognate gene is located between the *cob* operon (encoding genes required for cobalamin biosynthesis) and the *pdu* operon (encoding genes required for cobalamin-dependent degradation of propanediol).[31-33] The physical arrangement of the genes suggested this regulatory relationship, which was verified by mutations in the *pocR* gene. The *cob* and *pdu* operons are maximally induced anaerobically under carbon starvation conditions, effects which are mediated by the Fnr protein and the CAP/cAMP complex, respectively.[34] A similar physical juxtaposition of the cobalamin biosynthetic genes and the propanediol utilization genes is found in *Klebsiella aerogenes*, and similar regulatory patterns are also found (Kolko, Scott and Lawrence, unpublished observations).

More recently, it was found that the IHF protein is also likely involved in the maximal expression of these genes,[35] since cognate mutants have lower levels of PocR protein and fail to utilize propanediol as a carbon source. Mutations in the *mivA* gene (which regulates RpoS levels) also negatively affects expression of the *cob/pdu* regulon, but it is not clear is the effect is direct or indirect due to abnormally high levels of the RpoS protein.[36] These data demonstrate that while biochemical evidence provides a satisfying confirmation of strongly suspected direct interactions (like PocR binding to adjacent promoters under its control), it is absolutely requisite if one is to establish the mode of action of global regulatory proteins in gene expression.

The regulation of cobalamin biosynthesis in other organisms is less clear than in the enteric bacteria discussed so far. While the biosynthesis of cobalamin has been exhaustively studied in Pseudomonas denitrificans, little is known about its regulation. Preparations of cobamides for chemical analysis are made from cells growing in the presence of betaine and/or ethanolamine,[37] which were known to maximize yields, so one could suspect that their utilization could provide the physiological significance behind cobalamin biosynthesis in this organism. In other organisms, information is even more tenuous. For example, *Paracoccus denitrificans* requires cobalamin for anaerobic growth,[38] where it employs a B_{12}-dependent ethanolamine deaminase. Yet this requirement does not rule out the possibility that the required cobalamin is produced by baseline expression of these genes—as the B_{12} requirement for the MetH enzyme is satisfied in *Salmonella* growing anaerobically under noninducing conditions—and the cobalamin biosynthetic genes are induced under other conditions. Cyanobacteria have cobalamin-dependent nucleotide reductases,[39] but again it is not clear that their cofactor requirements induce the expression of B_{12} biosynthesis in these organisms. A similar nucleotide reductase was thought to be the enzyme selecting cobalamin biosynthesis in *Sinorhizobium meliloti*[40] until methionine biosynthesis was shown to require the cofactor as well.[41] Analysis of the genome sequence of *Sinorhizobium meliloti*[42] is inconclusive, since the large clusters of cobalamin biosynthetic genes are not adjacent to known B_{12}-requiring enzymes, thereby not providing an easy hypothesis to test as was the case with the juxtaposition of the *cob* and *pdu* operons in *Salmonella*. In some cases, even genetic data are inconclusive. For example, expression of a five-gene operon biosynthesis in *Rhodobacter capsulatus* whose products are likely involved in cobalamin (by sequence homology to genes in *Salmonella*) is unaffected by oxygen tension or the presence of cobalamin;[43] both factors affect the expression of the *Salmonella cob* genes dramatically. Mutations in these genes result in strains that fail to produce cobalamin but are corrected by the addition of exogenous cobalamin to the media, indicating that they are involved in cobalamin biosynthesis. The same mutations also disturb the formation of the photosynthetic apparatus; whether these genes are under similar regulation as the photosynthetic genes is unclear.

As expected from the dual role of the CobU protein in both de novo cobalamin biosynthesis and in the utilization of externally acquired cobinamide, the *cobUST* genes are expressed from a separate promoter in *Salmonella enterica*.[44] This is known because *lacZ* transcriptional fusions show significant levels of expression under noninducing conditions, even when polar insertions are placed upstream of the *cobU* gene. In *Salmonella*, it is not clear if the *cobUST* genes are expressed constitutively, or respond to outside signals.

In *Escherichia coli*, most of the *cob* operon is absent;[45] only the *cobUST* genes remain (Fig. 4B), allowing the synthesis of cobalamin from the complex intermediate cobinamide (which lacks the DMB lower axial ligand). *E. coli* also cannot degrade propanediol in a cobalamin-dependent fashion (among a survey of more than 100 isolates), and no genes homologous to the *Salmonella pdu*

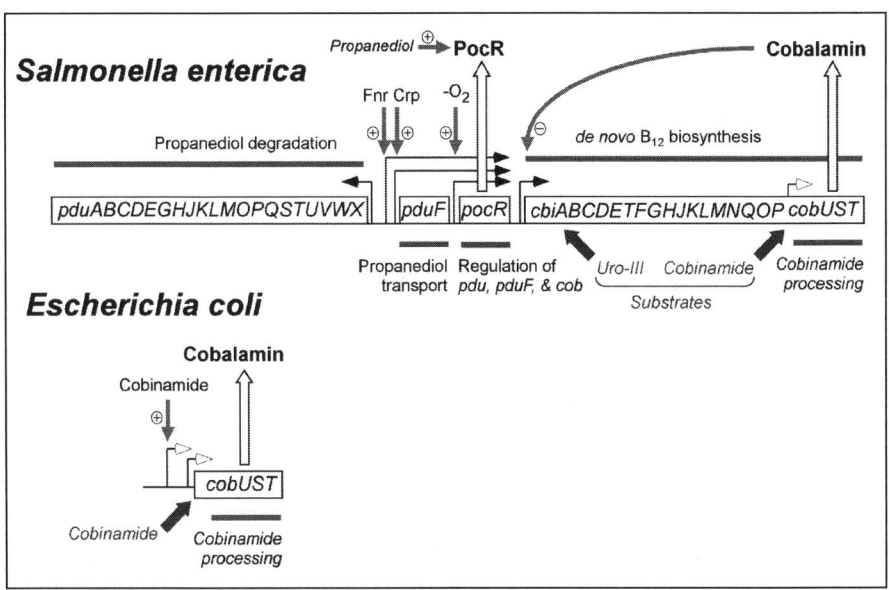

Figure 4. Regulation of the primary cobalamin biosynthetic regulons in *Salmonella enterica* and *Escherichia coli*. The five darkened arrow-heads in the *Salmonella* regulon represent promoters under the positive control of the PocR protein; the hollow arrow-head denotes a putatively constitutive promoter. The grey arrows direct substrates towards genes whose encoded enzymes act on those substances.

genes have been detected in any strain.[46-48] Therefore, it is reasonable to conclude that propanediol utilization does not provide selection for the retention of the *cobUST* genes in *E. coli*. Preliminary studies have found that the putative substrate for the *cobUST* genes (cobinamide) acts as an inducer,[45] allowing for cobalamin production from cobinamide when cobinamide is present. However, this activity was not influenced by methionine starvation or the need to utilize ethanolamine as a carbon or nitrogen source. Therefore, the physiological significance of cobalamin production in *E. coli* is unclear.

The regulation of the *Salmonella pdu* and *cob* operons by the PocR protein was predicted from the sequence of the *pocR* gene (it encodes an AraC-family DNA-binding protein with a C-terminal helix-turn-helix domain). Careful genetic studies revealed five promoters—one each for the *cob*, *pdu*, and *pocR* transcripts and 2 for the *pduF/pocR* transcript—were activated by the PocR protein.[31,33,49] A bioinformatic analysis of the promoter regions predicted eight strong and 2 weak potential binding sites for the PocR proteins in positions where plausible regulation could occur;[49] the half-sites were 30 bases in length with dyad symmetry expected of AraC-family proteins and three sites were proposed in the vicinity of the *cob* operon promoter. Biochemical footprinting studies verified that a GST-PocR fusion protein bound directly to two of these sites,[50] thereby establishing PocR as the protein directly responsible for induction of the *Salmonella cob* operon in response to propanediol.

Operon Repression and mRNA Binding

As indicated in Figure 4A, cobalamin also serves to repress its own synthesis;[51] the repression of a biosynthetic pathway by its end product is not uncommon for bacterial regulatory circuits. However, the mode of action by which cobalamin exerts its control in not thought to be as common. Numerous mutant hunts were performed to find the protein(s) responsible for B_{12}-mediated repression of the *Salmonella cob* operon; no protein target was ever identified despite exhaustive searches (J.R. Roth, personal communication). The *E. coli* and *Salmonella btuB* genes—encoding the outer-membrane cobamide transporter—are similarly repressed by vitamin

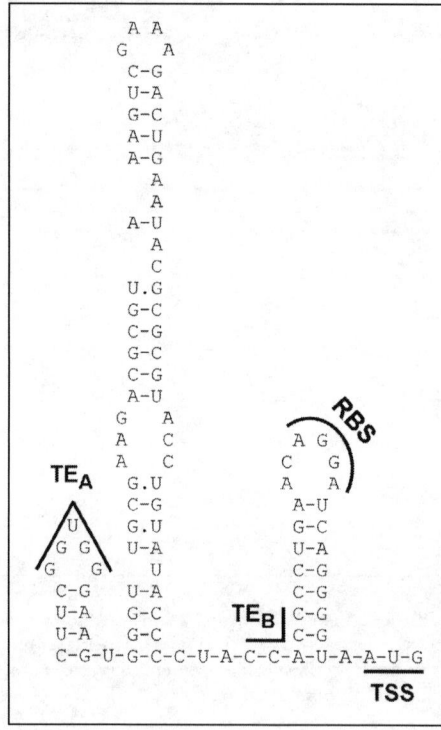

Figure 5. Leader region of the *Salmonella enterica* Typhimurium *cob* transcript, upstream of the *cbiA* gene. TSS: Translation Start Site, RBS: Ribosome Binding Site, TE$_A$, TE$_B$: Translational Enhancer, sites A and B, as defined by mutational analysis.[59]

B$_{12}$, cyanocobalamin;[52] similar mutant hunts uncovered the *E. coli btuR* gene (homologous to the *Salmonella cobA* gene), encoding a cobalamin adenosyl-transferase which adds the upper (Coβ) axial ligand to the central cobalt atom in the corrinoid ring.[53] In retrospect this result was not surprising, since only adenosyl-cobalamin represses the *btuB* gene or the *cob* operon, cobamides with other upper axial ligands would fail to repress.

There were similarities noted between the *btuB* and *cob* transcripts: both had promoters situated several hundred base-pairs upstream of the translation start site, and both shared some sequences in common, although the significance of these sequences was unclear.[2,54-56] All mutations which only affected repression of the *cob* operon by cobalamin lay in this "leader" region between the promoter and the translation start site of the *cbiA* gene (Fig. 5). Some mutations lay in the *pdu* operon which are thought to affect the levels of adenosyl-cobalamin in the cell, similar to mutations in the *cobA* gene.[57] While this result could reflect redundant proteins mediating repression, and the mutagenesis only affected the binding sites, it was tantalizing to speculate that there was a direct mRNA-B$_{12}$ interaction mediating this effect.

More recent studies have confirmed this direct mRNA-B$_{12}$ interaction, which stabilizes an mRNA secondary structure that occludes the ribosome loading site for both the *btuB* and *cob* transcripts.[58-61] In the case of the *Salmonella cob* operon, pseudoknots and long-distance mRNA secondary structures contribute towards stabilizing an RNA hairpin occluding the ribosome binding site (Fig. 5); the binding of cobalamin to this structure, mediated by a so-called translational enhancer (TE element), stabilizes this structure.[59] The control of gene expression via translational control by the direct binding of a small molecule to an mRNA is also seen in the so-called "S-Box" genes of *Bacillus*, where S-adenosyl-methionine has been demonstrated to bind directly to the mRNA of genes involved in methionine biosynthesis.[62,63] The control of gene expression via translational control—that is, manipulation of the mRNA by ribosome stalling, small molecule binding, antisense binding, and pseudoknots and other long-distance interactions is becoming an increasingly more frequent finding when the dissection of promoters which show complex regulation is performed.[64]

The Synergy of Cobalamin Transport and Synthesis

As indicated in Figure 4A, the genes for de novo cobalamin biosynthesis in *Salmonella* also serve for the conversion of cobinamide—presumably transported into the cell—into cobalamin. Cobalamin can be transported into the cell via the high-affinity BtuB outer membrane protein, which uses the power of proton-motive force mediated by the TonB protein to drive cobalamin influx into the periplasm.[65-67] There cobalamin is bound by the BtuF[68,69] and delivered to the BtuCD inner membrane transporter; here, ATP hydrolysis by the BtuD protein drives cobalamin influx into the cell.[70-72] Since *E. coli* can synthesize cobalamin only from cobinamide, we can safely assume that cobinamide is also transported into the cytoplasm where it is converted into cobalamin; as expected, cobinamide induces the expression of the *E. coli cobUST* genes,[45] which encode the proteins required for this conversion.

The troubling part of this scenario is a clear source of cobinamide in the environment. First, it is not a known biosynthetic intermediate in cobalamin biosynthesis (rather, cobinamide-phosphate is). Second, the pools of biosynthetic intermediates is quite low in cells, so even if cobinamide were formed from cobinamide phosphate upon cell lysis, only very small amounts would be present in the environment. It is more likely that cell lysis would release a complete cobamide, but perhaps one without 5,6-dimethylbenzinidazole as the lower (Coα) axial ligand; as noted above, many alternative forms are known from many different taxa. Therefore, it is reasonable to speculate that there cobamides could serve as a source of cobinamide if the "incorrect" lower axial ligand were cleaved at the phosphate bond, yielding cobinamide as a product.

While far from established, there is some evidence for this scenario. Organisms have been isolated that produce cobamides that require *cobUST* activity for them to serve as cofactors for the *E. coli* MetH methionine synthase [that is, *metE* mutants can utilize cell extracts as a source of cobalamin to grow, but *metE cobS* double mutants cannot (Halo, Somers and Lawrence, unpublished

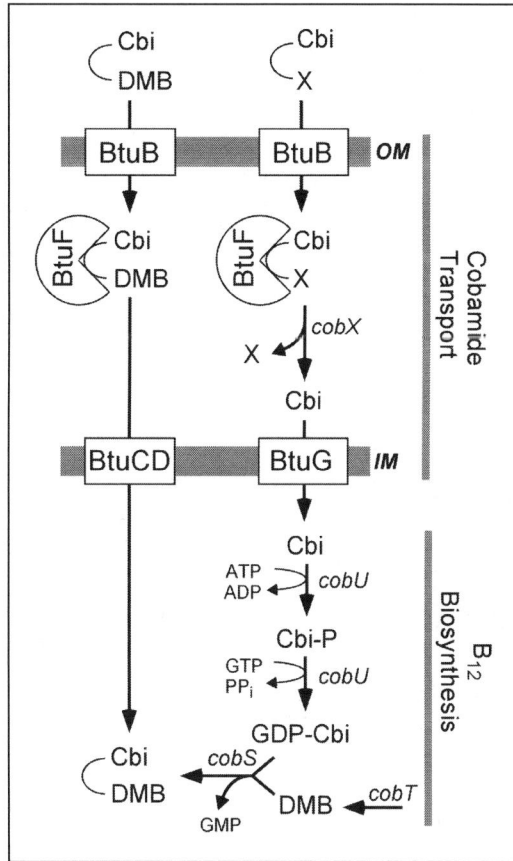

Figure 6. Scheme for transport and utilization of alternative cobamides in enteric bacteria. Corrinoids with Coα ligands significantly different from 5,6-dimethylbenzimidazole would be directed towards a cleavage enzyme, producing cobinamide. Cobinamide passes through the cell through the BtuG transporter (Halo, Somers and Lawrence, unpublished results) where it induces the *cobUST* genes.[45] There, the kinase activity of the CobU protein (which is not used in de novo cobalamin biosynthesis) makes cobinamide-phosphate, which enters the biosynthetic pathway.

results)]. Searches for mutants that cannot utilize these "alternative" cobamides have uncovered several novel genes involved in corrinoid transport and utilization (Fig. 6). One gene (preliminarily termed *btuG*) appears to encode an inner membrane transporter which appears to mediate cobinamide movement across the inner (Halo, Somers and Lawrence, unpublished results). Such a semi-redundant transporter could explain why *btuCD* mutants have such mild phenotypes, whereas cobalamin is a large molecule to make its way into the cell unassisted or via a transporter with little or no corrinoid specificity. Mutations in other genes appear to eliminate the proposed "cleavage" activity (termed *btuX*), although these genes have not been characterized fully (Halo, Somers and Lawrence, unpublished results).

A more complex "salvage" route for corrinoids was proposed for *Porphyromonas ginivalis*, which harbors all of the genes required for cobalamin biosynthesis using precorrin-2 as a substrate, but apparently cannot synthesize Uro-III.[73] It is not clear whether the observed cobalamin biosynthetic genes represent selfish operons[74] that were introduced by recent lateral transfer but are nonfunctional and will soon be lost to deletion,[75,76] or whether a cryptic pathway for Uro-III synthesis or acquisition has yet to be discovered. If these genes do represent a "salvage" pathway similar to the one proposed above, it is not clear what the substrate would be, but hemes or sirohemes would be likely candidates; their utilization would require a "reverse"-chelatase activity that would be a novel addition to the family of enzymes participating in the transport, biosynthesis and reuse of the complex small molecule.

Note Added in Proof

Earlier work pointed to direct interactions between cobalamin and nascent mRNA molecules as a possible mechanism mediating control of operons responsible for cobalamin synthesis or transport. The widespread discovery of riboswitches in many bacteria[77,78], including one in the leaders of the *btuB* genes of *Escherichia coli* and *Salmonella enterica*,[79] validates this hypothesis. Here, coenzyme B_{12} (adenosylcobalamin) binds directly to mRNA to facilitate translational repression.

References

1. Debussche L, Thibaut D, Danzer M et al. Biosynthesis of vitamin B_{12}: Structure of precorrin-3B, the trimethylated substrate of the enzyme catalyzing ring contraction. J Chem Soc Chem Commun 1993; 1993:1100-1103.
2. Roth JR, Lawrence JG, Bobik TA. Cobalamin (coenzyme B_{12}): Synthesis and biological significance. Annu Rev Microbiol 1996; 50:137-181.
3. Benner SA, Ellington AD, Tauer A. Modern metabolism as a palimsest of the RNA world. Proc Natl Acad Sci USA 1989; 86:7054-7058.
4. Eschenmoser A. Vitamin B_{12}: Experiments concerning the origin of its molecular structure. Angew Chem Int Ed Eng 1988; 27:5-39.
5. Stupperich E, Eisinger HJ, Schurr S. Corrinoids in anaerobic bacteria. FEMS Microbiol Rev 1990; 87:355-360.
6. Brushaber KR, GA OT, Escalante-Semerena JC. CobD, a novel enzyme with L-threonine -O-3-phosphate decarboxylase activity, is responsible for the synthesis of (R)-1-amino- 2-propanol O- 2-phosphate, a proposed new intermediate in cobalamin biosynthesis in Salmonella typhimurium LT2. J Biol Chem 1998; 273(5):2684-91.
7. Roth JR, Lawrence JG, Rubenfield M et al. Characterization of the cobalamin (vitamin B_{12}) biosynthetic genes of Salmonella typhimurium. J Bacteriol 1993; 175:3303-3316.
8. Battersby AR. How nature builds the pigments of life: The conquest of vitamin B_{12}. Science 1994; 264:1551-1557.
9. Raux E, Schubert HL, Warren MJ. Biosynthesis of cobalamin (vitamin B_{12}): A bacterial conundrum. Cell Mol Life Sci 2000; 57(13-14):1880-1893.
10. Roessner CA, Santander PJ, Scott AI. Multiple biosynthetic pathways for vitamin B_{12}: Variations on a central theme. Vitam Horm 2001; 61:267-297.
11. Smith DA, Childs JD. Methionine genes and enzymes of Salmonella typhimurium. Heredity 1966; 21:265-286.
12. Greene RC. Biosynthesis of methionine. In: Neidhardt FC, Curtiss IIIrd R, Ingraham JL, Lin ECC, Low KB, Magasanik B et al, eds. Escherichia coli and Salmonella typhimurium: Cellular and molecular biology. 2nd ed. Washington, DC: American Society for Microbiology, 1996:542-560.

13. Warren MJ, Roessner CA, Santeander PJ et al. The Escherichia coli cysG gene encodes S-adenosylmethionine-dependent uroporphyrinogen III methylase. Biochem J 1990; 265(3):725-729.
14. Warren MJ, Bolt EL, Roessner CA et al. Gene dissection demonstrates that the Escherichia coli cysG gene encodes a multifunctional protein. Biochem J 1994; 302(3):837-844.
15. Fazzio TG, Roth JR. Evidence that the CysG protein catalyzes the first reaction specific to B_{12} synthesis in Salmonella tyhimurium: Insertion of cobalt. J Bacteriol 1996; 178:6952-6959.
16. Raux E, Thermes C, Heathcote P et al. A role for Salmonella typhimurium cbiK in cobalamin (vitamin B_{12}) and siroheme biosynthesis. J Bacteriol 1997; 179(10):3202-3212.
17. Schubert HL, Raux E, Wilson KS et al. Common chelatase design in the branched tetrapyrrole pathways of heme and anaerobic cobalamin synthesis. Biochemistry 1999; 38(33):10660-10669.
18. Brindley AA, Raux E, Leech HK et al. A story of chelatase evolution: Identification and characterization of a small 13-15 kDa "ancestral" copbaltochelatase (CbiXS) in the Archaea. J Biol Chem 2003; 278(25):22388-22395.
19. Goldman BS, Roth JR. Genetic structure and regulation of the cysG gene in Salmonella typhimurium. J Bacteriol 1993; 175(5):1457-1466.
20. Leech HK, Raux-Deery E, Heathcote P et al. Production of cobalamin and sirohaem in Bacillus megaterium: An investigation into the role of the branchpoint chelatases sirohydrochlorin ferrochelatase (SirB) and sirohydrochlorin cobalt chelatase (CbiX). Biochem Soc Trans 2002; 30(4):610-613.
21. Raux E, Leech HK, Beck R et al. Identification and functional analysis of enzymes required for precorrin-2 dehydrogenation and metal ion insertion in the biosynthesis of sirohaem and cobalamin in Bacillus megaterium. Biochem J 2003; 370(2):505-516.
22. Blanche F, Debussche L, Thibaut D et al. Purification and characterization of S-adenosylmethionine: Uroporphyrinogen III methyltransferase from Pseudomonas denitrificans. J Bacteriol 1989; 171:4222-4231.
23. Anderson PJ, Entsch B, McKay DB. A gene, cobA + hemD, from Selenomonas ruminantium encodes a bifunctional enzyme involved in the synthesis of vitamin B_{12}. Gene 2001; 281(1-2):63-70.
24. Kolko MM, Kapetanovich LA, Lawrence JG. Alternative pathways for siroheme synthesis in Klebsiella aerogenes. J Bacteriol 2001; 183:328-335.
25. Childs JD, Smith DA. New methionine structural gene in Salmonella typhimurium. J Bacteriol 1969; 100:377-382.
26. Chang GW, Chang JT. Evidence for the B_{12}-dependent enzyme ethanolamine deaminase in Salmonella. Nature (London) 1975; 254:150-151.
27. Faust LP, Babior BM. Overexpression, purification, and some properties of the adocbl-dependent ethanolamine ammonia-lyase from Salmonella typhimurium. Arch Biochem Biophys 1992; 294(1):50-54.
28. Roof DM, Roth JR. Functions required for vitamin-B_{12} dependent ethanolamine utilization in Salmonella typhimurium. J Bacteriol 1989; 171:3316-3323.
29. Bobik TA, Xu Y, Jeter RM et al. Propanediol utilization genes (pdu) of Salmonella typhimurium: Three genes for the propanediol dehydratase. J Bacteriol 1997; 179(21):6633-9.
30. Jeter RM. Cobalamin dependent 1,2-propanediol utilization by Salmonella typhimurium. J Gen Microbiol 1990; 136:887-896.
31. Bobik TA, Ailion M, Roth JR. A single regulatory gene integrates control of vitamin B_{12} synthesis and propanediol degradation. J Bacteriol 1992; 174:2253-2266.
32. Rondon MR, Escalante-Semerena JC. The poc locus is required for 1,2-propanediol-dependent transcription of the cobalamin biosynthetic (cob) and propanediol utilization (pdu) genes of Salmonella typhimurium. J Bacteriol 1992; 174:2267-2272.
33. Chen P, Andersson DI, Roth JR. The control region of the pdu/cob regulon in Salmonella typhimurium. J Bacteriol 1994; 176:5474-5482.
34. Ailion M, Bobik TA, Roth JR. Two global regulatory systems (Crp and Arc) control the cobalamin/propanediol regulon of Salmonella typhimurium. J Bacteriol 1993; 175:7200-7208.
35. Rondon MR, Escalante-Semerena JC. Integration host factor is required for 1,2-propanediol-dependent transcription of the cob/pdu regulon in Salmonella typhimurium LT2. J Bacteriol 1997; 179(11):3797-3800.
36. Rondon MR, Escalante-Semerena JC. High levels of transcription factor RpoS (sigma S) in mviA mutants negatively affect 1,2-propanediol-dependent transcription of the cob/pdu regulon of Salmonella typhimurium LT2. FEMS Microbiol Lett 1998; 169(1):147-153.
37. Cameron B, Briggs K, Pridmore S et al. Cloning and analysis of genes involved in coenzyme B_{12} biosynthesis in Pseudomonas denitrificans. J Bacteriol 1989; 171:547-557.
38. Shearer N, Hinsley AP, Van Spanning RJ et al. Anaerobic growth of Paracoccus denitrificans requires cobalamin: Characterization of cobK and cobJ genes. J Bacteriol 1999; 181(22):6907-6913.

39. Gleason FK, Wood JM. Ribonucleotide reductase in blue-green algae: Dependence on adenosylcobalamin. Science 1976; 192(4246):1343-1344.
40. Cowles JR, Evans HJ. Some properties of the ribonucleotide reductase from Rhizobium meliloti. Arch Biochem Biophys 1968; 127(1):770-778.
41. Sato K, Inukai S, Shimizu S. Vitamin B_{12}-dependent methionine synthesis in Rhizobium meliloti. Biochem Biophys Res Commun 1974; 60(2):723-728.
42. Galibert F, Finan TM, Long SR et al. The composite genome of the legume symbiont Sinorhizobium meliloti. Science 2001; 293(5530):668-672.
43. Pollich M, Klug G. Identification and sequence analysis of genes involved in late steps in cobalamin (vitamin B_{12}) synthesis in Rhodobacter capsulatus. J Bacteriol 1995; 177(15):4481-4487.
44. Escalante-Semerena JC, Johnson MG, Roth JR. The CobII and CobIII regions of the cobalamin (vitamin B_{12}) biosynthetic operon of Salmonella typhimurium. J Bacteriol 1992; 174(1):24-29.
45. Lawrence JG, Roth JR. The cobalamin (coenzyme B_{12}) biosynthetic genes of Escherichia coli. J Bacteriol 1995; 177:6371-6380.
46. Cocks GT, Aguilar J, Lin ECC. Evolution of L-1,2 propanediol catabolism in Escherichia coli by recruitment of enzymes for L-fucose and L-lactate metabolism. J Bacteriol 1974; 118:83-88.
47. Hacking AJ, Aguilar J, Lin ECC. Evolution of propanediol utilization in Escherichia coli. Mutants with improved substrate scavenging power. J Bacteriol 1978; 136:522-530.
48. Lawrence JG, Roth JR. Evolution of coenzyme B_{12} synthesis among enteric bacteria: Evidence for loss and reacquisition of a multigene complex. Genetics 1996; 142:11-24.
49. Chen P, Ailion M, Bobik T et al. Five promoters integrate control of the cob/pdu regulon in Salmonella typhimurium. J Bacteriol 1995; 177(19):5401-5410.
50. Rondon MR, Escalante-Semerena JC. In vitro analysis of the interactions between the PocR regulatory protein and the promoter region of the cobalamin biosynthetic (cob) operon of Salmonella typhimurium LT2. J Bacteriol 1996; 178(8):2196-2203.
51. Andersson D. Kinetics of cobalamin repression in the cob operon in Salmonella typhimurium. FEMS Microbiol Lett 1995; 124:89-94.
52. Kadner RJ. Repression of synthesis of the vitamin B_{12} receptor in Escherichia coli. J Bacteriol 1978; 136(3):1050-1057.
53. Lundrigan MD, Kadner RJ. Altered cobalamin metabolism in Escherichia coli btuR mutants affects btuB regulation. J Bacteriol 1989; 171:154-161.
54. Lundrigan MD, Koster W, Kadner RJ. Transcribed sequences of the Escherichia coli btuB gene control its expression and regulation by vitamin B_{12}. Proc Natl Acad Sci USA 1991; 88:1479-1483.
55. Richter-Dahlfors AA. Cobalamin (vitamin B_{12}) repression of the cob operon in Salmonella typhimurium requires sequences within the leader and the first translated open reading frame. Mol Microbiol 1992; 6(6):743-749.
56. Richter-Dahlfors AA, Ravnum S, Andersson DI. Vitamin B_{12} repression of the cob operon in Salmonella typhimurium: Translational control of the cbiA gene. Mol Microbiol 1994; 13(3):541-553.
57. Ailion M, Roth JR. Repression of the cob operon of Salmonella typhimurium by adenosylcobalamin is influenced by mutations in the pdu operon. J Bacteriol 1997; 179(19):6084-6091.
58. Franklund CV, Kadner RJ. Multiple transcribed elements control expression of the Escherichia coli btuB gene. J Bacteriol 1997; 179(12):4039-4042.
59. Ravnum S, Andersson DI. An adenosyl-cobalamin (coenzyme-B_{12})-repressed translational enhancer in the cob mRNA of Salmonella typhimurium. Mol Microbiol 2001; 39(6):1585-1594.
60. Nahvi A, Sudarsan N, Ebert MS et al. Genetic control by a metabolite binding mRNA. Chem Biol 2002; 9(9):1043.
61. Ravnum S, Andersson DI. Vitamin B_{12} repression of the btuB gene in Salmonella typhimurium is mediated via a translational control which requires leader and coding sequences. Mol Microbiol 1997; 23(1):35-42.
62. Epshtein V, Mironov AS, Nudler E. The riboswitch-mediated control of sulfur metabolism in bacteria. Proc Natl Acad Sci USA 2003; 100(9):5052-5056.
63. McDaniel BA, Grundy FJ, Artsimovitch I et al. Transcription termination control of the S box system: Direct measurement of S-adenosylmethionine by the leader RNA. Proc Natl Acad Sci USA 2003; 100(6):3083-8.
64. Draper DE, Gluick C, Schlax PJ. Pseudoknots, RNA folding, and translational regulation. In: Simons RW, Grunberg-Manago M, eds. RNA Structure and Functions. Cold Spring Harbor: Cold Spring Harbor Laboratory Press, 1998:415-436.
65. Kadner RJ, Liggins GL. Transport of vitamin B_{12} in Escherichia coli: Genetic studies. J Bacteriol 1973; 115(2):514-521.

66. Bassford Jr PJ, Bradbeer C, Kadner RJ et al. Transport of vitamin B_{12} in tonB mutants of Escherichia coli. J Bacteriol 1976; 128(1):242-247.
67. Bassford Jr PJ, Kadner RJ. Genetic analysis of components involved in vitamin B_{12} uptake in Escherichia coli. J Bacteriol 1977; 132(3):96-105.
68. Cadieux N, Bradbeer C, Reeger-Schneider E et al. Identification of the periplasmic cobalamin-binding protein BtuF of Escherichia coli. J Bacteriol 2002; 184(3):706-717.
69. Van Bibber M, Bradbeer C, Clark N et al. A new class of cobalamin transport mutants (btuF) provides genetic evidence for a periplasmic binding protein in Salmonella typhimurium. J Bacteriol 1999; 181(17):5539-5541.
70. DeVeaux LC, Kadner RJ. Transport of vitamin B_{12} in Escherichia coli: Cloning of the btuCD region. J Bacteriol 1985; 162(3):888-896.
71. DeVeaux LC, Clevenson DS, Bradbeer C et al. Identification of the BtuCED polypeptides and evidence for their role in vitamin B_{12} transport in Escherichia coli. J Bacteriol 1986; 167(3):920-927.
72. Friedrich MJ, DeVeaux LC, Kadner RJ. Nucleotide sequence of the btuCED genes involved in vitamin B_{12} transport in Escherichia coli and homology with components of periplasmic-binding-protein-dependent transport systems. J Bacteriol 1986; 167(3):928-934.
73. Roper JM, Raux E, Brindley AA et al. The enigma of cobalamin (Vitamin B_{12}) biosynthesis in Porphyromonas gingivalis. Identification and characterization of a functional corrin pathway. J Biol Chem 2000; 275(51):40316-40323.
74. Lawrence JG, Roth JR. Selfish operons: Horizontal transfer may drive the evolution of gene clusters. Genetics 1996; 143:1843-1860.
75. Lawrence JG, Ochman H. Molecular archaeology of the Escherichia coli genome. Proc Natl Acad Sci USA 1998; 95:9413-9417.
76. Lawrence JG, Roth JR. Genomic flux: Genome evolution by gene loss and acquisition. In: Charlebois RL, ed. Organization of the Prokaryotic Genome. Washington, DC: ASM Press, 1999:263-289.
77. Vitreschak AG., Rodionov DA, Mironov AA, Gelfand MS. Regulation of the vitamin B_{12} metabolism and transport in bacteria by a conserved RNA structural element. RNA 2003; 9:1084-1097.
78. Nahvi A, Barrick JE, Breaker RR. Coenzyme B_{12} riboswitches are widespread genetic control elements in prokaryotes. Nucleic Acids Res 2004; 32:143-150.
79. Nahvi A, Sudarsan N, Ebert MS et al. Genetic control by a metabolite binding mRNA. Chem Biol 2002; 9:1043-1049.

Coenzyme B_{12}-Catalyzed Radical Isomerizations

Dominique Padovani and Ruma Banerjee*

Abstract

Cobalamins are complex organometallic cofactors essential for catalysis in three enzyme families, the isomerases, methyltransferases and reductive dehalogenases. This account focuses on the isomerases, which catalyze difficult and unusual 1,2-rearrangement reactions. These enzymes utilize AdoCbl as a radical reservoir. They induce homolytic cleavage of the Co-C bond of the cofactor thus generating a highly reactive 5′-deoxyadenosyl radical that initiates the radical rearrangement process. Over the past decade, structure-function and computational studies of several members of the isomerase family have provided interesting insights into the mechanism of AdoCbl-dependent radical isomerizations. In particular, these studies have revealed the critical role played by the protein in catalyzing the trillion-fold rate enhancement of Co-C bond homolysis, in shielding radical intermediates from adventitious side-reactions and in facilitating the rearrangement process itself.

Cobalamins, or B_{12}-cofactors, are used by three classes of enzymes and have long fascinated bioinorganic chemists because of their complex structure and their stable organometallic bond which is the only known biologically.[1] These tetrapyrrolic cofactors contain a central cobalt atom tethered by four equatorial nitrogen ligands from pyrroles A-D of the corrin ring (Fig. 1). One peculiarity of the corrin ring is the degree of flexibility, not enjoyed by the other tetrapyrrole cofactors (e.g., porphyrins), afforded by its fairly reduced state. In solution and at physiological pH, the lower trans-axial ligand of the cobalt atom is a nitrogen atom provided by the bulky base, 5,6-dimethylbenzimidazole (DMB),[1] that is appended via a nucleotide loop from the periphery of ring D of the corrin macrocycle. Diversity is thus present at the upper cis-axial ligand where cyano-, hydroxo-, -methyl- and deoxyadenosyl- groups are seen in vitamin B_{12}, hydroxocobalamin, methylcobalamin (MeCbl) and adenosylcobalamin (AdoCbl) respectively. Strictly speaking, diversity is also present in the trans position; however, technically these are described as cobamides rather than cobalamins.[1]

A key feature of the enzymes that utilize B_{12} as cofactors is their ability to control the reactivity of the stable Co-C bond to undergo either heterolytic or homolytic scission. Thus, biological methylation reactions carried out by MeCbl-dependent enzymes, operate through heterolytic cleavage of the Co-C bond generating cob(I)alamin and the methyl cation is transferred to an acceptor molecule.[1] On the other hand, biological rearrangement reactions catalyzed by AdoCbl-dependent isomerases, proceed via the homolytic fission of the Co-C bond, producing cob(II)alamin and a

---

Due to space limitations, the reference list is not exhaustive. A reference followed by an asterisk (e.g., 2*) refers to work cited in this chapter.

*Corresponding Author: Ruma Banerjee—Department of Biological Chemistry, University of Michigan, 3220B MSRBIII, 1150 W. Medical Center Dr., Ann Arbor, MI 48109-0606, USA. Email: rbanerje@umich.edu

Tetrapyrroles: Birth, Life and Death, edited by Martin J. Warren and Alison G. Smith.

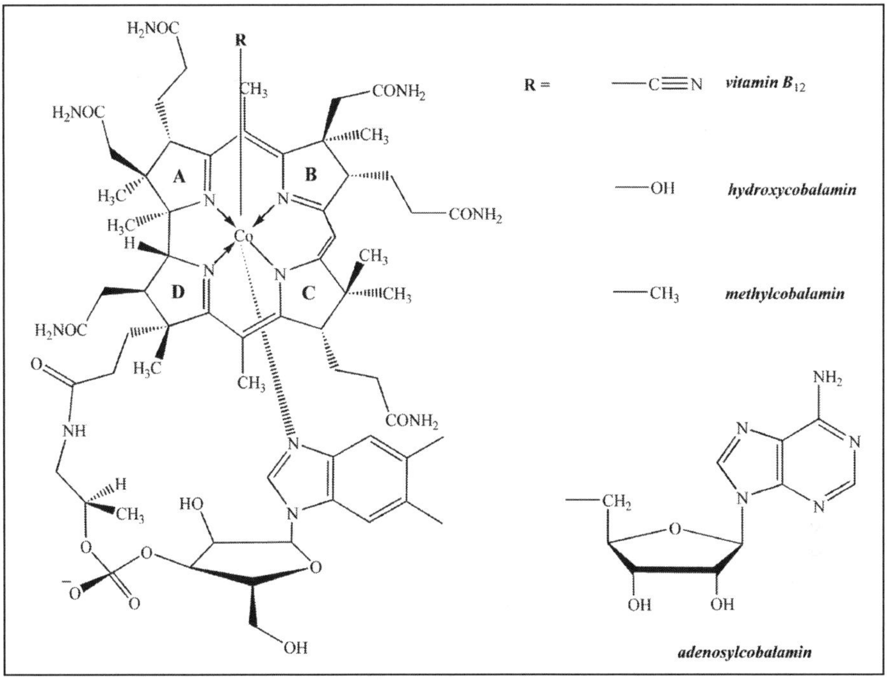

Figure 1. Structure of cobalamin derivatives. In solution and at physiological pH, the trans-axial ligand is the bulky base DMB.

highly reactive 5'-deoxyadenosyl radical that initiates radical chemistry.[2] In this chapter, we will focus on the B$_{12}$-dependent isomerases.

Structural Insights into the B$_{12}$-Dependent Isomerases

The AdoCbl-dependent isomerases are the largest subfamily of B$_{12}$-dependent enzymes. They are found in bacteria, where they participate in fermentative pathways. The exceptions are class II ribonucleotide reductase (RNR), that is important in de novo synthesis of deoxyribonucleotides, and mammalian methylmalonyl-CoA mutase, which is involved in the catabolism of odd-chain fatty acids, cholesterol and branched-chain amino acids.

The isomerases catalyze difficult and unusual rearrangement reactions involving the formal 1,2-interchange of a hydrogen atom on a carbon with a variable migrating group (-OH, -NH$_2$ or a carbon-containing fragment) on a vicinal carbon.[2] An exception to this rule is encountered by class II RNRs that catalyse the reduction of ribonucleotides to deoxyribonucleotides used by the cells for DNA replication and repair.[3] A common strategy developed by these enzymes to enable these chemically difficult reactions is the deployment of radical chemistry with AdoCbl serving as a radical reservoir. Thus, Co-C bond homolysis generates the radical pair, cob(II)alamin and the 5'-deoxyadenosyl radical. The latter abstracts a hydrogen atom and leads to a substrate radical (or a protein radical in RNR) that, in turn, rearranges to a product radical. Reabstraction of a hydrogen atom from 5'-deoxyadenosine generates product and the 5'-deoxyadenosyl radical. Recombination of the cofactor radicals completes the catalytic cycle (Fig. 2).[4]

During the last decade, X-ray analysis[5-8] and spectroscopic studies[9-11] have provided structural insights into B$_{12}$-dependent isomerases, leading in particular to the recognition of two subclasses of enzymes with respect to their mode of cofactor binding (Fig. 3).[2]

To promote catalysis, enzymes of the Class-I/His-on subfamily abstract a hydrogen atom from a carbon atom without substituents containing lone electron pairs. Enzymes in this family exhibit the consensus sequence Asp-X-His-X-X-Gly that provides the trans-axial histidine ligand to the cobalt

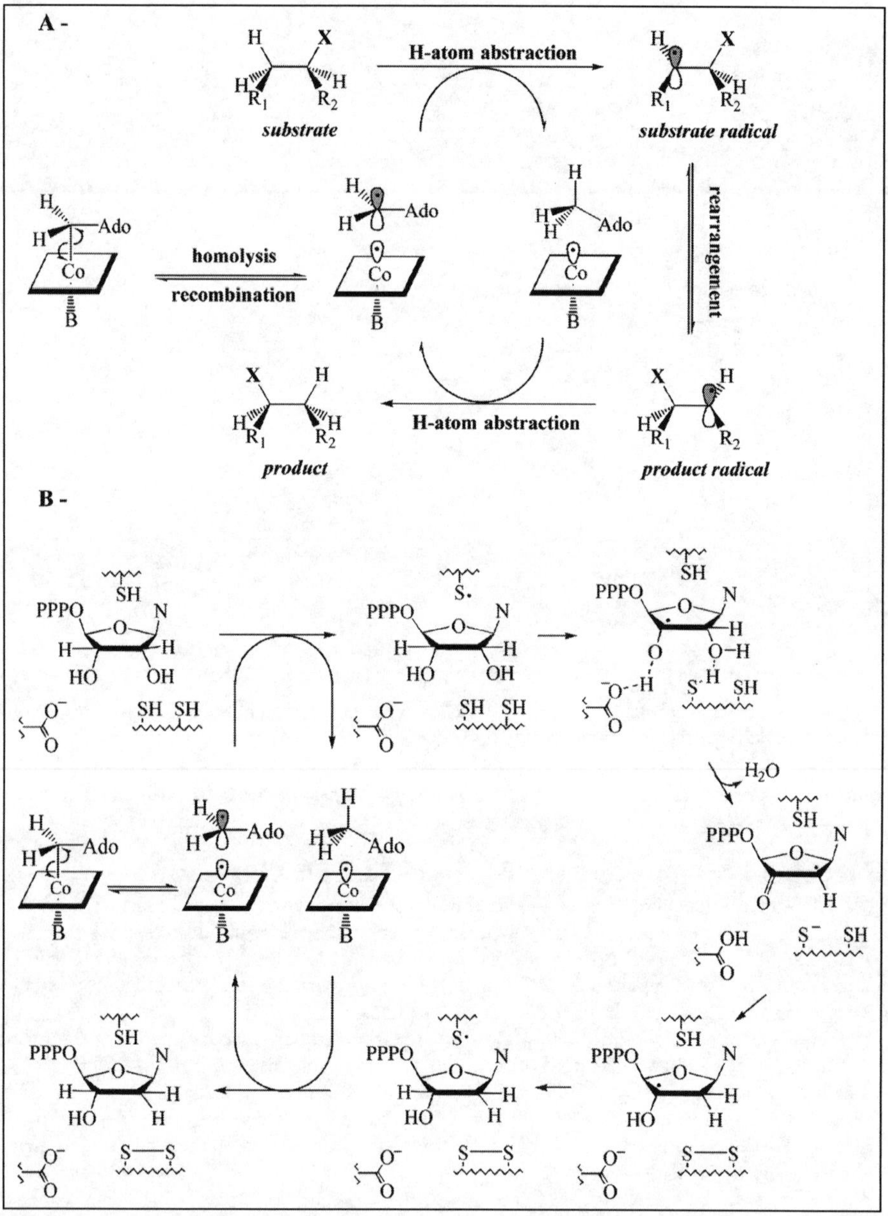

Figure 2. Postulated general reaction mechanism for AdoCbl-catalyzed radical isomerizations. The cofactor serves as a radical reservoir to generate a highly reactive 5'-deoxyadenosyl radical that initiates radical catalysis by abstraction of a hydrogen atom on the substrate (A) or on a cysteine residue as in RNR (B).

atom. Indeed, crystal structures of two members of the Class-I/His-on subfamily, methylmalonyl-CoA mutase[5] and glutamate mutase,[6] reveal that cobalamin undergoes a large conformational change upon binding. The corrin ring of cobalamin sits on an $\alpha_5\beta_5$ Rossman-fold, the trans-axial ligand in solution, 5,6-dimethylbenzimidazole, being replaced by a histidine residue donated by the protein matrix. In

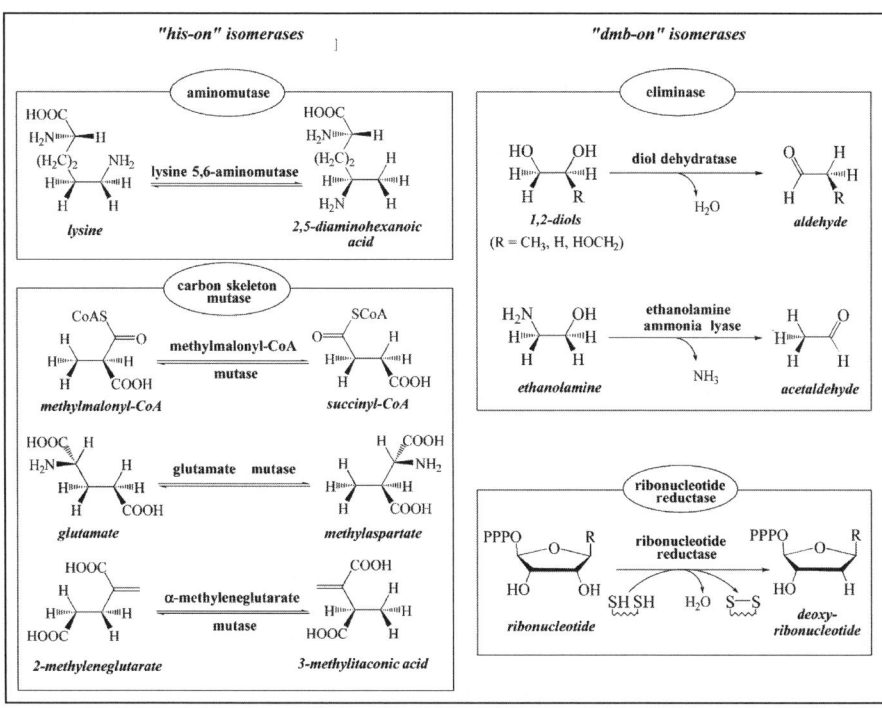

Figure 3. Representative reactions catalyzed by the two subclasses of AdoCbl-dependent isomerases.

this binding mode, the nucleotide tail is displaced from the cobalt and is bound in a deep pocket in an extended conformation with the DMB being moved 14 Å away from the cobalt atom (Fig. 4A). Based on EPR studies and/or sequence alignments, other isomerases belonging to this subfamily include lysine 5,6-aminomutase, α-methyleneglutarate mutase and isobutyryl-CoA mutase.[2,9]

In contrast, crystal structures of enzymes of the class-II/Base-on subfamily, which initiate catalysis by abstracting a hydrogen atom from a carbon bonded to a heteroatom, shows that the solution conformation of AdoCbl is retained upon binding (Fig. 4B). Members of this subfamily include the eliminases (e.g., diol dehydratase and ethanolamine ammonia lyase) and RNR.[7,8,10,11]

Despite the diversity in cofactor binding and the chemistry of the reactions catalyzed by B$_{12}$-dependent isomerases, the enzymes, with the exception of RNR, share similarities in their active site architecture (Fig. 5).[5-7] A (β/α)$_8$ TIM-barrel lies on the upper side of the cofactor with the corrin ring covering a lower cavity and isolating the active site and with the cis-axial ligand being directed toward the inside of the barrel. This structural design thus provides direct access to the active site cavity only via the substrate binding site. Such an architecture is well suited for radical chemistry with the deeply buried active site cavity minimizing unwanted side reactions of radical intermediates with solvent molecules. In contrast, the class II RNR contains a cobalamin-binding region and an active site that shares no structural similarities to other B$_{12}$-enzymes.[8] Instead, this enzyme retains all the key structural features shared by class I and class III RNRs, that is a global fold composed of a 10 stranded α/β-barrel core. This barrel is composed of two parallel five-stranded β-sheets oriented in an antiparrallel fashion and contains at its center, the RNR fingerloop consisting of two antiparrallel β-strands with the active site Cys located at the fingertip.[8]

Co-C Bond Activation in B$_{12}$-Dependent Isomerases

A subject of enduring debate in the bioinorganic field is how the AdoCbl-dependent isomerases overcome the inherent kinetic inertness of the Co-C bond. In solution, the Co-C bond dissociation

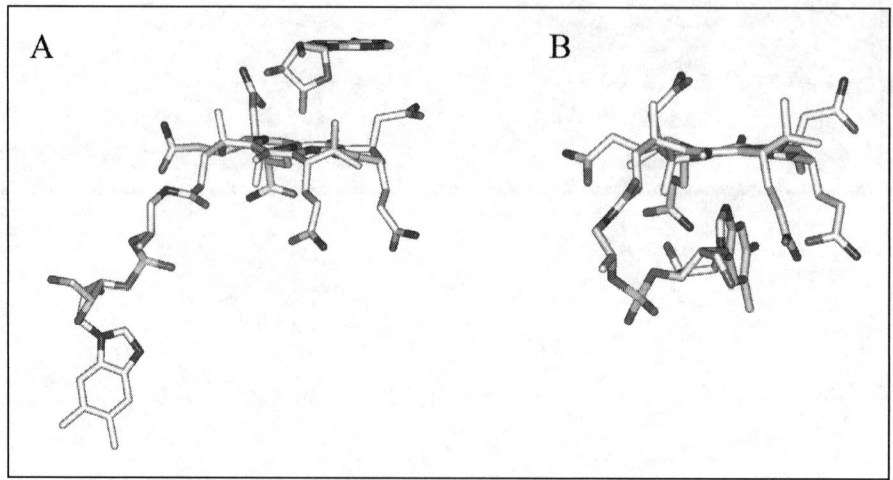

Figure 4. Two modes for binding AdoCbl by the isomerases. In the "his-on" subclass of isomerases, the nucleotide tail of DMB is removed from the cobalt and adopts an extended conformation (A). In the "DMB-on" subclass of isomerases, the solution conformation of the cofactor is maintained (B).

enthalpy is 31.4 ± 1.5 kcal mol^{-1} for AdoCbl and the thermal homolysis rate at 25°C is only $10^{-9 \pm 1}$ s^{-1}, representing a half-life of one year at 37°C for the free cofactor![12] However, in AdoCbl-dependent isomerases, presteady-state kinetic revealed that this rate is accelerated by a factor $10^{12 \pm 1}$ over the uncatalyzed reaction, implying an ~15 kcal mol^{-1} destabilization of the Co-C bond.[13] Several mechanisms have been considered so far to explain this trillion-fold rate enhancement of Co-C bond homolysis. Most of them involve steric effect, such as a trans-ligand effect where the bond strength to the upper axial ligand could be significantly weakened by an upward flexing of the corrin ring through steric interaction with the adenosyl moiety, an angular distortion of the Co-C bond or wedging in of an active site residue to pry apart the Co-C bond.[14*]

Stopped-flow spectroscopy performed on AdoCbl-dependent isomerases reveal that Co-C bond homolysis occurs at a rate that is much faster than catalytic turnover, indicating that this step is not rate determining.[2*,15] Moreover, in the absence of substrate, the homolysis products are not observed, indicating that the equilibrium for the homolysis reaction favors geminate recombination. In contrast, in the presence of substrate, Co-C bond homolysis is enhanced by ca. trillion-fold, suggesting that substrate binding energy triggers a change in the equilibrium in favor of homolysis.

Presteady state studies on methylmalonyl-CoA mutase,[16] RNR[17,18] and more recently, diol dehydratase,[15] have yielded kinetic and thermodynamic parameters for Co-C bond homolysis and the hydrogen transfer steps. In the reactions catalyzed by methylmalonyl-CoA mutase and RNR, the equilibrium constants for homolysis are close to unity and $\Delta G^{\ddagger}$ is ~15-17 kcal mol^{-1}, corresponding to a rate enhancement of ~10^{12}. In methylmalonyl-CoA mutase, lowering of the homolysis barrier is mainly due to a decrease in the enthalpy of dissociation ($\Delta H^{\ddagger}$ ~16 kcal mol^{-1}).[16] In RNR, conflicting reports of the relative importance of enthalpic versus entropic contributions have been published which may arise from differences in the concentrations and therefore the oligomeric states of the enzyme.[17,18] Resonance Raman studies, in which isotope editing was used to identify several vibrational modes of the adenosyl moiety, have been reported for methylmalonyl-CoA mutase.[19*] These studies led to the conclusion that the ν(Co-C) is minimally perturbed in the bound versus free cofactor with the vibrational frequency shifting from 430 cm^{-1} to 424 cm^{-1} upon cofactor binding. This 6 cm^{-1} downshift corresponds to a small destabilization of 0.5 kcal mol^{-1}. Resonance Raman spectra of the enzyme-substrate complex reveal that the ν(Co-C) shifts back to 430 cm^{-1}, corresponding to a slight strengthening of the Co-C bond upon substrate binding. In contrast, in diol dehydratase, the Co-C bond homolysis of the cofactor is achieved in two stages. A rate acceleration of $10^{6 \pm 1}$ occurs upon binding of AdoCbl to the

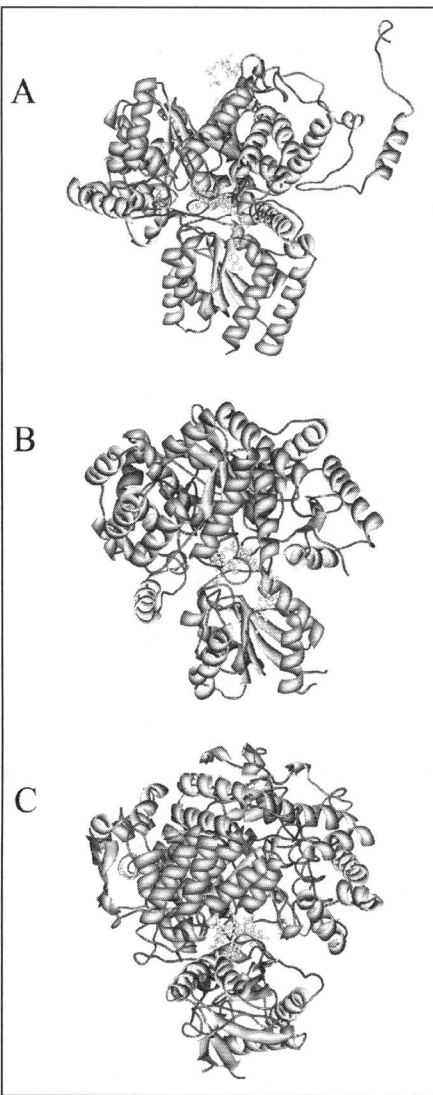

Figure 5. Protein architecture of three different B$_{12}$-dependent isomerases. The cofactor sits on the $\alpha_5\beta_5$-Rossman fold (lower part of the structure) and the $(\alpha/\beta)_8$ TIM-barrel (substrate binding domain) lies on the upper side. The cis-axial ligand, not shown in the structures, is directed toward the barrel. A) The α subunit of the heterodimeric $\alpha\beta$ methylmalonyl-CoA mutase from *Propionibacterium shermanii* represented in an open conformation (2REQ). The partial substrate analog, desulpho-CoA, is shown in ball-and-stick model in yellow (top side). B) An $\epsilon\alpha$ unit of the heterodimeric glutamate mutase from *Clostridium cochlearium* (1CB7) cocrystallized with MeCbl. C) An $\alpha\beta\gamma$ unit of the heterodimeric diol dehydratase from *Klebsiella oxytoca* (1DIO) in complex with CNCbl. Alpha helices are represented in sky blue, β-sheets in light green and loops in sienna. The cofactor is shown in ball-and-stick representation and is colored in yellow. A color version of this figure is available online at www.eurekah.com.

apoenzyme and a further rate enhancement of 10^6-fold ($\Delta G^\ddagger$ ~8.8 kcal mol^{-1}) is accomplished upon binding of substrate to the holoenzyme.[15]

Kinetic investigations of B$_{12}$-dependent enzymes present a clearer picture of how the isomerases may control the homolytic scission of the cofactor. Indeed, for all the enzymes studied so far, homolysis of the Co-C bond is coupled to hydrogen atom abstraction from the substrate or from an active site cysteine in RNR.[2*,15*] This coupling is reflected by observation of an isotope effect on the homolysis rate by isotopic substitution in the substrate. In methylmalonyl-CoA mutase and glutamate mutase, Co-C bond homolysis is rapid and very sensitive to substrate deuteration exhibiting a deceleration of ~20 when [CD$_3$]methyl-malonyl-CoA or [2,4,4-D$_3$]glutamate are respectively used instead of the corresponding protiated substrates.[2*] A plausible explanation for the isotopic sensitivity of the homolysis step is that it is kinetically coupled to the following step. Thus, the high energy 5'-deoxyadenosyl radical (dAdo$^\bullet$) intermediate abstracts a H atom to form a more stable substrate-centered radical, in the case of glutamate mutase (Fig. 6), which rearranges to a more stable product radical in the case of methylmalonyl-CoA mutase. More recently, similar experiments have been performed with diol dehydratase and ethanolamine ammonia lyase and an isotope effect, albeit of a lower magnitude was also observed. When [1,1-D$_2$]propanediol and [1,1-D$_2$]ethanolamine were employed as substrates k_H/k_D of ~10 and >10 respectively were measured.[15*]

In contrast to other B$_{12}$-dependent isomerases, the AdoCbl-dependent RNR catalyzes the exchange of hydrogen from the 5' position of AdoCbl with solvent in the presence of an allosteric effector and in the absence of substrate.[3*] This unusual exchange reaction can be explained by the cofactor-dependent formation of a thiyl radical, at Cys408 in the *Lactobacillus leichmannii* RNR. The Cys408Ser mutant renders the Co-C bond cleavage reaction unfavorable suggesting that the homolysis step is coupled to the generation of the thiyl radical.[3*] Indeed, this coupling would be expected to significantly reduce (by ~10 kcal mol^{-1}) the enthalpic cost of Co-C bond homolysis. By the same token, coupling of the unfavorable H atom reabstraction step (from 5'-deoxyadenosine by a thiyl radical) to the energetically favorable Co-C bond reformation would result in a net negative enthalpic change.

The observed kinetic isotope effects for Co-C bond homolysis in methylmalonyl-CoA mutase and glutamate mutase are much larger than expected from semi-classical considerations.[2*] These results suggest the involvement of quantum chemical tunneling in the H-atom transfer from substrate to the 5'-deoxyadenosyl radical. Indeed, experimental evidence for tunneling has been reported for methylmalonyl-CoA mutase where the temperature dependence of the isotope effect was studied.[20] Arrhenius analysis of the data revealed large deviations for the parameters A_H/A_D and $E_{aD}-E_{aH}$ (0.078 ± 0.009 and 3.41 ± 0.07 kcal mol^{-1} respectively) from the values predicted from semi classical theory (0.7-1 and 1.15 kcal mol^{-1}).

As discussed previously, X-ray structure determinations of B$_{12}$-dependent isomerases revealed two different AdoCbl conformations and raises the possibility that trans ligand effects may be more important in one subset of enzymes than in the other. An extensive study on the Co-C bond thermolysis in cobinamide derivatives with different axial bases showed that an increase in the basicity of the trans-axial ligand enhances the rate of Co-C bond homolysis but has a more pronounced effect on enhancing the competing heterolysis reaction.[21*] Recent ab initio DFT calculations on cobamide models revealed that the homolytic Co-C bond scission has a very small dependence on the nature of the trans-axial ligand.[21] Neither the thermodynamic nor the kinetic parameters are affected when the Co-N$_{ax}$ distance is changed ± 0.3 Å. Taken together, these results rule out a significant role for the trans-axial ligand in accelerating the Co-C bond homolysis.[21]

Recent attention has focused on the role of protein conformational changes when substrate binds to the enzyme to explain the large increase in the Co-C bond homolysis rate. In methylmalonyl-CoA mutase, the substrate binding domain is a TIM barrel perched above the corrin ring of cobalamin.[5] Binding of substrate results in a dramatic conformational change that closes up the active site and apparently destroys the binding pocket for the deoxyadenosyl group by pushing a key active site residue, Tyr89, into the area previously occupied by the deoxyadenosine moiety.[22] These crystallographic observations led to the hypothesis that the substrate-driven barrel closure triggers homolytic cleavage and that steric crowding by Tyr89 labilizes the Co-C bond. Kinetic studies on the Tyr89Phe mutant revealed a 10^3-fold and a 580-fold decrease in the k$_{cat}$ in the forward and reverse directions respectively and a suppression of the overall kinetic isotope effect in the forward but not the reverse direction.[23*] The homolysis equilibrium in the Tyr89Phe mutant was perturbed in favor of geminate recombination, consistent with the Co-C bond homolysis step becoming rate determining in the forward direction.[23] Moreover, the crystal structure of the Tyr89Phe mutant revealed that it is essentially superimposable on the structure of the wild type enzyme except at the position of the mutation which shows enhanced flexibility further consistent with a role for Tyr89 as a molecular wedge that labilizes the Co-C bond.[23*]

Crystallographic studies on diol dehydratase using the B$_{12}$-analog adeninylpentylcobalamin revealed that the adeninylpentyl group is located in the TIM barrel above the corrin ring in a nucleotide binding pocket.[15*] It should be noted that the existence of such a pocket provides a rationale for the specificity of the enzyme for the adenosyl group of the cofactor. Crystallographic and modeling studies reveal that distortion of the Co-C bond is already visible in the absence of substrate and that binding of substrate shifts the equilibrium in favor of Co-C bond dissociation.[15*,24] It was concluded that an "adenine-attracting effect", that is strong enzyme-cofactor interactions at both the corrin and adenine moieties, produces angular strain and tensile forces contributing to Co-C bond labilization.

Radical Flights: Conformational Changes at Play

The first step in the isomerase-catalyzed reactions is postulated to be homolytic fission of the Co-C bond of the cofactor to generate cob(II)alamin and dAdo$^{\bullet}$. The latter radical then abstracts a hydrogen atom from the substrate (or the protein in the case of RNR), leading to formation of a radical-pair intermediate.[4] The existence of biradical species has been corroborated by EPR spectroscopy which reveals the presence of paramagnetic intermediates having a cob(II)alamin-centered radical coupled to an organic radical.[2*,3*,15,25,26,30*] In methylmalonyl-CoA mutase,[2*]

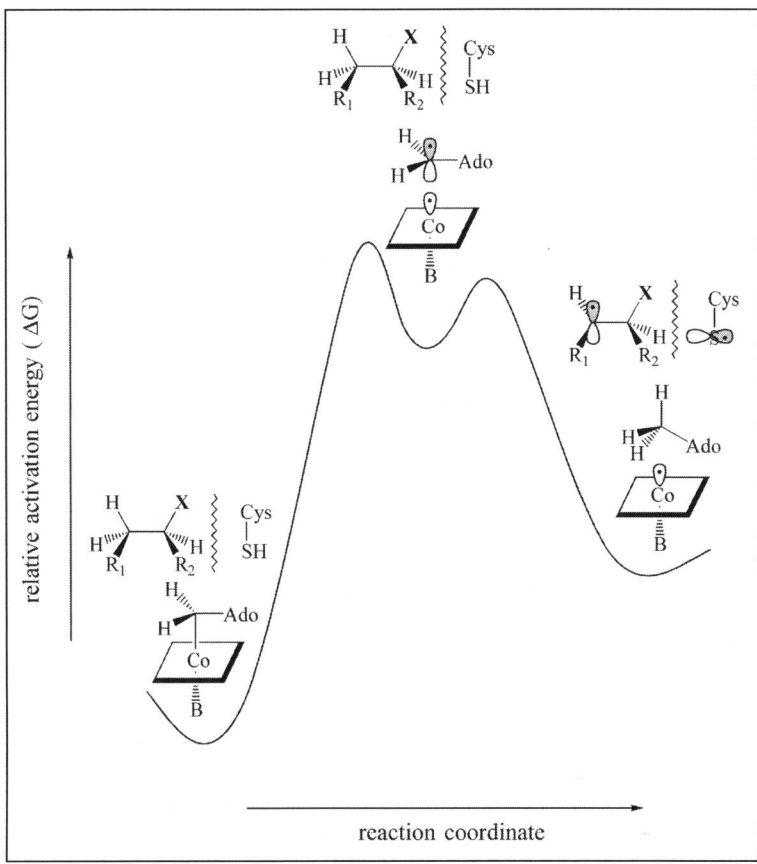

Figure 6. Qualitative free energy profile for Co-C bond homolysis coupled to the following chemical step. The unfavorable Co-C bond homolysis step is rendered more favorable by kinetic coupling to a hydrogen abstraction step.

glutamate mutase,[25] α-methyleneglutarate mutase[30*] and RNR,[3*] a broad axial spectrum with g values of 2.11 and 2.0 have been observed. Simulations of the spectra obtained with RNR labeled with deuterated cysteine supports assignment of the EPR spectrum resulting from a thiyl radical in dipolar and exchange coupling interactions with cob(II)alamin.[3*] From the analysis of the dipolar interactions, an estimate of the interradical distance of 5-7 Å was obtained.[3*] In glutamate mutase, the EPR signal has been assigned to the 4-glutamyl radical interacting with cob(II)alamin with the two separated by 6.6 Å.[25] In contrast, in diol dehydratase[15*] and in ethanolamine ammonia lyase,[26] the EPR spectra reveal the presence of a high-field doublet that arises from an organic radical weakly coupled to low-spin cob(II)alamin. Simulations including both exchange and dipolar coupling in ethanolamine ammonia-lyase yield to an inter-radical distance ≥10 Å.[26]

These results suggest that dAdo• moves over several angstroms from its original position, proximal to cob(II)alamin so as to abstract a hydrogen atom from the substrate. With the exception of RNR, which involves a protein radical intermediate, the dAdo• journeys directly to the substrate.

Structural studies have provided fascinating insights into how conformational changes might enable radical propagation in the case of diol dehydratase[15*] and glutamate mutase.[27] In diol dehydratase, where the distance that needs to be bridged between the C1 of substrate and the cobalt is very large, rotation of the ribosyl group around the glycosidic bond swings the 5'-methylene group of dAdo• to within 2 Å of C1 of the substrate.[15*] This "ribosyl rotation" enables not only the radical transfer but ensures that the substrate hydrogen atom abstraction is stereospecific. In glutamate mutase, where the radical journeys over a much shorter distance, a "ribose pseudorotation", from a C2'-*endo* to a C3'-*endo* conformation, switches the orientation of the C5' of 5'-dAdo• from pointing toward cobalt to pointing toward the substrate hydrogen atom that will be abstracted.[27] Such a "conformational toggle mechanism" could also occur in methylmalonyl-CoA mutase where the interradical distance between the substrate and the cofactor is similar to that in glutamate mutase.

The mechanism of radical translation in ethanolamine ammonia lyase has been probed by pulsed ENDOR spectroscopy.[28] The distance between the C1 of the substrate and the C5'-methylene group of 5'-deoxyadenosine is estimated to be 3.4 ± 0.2 Å, revealing that the C5'-dAdo• moves ~7 Å from its original position to one where it is in van der Waals contact with the C1 of substrate. This was independently confirmed by pulsed ESEEM spectroscopy which revealed that the unpaired electron on the C1 of substrate and the C5'-dAdo• are 3.2 Å apart consistent with a direct abstraction of a hydrogen atom by the dAdo• from the C1 position of the substrate.[29]

Rearrangement Reactions Catalyzed by B_{12}-Dependent Enzymes

The mechanisms by which rearrangement reactions catalyzed by B_{12}-dependent isomerases occur are of interest considering the limited analogies in organic radical chemistry. EPR spectroscopy has provided compelling evidence that these reactions involve free radical chemistry (see above). The availability of crystal structures has spurred computational studies ab initio molecular theory calculations. Nevertheless, the rearrangement mechanism remains one the least well understood aspects of AdoCbl-dependent reactions. While a hydrogen atom is transferred intermolecularly from the substrate to dAdo• (or to the active site Cys• in RNR) to generate a substrate radical, a variable group migrates intramolecularly to an adjacent carbon atom. This migration occurs with inversion (e.g., in glutamate mutase and α-methyleneglutarate mutase) or with retention (e.g., in methylmalonyl-CoA mutase) of configuration. Due to these differences and to the variability in the reactivity of the groups undergoing rearrangement, enzymes have developed different strategies to facilitate the interconversion step (Fig. 7).

In methylmalonyl-CoA mutase and α-methyleneglutarate mutase, the carbon-centered migrating group is an sp^2 hybridized carbon (thioacyl carbon and vinylic carbon respectively), and rearrangement via a cyclopropyl radical intermediate is a plausible mechanism.

High level ab initio calculations predicted that partial protonation would substantially reduce the energy barrier for rearrangement by ~6.2 kcal mol^{-1} in methylmalonylCoA-mutase.[2*] The crystal structure of the enzyme revealed the presence of an ideally positioned histidine residue (His244) that could assist the partial protonation of the carbonyl oxygen atom of the substrate.[5] Mutation of His244 to Gly, Ala or Gln results in a 10^2 to 10^3-fold decrease in the k_{cat} compared to wild type enzyme.[2*] Moreover, the His244Ala mutant lost one of the two titrable pKa groups that govern activity in the wild type enzyme. These studies also revealed that His244 plays a major role in shielding the radical site from access to oxygen.[2*]

Although ab initio calculations predict that a rearrangement pathway via a cyclopropyl radical intermediate would be more favorable in α-methyleneglutarate mutase than a fragmentation-recombination pathway ($\Delta\Delta G^{\ddagger}$ ~21.8 kcal mol^{-1}),[51] experimental results support the latter.[30*,31] The enzyme is inhibited by acrylate, one of the fragmentation products of 2-methyleneglutarate, and an EPR spectrum is generated that resembles the spectrum observed in the presence of the substrate.[30*] Furthermore, studies with the labeled (Z)-isomers of the product, methylitaconate, revealed an "E-overshoot" in the initial phase of the enzyme-catalyzed equilibration. This was interpreted as evidence for the rotation of the exo-methylene group that could only take place in a fragmentation-recombination pathway.[30] Finally, studies on the recombination pathway of α-ester radicals produced by flash photolysis of α-phenylselenyl ester derivatives indicated that the cylcopropylcarbinyl radical is unlikely to be formed during the rearrangement reaction. Instead, it was pointed out that the conversion of the substrate to product radical is likely to involve a polar radical reaction mechanism, that is a heterolytic fragmentation-recombination pathway involving the participation of radical anions.[31]

The interconversion of glutamate into 3-methylaspartate catalyzed by glutamate mutase involves the 1,2-intershift of a sp^3 hybridized carbon, thus ruling out the formation of a cyclo-propylcarbinyl

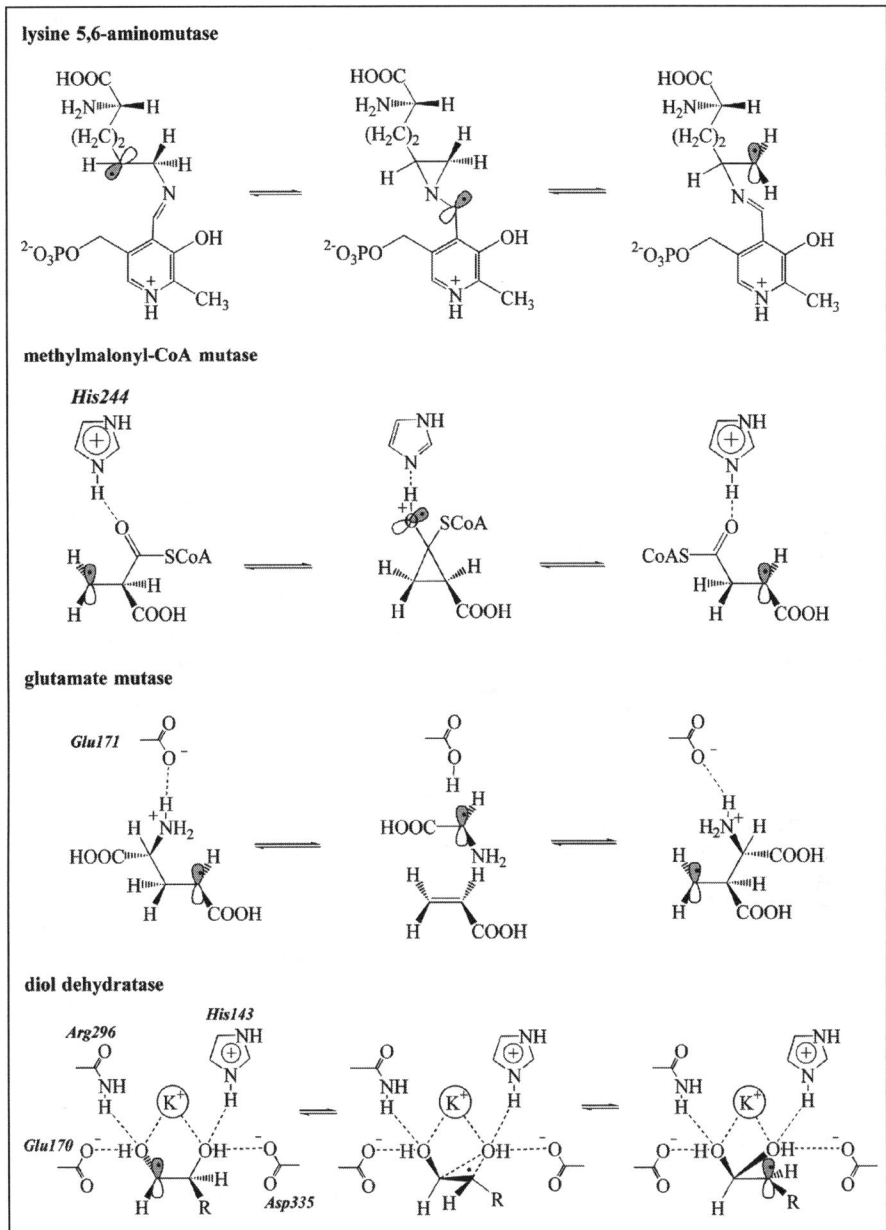

Figure 7. Plausible rearrangement pathways for representative members of the AdoCbl-dependent isomerases.

radical intermediate. A plausible mechanism is a fragmentation-recombination pathway. Evidence in the support of this has been obtained from rapid quench studies which demonstrated the formation of acrylate, a fragmentation product of the 4-glutamyl radical, at a kinetically competent rate.[32] Computational studies indicate glutamate mutase could assist the fragmentation-recombination pathway by partial proton-transfer processes to the amino and α-carbonyl moiety of the substrate.[33*] The crystal structure of the enzyme reveals a suitably positioned glutamate residue (Glu171) in hydrogen bonding distance with the amino group of glutamate.[6] Mutation of Glu171 to Gln, Ala or Asn, resulted in a >50-fold reduction in the k_{cat} and in contrast to the wild type enzyme the activity of the Glu171Gln mutant is pH independent.[33] These results are consistent with a role for Glu171 as a general base in the mutase reaction facilitating formation of the glycyl radical intermediate.

In diol dehydratase, computational studies suggest that the shift of the hydroxyl group from adjacent carbon atoms C2 to C1 would be feasible via a concerted pathway through a cyclic transition state and facilitated by partial protonation of the migrating hydroxyl group.[15*] Moreover, the active site Lewis acid K^+, that contacts both hydroxyl groups of the substrate, would modestly participate in this process by lowering the barrier of the transition state by ~2.3 kcal mol^{-1}.[15*] Nevertheless, calculations show that two active site residues, His143 and Glu170, could support a "synergistic retro-push-pull catalysis" where partial protonation of the migrating OH group and partial deprotonation of the spectator OH group would accelerate the rearrangement step.[15*] Recent kinetic studies have confirmed the importance of both residues in the elimination reaction. Indeed, the His143Leu and Glu170Ala mutant enzymes demonstrate a 524-fold and >3 × 10^4-fold decrease in k_{cat} respectively and in the former mutant, hydrogen atom transfer becomes rate determining.[15]

Rearrangement reaction catalyzed by ethanolamine ammonia-lyase remains one of the less understood mechanisms in the field. Although recent ab initio molecular orbital calculations failed to provide a transition state solution for the catalyzed-reaction, they suggest that an intramolecular rearrangement reaction is more likely than a direct elimination of the amino group.[34] Moreover, a full protonation of the migrating group would be synergistic with the spectator hydroxyl group interacting with a basic catalyst, would promote efficiency of the rearrangement process.[34]

In lysine 5,6-aminomutase, the 1,2-intershift of the amino group is favored by the formation of an external aldimine linkage with pyridoxal phosphate (PLP). It renders the nitrogen sp^2-hybridized and affords the substrate radical a low-energy intramolecular rearrangement pathway via an azacyclopropylcarbinyl intermediate radical in which the unpaired electron is delocalized through a captodative effect.[35*] Furthermore, computational studies indicate that partial protonation of PLP would facilitate the 1,2-amino shift by preventing overstabilization of the cyclic radical intermediate.[35]

In RNR, the thiyl radical is essential for activity as it triggers the first committed step of the reduction process that is hydrogen atom abstraction from C3' of the ribose. The substrate radical is then reduced to the deoxyribonucleotide, via elimination of the C2' hydroxyl group and the formation of a ketyl radical intermediate, by three reducing equivalents provided by NADPH and the Cys408 residue which returns to its radical form.[3]

In summary, X-ray crystal structures and the kinetic and computational studies described above have provided fascinating insights into the mechanisms of B$_{12}$-dependent radical isomerizations. In particular, these investigations have begun to reveal the critical role of individual active site residues in facilitating Co-C bond homolysis and in stabilizing radical intermediates. Nevertheless, much has still to be learnt to resolve intriguing properties of these enzymes. The availability of a range of spectroscopic methods combined with computational studies will further unravel unresolved mechanistic issues in the future.

Acknowledgements

This work was supported by a grant from the National Institutes of Health (DK45776). R. Banerjee is an Established Investigator of the American Heart Association.

References

1. Banerjee R, Ragsdale SW. The many faces of vitamin B$_{12}$: Catalysis by cobalamin-dependent enzymes. Annu Rev Biochem 2003; 72:209-247.
2. Banerjee R. Radical carbon skeleton rearrangements: Catalysis by coenzyme B$_{12}$-dependent mutases. Chem Rev 2003; 103:2083-2094.
3. Stubbe J, van der Donk WA. Protein radicals in enzyme catalysis. Chem Rev 1998; 98:705-762.
4. Halpern J. Mechanisms of coenzyme B$_{12}$-dependent rearrangements. Science 1985; 227:869-875.
5. Mancia F, Keep NH, Nakagawa A et al. How coenzyme B$_{12}$ radicals are generated: The crystal structure of methylmalonyl-CoA mutase at 2 Å resolution. Structure 1996; 4:339-350.

6. Reitzer R, Gruber K, Jogl G et al. Structure of coenzyme B$_{12}$ dependent enzyme glutamate mutase from Clostridium cochlearium. Structure 1999; 7:891-902.
7. Shibata N, Masuda J, Tobimatsu T et al. A new mode of B$_{12}$ binding and the direct participation of a potassium ion in enzyme catalysis: X-ray structure of diol-dehydratase. Structure 1999; 7:997-1008.
8. Sintchak MD, Arjara G, Kellogg BA et al. The crystal structure of class II ribonucleotide reductase reveals how an allosterically regulated monomer mimics a dimer. Nat Struct Biol 2002; 9:293-300.
9. Michel C, Albracht SP, Buckel W. Adenosylcobalamin and cob(II)alamin as prosthetic groups of 2-methyleneglutarate mutase from Clostridium barkeri. Eur J Biochem 1992; 205:767-773.
10. Abend A, Bandarian V, Nitsche R et al. Ethanolamine ammonia-lyase has a "base-on" binding mode for coenzyme B$_{12}$. Arch Biochem Biophys 1999; 370:138-141.
11. Lawrence CC, Gerfen GJ, Samano V et al. Binding of cob(II)alamin to the adenosyl-cobalamin-dependent ribonucleotide reductase from Lactobacillus leichmannii. Identification of dimethylbenzimidazole as the axial ligand. J Biol Chem 1999; 274:7039-7042.
12. Hay BP, Finke RG. Thermolysis of the Co-C bond in adenosylcobalamin. 2. Products, kinetics, and Co-C bond dissociation energy in aqueous solution. J Am Chem Soc 1986; 108:4820-4829.
13. Hay BP, Finke RG. Thermolysis of the Co-C bond in adenosylcorrins. 3. Quantification of the axial base effect in adenosylcobalamin by the synthesis and thermolysis of axial base-free adenosylcobinamide. Insights into the energetics of enzyme-assisted cobalt-carbon bond homolysis. J Am Chem Soc 1987; 109:8012-8018.
14. Kozlowski PM. Quantum chemical modeling of Co-C bond activation in B$_{12}$-dependent enzymes. Curr Opin Chem Biol 2001; 5:736-743.
15. Toraya T. Radical catalysis in coenzyme B$_{12}$-dependent isomerization (eliminating) reactions. Chem Rev 2003; 103:2095-2127.
16. Chowdhury S, Banerjee R. Thermodynamic and kinetic characterization of Co-C bond homolysis catalyzed by coenzyme B$_{12}$-dependent methylmalonyl-CoA mutase. Biochemistry 2000; 39:7998-8006.
17. Brown KL, Li J. Activation parameters for the carbon-cobalt bond homolysis of coenzyme B$_{12}$ induced by the B$_{12}$-dependent Ribonucleotide Reductase from Lactobacillus leichmanii. J Am Chem Soc 1998; 120:9466-9474.
18. Licht SS, Lawrence CC, Stubbe J. Thermodynamic and kinetic studies on carbon-cobalt bond homolysis by ribonucleoside triphosphate reductase: The importance of entropy in catalysis. Biochemistry 1999; 38:1234-1242.
19. Dong S, Padmakumar R, Banerjee R et al. Co-C bond activation in B$_{12}$-dependent enzymes: Cryogenic resonance Raman studies of methylmalonyl-CoA mutase. J Am Chem Soc 1999; 121:7063-7070.
20. Chowdhury S, Banerjee R. Evidence for quantum mechanical tunneling in the coupled cobalt-carbon bond homolysis-substrate radical generation reaction catalyzed by methylmalonyl-CoA mutase. J Am Chem Soc 2000; 122:5417-5418.
21. Dolker N, Maseras F, LledÓs A. A density functional study on the effect of the trans axial ligand of cobalamin on the homolytic cleavage of the Co-C bond. J Phys Chem B 2001; 105:7564-7571.
22. Mancia F, Smith GA, Evans PR. Crystal structure of substrate complexes of methylmalonyl-CoA mutase. Biochemistry 1999; 38:7999-8005.
23. Vlasie M, Banerjee R. Tyrosine 89 accelerates cocarbon bond homolysis in methylmalonyl-CoA mutase. J Am Chem Soc 2003; 125:5431-5435.
24. Shibata N, Masuda J, Morimoto Y et al. Substrate-induced conformational change of a coenzyme B$_{12}$-dependent enzyme: Crystal structure of the substrate-free form of diol dehydratase. Biochemistry 2002; 41:12607-12617.
25. Bothe H, Darley DJ, Albracht SP et al. Identification of the 4-glutamyl radical as an intermediate in the carbon skeleton rearrangement catalyzed by coenzyme B$_{12}$-dependent glutamate mutase from Clostridium cochlearium. Biochemistry 1998; 37:4105-4113.
26. Bandarian V, Reed GH. Analysis of the electron paramagnetic resonance spectrum of a radical intermediate in the coenzyme B$_{12}$-dependent ethanolamine ammonia-lyase catalyzed reaction of S-2-aminopropanol. Biochemistry 2002; 41:8580-8588.
27. Gruber K, Reitzer R, Kratky C. Radical shuttling in a protein: Ribose pseudorotation controls alkyl-radical transfer in the coenzyme B$_{12}$ dependent enzyme glutamate mutase. Angew Chem Int Ed Engl 2001; 40:3377-3380.
28. LoBrutto R, Bandarian V, Magnusson OT et al. 5'-Deoxyadenosine contacts the substrate radical intermediate in the active site of ethanolamine ammonia-lyase: ^{2}H and ^{13}C electron nuclear double resonance studies. Biochemistry 2001; 40:9-14.

29. Warncke K, Utada AS. Interaction of the substrate radical and the 5'-deoxyadenosine-5'-methyl group in vitamin B_{12} coenzyme-dependent ethanolamine deaminase. J Am Chem Soc 2001; 123:8564-8572.

30. Pierik AJ, Ciceri D, Broker G et al. Rotation of the exo-methylene group of (R)-3-methylitaconate catalyzed by coenzyme B_{12}-dependent 2-methyleneglutarate mutase from Eubacterium barkeri. J Am Chem Soc 2002; 124:14039-14048.

31. Newcomb M, Miranda N. Kinetic results implicating a polar radical reaction pathway in the rearrangement catalyzed by α-methyleneglutarate mutase. J Am Chem Soc 2003; 125:4080-4086.

32. Chih HW, Marsh ENG. Mechanism of glutamate mutase: Identification and kinetic competence of acrylate and glycyl radical as intermediates in the rearrangement of glutamate to methylaspartate. J Am Chem Soc 2000; 122:10732-10733.

33. Madhavapeddi P, Marsh ENG. The role of the active site glutamate in the rearrangement of glutamate to 3-methylaspartate catalyzed by adenosylcobalamin-dependent glutamate mutase. Chem Biol 2001; 8:1143-1149.

34. Wetmore SD, Smith DM, Bennett JT et al. Understanding the mechanism of action of B_{12}-dependent ethanolamine ammonia-lyase: Synergistic interactions at play. J Am Chem Soc 2002; 124:14054-14065.

35. Wetmore SD, Smith DM, Radom L. Enzyme catalysis of 1,2-amino shifts: The cooperative action of B_6, B_{12}, and aminomutases. J Am Chem Soc 2001; 123:8678-8689.

CHAPTER 22

Biosynthesis of Siroheme and Coenzyme F$_{430}$

Martin J. Warren,* Evelyne Deery and Ruth-Sarah Rose

Abstract

The biosynthesis of siroheme from uropoprhyirnogen III in bacteria, yeasts and plants is described. The pathway requires the *bis*-methylation of uroporphyrinogen III to generate precorrin-2, which is then oxidised to sirohydrochlorin prior to its ferrochelation. A number of structures of the various biosynthetic enzymes have been elucidated and thus the overall process is known in molecular detail. In contrast, the biosynthesis of coenzyme F$_{430}$, which is synthesized soley by methanogenic bacteria, is poorly understood. It is estimated that between 6 and 8 steps are required for the transformation of uroporphyrinogen III into coenzyme F$_{430}$, yet none of the biosynthetic enzymes have been identified and only one potential intermediate has been isolated.

Introduction

Other chapters in this book cover the biosynthesis of heme, chlorophyll, bilins and vitamin B$_{12}$. In this chapter we will focus on the biosynthesis of siroheme and coenzyme F$_{430}$, two modified tetrapyrroles that are involved in the reduction of sulphite and nitrite (see Chapter 24) and the process of methanogenesis respectively (see Chapter 23). The relationship of these compounds is outlined in the summary of the branched biosynthetic pathway shown in Figure 1. The biosynthesis of siroheme from uroporphyrinogen III requires three biosynthetic steps and is well understood, whereas the biosynthesis of F$_{430}$ requires at least five enzyme mediated steps in a process that has not yet been characterised in any significant detail.

Siroheme Biosynthesis

Siroheme was first identified as a prosthetic group in sulphite reductase[1,2] and subsequently in nitrite reductases.[3] The structure of siroheme was determined to be that of an iron-containing isobacteriochlorin, derived from the common uroporphyrinogen III template that is characteristic of all modified tetrapyrrole.[2] The transformation of uroporphyrinogen III into siroheme requires three steps, namely the *bis*-methylation of uroporphyrinogen III into precorrin-2, the oxidation of precorrin-2 into sirohydrochlorin and the chelation of sirohydochlorin with ferrous iron to give siroheme (Fig. 2).

The biosynthesis of siroheme was first investigated in *Escherichia coli* where it was found that lesions in the *cysG* gene resulted in the loss of NADH-dependent nitrite reductase activity.[4] As the main structural genes for nitrite reductase had already been identified, it was suggested that *cysG* encoded an enzyme for the synthesis of siroheme, a known prosthetic group for this enzyme.[5] Subsequently, the gene was sequenced and shown to encode a 457 amino acid protein.[6] Sequence comparisons revealed that CysG shared similarity with the cobalamin biosynthetic methyltransferases, including CobA, CobI, CobJ, CobM, CobF and CobL (see Chapter 18) (Fig. 2).[7,8] However, the sequence similarity was shared only over the C-terminal region of CysG, from amino acids 202-457. The N-terminal region of CysG did not show sequence similarity to any other protein on the data

*Corresponding Author: Martin J. Warren—Protein Science Group, Department of Biosciences, University of Kent, Canterbury, Kent CT2 7, UK. Email: m.j.warren@kent.ac.uk

Tetrapyrroles: Birth, Life and Death, edited by Martin J. Warren and Alison G. Smith.
©2009 Landes Bioscience and Springer Science+Business Media.

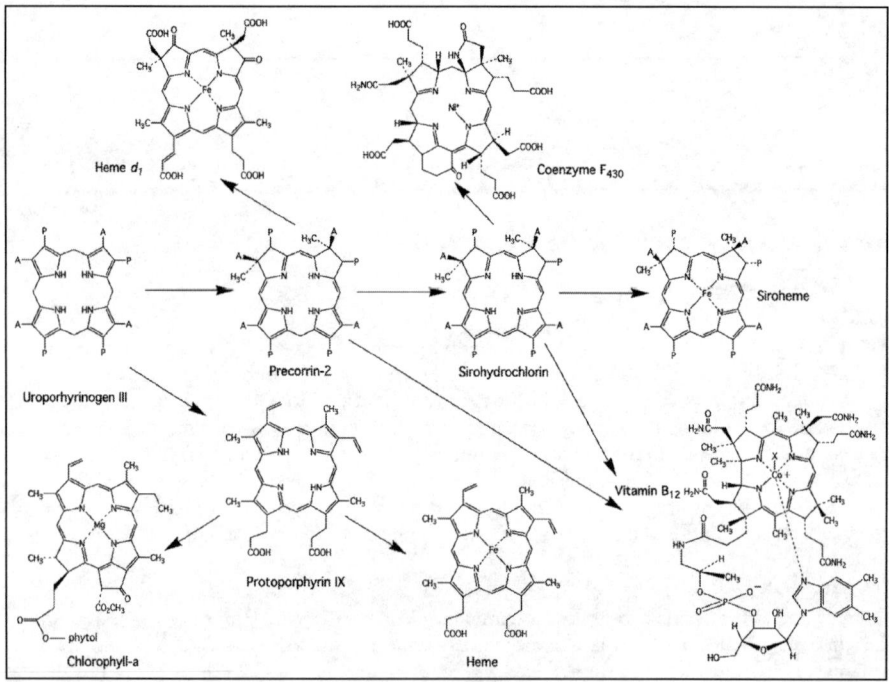

Figure 1. The branched tetrapyrrole biosynthetic pathway, highlighting how siroheme and coenzyme F_{430} are synthesised from uroporphyrinogen III via precorrin-2. (A = acetic acid side chain, P = propionic acid side chain).

bases (Fig. 2). The high level of sequence identity of the C-terminus with the cobalamin biosynthetic enzymes, and especially with CobA, the uroporphrinogen III methyltransferase that methylates at positions C2 and C7, strongly implied that *cysG* was able to catalyse a very similar reaction.[9] Indeed, it was very shortly shown that CysG did catalyse the transformation of uroporphyrinogen III into precorrin-2, by the addition of two S-adenosyl-L-methionine derived methyl groups to C2 and C7.[10] Strangely, the enzyme was able to catalyse the addition of a third methyl group to C12 to generate a trimethylpyrrocorphin.[11] However, this third methylation only occurred under conditions where a high concentration of enzyme was present, and is not thought to represent a physiological process.

Although these experiments demonstrated that CysG was able to synthesize precorrin-2, the dehydrogenase and ferrochelatase activities required to convert precorrin-2 into siroheme still had to be found. Significantly, though, no other mutants in siroheme deficiency had been located and it was thus suggested that at least one of these functions may be resident within the N-terminal region of CysG. In fact, gene dissection experiments reinforced this idea, as production of the C-terminal region of CysG gave a peptide that was still able to catalyse the synthesis of precorrin-2 from uroporphyrinogen III even though the corresponding truncated gene was not able to complement an *E. coli cysG* phenotype.[12] Moreover, within the N-terminal region of CysG a putative NAD$^+$ binding site (GXGXXA) was detected.[12] Finally, purified CysG was shown to catalyse the synthesis of siroheme not only by methylating uroporphyrinogen III, but also by performing an NAD$^+$-dependent dehydrogenation to generate sirohydrochlorin and a ferrochelation to give siroheme.[13] Thus, CysG was shown to be a multifunctional enzyme, whereby the C-terminal region of the protein is responsible for the methylation reactions and the N-terminal region is required for oxidation and metal insertion activities.

The structure of CysG was determined by X-ray diffraction studies after the *Salmonella enterica* protein had been overproduced and crystallised.[14] The protein was shown to exist as a homodimer

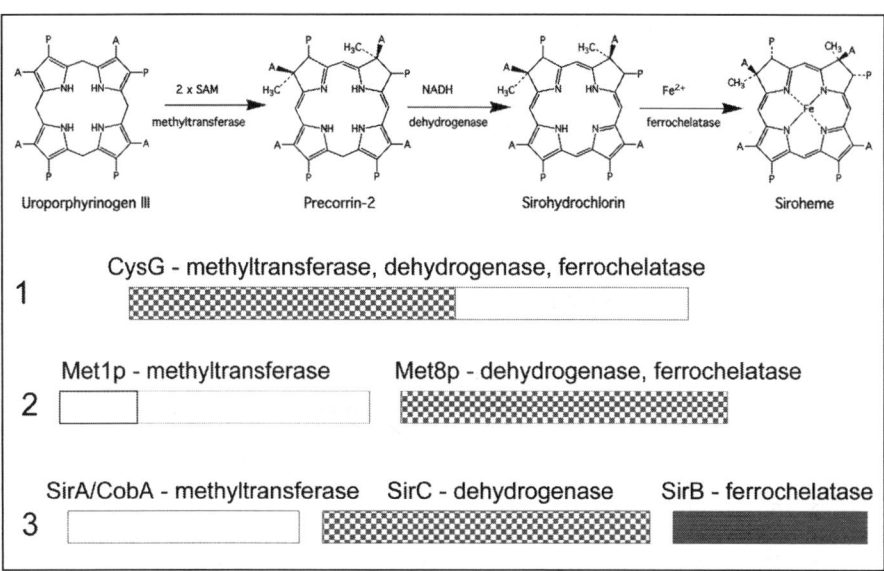

Figure 2. Synthesis of siroheme and the enzyme modules that are responsible for its synthesis. Siroheme is synthesised from uroporphyrinogen III in three steps. Initially, uorporphyrinogen III is methylated at positions 2 and 7 to yield precorrin-2. Two electrons and two protons are removed from preocrrin-2 to give sirohydrochlorin and finally sirohydrochlorin is chelated with ferrous iron to yield siroheme. Strategies to make siroheme: (1) In some enteric bacteria, siroheme is made from uroporphyrinogen III by the action of a single multifunctional enzyme called CysG. This enzyme is composed of a C-terminal methyltransferase region (diagonal shaded area) and an N-terminal dehydrogenase/chelatase region (checked area). (2) In yeast, siroheme is made by the action of two enzymes, one which carries out the methylations for the synthesis of precorrin-2 (Met1p - which displays similarity to the C-terminal region of CysG) and the other which performs the dehydrogenation and ferrochelation reactions (Met8p - and displays similarity to the N-terminal region of CysG). Finally, some bacteria have individual enzymes for each of the reactions required for siroheme synthesis. Thus, some bacteria have a separate uroporphyrinogen methyltransferase (SirA), a dehydrogenase (SirC) and a ferrochelatase (SirB).

with two structurally independent modules. The C-terminal region has a topology that is, as expected, similar to the class III methyltransferases of cobalamin biosynthesis.[14,15] The N-terminal region forms a large active site between a Rossmann fold domain and a smaller α/β domain.[14] This region of the protein contains one key catalytic residue, an aspartic acid. Although this region of the protein is able to catalyse both the dehydrogenation and chelation reactions within the one active site, it is not clear how this is achieved.[14]

The crystal structure of CysG revealed one more fascinating finding in that it was shown to be a phosphoprotein. An analysis of mutant variants of the serine residue that is post-translationally modified suggests that phosphorylation may prevent dehydrogenase activity.[14] This is significant since in many enteric bacteria, CysG acts as the source of precorrin-2 for cobalamin (vitamin B$_{12}$) biosynthesis as well as siroheme. Thus inactivation of the dehydrogenase activity would result in elevated precorrin-2 levels that would then be incorporated into cobalamin and represents a way that flux is controlled along a branched pathway.

In yeast, mutagenesis studies suggested that siroheme synthesis required two gene products, Met1p and Met8p (Fig. 2).[16] Sequencing of *MET1* revealed that that it encodes a protein product of 526 amino acids.[16] The first 325 amino acids of Met1p do not reveal similarity to any other protein, whereas the C-terminal 220 amino acids share a high degree of similarity with the uroporphyrinogen III methyltransferases, including the C-terminal region of CysG (Fig. 2).[16] No function for the N-terminal region of Met1p has been demonstrated but there is always the possibility that this part of the protein helps to control the activity of the enzyme. Met8p is a 274 amino acid protein that has a high degree of similarity with the N-terminal region of CysG (Fig. 2), which is known to house the dehydrogenase and ferrochelatase activities.[17] As with CysG, Met8p also contains a putative nucleotide binding sequence in the form of a GXGXXG motif. On the basis of the sequence data, it thus appears that Met1p is a uroporphyrinogen III methyltransferase and Met8p is a precorrin-2 dehydrogenase and sirohydrochlorin ferrochelatase (Fig. 2). Indeed, yeast mutants in both *MET1* and *MET8* can be complemented by *cysG*.[16,17] Finally, the activity of Met8p was shown in vitro after the *MET8* gene was cloned and the encoded protein overproduced as a recombinant enzyme.[17] Met8p was also crystallised and its structure determined to 2.2 Å resolution.[18] The protein, as expected, has a similar topology to the N-terminal region of CysG.[14,18] Both the dehydrogenase and ferrochelatase activities are housed within a single active site, where Asp141 is thought to play an essential role in both catalytic processes.[18]

In the enteric bacteria and a range of other eubacteria, as mentioned above, siroheme is made by the action of the multifunctional enzyme CysG. However, in the bacilli, such as *Bacillus subtilis* and *Bacillus megaterium*, siroheme is synthesised by the action of three separate enzymes, which are encoded by *sirA*, *B* and *C* (Fig. 2).[19,20] These encode a uroporphyrinogen III methyltransferase (SirA, which is also known as CobA), a precorrin-2 dehydrogenase (SirC) and a sirohydrochlorin ferrochelatase (SirB) (Fig. 2). The SirA protein has a high level of sequence identity with other uroporphyrinogen III methyltransferases such as CobA and the C-terminal region of CysG (Fig. 2). SirC has sequence identity with Met8p and the N-terminal region of CysG (Fig. 2), although it has no ferrochelatase activity. It would seem likely that enzymes such as Met8p started off as straightforward dehydrogenases like SirC and then acquired chelatase activity. It is not apparent from the study of Met8p and SirC why Met8p is able to chelate metal ions into sirohydrochlorin. The final member of this bacterial siroheme pathway branch trilogy is SirB, an enzyme that has some sequence identity with the cobaltochelatases found in the anaerobic route of cobalamin (vitamin B_{12}) biosynthesis such as CbiX and CbiK.[21]

In higher plants, siroheme is made in the plastid. Uroporphyrinogen III methyltransferases have been described in both maize and *Arabidopsis* and their function determined.[22,23] The enzyme is made in the cytosol with an N-terminal extension that acts as a chloroplast transit peptide, and which is cleaved after translocation into the plastid. In maize the expression of the gene appeared to be coregulated with genes associated with nitrate assimilation, consistent with siroheme playing a role in nitrite reduction.[22] Although no precorrin-2 dehydrogenase has yet been identified in higher plants, the terminal enzyme of the siroheme pathway has recently been discovered in *Arabidopsis thaliana*.[24] The enzyme, a SirB homologue, is made as a precursor protein of 225 amino acids. The mature form of the enzyme consists of 150 amino acids, giving it a size of only about half that of the SirB orthologues found in the bacilli. In fact, the large SirB proteins appear to be made from a protein fusion of two such smaller gene products, presumably after a gene duplication event, as the N- and C-terminal region of the larger SirB proteins display a level of similarity to each other. Green fluorescent protein (GFP) tagging of the plant SirB confirmed that the protein is localised to the plastid. Surprisingly, the recombinant enzyme was found to contain a Fe_2-S_2 centre, which is very unstable. The Fe-S centre is not essential for activity and is likely to act as a redox sensor. Fe-S centres have been reported on some protoporphyrin ferrochelatases in both prokaryotes and eukaryotes, but the plant SirB is the only sirohydrochlorin ferrochelatase known to contain such a redox group. The Fe-S centre is housed in the C-terminal region of the *Arabidopsis* SirB.

Coenzyme F_{430} Biosynthesis

It has been estimated that methanogens generate about a trillion tons of methane gas per annum, of which about one third escapes into the atmosphere where it is photochemically converted into CO_2.[25-27] The concentration of methane in the atmosphere has been rising steadily over the past 300 hundred years, a highly significant finding since methane is about 50 times more potent than CO_2 as a greenhouse gas. The concentration of methane in the atmosphere is likely to increase further with the employment of intensive farming of crops such as rice and ruminant livestock. As a combustible gas, methane is also of commercial importance and through the employment of methanogens on the decomposition of organic material it represents a cheap source of energy. There are thus strong environmental and strategic arguments for a thorough understanding of the process of methanogenesis and this is covered in Chapter 23.

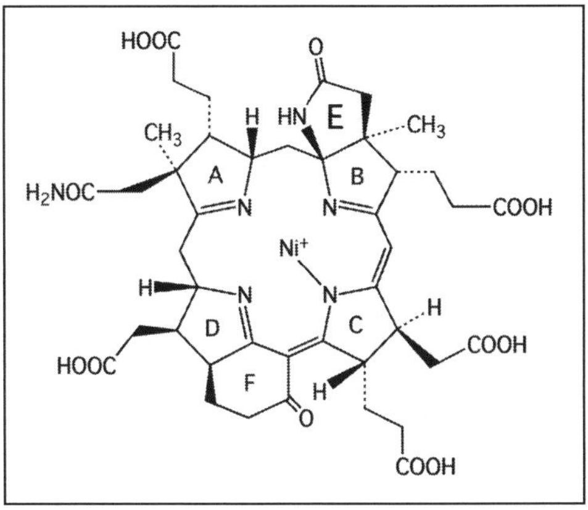

Figure 3. The structure of coenzyme F430 with the extra rings highlighted.

Methanogens are members of the *Archaea* that are highly specialised in terms of their biochemistry and are unique in being able to produce methane. Classical experiments by a number of eminent scientists have allowed much of the biochemistry of methane formation from C1 compounds such as CO_2, methanol, methylthiols and methylamines as well as C2 acetate to be elucidated.[27] Key to this process is the role of methyl-coenzyme M, which is the penultimate intermediate in methane production.[25] It is reduced by the enzyme methyl CoM reductase to give methane with the formation of a heterodisulphide between CoM and another coenzyme called coenzyme B. It is the reduction of this heterodisulphide that is coupled to ATP formation. The ability of methyl CoM reductase to catalyse the formation of methane lies in the use of a novel cofactor called coenzyme F_{430} (Fig. 3).[25] This is a modified tetrapyrrole with a centrally chelated nickel ion. Analogies with the role of the cobalt ion in vitamin B_{12} can be drawn with the change in oxidation state that nickel undergoes in the role of coenzyme F_{430}. In its active form the nickel ion at the centre of F_{430} is in a Ni(I) form, which is methylated by methyl CoM to generate a Ni(III)-CH_3 species, a transient unstable intermediate that readily reduces to Ni(II)-CH_3. This then spontaneously protonlyses to give CH_4 and Ni(II).[25]

Despite the indispensable role played by F_{430} in the process of methanogenesis and its global importance, little is known about how this remarkable cofactor is made.[26] As a modified tetrapyrrole, the synthesis of coenzyme F_{430} is based on the macrocyclic template design also observed in

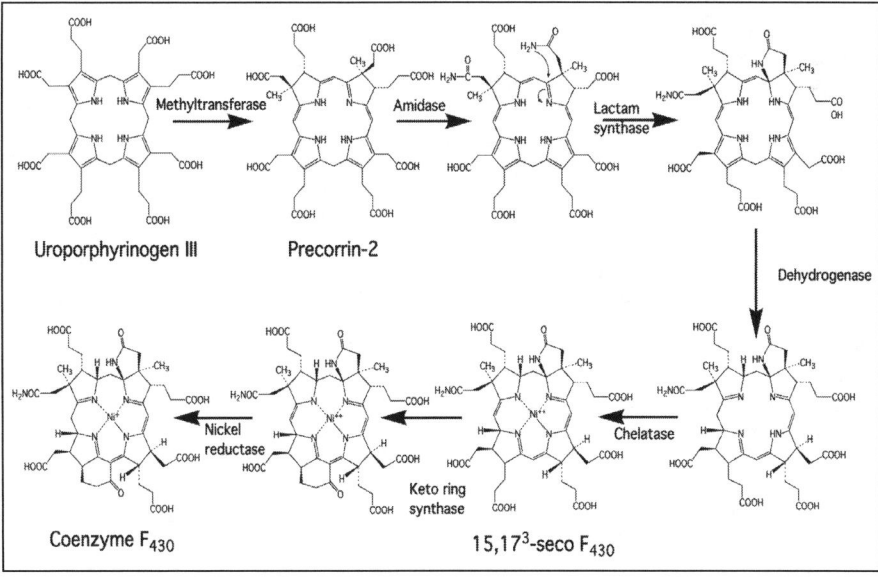

Figure 4. An outline of the biosynthesis of coenzyme F_{430} based on the synthesis from precorrin-2 and with late nickel insertion.

the hemes, chlorophylls, sirohemes, corrins and heme d_1.[26,28] However, it differs from these other modified tetrapyrroles in the nature of the centrally chelated metal ion and in the oxidation state of the macrocycle as it is a tetrahydroporphyrinogen and is therefore the most reduced member of the family. As well as the 4 pyrrole-derived rings found in all modified tetrapyrroles (labelled A-D; Fig. 3), F_{430} also contains two extra rings (E and F; Fig. 3). E is a lactam derived from the amidated acetic acid side chain attached to ring B whilst F is a keto ring derived from the propionic acid side chain on ring D. The two methyl groups found at positions 2 and 7 also suggest that F_{430} is derived from precorrin-2, an intermediate of the synthesis of cobalamin, siroheme and heme d_1. Indeed, labelling experiments clearly indicate that the biosynthesis of coenzyme F_{430} proceeds via either precorrin-2 or its oxidised relative sirohydrochlorin.[29] Moreover, under certain growth conditions, a $15,17^3$-*seco* intermediate of coenzyme F_{430}, missing ring F, can be isolated and converted into coenzyme F_{430} in the presence of ATP.[6] This is the only intermediate in the biosynthesis of F_{430} that has been isolated and characterised. Based on the observation that coenzyme F_{430} is derived from either precorrin-2 or sirohydrochlorin and proceeds via the *seco* intermediate, a biosynthetic pathway has been postulated and this is outlined in Figure 4.[26,30] There are likely to be between 6 and 8 enzymes required in the transformation of precorrin-2 into F_{430}. The steps include amidation, lactam synthesis, macrocyclic ring reduction, nickel insertion, nickel reduction, propionic acid side chain activation and ring F cyclisation.

Coenzyme F_{430} biosynthesis in a disrupted crude cell extract has been investigated more recently with the application of mass spectrometry to help in the identification of potential pathway intermediates.[31] Here, either uroporphyrinogen III or precorrin-2 was incubated with crude cell extracts from *Methanothermobacter thermoautotrophicus*. This involved the construction of a system for the rapid generation of both uroporphyrinogen III and precorrin-2, which was achieved by making plasmids encoding *hemB, C, D*, and *hemB, C, D, -cobA*. Despite the presence of these substrates in the crude cell incubation mixture, there was so significant increase in the production of coenzyme F_{430}. One reason to explain the lack of de novo synthesis of F_{430} could be the absence of a coenzyme or lack of appropriate conditions to allow relevant reactions to take place. Alternatively it may be due to the presence of a bottleneck in the pathway, but in all cases these possibilities would result in the accumulation of an earlier metabolite. A variety of methods were employed to determine the nature of any molecules that were accumulating, including an hplc method for F_{430} derivatives, isolated as their free acids. The use of this hplc method coupled to mass spectrometry provided a powerful tool for analysis of pathway intermediates. Incubation of uroporphyrinogen III and precorrin-2 with a cell free extract resulted in the isolation and identification of a range of compounds including sirohydrochlorin and nickel-sirohydrochlorin. A further compound with a mass of two units less that nickel sirohydrochlorin was also observed. It had a UV-visible spectrum similar to nickel-sirohydrochlorin although the retention time of this compound is considerably greater than nickel-sirohydrochlorin on a reverse phase column, indicating that the peripheral groups had undergone modification. Amidation of carboxyl peripheral groups of a tetrapyrrole is known to cause a dramatic increase in the retention time of a molecule. Consequently, amidation of nickel-sirohydrochlorin is the most likely cause of the shift in retention time. Mass spectral analysis of intermediates containing nickel is easy due to the isotope profile of the metal. A further compound was found in low abundance that contained nickel, with a m/z ratio of 949 that could represent a later intermediate. Based on these results it is possible that coenzyme F_{430} synthesis could proceed via sirohydrochlorin, nickel sirohydrochlorin and nickel sirohydrochlorin *a,c*-diamide as suggested in Figure 5. However, these intermediates could also be the result of mis-incoporation of nickel into the cobalamin biosynthetic pathway, which is also known to proceed via sirohydrochlorin, and thus the observed intermediates could be artifactual.

There is thus a great deal still to be learnt about coenzyme F_{430} biosynthesis, including the identification of the pathway intermediates and biosynthetic enzymes. With more archeal genomes becoming available and with the development of molecular tools for archeal genetic manipulation there is hope that some progress will be made on this very interesting biochemical pathway. What is clear from both coenzyme F_{430} and siroheme biosynthesis is that precorrin-2 plays a pivotal role as an intermediate and that this branch of the pathway is likely to represent the primordial route for modified tetrapyrrole synthesis.

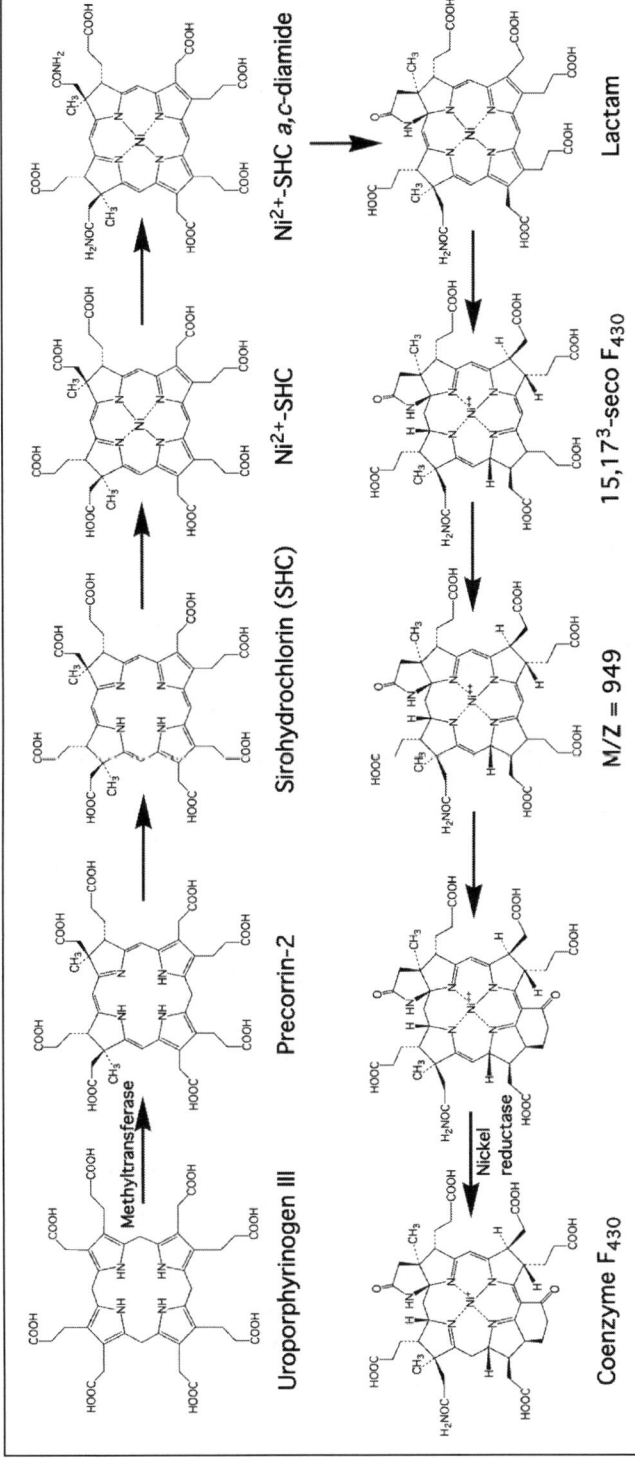

Figure 5. An alternative proposed biosynthesis of coenzyme F430, via sirohydrochlorin and with early nickel insertion.

Acknowledgements

The authors thank the BBSRC for financial support and Prof Rolf Thauer (Marburg) for helpful discussions and the provision of *M. thermoautotrophicus* cell pastes.

References

1. Murphy MJ, Siegel LM. Siroheme and sirohydrochlorin: The basis for a new type of porphyrin-related prosthetic group common to both assimilatory and dissimilatory sulfite reductases. J Biol Chem 1973; 248:6911-6919.

2. Murphy MJ, Siegel LM, Tove SR et al. Siroheme: A new prosthetic group participating in six-electron reduction reactions catalyzed by both sulfite and nitrite reductases. Proc Natl Acad Sci USA 1974; 71:612-616.

3. Vega JM, Garrett RH. Siroheme: A prosthetic group of the Neurospora crassa assimilatory nitrite reductase. J Biol Chem 1975; 250:7980-7989.

4. Ostrowski J, Hulanicka D. Constitutive mutation of cysJIH operon in a cysB deletion strain of Salmonella typhimurium. Mol Gen Genet 1979; 175:145-149.

5. Cole JA, Newman BM, White P. Biochemical and genetic characterization of nirB mutants of Escherichia coli K12 pleiotropically defective in nitrite and sulphite reduction. J Gen Microbiol 1980; 120:475-483.

6. Peakman T, Crouzet J, Mayaux JF et al. Nucleotide sequence, organisation and structural analysis of the products of genes in the nirB-cysG region of the Escherichia coli K-12 chromosome. Eur J Biochem 1990; 191:315-323.

7. Crouzet J, Cameron B, Cauchois L et al. Genetic and sequence analysis of an 8.7-kilobase Pseudomonas denitrificans fragment carrying eight genes involved in transformation of precorrin-2 to cobyrinic acid. J Bacteriol 1990; 172:5980-5990.

8. Crouzet J, Cauchois L, Blanche F et al. Nucleotide sequence of a Pseudomonas denitrificans 5.4-kilobase DNA fragment containing five cob genes and identification of structural genes encoding S-adenosyl-L-methionine: Uroporphyrinogen III methyltransferase and cobyrinic acid a,c-diamide synthase. J Bacteriol 1990; 172:5968-5979.

9. Blanche F, Robin C, Couder M et al. Purification, characterization, and molecular cloning of S-adenosyl-L-methionine: Uroporphyrinogen III methyltransferase from Methanobacterium ivanovii. J Bacteriol 1991; 173:4637-4645.

10. Warren MJ, Roessner CA, Santander PJ et al. The Escherichia coli cysG gene encodes S-adenosylmethionine-dependent uroporphyrinogen III methylase. Biochem J 1990; 265:725-729.

11. Warren MJ, Stolowich NJ, Santander PJ et al. Enzymatic synthesis of dihydrosirohydrochlorin (precorrin-2) and of a novel pyrrocorphin by uroporphyrinogen III methylase. FEBS Lett 1990; 261:76-80.

12. Warren MJ, Bolt EL, Roessner CA et al. Gene dissection demonstrates that the Escherichia coli cysG gene encodes a multifunctional protein. Biochem J 1994; 302(Pt 3):837-844.

13. Spencer JB, Stolowich NJ, Roessner CA et al. The Escherichia coli cysG gene encodes the multifunctional protein, siroheme synthase. FEBS Lett 1993; 335:57-60.

14. Stroupe ME, Leech HK, Daniels DS et al. CysG structure reveals tetrapyrrole-binding features and novel regulation of siroheme biosynthesis. Nat Struct Biol 2003; 10:1064-1073.

15. Schubert HL, Blumenthal RM, Cheng X. Many paths to methyltransfer: A chronicle of convergence. Trends Biochem Sci 2003; 28:329-335.

16. Hansen J, Muldbjerg M, Cherest H et al. Siroheme biosynthesis in Saccharomyces cerevisiae requires the products of both the MET1 and MET8 genes. FEBS Lett 1997; 401:20-24.

17. Raux E, McVeigh T, Peters SE et al. The role of Saccharomyces cerevisiae Met1p and Met8p in sirohaem and cobalamin biosynthesis. Biochem J 1999; 338(Pt 3):701-708.

18. Schubert HL, Raux E, Brindley AA et al. The structure of Saccharomyces cerevisiae Met8p, a bifunctional dehydrogenase and ferrochelatase. EMBO J 2002; 21:2068-2075.

19. Johansson P, Hederstedt L. Organization of genes for tetrapyrrole biosynthesis in gram—positive bacteria. Microbiology 1999; 145(Pt 3):529-538.

20. Raux E, Leech HK, Beck R et al. Identification and functional analysis of enzymes required for precorrin-2 dehydrogenation and metal ion insertion in the biosynthesis of sirohaem and cobalamin in Bacillus megaterium. Biochem J 2003; 370:505-516.

21. Schubert HL, Raux E, Matthews MA et al. Structural diversity in metal ion chelation and the structure of uroporphyrinogen III synthase. Biochem Soc Trans 2002; 30:595-600.

22. Leustek T, Smith M, Murillo M et al. Siroheme biosynthesis in higher plants: Analysis of an S-adenosyl-L-methionine-dependent uroporphyrinogen III methyltransferase from Arabidopsis thaliana. J Biol Chem 1997; 272:2744-2752.

23. Sakakibara H, Takei K, Sugiyama T. Isolation and characterization of a cDNA that encodes maize uroporphyrinogen III methyltransferase, an enzyme involved in the synthesis of siroheme, which is prosthetic group of nitrite reductase. Plant J 1996; 10:883-892.
24. Raux-Deery E, Leech HK, Nakrieko KA et al. Identification and characterization of the terminal enzyme of siroheme biosynthesis from Arabidopsis thaliana: A plastid-located sirohydrochlorin ferrochelatase containing a 2FE-2S center. J Biol Chem 2005; 280:4713-4721.
25. Thauer RK. Biochemistry of methanogenesis: A tribute to Marjory Stephenson. 1998 Marjory Stephenson Prize Lecture. Microbiology 1998; 144(Pt 9):2377-2406.
26. Thauer RK, Bonacker LG. Biosynthesis of coenzyme F430, a nickel porphinoid involved in methanogenesis. Ciba Found Symp 1994; 180:210-222, (discussion 222-217).
27. Wolfe RS. An historical overview of methanogenesis. In: Ferry JG, ed. Methanogenesis. New York and London: Chapman and Hall, 1993:1-32.
28. Raux E, Schubert HL, Roper JM et al. Vitamin B12; insights into biosynthesis's Mount improbable. Bioorganic Chem 1999; 27:100-118.
29. Mucha H, Keller E, Weber H et al. Sirohydrochlorin, a precursor of factor F430 biosynthesis in Methanobacterium thermoautotrophicum. FEBS Letts 1985; 190:169-171.
30. Pfaltz A, Kobelt A, Huster R et al. Biosynthesis of coenzyme F430 in methanogenic bacteria: Identification of 15,17(3)-seco-F430-17(3)-acid as an intermediate. Eur J Biochem 1987; 170:459-467.
31. Rose RS. The biosynthesis of coenzyme F430 in Methanothermobacter thermoautotrophicus. University of London 2005.

CHAPTER 23

Role of Coenzyme F$_{430}$ in Methanogenesis

Evert C. Duin*

Abstract

Methane production by methanogenic archaea takes place in many anaerobic microbial habitats such as swamps, rice paddies, fresh water sediments and the intestinal tract of animals and insects. Almost all species of methanogenic archaea are able to oxidize H$_2$ and to use CO$_2$ as the electron acceptor. Species from several families are also able to use formate, methanol, methylamine, and acetate. Although every pathway starts out different, they all end with the same step, the reduction of methyl-coenzyme M (CH$_3$-S-CoM) with coenzyme B (HS-CoB) to CH$_4$ and the mixed disulfide of coenzyme M and coenzyme B, CoM-S-S-CoB. This step is catalyzed by methyl-coenzyme M reductase (MCR). The enzyme contains a tightly bound nickel porphinoid, Factor 430 (F$_{430}$). For the enzyme to be active the nickel has to be in the Ni(I) state. Although the enzyme was first isolated in 1981, only in recent years have we seen the development of methods to purify highly active MCR. This opened up the way to study the different states of MCR with spectroscopic methods including electron paramagnetic resonance, electron nuclear double resonance, resonance Raman, X-ray absorption, and magnetic circular dichroism spectroscopies. Here, I will present an overview of the most important findings and how these relate to the proposed reaction mechanisms as discussed in the literature.

Introduction

I grew up in a city called Haarlem in the western part of the Netherlands. The street we lived on was on the outskirts of the city and only one row of houses separated us from the farmlands behind it. To the horror of our mothers, my friends and I would spend most of our summer holiday trying to jump over the small canals that crisscrossed the meadows, more than once ending up in the middle of a canal. Of course all kids were familiar with the fact that if you poked the mud at the bottom of the canals and collected the gas bubbles that emerged in a jam jar you could make the gas explode by igniting it with a match taken from your mom's kitchen. In our play we were in fact repeating an experiment described by Alessandro Volta in 1776.[1]

It took several other great scientists to identify the gas bubbles as methane and to show that this is produced by microbes living in the anaerobic layer of the sediment. A whole series of enzymes has been characterized being involved in methanogenesis, the pathways involved in the production of methane from several substrates.[2] The enzyme directly involved in the methane production, methyl-coenzyme M reductase (MCR), was first purified by Ellefson and Wolfe in 1981.[3] At that time MCR was known as component C of the methylreductase system. Two more components were needed for this system to work in vitro, component A and component B. Component B, which was later renamed into coenzyme B, turned out to be a cosubstrate in the reaction: when methyl-coenzyme M (CH$_3$-S-CoM) is converted to CH$_4$, coenzyme B (HS-CoB) forms a mixed disulfide with coenzyme M (HS-CoM).[4,5]

$$CH_3\text{-S-CoM} + HS\text{-CoB} \rightarrow CH_4 + CoM\text{-S-S-CoB}$$

*Evert C. Duin—Department of Chemistry and Biochemistry, 179 Chemistry Building, Auburn University, Auburn, Alabama 36849-5312, USA. Email: duinedu@auburn.edu

Tetrapyrroles: Birth, Life and Death, edited by Martin J. Warren and Alison G. Smith.
©2009 Landes Bioscience and Springer Science+Business Media.

Component A, on the other hand, turned out to be a set of proteins, A1, A2, and A3 that are involved in the ATP dependent reductive activation of MCR.[6,7]

Component C had a characteristic nonfluorescent yellow color, giving rise to an absorption band at around 425 nm with a shoulder at 450 nm. It was immediately realized that this color had to be due to the nickel-containing factor 430 (F_{430}) that had been purified earlier from cell extracts.[3] Unfortunately MCR turned out to be a very labile enzyme. Later studies with the enzyme from *Methanothermobacter marburgensis* (previously known as *Methanobacterium thermoautotrophicum*, strain Marburg), showed that already upon breakage of the cells most activity was lost. Based on the MCR content of cells and methane formation in cell suspensions it was estimated that in purified protein only 1 to 4% of the protein molecules were expected to be active and participate in catalysis.[8] As a result we know more about free F_{430} in solution and F_{430} bound to inactive enzyme than we do about the cofactor in active MCR. In recent years new methods have been developed to purify highly active MCR. This opened up the way for spectroscopic investigation of F_{430} in the active protein.

This Chapter focuses on these recent developments. It turns out that some of the properties of free F_{430} are very different from those of protein-bound F_{430}. It has even been proposed that F_{430} might have a different structure in active MCR.[9] In addition we will discuss the different reaction mechanisms that have been proposed for methane production and see how these are in agreement or disagreement with the recent spectroscopic data.

Methanogenesis

The production of CH_4 by archaea has an important influence on life on this planet. First of all, methanogenesis is the last step in the anaerobic breakdown of biopolymers and takes place in many anaerobic microbial habitats such as swamps, rice paddies, fresh water sediments and the intestinal tract of animals and insects. The complex polymers, like cellulose, starch and proteins, are first broken down into their respective building blocks, which in turn can be fermented into propionate and butyrate and further into H_2, CO_2, formate and acetate. These last four compounds are typical substrates for the methanogens that convert them into CH_4. Other substrates are methanol and methylamines. The second influence on life is due to the fact that CH_4 is a potent greenhouse gas. Although part of the CH_4 produced by methanogens is reoxidized by other organisms, more and more CH_4 escapes to the atmosphere mainly due to increasing areas for rice production and the increase in amounts of livestock.[10]

Methanogens contain different metabolic pathways for the conversion of the different substrates into methane. The two major metabolic pathways of methanogenesis are shown in Figure 1. The CO_2-reducing pathway or hydrogenothropic pathway uses H_2, CO_2 and/or formate as substrates, and the aceticlastic pathway utilizes acetate. The entrance of the methylothrophic pathway (methanol and methylamine) into these pathways is also indicated. The reduction of CO_2 to CH_4 proceeds via coenzyme-bound C_1-intermediates, methanofuran, tetrahydromethanopterin, and coenzyme M (see also Fig. 2). In the first step CO_2 is reduced to the level of formate, formylmethanofuran. From formylmethanofuran the formyl group is transferred to tetrahydromethanopterin, which serves as the carrier of the C_1 unit during its reduction to methylene- and methyltetrahydromethanopterin. The fourth and last reduction step in the pathway utilizes the structurally simple substrate, methyl-coenzyme M (see also Fig. 2). For the reduction steps electrons from the oxidation of H_2 are used. This input is either directly via hydrogenase or via coenzyme F_{420}. In the aceticlastic pathway, acetate is first activated by phosphorylation and subsequent formation of acetyl-coenzyme A. Carbon monoxide dehydrogenase cleaves acetyl-coenzyme A into two one-carbon units. The carbon monoxide unit is oxidized to carbon dioxide at the same enzyme complex, whereas the methyl group is transferred to tetrahydromethanopterin. Two reducing equivalents are generated in this process that can be used for the reduction of the heterodisulfide and ATP synthesis.

As you can see in Figure 1, all pathways use the same last step, the conversion of methyl-coenzyme M into CH_4 by MCR, making MCR the central protein in methanogenesis. Methyl-coenzyme M is converted to CH_4 and the mixed disulfide of coenzyme M and coenzyme B. From our point of view the production of methane is the most interesting aspect of methanogenesis. The production of the mixed disulfide (also called heterodisulfide) is, however, much more important for the cell.[11] The CH_4 is only a waste product. Most methanogens couple the reduction of the disulfide bond of the heterodisulfide indirectly to ATP synthesis.

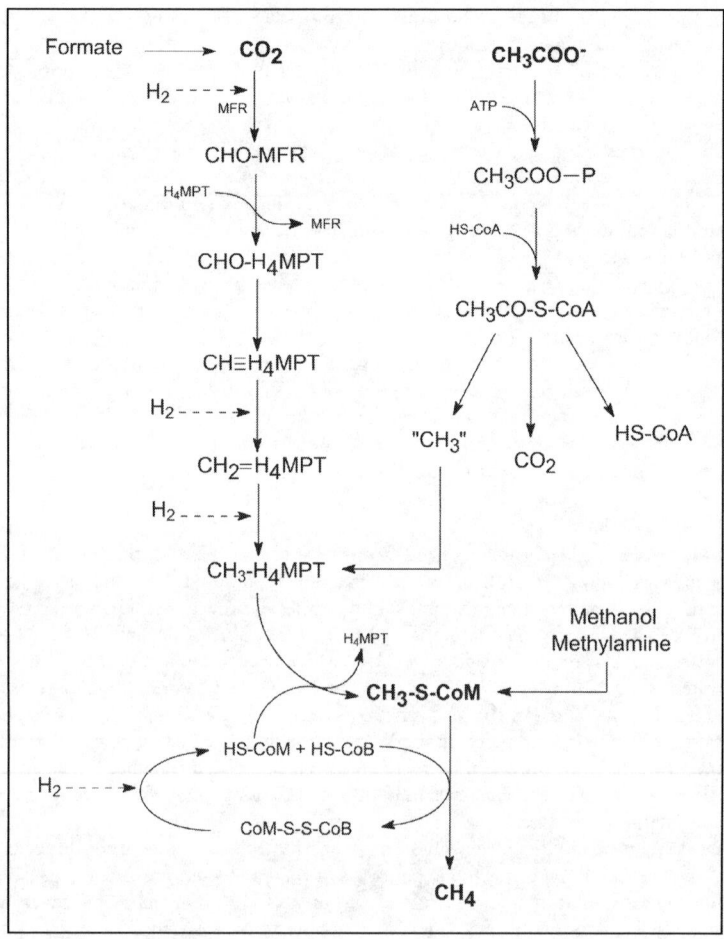

Figure 1. Metabolic pathway of methanogenesis from CO_2 and Acetate. CHO-MFR, N-formylmethanofuran; CHO-H_4MPT, N^5-formyltetrahydromethanopterin; CH=H_4MPT$^+$, N^5,N^{10}-methenyltetrahydromethanopterin; CH_2=H_4MPT, N^5,N^{10}-methylenetetrahydromethanopterin; CH_3-H_4MPT$^+$, N^5-methyltetrahydromethanopterin; CH_3-S-CoM, methyl-coenzyme M; CH_3COO^-: Acetate; P, phosphate; HS-CoA, coenzyme A; CH_3COS-CoA; acetyl-coenzyme A.

Free Factor 430

Coenzyme F_{430} is a nickel porphinoid of unique structure[12,13] (Fig. 2). The π chromophore extends only over three of the four nitrogens, making F_{430} the most extensively reduced tetrapyrrole found in nature. F_{430} is a tetrahydrocorphin, 'corphin' being the name proposed by Eschenmoser[14] for this class of tetrapyrroles in which the carbon framework of a porphin is combined with the linear chromophore typical for corrins. Two additional rings are found in F_{430}, a lactam ring fused to ring B and a 6-membered carbocycle formed through intramolecular acylation of C_{15} by the propionic acid side chain at ring D.

Native, as-isolated, F_{430} is thermally unstable to epimerization of the acid side chains at β-carbon positions 12 and 13 of pyrrole ring C to form the 13-monoepimer and ultimately the 12,13-diepimer.[15] It is also oxidatively unstable, being slowly oxidized in air to 12,13-didehydro-F_{430}, which is termed F_{560} because it exhibits an absorption maximum at 560 nm.[15]

Native F$_{430}$ is silent in electron paramagnetic resonance (EPR) spectroscopy. Figure 3, panel B, shows the EPR spectra for the free cofactor in three different oxidation states. F$_{430}$ is present in cells in the pentaacid form (Fig. 2). This form is present in a protein bound and in a free form.[16,17] F$_{430}$ was also prepared as a pentamethyl ester[12] and a pentaamide.[18] The structure of F$_{430}$ was initially solved by nuclear magnetic resonance (NMR) spectroscopy on the pentamethyl ester (F$_{430}$M). In contrast to F$_{430}$, F$_{430}$M is soluble in noncoordinating organic solvents. This was crucial for the NMR studies since in the presence of donors the nickel ion in both F$_{430}$M and F$_{430}$ has the tendency to form 5- and 6-coordinated, paramagnetic complexes. These could not be studied by NMR spectroscopy. The spectra shown in Figure 3 are for the F$_{430}$M form. Both Ni(II)F$_{430}$ and Ni(II)F$_{430}$M could be reduced to Ni(I).[19-21] Oxidation of Ni(II)F$_{430}$M in acetonitrile leads to Ni(III)F$_{430}$M.[21]

Dependent on the temperature, the divalent nickel ion can be found in two forms in aqueous solution, a tetragonally coordinated paramagnetic (S = 1) form with two water molecules as the axial ligands at 10 K or a 4-coordinated diamagnetic (S = 0) form at room temperature.[22] The high amount of saturation makes the corphin ring quite flexible. To coordinate the small low-spin Ni(II) form, F$_{430}$ deforms into a saddle shape[14,23] which results in a shortening of the Ni-N distances. It has even been shown by theoretical calculations that F$_{430}$ can in principle accommodate a trigonal-bipyramidal coordination geometry about the nickel.[24]

Name That Signal

Several different forms of MCR have been prepared and described. Some of these forms, however, are produced under very artificial conditions in vitro and will not be discussed here. Table 1 gives an overview of the properties of the different MCR and F$_{430}$ forms discussed in this paper. The names of the MCR forms are based on the initial detection of the respective EPR signals in whole cells and cell extracts under more reducing conditions, like MCRred1 and MCRred2, or more oxidizing conditions, like for example MCRox1 (see below).[25] In the absence of an EPR signal the form was dubbed MCRsilent for EPR silent. Now we know, however, that there is more than one silent form, based on crystallization experiments and X-ray absorption measurements. Three silent forms can be discerned: MCRsilent, MCRred1-silent and MCRox1-silent. The differences between these forms will be discussed below. Figure 3, panel A, shows the EPR spectra of some of the MCR forms.

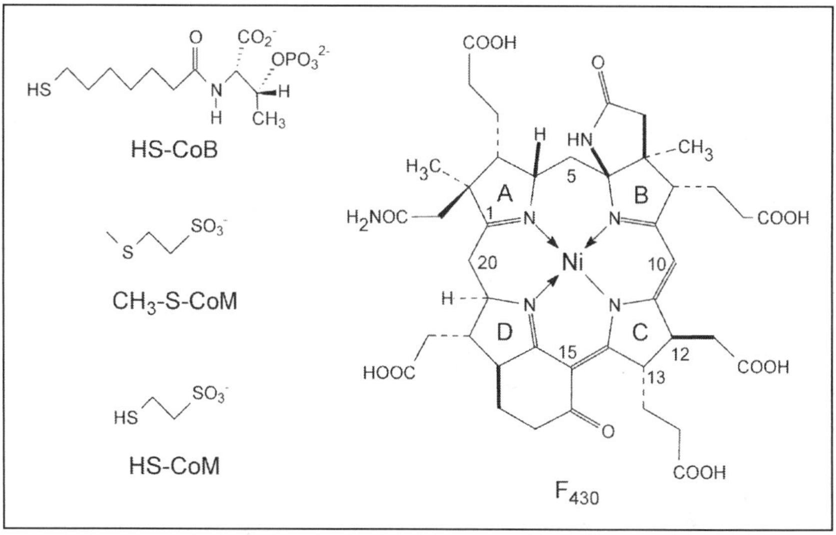

Figure 2. Structures of coenzyme B (HS-CoB, *N*-7-mercaptoheptanoyl-*O*-phospho-L-threonine), methyl-coenzyme M (CH$_3$-S-CoM, 2-(methylthio)ethanesulfonate), coenzyme M (HS-CoM, 2-mercaptoethanesulfonate) and Factor 430.

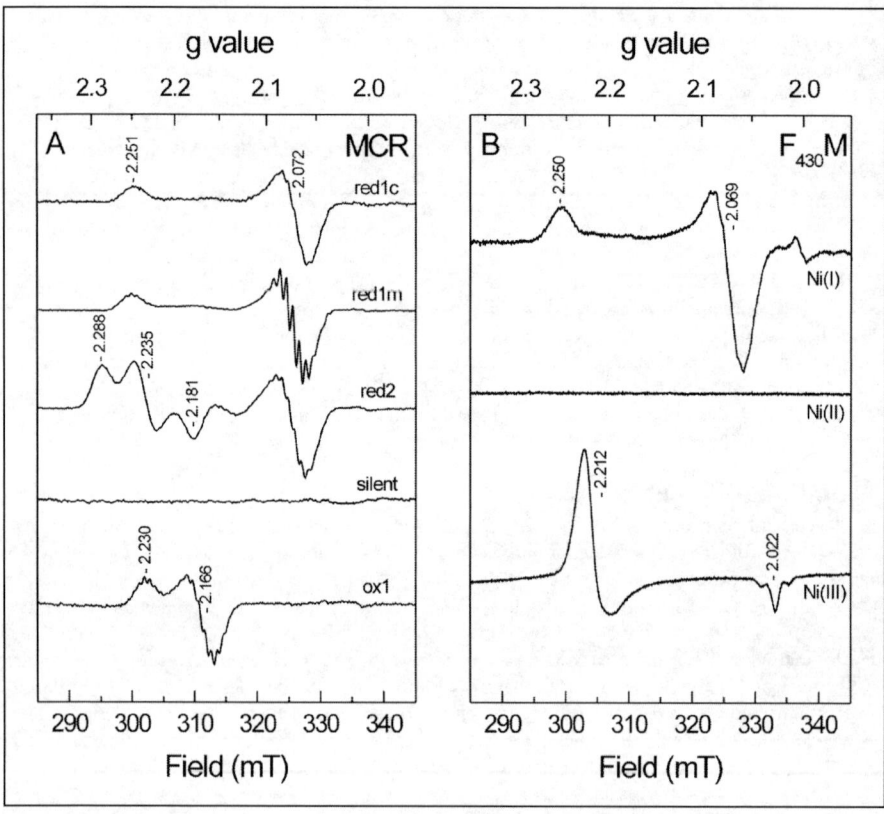

Figure 3. EPR spectra of the methyl-coenzyme M reductase (A) and coenzyme $F_{430}M$ (B) in different forms and oxidation states. $F_{430}M$ is the pentamethyl ester form of F_{430}. Data collected at 77 K.

Reviewing the spectra in Figure 3, there are some clear similarities and differences between the different EPR spectra. The two top spectra in panel A are from MCR in the active form which is named the MCRred1 form. The properties of this form can show more subtle changes dependent on the presence of methyl-coenzyme M or coenzyme M in the buffer solution. MCRred1 in the *a*bsence of either compound, designated as MCRred1*a*, is very labile and the activity and EPR signal have a half life of only a couple of hours. In the presence of one of these compounds the half life increases significantly. MCRred1 in the presence of *m*ethyl-coenzyme M, designated as MCRred1*m*, has a half life of a couple of days. MCRred1 in the presence of *c*oenzyme M, designated as MCRred1*c* has a half life of a whole week. So although one would prefer to study the red1a form this is not possible due to the low stability and the red1c and red1m forms, therefore, have been studied more extensively. Throughout the text the name MCRred1 will be used when no specific information is known about the subtype. In addition to the differences in stability more differences can be found for red1c and red1m. Although both are highly active, the red1m form is the more active. Also the absorption spectra (Table 1) and EPR spectra (Fig. 3) are similar but not the same. The EPR spectra in Figure 3 show that the superhyperfine coupling due to the nuclear spin (I = 1) of the four nitrogen nuclei coordinated to the nickel, is hardly resolved on the MCRred1c signal but is highly resolved on the MCRred1m signal. This difference was proposed to be due to binding of the methyl-coenzyme M substrate to the nickel,[26] but X-ray absorption spectroscopy (XAS) measurements showed that the coordination around the nickel is similar in both cases and only involves five ligands from the enzyme (see below).

Table 1. Properties of coenzyme F_{430} and of the methyl-coenzyme M reductase in different oxidation states

	EPR Parameters			Absorption Band (nm) (ε, M⁻¹ cm⁻¹)		Axial Ligand	Nickel Charge	Reference
	g_z	g_y	g_x					
MCRsilent	-	-	-	423 (22 000)	445 (shoulder)	-O-SO₂-CoM-S-S-CoB	2+	56
MCRox1	2.2310	2.1667	2.1532	420	650	-S-CoM	3+	26,35,46
MCRox1-silent	-	-	-	423	445 (shoulder)	-S-CoM	2+	46
MCRred1c	2.2500	2.0710	2.0605	387	720	-	1+	26
MCRred1m	2.2515	2.0730	2.0635	387	748	-	1+	26
MCRred1-silent	-	-	-	423 (shoulder)	445	(-S-CoM)[a]	2+	46
MCRred2	2.2869	2.2313	2.1753	416	600, 690	-S-CoM	1+	26,43
Ni(III)F₄₃₀M[b]	2.020	2.211	2.211	368	595, 890, 1020	-NCMe	3+	24
Ni(II)F₄₃₀M[c]	-			431 (19 900) 439[b]	-	(H₂O)[d] -	2+	12 24
Ni(II)F₄₃₀[c]	-			430 (23 000) 432 (20 300)	-	(H₂O)[d] (H₂O)[d]	2+	67 22
Ni(I)F₄₃₀M[e]	2.250	2.074	2.065	382 (29 600)	754 (2500)	-	1+	23
Ni(I)F₄₃₀[c]	2.244	2.061	2.061	376 (27 600)	710 (1600)	-	1+	22

a) The amount of coordination depends on the crystallization conditions. b: In MeCN. c) In H₂O. d) The amount of coordination depends on the temperature. e) In THF. F₄₃₀M, pentamethyl-F₄₃₀.

The spectrum of MCRred2 is more complex. In this case we see an overlap of an axial red1 signal (two apparent g values) and a rhombic signal (three apparent g values), the MCRred2 signal. The fact that only up to 50% of the red2 signal can be induced has been puzzling for a long time but recently we found that this is probably an artifact of freezing. The MCRred2 form can be induced by adding coenzyme M and coenzyme B to MCR in the red1 form.[26] Temperature studies showed that in fact there is equilibrium between the red2 and the red1 forms, the protein being mainly red2 at higher temperatures, > 40°C, and mainly red1 at lower temperatures, <20°C.[27] Upon freezing part of the enzyme in the red2 form is converted into the red1 form, even when the sample is flash frozen in cold ethanol (200 K).

Native F_{430} is EPR-silent and therefore the nickel is in the 2+ oxidation state. Both $Ni(II)F_{430}$ and $Ni(II)F_{430}M$ could be reduced to $Ni(I)$,[19-21] forming 4-coordinate complexes.[28] Their EPR spectra are consistent with a S = 1/2 species with its spin density residing predominantly in the $d_{x^2-y^2}$ orbital of the central nickel ion (Fig. 3, panel B). Although not very well resolved, the $Ni(I)F_{430}M$ spectrum shows superhyperfine structure on the g_{xy} line (g = 2.069) due to the interaction of the free electron on the nickel with the nuclear spin of the four nitrogen ligands from F_{430}. Oxidation of $Ni(II)F_{430}M$ in acetonitrile leads to $Ni(III)F_{430}M$.[21] The EPR spectrum of $Ni(III)F_{430}M$ consists of an axial type signal with no detectable fine structure on the g_{xy} line (g = 2.212) and resolved fine structure on the g_z part (g = 2.022, Fig. 3, panel B). The superhyperfine structure is due to the coupling with the nitrogen nuclei of two acetonitrile solvent molecules coordinated in the axial positions. The EPR spectrum is typical for an S = 1/2 species with the unpaired electron in a nickel d_{z^2} orbital.[21]

It might seem strange to show the EPR 'spectrum' for MCRsilent and $Ni(II)F_{430}M$, but not everybody might be familiar with this technique. The absence of a signal in MCRsilent tells us that the nickel has to be in the 2+ oxidation state, since $Ni(II)F_{430}$ does not show a spectrum either. In EPR spectrometry only paramagnetic species with spin 1/2, 3/2, 5/2, etc. show up under the standard (perpendicular mode) measuring conditions. This was further proven by magnetic circular dichroism measurements that showed that the nickel is high spin (S = 1) $Ni(II)$.[29,30] Based on the similar absorption and EPR spectra of MCRred1 and $Ni(I)F_{430}M$ and $Ni(I)F_{430}$ (Table 1) it was proposed that the nickel in the red1 state is 1+. Since red2 can be formed from red1 by the simple addition of coenzyme M and coenzyme B and the fact that this conversion is reversible an oxidation state of 1+ has been assigned to this form.

The oxidation state of the nickel in the MCRox1 form has always been a point of discussion. Due to the high stability against oxygen and the fact that the signal could be induced in cells under relatively oxidizing conditions (see below) the nickel was proposed to be 3+. Looking at the EPR spectra, however, the ox1 EPR signal shows more similarities with the signals of $Ni(I)F_{430}M$ and MCRred1 than with that of $Ni(III)F_{430}M$ (Fig. 3). Although model studies[31] and density functional theory calculations[32] could not rule out that the signal was not due to $Ni(III)$, electron nuclear double resonance (ENDOR) studies were more in favor of the nickel being $Ni(I)$.[33] Several techniques have been applied to address this issue, which will be discussed below in detail, but the emerging picture is that the MCRox1 form is best described as a coupled $Ni(II)$-thiyl radical species with the unpaired electron in the $d_{x^2-y^2}$ orbital.

EPR Signals in Whole Cells

To be able to study the MCR enzyme one must obtain either the stable inactive MCRox1 form which can be converted into the active MCRred1 form,[34] or to obtain the MCRred1 form directly. Different methods can be used, but they all start with inducing the desired form in whole cells. This can be done for example by changing the gas phase before the cells are harvested. Figure 4, panel A, shows an experiment where the MCRox1 signal is induced in whole cells. In this experiment, aliquots of 50 ml were taken under exclusion of oxygen from a batch fermenter containing 10 L of cell suspension growing at 65°C. The 50 ml was immediately cooled to 4°C and the cells were spun down within 5 min. The cell pellet was resuspended in 1 ml buffer and a 350 μl sample was frozen in an EPR tube and stored in liquid nitrogen. The EPR spectrum taken at t = 0, shows the EPR signals that are detectable in cells in the late exponential phase growing on a mixture of 80% H_2 and 20% CO_2. Due to the fact that 10% of all protein in these cells is MCR we can clearly discern the MCR EPR signals. A small amount of the MCR enzymes must be in the MCRsilent form at t = 0 since the overall signal intensity of the EPR signals increases during this experiment. The major EPR detectable form at t = 0 is the MCRred1 form. A small percentage is present in the MCRox1 state. On t = 0 the gas phase was changed to 80% N$_2$ and 20% CO$_2$. As indicated in Figure 1, in the hydrogenothropic pathway used by *M. marburgensis*, H$_2$ can be considered the reductant and CO$_2$ the oxidant. By removing H$_2$ we create a relatively oxidizing condition for the cell. Within 5 min we see a decrease in the MCRred1 signal intensity and an increase in the MCRox1 signal intensity. After about 30 min most of the MCR is now in the MCRox1 state. This is the point where the cells are cooled down to stop any further changes and the cells are harvested. Upon further purification a pure MCR preparation can be obtained that contains 50 to 70% of the MCRox1 form, the remaining part being MCRsilent.[26,35]

By changing the gas phase to 100% H$_2$ a more reducing environment is created in the cells, not only increasing the intensity of the MCRred1 signal but also inducing a new signal, MCRred2 (Fig. 4, panel B). The MCRred2 signal is already detectable after 5 min but overlaps with signals from different hydrogenases present in the cells. Again after 30 min the cells are ready to be harvested. Since we know that the red2 signal can be induced by adding coenzyme M and coenzyme B to MCR in the red1 form, it is easy to explain what is happening in whole cells. The heterodisulfide reductase in the cells must be using the H$_2$ to split the heterodisulfide into coenzyme M and coenzyme B. Apparently these coenzymes are produced in high enough amounts to induce the red2 signal. The MCRred2 form is lost during purification (due to the loss of coenzyme B and is converted into MCRred1) and the final purified enzyme normally has 70 to 90% in a red1 form, in this case the MCRred1c form. Only when all buffers are supplemented with coenzyme M can MCR directly be purified from whole cells.[26,35]

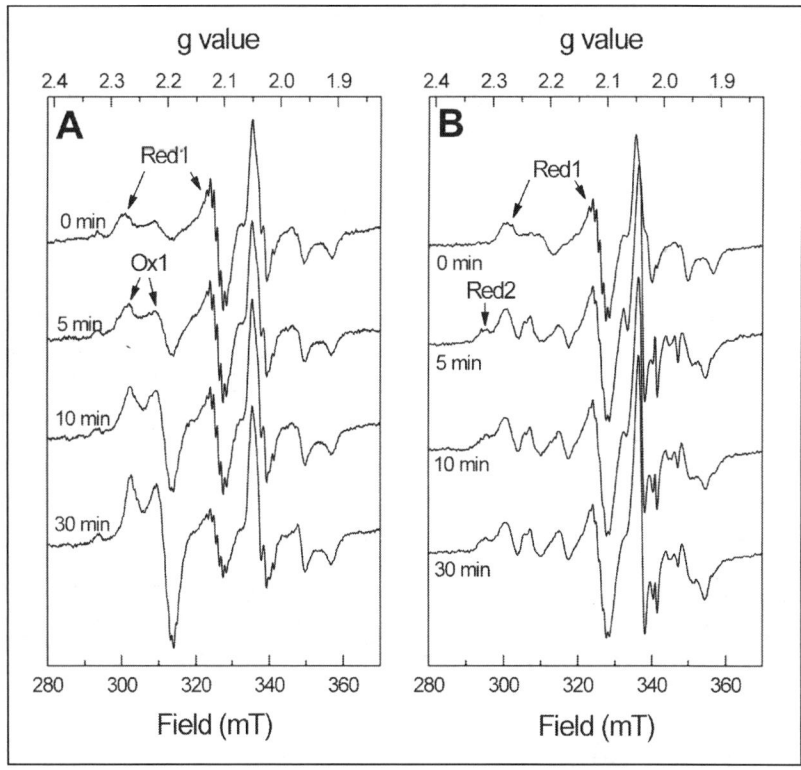

Figure 4. Induction of the methyl-coenzyme M reductase states MCRox1 (A) and MCRred1 and MCRred2 (B) in whole cells of *Methanothermobacter marburgensis*. Data collected at 77 K.

Structure of MCR and the Nickel Site

Crystal structures of MCR from *M. marburgensis* in various enzymatic inactive, EPR-silent Ni(II) states have been resolved to 1.16 Å.[36-39] The 300-kDa enzyme is a functional dimer of two $\alpha\beta\gamma$ heterotrimers with two independent F_{430}-harboring active sites. Each F_{430} is deeply buried within the protein and accessible from the outside only by a 50 Å channel through which methyl-coenzyme M can reach F_{430}. Two forms of particular interest are shown in schematic form in Figure 5. Structure A has been prepared from MCR in the ox1 form. Since all MCR forms turn silent upon crystallization via autooxidation of the nickel, this form was called MCRox1-silent. In this form, coenzyme M coordinates axially with its thiol group to the nickel from the proximal side of F_{430} (Fig. 2, reduced pyrrole rings A, B, C and D clockwise). From the distal side the oxygen of a glutamine residue axially coordinates the nickel, as it does in all crystal structures. Coenzyme B can reach into its channel with its thiol group-containing arm only up to a distance of 8 Å from the nickel. Structure B has been prepared from cells without changing the gas phase before harvesting and without adding coenzyme M to the buffers during purification. In this case a completely silent form is obtained, called MCRsilent. CoM-S-S-CoB was found to be coordinated with its sulfonate group to the Ni(II) from the proximal side. If we would imagine methyl-coenzyme M on the position where we find coenzyme M in structure A, this would represent a possible starting point of the catalytic cycle. Structure B would represent the end of the reaction cycle.

The crystal structures showed some more surprises. They revealed the presence of five modified amino acids near the active-site region. Four of these modifications are methylations.[36] A fifth modification is the insertion of sulfur into a glycine residue forming a thioglycin (indicated as S=C< in Fig. 5).[36] Three of the four methylated amino acid residues were also present in the crystal structure of MCR from *Methanosarcina barkeri*.[37] The fact that these residues are completely conserved among the known MCR amino acid sequences already indicates that these modifications are put there on purpose and are not a side effect of the formation of CH4. Labeling studies showed that the methyl group of all four amino acids in *M. marburgensis* are posttranslational modifications and are derived from the methyl group of methionine, most likely via S-adenosyl methionine.[40] The glycine residue used to form the thioglycine is also highly conserved. The exact function of these five modifications is not known.

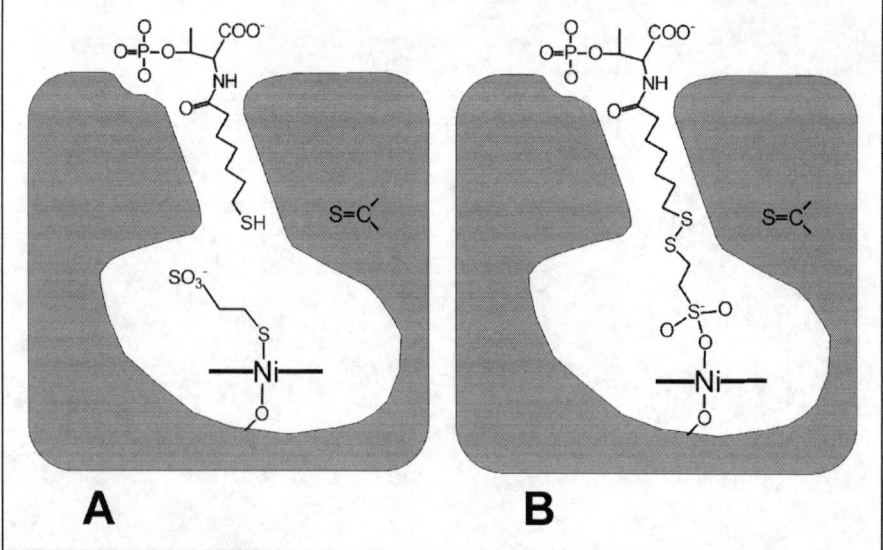

Figure 5. Models of the crystal structure from methyl-coenzyme M reductase in the MCRox1-silent state (A) and the MCRsilent state (B).

All crystal structures of methyl-coenzyme M reductase obtained to date are of an inactive enzyme in the Ni(II) oxidation state. The active enzyme has the nickel in the 1+ oxidation state. To study the nickel in active MCR other techniques have been used: XAS and ENDOR spectroscopy. Two major XAS studies have been described in the literature.[9,41] These showed that the nickel in MCR always contains at least five ligands, the four nitrogen ligands from F$_{430}$ and the distal oxygen ligand from the glutamine residue. As already proven by the crystallization studies all silent forms were found to have a sixth axial ligand, being sulfur in the case of MCRox1-silent and MCRred1-silent, and oxygen in the case of MCRsilent (Table 1). Also the nickel in the MCRox1 state was found to be 6-coordinate, having a sixth sulfur ligand. This was a confirmation of cryoreduction experiments where the MCRox1-silent form was directly converted into the MCRox1 form at 77 K indicating that the nickel in both states must have the same coordination.[42] So the sulfur ligand must be due to the thiolate sulfur of coenzyme M. The two XAS studies disagree on the coordination and geometry around the nickel in the MCRred1 states. The study done by Duin et al[41] showed that the MCRred1 form is 5-coordinate, leaving the sixth position open for the nickel to bind the substrate methyl-coenzyme M. Tang et al[9] found that the MCRred1 state contained a sixth, oxygen ligand, but proposed that this was bound very weakly to explain for the difference found in reactivity of this state in comparison to the other inactive states that are 6-coordinate. The differences between these two studies can be easily explained. In the case of Tang et al the samples only contain up to 60% of the relevant MCR state in comparison to 80 to 90% in the Duin paper. Since the contaminating silent forms are 6-coordinate and XAS detects all nickel forms it is possible that the presence of 40% of a silent form resulted in an overestimation of the amount of ligands around the nickel in the Tang paper.

Additional evidence for the absence of a sixth ligand in MCRred1 comes from crystallization studies.[39] When the MCRred1 form was crystallized the so-called MCRred1-silent form was obtained. Several crystals have been made and these showed that this enzyme state is only partially occupied by coenzyme M. The coenzyme M was found to be bound via the thiolate sulfur, and the occupancy varied between 0 and 90%. This is not unexpected since the red1 form was purified in the presence of coenzyme M. When the enzyme was washed extensively lower amounts of coenzyme M were found in the crystal structure. When the protein was crystallized in the presence of coenzyme M an occupancy of 100% was found. It was proposed that in the red1 state the nickel is 5-coordinate and binds coenzyme M when this is present to become 6-coordinate in the red1-silent state. The fact that no other, oxygen-based, ligands were detected in the MCRred1-silent crystal structure argues against the presence of such a ligand in the Ni(I) state. It is not very likely that the nickel would lose a ligand that is already bound upon oxidation to the Ni(II) state.

As pointed out, XAS measurements cannot be used when more than one enzyme form is present in the sample. To get more insight into the geometry and ligands around the nickel in the MCRred2 state, ENDOR studies were undertaken. Although there is a mixture of two EPR signals in frozen samples that show partial overlap, by observation of the red2 region that does not overlap and comparisons with samples that only contain the MCRred1 signal, it is possible to assign the observed peaks in the MCRred2 sample to the respective red1 or red2 components.

Upon discovery of the MCRred2 signal in whole cells and cell extracts it was already noted that in contrast to the MCRred1 signal no resolved superhyperfine structure was detected. However, the broadness of the peaks of the EPR signal was proposed to indicate the presence of unresolved hyperfine splitting. Therefore this signal was assigned to MCR.[25] ENDOR measurements[43] confirmed that the nitrogen superhyperfine couplings in MCRred2 are smaller than those in MCRred1. The nitrogens can be divided into two sets with a superhyperfine coupling differing by about a factor of two.[43] This might indicate that the macrocycle in the MCRred2 state is significantly distorted. It was proposed that at least pyrrole ring A must be slightly bent out of the plane of the macrocycle. This together with the influence of the axial ligands would lower the symmetry around the nickel and cause a strong mixing of the d orbitals of the ground state. This would explain the unusual g values for this state.

In addition to the nitrogen coupling, isotropic proton superhyperfine couplings were detected.[43] These were similar in character to proton superhyperfine couplings observed in type II blue copper proteins for cysteine β-protons and in iron-sulfur clusters. Since the red2 state is induced from the MCRred1 state by the simple addition of coenzyme M and coenzyme B, and the fact that coenzyme

B cannot reach far enough into the active-site channel, it was proposed that these couplings were due to coenzyme M bound to nickel via the thiolate sulfur in axial position.[43] To prove this, studies were performed with ^{33}S-labeled coenzyme M.[44] In Figure 6 the outcome of this experiment is shown. When H^{33}S-CoM was used to induce the red2 state, the MCRred2 EPR signal was clearly broadened due to the interaction of the free electron on the nickel with the nuclear spin of the sulfur nucleus (I = 3/2). This result in combination with hyperfine sublevel correlation (HYSCORE) data[44] indicates that the thiolate sulfur of coenzyme M is directly bound to the nickel. For clarity the residual red1 signals present in the samples have been subtracted from the EPR spectra in Figure 6.

So in summary we have proven that the nickel in the MCRred2 state is also 6-coordinate, the sixth ligand being the thiolate sulfur of coenzyme M, and that the macrocycle underwent a conformational change, explaining the observed differences in g values in comparison to the other MCR forms with 6-coordinated nickel.

All of this structural information shows how the protein affects the properties of F_{430}. F_{430} is rigidly bound to the protein by 21 hydrogen bonds and the active site cavity does not allow a large amount of bending, so instead of being bent like the free cofactor in solution, the structure of F_{430} in the protein has a relatively flat geometry around the nickel. Recent calculations showed that outside of the protein the conformation of the 12,13-diepi-F_{430} is energetically more favorable, but the protein restraints are causing F_{430} to have a lower energy conformation than the diepimer in MCR.[45] The diepimer has different chemical properties from F_{430}. Unlike F_{430} the diepimer does not bind axial ligands in the Ni(II) state. So the protein forces the nickel in F_{430} to accept axial ligands. This is further evidenced by the fact that free Ni(I)F_{430} is square planar and does not bind ligands in the axial positions. Inside the protein, however, the nickel in the 1+ oxidation state is 5-coordinate in the MCRred1 form or even 6-coordinate in the MCRred2 form.

Oxidation State of Nickel in the MCRox1 Form

Evaluation of the EPR spectra and ENDOR data obtained for MCRox1 left room for doubt about the oxidation state of the nickel in this form. Therefore several other techniques were used in an attempt to solve this question. Cryoreduction experiments were the first in this series.[42] Cryoreduction turned out to be very useful to study EPR silent forms of metalloenzymes. Upon exposure of a sample to X-ray radiation water molecules in the sample get oxidized, creating free-electrons that in turn can reduce the metal site resulting in the detection of an EPR signal. This experiment is performed at 77 K and it is assumed that no ligand rearrangements take place at this temperature upon reduction of the metal. Therefore the induced EPR spectrum can be used to draw conclusions about the coordination of the metal in the previous EPR-silent state. In addition annealing of the sample might result in ligand rearrangements and changes in the EPR spectrum. This experiment was done on a MCRox1-silent sample that was prepared from MCRox1. Upon exposure to X-rays at 77 K part of the silent form, about 10 to 15%, was directly converted back into MCRox1 without annealing, indicating that this state is more reduced than the silent form with Ni(II). However, it cannot be excluded that the nickel was oxidized to the 3+ oxidation state, since also the water in the sample gets oxidized. In a parallel experiment, it was shown that exposure of the MCRred1-silent form to X-ray radiation resulted in the formation of a new red1-like EPR signal that when the temperature of the sample was increased slowly converted into the red1 signal. This results could clearly be interpreted as a reduction of the Ni(II) in MCRred1-silent to Ni(I) in MCRred1. Therefore it was concluded that the conversion of MCRox1-silent into MCRox1 was also due to reduction.[42]

Although this conclusion seemed very solid, new evidence indicated that in the case of MCRox1 the X-ray radiation might have oxidized the nickel. The first evidence came from X-ray absorption spectroscopy. XAS is not only used to determine the ligands and geometry around a specific metal but can also be used to determine the oxidation state of that metal since the edge position is sensitive to the charge on the metal. XAS studies revealed that the energy of the nickel K-edge in the MCRox1 sample was the same as those observed for EPR-silent MCR forms with the nickel in the 2+ oxidation state and clearly different from the energy of the nickel K-edge of the MCRred1 forms with the nickel in the 1+ oxidation state.[41] Therefore it was proposed that the MCRox1 state might corresponds to high spin Ni(II) (S = 1) coupled to a radical on the axial sulfur ligand to explain the fact that this species is EPR active. Of the two theoretical cases of an Ni(II)-thiolate dianion or an Ni(II)-thiyl radical the second one would be more realistic. This meant that the

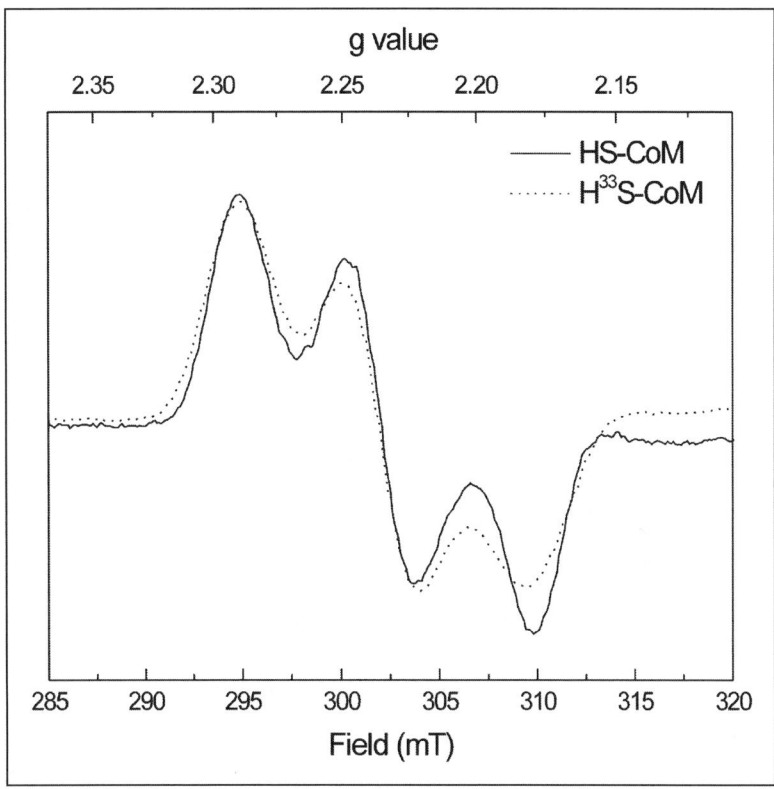

Figure 6. EPR spectra of methyl-coenzyme M reductase in the MCRred2 state. Solid line: induction of the red2 signal with coenzyme M (HS-CoM), dotted line: induction of the red2 signal with ^{33}S-labeled coenzyme M (H^{33}S-CoM). Data collected at 77 K. Reproduced from: Finazzo C et al. J Am Chem Soc 125:4988-4989; ©2003 with permission from The American Chemistry Society.[44]

whole coenzyme M/NiF$_{430}$ system is an electron more oxidized than the silent Ni(II) forms. The formal charge on the nickel would therefore be 3+. However, the correct description of the MCRox1 state would be somewhere between a pure Ni(III) species and a Ni(II)-thiyl radical. Additional evidence came from magnetic circular dichroism (MCD) measurements. The energies of the nickel d-d transitions of the different MCR forms were determined and compared with those of F$_{430}$ in the 1+, 2+, and 3+ oxidation state. The position of the d-d transitions in the MCD spectra is related to the oxidations state of the nickel. Again it was found that the overall shape of the MCD spectra of MCRox1 were very similar to those of the MCRsilent forms and Ni(II)F$_{430}$M with the nickel in the 2+ oxidation state.[46-48] The position of the d-d transitions, however, was in line with a more oxidized state of the Ni, confirming the Ni(III)/Ni(II)-thiyl radical description. This model was further validated by Density functional theory (DFT) calculations.[48]

Activation and Inactivation of Methyl-Coenzyme M Reductase

Now that all the MCR forms have been introduced and their structures are known we can go up one level and see how these forms are related. A very important relationship between the MCRred1, MCRred2 and MCRox1 forms is shown in Figure 7.

The MCRox1 form can be activated and converted into the MCRred1 form by incubation with the reductant titanium(III)citrate at pH 9.0-10.0 and 65°C.[34] Subsequently MCRred1 can be converted into MCRred2 by the addition of both coenzyme M and coenzyme B.[26] The MCRred2 form, in return, can be converted back into the MCRox1 form by the addition of polysulfide.[35]

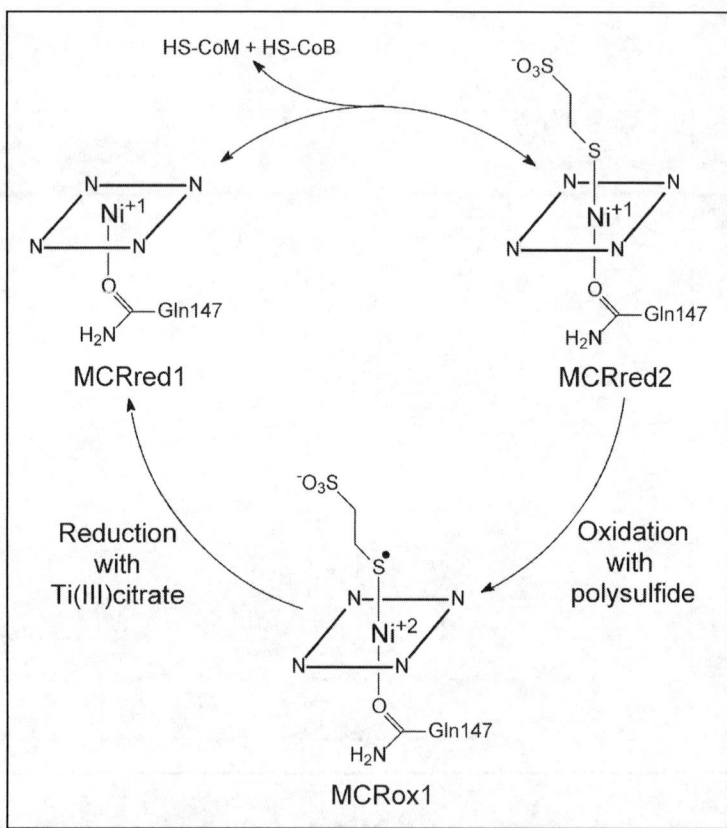

Figure 7. Relationship of three forms from methyl-coenzyme M reductase, MCRred1, MCRred2, and MCRox1, and the structure of the nickel site.

There are some very important aspects of this cycle. First of all these forms are present in whole cells and one can ask if such a cycle might serve a purpose in the cell. The answer is probably yes. Some methanogens encounter microaerobic conditions during their daily cycle. The active MCRred1 form is highly oxygen sensitive. Upon exposure to oxygen or other oxidants, the red1 form is quenched and converted into the silent form. Till now, no methods have been found that can convert the silent- or red1silent form into the red1 form in vitro. The cell probably contains an activating system to do this that requires ATP.[7] Still, it is not sure if this systems works for both the MCRsilent and MCRred1-silent forms. In addition, the cycle in Figure 6 might provide the cells with a method that does not require ATP.

To protect MCR from damage by oxygen the cell has to find a way to convert MCRred1 into MCRred2. This can be done by gassing the cells with 100% H_2 (Fig. 4). Of course in nature the conditions would be completely opposite upon exposure of the anaerobic microbial community towards oxygen or higher redox potentials. It does not seem unreasonable to assume that under microaerobic conditions the production of H_2 due to fermentation is slowed down or that the hydrogenases in the methanogens would be inactivated under these conditions. The production of methyl-coenzyme M would stop and the concentration of coenzyme M would increase. Dependent on the relative amount of coenzyme B present in the cell this could induce the red2 state that would subsequently react with an oxidant like polysulfide causing the conversion into the MCRox1 form.

We think polysulfide works as an oxidant since addition of polysulfide to the MCRred1 form results in quenching of the signal due to the oxidation of Ni(I) to Ni(II). It is not clear what oxidant is used by the cells. Both high concentration of sulfide and low concentrations of polysulfide have been used to convert MCRred2 into MCRox1 in vitro.[35] The effect of sulfide can probably be explained by the presence of traces of polysulfide. Both sulfide and polysulfide might be present in the cell, but it is much more likely that a different disulfide or other oxidant is involved in this conversion. This oxidant, however, has to be very specific. When the MCRred2 EPR signal was titrated with ferricyanide, the EPR signal disappeared due to the oxidation of Ni(I) to Ni(II).[35] The plot of the EPR signal intensity against the potential can be fitted with a one-electron Nernst curve with E_m = -440 mV (pH 7.2, vs NHE). This process, however, is not reversible and the addition of titanium(III)citrate does not result in rereduction to MCRred2 even at higher temperatures. Addition of other compounds like methylviologen, FAD and FMN also resulted in quenching of the EPR signal. Both NAD$^+$ and NADP$^+$ do not have an effect on the signal intensity (Mahlert, Thauer, Duin, unpublished results).

The second important aspect of this cycle has to do with the uncertainty of the oxidation state of the nickel in the MCRox1 form. Based on their cryoreduction experiments Ragsdale and coworkers assumed that the nickel in the MCRox1 form was in the 1+ oxidation state.[42] This meant that it was necessary to propose the existence of a second redox active group in MCR to explain the observed redox processes. Based on resonance Raman (RR) studies coupled with reductive titrations, Ragsdale and coworkers proposed that reductive activation of MCRox1 to MCRred1 involved a two-electron reduction of a C=N bond in ring B or D of the F$_{430}$ macrocycle (Fig. 2).[9] They studied both the free F$_{430}$ and different MCR forms. However, these studies are not as straightforward as they seem. The reductive titrations were performed with titanium(III)citrate and as already noted by the authors this is a tricky substance due to its instability at high pH and 65°C, the conditions used to activate MCR. The reduced MCR itself is also unstable. Therefore the amount of electrons needed to convert ox1 into red1 was found to be either 1, 2 or 3. The titrations for F$_{430}$ were performed at both pH 10.0 and 7.0. At pH 7.0, F$_{430}$ cannot be fully reduced, but titanium(III)citrate is more stable. Extrapolation of the data resulted in a total of 2.7 electrons for the reduction of Ni(II)F$_{430}$ to Ni(I)F$_{430}$. The authors concluded from these results that for the reduction of Ni(II)F$_{430}$ to Ni(I)F$_{430}$ one electron is needed for the nickel and two electrons for a second redox active group. This is the same group that is reduced in the conversion of MCRox1 into MCRred1 which was proposed to involve two electrons. In addition they showed that the RR spectra of the ox1 form and the red1 form are very different, especially in the hydrocorphin ring modes with C=N stretching character. This assignment was based on shifts observed in the RR spectra after labeling with ^{15}N. Taken all this data together it was proposed that one of the C=N bonds in the macrocycle gets reduced in the reductive titrations.

Aside from the fact that the nickel can be better described as a Ni(III)/Ni(II)-thiyl radical species, the titration results are not in line with previous voltammetric studies on F$_{430}$ and F$_{430}$M. Holliger et al[19] have performed a titration on F$_{430}$ under similar conditions. They found that only 1 electron was needed to reduce Ni(II)F$_{430}$ to Ni(I)F$_{430}$. Cyclic voltammetric studies on F$_{430}$M showed that one electron was needed for the reduction to Ni(I) or the oxidation to Ni(III).[49] Ragsdale and coworkers argued that in this latter case the spectroscopic characterization of the reduced Ni(I)F$_{430}$M was not on the species reduced in the voltammetric experiment but was on a species separately prepared by chemical reduction. In this case the amount of reduction was not controlled and more than 1 electron might have been taken up. Therefore these experiments were repeated recently.[50] It was shown that only one electron is involved in both the bulk electrolysis and chemical reduction of Ni(II)F$_{430}$M to Ni(I)F$_{430}$M. The spectral properties of the Ni(I)F$_{430}$M forms obtained with both methods were identical showing that only one electron is needed for the reduction and that this electron goes to the nickel.[50] Reduction of the ring probably would have resulted in completely different absorption spectra since the most intense band is due to the π-π^* transition. It can be concluded that for the conversion of MCRred2 into MCRox1 and MCRox1 into MCRred1 no ring reduction or oxidation has to be invoked. The changes observed in the RR spectra for the different MCR forms[9] can easily be explained by proposing a change in conformation of the macrocycle or a change in axial conformation.

There is another experiment that is related to this activation cycle. This is a labeling study done on whole cells with radioactive ^{35}S.[51] An alternative method to induce MCRox1 in whole cells is to add sodium sulfide to a growing cell culture prior to harvesting the cells. By using ^{35}S labeled sulfide it was found that the sulfur gets incorporated into the protein and only gets released upon activation of the MCRox1 state to the MCRred1 state. At the time these experiments were done it was not clear how MCRred2 was related to the other states, but now we know that the addition of sulfide with probably traces of polysulfide can induce the MCRox1 state from the red2 state present in cells. Although the red2 state is normally not present in growing cells, in this labeling experiment the cells were incubated for 20 to 50 min with sodium sulfide without gassing the cells. Under these conditions it is possible that the red2 state is induced, which subsequently reacts with the sulfide (or polysulfide). In the labeling studies it was proposed that the sulfur from the sodium sulfide forms a Ni-S adduct in the ox1 state that is reduced upon activation resulting in the observed release of radioactive sulfur. However, from the in vitro experiments it is clear that treatment of MCRred2 with polysulfide does not result in a new Ni-S adduct. In both the red2 and the resulting ox1 state coenzyme M is bound to the nickel. Unless we are talking about two different ox1 states, which is not impossible but not very likely due to the spectral similarities of the different ox1 preparations, we propose that the sulfur from sodium sulfide or polysulfide binds somewhere to the protein, maybe in the active site, but not to the nickel. Upon activation this sulfur is split off and two electrons are needed to break a covalent X-S bond. It is clear that the ^{35}S labeling experiment has to be repeated with the MCR in the MCRred2 form in vitro to learn more about the activation and inactivation of MCR.

Prelude to the Catalytic Mechanism

Several different reaction mechanisms have been proposed for the formation of CH_4 by MCR. These, however, had to be adjusted every time new evidence was found related to the catalytic mechanism. So before discussing possible reaction mechanisms let us first go over the evidence that make up the rules every proposal must follow:

    i.   The catalytic cycle starts with the enzyme in the Ni(I) oxidation state. The only active form of MCR is MCRred1, in which the nickel has the 1+ oxidation state. This also means that the starting point is 5-coordinate nickel with one open coordination site for substrate binding.

    ii. Of the two substrates, methyl-coenzyme M has to bind first to the enzyme. From the crystal structures it is clear that methyl-coenzyme M has to enter the active site first, since binding of coenzyme B closes off the entrance of the active site, stopping other molecules from entering.[36]

    iii. The enzyme exhibits a ternary complex involving methyl-coenzyme M and coenzyme B. The reaction does not start till both substrates are bound. Hypothetically it would be possible that CH_4 is immediately formed when methyl-coenzyme M enters the active site channel in the absence of any coenzyme B. Coenzyme B would then be needed to regenerate active MCR after CH_4 is released. Steady-state[52] and pre steady-state kinetic experiments,[53] however, showed that coenzyme B is integrally involved in methane formation and methane is not produced until coenzyme B is bound.

    iv. MCR is very specific for the substrate methyl-coenzyme M and coenzyme B. MCR can only convert a few other compounds that are structurally very similar to the natural substrates. Ethyl-coenzyme M (2-(ethylthio)ethanesulfonate) and difluoromethyl-coenzyme M (2-(difluoromethylthio)ethanesulfonate) are substrate analogs. Interestingly, trifluoromethyl-coenzyme M (2-(trifluoromethylthio)ethanesulfonate) was not reduced in activity assays and turned out to be an inhibitor.[54,55] Replacing the thioether sulfur with oxygen (2-methoxyethanesulfonate) resulted in the creation of an inhibitor, but replacing the sulfur with selenium (2-(methylseleno)ethanesulfonate) gave a substrate with a even higher V_{max} than found for methyl-coenzyme M.[54] The aliphatic arm length of coenzyme B (N-7-(mercaptohexanoyl)-L-threonine phosphate) turned out to be very important. In the case of C_6-coenzyme B (N-6-(mercaptohexanoyl)-L-threonine phosphate) only 1% of the maximal activity was found. No activity was found with C_8-coenzyme B (N-8-(mercaptooctanoyl)-L-threonine phosphate). Both compounds are inhibitors of the reaction.[56,57]

    v. Upon conversion of methyl-coenzyme M into CH_4, an inversion of configuration takes place around the C atom of the methyl group.[58] This was found with activity studies using the labeled substrate analogs (R)- or (S)-[1-^{2}H,^{3}H]ethyl-coenzyme M. The produced ethane was subsequently converted into acetate. Configurational analysis of the acetate showed that inversion of configuration had taken place.[58]

In addition to these five 'rules' there are a few further observations that may be of importance: Native Ni(I)F$_{430}$M does not react with thioethers, including methyl-coenzyme M.[49] Methane formation was shown, however, with the more reactive electrophile methyl iodide (CH$_3$I). When CH$_3$I was allowed to react with Ni(I)F$_{430}$M first a Ni(II) complex was formed without the formation of CH$_4$. Only after the addition of acid was CH$_4$ formed. Addition of deuterated acid gave over 85% CH$_3$D.[59] The intermediate form was proposed to be Ni(II)-CH$_3$. Ni(II)F$_{430}$M reacted with (CD$_3$)$_2$Mg to give Ni(II)-CD$_3$ which could be detected in ^{2}H-NMR. The signal disappeared after the addition of acid.[60]

As already mentioned the properties of F$_{430}$ change upon binding to MCR. Although the nickel is more prone to coordinate ligands in its axial positions, it is clear from rule (iii) that methyl-coenzyme M does not interact with the nickel till coenzyme B is bound. This is also evidenced from the studies on the MCRred2 form. From the difference in stability found between MCRred1 in the absence and presence of coenzyme M it is clear that coenzyme M has to bind to the enzyme. The place of binding is not known but it might be somewhere in the 50 Å-long active-site channel. From XAS studies we know that coenzyme M is not bound to the nickel in the red1 state. If coenzyme B is now added to the enzyme solution, the MCRred2 form is induced. ENDOR studies have proven that in this form coenzyme M is bound to the nickel. So the binding of coenzyme B to MCR might induce a conformational change forcing the nickel (or coenzyme M) to be more reactive. Since methyl-coenzyme M also stabilizes the protein activity and EPR signal in the MCRred1m form, it can be proposed that the substrate is already present in the active site without interacting with the nickel and that upon binding of coenzyme B a conformational change makes the nickel (or methyl-coenzyme M) more reactive resulting in the binding of methyl-coenzyme M to the nickel. So based on this it might be necessary to include in a proposal for the catalytic cycle some kind of activation step.

Based on the work with the free cofactor 430, most proposals include a CH$_3$-Ni-F$_{430}$ as an intermediate in the catalytic cycle. Recent density functional theory calculations, however, showed that it is possible to come up with a reaction scheme that does not involve such an intermediate.[61]

Catalytic Mechanism

Figure 8 shows a reaction mechanism (mechanism I) where binding of coenzyme B to coenzyme M is used to activate methyl-coenzyme M which results in the formation of a CH$_3$-Ni(III)-F$_{430}$ intermediate. This mechanism was recently proposed by Ragsdale and coworkers[53] and is similar to an older proposal by Berkessel.[62] After both substrates are bound, a coenzyme B radical is generated with the concomitant formation of a radical R· on the protein (step 1). The coenzyme B radical reacts with methyl-coenzyme M to form a sulfuranyl radical (step 2), making the methyl-S bond more prone to homocatalytic cleavage by Ni(I) to form CH$_3$-Ni(III)-F$_{430}$ and a disulfide anion radical (step 3). The CH$_3$-Ni(III)-F$_{430}$ is reduced by the R· radical to CH$_3$-Ni(II)-F$_{430}$ (step 4) and subsequent protonation leads to the formation of CH$_4$ (step 5). Ni(II) is regenerated to Ni(I) by taking up the free electron from the disulfide-anion radical (step 6). In principle this proposal follows all five rules. In regard to rule (v) it has to be assumed that displacement of the sulfur from the methyl group by Ni(I) proceeds with inversion of configuration. This is important since the subsequent protonolytic cleavage of the alkyl-metal bond would be expected to proceed in a retention mode.[58] The model has one weak aspect. In the crystallization studies it was found that coenzyme B cannot enter the active site channel all the way. With the thiolate sulfur of coenzyme B being 8 Å away from the nickel, the methyl carbon of the sulfuranyl radical species would be about 4 Å away from the nickel, too far for a direct interaction with the nickel. The model would predict that the C$_8$-coenzyme B analog would be a better substrate, since it would place the methyl-S bond closer to the nickel. This, however, is not the case.

A different reaction mechanism has been proposed by Thauer and coworkers (Mechanism II, Fig. 9).[38,39] This proposal uses the fact that MCR catalyses a reaction that is somewhat similar to that catalyzed by ribonucleotide reductase; a hydroxyl group (in the case of MCR a thioether group) is reduced at the expense of the formation of a disulfide. Therefore it was proposed that the

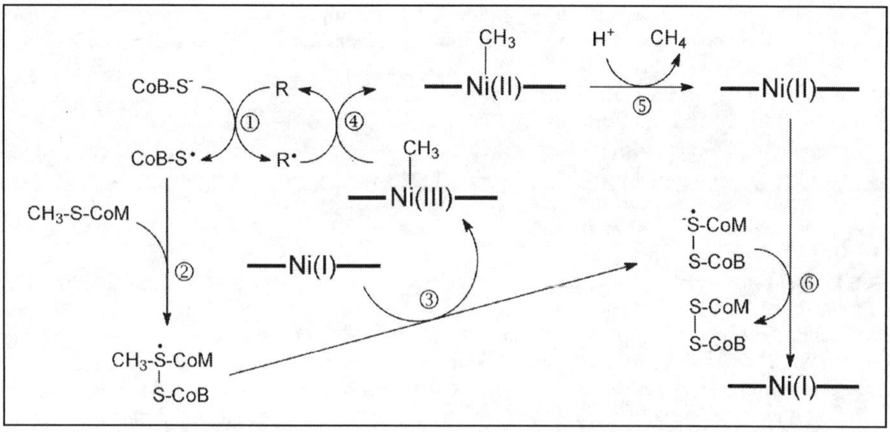

Figure 8. Hypothetical reaction mechanism I. Adapted from reference 53. See text for description.

formation of CH_4 proceeds in part via similar intermediates and uses similar principles. So in both proposed reaction mechanisms a thiyl radical is formed in an oxidation reaction that subsequently reacts with a thiolate to form a disulfide anion radical. In the hypothetical mechanism for MCR in Figure 9 it is shown that after the sequential binding of the two substrates the methyl group of methyl-coenzyme M is transferred onto the Ni(I) yielding a CH_3-Ni(III)-F_{430} intermediate (step 1). The CH_3-Ni(III)-F_{430} immediately oxidizes coenzyme M to the thiyl radical concomitantly forming CH_3-Ni(II)-F_{430} (step 2). After protonation CH_4 and Ni(II)-F_{430} are formed. Subsequently the coenzyme M thiyl radical forms a heterodisulfide anion radical with the coenzyme B thiolate (step 3). The heterodisulfide anion radical reduces the Ni(II)-F_{430} back to Ni(I)-F_{430} (step 4). This proposal is in line with the five rules. In step 1 it was proposed that the binding of coenzyme B induces a conformational change that makes the nickel more reactive. Also the inversion of configuration is taken into account. A weakness of the proposal is that some proton shuffling is needed. It was proposed that one of the two tyrosine residues in the active site is involved in the process. In step 1 a proton is needed to make HS-CoM. This proton is released upon formation of the coenzyme-M-thiyl radical. In step 2 and 3 the CH_3-Ni(II)-F_{430} intermediate is protonated by one of the tyrosine residues. Subsequently, a proton is abstracted from coenzyme B and accepted by the thiyl radical which in turn reprotonates the tyrosine residue. The crystal structure, however, shows that the position and orientation of one tyrosine residue and the absence of a hydrogen bond to the other residue, making the phenol oxygen less acidic, are not in support of such a role in the catalytic mechanism.[38] However, the crystal structures show the enzyme in the inactive form and upon activation the expected conformational change might place one of these tyrosine residues in a more favorable position to be able to take part in the catalytic mechanism.

The model explains why some of the analogs are inhibitors. The fact that trifluoromethyl-coenzyme M is an inhibitor and the difluoro compound is not can be explained by the fact that the bulky fluorine atoms prevent the protonation of the CF_3-Ni-F_{430} intermediate while this is still possible for CHF_2-Ni-F_{430}.[63] That with ethyl-coenzyme M a lower activity is found is due to the fact that a nucleophilic attack by the nickel is more difficult due to the more bulky ethyl group. The model does not give an obvious reason why changing the length of the aliphatic chain of coenzyme B results in the observed absence of activity.

So altogether it seems that mechanism II is inline with most of the available data. A recent theoretical paper by Pelmenschikov et al,[61,64] however, showed that some of the assumptions made in this model might not be correct. In mechanism II it is assumed that the CH_3-Ni(III)-F_{430} compound formed in step 1 is a strong oxidant which immediately oxidizes coenzyme M forming the thiyl radical (step 2). DFT calculations, on the other hand, showed that the CH_3-Ni bond is very weak for both Ni(III) and Ni(II) and the thiyl radical formation turned out to be highly endothermic, making the whole reaction scheme energetically very unrealistic.

In addition to the evaluation of mechanism II as proposed by Thauer and coworkers, a new model was proposed completely based on DFT calculations (Mechanism III, Fig. 10). At the beginning of the reaction cycle the two substrates are bound with a distance of 5.9 Å between the thioether sulfur of methyl-coenzyme M and the thiolate sulfur of coenzyme B. In the first step (step 1) the nickel interacts with the thioether sulfur of methyl-coenzyme M resulting in the formation of a methyl radical and a Ni(II)-thiolate complex. The methyl radical then abstracts a hydrogen atom from coenzyme B forming CH$_4$ and the coenzyme B radical (step 2), leading to the stereo inversion at the carbon. In step 3, the coenzyme M anion is released and reacts with the coenzyme B radical to form the disulfide anion radical that rereduces Ni(II) to Ni(I) (step 4). The proposed model seems to be in line with the five rules but does not contain an activation step. It was proposed that methyl-coenzyme M would not access the nonpolar active site channel by itself but would bind to the coenzyme B binding place. With the arrival of coenzyme B, methyl-coenzyme M is pushed into the active site where it reacts with the nickel. The tyrosine residues do play a role in this model, but are only involved in stabilizing some of the reaction intermediates via hydrogen bonds.

For the calculations two simplified models of F$_{430}$ were used. One model, F$_{430}{}^B$, contains only the nickel, the four nitrogen ligands and the bonds connecting the nitrogens, like a tetraaza complex. A second model, F$_{430}{}^A$, used for the final calculations contained most of the typical structural features of F$_{430}$. To reduce the size of the system, the acetamide group, the two methyl groups and all five acid side chains were omitted or replaced by hydrogen. In regard to the different properties of F$_{430}$ and diepimer F$_{430}$ it is clear that the side chains have an important influence on the properties of the nickel. In addition to the influence of the acid side chains at β-carbon position 12 and 13 it is known that removal of the carbonyl group of the 6-membered carbocycle moves the redox potential for reduction of the nickel up by 300 mV. This group is present in F$_{430}{}^B$ and not in F$_{430}{}^A$. Still the DFT calculations showed that nickel-sulfur and nickel-carbon bond strengths are the same for both models. One would expect these to be different for both models and even more different for the complete F$_{430}$.

In respect to the substrate analogs and inhibitors it was found that the relative long distance between the two thiol sulfurs and coenzyme M and coenzyme B is actually optimal for methyl release. Making this distance longer or shorter would make the mechanism energetic less favorable, which is in line with the absence of activity with C$_8$-coenzyme B and the almost total absence with C$_6$-coenzyme B. In addition it was proposed why trifluoro-coenzyme M is not a substrate. For a ·CF$_3$ radical a stereo inversion barrier of 24.3 kcal/mol was calculated, compared to a value of 6.1

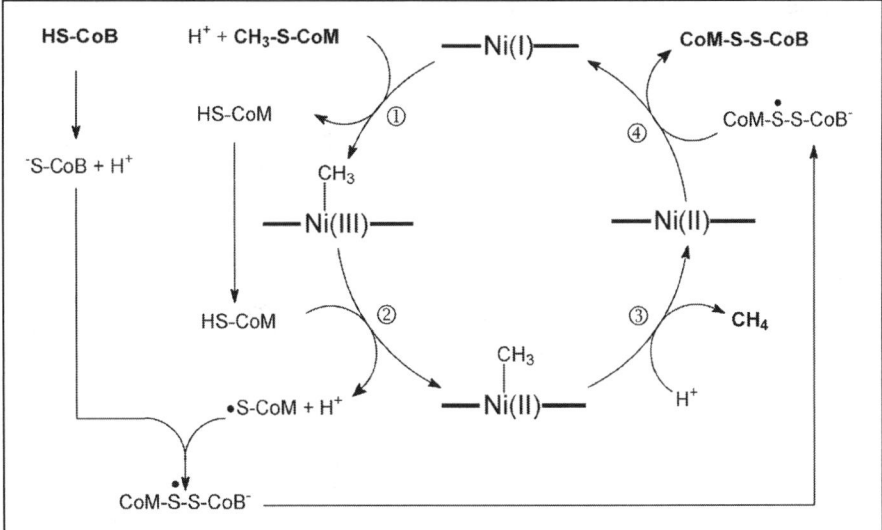

Figure 9. Hypothetical reaction mechanism II. Adapted from reference 39. See text for description.

kcal/mol for a $\cdot$CHF$_2$ radical. Apparently the inversion barrier is too high for a $\cdot$CF$_3$ radical and there is no possibility for the radical to rotate in the pit. This explanation, however, is not in line with the fact that ethyl-coenzyme M is a substrate. Dependent on the orientation of the ethyl group the ethyl radical formed might have to rotate in the pit, explaining why a lower activity is observed. If, however, the ethyl group has already the correct orientation (the methyl moiety of the ethyl group would be lying just above the macrocycle, a position in principle allowed by the crystal structure) no rotation is needed and it cannot be explained why a lower activity is found.

Although mechanism II and III seem to be in line with our rules, at the moment it is too early to predict which model comes close to the actual mechanism. The key evidence lacking is whether the actual mechanism involves a CH$_3$-Ni or a Ni-thiolate complex. Stop-flow and freeze-quench techniques would be needed to trap possible intermediates. Unfortunately both types of intermediates are expected to be EPR silent. The nickel-carbon bond of a Ni(III)-CH$_3$ complex would be expected to be very electron-rich and it would not be possible to trap it in the enzyme due to the availability of hydrogen donors in the active site channel.[49] A combination of freeze-quench techniques, cryoreduction of the samples and isotopic labeling might be able to give the answer to this problem. Due to the fact that there will be different forms present in these samples it might not be straight forward to explain the data obtained in such an experiment.

Anaerobic CH$_4$ Oxidation

There is another interesting development that might be important for the understanding of the formation of methane by MCR. Geochemical evidence, radiotracer experiments and stable isotope measurements show that methane is converted to CO$_2$ in anoxic layers of sediments.[65,66] Based on the fact that methanogens can oxidize small amounts of methane during growth on other substrates it was proposed that methanogens are responsible for this anaerobic methane oxidation. In addition it was proposed that sulfate-reducing bacteria were needed to provide the appropriate environmental conditions making the methane oxidation thermodynamically feasible.[65] Studies of the effects of adding sulfate to methanogenic sediments were in line with this so-called Consortium Hypothesis. Recently the first microbial consortium capable of anaerobic methane oxidation has been isolated from the Cascadia convergent margin off the coast of Oregon where discrete methane hydrate layers are exposed at the sea floor.[67] The consortium consists of archaea that grow in dense aggregates of approximately 100 cells that are surrounded by sulfate reducing bacteria. The remaining question now is how these archaea oxidize methane anaerobically. The steps in methanogenesis from CO$_2$ to methyl-coenzyme M are reversible[68] (Fig. 1) and methylotropic methanogens use directly analogous reactions to oxidize a fraction of the methyl substrate to CO$_2$. Since MCR has been detected in these consortia, it has been proposed that MCR might be involved in the anaerobic oxidation of CH$_4$.[69] In the simplest model that has been proposed the MCR reaction takes place in reverse direction. If this is true we have to add another rule that it is a prerequisite that the reaction can go in both directions. In this case none of the proposed mechanisms can be envisioned to proceed in both directions.

MCR Is Still a Mystery

The recent development of methods to obtain highly active protein has sparked new interest in methane formation by MCR. This is clearly evidenced by the amount of recent papers published by different groups, using very different approaches. We are now at a stage where we know the geometry and coordination around the nickel in all relevant MCR forms. We are still in the dark, however, about structural changes at other positions in MCR that are not sensed by the nickel. The only way to solve this is to obtain crystals of MCR in the active MCRred1 form, in addition to the MCRred2 and MCRox1 forms. This will not be easy but is definitely possible. With current methods, crystals can be obtained within 6 hours. If care is taken that all handling is done under exclusion of oxygen it should be possible to obtain structures of the oxygen sensitive forms. In addition we have to be careful comparing the properties of free F$_{430}$ with those of F$_{430}$ bound to the protein. DFT calculations showed that the CH$_3$-Ni intermediate found in the reaction of F$_{430}$M with ICH$_3$ might not be an intermediate in the formation of methane by MCR. Therefore we should be open to different possible reaction intermediates. In principle the reaction intermediates would have to be

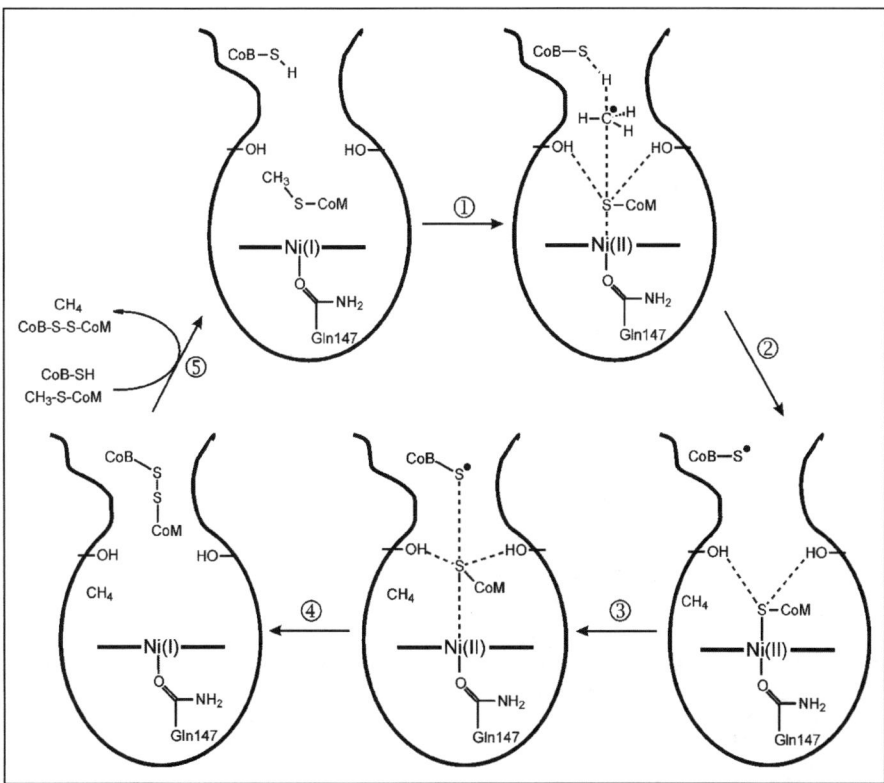

Figure 10. Hypothetical reaction mechanism III. Adapted from reference 61. See text for description.

trapped and characterized. This, however, might be very difficult if mixtures of forms are obtained. Since the intermediate will probably be EPR silent we have to use other techniques for the characterization like cryoreduction or even XAS. Both techniques, however, do not work very well for samples containing mixtures of different forms. Another path that is being followed at the moment is a reevaluation of the different substrate analogs and inhibitors of MCR. This list is much longer than the analogs mentioned here in this Chapter. Some of the inhibitors have only been tested in cell extracts. Retesting of these inhibitors on purified protein showed that some of these result in irreversible inhibition.[70] Characterization of MCR inhibited by these complexes and the inhibition mechanism might give clues about the actual reaction mechanism of MCR.

Acknowledgement

I would like to thank Dr. Rudolf Thauer at the Max Planck Institute in Marburg, Germany, for critically reading the manuscript. I would also like to thank the groups of Dr. Bernhard Jaun and Dr. Arthur Schweiger from the ETH-Zurich in Switzerland for many stimulating and helpful discussions during our yearly brain storm sessions on MCR.

References

1. Wolfe RS. 1776-1996: Alessandro Volta's combustible air. ASM News 1996; 62(10):529-534.
2. Thauer RK. Biochemistry of methanogenesis: A tribute to Marjory Stephenson. Microbiology 1998; 144:2377-2406.
3. Ellefson WL, Wolfe RS. Component C of the methylreductase system of Methanobacterium. J Biol Chem 1981; 256(9):4259-4262.

4. Noll KM, Wolfe RS. Component C of the methylcoenzyme M methylreductase system contains bound 7-mercaptoheptanoylthreonine phosphate (HS-HTP). Biochem Biophys Res Com 1986; 139(3):889-895.

5. Ellermann J, Kobelt A, Pfaltz A et al. On the role of N-7-mercaptoheptanoyl-O-phospho-L-threonine (component B) in the enzymatic reduction of methyl-coenzyme M to methane. FEBS Lett 1987; 220(2):358-362.

6. Rouvière PE, Wolfe RS. Component A3 of the methylcoenzyme M methylreductase system of Methanobacterium thermoautotrophicum ΔH: Resolution into two components. J Bacteriol 1989; 171(9):4556-4562.

7. Kuhner CH, Lindenbach BD, Wolfe RS. Component A2 of methylcoenzyme M reductase system from Methanobacterium thermoautotrophicum ΔH: Nucleotide sequence and functional expression by Escherichia coli. J Bacteriol 1993; 175(10):3195-3203.

8. Brenner MC, Zhang H, Scott RA. Nature of the low activity of S-methyl-coenzyme M reductase as determined by active site titrations. J Biol Chem 1993; 268(25):18491-18495.

9. Tang Q, Carrington PE, Horng YC et al. X-ray absorption and resonance Raman studies of methyl-coenzyme M reductase indicating that ligand exchange and macrocycle reduction accompany reductive activation. J Am Chem Soc 2002; 124(44):13242-13256.

10. Conrad R. Soil microorganisms as controllers of atmospheric trace gases (H_2, CO, CH_4, OCS, N_2O and NO). Microbiol Rev 1996; 60:609-640.

11. Hedderich R, Klimmek O, Kröger A et al. Anaerobic respiration with elemental sulfur and with disulfides. FEMS Microbiol Rev 1998; 22:353-381.

12. Pfaltz A, Jaun B, Fässler A et al. Zur Kenntnis des Faktors F430 aus methanogenen Bakterien: Struktur des porphinoiden Ligandsystems. Helv Chim Acta 1982; 65(3):828-865.

13. Färber G, Keller W, Kratky C et al. Coenzyme F430 from methanogenic bacteria: Complete assignment of configuration based on an X-ray analysis of 12,13-diepi-F430 pentamethyl ester and on NMR spectroscopy. Helv Chim Acta 1991; 74:697-716.

14. Eschenmoser A. Chemistry of corphinoids. Ann NY Acad Sci 1986; 471:108-129.

15. Pfaltz A, Livingston DA, Jaun B et al. Zur Kenntnis des Faktors F430 aus methanogenen Bakterien: Über die Natur der Isolierungsartefakte von F430, ein Beitrag zur Chemie von F430 und zur konformationellen Stereochemie der Ligandperipherie von hydroporphinoiden Nickel(II)-Komplexen. Helv Chim Acta 1985; 68:1338-1358.

16. Livingston DA, Pfaltz A, Schreiber J et al. Zur Kenntnis des Faktors F430 aus methanogenen Bakterien: Struktur des proteinfreien Faktors. Helv Chim Acta 1984; 67(1):334-351.

17. Hausinger RP, Orme-Johnson WH, Walsh C. Nickel tetrapyrrole cofactor F_{430}: Comparison of the forms bound to methyl coenzyme M reductase and protein free in cells of Methanobacterium thermoautotrophicum ΔH. Biochemistry 1984; 23:801-804.

18. Hamilton CL, Ma L, Renner MW et al. Ni(II) and Ni(I) forms of pentaalkylamide derivatives of cofactor F_{430} of Methanobacterium thermoautotrophicum. Biochim Biophys Acta 1991; 1074:312-319.

19. Holliger C, Pierik AJ, Reijerse EJ et al. A spectroelectrochemical study of factor F_{430} nickel(II/I) from methanogenic bacteria in aqueous solution. J Am Chem Soc 1993; 115:5651-5656.

20. Jaun B, Pfaltz A. Coenzyme F430 from methanogenic bacteria: Reversible one-electron reduction of F430 pentamethyl ester to the nickel(I) form. J Chem Soc Chem Commun 1986; 1327-1329.

21. Jaun B. Coenzyme F430 from methanogenic bacteria: Oxidation of F430 pentamethyl ester to the Ni(III) form. Helv Chim Acta 1990; 73:2209-2217.

22. Shiemke AK, Shelnutt JA, Scott RA. Coordination chemistry of F_{430}. J Biol Chem 1989; 264(19):11236-11245.

23. Kratky C, Waditschatka R, Angst C et al. Die Sattelkonformation der hydroporphinoiden Nickel(II)-Komplexe: Struktur, Ursprung und stereochemische Konsequenzen. Helv Chim Acta 1985; 68:1312-1337.

24. Zimmer M, Crabtree RH. Bending of the reduced porphyrin of factor F430 can accommodate a trigonal-bipyramidal geometry at nickel: A conformational analysis of this nickel-containing tetrapyrrole, in relation to archaebacterial methanogenesis. J Am Chem Soc 1990; 112:1062-1066.

25. Albracht SPJ, Ankel-Fuchs D, Böcher R et al. Five new EPR signals assigned to nickel in methyl-coenzyme M reductase from Methanobacterium thermoautotrophicum, strain Marburg. Biochim Biophys Acta 1988; 955:86-102.

26. Mahlert F, Grabarse W, Kahnt J et al. The nickel enzyme methyl-coenzyme M reductase from methanogenic archaea: In vitro interconversions among the EPR detectable MCR-red1 and MCR-red2 states. J Biol Inorg Chem 2002; 7:101-112, (2002; 7:151).

27. Goenrich M, Duin EC, Mahlert F et al. Temperature dependence of methyl-coenzyme M reductase activity and of the formation of the methyl-coenzyme M reductase red2 state induced by coenzyme B. J Biol Inorg Chem 2005; 10:333-342.

28. Furenlid LR, Renner MW, Fajer J. EXAFS studies of nickel(II) and nickel(I) factor 430 M. J Am Chem Soc 1990; 112:8987-8989.

29. Cheesman MR, Ankel-Fuchs D, Thauer RK et al. The magnetic properties of the nickel cofactor F430 in the enzyme methyl-coenzyme M reductase of Methanobacterium thermoautotrophicum. Biochem J 1989; 260:613-616.

30. Hamilton CL, Scott RA, Johnson MK. The magnetic and electronic properties of Methanobacterium thermoautotrophicum (strain ΔH) methyl coenzyme M reductase and its nickel tetrapyrrole cofactor F$_{430}$. J Biol Chem 1989; 264(20):11605-11613.

31. Renner MW, Fajer J. Oxidative chemistry of nickel porphyrins. J Biol Inorg Chem 2001; 6:823-830, (2002; 7:352).

32. Wondimagegn T, Ghosh A. Theoretical modeling of putative Ni(III)-F$_{430}$ intermediates of methylcoenzyme M reductase. Can nickel(III) masquerade as nickel(I) with respect to electron spin density distribution. J Am Chem Soc 2001; 123:1543-1544.

33. Telser J, Horng YC, Becker DF et al. On the assignment of nickel oxidation states of the ox1, ox2 forms of methyl-coenzyme M reductase. J Am Chem Soc 2000; 122:182-183.

34. Goubeaud M, Schreiner G, Thauer RK. Purified methyl-coenzyme-M reductase is activated when the enzyme-bound coenzyme F430 is reduced to the nickel(I) oxidation state by titanium(III) citrate. Eur J Biochem 1997; 243:110-114.

35. Mahlert F, Bauer C, Jaun B et al. The nickel enzyme methyl-coenzyme M reductase from methanogenic archaea: In vitro induction of the nickel-based MCR-ox EPR signals from MCR-red2. J Biol Inorg Chem 2002; 7:500-513.

36. Ermler U, Grabarse W, Shima S et al. Crystal structure of a 300 kDa methyl-coenzyme M reductase, the key enzyme of biological methane formation, at 1.45 Å resolution. Science 1997; 278:1457-1462.

37. Grabarse W, Mahlert F, Shima S et al. Comparison of three methyl-coenzyme M reductases from phylogenetically distant organisms: Unusual amino acid modification, conservation and adaptation. J Mol Biol 2000; 303:329-344.

38. Grabarse W, Mahlert F, Duin EC et al. On the mechanism of biological methane formation: Structural evidence for conformational changes in methyl-coenzyme M reductase upon substrate binding. J Mol Biol 2001; 309:315-330.

39. Grabarse W, Shima S, Mahlert F et al. Methyl-coenzyme M reductase. In: Messerschmidt A, Huber R, Poulos T, Wieghardt K, eds. Handbook of Metalloproteins. Chichester: John Wiley and Sons Ltd., 2001:897-914.

40. Selmer T, Kahnt J, Goubeaud M et al. The biosynthesis of methylated amino acids in the active site region of methyl-coenzyme M reductase. J Biol Chem 2000; 275(6):3755-3760.

41. Duin EC, Cosper NJ, Mahlert F et al. Coordination and geometry of the nickel atom in active methyl-coenzyme M reductase from Methanothermobacter marburgensis as detected by X-ray absorption spectroscopy. J Biol Inorg Chem 2003; 8:141-148.

42. Telser J, Davydov R, Horng YC et al. Cryoreduction of methyl-coenzyme M reductase: EPR characterization of forms, MCR$_{ox1}$ and MCR$_{red1}$. J Am Chem Soc 2001; 123(25):5853-5860.

43. Finazzo C, Harmer J, Jaun B et al. Characterization of the MCR$_{red2}$ form of methyl-coenzyme M reductase, a pulse EPR and ENDOR study. J Biol Inorg Chem 2003; 8:586-593.

44. Finazzo C, Harmer J, Bauer C et al. Coenzyme B induced coordination of coenzyme M via its thiol group to Ni(I) of F$_{430}$ in active methyl-coenzyme M reductase. J Am Chem Soc 2003; 125:4988-4989.

45. Todd LN, Zimmer M. Moderating influence of proteins on nonplanar tetrapyrrole deformations: Coenzyme F430 in methyl-coenzyme-M reductase. Inorg Chem 2002; 41:6831-6837.

46. Duin EC, Signor L, Piskorski R et al. Spectroscopic investigation of the nickel-containing porphinoid cofactor F$_{430}$: Comparison of the free cofactor in the +1, +2 and +3 oxidation states with the cofactor bound to methyl-coenzyme M reductase in the silent, red and ox forms. J Biol Inorg Chem 2004; 9:563-576.

47. Craft JL, Horng YC, Ragsdale SW et al. Spectroscopic and computational characterization of the nickel-containing F$_{430}$ cofactor of methyl-coenzyme M reductase. J Biol Inorg Chem 2004; 9(1):77-89.

48. Craft JL, Horng YC, Ragsdale SW et al. Nickel oxidation states of F$_{430}$ cofactor in methyl-coenzyme M reductase. J Am Chem Soc 2004; 126:4068-4069.

49. Jaun B. Methane formation by methanogenic bacteria: Redox chemistry of coenzyme F430. In: Sigel H, Sigel A, eds. Metal ions in biological systems. New York: Marcel Dekker Inc., 1993:287-337.

50. Piskorski R, Jaun B. Direct determination of the number of electrons needed to reduce coenzyme F430 pentamethyl ester to the Ni(I) species exhibiting the electron paramagnetic resonace and ultraviolet-visible spectra characteristic for the MCR$_{red1}$ state of methyl-coenzyme M reductase. J Am Chem Soc 2003; 125:13120-13125.

51. Becker DF, Ragsdale SW. Activation of methyl-SCoM reductase to high specific activity after treatment of whole cells with sodium sulfide. Biochemistry 1998; 37:2639-2647.

52. Bonacker LG, Baudner S, Mörschel E et al. Properties of the two isoenzyme of methyl-coenzyme M reductase in Methanobacterium thermoautotrophicum. Eur J Biochem 1993; 217:587-595.
53. Horng YH, Becker DF, Ragsdale SW. Mechanistic studies of methane biogenesis by methyl-coenzyme M reductase: Evidence that coenzyme B participates in cleaving the C-S bond of methyl-coenzyme M. Biochemistry 2002; 40:12875-12885.
54. Wackett LP, Honek JF, Begley TP et al. Substrate analogues as mechanistic probes of methyl-S-coenzyme M reductase. Biochemistry 1987; 26:6012-6018.
55. Gunsalus RP, Wolfe RS. ATP activation and properties of the methyl coenzyme M reductase system in Methanobacterium thermoautotrophicum. J Bacteriol 1978; 135(3):851-857.
56. Olson KD, Chmurkowska-Cichowlas L, McMahon CW et al. Structural modifications and kinetic studies of the substrates involved in the final step of methane formation in Methanobacterium thermoautotrophicum. J Bacteriol 1992; 174(3):1007-1012.
57. Ellermann J, Hedderich R, Böcher R et al. The final step in methane formation. Eur J Biochem 1988; 172:669-677.
58. Ahn Y, Krzycki JA, Floss HG. Steric course of the reduction of ethyl coenzyme M to ethane catalyzed by methyl coenzyme M reductase from Methanosarcina barkeri. J Am Chem Soc 1991; 113:4700-4701.
59. Jaun B, Pfaltz A. Coenzyme F430 from methanogenic bacteria: Methane formation by reductive carbon-sulphur bond cleavage of methyl sulphonium ions catalysed by F430 pentamethyl ester. J Chem Soc Chem Commun 1988; 293-294.
60. Lin SK, Jaun B. Coenzyme F430 from methanogenic bacteria: Detection of a paramagnetic methylnickel(II) derivative of the pentamethyl ester by ^{2}H-NMR spectroscopy. Helv Chim Acta 1991; 74:1725-1738.
61. Pelmenschikov V, Blomberg MRA, Siegbahn PEM et al. A mechanism from quantum chemical studies for methane formation in methanogenesis. J Am Chem Soc 2002; 124(15):4039-4049.
62. Berkessel A. Methyl-coenzyme M reductase: Model studies on pentadentate nickel complexes and a hypothetical mechanism. Bioorg Chem 1991; 19:101-115.
63. Wackett LP, Honek JF, Begley TP et al. Methyl-S-coenzyme-M reductase: A nickel-dependent enzyme catalyzing the terminal redox step in methane biogenesis. In: Lancaster Jr JR, ed. The Bioinorganic Chemistry of Nickel. Weinheim: VCH Verlagsgesellschaft mbH, 1988:249-274.
64. Pelmenschikov V, Siegbahn PEM. Catalysis by methyl-coenzyme M reductase: A theoretical study for heterodisulfide product formation. J Biol Inorg Chem 2003; 8(6):653-662.
65. Hoehler TM, Alperin MJ. Anaerobic methane oxidation by a methanogen-sulfate reducer consortium: Geochemical evidence and biochemical considerations. In: Lidstrom ME, Tabita FR, eds. Microbiol Growth on C1 Compounds. Dordrecht: Kluwer Academic Publishers, 1996:326-333.
66. Valentine DL, Reeburgh WS. New perspectives on anaerobic methane oxidation. Environmental Microbiology 2000; 2(5):477-484.
67. Boetius A, Ravenschlag K, Schubert CJ et al. A marine microbial consortium apparently mediating anaerobic oxidation of methane. Nature 2000; 405:623-626.
68. Thauer RK, Hedderich R, Fischer R. Reactions and enzymes involved in methanogenesis from CO_2 and H_2. In: Ferry JG, ed. Methanogenesis. New York: Chapman and Hall, 1993:209-252.
69. Krüger M, Meyerdierks A, Glöckner FO et al. A conspicuous nickel protein in microbial mats that oxidize methane anaerobically. Nature 2003; 426:878-881.
70. Goenrich M, Mahlert F, Duin EC et al. Probing the reactivity of Ni in the active site of methyl-coenzyme M reductase with substrate analogues. J Biol Inorg Chem 2004; 9:691-705.

CHAPTER 24

The Role of Siroheme in Sulfite and Nitrite Reductases

M. Elizabeth Stroupe and Elizabeth D. Getzoff*

Abstract

Sulfite and nitrite reductases (SiRs/NiRs) use siroheme, an iron-containing isobacteriochlorin, alongside a [4Fe-4S] cluster to perform the six-electron reduction of sulfite to sulfide or nitrite to ammonia. X-ray crystallographic structures of the catalytic siroheme-containing subunit of *Escherichia coli* assimilatory SiR provide clues about the relationship between the SiRs and the NiRs, allowing us to use the *E. coli* enzyme as a model system for other siroheme-containing SiRs and NiRs.[1-3] These structures also provide insight into the role of the siroheme in this powerful redox reaction, both as an anchor for the acid/base chemistry that directs substrate formation and as an electronically-flexible cofactor that drives the electron transfer reaction.

Introduction

To date, scientists have identified only one class of enzymes that uses a siroheme cofactor: the sulfite and nitrite reductases (SiRs and NiRs). Siroheme, which is equivalent to iron-sirohydrochlorin, is an isobacteriochlorin; that is, the central macrocycle is more reduced than the canonical protoporphyrin IX-derived macrocycles like heme or chlorophyll. Furthermore, the siroheme biosynthetic enzymes are more closely related to the enzymes that make the porphyrinoid component of vitamin B12, suggesting that siroheme is of antique origin, perhaps evolving before the predominance of an aerobic environment.[4] To appreciate the role of siroheme in SiRs and NiRs, we must first define the relationship between the sulfur-reducing and the nitrogen-reducing components of this enzyme class. From there, we can explore the structure-function relationship of the siroheme based on X-ray crystallographic structural analysis of the *E. coli* SiR. The first component of this relationship is the role of the enzyme/cofactor ensemble in directing proton transfer for accurate catalysis. The second is the effect of siroheme's unique electronic properties on the six-electron reduction of sulfite to sulfide or nitrite to ammonia.

Siroheme-Containing Sulfite and Nitrite Reductases Represent a Single Enzyme Class

Several different types of SiRs and NiRs fall into this enzyme class, including the nicotinamide adenine dinucleotide phosphate (NADPH)-dependent assimilatory sulfite reductases (aSiRs) and nitrite reductases (aNiRs);[5,6] the ferredoxin-dependent aSiRs and aNiRs;[7] the cytochrome c3-dependent dissimilatory sulfite reductases (dSiRs);[8; and references therein] the NAD(P)H-dependent dissimilatory nitrite reductases (dNiRs);[9] and a low-molecular weight aSiR (Fig. 1).[10] Of all these siroheme-containing SiRs and NiRs, only the dNiRs subclass has nonsiroheme-containing counterparts, representing an extended group of metalloenzymes.[11,12]

*Corresponding Author: Elizabeth D. Getzoff—The Scripps Research Institute, 10550 North Torrey Pines Road, La Jolla, California 92037, USA. Email: edg@scripps.edu

Tetrapyrroles: Birth, Life and Death, edited by Martin J. Warren and Alison G. Smith.
©2009 Landes Bioscience and Springer Science+Business Media.

Figure 1. Sequence alignments of the core regions of the SiRs and NiRs and their relationship to the pseudo two-fold symmetry in the *E. coli* aSiRHP. A) Sequence alignments between the N- and C-terminal halves of *E. coli* aSiRHP, *A. thalinia* aSiR, and *S. oleracea* dNiR, as well as the corresponding homology regions of the α and β subunits from *A. fulgidus* dSiR, and *D. vulgaris* dSiR. The homology regions H1-H4 are underlined and colored according to B. Conserved residues are in bold and are labeled below with their function: S = siroheme binding, A = active site residue, C = cysteine ligands of the [4Fe-4S] cluster, and + = structurally important. Figure is continued on next page.

```
1)  N-terminus (62-333)   Escherichia coli NADPH-dependent aSiRHP
2)  N-terminus (118-399)  Arabidopsis thalinia ferredoxin-dependent aSiR
3)  N-terminus (131-369)  Spinacia oleracea ferredoxin-dependent dNiR
4)  C-terminus (347-570)  Escherichia coli NADPH-dependent aSiRHP
5)  C-terminus (400-640)  Arabidopsis thalinia ferredoxin-dependent aSiR
6)  C-terminus (370-594)  Spinacia oleracea ferredoxin-dependent dNiR
7)  Archaeoglobus fulgidus dSiR - subunit α   (H1, H2, H3, H4)
8)  Archaeoglobus fulgidus dSiR - subunit β   (H1, H2, H3, H4)
9)  Desulfovibrio vulgaris dSiR - subunit α   (H1, H2, H3, H4)
10) Desulfovibrio vulgaris dSiR - subunit β   (H1, H2, H3, H4)
```

```
                                                                                      .200
1)  DRDIRAERAEQKLEPRHAMLLRCRLPGGVITTKQWQAIDKFAGENTIYGSIRLTNRQ-TFQFHGILKKNVKPVHQMLHSVGLDALATANDMNRN-VLCTS  159
2)  NREERGGRS----------YSFMLRTKNPSGEVPNQLYLIMDDLADEFG-IGTIRLTTRQ-TFQLHGVLKQNLKTVMSSIIKNMGSTLGACGDLNRN-VLAPA  198
3)  RKHHY----------GRFMMRLKLPNGIVTSEQTRVLASVIKKYGDGCADVTTRQ-NWQIRGVVLPDVPEIIKGLESVGLTSLQSGMDNVRNPVGNPL  218
4)  ----GW-VKGIDRWHLFLFERITANGIDYPARPLKTGLEIAKIIHKGQFRITANG-NLIIAGVPESEKAKIEKIAKESGLMN-AVTPQ-RENSNACV  429
5)  ----LGWHEQGD-GTWFCGLHVDSGRV--GGIMKKTLREVIEKY-KIDVRITPNQ-NIVLCDIKTEWKRPITTVLAAGLLQPEFVDPI-NQTAMACP  501
6)  ----GVHPQKQQGLSFVGLHIPVGRL--QADEMEELARIADVYGSGELRLTVEQ-NIIIPNVENSKIDSLLNEPLLKERYSPE--PPILMKGLVACT  474
7)                                      GLTNFHGSTGDIIFLG-------------------RTPS-ACM
8)                                      GYLRWTSRN-NVEFFV-------------------SNIV-HTQ
9)                                      GLTNMHGSTGDIVLLG-------------------RTPE-SCL
10)                                     GHLRFTTRN-NVEFMV-------------------SNIV-HTQ
        A                              + S SS SS S+                        A   C
                                              H1                               H2
```

```
1)  NPYESQL-HAEAYEWAKKISEHLLPRTRAYAEIWLDQEKVATTD---------------EEPILGQTYLPRKFKFKTVVIPPQNDIDLHAN-DMN  236
2)  APYVKKD-YLFFAQETADNIAALLSPQSGFYDMWVDGEQFMTAEPPEVVKARNDNSHGTNFVDS-PEPIYGTQFLPRKFKFKVAVTVPTDNSVDLLTN-DIG  304
3)  AGIDPHE-IVUTRPFTNLISQFVTANSRGNLSI----------------------------------TNLPRKFWNPCVIGSHDLYEHPHIN-DLA  277
4)  SFPTCPLAMEEAERFLPSFIDNIMAKHGVS-----------------------------------DEHIVNRVTGCPNGCGRAMLA-EVG  492
5)  AFPLCPLAITEAERGIPSILKRVRAMFEVGLDY--------------------------------DESVVIRVTGCPNGCARPYMA-ELG  560
6)  GSQFCGQAIIETKARALKVTEEVQRL---------------------------------VSVTRFVR--NHWTGCPNSCGQVQVA-DIG  527
7)                      GPALCEFACYDT----------------------------PYKFKIKCAGCPNDCVASKARSD---
8)                      GWIHCHIPAIDA----------------------------PAMCRJSLACCNMCGAVHA-SD---
9)                      GKSRCEFACYDS----------------------------PYKFKFDACPNGCVASIARSD---
10)                     GWVRCHTPAIDA----------------------------PAPVRISLACCINMCGAVHC-SD---  +
        C                                               +SA+A           H3   C C S
   H2 (con't)
```

```
                                                                                      .400
1)  -FVAIA-ENGKL-VGFNLLVGGGLSIEHGNKKTYARTASEFGYLPLEHTLAVAEAVVTTQRDWGNRTDRKNAKTKYTLERVGVETFKAEVERRA-------GIK  330
2)  --VVVVSDENGEP-QGFNIYVGGGMGRTHRMESTFARLAEPIGYVPKEDILXAVKAIVVTQREHGRDDRKYSRMKYLISSWGIEKFRDVVE------QYYGKK  399
3)  --YMPATKNG--KFGFNLLVGGFFSIKKPCEEAI-----PLDAWVSAEDVVVPVCKAMLEAFRDLGFRGNRQGCRMQWLIDELGMEAFRGEVEKRMEQVLERA  369
4)  --LVGKAP--GR--YNLHLGG---NRIGTRIPR--MYKENITEPEILASLDELIGRWAKEREAGEQFGDFTVRAGI-IRPVLDPARDLWD-----------  570
5)  -LVGD-----G--PNSYQVWLGGTFNLTC----IAR  SFMDKVKVHDLEKVCEPLFYHWKLERQTKESFGEYTTRMGFEKLKELIDTYKGVSQ---------  640
6)  FMGCMTRDENGKPCEGADVFVGGRIGSDSHLGDIYKKAVPCKDLVFVVAEILINGQFGAVPREREEAE-----------------------------------  594
7)       GK-APFVEGAVIG
8)       G--AAIMVGG
9)       GAKAPIIDGAQMG
10)      G--DGVV IMVGG
             +   ++                                              +
     H4
```

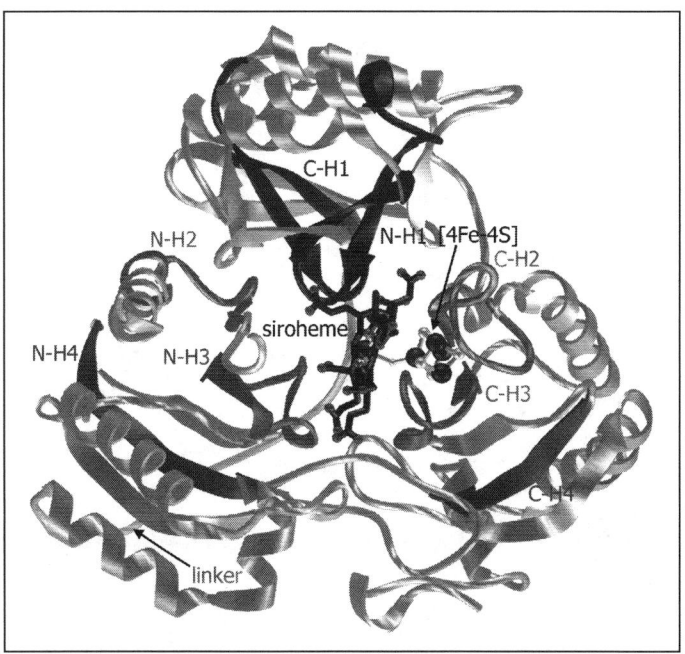

Figure 1, continued. B) A ribbon diagram of *E. coli* aSiRHP that highlights the pseudo two-fold structural symmetry that also relates the homology regions that are defined above. The homology regions (defined in A) are labeled N-H1 to N-H4 or C-H1 to C-H4 according to their positions in the N-terminal or C-terminal half of aSiRHP. H1 is blue, H2 is gold, H3 is green, and H4 is red. The siroheme (dark gray bonds) binds in a deep groove at the junction of the three domains. Oxygens are red balls, nitrogens are blue balls, sulfurs are yellow balls, and irons are green spheres. The linker (cyan) is the two-fold symmetric partner of the siroheme. The distally-bound phosphate has been removed from the figure for clarity. A color version of this figure is available online at www.eurekah.com.

Diversity between the SiR and NiR Enzymes

Assimilatory and dissimilatory SiRs and NiRs differ in their biological function, oligomeric state, and whether or not they release partially-reduced intermediates during catalysis. Both enzymes react with either substrate (sulfite or nitrite), however each shows binding selectivity for its namesake.[7,13] NADPH-dependent aSiRs are found in prokaryotes and plant eukaryotes where they reduce sulfur to the proper redox state for incorporation into sulfur-containing amino acids and cofactors.[5] aSiRs are large (> 700 kDa) oligomers that contain eight copies of a cytochrome P450-reductase-like, flavin-containing protein (SiRFP; 60 kDa) and either four or eight copies of a siroheme-containing hemoprotein (SiRHP; 64 kDa).[14,15] SiRFP interacts with a molecule of NADPH and passes reducing equivalents through a flavin adenine dinucleotide (FAD) and a flavin mononucleotide (FMN) to SiRHP, which binds substrate on the distal face of the siroheme.[14,15] When urea-denatured, SiRFP dissociates as an octomer whereas SiRHP dissociates as a monomer.[14] SiRHP is active when reducing equivalents are supplied chemically and can be evaluated as a monomeric hemoprotein.[16] aSiRs perform the six-electron reduction without releasing partially-reduced biproducts and without any discernable spectroscopic intermediates.[17] The ferredoxin-dependent aSiRs and aNiRs are 60 kDa, monomeric enzymes with one siroheme and one [4Fe-4S] cluster per molecule.[7] In contrast, most dSiRs are $\alpha_2\beta_2$ heterotetramers found in sulfur-reducing eubacteria and archae.[8] The α subunits are typically about 50 kDa whereas the β subunits are typically about 40 kDa.[8] Some variations exist in both the subunit composition (some dSiRs have an additional two γ subunits) and the number of metal centers in the heterotetramers, however the hallmarks of the dissimilatory enzyme are its role in facilitating the final step of sulfur-based, anaerobic respiration and the release of partially-reduced intermediates during catalysis.[8,17] NAD(P)H-dependent dNiRs are dimeric enzymes involved in nitrate-based respiration in which the major subunit is the siroheme-containing hemoprotein and the smaller subunit is the NAD(P)H-dependent reductase.[9]

Symmetry Defines Homology between the Assimilatory and Dissimilatory Enzymes

In 1995, Crane et al solved the X-ray crystallographic structure of the *E. coli* aSiRHP (Fig. 1B). Their structure reveals a pseudo two-fold axis of symmetry that relates the N-terminal and C-terminal halves of *E. coli* aSiRHP (Fig. 1). Structural homology between the N- and C-terminal halves of the monomeric hemoprotein identifies three core regions that define the siroheme binding site and the [4Fe-4S] cluster coordination (Fig. 1). These structural units are called homology regions 1-4 (H1-H4). The homology extends to other monomeric SiRs and NiRs, as well as the α and β subunits of the tetrameric dSiRs (Fig. 1A).[1] Both the structural symmetry in aSiRHP and the sequence homology between each half of aSiRHP with the α and β subunits of the dSiRs suggest that these enzymes evolved through a gene duplication event, in which each homologous half is needed to form a two-component active site.[8] Whereas aSiRHP houses the siroheme, [4Fe-4S] cluster, and active site within a single 60 kDa monomer, dSiRs likely assemble an α and a β subunit, which are together about 90 kDa in size, to bind coupled cofactors and form the active site.

SiRs and NiRs Have Multiple Redox Centers and Intricate Spectroscopic Features

In the siroheme-containing SiRs and NiRs, siroheme works with a [4Fe-4S] cluster, reducing sulfite to sulfide (SO_3^{-2} to S^{-2}) or nitrite to ammonia (NO_2^- to NH_3). Together, this ensemble of iron-containing cofactors acts as a "metallic battery," passing six electrons from a molecule of NADPH, ferredoxin, or cytochrome *c* to the central atom of the anionic substrate.

The earliest studies on siroheme-containing SiRs and NiRs from *E. coli* and *S. oleracea* probed the enzymes' diverse ultra-violet/visible (UV/vis), electron paramagnetic resonance (EPR), resonance Raman (RR), and Mössbauer spectra.[13,18-29] In brief, aSiR and aNiR can access at least four distinct oxidation states: oxidized (in which the siroheme iron is formally +3 and the [4Fe-4S] cluster is formally +2), one-electron reduced (in which the siroheme iron is formally +2 and the [4Fe-4S] cluster is formally +2), two-electron reduced (in which the siroheme iron is formally +2 and the [4Fe-4S] cluster is formally +1), and "super-oxidized" (in which the siroheme is formally +4 and the [4Fe-4S] cluster is formally +2). Each state is characterized by a unique UV/vis, EPR, or Mössbauer spectra. The *E. coli* aSiRHP is the most extensively studied of these enzymes and details about its spectroscopy have been reviewed elsewhere.[30] By using the distinct EPR signals for the +3/+2, +2/+2, and +2/+1 states, redox potentials were measured for *E. coli* aSiRHP of -340 mV for the siroheme[13] and -405 mV for the [4Fe-4S] cluster;[23] addition of anions (CO and CN⁻) affects the redox potentials of each cofactor.[22] These extensive studies on the *E. coli* aSiRHP provided an indirect picture of the enzyme's active site and were the basis of the hypothesis that the five iron atoms are electronically coupled throughout the whole catalytic cycle by an endogenous cysteine ligand.

Later studies on dSiR from *Desulfovibrio vulgaris* (Hildenborough) suggest that the relationship between the siroheme and the [4Fe-4S] cluster may be slightly different in other systems. In particular, inorganic sulfur has been implicated in bridging the cofactors in the dSiR system, despite conservation of four potential cluster-ligating cysteine residues (Fig. 1A).[31] Further, redox potentials of -298 mV and -620 mV were measured for the dSiR's siroheme iron and [4Fe-4S] cluster.[32] The exogenous inorganic sulfur remained bound to the enzyme throughout turnover, supporting the hypothesis that the cofactors are coupled throughout the catalytic cycle.[33] Despite the differences between aSiRs and dSiRs, however, the overall relationship between the siroheme and cluster is similar in both systems.

X-Ray Crystallographic Structures Support the Spectroscopic Data

E. coli aSiRHP has an elegant active site in which a single cysteine ligand serves both as one of the four ligands to the [4Fe-4S] cluster and as the proximal ligand to the iron of the siroheme (Fig. 2). Based on Crane et al's subsequent structures of the enzyme in various redox states and in complex

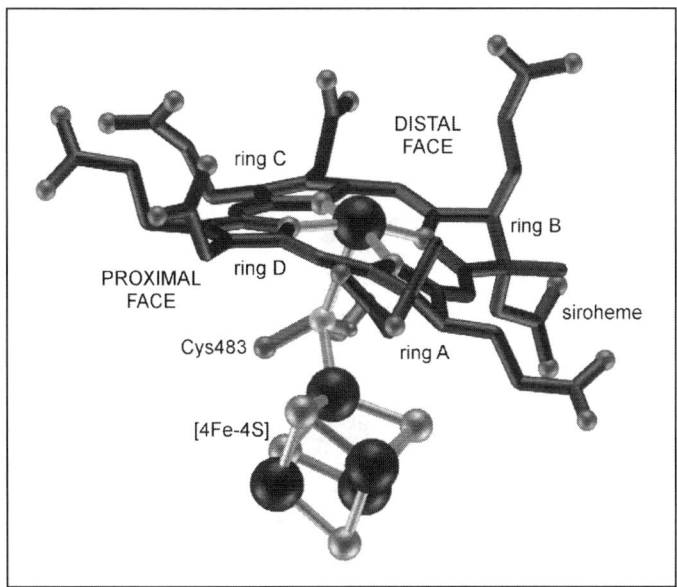

Figure 2. Siroheme and [4Fe-4S] cluster cofactors from the X-ray crystallographic structure of the *E. coli* assimilatory sulfite reductase hemoprotein. The siroheme (dark gray tubes, top) from the X-ray crystallographic structure of the oxidized aSiRHP with the proximal cysteine ligand (green, center) and the [4Fe-4S] cluster (gold and dark green spheres with light gray bonds, bottom). The distally-bound phosphate anion has been removed for simplicity. Irons are dark green spheres, oxygens are red balls, nitrogens are blue balls, inorganic sulfites are gold spheres, sulfurs are yellow balls, metal – ligand and covalent bonds between cofactors are light gray. A color version of this figure is available online at www.eurekah.com.

with a series of distally-bound anions, we can assign structures that correlate to a proposed mechanism in which the six-electron reduction occurs by three additions of two electrons (Fig. 3).[2,3] These structures, alongside careful spectroscopic characterization of the species present in the protein crystals, provide a structural basis for a wealth of biochemical data that has been collected on the aSiRs. Together with recently obtained ultra-high resolution X-ray crystallographic data on the *E. coli* aSiRHP (Stroupe and Getzoff, unpublished data), the many structures of aSiRHP provide a detailed view into the life of the siroheme cofactor as it serves alongside a [4Fe-4S] cluster to perform a powerful six-electron reduction reaction.

Siroheme-containing SiRs and NiRs represent a large class of fairly diverse enzymes with two defining features. First, they use a coupled siroheme/[4Fe-4S] cluster as the active-site cofactors. Second, these coupled cofactors reside within a conserved, two-fold symmetric assembly, whether a monomer or an arrangement of α and β subunits. Given these class characteristics, the structure of *E. coli* aSiRHP can serve as a model for the siroheme-dependent sulfite and nitrite reductases.

Siroheme Is at the Heart of the Six-Electron Reduction of Sulfite to Sulfide or Nitrite to Ammonia

Two characteristic structural features of the siroheme contribute to its distinct conformation in the context of aSiRHP (Fig. 4). First, the eight porphobilinogen-derived carboxylates remain intact on siroheme, distributed according to the asymmetry of the tetrapyrrolic precursor uroporphyrinogen III (starting with ring A, the macrocycle's side chains follow an AP AP AP PA pattern, where A = acetate and P = propionate) (Fig. 4B). Second, the two *S*-adenosyl-L-methionine (SAM)-derived methyl groups on C2 and C7 of rings A and B break the four pyrrole rings' conjugation by saturating C2, C3, C7, and C8 (Fig. 4B). In the context of the protein, siroheme's unique structural features help explain how the enzyme binds geometrically diverse anionic substrates and how distortions in

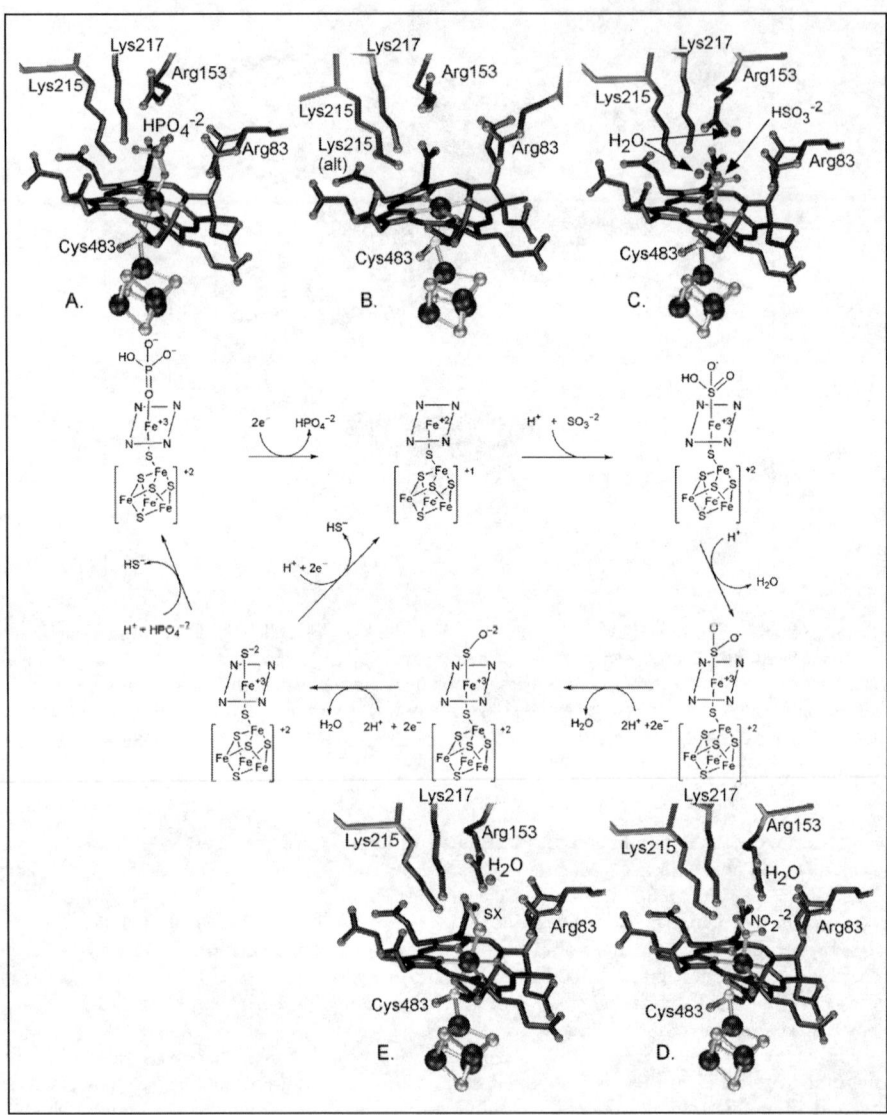

Figure 3. A series of X-ray crystallographic structures illustrate the changes in the active site as aSiRHP binds to various anions, suggesting a means by which the enzyme binds to geometrically diverse anions. The following structures are complexes of *E. coli* aSiRHP with HPO_4^{-2} (A), reduced chemically by two electron equivalents (B), bound to HSO_3^{-2} (C), bound to NO_2^- (D), and bound to a reduced sulfur intermediate, modeled as SO^{-2} (E).[2,3] Each substrate and intermediate complex (B-E) was generated in the crystal by chemically reducing the enzyme and, in C-E, the small molecule was soaked into the crystal.[2,3] Whereas the anion geometry correlates with the species in the proposed mechanism, the structures do not necessarily represent on-path intermediates.

the siroheme facilitate coupling between the two metal centers and the bound substrate throughout the catalytic cycle. By expanding on the themes of charge in the active site and the coupled metallic cofactors, we can explore the role of siroheme as the anchor for the six-electron reduction of sulfite to sulfide or nitrite to ammonia.

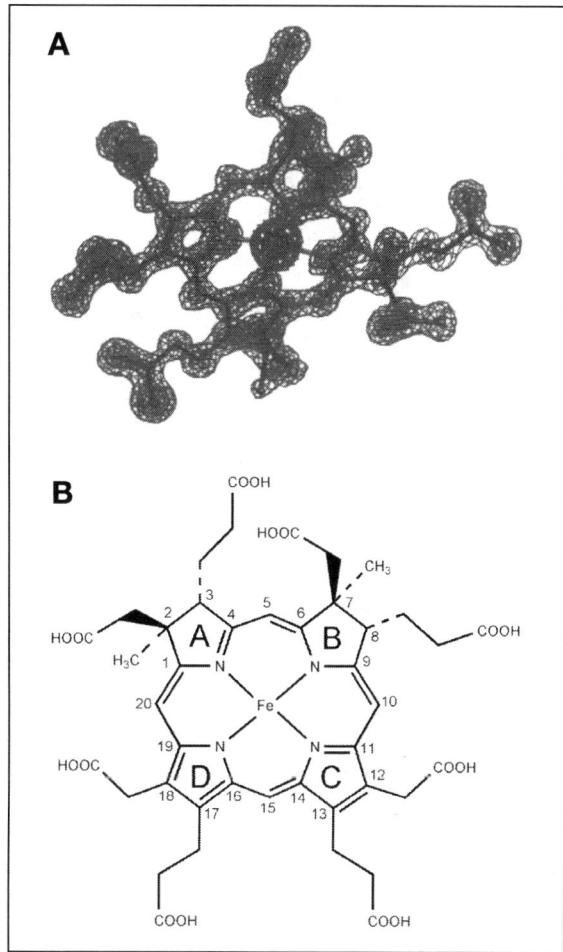

Figure 4. Siroheme—the X-ray crystallographic density and the chemical structure of the iron-containing isobacteriochlorin. A) The $2F_o$-F_c electron density of the siroheme calculated at 1.1 Å-resolution. Blue cages represent the map contoured at 3 sigma, purple contours at 5 sigma, and red contours are at 10 sigma. B) Siroheme's chemical structure.

Siroheme Anchors the Transformation of Sulfite to Sulfide or Nitrite to Ammonia

Siroheme's eight carboxylates contribute an overwhelmingly negative charge that is neutralized in the protein framework by positively-charged residues and hydrogen-bonding partners lining the active site cavity. Arg113, Thr115, Arg117, His123, Arg214, Arg302, Lys306, Gln396, and Arg485 interact with the siroheme side chains, either through hydrogen bonds or salt bridges (Fig. 5A). These residues form a dome around on the distal side of the siroheme, positioning all eight of the siroheme side chains on this face to make room for the proximally-bound [4Fe-4S] cluster that sits directly beneath the siroheme ring A.[1]

Four other positively-charged residues, Arg83, Arg153, Lys215, and Lys217, interact with the distally-bound anion and likely serve as general acids during catalysis, providing the protons needed to release oxygen atoms from sulfite and nitrite as water molecules (Fig. 5B).[1,2] In the absence of a protein, but with an appropriate reductant, free siroheme functions as an efficient catalyst in incompletely reducing sulfite or nitrite (the major product from the sulfur reduction reaction is hydrosulfite, $S_2O_4^{-2}$).[34-36] Therefore, the protein scaffold likely plays an important role in directing proton transfer to keep the reaction on pathway, but need not contribute significantly to the energetics of the reaction. Of these four residues, Lys217 is the only one that is not strictly conserved between aSiRHP and dNiRs (Fig. 1A) and has been proposed to be significant in discriminating between sulfite and nitrite binding.[1] Further, these four residues are not strictly conserved with the heterodimeric dSiRs (Fig. 1A), suggesting that the active site's protein environment may change with the varying subunit assembly. These differences in the active site may be related to the dissimilatory enzyme's ability to release partially reduced species.

As purified and crystallized in the oxidized state, aSiRHP has a phosphate molecule in its active site, with one of its oxygens bound to the siroheme iron.[1] Upon reduction, the phosphate is released (Fig. 3B).[3] This "redox gating" phenomenon offers an explanation for the observation that the oxidized enzyme experiences a long lag time in binding substrate, whereas the reduced enzyme shows no such lag.[3,22] Interactions between the phosphate anion and the surrounding side chains are extensive (Figs. 5B, 6A). A second phosphate oxygen interacts indirectly with the siroheme by bridging hydrogen bonds through Lys215Nζ to the C18 acetate and through Lys217Nζ to both the C12 acetate and the C13 propionate. Another phosphate oxygen makes hydrogen bonds with Arg83Nη, Arg153Nε, and an ordered water molecule. The final oxygen makes a direct hydrogen bond to the C8 propionate. This P-O bond is longer than the others (1.60 ± 0.01 Å, compared to 1.51 ± 0.01 Å, 1.53 ± 0.01 Å or 1.53 ± 0.01 Å [Stroupe and Getzoff, unpublished data]), supporting the notion that the enzyme specifically favors protonation of this oxygen.[1]

Unlike the phosphate anion, which binds SiR's siroheme iron through one of its four oxygen atoms, substrate (sulfite or nitrite) binds through its central sulfur or nitrogen atom.[2] In the case of sulfite, the central atom (sulfur) is about 1.7 Å closer to the siroheme iron than is the phosphorous. As a result, the anion is no longer in direct hydrogen-bonding distance to the siroheme carboxylates (Fig. 6B). The lysine-mediated hydrogen bonds remain between two of the three sulfite oxygens and the C12, C13, and C18 side chain carboxylates. One of the sulfite oxygens is believed to be protonated because in the X-ray crystal structure, the S-O bond is about 0.3 Å longer than the other two S-O bonds; this protonated oxygen also interacts with an ordered water molecule.[2] A second water molecule bridges the hydrogen bond between the unprotonated oxygen and the C8 propionate. The same water molecule also makes a hydrogen bond to Arg153Nη, which has flipped over to improve interactions with the smaller substrate through a direct hydrogen bond to the sulfite's protonated oxygen. Arg153Nη's guanidinium group is likely the source of the first proton in the six-electron reduction (Figs. 3C, 6). In participating in the interactions between the substrate and the proton source, the siroheme anchors a fundamental hydrogen bond network that facilitates the proton transport needed for accurate catalysis.

When the substrate nitrite binds, a water-mediated hydrogen bond links one nitrite oxygen to both the C8 propionate and Arg153Nε (Fig. 6C).[2] The two sets of lysine-mediated hydrogen bonds are maintained, linking the same substrate oxygen with the C12, C13, and C18 siroheme side chains (Fig. 6C). In the NO-bound structure, only the water-mediated hydrogen bond to the C8 propionate group and a direct hydrogen bond to Arg153Nη remain (Fig. 6D). In the SX-bound structure, both hydrogen bonds to the C8 propionate group and Arg153Nε are water-mediated (Fig. 6E). For the last step of the reaction, Arg153 is the likely source of the final proton (Fig. 3).

In the structures described above, we see aSiRHP as it binds to geometrically diverse anions (PO_4^{-2}, SO_3^{-2}, NO_2^{-2}, NO, and SX), reflecting the structural changes that likely occur during catalysis.[2] During turn-over, six electrons and six protons are added to the substrate and three water molecules are released. In each step, the siroheme side chains play an important role in directing proton transfer by anchoring hydrogen-bonding networks to ordered water molecules, protein side chains, and substrate oxygens.

The Siroheme Tetrapyrrole Shows Significant Departure from Planarity

One of the most striking features in the structure of aSiRHP is the siroheme's pronounced saddle shape.[1] Similar deviations are described for some protoporphyrin IX hemes and are attributed to the influence of the surrounding protein residues in bending the typically planar molecule.[37] Unlike

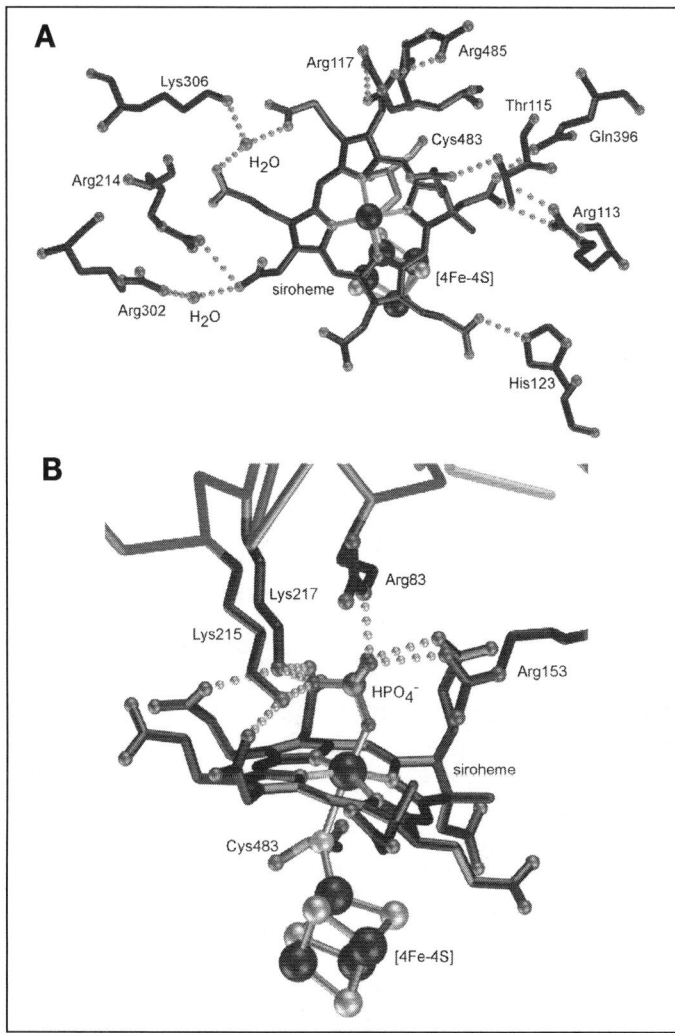

Figure 5. Protein and water interactions with the siroheme. A) A view of the siroheme (dark gray bonds)-[4Fe-4S] cluster (gold and dark green spheres) ensemble, viewed from the top. The distally-bound phosphate has been removed for clarity. Protein side chains (purple) interact with the siroheme side chains and hold them all on the distal face of the isobacteriochlorin. Oxygens and ordered water molecules are red balls, nitrogens are blue balls, hydrogen bonds are gray balls, and covalent bonds between cofactors are light gray. B) The four proton-supplying residues, that interact with the distally-bound phosphate: Arg83, Arg153, Lys215, and Lys217. A color version of this figure is available online at www.eurekah.com.

protoporphyrin IX-derived hemes, however, siroheme's flexibility stems in part from the puckers in the pyrroline rings A and B. These puckers cause significant deviation from planarity (Figs. 2, 4A). C3 of ring A is about 0.15 Å below (exo) the plane formed by C1, C4, and the two pyrroline nitrogens A and B, whereas C2 is about 0.40 Å above (endo) the same plane. In ring B, C7 and C8 are 0.26 Å and 0.78 Å below and above, respectively, the plane define by C6, C9, and the two pyrroline nitrogens A and B. The endo/exo patterns of C2, C3, C7, and C8 contribute to siroheme's asymmetry by accentuating the nonplanarity of rings A and B. Additionally, the bridging meso carbon between rings A and B sits almost 2 Å below the plane formed by the 11 central carbons

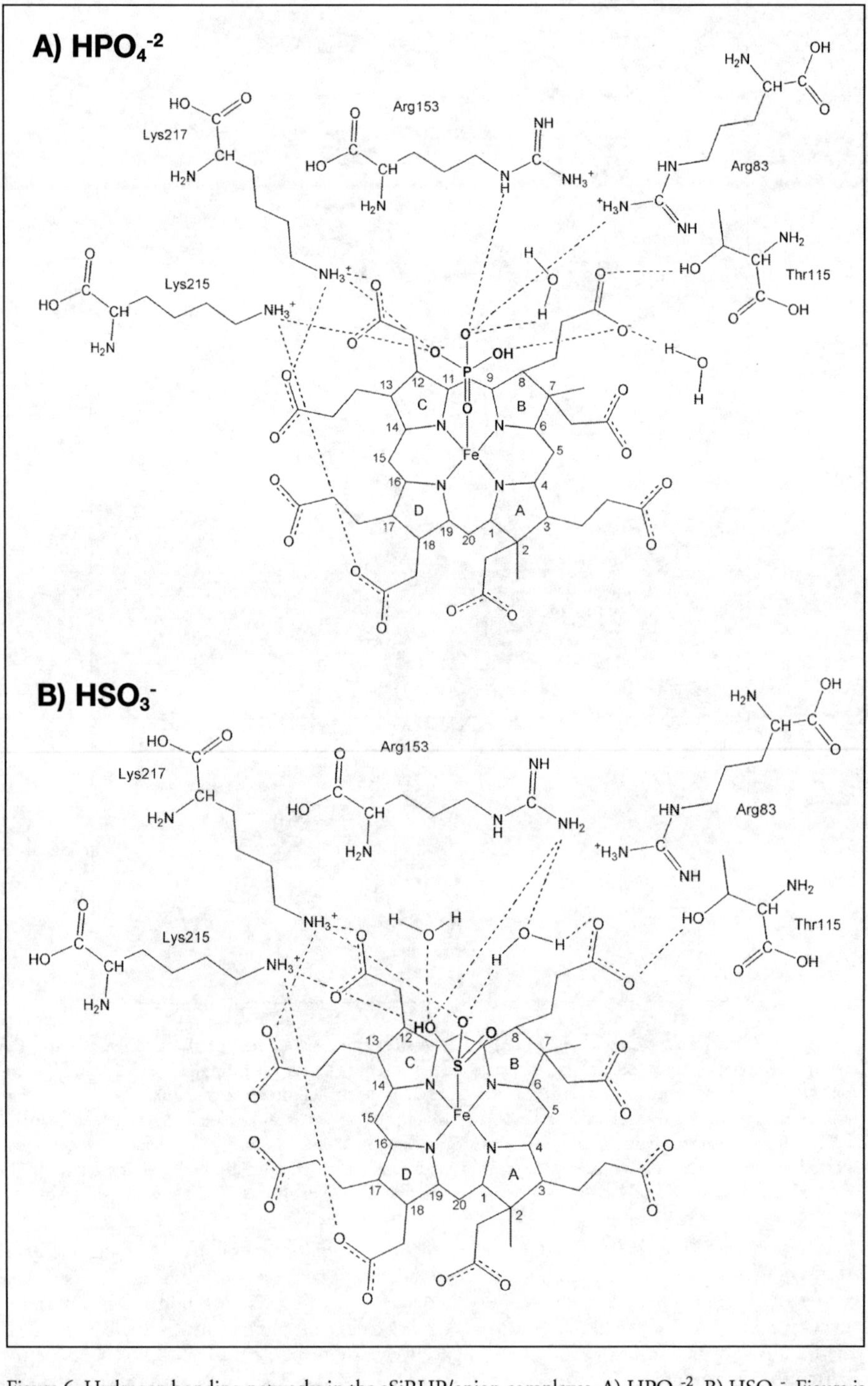

Figure 6. Hydrogen bonding networks in the aSiRHP/anion complexes. A) HPO$_4^{-2}$. B) HSO$_3^-$. Figure is continued on next pages.

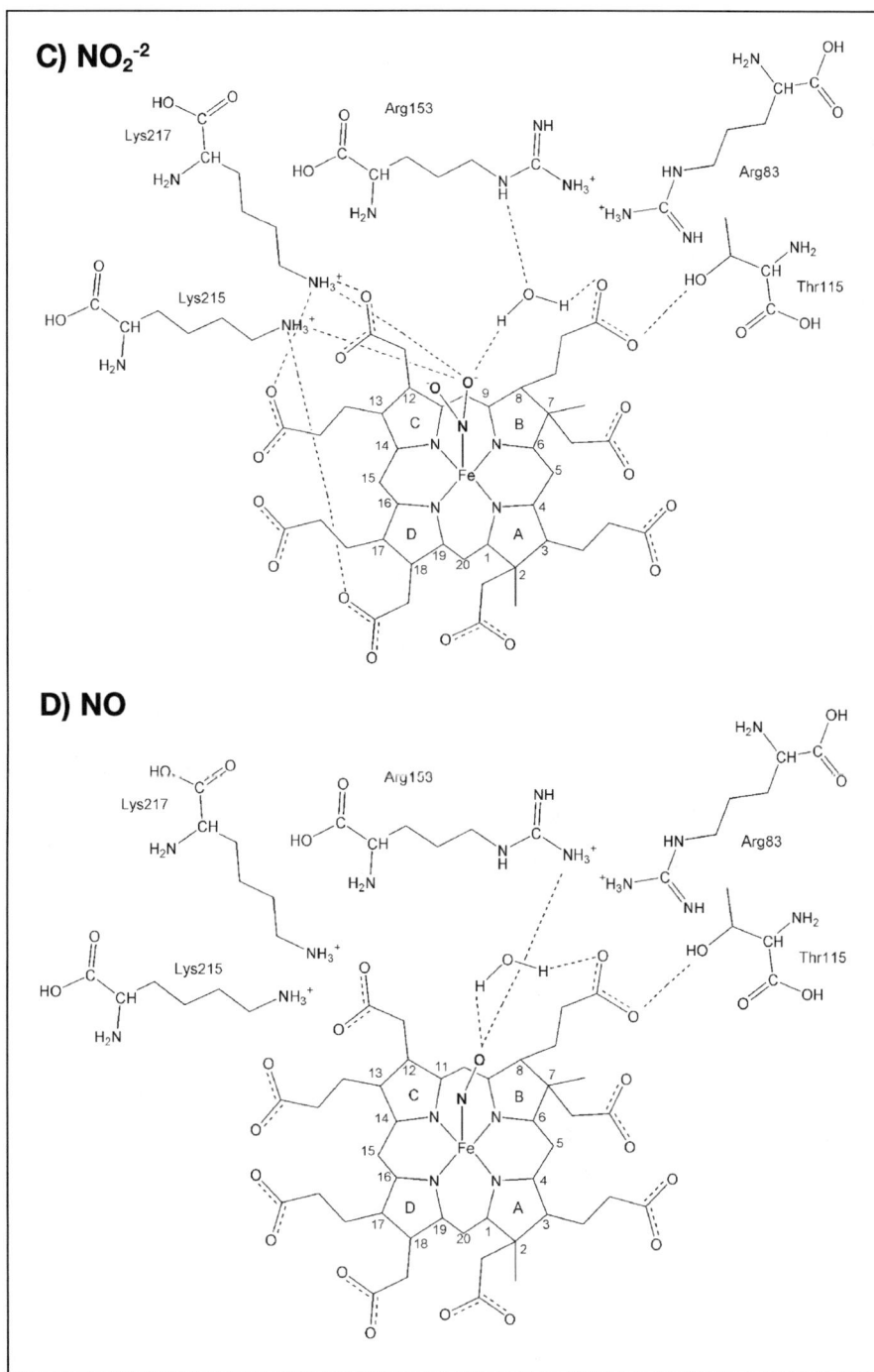

Figure 6, continued. C) NO_2^{-2}. D) NO. Figure is continued on next page.

Figure 6, continued. E) SX.

(C10-C20) of the planar rings C and D (Fig. 2) (Stroupe and Getzoff, unpublished data), almost a whole Ångstrom more than the most pronounced of the protoporphyrin IX hemes analyzed computationally by Jentzen, et al.[37,38] As a result of the saddle-shaped isobacteriochlorin and the asymmetric placement of the [4Fe-4S] cluster beneath ring A (Fig. 5), the bridging carbon between rings A and B is within van der Waal's contacts of one of the cluster's inorganic sulfur atoms and may provide another pathway for the electrons during catalysis.[3]

Siroheme's Structural and Electronic Characteristics Control Anion Interactions

The saddle-shaped deformation of the isobacteriochlorin also defines the interaction between the macrocycle and the siroheme iron which, in turn, affects bonding to the proximal cysteine ligand and binding of the distally-bound anion. As a result of the bowing of ring A and B, the iron-binding central cavity formed by the pyrrole/pyrroline nitrogens is larger in this isobacteriochlorin than in typical porphyrins. Consequently, the oxidized metal sits only about 0.26 Å above the plane formed by the tetrapyrrole nitrogens, closer than in most ferric-porphyrin complexes (the typical distance for the oxidized iron porphyrins is greater than 0.3 Å).[1] Upon reduction, the iron radius shrinks as the iron goes from high-spin Fe^{+3} to low spin Fe^{+2}; the iron moves essentially into the plane of the pyrrole/pyrroline nitrogens at a distance of only 0.06 Å.[3] When the iron moves into the plane of the pyrrole nitrogens, the Fe-Cys483Sγ distance shortens by about 0.2 Å, suggesting that the coupling of the cofactors is strengthened by reduction of the system.[3] This coupling extends to the distally-bound anion, as shown through extensive spectroscopic work in which the redox potential of the cluster is modulated by the nature of the distally bound anion.[21-23]

The cysteine thiolate proximal ligand in *E. coli* aSiRHP has a direct affect on the length of the anion-siroheme iron bond. Cys483, the bridging cysteine ligand, has unusual stereochemistry in which the Sγ-Cβ bond is eclipsed with the Cα-N bond (Fig. 2). This noncanonical side-chain configuration staggers the Sγ-Cβ bond with the Fe-NB and Fe-NC bonds (Fig. 5). As a ligand to both the siroheme iron and to the [4Fe-4S] cluster, Cys483Sγ approximates sp^3 hybridization and the sulfur lone pair projects opposite the vector formed by the Sγ-Cβ bond, staggered between the Fe-NA and Fe-ND bonds.[3] On the other hand, the canonical d-orbital symmetry for a heme iron is O_h, in which the d_{xy} orbital projects between the iron-pyrrole nitrogen bonds and the d_{xz} and d_{yz} orbitals project below and above the pyrrole nitrogens. Given this configuration, there would be very little π-interaction between the sp^3-hybridized Cys483Sγ and the siroheme Fe^{+3} because the sulfur lone pair would be parallel to the d_{xy} orbital and would pass between the d_{xz} and d_{yz} orbitals.[2] The same does not hold for the distally-bound anion, however, and the lack of macrocyclic conjugation in rings A and B improves the sirohemes ability to backbond with a distally-bound anion.[2] As a result of both the lack of backbonding with the proximal ligand and siroheme's improved ability to share electrons with a distally-bound anion, the factors that determine anion-to-siroheme iron bond length are different in *E. coli* aSiRHP than in most porphyrin systems. In most metalloporphyrin complexes with a backbonding proximal ligand, when a distal ligand that makes strong π-interactions with the metal binds, the bond length between the proximal ligand and the metal increases. This occurs because electron density moves into the distally-bound ligand, decreasing the metal's ability to bond with the proximal ligand. With the staggered cysteine ligand whose interactions with the siroheme depend very little on backbonding, a strongly-interacting distal ligand pulls electrons from the metal, improving the metal's ability to interact with the negatively-charged cysteine ligand. Consequently, the siroheme iron-Cys483Sγ bond shortens. This trend is seen a series of structures in which *E. coli* aSiRHP is bound to CO, CN$^-$, SO$_3^{-2}$, NO$_2^{-2}$, NO, and S^{-2} (Fig. 3).[2]

A Possible π Cation Radical Intermediate

In addition to the structural distortions caused by saturation of rings A and B, there are significant electronic characteristics that control the reactivity of the macrocyclic ring. Decreased conjugation destabilizes the highest occupied molecular orbital (HOMO) of the macrocycle, allowing it to be more easily oxidized than the fully conjugated porphyrin.[39] Such an oxidation would result in formation of a π cation radical on the ring. Experimentally, a siroheme π cationic state has been isolated, both in free siroheme and within the aSiRHP enzyme.[19,40] Computationally, Chang et al. propose that this unique oxidation event occurs because the energy of the macrocycle's HOMO is raised above that of the reduced metal's HOMO, promoting the removal of an electron from the ring before metal oxidation.[39] Mechanistically, if such a π cationic state were to exist, the siroheme/ [4Fe-4S] cluster could perform the six-electron reduction in two steps of three electrons each (equivalent to going from SO$_3^{-2}$ to SO$^-$/SO$_2^{-3}$ and then from SO$^-$/SO$_2^{-3}$ to S^{-2}) rather than in three steps of two electrons each (as depicted in Fig. 3). Attempts to differentiate the two mechanisms are inconclusive. A single ferroheme-NO intermediate has been trapped during nitrite reduction by aNiR or aSiRHP. This species correlates to a two-electron reduced intermediate with a reduced siroheme iron, accounting for three electron equivalents.[18,21,41] On the other hand, both SiR and NiR can react with NH$_3$OH, which is four-electrons more reduced that NO$_2^-$, supporting the mechanism in which reduction occurs in three steps of two electrons where this substrate correlates to the second intermediate (Fig. 3). Crane, et al suggest that the enzyme might avoid being trapped in this stable intermediate by passing three electrons in a concerted fashion.[3] No electronic intermediates have been characterized during the sulfur reduction process, thus the mechanism by which aSiRHP performs its six-electron reduction reaction is still open for debate and study.

Conclusions

Siroheme is an ancient, unique cofactor whose utility is harnessed by SiRs and NiRs for the six-electron reduction reactions of sulfite to sulfide and nitrite to ammonia. Although this reaction can also be catalyzed by other metallic cofactors, the siroheme/[4Fe-4S] cluster is an elegant ensemble that can pass the electrons efficiently without release of partially reduced intermediates. The siroheme's importance in this reaction is two-fold. First, the cofactor's carboxylate side chains anchor the acid/base chemistry that occurs during catalysis as six protons are delivered and three water molecules released. Second, the unique electronic nature of the isobacteriochlorin allows its saddle-shaped structure. The extreme distortions of the siroheme macrocycle facilitates anion binding by allowing close contact among the siroheme, its central iron, and the distally-bound anion. The macrocycle's accentuated asymmetry also allows for tight interactions between the siroheme and its [4Fe-4S] partner, both through the shared cysteine ligand and by through-space interactions of a cluster sulfide and the meso carbon that bridges rings A and B. Such unique features raise questions about the enzyme's electron transfer mechanism that have yet to be answered by the biochemical and structural studies on the enzyme.

Note Added in Proof

Since the writing of this chapter, the structure of the heterotetrameric dissimilatory sulfite reductase _Archaeoglobus fulgidus_ was reported by Schiffer et al.[42] Their structure support the hypothesis put forward by Crane et al that aSiR is the product of gene duplication event: a second siroheme-[4Fe4S] cluster is found in dSiR sitting in the pocket that is occupied by aSiR's linker. A bulky side chain blocks access to the active site on the distal face of the siroheme so this second set of cofactors are believed to be structural, rather than catalytic.

Acknowledgements

Special thanks to D. Barondeau and D. Daniels for reading of the manuscript. Funding was provided by National Institues of Health grant GM37684 (to EDG) and a National Science Foundation Predoctoral Fellowhip (to MES).

References

1. Crane BR, Siegel LM, Getzoff ED. Sulfite reductase structure at 1.6 A: Evolution and catalysis for reduction of inorganic anions. Science 1995; 270(5233):59-67.
2. Crane BR, Siegel LM, Getzoff ED. Probing the catalytic mechanism of sulfite reductase by X-ray crystallography: Structures of the Escherichia coli hemoprotein in complex with substrates, inhibitors, intermediates, and products. Biochemistry 1997; 36(40):12120-12137.
3. Crane BR, Siegel LM, Getzoff ED. Structures of the siroheme- and Fe4S4-containing active center of sulfite reductase in different states of oxidation: Heme activation via reduction-gated exogenous ligand exchange. Biochemistry 1997; 36(40):12101-12119.
4. Martens JH, Barg H, Warren MJ et al. Microbial production of vitamin B12. Appl Microbiol Biotechnol 2002; 58(3):275-285.
5. Siegel LM, Murphy MJ, Kamin H. Reduced nicotinamide adenine dinucleotide phosphate-sulfite reductase of enterobacteria. I. The Escherichia coli hemoflavoprotein: Molecular parameters and prosthetic groups. J Biol Chem 1973; 248(1):251-264.
6. Vega JM, Garrett RH. Siroheme: A prosthetic group of the Neurospora crassa assimilatory nitrite reductase. J Biol Chem 1975; 250(20):7980-7989.
7. Krueger RJ, Siegel LM. Spinach siroheme enzymes: Isolation and characterization of ferredoxin-sulfite reductase and comparison of properties with ferredoxin-nitrite reductase. Biochemistry 1982; 21(12):2892-2904.
8. Crane BR, Getzoff ED. The relationship between structure and function for the sulfite reductases. Curr Opin Struct Biol 1996; 6(6):744-756.
9. Harborne NR, Griffiths L, Busby SJ et al. Transcriptional control, translation and function of the products of the five open reading frames of the Escherichia coli nir operon. Mol Microbiol 1992; 6(19):2805-2813.
10. Huynh BH, Kang L, DerVartanian DV et al. Characterization of a sulfite reductase from Desulfovibrio vulgaris. Evidence for the presence of a low-spin siroheme and an exchange- coupled siroheme-[4Fe-4S] unit. J Biol Chem 1984; 259(24):15373-15376.
11. Liu MC, Peck Jr HD. The isolation of a hexaheme cytochrome from Desulfovibrio desulfuricans and its identification as a new type of nitrite reductase. J Biol Chem 1981; 256(24):13159-13164.
12. Godden JW, Turley S, Teller DC et al. The 2.3 angstrom X-ray structure of nitrite reductase from Achromobacter cycloclastes. Science 1991; 253(5018):438-442.
13. Siegel LM, Rueger DC, Barber MJ et al. Escherichia coli sulfite reductase hemoprotein subunit. Prosthetic groups, catalytic parameters, and ligand complexes. J Biol Chem 1982; 257(11):6343-6350.
14. Siegel LM, Davis PS. Reduced nicotinamide adenine dinucleotide phosphate-sulfite reductase of enterobacteria. IV. The Escherichia coli hemoflavoprotein: Subunit structure and dissociation into hemoprotein and flavoprotein components. J Biol Chem 1974; 249(5):1587-1598.
15. Zeghouf M, Fontecave M, Coves J. A simplifed functional version of the Escherichia coli sulfite reductase. J Biol Chem 2000; 275(48):37651-37656.
16. Siegel LM, Davis PS, Kamin H. Reduced nicotinamide adenine dinucleotide phosphate-sulfite re-ductase of enterobacteria. 3. The Escherichia coli hemoflavoprotein: Catalytic parameters and the sequence of electron flow. J Biol Chem 1974; 249(5):1572-1586.

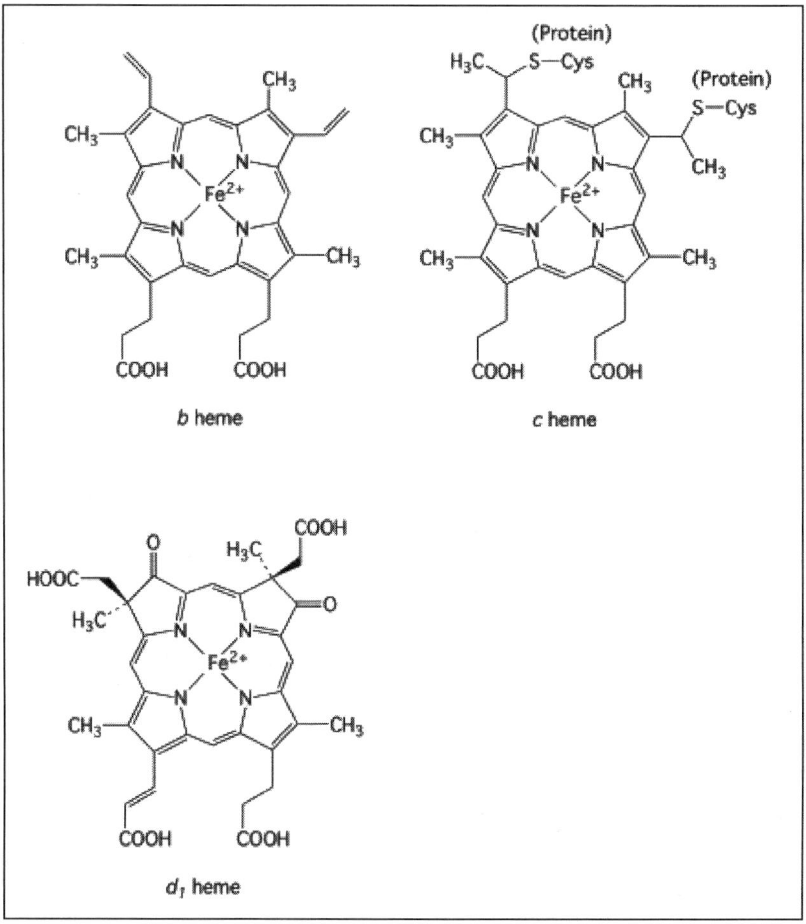

Figure 1. Structural formulae for *b*, *c* and d_1 type hemes.

The reduction of nitrite to nitric oxide requires several things of a heme protein. First the anionic nitrite must bind with sufficient affinity to the heme. One might expect that the ferric state of the heme would form the initial complex with the substrate nitrite, but an optimal mechanism might require the ferrous heme to form the initial Michaelis complex with the nitrite. Second, the nitrite must be reduced only as far as nitric oxide, and not further to ammonium via hydroxylamine as some nitrite reductases are able to do. Third, the product nitric oxide must not be reduced further to nitrous oxide. Finally, the heme must release nitric oxide which usually has high affinity for heme groups, especially in their ferrous oxidation state. We will consider how the unusual features of the cd_1 heme might have been selected so as to meet at least some of these requirements. This has to be done in the context of the three dimensional structure of cytochrome cd_1.

Structure of Cytochrome cd_1

Crystal structures have been obtained for the enzymes from *Paracoccus pantotrophus* (formerly *Thiosphaera pantotropha*) and *Pseudomonas aeruginosa*. In each case, the molecule is a homodimer, with each polypeptide being folded into two domains. An alpha helical domain comprises the *c*-type cytochrome part of the molecule. The d_1 heme is held in an eight bladed beta propeller structure within a pocket which provides many hydrogen bond partners for the polar substituents.[5-7] The overall fold of the *P. pantotrophus* protein is shown in Figure 2. The hemes on a monomer are sufficiently close to permit electron transfer on a timescale compatible with steady state turnover, but electron transfer across the dimer interface between any pair of hemes is thought to be kinetically insignificant. Several reviews dealing with cytochrome cd_1 as a whole rather than just the d_1 heme have been published recently.[6,8-10] In addition, another recent review provides complementary perspective on both d_1 heme and its role in the nitrite reductase.[11]

Surprisingly, the structures of the 'as prepared' oxidised enzymes from *P. pantotrophus* and *P. aeruginosa* are not identical. For the *P. pantotrophus* enzyme the d_1 heme has two axial ligands; on the proximal side a histidine, as expected, but on the distal side there is a tyrosine (residue 25) from the cytochrome *c* domain. Thus the 6-coordinated iron atom of the d_1 heme has no vacant coordination site for the substrate, nitrite. However, reduction of the crystals showed that the tyrosine residue had 'backed out' of the active site and thus generated a vacant coordination position at the Fe of the d_1 heme. Soaking of such crystals with nitrite permitted crystallographic observation of nitrite bound via its nitrogen atom to the d_1 heme iron.[12] One of the oxygen atoms of nitrite, presumably that destined to become product water, was hydrogen bonded to two histidine residues that sit in the active site above the d_1 heme. In some crystals it was possible to observe nitric oxide bound to the d_1 heme,[12] presumably as a result of electron transfer to the substrate nitrite. Although it is generally assumed that a d_1 heme iron to oxygen bond does not form during nitrite reduction, a recent theoretical study argues that oxygen-linked binding of nitrite may play a role in the enzyme reaction mechanism.[13] However, there is no experimental proof for this, but nor it can be definitely disproved by existing experimental data; it is, nevertheless, the least likely mechanism.

Tyrosine movement away from the d_1 heme iron upon reduction of the *P. pantotrophus* cytochrome cd_1 was accompanied by a striking ligand switching in the cytochrome c domain.[12] In the original oxidised state the heme iron in the cytochrome c domain had bis-histidinyl coordination, but reduction caused replacement of one of the histidines ligands by a methionine residue which had previously been some distance away from the heme. Currently it is believed that the crystal structure of the oxidised as prepared enzyme does not relate to conformational states of the enzyme that occur during steady state catalysis. There are solution spectroscopic studies that show the catalytically active form of the oxidised state of the enzyme to have histidine/methionine coordination at the heme in the c-type cytochrome domain centre.[14-15] Thus the enzyme needs to be activated by reduction. A straightforward conclusion would be that the structure seen for the reduced form of the *P. pantotrophus* enzyme also applies to the oxidised state during catalytic turnover. However, in more recent work the enzyme has been reduced before crystallisation. In this case the resulting crystal structure showed differences from the structure obtained following

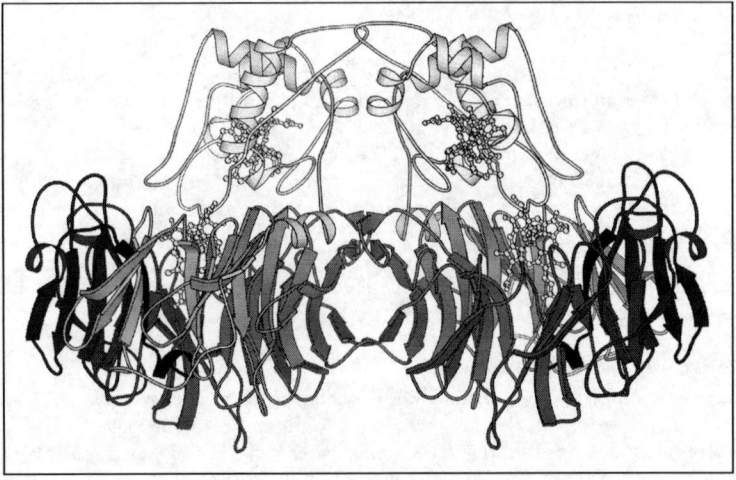

Figure 2. The structure of cytochrome cd_1 from *Paracoccus pantotrophus*.

reduction of crystals of oxidised protein.[16] The heme iron ligands were still histidine/methionine at the c and histidine/vacant at the d_1, but the relative positions of the two (c and d_1) protein domains had changed. Oxidation of the new crystal form by the alternative substrate oxygen did not change the relative orientations of the domains; the c-heme ligands remained the same but an oxygen molecule, or reduced species derived from it, could be seen at the d_1 heme. There is currently some uncertainty about the factors and conformations that open up the d_1 heme active site to the approach of the substrate nitrite. Nor is the driving force behind ligand switching from the bis-histidinyl at the c heme and histidine/tyrosine at the d_1 heme to the histidine/methionine at the former and histidine/vacant at the latter yet understood. Production of enzyme carrying site-directed mutations of critical residues has not been a simple matter because the specialised d_1 heme is not synthesised by the common host organisms such as *Escherichia coli*. Recently, it has proved possible to obtain active variants of cytochrome cd_1 of *P. pantotrophus*. The first such mutant was Y25S. This was active and thus established that there is no obligatory role for the tyrosine residue 25 that is normally ligated to the d_1 heme iron.[17]

The crystal structure of the oxidised Y25S protein shows that the bis-histidinyl coordination is retained at the c-type centre but that the d_1 heme iron is coordinated by the same histidine ligand as seen in the wild type structure. The absent tyrosine side chain is replaced by sulphate from the crystallisation liquor and the hydroxy group of the serine side chain is some distance away. Rather surprisingly this variant Y25S protein did not, in contrast to the wild-type form, first require switching into an active form by reduction.[17] In an unexpected development it has emerged that in solution the Y25S variant has his/met coordination at the c-type heme center.[18] This explains why this protein has no requirement for reductive activation, a process that switches the c heme from his/his to his/met in the wild type.

The significance of tyrosine coordination to the ferric d_1 heme iron comes under further scrutiny when the structure of the oxidised cytochrome cd_1 from *Pseudomonas aeruginosa* is considered. In this case the iron of the cytochrome c domain had histidine/methionine coordination and the d_1 heme had the equivalent histidine ligand as in the *P. pantotrophus* enzyme but hydroxide was the other axial ligand.[19] The structure of the *P. aeruginosa* enzyme also did not show an obvious route for substrate into the active site. There is also a further difference with the *P. pantotrophus* structure. In the *P. aeruginosa* enzyme there is a crossing over between the c domains of each monomer.[10,19] The significance of this is not known. In contrast to the *P. pantotrophus* enzyme, the cytochrome cd_1 from *P. aeruginosa* does not undergo such a large conformational change upon reduction,[10,20] although access to the active site must be opened up. A structural investigation suggests that it is the conversion of ferric d_1 heme to the ferrous state that triggers the opening of the d_1 heme distal pocket.

An important similarity between the structures of the enzymes from the two sources is that in each case there are two histidines which sit above the nitrite binding site on the d_1 heme. These histidines are regarded as proton donors to the substrate[5,21] and their importance for the nitrite reductase reaction can be appreciated by the finding that their loss through mutagenesis in the case of the *P. aeruginosa* enzyme results in loss of activity.[21] Furthermore, the absence of one of these histidines causes the c and d_1 domains of the enzyme to move with respect to each other.[22] The meaning of this relative domain movement and its relation to the different domain orientations seen for the *P. pantotrophus* enzyme remains to be determined. It is possible that during each turnover of substrate there is a reordering of the protein such that each of these overall conformations is sequentially adopted.

As will be discussed later, the d_1 heme requires a complex series of biosynthesis steps. Many hemes of periplasmic proteins are covalently bound, ie the c-type cytochromes, but the d_1 is not. It may be that forms of cytochrome cd_1 nitrite reductase in which the polypeptide chain completely surrounds the d_1 heme are important in order to avoid possible heme loss. Some support for this suggestion comes from the finding that during purification d_1 heme is lost more readily from Y25S variant compared with the wild type.[17]

Crystal structures of cytochromes cd_1 from *P. aeruginosa* or *P. pantotrophus* with not only nitrite or an oxygen-derived species but also nitric oxide, carbon monoxide, a partially reduced form of oxygen and cyanide have been reported.[4,12,23-26] The nitric oxide and carbon monoxide complexes of the ferrous d_1 heme in the *P. aeruginosa* enzyme have been investigated by resonance Raman spectroscopy. The iron to ligand bonds have characteristics distinct from those found in other hemoproteins. These distinct features were thought in part to reflect the influence of a highly ruffled structure of the d_1 heme on its electronic properties.[27]

Mechanism of Nitrite Reduction

The mechanism of nitrite reduction reaction catalysed by cytochrome cd_1 is far from certain. It is generally assumed that nitrite binds to the ferrous state of the d_1 heme, an event that is followed by electron transfer to the substrate and loss of one oxygen atom as a water molecule. This leaves a bound nitrosyl species which can rehydrate to nitrite, as evidenced by exchange studies with ^{18}O labelled water.

Bound nitrosyl has to leave the active site as nitric oxide. This is most likely to be promoted by the heme remaining in the ferric oxidation states as ferrous hemes in general have the higher affinity for nitric oxide. The factors that could prevent the rapid transfer of an electron from the c heme to form a ferrous-nitrosyl complex are not understood. Apart from this issue, one might envisage that the d_1 heme is optimised to release nitric oxide. In this context it may be relevant that the energy levels of the Fe d orbitals are, for the ferric state, inverted relative to 'normal' heme.[28] Ferric and ferrous d_1 heme nitrosyl complexes have been observed[10,29] but there is as yet no kinetic study that establishes the kinetic competence of one of these species.

Studies of cyanide binding to the cytochrome cd_1 of *Paracoccus pantotrophus* showed that the ferrous state has an unusually high affinity for cyanide.[24] A crystal structure was obtained for the cyanide complex of the enzyme in this oxidation state. This showed that the cyanide was not bound in any particularly stabilised form with respect to hydrogen bonds or other noncovalent interactions. The oxidised state of the cytochrome cd_1 from *P. pantotrophus* did not bind cyanide, in part at least because there is a powerful driving force for the phenolate oxygen of a tyrosine side chain to bind to the ferric iron of the d_1 heme and thus displace the cyanide. Consequently, and because the tyrosine does not bind to the ferrous d_1 heme, these studies do not provide direct insight into the relative affinities of the ferrous and ferric states. However, some insight into this has come from a study of cyanide binding to d_1 heme held in the heme-binding pocket of myoglobin.[30] In this system the affinity for cyanide was relatively independent of the iron oxidation state, again indicating the unusual affinity of the ferrous state for the anionic cyanide.

The binding of cyanide to *P. aeruginosa* cd_1 has also been studied both in solution and by crystallography.[25] In common with the *P. pantotrophus* enzyme, this enzyme binds cyanide in the ferric state, but in contrast to the *P. pantrophus* enzyme a comparable affinity is seen also for the ferrous state. This difference reflects the absence of the 'rebinding tyrosine' that displaces the cyanide from the *P. pantotrophus* enzyme. In the enzymes from both sources the cyanide is bound similarly with one of the two conserved active site histidines that lie in the nitrite binding site being with hydrogen bond distance.[24-25] It is reasonable to envisage that these histidines provide positive charge as one of the factors that contributes to anion binding, in particular the substrate nitrite to the heme. Studies with a varant of the *P. aeruginosa* enzyme that lacks one of these histidines shows the importance of this residue for the binding of both cyanide and nitrite.[25]

The d_1 heme may be fine tuned to bind nitrite rather than cyanide even though these two ions are close in the spectrochemical series. Direct evidence that the ferric state of cytochrome cd_1 from *P. pantotrophus* has a preference for nitrite over cyanide has been obtained.[31] The enzyme from this source requires activation by switching into a conformation for oxidised enzyme in which the c heme has switched to histidine/methionine coordination from bishistidinyl and the d_1 heme has become five coordinate as a consequence of the loss of a tyrosine ligand. Nitrite will bind to this 'activated' oxidised state to form a very long lived nitrite - ferric enzyme complex.[31] In effect, the binding of nitrite appears to stabilise the activated oxidised state of the protein. In contrast, cyanide is unable to achieve this stabilisation; it is displaced by the tyrosinate side chain, thus strongly suggesting that the special features of the d_1 heme ring are indeed tailored to optimise the binding and reduction of nitrite. Further support for this comes from studies of the Y25S *P. pantotrophus* variant protein. Cyanide binds to both the oxidised and reduced forms of this protein but more strongly to the latter. The comparison of the affinity of Y25S variant in its two redox states for cyanide is not complicated by the tendency of Y25 to bind to the wild type oxidised enzyme. Y25S binds nitrite in its oxidised form more strongly than the wild type (dissociation constant c

7×10^{-5} M compared with 2×10^{-3} M) and if we assume that the ten fold higher affinity of the ferrous state for cyanide will also apply to nitrite we can estimate a dissociation constant of c 7×10^{-6} M for nitrite binding to the ferrous state of the d_1 heme.

In general terms, the special features of the d_1 ring can be seen as contributing to the withdrawal of electrons from the ring. The contribution of each feature could in principle be delineated through preparing, by synthetic methods, variations of the d_1 ring. The only studies in this direction are those by Chan and coworkers who were able to insert the d_1 heme variants into a semi-apo preparation (ie with d_1 heme lost but c-type cytochrome centre retained) of cytochrome cd_1 from *Pseudomonas stutzeri*. Deletion of one carbonyl group could be tolerated but other changes that made the molecule more like 'standard' heme could not support enzyme activity.[32] More needs to be done in this direction if we are to understand the precise tailoring of the d_1 heme ring.

Beyond providing selectivity for nitrite, a second feature required of the d_1 heme ring is that it must not allow the reduction of nitrite to proceed beyond the nitric oxide oxidation state. This may indeed be a very critical design feature of the d_1 heme ring as those other nitrite reductases, which can reduce nitrite to ammonium via formal intermediates of nitric oxide and hydroxylamine.[24] These include a siroheme-containing enzyme and a pertaheme c-type cytochrome known as NrFA.

A critical reason why the d_1 ring is not involved in the onward reduction of nitric oxide to ammonium relates to the reduction potential required for the latter reaction. Unlike reduction of nitrite to nitric oxide and of hydroxylamine to ammonium, which have mid point potentials of +300mV and +800mV respectively, the reduction of nitric oxide to hydroxylamine requires a stronger reducing potential, in the region of 0mV. Determination of the reduction potential of the d_1 heme within cytochrome cd_1 has proved problematic, but estimates are in the range of +300mV. Similar values have been reported for isolated d_1 heme. These values correlate well with the position of cytochrome cd_1 in electron transport systems.[24] As pointed out earlier, electrons reach this enzyme via the cytochrome bc$_1$ complex and soluble c-type cytochromes and cupredoxins which operate at a mid point potential close to +250mV. In contrast, the siroheme type of nitrite reductase operates with NADH (-320mV) as its direct electron donor and the c-type cytochrome NrfA nitrite reductase receives electrons directly from the ubiquinol or menaquinol pools, with respective approximate redox potentials of +60mV and -50mV respectively; ammonia is the reaction product for both these enzymes.

In summary, the behaviour of the d_1 heme in nitrite reductase is modulated to at least some extent by the environment provided by the protein. Separating the intrinsic contribution of the d_1 heme ring from that provided by the protein is difficult. Thus the ability to bind anions in the Fe^{2+} state is thought to be a consequence of both the specialised features of the modified pyrrole and key amino acid residues, for instance the two positively charged histidines which are at the active site. A theoretical analysis by Ranghino et al suggests that the redox state of the d_1 heme and the protonation states of the two histidines together provide critical advantageous features for catalysis.[33]

The value of the reduction potential of the d_1 heme in cytochrome cd_1 is a puzzle. Redox titration of the enzyme from *P. pantotrophus* generates a remarkable hysteretic and cooperative titration. In the reductive direction a potential of 60 mv is observed but in the oxidative direction the potential is for both types of heme c. 250 mV.[34] Such complexities have not been seen for the *P. aeruginosa* enzyme but nevertheless there is a puzzling variation, from +190 mV to +280 mV, in the value of the reduction potentials reported for the d_1 heme.[6] Cytochrome cd_1 from *Pseudomonas nautica* gave the simplest oxidation/reduction titrations yet reported, allowing a clear determination of the c and d_1 redox potentials as +234 mV and +199 mV respectively.[35] Furthermore, it should be recalled that the redox potential has to fit in with the redox potential at which the enzyme works in the respiratory chain. There seems little doubt that the reduction potentials of the heme in the cytochrome c domain is 250 mV, consistent with it receiving electrons from a cupredoxin or c-type cytochrome with comparable potentials. As explained earlier, these electron donors are in turn reduced by the bc$_1$ complex.

Insights into d_1 heme Chemistry from Model Compound Studies

Analysis of model compounds has shown that the ring oxidation potential for the free base form of a dioxoisobacteriochlorin are almost identical to that for porphyrin itself.[36] Other model compound studies have suggested that a dioxoisobacteriochlorin such as d_1 is neither easier nor harder to oxidise than the standard Fe protoporphyrin IX,[37] but the absence of the two keto groups has been shown to make the ring much harder by c 400 mV to reduce. Thus on this basis the biosynthetic effort in making d_1 heme appears to achieve nothing relative to the porphyrins themselves,. However, a consequence of the two oxo groups in the d_1 heme is increased charge on the metal, which, for model compounds, shifts the Fe(II)/Fe(III) couple to less (c 200 mV) negative values with respect to a porphyrin. As explained earlier, an oxidation/ reduction potential of c 250 mV is required for the active site of cd_1. However, whilst such a value is towards the positive end of the range of values for the 'standard' porphyrins in b-type cytochromes, such potentials can be readily achieved by tuning the environment provided by a protein for the heme pocket. Thus reasons other than tuning redox potential are indicated for the presence of the keto groups. Increased charge on the ferrous state of the d_1 heme could be a contributory factor in the ability of this species to bind anions (see earlier), a property not normally associated with the ferrous state of b-type cytochromes.

Amongst other model compound work Ozawa et al studied the oxidation of the nitrosyl-iron (II) complex of dioxoisobacteriochlorin.[38] The product was shown to be an Fe(II)-NO π-cation radical complex with a bent Fe-NO bond. It was argued that the π cation radical was formed in preference to the FeIII-NO complex because the Fe-NO hybrid orbitals were lowered in energy, relative to a standard porphyrin, by the presence of the ring oxo groups.[38] The relevance of this work is that a nitrosyl complex (FeIINO)$^+$ is often postulated to be a reaction intermediate in the cytochrome cd_1 reaction that follows the dehydration of an initial ferrous d_1 heme-nitrite complex.[39]

In a development of their initial work, Ozawa et al showed that the presence of an axial imidazole ligand resulted in valence isomerisation such that the π cation was lost and the FeII-NO speices became FeII-NO$^+$ imidazole.[40] Addition of a further mole of imidazole displaced the NO; through comparative studies with other model compounds the ease of this displacement was judged to correlate with the presence of the electron withdrawing oxo groups. The latter would cause increased positive charge on the Fe which would, via an effect on Fe (dz) ->NO (π*) backbonding weaken the Fe-NO bond. By analogy the the axial proximal histidine ligand may have a role in contributing to the release of NO from the d_1 heme. On the basis of the model compound work it was suggested that a distal histidine or other nucleophile might also contribute to the release of NO from the d_1 heme active site. Comparison of several model compounds certainly suggested that the oxo groups in d_1 heme would contribute to the dissociation of NO from the d_1 heme in cytochrome cd_1.

Biosynthesis of d_1 heme

Lack of understanding of the advantageous features for catalysis of the d_1 heme ring is mirrored by a similar paucity of information as to how formation of these features is achieved.

Sequence analysis of genes adjacent to the structural gene for cytochrome cd_1, *nirS*, has identified seven genes which, on the basis of mutational analysis, are believed to be required for d_1 biogenesis. Of these, a clear role can be assigned for only one, *nirE*. This codes for an S-adenosyl methionine transferase which is believed to be responsible for converting uroporphyrinogen III into precorrin-2, a reaction step that is common to several other porphyrin biosyntheses (Chapter 18). Of the other genes, *nirD, nirF, nirG, nirH, nirJ,* and *nirL*, only *nirJ* has any clear resemblance to other genes in the databases. The presence of a predicted CXXCXYC sequence in NirJ suggests an Fe/S protein with some similarity to other cofactor assembly factors (e.g., NifB and PqqE) which are believed to convert substrate S-adenosyl methionine into a radical form, but, as the functions of the latter have also not been assigned, the comparison does not contribute to understanding function. By analogy with heme and siroheme synthesis it is likely that metallochelation will be the last step in synthesis of the d_1 heme per se but this step may have to be followed by transport of the completed d_1 heme to the periplasm. The apo form of cytochrome cd_1 is thought to be delivered to the periplasm as an unfolded polypeptide and therefore is unlikely to carry the d_1 heme with it across the membrane. Simple diffusion of the expensively assembled d_1 heme moiety does not seem to be a sensible strategy for delivery of this cofactor to its protein. There are several 'double' reactions, e.g., decarboxylation that have to be achieved to generate a d_1 heme (Fig. 1) molecule but quite how these are achieved by the Nir proteins is as mysterious as the reason why biology has evolved the unusual d_1 heme. Several possibilities for a biosynthetic pathway are discussed elsewhere.[11] In a recent development, the expression of *nirFDLGH* from *P. stutzeri* in *P. aeruginosa* has been found to cause production of a pigment, termed P800, that in vitro can be converted to the d_1 heme ring. Pigment 800 has several unusual features, including an epoxide ring and it remains to be seen as to what extent this compound 800 is related to a true intermediate on the pathway.[41]

Note Added in Proof

Two recent papers have suggested that nitric oxide can be released from the ferrous state of the d_1 heme.[42-43] Other mechanistic developments include the demonstration that interaction with an electron donor protein, pseudoazurin, facilitates product release,[44] the reaction studied at pH 6 differs from at pH 7[45] and a very rapid reaction study that has identified some new intermediate species including a putative Michaelis complex.[46] There has been provision of evidence that binding of nitrite to the active site of the *P. pantotrophus* enzyme can drive a lignad switch from histidine to methionine at the c-type heme[47] and that a weakly coupled heme-radical pair may occur at the active site.[48] Production of a variant of *P. pantotrophus* cytochrome *cd$_1$* in which the c-heme is bishistidinyl coordinated in both oxidation states has allowed the oxidation/reduction potential of the d_1 heme in the enzyme to be resolved at +175 mV because the c-heme was shifted to -60 mV.[49] In the same paper[49] it was shown that transcription of the the d_1 biosynthesis gene cluster requires NO, probably mediated by the Nnr transcription factor. It has been suggested that two of the biosynthesis genes in the *nir* cluster, *nirD* and *nirL*, are transcription factors and that possibly only *nirJ* is needed to catalyse a reaction step that is specific for the d_1 biosynthesis.[50]

Acknowledgements

Work from the author's laboratory owes everything to many colleagues who names appear in the bibliography. It was supported by BBSRC.

References

1. Zumft WG. Cell biology and molecular basis of denitrification. Microbiol Mol Biol Rev 1997; 61:533-616.
2. Pearson IV, Page MD, van Spanning RJM et al. A mutant of Paracoccus denitrificans with disrupted genes coding for cytochrome c550 and pseudoazurin establishes these two proteins as the in vivo electron donors to cytochrome cd$_1$ nitrite reductas. J Bacteriol 2003; 185:6308-6315.
3. Richter CD, Allen JWA, Higham CW et al. Cytochrome cd$_1$: Reductive activation and kinetic analysis of a multi-functional respiratory enzyme. J Biol Chem 2002; 277:3093-3100.
4. Sjogren T, Hadju J. Structure of the bound dioxygen species in the cyotchome oxidase reaction of cytochrome cd1 nitrite reductase. J Biol Chem 2001; 276:13072-13076.
5. Fulop V, Moir JWB, Ferguson SJ et al. The anatomy of a bifunctional enzyme structural basis for reduction of oxygen to water and synthesis of nitric oxide by cytochrome cd$_1$. Cell 1995; 81:369-377.
6. Allen JWA, Ferguson SJ, Fulop V. Cytochrome cd$_1$ nitrite reductase. In: Messerschmidt A, Huber R, Poulos T, Wieghardt K, eds. Handbook of Metalloporoteins. Chichester: J Wiley and Sons, 2001:424-439.
7. Baker SC, Saunders NF, Willis AC et al. Cytochrome cd$_1$ structure: Unusual heme environments in a nitrite reductase and analysis of factors contributing to beta-propeller folds. J Mol Biol 1997; 269:440-455.
8. Ferguson SJ, Fulop V. Cytochrome cd$_1$ nitrite reductase structure raises interesting mechanistic questions. Subcellular Biochem 2000; 35:519-540.
9. Fulop V, Watmough NJ, Ferguson SJ. Structure and enzymology of two bacterial diheme enzymes: Cytochrome cd$_1$ nitrite reductase and cytochrome c peroxidase. Adv Inorgan Chem 2001; 51:163-204.
10. Cutruzzola F, Rinaldo S, Centola F et al. NO Production by Pseudomonas aeruginosa cytochrome cd$_1$ nitrite reductase. IUBMB Life 2003; 55:617-621.
11. Timkovich R. The family of d-type hemes: Tetrapyrroles with unusual substituents. In: Kadish K, Smith K, Guilard R, eds. Porphyrin Handbook. Part II. San Diego: Academic Press, 123-156.
12. Williams PA, Fulop V, Garman EF et al. Haem-ligand switiching during catalysis in crystals of a nitrogen cycle enzyme. Nature 1997; 389:406-412.
13. Silaghi-Dumitrescu R. Linkage isomerisation in nitrite reduction bycytochrmoe cd$_1$ nitrite reductase. Inorganic Chemistry 2004; 43:3715-3718.
14. Allen JWA, Watmough NJ, Ferguson SJ. A switch in heme axial ligation prepares Paracoccus pantrotophus cytochrome cd$_1$ for catalysis. Nature Struct Biol 2000; 7:885-888.

15. Allen JWA, Cheesman MR, Higham CW et al. A novel conformer of oxidized Paracoccus pantotrophus cytochrome cd1 observed by freeze-quench NIR-spectroscopy. Biochem Biophys Res Commun 2000; 279:674-677.
16. Sjogren T, Hajdu J. The structure of an alternative form of Paracoccus pantotrophus cytochrome cd$_1$ nitrite reductase. J Biol Chem 2001; 276:29450-29455.
17. Gordon EHJ, Sjogren T, Lofqvist M et al. Structure and kinetic properties of Paracoccus pantotrophus cytochrome cd1 nitrite reductase with the d1 heme active site ligand tyrosine 25 replaced by serine. J Biol Chem 2003; 278:11773-11781.
18. Zajicek RS, Cheesman MR, Gordon EHJ et al. Y25S variant of Paracoccus pantotrophus Cytochrome cd$_1$: Provides insight into anion binding by d$_1$ heme and a rare example of a critical difference between solution and crystal structures. J Biol Chem 2005; 280:26073-26079.
19. Nurizzo D, Silivestrini MC, Mathieu M et al. N-terminal arm exchange is observed in the 2.15A crystal structure of oxidized nitrite reductase from Pseudomonas aeruginosa. Structure 1997; 5:1157-1171.
20. Nurizzo D, Cutruzzola F, Arese M et al. Does reduction of c heme trigger the conformational change of crystalline nitrite reductase? J Biol Chem 1999; 274:14997-15004.
21. Cutruzzola F, Brown K, Wilson EK et al. The nitrite reductase from Pseudomonas aeruginosa: Essential role of two active site-histidines in the catalytic and structural properties. Proc Natl Acad Sci US 2001; 98:2232-2237.
22. Brown K, Roig-Zamboni V, Cutruzzola F et al. Domain swing upon his to ala mutation in nitrite reductase of Pseudomonas aeruginosa. J Mol Biol 2001; 312:541-554.
23. Nurizzo D, Cutruzzola F, Arese M et al. Conformational changes occurring upon reduction and NO binding in nitrite reductase from Pseudomonas aeruginosa. Biochemistry 1998; 37:13987-13996.
24. Jafferi A, Allen JWA, Ferguson SJ et al. X-ray crystallographic study of cyanide binding provides insights into the structure-function relationship for cytochrome cd$_1$ nitrite reductase from Paracoccus pantotrophus. J Biol Chem 2000; 275:25089-25094.
25. Sun WL, Arese M, Brunori M et al. Cyanide binding to cd$_1$ nitrite reductase from Pseudomonas aeruginosa role of the active site his 369 in ligand stabilisation. Biochem Biophys Res Commn 2002; 291:1-7.
26. Sjogren T, Svensson-Ek M, Hajdu J et al. Proton coupled structural changes upon binding of carbon monoxide to cytochrome cd1; a combined flash photolysis and X-ray crystallography study. Biochemistry 2000; 39:10967-10974.
27. Das TK, Wilson EK, Cutruzzola F et al. Binding of NO and CO to the d$_1$ heme of cd$_1$ nitrite reductase from Pseudomonas aeruginosa. Biochemistry 2001; 40:10774-10781.
28. Cheesman MR, Ferguson SJ, Moir JWB et al. Two enzymes with a common function but different heme ligands in the forms as isolated. Optical and magnetic properties of the heme groups in the oxidized forms of nitrite reductase, cytochrome cd$_1$ from Pseudomonas stutzeri and Thiosphaera pantotropha. Biochemistry 1997; 36:16267-16276.
29. George SJ, Allen JWA, Ferguson SJ et al. Time resolved infra-red spectroscopy reveals a stable Fe(III)-NO intermediate in the reaction of Paracoccus pantotrophus cytochrome cd1 nitrite reductase with nitrite. J Biol Chem 2000; 275:33231-33237.
30. Steup MB, Muhoberac BB. Preparation and spectral characterization of the heme d$_1$ apomyoglobin complex; An unusual protein environment for the substrate binding heme of pseudomonas cytochrome oxidase. J of Inorganic Biochemistry 1989; 37:233-258.
31. Allen JWA, Higham CW, Zajicek RS et al. A novel, kinetically stable, catalytically active, all ferric, nitrite-bound complex of Paracoccus pantotrophus cytochrome cd$_1$. Biochem J 2002; 366:883-888.
32. Chang CK. The Biosynthesis of the tetrapyrrole pigments. Wiley, Chichester: CIBA Foundation symposium 180, 1994:228-246.
33. Ranghino G, Scorza E, Sjogren T et al. Quantum mechanical interpretation of nitrite reduction by cytochrome cd1 nitrite reductase from Paracoccus pantotrophus. Biochemistry 2000; 39:10958-10966.
34. Koppenhofer A, Turner KL, Allen WA et al. Cytochrome cd$_1$ from Paracoccus pantotrophus exhibits kinetically gated, conformationally dependent, highly cooperative two-electron redox behaviour. Biochemistry 2000; 39:4243-4249.
35. Besson S, Carneiro C, Moura JJG et al. Cytochrmoe cd$_1$-type nitrite reductase isolated from the marine denitrifier Pseudomonas nautical 617 purification and characterization. Anaerobe 1995; 1:219-226.
36. Barkigia KM, Chang CK, Fajer J et al. Models of Heme d1. Characterization of an Iron (III) dioxoisobacteriochlorin (Porphyridione). J Am Chem Soc 1992; 114:1701-1707.
37. Chang CK, Barkigia LK, Fajer J. Models of heme d$_1$. structure and redox chemistry of dioxoisobacteriochlorins. J Am Chem Soc 1986; 108:1352-1354.

38. Ozawa S, Sakamoto E, Watanabe Y et al. Formation of nitrosyl-iron(II) iron(II) beta-oxoporphyrin pi-cation radical complexes—Models for a reaction intermediate of dissimilatory nitrite reductases. Chem Commun 1994; 935-936.
39. Ye RW, Toro-Suarez I, Tiedje JM et al. H$_2$^{18}O isotope exchange studies on the mechanism of reduction of nitrite to nitrous oxide during reduction of nitric oxide. J Biol Chem 1991; 266:12848-12851.
40. Ozawa S, Sakamoto E, Ichikawa T et al. Model studies of nitrosyl intermediates in the catalytic cycle of dissimilatory nitrite reductases. Inorganic Chem 1995; 34:6362-6370.
41. Youn HS, Liang Q, Cha JK et al. Compound 800, a natural product isolated from genetically engineered Pseudomonas: Proposed structure reactivity and putative relation to heme d$_1$. Biochemistry 2004; 43:10730-10738.
42. Rinaldo S, Arcovito A, Brunori M, Cutruzzola F. Fast dissociation of nitric oxide from ferrous Pseudomnas aeruginosa cytochrome cd$_1$ nitrite reductase—a novel outlook on the reaction mechanism. J Biol Chem 2007; 282:14761-14767.
43. Rinaldo S, Brunori M, Cutruzzola F. Nitrite controls the release of nitric oxide in Pseudomnas aeruginosa cytochrome cd$_1$ nitrite reductase. Biochem Biophys Res Comm 2007; 363:662-666.
44. Sam KA, Fairhurst SA, Thorneley RNF et al. Pseudoazurin dramatically enhances the reaction profile of nitrite reduction by Paracoccus pantotrophus cytochrome cd$_1$ and facilitates release of product nitric oxide. J Biol Chem 2008; 283:12555-12563.
45. Sam KA, Tolland JD, Fairhurst SA et al. Unexpected dependence on pH of NO release from Paracoccus pantotrophus cytochrome cd$_1$. Biochem Biophys Res Commun 2008; 371:719-723.
46. Sam KA, Strampraad MJF, de Vries S, Ferguson SJ. Very early reaction intermediates detected by microsecond time scale kinetics of cytochrome cd$_1$-catalyzed reduction of nitrite. J Biol Chem 2008; 283:27403-27409.
47. van Wonderen JH, Knight C, Oganesyan VS et al. Activation of the cytochrome cd$_1$ nitrite reductase from Paracoccus pantotrophus – reaction of oxidized enzyme with substrate drives a ligand switch at heme c. J Biol Chem 2007; 282:28207-28215.
48. Oganesyan VS, Cheesman MR, Thomson AJ. Magnetic circular dichroism evidence for a weakly coupled heme-radical pair at the active site of cytochrome cd1, a nitrite reductase. Inorganic Chemistry 2007; 46:10950-10952.
49. Zajicek RS, Cartron ML, Ferguson SJ. Probing the unusual oxidation/reduction behavior of Paracoccus pantotrophus cytochrome cd$_1$ nitrite reductase by replacing a switchable methionine heme iron ligand with histidine. Biochemistry 2006; 45:11208-11216.
50. Xiong J, Bauer CE, Pancholy A. Insight into the haem d$_1$ biosynthesis pathway in heliobacteia through bioinformatic analysis. Microbiology 2007; 153:3548-3562.

Index

A

ABC transporter 152, 153, 155, 157, 193, 199, 200
Acute porphyria 4-7, 10-17, 21, 22, 28, 89, 90, 93, 94, 96, 97, 99
Adenosylcobinamide (AdoCbi) 304-308, 311
Adenosylcobinamide-phosphate (AdoCbi-P) 304-308
AdoCbi-GDP 305, 308, 311
Aerobic metabolism 74, 82, 83, 149, 162, 163, 185, 189, 201, 240, 256, 286-296, 300, 318-321, 375
Alpha-5,6-dimethylbenzimidazole adenine dinucleotide (α-DAD) 307-309
Amino acid ligand 168, 175
5-Aminolaevulinic acid (ALA) 6-8, 10, 13, 14, 16, 17, 19, 20, 22, 29, 30, 32, 33, 35, 37, 40, 43, 46, 51, 52, 54-57, 66, 70, 73, 84, 85, 89-93, 95-98, 116, 117, 122, 140-142, 147, 148, 151, 160, 161, 175, 185, 186, 196-201, 230, 250-255, 258, 259, 263, 268, 270
  ALA ester 141, 142
  ALA synthesis 7, 29, 116, 122, 230, 254
5-Aminolaevulinic acid dehydratase 43
5-Aminolaevulinic acid synthase (ALAS) 29-31, 90, 116-123, 125, 171, 186, 196, 199, 201
  ALAS-1 30, 90, 96, 116, 117, 119-121, 125, 196, 171
  ALAS-2 30, 90, 95, 116-123, 125, 186, 196, 199
Anaerobic metabolism 8, 37, 74-77, 82, 108, 168, 201, 213, 240, 241, 286-289, 291, 293-296, 300, 304, 310, 319-322, 346, 352, 353, 364, 370, 374, 377
AP1 122
Arabidopsis thaliana 37, 76, 78, 83, 211, 212, 213, 215, 216, 217, 221, 222, 225, 227, 228, 229, 230, 238, 252, 253, 254, 255, 256, 257, 258, 259, 267, 268, 271
Archaeoglobus fulgidus 376, 388
Arsenic 22
aSiR 375-379, 388
aSiRHP 376-380, 382, 384, 386, 387
Atherosclerosis 114, 140

B

B_{12}-dependent isomerase 331, 333, 335, 336, 338
Bach1 171, 176, 184, 193, 195, 196, 199
Bacillus megaterium 172, 294, 320, 346
Bacillus subtilis (*B. subtilis*) 32, 39, 77-82, 85, 162, 178, 180, 181, 186, 346
Bacteriochlorophyll 74, 80, 137, 235, 237, 239-241, 243-247
Bacteriochlorophyll synthase 246
Bilin 74, 208-210, 212, 213, 215-218, 221-224, 263, 343
Bilin reductase 208, 212, 213, 215-217
Bilirubin IXα 101, 113, 114
Biliverdin 101-105, 107, 109, 110, 112-114, 176, 187, 188, 193, 208, 210, 212, 214, 216, 222, 225, 259
  IXα 101, 102, 110, 113, 208, 210, 214, 222, 225
  IXα reductase 101, 102, 110, 113
  IXβ 113
Biosynthesis 5-7, 9-12, 22, 29, 30, 32, 33, 43, 58, 61, 76, 77, 82, 84, 85, 90, 103, 116, 117, 149, 152, 153, 155, 160, 185, 186, 196, 200, 201, 208, 212-214, 216-218, 230, 235, 237, 239, 240, 243-245, 247, 252, 256, 263-267, 270, 286, 287, 289, 293, 295, 296, 300, 304, 306-308, 311, 317-322, 324-326, 343, 345-349, 393, 396, 397
Branchpoint regulation 256, 258, 259

C

Carbon monoxide (CO) 101-103, 105-107, 109, 114, 167, 168, 170-172, 176, 185-188, 193, 197, 198, 201, 210, 353, 378, 387, 393
Catabolism 114, 184-186, 188-190, 193, 195, 197, 199, 201, 274, 282, 283, 331
Catalytic mechanism 30, 34, 37, 43, 52, 54, 61, 63-65, 77, 242, 243, 247, 366-368
Channel 40, 45, 80, 85, 176, 184, 187, 188, 258, 360, 362, 366, 367, 369, 370
Chemical structure 301, 303, 381

Chlamydomonas reinhardtii 30, 39, 78, 79, 187, 240, 242, 244, 252, 268, 269

Chlorophyll (Chl) 1, 4, 5, 7-9, 18, 29, 32, 43, 61, 74, 75, 80, 82, 83, 85, 103, 137, 138, 162, 208, 213, 216, 217, 227, 230, 235-247, 250-256, 258, 259, 263-270, 274-283, 286, 293, 317, 343, 348, 375
  a-binding protein 263
  a oxygenase CAO) 230, 258, 259, 267-269
  b formation 263
  cycle 267
  synthase 246, 251, 267
  synthesis 253

Chloroplast development 221, 227, 229, 255

Circadian 37, 184, 185, 196, 197, 199, 201, 227, 230, 254, 255, 258, 259

Clinical trial 98, 128, 131, 133-136, 138, 139, 141, 142

Co-C bond 300, 330, 331, 333-337, 340

Cob 300, 319, 324

Cob(II)alamin 302, 303, 330, 331, 336, 337

Cobalamin 43, 61, 240, 286-289, 291, 293-296, 300, 302, 304-307, 311, 317-326, 330-333, 336, 343-346, 348
  biosynthetic methyltransferase 296
  transport 324

Cobalt 8, 286, 287, 292-296, 301, 302, 304, 319, 320, 324, 330, 331, 333, 334, 337, 338, 347

Cobinamide (Cbi) 300, 301, 304, 306, 307, 317, 319, 322-326, 336

CobU kinase 319

Cobyric acid 288, 293, 295, 296, 301, 304, 306

Coenzyme B 347, 352, 353, 355, 358-361, 363, 364, 366-369

Coenzyme B$_{12}$ 300, 301, 303, 307, 308, 310, 317, 319, 326, 330
  biosynthesis 308

Coenzyme F$_{430}$ 29, 61, 74, 287, 343, 344, 346-349, 352, 354, 357

Coenzyme M (CoM) 347, 352-364, 366-370
  reductase 347

Cofactor 8, 30, 34, 38, 39, 58-61, 74, 76, 79, 81, 83, 85, 110, 119, 120, 122, 125, 149, 156, 160-163, 165, 167-169, 171, 174-176, 184, 235, 240-244, 251, 252, 256, 259, 263, 264, 277, 302, 304, 317, 319, 321, 322, 325, 330-336, 338, 347, 353, 355, 362, 367, 375, 377-380, 383, 386-388, 396

Congenital porphyria 6, 9, 12, 17, 18, 21

Coproporphyrinogen oxidase 7, 8, 14, 90, 122, 123, 161, 162, 201, 252

Corrin 8, 74, 76, 77, 286, 292-296, 300, 301, 304, 306, 307, 330, 332-334, 336, 348, 354

Corrinoid 300-302, 304, 306, 317-319, 324-326
  adenosylation 300-302, 306

Corynebacterium diptheriae 103

Covalent modification 190

CpG island 124

Crigler-Najjar syndrome 102, *see also* Neonatal jaundice

Cubic membrane 266

Cyanobacteria 76, 77, 101, 103, 156, 164, 211-214, 217, 221, 222, 224-226, 231, 237, 242-244, 253, 258, 264, 268, 269, 322

Cyclase 103, 167, 170, 174, 186, 187, 230, 239-241, 259, 270

cysFDNC operon 320

CysG 319-321, 343-346

Cysteine 34, 35, 37, 44-49, 52, 53, 57-60, 78, 82, 129, 150, 155, 156, 161-168, 170-172, 175, 187, 195, 196, 223-225, 319, 332, 335, 337, 361, 376, 378, 379, 386-388

Cytochrome *cd*$_1$ 162, 166, 169, 170, 390-397

Cytochrome P450 12, 90, 101, 107-110, 112, 119, 120, 160, 162, 164-169, 171, 172, 174-176, 185, 188, 210, 377

D

d$_1$ heme   *see* Heme *d*$_1$

Deaminase 7, 8, 14, 16, 43, 58-61, 63, 66, 91, 96, 121, 140, 161, 252, 321, 322

Death 10, 117, 130, 131

Degradation 1, 9, 37, 101, 107, 109, 114, 152, 171, 176, 184, 186, 195, 197, 199-201, 213, 217, 259, 266, 268, 269, 274-276, 280-282, 322

Denitrification 166, 390

Dipyrromethane cofactor 58-60, 161

Drugs and chemical porphyria 11

dSiR 375-378, 382, 388

E

Electron paramagnetic resonance (EPR) 50, 107, 108, 172, 174, 188, 333, 336-338, 352, 355-363, 365, 367, 370, 371, 378
Enzymology 217
Erythroid cell 60, 90, 116-118, 122, 175, 186, 196, 199, 200
Erythroid Krüppel-like factor (EKLF) 118
Erythroid-specific 5-aminolevulinate synthase 90, 116
Erythropoietic protoporphyria 6-8, 13, 14, 18, 89
Escherichia coli 30, 32, 33, 34, 36-38, 44-50, 52-55, 57, 58, 60, 61, 76, 77, 79, 80, 82, 83, 151, 152, 155, 156, 162, 164, 166-168, 195, 211, 216, 224, 237-239, 246, 300-302, 307, 319-326, 343, 344, 375-380, 386, 387, 393
Etioplast 84, 242, 243, 258, 266, 269

F

Feline leukemia virus group C (FLVCR) 184, 193, 195, 200, 201
Ferrochelatase 6-8, 12, 14, 74, 75, 77, 80-86, 90, 95, 119, 122, 124, 140, 149, 150, 153, 154, 161, 162, 185, 186, 193, 197, 199, 237, 250, 251, 253, 256-259, 295, 344-346

G

Gas sensor 168
GATA-1 67, 118, 121-124
Geranylgeraniol-bacteriochlorophyll reductase 247
Geranylgeranyl diphosphate (GGPP) 265-267, 269
GF-1 121
Glutamate-1-semialdehyde-2,1-aminomutase (GSAM) 29, 32, 33, 35, 37-40
Glutamyl-tRNA reductase 29, 33, 230, 252, 254
Glutamyl-tRNA synthetase 29, 30, 33, 34, 35, 255
Guanylyltransferase 305, 307

H

Haem (Heme) 1, 4, 5, 7-14, 20, 22, 29, 30, 34, 37, 43, 58, 61, 63, 74-77, 80-86, 89-93, 96, 97, 101-110, 112-114, 116-125, 140, 141, 149-157, 160-172, 174-176, 184-202, 208, 210, 211, 213, 216, 217, 221, 235-237, 240, 250-259, 263, 282, 286, 290, 292, 293, 295, 317-320, 326, 343, 348, 375, 382, 383, 386, 387, 390-397
  biosynthesis 7, 10-12, 22, 58, 77, 84, 85, 90, 116, 117, 149, 153, 185, 186, 196, 200, 201, 320, 396
  coordination 166, 167, 174, 188
  oxygenase (HO) 101-110, 114, 112, 115, 117, 119, 176, 184, 185, 187-190, 192, 193, 195, 197-201, 208, 210-213, 217, 251-253, 255, 267
  synthesis 12, 74-76, 80, 82-85, 89, 90, 116-120, 123-125, 152, 160, 161, 175, 184, 186, 196, 197, 199-201, 235-237, 250, 253, 256, 295, 345, 396
Haematoporphyrin derivative (HpD) 19, 131-134, 142
Harderian gland 5, 6, 19
Heme carrier protein 1 (HCP1) 192, 193, 195
Heme d_1 162, 166, 172, 175, 348, 390-397
Hemoglobin (Hb) 5-7, 30, 101, 113, 114, 116, 117, 150-155, 162, 163, 166, 168, 175, 186-189, 193, 196, 199
Hemopexin (HPX) 7, 153-155, 184, 185, 188-195, 202
Hemoprotein 1, 117, 150, 155, 160, 165, 167-172, 174-176, 191, 377-379, 394
Hepatocyte 117, 124, 154, 191, 192
Hepatocyte nuclear factor 1 alpha (HNF-1α) 124
History 4, 8, 9, 17, 20, 21, 92, 128, 216, 221, 227
Housekeeping 5-aminolevulinate synthase 30, 116, 117, 186, 196
Hydrogenobyrinic acid 290-293
Hydroxymethylbilane synthase 7, 43, 58, 91, 121, 161
Hyperbilirubinemia 21, 114

I

Inhibition 7, 16, 34, 38, 44, 57, 79, 80, 96,
        103, 116, 118, 122, 140, 160, 171, 176,
        186, 188, 196, 199, 200, 227-229, 239,
        240, 250, 251, 253, 255, 268, 270, 287,
        303, 319, 371
Iron 4-6, 8, 9, 75, 81, 82, 85, 90, 93, 95-97,
        101-103, 105, 107-109, 116-121, 124,
        125, 139, 140, 149-152, 154, 160, 162,
        164-168, 170-172, 174-176, 184,
        186-197, 199-201, 211, 217, 237, 240,
        256, 257, 263, 268, 276, 290, 319, 320,
        343, 345, 361, 375, 378, 381, 382, 386,
        387, 392-394, 396
Iron delivery 81, 85
Iron regulatory protein (IRP) 118, 171, 199
Iron responsive element (IRE) 117-119, 121,
        124, 196

K

Klebsiella aerogenes 320, 322
Krüppel-like transcriptional factor 124

L

Lactobacillus leichmannii 335
Lamellar membrane 267
Lead 5-7, 16, 21, 22, 44, 57, 93, 96
Ligand binding 56, 184, 191
Locus control region (LCR) 118
Lycopene cyclase 270

M

Macrocycle 4, 8, 74, 75, 81, 103, 107, 108,
        139, 162-165, 174, 175, 211, 213, 237,
        238, 245, 274-276, 282, 283, 295, 300,
        304, 330, 348, 361, 362, 365, 370, 375,
        386, 387
Magnesium chelatase 230, 237, 238, 293
Magnetic circular dichroism (MCD) 172,
        174, 352, 358, 363
Maize 253, 346
Mechanism 7, 8, 17, 20, 29-32, 34, 37, 43,
        52, 54, 55, 61-65, 74, 76, 77, 79-81, 85,
        86, 89, 95-97, 101, 103, 107, 109-114,
        116-118, 125, 129-131, 135, 141, 149,

        151, 152, 154-156, 162, 171, 174-176,
        184, 186, 187, 189, 191-193, 195, 197,
        200, 222-224, 229, 230, 236, 237,
        239-245, 247, 250, 255, 256, 258, 259,
        264-266, 269, 287, 290, 295, 317, 326,
        330, 332, 334, 338-340, 366-371, 379,
        380, 387, 388, 391, 392, 394, 397
Met1p 345, 346
Met8p 345, 346
Metal insertion 293, 344
Methanogenesis 300, 343, 346, 347, 352-354,
        370
Methanogenic archaea 300, 352
Methanopyrus kandleri 34, 36
Methanothermobacter thermoautotrophicus 348,
        350
Methyl-coenzyme M (methyl CoM) 347,
        352-357, 359-361, 363, 364, 366-370
Methyl-coenzyme M reductase 347, 352, 356,
        357, 359-361, 363, 364
Methylmalonyl-CoA mutase 304, 331, 332,
        334-336, 338
Methyltransferase 238, 239, 251, 256, 287,
        288, 290, 291, 295, 296, 330, 343-346
Model compound studies 390, 395
Monopyrrole 10, 11, 16, 58

N

Neisseria meningitides (nm-HO) 103-107
Neonatal jaundice 102, 114, *see also*
        Crigler-Najjar syndrome
Nickel 139, 347-349, 352-358, 360-370
Nitric oxide 160, 166-170, 185-187, 390,
        391-395, 397
Nitrite reductase 162, 166, 168, 170, 176,
        251, 319, 321, 343, 375, 379, 390-393,
        395
Nonphototransformable 270
NPAS 197
NPAS2 184, 185, 197, 198
Nuclear factor erythroid-derived 2 (NF-E2)
        118, 121, 124, 200
Nuclear factor kappa B (NFκB) 119, 120,
        123
Nuclear respiratory factor-1 (NRF-1) 119,
        120
Nucleotide loop assembly 287, 300, 301,
        304-307

O

Oct1 122
Operon induction 321
Operon repression 323
Organ transplant rejection 114
Organometallic bond 330
Oxygen (O₂) 4, 8, 34, 35, 37, 39, 49, 76-80,
    82, 83, 89, 91, 96, 101-110, 114, 117,
    119, 128-130, 132, 136-139, 151, 153,
    160, 162, 166-170, 176, 184-190, 195,
    197, 199, 201, 208, 210, 211, 213, 215,
    230, 237, 240, 243, 244, 251-253, 255,
    259, 263, 264, 267, 269, 275-278, 282,
    283, 286, 290, 292-294, 310, 322, 338,
    358, 360, 361, 364, 366, 368, 370, 377,
    379, 381-383, 390, 392-394
    activation 107, 168

P

Paracoccus denitrificans 168, 293, 302-304,
    307, 310, 311, 322
Paracoccus pantotrophus 166, 169, 170,
    391-395, 397
Phosphoribosyltransferase (PRTase) 305, 307,
    309, 310
Photodynamic therapy 20, 80, 128-131, 137,
    138, 140, 142
Photomorphogenesis 227, 228
Photoreceptor 208, 209, 212, 221-224, 227,
    228, 230, 241, 255
Photosynthesis 5, 7, 8, 185, 227, 235, 237,
    263, 264, 268
Phycocyanobilin (PCB) 208, 209, 211-214,
    216, 217, 222-226
Phycoerythrobilin 208, 210, 212-214
Phytochrome 37, 103, 208, 209, 211-217,
    221-231, 252-255, 268, 269
Phytochromobilin 83, 103, 208, 211-214,
    221, 225, 250-253
Phytol 245-247, 265, 275
Plant pigment 274
Plastid factor 268
Plastid signaling 254-256, 258
Phytocholorophyllide reductase (POR) 230,
    241-244, 251, 252, 258, 259, 264-267,
    269
Porphobilinogen 6, 7, 10, 11, 14-16, 43, 44,
    58-61, 63, 66, 91, 93, 96, 121, 140, 161,
    252, 379

Porphobilinogen deaminase (PBD) 7, 14, 43,
    58-61, 66, 91, 96, 121, 140, 161, 252
Porphyria 1-18, 20-23, 55, 56, 58, 60, 66, 75,
    82, 89, 90, 92-99, 122, 128, 186, 199,
    217, 252
Porphyria cutanea tarda (PCT) 6, 7, 13, 14,
    16, 17, 89-92, 94, 95, 97
Precorrin 77, 286-288, 290-296, 304, 319,
    320, 326, 343-348, 396
Precorrin-2 286-288, 290, 291, 293-295,
    304, 319, 320, 326, 343-348, 396
Prolamellar body 242, 243, 266
Protochlorophyllide 230, 239, 251-254, 263,
    264, 265
Protochlorophyllide oxidoreductase 251, 264
Protoporphyrinogen 7, 8, 14, 16, 21, 74-77,
    79, 80, 83, 84, 90, 95, 96, 122, 161, 162,
    250, 251
Protoporphyrinogen IX oxidase (PPO) 74, 75,
    77-85, 122-124, 162, 250, 251
Pseudoazurin 390, 397
Pseudomonas aeruginosa (P. aeruginosa, pa-HO)
    37, 39, 48, 50, 53, 103, 104, 105, 210,
    217, 224, 293, 391, 392, 393, 394, 395,
    397
Pseudoporphyria 22
Purple bacteria 221, 222, 237

R

Receptor 103, 131, 135, 151, 152, 154,
    184-187, 189-193, 195, 196, 199-201
Red blood cell (RBC) 92, 97
Red chlorophyll catabolite (RCC) 213,
    276-278
Red chlorophyll catabolite reductase
    (RCC-reductase) 213, 277, 278
Redox potential 166, 168, 175, 185, 188,
    369, 386, 395, 396
Redox sensing 168
Regulation of tetrapyrrole synthesis 80, 117,
    229, 250, 253
Resonance Raman 81, 107, 174, 334, 352,
    365, 378, 393
Respiration 168, 319, 377, 378
Reticulocyte 117
Rhodobacter 19, 29, 30, 116, 160, 162, 168,
    235, 322
Rhodobacter capsulatus 235, 237, 238, 239,
    240, 243, 244, 245, 246, 290, 293, 322

Rhodobacter sphaeroides 19, 30, 162, 168, 235, 237-240, 244, 245, 247
Ribosyl rotation 337
Ring contraction 82, 286, 287, 290, 295
Royal malady 7, 20
RpoS protein 322

S

Salmonella 82, 294, 307, 319-324, 326, 344
Salmonella cob 320-324
Salmonella enterica 294, 307, 319, 321-324, 326, 344
Salmonella typhimurium 37, 82
Seco intermediate 348
Senescence 227, 274, 275, 281, 283
Shade avoidance 227-229
Shemin pathway 29
Shibata shift 242, 265, 266
Sinorhizobium meliloti 322
Siroheme 61, 165, 167, 317, 319-321, 326, 343-346, 348, 375-383, 386-388, 395, 396
Siroheme-[4Fe4S] cluster 388
Sirohydrochlorin 251, 256, 257, 293, 294, 343-346, 348, 349, 375
Sp1 118-124
Structure-function relationship 176, 375
Succinyl coenzyme A (succinyl CoA) 29, 30, 84, 160, 196, 201
Sulfite reductase 167, 319, 321, 375, 379, 388
Synechocystis 30, 76, 175, 209, 211, 215-217, 221, 223-227, 237-240, 242, 243, 246, 247, 258, 268

T

TATA box 120, 122, 123
Tetrapyrrole 1-3, 5, 8, 9, 11, 29, 30, 33, 43, 57, 58, 60-62, 66, 74, 77, 80, 83, 90, 116, 117, 128, 129, 131, 133, 135, 136, 139, 140, 142, 154, 160, 161, 165, 187, 193, 201, 208, 212-214, 217, 221-223, 229, 230, 239, 240, 245, 246, 250-259, 263, 268, 269, 276, 277, 280-283, 286-288, 290, 292, 295, 300, 317-320, 330, 343, 344, 347, 348, 354, 382, 386

Tetrapyrrole Discussion Group (TPDG) 1-4
Tetrapyrrole synthesis 80, 116, 117, 229, 250-256, 268, 288, 348
Thermus thermophilus (Tt) 33, 64
Thiosphaera pantotropha 391
Transcription 37, 39, 116-124, 153, 167, 168, 170, 171, 176, 184-186, 190, 193, 195-199, 217, 227, 230, 250, 254, 258, 259, 319, 322, 397
Transcriptional activation 190, 197
Transport 80, 84, 85, 119, 149, 150-157, 160, 162, 168, 184-187, 189-193, 195, 199-201, 216, 259, 268, 279, 301, 319, 320, 323-326, 382, 395, 396
tRNA 29, 30, 32-37, 39, 40, 160, 230, 251, 252, 254, 255, 259
tRNA^{Glu} 30, 32-37, 39, 40, 160

U

Uroporphyrinogen cosynthase 122
Uroporphyrinogen decarboxylase 8, 14, 16, 22, 90, 122, 161, 162, 253
Uroporphyrinogen III 7, 8, 14, 19, 43, 61, 62, 63, 66, 74, 75, 76, 77, 82, 83, 85, 90, 122, 161, 162, 165, 251, 256, 286, 287, 288, 290, 317, 343, 344, 345, 346, 348, 379, 396
Uroporphyrinogen III synthase (UROS) 43, 61-63, 66, 90, 94, 96, 98, 122, 123, 161

V

Verdoheme 107-109, 111, 112
Vinyl reductase 240
Vitamin B$_{12}$ 6, 7, 29, 74, 82, 152, 237, 240, 256, 286-288, 292, 295, 296, 319, 323, 330, 343, 345-347, 375

W

Werewolves 7, 21